*Linear Circuit Design Handbook*

# Linear Circuit Design Handbook

*Hank Zumbahlen*

*with the engineering staff of Analog Devices*

ELSEVIER

AMSTERDAM • BOSTON • HEIDELBERG • LONDON • NEW YORK • OXFORD
PARIS • SAN DIEGO • SAN FRANCISCO • SINGAPORE • SYDNEY • TOKYO

Newnes is an imprint of Elsevier

Newnes

Cover images courtesy of Analog Devices

Newnes is an imprint of Elsevier

30 Corporate Drive, Suite 400, Burlington, MA 01803, USA

Linacre House, Jordan Hill, Oxford OX2 8DP, UK

 Recognizing the importance of preserving what has been written, Elsevier prints its books on acid-free paper whenever possible.

**Library of Congress Cataloging-in-Publication Data**

Linear circuit design handbook / edited by Hank Zumbahlen ; with the engineering staff of Analog Devices.
    p. cm.
  ISBN 978-0-7506-8703-4
  1. Electronic circuits. 2. Analog electronic systems. 3. Operational amplifiers. I. Zumbahlen, Hank.
II. Analog Devices, inc.
  TK7867.L57  2008
  627.39'5--dc22
                                                                  2007053012

**British Library Cataloguing-in-Publication Data**

A catalogue record for this book is available from the British Library.

ISBN: 978-0-7506-8703-4

> For information on all Newnes publications
> visit our Web site at www.books.elsevier.com

Typeset by Charon Tec Ltd (A Macmillan Company), Chennai, India
www.charontec.com

Printed and bound by CPI Group (UK) Ltd, Croydon, CR0 4YY

Transferred to Digital Print 2011

# Contents

# *Preface*

This work is based on the work of many other individuals who have been involved with applications and Analog Devices since the company started in 1965. Much of the material that appears in this work is based on work that has appeared in other forms. My major job function in this case was one of editor. The list of people I would like to credit for doing the pioneering work include: Walt Kester, Walt Jung, Paul Brokaw, James Bryant, Chuck Kitchen, and many other members of Analog Devices technical community.

In addition many others contributed to the production of this edition by helping out with the production of this book by providing invaluable assistance by proofreading and providing commentary. I especially want to thank Walt Kester, Bob Marwin, and Judith Douville, who also did the indexing.

Again, many thanks to those involved in this project.

Hank Zumbahlen
Senior Staff Applications Engineer

# CHAPTER 1

# *The Op Amp*

## Chapter Introduction

In this chapter we will discuss the basic operation of the op amp, one of the most common linear design building blocks.

In Section 1-1 the basic operation of the op amp will be discussed. We will concentrate on the op amp from the black box point of view. There are a good many texts that describe the internal workings of an op amp, so in this work a more macro view will be taken. There are a couple of times, however, that we will talk about the insides of the op amp. It is unavoidable.

In Section 1-2 the basic specifications will be discussed. Some techniques to compensate for some of the op amps limitations will also be given.

Section 1-3 will discuss how to read a data sheet. The various sections of the data sheet and how to interpret what is written will be discussed.

Section 1-4 will discuss how to select an op amp for a given application.

# Op Amp Operation

## Introduction

The op amp is one of the basic building blocks of linear design. In its classic form it consists of two input terminals—one of which inverts the phase of the signal, the other preserves the phase—and an output terminal. The standard symbol for the op amp is given in Figure 1-1. This ignores the power supply terminals, which are obviously required for operation.

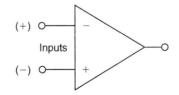

**Figure 1-1: Standard op amp symbol**

The name "op amp" is the standard abbreviation for operational amplifier. This name comes from the early days of amplifier design, when the op amp was used in analog computers. (Yes, the first computers were analog in nature, rather than digital.) When the basic amplifier was used with a few external components, various mathematical "operations" could be performed. One of the primary uses of analog computers was during World War II, when they were used for plotting ordinance trajectories.

## Voltage Feedback Model

The classic model of the voltage feedback (VFB) op amp incorporates the following characteristics:

1.  Infinite input impedance
2.  Infinite bandwidth
3.  Infinite gain
4.  Zero output impedance
5.  Zero power consumption

None of these can be actually realized, of course. How close we come to these ideals determines the quality of the op amp.

This is referred to as the VFB model. This type of op amp comprises nearly all op amps below 10 MHz bandwidth and on the order of 90% of those with higher bandwidths (Figure 1-2).

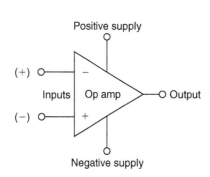

- Ideal op amp attributes
  - Infinite differential gain
  - Zero common mode gain
  - Zero offset voltage
  - Zero bias current
  - Infinite bandwidth

- Op amp input attributes
  - Infinite impedance
  - Zero bias current
  - Respond to differential voltages
  - Do not respond to common mode voltages

- Op amp output attributes
  - Zero impedance

**Figure 1-2: The attributes of an ideal op amp**

### Basic Operation

The basic operation of the op amp can be easily summarized. First we assume that there is a portion of the output that is feedback to the inverting terminal to establish the fixed gain for the amplifier. This is negative feedback. Any differential voltage across the input terminals of the op amp is multiplied by the amplifier's open-loop gain. If the magnitude of this differential voltage is more positive on the inverting (−) terminal than on the non-inverting (+) terminal, the output will go more negative. If the magnitude of the differential voltage is more positive on the non-inverting (+) terminal than on the inverting (−) terminal, the output voltage will become more positive. The open-loop gain of the amplifier will attempt to force the differential voltage to zero. As long as the inouts and output stays in the operational range of the amplifier, it will keep the differential voltage at zero and the output will be the input voltage multiplied by the gain set by the feedback. Note from this that the inputs respond to differential-mode not common-mode input voltage:

$$A = -\frac{R_{FB}}{R_{IN}} \qquad (1\text{-}1)$$

## Inverting and Non-Inverting Configurations

There are two basic ways configure the VFB op amp as an amplifier. These are shown in Figure 1-3 and Figure 1-4.

Figure 1-3 shows what is known as the inverting configuration. With this circuit, the output is out of phase with the input. The gain of this circuit is determined by the ratio of the resistors used and is given by:

$$A = -\frac{R_{FB}}{R_{IN}} \qquad (1\text{-}2a)$$

Figure 1-4 shows what is know as the non-inverting configuration. With this circuit, the output is in phase with the input. The gain of the circuit is also determined by the ratio of the resistors used and is given by:

$$A = 1 + \frac{R_{FB}}{R_{IN}} \qquad (1\text{-}2b)$$

Note that since the output drives a voltage divider (the gain setting network) the maximum voltage available at the inverting terminal is the full output voltage, which yields a minimum gain of 1.

Also note that in both cases the feedback is from the output to the inverting terminal. This is negative feedback and has many advantages for the designer. These will be discussed in more detail further in this chapter.

It should also be noted that the gain is based on the ratio of the resistors, not their actual values. This means that the designer can choose just about any value he or she wishes within practical limits.

If the value of the resistors is too low, a great deal of current would be required from the op amps output for operation. This causes excessive dissipation in the op amp itself, which has many disadvantages. The increased dissipation leads to self-heating of the chip, which could cause a change in the DC characteristics of the op amp itself. Also the heat generated by the dissipation could eventually cause the junction temperature to rise above the 150°C, the commonly accepted maximum limit for most semiconductors.

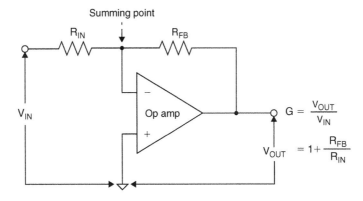

**Figure 1-3: Inverting-mode op amp stage**

5

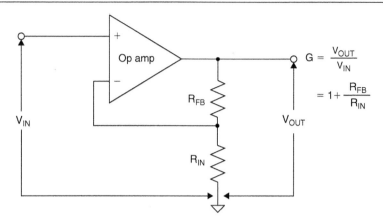

**Figure 1-4: Non-inverting-mode op amp stage**

The junction temperature is the temperature at the silicon chip itself. On the other end of the spectrum, if the resistor values are too high, there is an increase in noise and the susceptibility to parasitic capacitances, which could also limit bandwidth and possibly cause instability and oscillation.

From a practical sense, resistors below $10\,\Omega$ and above $1\,M\Omega$ become increasingly difficult to purchase especially if precision resistors are required.

Let us look at the case of an inverting amp in a little more detail. Referring to Figure 1-5, the non-inverting terminal is connected to ground. (We are assuming a bipolar (+ and −) power supply.) Since the op amp will force the differential voltage across the inputs to zero, the inverting input will also appear to be at ground. In fact, this node is often referred to as a "virtual ground."

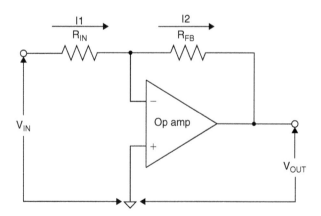

**Figure 1-5: Inverting amplifier gain**

If there is a voltage ($V_{IN}$) applied to the input resistor, it will set up a current (I1) through the resistor ($R_{IN}$) so that:

$$I1 = \frac{V_{IN}}{R_{IN}}$$

(1-3)

Since the input impedance of the op amp is infinite, no current will flow into the inverting input. Therefore, this same current (I1) must flow through the feedback resistor ($R_{FB}$). Since the amplifier will force the inverting terminal to ground, the output will assume a voltage ($V_{OUT}$) such that:

$$V_{OUT} = I1 \times R_{FB} \qquad (1\text{-}4)$$

Doing a little simple arithmetic we then can come to the conclusion of Eq. (1-1):

$$\frac{V_{OUT}}{V_{IN}} = A = -\frac{R_{FB}}{R_{IN}} \qquad (1\text{-}5)$$

Now we examine the non-inverting case in more detail. Referring to Figure 1-6, the input voltage is applied to the non-inverting terminal. The output voltage drives a voltage divider consisting of $R_{FB}$ and $R_{IN}$. The name "$R_{IN}$," in this instance, is somewhat misleading since the resistor is not technically connected to the input, but we keep the same designation since it matches the inverting configuration, has become a de facto standard, anyway. The voltage at the inverting terminal ($V_a$), which is at the junction of the two resistors, is:

$$V_a = \frac{R_{IN}}{R_{IN} + R_{FB}} V_{OUT} \qquad (1\text{-}6)$$

The negative feedback action of the op amp will force the differential voltage to 0 so:

$$V_a = V_{IN} \qquad (1\text{-}7)$$

Again applying a little simple arithmetic we end up with:

$$\frac{V_{OUT}}{V_{IN}} = \frac{R_{FB} + R_{IN}}{R_{IN}} = 1 + \frac{R_{FB}}{R_{IN}} \qquad (1\text{-}8)$$

Which is what we specified in Eq. (1-2).

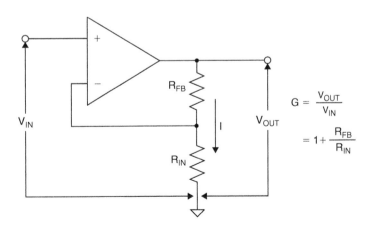

**Figure 1-6: Non-inverting amplifier gain**

In all of the discussions above, we referred to the gain setting components as resistors. In fact, they are impedances, not just resistances. This allows us to build frequency dependent amplifiers. This will be covered in more detail in a later section.

## Open-Loop Gain

The open-loop gain (usually referred to as $A_{VOL}$) is the gain of the amplifier without the feedback loop being closed, hence the name "open loop." For a precision op amp this gain can be vary high, on the order of 160 dB or more. This is a gain of 100 million. This gain is flat from DC to what is referred to as the dominant pole. From there it falls off at 6 dB/octave or 20 dB/decade. (An octave is a doubling in frequency and a decade is ×10 in frequency.) This is referred to as a single pole response. It will continue to fall at this rate until it hits another pole in the response. This second pole will double the rate at which the open-loop gain falls, i.e., to 12 dB/octave or 40 dB/decade. If the open-loop gain has dropped below 0 dB (unity gain) before it hits the second pole, the op amp will be unconditionally stable at any gain. This will be typically referred to as unity gain stable on the data sheet. If the second pole is reached while the loop gain is greater than 1 (0 dB), then the amplifier may not be stable under some conditions (Figure 1-7).

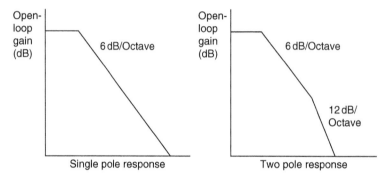

**Figure 1-7: Open-loop gain (Bode plot)**

It is important to understand the differences between open-loop gain, closed-loop gain, loop gain, signal gain, and noise gain (Figures 1-8 and 1-9). They are similar in nature, interrelated, but different. We will discuss them all in detail.

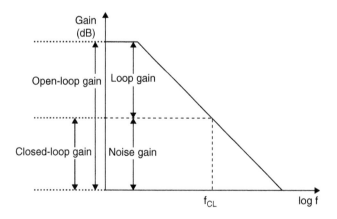

**Figure 1-8: Gain definition**

The open-loop gain is not a precisely controlled specification. It can, and does, have a relatively large range and will be given in the specifications as a typical number rather than a min/max number, in most cases. In some cases, typically high precision op amps, the specification will be a minimum.

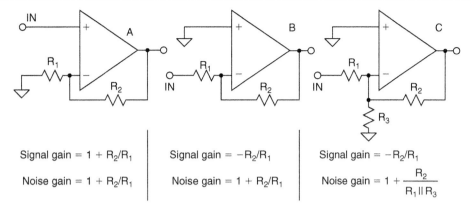

Signal gain $= 1 + R_2/R_1$

Noise gain $= 1 + R_2/R_1$

Signal gain $= -R_2/R_1$

Noise gain $= 1 + R_2/R_1$

Signal gain $= -R_2/R_1$

Noise gain $= 1 + \dfrac{R_2}{R_1 \| R_3}$

- Voltage noise and offset voltage of the op amp are reflected to the output by the noise gain.
- Noise gain, not signal gain, is relevant in assessing stability.
- Circuit C has unchanged signal gain, but higher noise gain, thus better stability, worse noise, and higher output offset voltage.

**Figure 1-9: Noise gain**

In addition, the open-loop gain can change due to output voltage levels and loading. There is also some dependency on temperature. In general, these effects are of a very minor degree and can, in most cases, be ignored. In fact this nonlinearity is not always included in the data sheet for the part.

## Gain-Bandwidth Product

The open-loop gain falls at 6 dB/octave. This means that if we double the frequency, the gain falls to half of what it was. Conversely, if the frequency is halved, the open-loop gain will double, as shown in Figure 1-8. This gives rise to what is known as the Gain-Bandwidth Product. If we multiply the open-loop gain by the frequency, the product is always a constant. The caveat for this is that we have to be in the part of the curve that is falling at 6 dB/octave. This gives us a convenient figure of merit with which to determine if a particular op amp is useable in a particular application (Figure 1-10).

For example, if we have an application with which we require a gain of 10 and a bandwidth of 100 kHz, we require an op amp with, at least, a gain-bandwidth product of 1 MHz. This is a slight oversimplification. Because of the variability of the gain-bandwidth product, and the fact that at the location where the closed-loop gain intersects the open-loop gain the response is actually down 3 dB, a little margin should be included. In the application described above, an op amp with a gain-bandwidth product of 1 MHz would be marginal. A safety factor of at least 5 would be better insurance that the expected performance is achieved.

## Stability Criteria

Feedback theory states that the closed-loop gain must intersect the open-loop gain at a rate of 6 dB/octave (single pole response) for the system to be stable. If the response is 12 dB/octave (two pole response) the op amp will oscillate. The easiest way to think of this is that each pole adds 90° of phase shift. Two poles then means 180°, and 180° of phase shift turns negative feedback into positive feedback, which means oscillations.

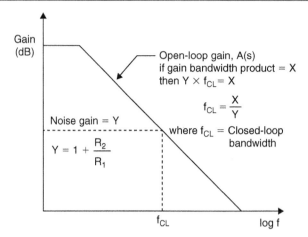

**Figure 1-10: Gain-bandwidth product**

The question could be then, why would you want an amplifier that is not unity gain stable. The answer is that for a given amplifier, the bandwidth can be increased if the amplifier is not unity gain stable. This is sometimes referred to as decompensated, but the gain criteria must be met. This criteria is that the closed-loop gain must intercept the open-loop gain at a slope of 6 dB/octave (single pole response). If not, the amplifier will oscillate.

As an example, compare the open-loop gain graphs in Figures 1-11, 1-12, 1-13. The three parts shown, the AD847, AD848, and AD849, are basically the same part. The AD847 is unity gain stable. The AD848 is stable for gains of two or more. The AD849 is stable for a gain of 10 or more.

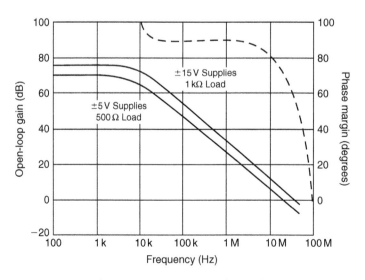

**Figure 1-11: AD847 open-loop gain**

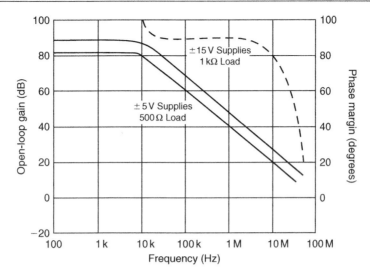

**Figure 1-12: AD848 open-loop gain**

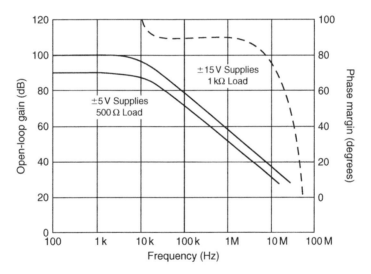

**Figure 1-13: AD849 open-loop gain**

## Phase Margin

One measure of stability is phase margin. Just as the amplitude response does not stay flat and then change instantaneously, the phase will also change gradually, starting as much as a decade back from the corner frequency. Phase margin is the amount of phase shift that is left until you hit 180° measured at the unity gain point.

The manifestation of low phase margin is an increase in the peaking of the output just before the close loop gain intersects the open-loop gain (see Figure 1-14).

*11*

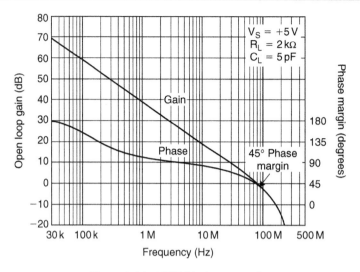

**Figure 1-14: AD8051 phase margin**

## Closed-Loop Gain

This, of course, is the gain of the amplifier with the feedback loop closed, as opposed the open-loop gain, which is the gain with the feedback loop opened. It has two forms, signal gain and noise gain. These are described and differentiated below.

The expression for the gain of a closed-loop amplifier involves the open-loop gain. If G is the actual gain, $N_G$ is the noise gain (see below), and $A_{VOL}$ is the open-loop gain of the amplifier, then:

$$G = N_G - \frac{N_G^2}{N_G + A_{VOL}} = \frac{N_G}{\dfrac{N_G}{A_{VOL}} + 1} \qquad (1\text{-}9)$$

From this you can see that if the open-loop gain is very high, which it typically is, the closed-loop gain of the circuit is simply the noise gain.

## Signal Gain

This is the gain applied to the input signal, with the feedback loop connected. In the basic operation section above, when we talked about the gain of the inverting and non-inverting circuits, we were actually more correctly talking about the closed-loop signal gain. It can be inverting or non-inverting. It can even be less than unity for the inverting case. Signal gain is the gain that we are primarily interested in when designing circuits.

The signal gain for an inverting amplifier stage is:

$$A = -\frac{R_{FB}}{R_{IN}} \qquad (1\text{-}10)$$

and for a non-inverting amplifier it is:

$$A = 1 + \frac{R_{FB}}{R_{IN}} \qquad (1\text{-}11)$$

## Noise Gain

Noise gain is the gain applied to a noise source in series with an op amp input. It is also the gain applied to an offset voltage. The noise gain is equal to:

$$A = 1 + \frac{R_{FB}}{R_{IN}} \qquad (1\text{-}12)$$

Noise gain is equal to the signal gain of a non-inverting amp. It is the same for either an inverting or non-inverting stage.

It is the noise gain that is used to determine stability. It is also the closed-loop gain that is used in Bode plots. Remember that even though we used resistances in the equation for noise gain, they are actually impedances (see Figure 1-9).

## Loop Gain

The difference between the open- and the closed-loop gain is known as the loop gain. This is useful information because it gives you the amount of negative feedback that can apply to the amplifier system (see Figure 1-8).

## Bode Plot

The plotting of open-loop gain versus frequency on a log–log scale gives is what is known as a Bode (pronounced *boh dee*) plot. It is one of the primary tools in evaluating whether a particular op amp is suitable for a particular application.

If you plot the open-loop gain and then the noise gain on a Bode plot, the point where they intersect will determine the maximum closed-loop bandwidth of the amplifier system. This is commonly referred to as the closed-loop frequency ($F_{CL}$). Remember that the true response at the intersection is actually 3 dB down. One octave above and one octave below $F_{CL}$, the difference between the asymptotic response and the real response will be less than 1 dB (Figure 1-15).

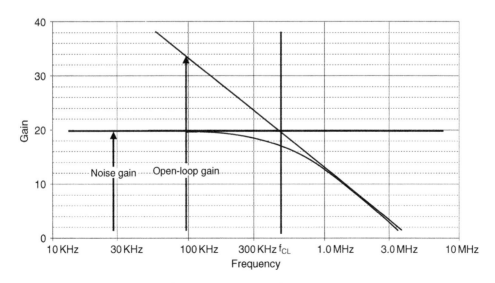

**Figure 1-15: Asymptotic response**

The Bode plot is also useful in determining stability. As stated above, if the closed-loop gain (noise gain) intersects the open-loop gain at a slope of greater than 6 dB/octave (20 dB/decade), the amplifier may be unstable (depending on the phase margin).

## Current Feedback Model

There is a type of amplifiers that have several advantages over the standard VFB amplifier at high frequencies. They are called current feedback (CFB) or sometimes transimpedance amps. There is a possible point of confusion since the current-to-voltage (I/V) converters commonly found in photo-diode applications are also referred to as transimpedance amps. Schematically CFB op amps look similar to standard VFB amps, but there are several key differences.

The input structure of the CFB is different from the VFB. While we are trying not to get into the internal structures of the op amps, in this case a simple diagram is in order (see Figure 1-16). The mechanism of feedback is also different, hence the names. But again, the exact mechanism is beyond what we want to cover here. In most cases if the differences are noted, and the attendant limitations observed, the basic operation of both types of amplifiers can be thought of as the same. The gain equations are the same as for a VFB amp, with an important limitation as noted in the next section.

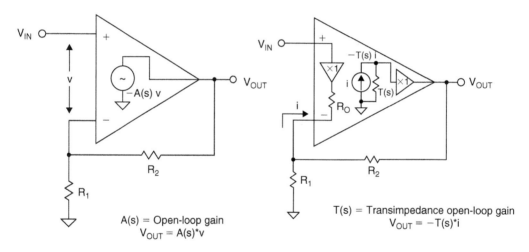

**Figure 1-16: VFB and CFB amplifiers**

### Difference from VFB

One primary difference between the CFB and VFB amps is that there is not a gain-bandwidth product. While there is a change in bandwidth with gain, it is not even close to the 6 dB/octave that we see with VFB (see Figure 1-17). Also, a major limitation is that the value of the feedback resistor determines the bandwidth, working with the internal capacitance of the op amp. For every CFB op amp there is a recommended value of feedback resistor for maximum bandwidth. If you increase the value of the resistor, reduce the bandwidth. If you use a lower value of resistor, the phase margin is reduced and the amplifier could become unstable. This optimum value of resistor is different for different operational conditions. For instance, the value will change for different packages, e.g., SOIC versus DIP (see Figure 1-18).

Also, a CFB amplifier should not have a capacitor in the feedback loop. If a capacitor is used in the feedback loop, it reduces the feedback impedance as frequency is increased, which will cause the op amp to

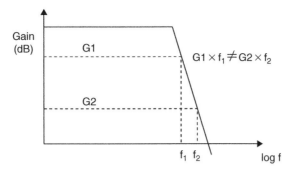

- Feedback resistor fixed for optimum performance. Larger values reduce bandwidth; smaller values may cause instability.

- For fixed feedback resistor, changing gain has little effect on bandwidth.

- Current feedback op amps do not have a fixed gain-bandwidth product.

**Figure 1-17: CFB amplifier frequency response**

| | AD8001AN (PDIP) Gain | | | | | AD8001AR (SOIC) Gain | | | | | AD8001ART (SOT-23-5) Gain | | | | |
|---|---|---|---|---|---|---|---|---|---|---|---|---|---|---|---|
| Component | −1 | +1 | +2 | +10 | +100 | −1 | +1 | +2 | +10 | +100 | −1 | +1 | +2 | +10 | +100 |
| $R_F$ (Ω) | 649 | 1050 | 750 | 470 | 1000 | 604 | 953 | 681 | 470 | 1000 | 845 | 1000 | 768 | 470 | 1000 |
| $R_G$ (Ω) | 649 | | 750 | 51 | 10 | 604 | | 681 | 51 | 10 | 845 | | 768 | 51 | 10 |
| $R_O$ (Nominal) (Ω) | 49.9 | 49.9 | 49.9 | 49.9 | 49.9 | 49.9 | 49.9 | 49.9 | 49.9 | 49.9 | 49.9 | 49.9 | 49.9 | 49.9 | |
| $R_S$ (Ω) | 0 | | | | | 0 | | | | | 0 | | | | |
| $R_T$ (Nominal) (Ω) | 54.9 | 49.9 | 49.9 | 49.9 | 49.9 | 54.9 | 49.9 | 49.9 | 49.9 | 49.9 | 54.9 | 49.9 | 49.9 | 49.9 | 49.9 |
| Small signal BW (MHz) | 340 | 880 | 460 | 260 | 20 | 370 | 710 | 440 | 260 | 20 | 240 | 795 | 380 | 260 | 20 |
| 0.1 dB Flatness (MHz) | 105 | 70 | 105 | | | 130 | 100 | 120 | | | 110 | 300 | 145 | | |

**Figure 1-18: AD8001 optimum feedback resistor versus package**

oscillate. You need to be careful of stray capacitances around the inverting input of the op amp for the same reason.

A common error in using a CFB op amp is to short the inverting input directly to the output in an attempt to build a unity gain voltage follower (buffer). This circuit will oscillate. Obviously, in this case, the feedback resistor value will be less than the recommended value. The circuit is perfectly stable if the recommended feedback resistor of the correct value is used in place of the short.

Another difference between the VFB and CFB amplifiers is that the inverting input of the CFB amp is low impedance. By low we mean typically 50–100 Ω. Therefore, there is not the inherent balance between the inputs that the VFB circuit shows.

Slew-rate performance is also enhanced by the CFB topology. The current that is available to charge the internal compensation capacitor is dynamic. It is not limited to any fixed value as is often the case in VFB topologies. With a step input or overload condition, the current is increased (current-on-demand) until the overdriven condition is removed. The basic CFB amplifier has no fundamental slew-rate limit. Limits only come about from parasitic internal capacitances and many strides have been made to reduce their effects.

The combination of higher bandwidths and slew rate allows CFB devices to have good distortion performance while doing so at a lower power.

The distortion of an amplifier is impacted by the open-loop distortion of the amplifier and the loop gain of the closed-loop circuit. The amount of open-loop distortion contributed by a CFB amplifier is small due to the basic symmetry of the internal topology. Speed is the other main contributor to distortion. In most configurations, a CFB amplifier has a greater bandwidth than its VFB counterpart. So at a given signal frequency, the faster part has greater loop gain and therefore lower distortion.

## How to Choose Between CFB and VFB

The application advantages of CFB and VFB differ. In many applications, the differences between CFB and VFB are not readily apparent. Today's CFB and VFB amplifiers have comparable performance, but there are certain unique advantages associated with each topology. VFB allows freedom of choice of the feedback resistor (or impedance) at the expense of sacrificing bandwidth for gain. CFB maintains high bandwidth over a wide range of gains at the cost of limiting the choices in the feedback impedance.

In general, VFB amplifiers offer:

• Lower noise

• Better DC performance

• Feedback component freedom

while CFB amplifiers offer:

• Faster slew rates

• Lower distortion

• Feedback component restrictions

## Supply Voltages

Historically the supply voltage for op amps was typically $\pm 15\,$V. The operational input and output range was on the order of $\pm 10\,$V. But there was no hard requirement for these levels. Typically the maximum supply was $\pm 18\,$V. The lower limit was set by the internal structures. You could typically go within 1.5 or 2 V of either supply rail, so you could reasonably go down to $\pm 8\,$V supplies or so and still have a reasonable dynamic range.

Lately though, there has been a trend toward lower supply voltages. This has happened for a couple of reasons.

First, high speed circuits typically have a lower full-scale range. The principal reason for this is the amplifier's ability to swing large voltages. All amplifiers have a slew-rate limit, which is expressed as so many volts per microsecond. So if you want to go faster, your voltage range must be reduced, all other things being equal. Another reason is that to limit the effects of stray capacitance on the circuits, you need to reduce their impedance levels. Driving lower impedances increases the demands on the output stage, and on the power dissipation abilities of the amplifier package. Lower voltage swings require lower currents to be supplied, thereby lowering the dissipation of the package.

A second reason is that as the speed of the devices inside the amplifier increased, the geometries of these devices tend to become smaller. The smaller geometries typically mean reduced breakdown voltages for these parts. Since the breakdown voltages were getting lower, the supply voltages had to follow. Today high speed op amps typically have breakdown voltage of ±7V, and so the supplies are typically ±5V, or even lower.

In some cases, operation on batteries established a requirement for lower supply voltages. Lower supplies would then lessen the number of batteries, which, in turn, reduced the size, weight, and cost of the end product.

At the same time there was a movement towards single supply systems. Instead of the typical plus and minus supplies, the op amps operate on a single positive supply and ground, the ground then becoming the negative supply.

## Single Supply Considerations

There is nothing in the circuitry of the op amp that requires ground. In fact, instead of a bipolar (+ and −) supply of ±15V you could just as easily use a single supply of +30V (ground being the negative supply), as long as the rest of the circuit was biased correctly so that the signal was within the common-mode range of op amp. Or, for that matter, the supply could just as easily be −30V (ground being the most positive supply).

When you combine the single supply operation with reduced supply voltages, you can run into problems. The standard topology for op amps uses a NPN differential pair (see Figure 1-19) for the input and emitter followers (see Figure 1-22) for the output stage. Neither of these circuits will let you run "rail-to-rail", i.e., from one supply to the other. Some circuit modifications are required.

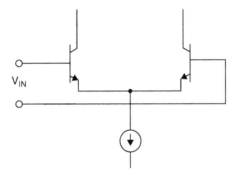

**Figure 1-19: Standard input stage (differential pair)**

The first of these modifications was the use of a PNP differential input (see Figure 1-20). One of the first examples of this input configuration was the LM324. This configuration allowed the input to get close to the negative rail (ground). It could not, however, go to the positive rail, But in many systems, especially mixed signal systems that were predominately digital, this was enough. In terms of precision, the 324 is not a stellar performer.

The NPN input cannot swing to ground. The PNP input cannot swing to the positive rail. The next modification was to use a dual input. Here a NPN differential pair is combined with a PNP differential pair (see Figure 1-21). Over most of the common-mode range of the input both pairs are active. As one rail or the other is approached, one of the inputs turns off. The NPN pair swings to the upper rail and the PNP pair swings to the lower rail.

It should be noted here that the op amp parameters which primarily depend on the input structure (bias current, for instance) will vary with the common-mode voltage on the inputs. The bias currents will even change direction as the front end transitions from the NPN stage to the PNP stage.

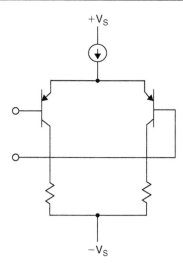

**Figure 1-20: PNP input stage**

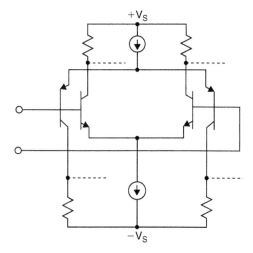

**Figure 1-21: Compound input stage**

Another difference is the output stage. The standard output stage, which is a complimentary emitter follower (common collector) configuration, is typically replaced by a common emitter circuit (Figure 1-22). This allows the output to swing close to the rails. The exact level is set by the $V_{CEsat}$ of the output transistors, which is, in turn, dependent on the output current levels. The only real disadvantage to this arrangement is that the output impedance of the common emitter circuit is higher than the common collector circuit. Most of the time this is not really an issue, since negative feedback reduces the output impedance proportional to the amount of loop gain. Where it becomes an issue is that as the loop gain falls this higher output impedance is more susceptible to the effects of capacitive loading.

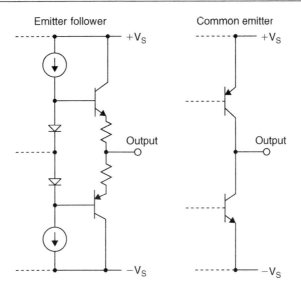

**Figure 1-22: Output stages: Emitter follower for standard configuration and common emitter for "rail-to-rail" configuration**

## Circuit Design Considerations for Single Supply Systems

Many waveforms are bipolar in nature. This means that the signal naturally swings around the reference level, which is typically ground. This obviously will not work in a single supply environment. What is required is to AC couple the signals.

AC coupling is simply applying a high pass filter and establishing a new reference level typically somewhere around the center of the supply voltage range (see Figure 1-23). The series capacitor will block the DC component of the input signal. The corner frequency (the frequency at which the response is 3 dB down from the midband level) is determined by the value of the components:

$$f_c = \frac{1}{2\pi \, R_{EQ} \, C} \qquad (1\text{-}13)$$

where:

$$R_{EQ} = \frac{R_4 \, R_5}{R_4 + R_5} \qquad (1\text{-}14)$$

It should be noted that if multiple sections are AC coupled, each section will be 3 dB down at the corner frequency. So if there are two sections with the same corner frequency, the total response will be 6 dB down; three sections would be 9 dB down, etc. This should be taken into account so that the overall response of the system will be adequate. Also keep in mind that the amplitude response starts to roll off a decade, or more, from the corner frequency.

The AC coupling of arbitrary waveforms can actually introduce problems which do not exist at all in DC coupled systems. These problems have to do with the waveform duty cycle, and are particularly acute with signals which approach the rails, as they can in low supply voltage systems which are AC coupled.

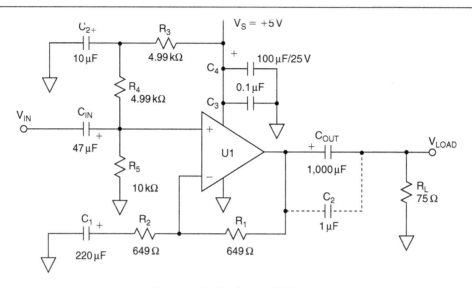

**Figure 1-23: Single supply biasing**

In an amplifier circuit such as that of Figure 1-23, the output bias point will be equal to the DC bias as applied to the op amp's (+) input. For a symmetric (50% duty cycle) waveform of a 2 Vp-p output level, the output signal will swing symmetrically about the bias point, or nominally 2.5 ± 1 V (using the values give in Figure 1-23). If however the pulsed waveform is of a very high (or low) duty cycle, the AC averaging effect of $C_{IN}$ and $R_4 \| R_5$ will shift the effective peak level either high or low, dependent on the duty cycle. This phenomenon has the net effect of reducing the working headroom of the amplifier, and is illustrated in Figure 1-24.

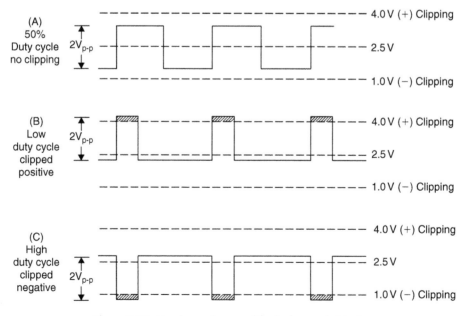

**Figure 1-24: Headroom issues with single supply biasing**

In Figure 1-24(A), an example of a 50% duty cycle square wave of about 2 Vp-p level is shown, with the signal swing biased symmetrically between the upper and lower clip points of a 5 V supply amplifier. This amplifier, for example, (an AD817 biased similarly to Figure 1-23) can only swing to the limited DC levels as marked, about 1 V from either rail. In cases (B) and (C), the duty cycle of the input waveform is adjusted to both low and high duty cycle extremes *while maintaining the same peak-to-peak input level*. At the amplifier output, the waveform is seen to clip either negative or positive, in (B) and (C), respectively.

## Rail-to-Rail

When the input and/or the output can swing very close to the supply rails, it is referred to as "rail-to-rail." There is no industry standard definition for this. At Analog Devices (ADI) we have defined this at swinging to within 100 mV of either rail. For the output this is driving a standard load, since the actual maximum output level will depend on the output current. Note that not all amplifiers that are touted as single supply are rail-to-rail. And not all rail-to-rail amplifiers are rail-to-rail on input and output. It could be one or the other, both, or neither. The bottom line is that you must read the data sheet. In no case can the output actually swing completely to the rails.

## Phase Reversal

There is an interesting phenomenon that can occur when the common-mode range of the op amp is exceeded. Some internal nodes can turn off and the output will be pulled to the opposite rail until the input comes back into the operational range (see Figure 1-25). Many modern designs take steps to eliminate this problem. Many times this is called out in the bullets on the cover page. Phase reversal is most common when the amplifier is in the follower mode.

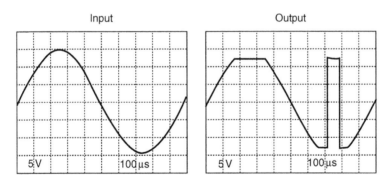

Vertical scale: 5 V / division
Horizontal scale: 100 µs / division

**Figure 1-25: Phase reversal**

## Low Power and Micropower

Along with the trend toward single supplies is the trend toward lower quiescent power. This is the power used by the amp itself. We have arrived at the point where there are whole amplifiers that can operate on the bias current of the 741.

However, low power involves some tradeoffs.

One way to lower the quiescent power is to lower the bias current in the output stage. This amounts to moving more toward class B operation (and away from class A). The result of this is that the distortion of the output stage will tend to rise.

Another approach to lower power is to lower the standing current of the input stage. The result of this is to reduce the bandwidth and to increase the noise.

While the term "low power" can mean vastly different things depending on the application, at ADI we have set a definition for op amps. Low power means the quiescent current is less than 1 mA per amplifier. Micropower is defined as having a quiescent current less than 100 μA per amplifier. As was the case with "rail-to-rail," this is not an industry wide definition.

## Processes

The vast majority of modern op amps are built using bipolar transistors.

Occasionally a junction FET is used for the input stage. This is commonly referred to as a Bi-Fet (for **Bi**polar-**FET**). This is typically done to increase the input impedance of the op amp, or conversely, to lower the input bias currents. The FET devices are typically used only in the input stage. For single supply applications, the FETs can be either N-channel or P-channel. This allows input ranges extending to the negative rail and positive rail, respectively.

Complementary-MOS processing (CMOS) is also used for op amps. While historically CMOS has not been that attractive a process for linear amplifiers, process and circuit design make progressed to the point that quite reasonable performance can be obtained from CMOS op amps.

One particularly attractive aspect of using CMOS is that it lends itself easily to mixed mode (analog and digital) applications. Some examples of this are the DigiTrim and chopper stabilized op amps.

"DigiTrim" is a technique that allows the offset voltage of op amps to be adjusted out at final test. This replaces the more common techniques of zener zapping or laser trimming, which must be done at the wafer level. The problem with trimming at the wafer level is that there are certain shifts in parameters due to packaging, etc. that take place after the trimming is done. While the shift in parameters is fairly well understood and some of the shift can be anticipated, trimming at final test is a very attractive alternative. The DigiTrim amplifiers basically incorporate a small digital-to-analog converter (DAC) used to adjust the offset.

Chopper stabilized amplifiers use techniques to adjust out the offset continuously. This is accomplished by using a DC precision amp to adjust the offset of a wider bandwidth amp. The DC precision amp is switched between a reference node (usually ground) and the input. This then is used to adjust the offset of the "main" amp.

DigiTrim and chopper stabilized amplifiers are covered in more detail in Chapter 2.

## Effects of Overdrive on Op Amp Inputs

There are several important points to be considered about the effects of overdrive on op amp inputs. The first is, obviously, damage. The data sheet of an op amp will give "absolute maximum" input ratings for the device. These are typically expressed in terms of the supply voltage, but, unless the data sheet expressly says otherwise, maximum ratings apply only when the supplies are present, and the input voltages should be held near zero in the absence of supplies.

A common type of rating expresses the maximum input voltage in terms of the supply, $V_{ss} \pm 0.3\,V$. In effect, neither input may go more than 0.3 V outside the supply rails, whether they are on or off. If current is limited to 5 mA or less, it generally does not matter if inputs do go outside $\pm 0.3\,V$ *when the supply is off* (provided that no base–emitter reverse breakdown occurs). Problems may arise if the input is outside this range when the supplies are turned on as this can turn on parasitic silicon controlled rectifiers (SCRs) in the device structure and destroy it within microseconds. This condition is called *latch-up*, and is much more common in digital CMOS than in linear processes used for op amps. If a device is known to be sensitive to latch-up, avoid the possibility of signals appearing before supplies are established. (When signals come from other circuitry using the same supply there is rarely, if ever, a problem.) Fortunately, most modern integrated circuit (IC) op amps are relatively insensitive to latch-up.

Input stage damage will be limited if the input current is limited. The standard rule-of-thumb is to limit the current to 5 mA. Reverse bias junction breakdown should be avoided at all cost. Note that the common—and differential—mode specifications may be different. Also, not all overvoltage damage is catastrophic. Small degradation of some of the specifications can occur with constant abuse by overvoltaging the op amp.

A common method of keeping the signal within the supplies is to clamp the signal to the supplies with Schottky diodes as shown in Figure 1-26. This does not, in fact, limit the signal to ±0.3 V at all temperatures, but if the Schottky diodes are at the same temperature as the op amp, they will limit the voltage to a safe level, even if they do not limit it at all times to within the data sheet rating. This is easily accomplished if overvoltage is only possible at turn-on, and diodes and op amp will always be at the same temperature then. If the op amp may still be warm when it is repowered, however, steps must be taken to ensure that diodes and op amp are at the same temperature when this occurs.

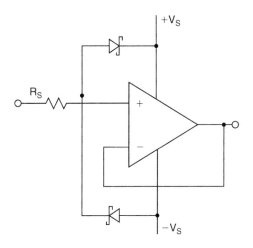

Figure 1-26: Input overvoltage protection

Many op amps have limited common-mode or differential input voltage ratings. Limits on common-mode are usually due to complex structures in very fast op amps and vary from device to device. Limits on differential input avoid a damaging reverse breakdown of the input transistors (especially super-beta transistors). This damage can occur even at very low current levels. Limits on differential inputs may also be needed to prevent internal protective circuitry from overheating at high current levels when it is conducting to prevent breakdowns—in this case, a few hundred microseconds of overvoltage may do no harm. One should never exceed any "absolute maximum" rating, but engineers should understand the reasons for the rating so that they can make realistic assessments of the risk of permanent damage should the unexpected occur.

If an op amp is overdriven *within* its ratings, no permanent damage should occur, but some of the internal stages may saturate. Recovery from saturation is generally slow, except for certain "clamped" op amps specifically designed for fast overdrive recovery. Overdriven amplifiers may therefore be unexpectedly slow.

Because of this reduction in speed with saturation (and also output stages unsuited to driving logic), it is generally unwise to use an op amp as a comparator. Nevertheless, there are sometimes reasons why op amps may be used as comparators. The subject is discussed in Reference 3 and Chapter 2.

# Op Amp Specifications

## Introduction

In this section, we will discuss basic op amp specifications. The importance of any of these specifications depends, of course, on the application. For instance, offset voltage, offset voltage drift, and open-loop gain (DC specifications) are very critical in precision sensor signal conditioning circuits, but may not be as important in high speed applications where bandwidth, slew rate, and distortion (AC specifications) are typically the key specifications.

Most op amp specifications are largely topology independent. However, although VFB and CFB op amps have similar error terms and specifications, the application of each part warrants discussing some of the specifications separately. In the following discussions, this will be done where significant differences exist.

It should be noted that not all of these specifications will necessarily appear on all data sheets. As the performance of the op amp increases, the more specifications it has and the tighter the specifications become. Also keep in mind the difference between typical and min/max. At ADI, a specification that is min/max is guaranteed by test. Typical specifications are generally not tested.

## DC Specifications

### Open-Loop Gain

The open-loop gain is the gain of the amplifier when the feedback loop is not closed. It is generally measured, however, with the feedback loop closed, although at a very large gain. In an ideal op amp, it is infinite with infinite bandwidth. In practice, it is very large (up to 160 dB) at DC. At some frequency (the dominant pole) it starts to fall at 6 dB/octave or 20 dB/decade. (An octave is a doubling in frequency and a decade is $\times 10$ in frequency.) This is referred to as a single pole response. The dominant pole frequency will range from in the neighborhood of 10 Hz for some high precision amps to several kHz for some high speed amps. It will continue to fall at this rate until it reaches another pole in the response. This second pole will double the rate at which the open-loop gain falls, that is to 12 dB/octave or 40 dB/decade. If the open-loop gain has gone below 0 dB (unity gain) before the amp hits the second pole, the op amp will be unconditionally stable at any gain. This will be referred to as unity gain stable on the data sheet. If the second pole is reached while the loop gain is greater than 1 (0 dB), then the amplifier may not be stable under some conditions (Figure 1-27).

Since the open-loop gain falls by half with a doubling of frequency with a single pole response, there is what is called a constant gain-bandwidth product. At any point along the curve, if the frequency is multiplied by the gain at that frequency, the product is a constant. For example, if an amplifier has a 1 MHz gain-bandwidth product, the open-loop gain will be 10 (20 dB) at 100 kHz, 100 (40 dB) at 10 kHz, etc. This is readily apparent on a Bode plot, which plots gain versus frequency on a log–log scale.

Since a VFB op amp operates as a voltage in/voltage out device, its open-loop gain is a dimensionless ratio, so no unit is necessary. Data sheets sometimes express gain in V/mV or V/$\mu$V instead of V/V, for the convenience of using smaller numbers. Or voltage gain can also be expressed in dB terms, as gain in dB $= 20 \times \log A_{VOL}$. Thus an open-loop gain of 1 V/$\mu$V (or 1000 V/mV or 1,000,000 V/V) is equivalent to 120 dB, and so on (Figure 1-28).

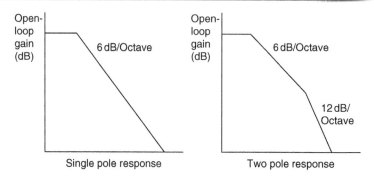

Figure 1-27: Open-loop gain

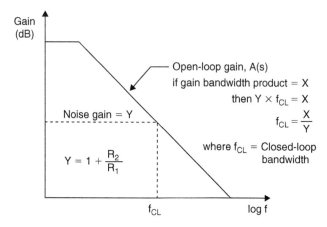

Figure 1-28: Bode plot (for VFB amps)

For very high precision work, the nonlinearity of the open-loop gain must be considered. Changes in the output voltage level and output loading are the most common causes of changes in the open-loop gain of op amps. A change in open-loop gain with signal level produces a *nonlinearity* in the closed-loop gain transfer function, which cannot be removed during system calibration. Most op amps have fixed loads, so $A_{VOL}$ changes with load are not generally important. However, the sensitivity of $A_{VOL}$ to output signal level may increase for higher load currents (see Figure 1-29).

The severity of this nonlinearity varies widely from one device type to another, and generally is not specified on the data sheet. The minimum $A_{VOL}$ is always specified, and choosing an op amp with a high $A_{VOL}$ will minimize the probability of gain nonlinearity errors. There is no way to compensate for $A_{VOL}$ nonlinearity.

## Open-Loop Transresistance of a CFB Op Amp

For CFB amplifiers, the open-loop response is voltage out for a current in, so it is a *transresistance* (expressed in ohms) rather than a gain. This is generally referred to as a *transimpedance*, since there is an AC component as well as a DC term. The *transimpedance* of a CFB amp will usually be in the range of $500 \, k\Omega$ to $1 \, M\Omega$.

A CFB op amp open-loop transimpedance does not vary in the same way as a VFB open-loop gain. Therefore, a CFB op amp will not have the same gain-bandwidth product as in VFB amps. While there is

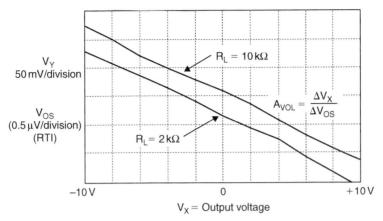

A<sub>VOL</sub> (average) ≈ 8 million

$A_{VOL,Max} \approx 9.1$ million, $A_{VOL,Min} \approx 5.7$ million

Open-loop gain nonlinearity ≈ 0.07 ppm

Closed-loop gain nonlinearity ≈ NG × 0.07 ppm

**Figure 1-29: Open-loop nonlinearity**

some variation of frequency response with frequency with a CFB amp, it is nowhere near 6 dB/octave (see Figure 1-30).

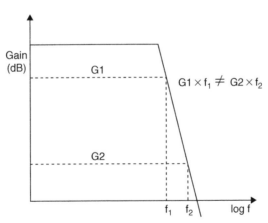

**Figure 1-30: Open-loop gain of a CFB op amp**

When using the term *transimpedance amplifier*, there can be some confusion. An amplifier configured as a current to voltage (I/V) converter, typically in photodiode circuits, is also referred to as a transimpedance amplifier. But the photodiode application will generally use a FET input VFB amp rather than a CFB amp. This is because the current levels in the photodiode applications will be very low, not the most compatible with the low impedance input of a CFB op amp.

## Offset Voltage

If both inputs of an op amp are at exactly the same voltage, then the output should be at zero volts, since a differential of 0 V should produce an output of 0 V. In practice, however, there will typically be some voltage at the output. This is known as the *offset voltage* or $V_{OS}$. The typical way to specify offset voltage is as the amount of voltage that must be added to the input to force 0 V out. This voltage, divided by the noise gain of the circuit, is the *input offset voltage* or *input referred offset voltage*. The offset voltage is usually input referred to eliminate the effect of circuit gain, which makes comparisons easier. The offset voltage is modeled as a voltage source, $V_{OS}$, in series with the inverting input of the op amp as shown in Figure 1-31.

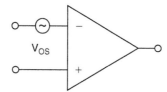

**Figure 1-31: Offset voltage**

## Offset Voltage Drift

The input offset voltage varies with temperature. Its temperature coefficient is known as $TCV_{OS}$, or more commonly, *drift*. Offset drift may be as low as $0.1\,\mu V/^{\circ}C$ (typical value for OP-177F, a very high precision op amp). More typical drift values for a range of general purpose precision op amps lie in the range $1–10\,\mu V/^{\circ}C$. Most op amps have a specified value of $TCV_{OS}$, but some, instead, have a second value of maximum $V_{OS}$ that is guaranteed over the operating temperature range. Such a specification is less useful, because there is no guarantee that $TCV_{OS}$ is constant or monotonic.

## Drift with Time

The offset voltage also changes as time passes, or *ages*. Aging is generally specified in $\mu V/month$ or $\mu V/1000$ hours, but this can be misleading. Aging is not linear, but instead a nonlinear phenomenon that is proportional to the *square root* of the elapsed time. A drift rate of $1\,\mu V/1000$ hours therefore becomes about $3\,\mu V/year$ (not $9\,\mu V/year$). *Long-term drift* of the OP-177F is approximately $0.3\,\mu V/month$. This refers to a time period *after* the first 30 days of operation. Excluding the initial hour of operation, changes in the offset voltage of these devices during the first 30 days of operation are typically less than $2\,\mu V$. The long-term drift of offset voltage with time is not always specified, even for precision op amps.

## Correction for Offset Voltage

Early op amps typically had pins available for nulling out offset voltages. A potentiometer connected to these pins, and the wiper connected to one or the other of the supply voltages, allowed balancing the input stage, which, in turn, nulled out the offset voltage (see Figure 1-32).

Makers of high precision op amps, such as Analog Devices (ADI) and Precision Monolithics (PMI) employed circuit design tricks to internally balance the input structures. ADI used laser trimming of the input stage load resistors to achieve balance. PMI used a technique called zener zapping to accomplish basically the same thing.

Laser trimming used lasers to eat away part of the collector resistors to adjust their value. Zener zapping involved having a string of resistors, each bypassed by a semiconductor structure that is basically a zener diode. By applying a pulse of voltage these zener diodes would be shorted out (zapped). This adjusts the value of the resistor string.

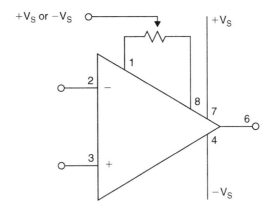

Figure 1-32: Offset adjustment pins

## DigiTrim™ Technology

DigiTrim is a technique which adjusts circuit offset performance by programming digitally weighted current sources, in essence a DAC. This technique makes use of the mixed signal capabilities of the CMOS process. While, historically, CMOS would not be the first choice for precision amplifiers, recent process improvements combined with the DigiTrim technology result in a very reasonable precision performance. In this patented new trim method, the trim information is entered through existing analog pins using a special digital keyword sequence. The adjustment values can be temporarily programmed, evaluated, and readjusted for optimum accuracy before permanent adjustment is performed. After the trim is completed, the trim circuit is locked out to prevent the possibility of any accidental re-trimming by the end user.

A unique feature of this technique is that the adjustment is done after the chip is packaged. With zener zapping and laser trimming, the offset must be adjusted at the die level. Subsequent processing, mounting the chip on a header and encapsulating in plastic cause a shift in the offset. This is due to both the mechanical stress of the mounting (strain gauge effect) and the heat of molding the package. While the amount of the shift is well profiled, the ability to trim at the package level versus the chip level is a distinct advantage.

The physical trimming, achieved by blowing polysilicon fuses, is very reliable. No extra pads or pins are required for this trim method and no special test equipment is needed to perform the trimming. The trimming is done through the input pins. A simplified representation of an amplifier with DigiTrim™ is shown in Figure 1-33. No testing is required at the wafer level assuming reasonable die yields. No special wafer fabrication process is required and circuits can even be produced by our foundry partners. All of the trim circuitry tend to scale with the process features so that as the process and the amplifier circuit shrink, the trim circuit also shrinks proportionally. The trim circuits are considerably smaller than normal amplifier circuits so that they contribute minimally to die cost. The trims are discrete as in link trimming and zener zapping, but the required accuracy is easily achieved at a very small cost increase over an untrimmed part.

The DigiTrim approach could also support user trimming of system offsets with a different amplifier design. This has not yet been implemented in a production part, but it remains a possibility.

### External Trim

The offset adjustment pins started to disappear with the advent of dual op amps since there were not enough pins left for them in the 8-pin package. Therefore external adjustment techniques were required.

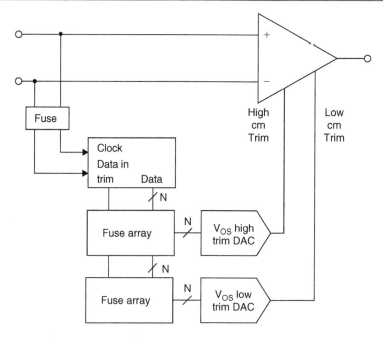

**Figure 1-33: Simplified schematic of the DigiTrim technology**

External trimming out offset involves basically adding a small voltage to the input to counteract the offset (see Figure 1-34). The polarity of the voltage applied to the offset pot will depend on the process used to manufacture the part as well as the polarity of the input devices (NPN or PNP). The offset can be accomplished with potentiometers, digital pots, or DACs. The major problem with external trimming is that the temperature coefficients of the internal and external components will probably not match. This will limit the effectiveness of the adjustment over temperature.

In addition, the mechanical pot is subject to aging and mechanical vibration.

There is an increase in noise gain due to the added resistance and the potentiometer resistance. The resulting increase in noise gain may be reduced by making $R_3$ much greater than $R_1$. Note that otherwise, the signal gain might be affected as the offset potentiometer is adjusted. The gain may be stabilized, however, if $R_3$ is connected to a fixed low impedance reference voltage sources, $\pm VR$.

The digital pot and DAC, however, can be adjusted in circuit, under control of a microprocessor or microcontroller, which could mitigate aging and temperature effects (Figure 1-35).

If response to DC is not required, an alternative approach would be to use a circuit called a servo (see Figure 1-36). This circuit is basically an integrator, which is placed in a feedback loop around the main amplifier. A precision amplifier should be used for the integrator; it need not be fast enough to pass the full frequency spectrum that the main amplifier must. The circuit operates by taking the average DC level of the output and feeding it back to the main amplifier, in effect subtracting it from the signal.

### Input Bias Current

In the ideal model of the op amp the inputs have infinite impedance and so no current flows into the input terminals. But since the most common input structure uses bipolar junction transistors (BJTs), there is always some current required for operation, since the BJT is a current controlled device. This is referred to

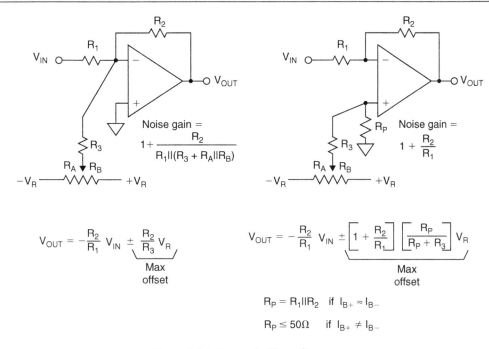

$$V_{OUT} = -\frac{R_2}{R_1} V_{IN} \pm \frac{R_2}{R_3} V_R$$

Max offset

$$V_{OUT} = -\frac{R_2}{R_1} V_{IN} \pm \left[1 + \frac{R_2}{R_1}\right]\left[\frac{R_P}{R_P + R_3}\right] V_R$$

Max offset

$R_P = R_1 \| R_2$   if $I_{B+} \approx I_{B-}$

$R_P \le 50\Omega$   if $I_{B+} \neq I_{B-}$

**Figure 1-34: External offset adjustment**

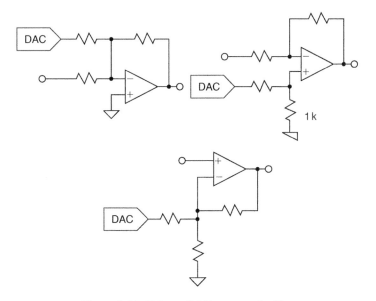

**Figure 1-35: Using a DAC to control offset**

as *bias current* ($I_B$) or *input bias current*. In practice, there are always two *input bias currents*, $I_{B+}$ and $I_{B-}$ (see Figure 1-37), one for each of the inputs. Values of $I_B$ range from 60 fA (about one electron every three microseconds) in the AD549 electrometer, to tens of microamperes in some high speed op amps. Due to the inherent nature of monolithic op amp fabrication processing, these bias currents tend to be equal, but

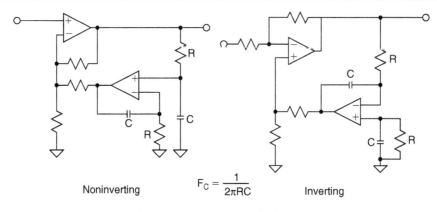

Noninverting $\qquad F_C = \dfrac{1}{2\pi RC} \qquad$ Inverting

**Figure 1-36: Servo controlled offset**

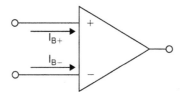

**Figure 1-37: Input bias current**

this is not guaranteed to be case. And in the case of CFB amplifiers, the non-symmetric nature of the inputs guarantees that the bias currents are different.

Input bias current is a problem to the op amp user because it flows in external impedances and produces offset voltages, which add to system errors. Consider a non-inverting unity gain buffer driven from a source impedance of $1\,\text{M}\Omega$. If $I_B$ is $10\,\text{nA}$, it will introduce an additional $10\,\text{mV}$ of error. Or, if the designer simply forgets about $I_B$ and uses capacitive coupling, the circuit will not work at all! This is because the bias currents need a DC return path to ground. If the DC return path is not there, the input of the op amp will drift to one of the rails. Or, if $I_B$ is low enough, it may work momentarily while the capacitor charges, giving even more misleading results. The moral here is not to neglect the effects of $I_B$ in any op amp circuit.

### Input Offset Current

The difference in the bias currents is the input offset current. Normally the difference between the bias currents is small, so that the offset current is also small. In bias-compensated op amps (see next section) the offset current is approximately equal to the bias current.

### Compensating for Input Bias Current

There are several ways to compensate for bias currents. It can be addressed by the manufacturer, or external techniques can be employed.

There are basically two different ways that an IC manufacturer can deal with bias currents.

The first is to use a "super-beta" transistors for the input stage. Super-beta transistors are specially processed devices with a very narrow base region. They typically have a current gain ($\beta$) of thousands or tens of thousands (rather than the more usual hundreds for standard BJT transistors). Op amps with super-beta input stages have much lower bias currents, but they also have more limited frequency response. Since the breakdown voltages of super-beta devices are typically quite low, they also require additional circuitry to protect the input stage from damage caused by overvoltage on the input.

The second method of dealing with bias currents is to use a bias-compensated input structure (see Figure 1-38). With a bias current compensated input, small current sources are added to the bases of the input devices. The idea is that the bias currents required by the input devices are provided by the current sources so that the net current seen by the external circuit is reduced considerably.

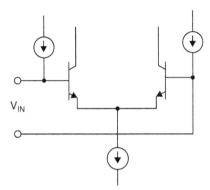

**Figure 1-38: Input bias current compensation**

Bias current compensated input stages have many of the good features of the simple bipolar input stage, namely: low voltage noise, low offset, and low drift. Additionally, they have low bias current which is fairly stable with temperature. However, their current noise is not very good, since current sources are added to the input. And their bias current matching is poor. These latter two undesired side effects result from the external bias current being the *difference* between the compensating current source and the input transistor base current. Both of these currents inevitably have noise. Since they are uncorrelated, the two noises add in a root-sum-of-squares fashion (even though the DC currents subtract).

Note that this can easily be verified, by examining the *offset current* specification (the difference in the bias currents). If internal bias current compensation exists, the offset current will be of the same magnitude as the bias current. Without bias current compensation, the offset current will generally be at least a factor of 10 smaller than the bias current. Note that these relationships generally hold, regardless of the exact magnitude of the bias currents.

Since the resulting external bias current is the difference between two nearly equal currents, there is no reason why the net current should have a defined polarity. As a result, the bias currents of a bias-compensated op amp may not only be mismatched, they can actually flow in opposite directions! In most applications this is not important, but in some it can have unexpected effects (e.g., the droop (change of voltage in the hold mode) of a sample-and-hold amplifier (SHA) built with a bias-compensated op amp may have either polarity).

In many cases, the bias current compensation feature is not mentioned on an op amp data sheet. It is easy to determine if bias current compensation is being used by examining the bias current specification. If the bias current is specified as a "±" value, the op amp is most likely compensated for bias current.

The designer can compensate for the effects of the bias current by equalizing the impedances seen by the two inputs (see Figure 1-39). If the impedances are equal, then the bias currents (which will tend to also be equal) flowing through them will produce the same offset voltage, which will appear as a common-mode signal. Since it is a common-mode signal it would tend not to add to the error due to the common-mode rejection (CMRR, to be discussed later in this section) of the amplifier.

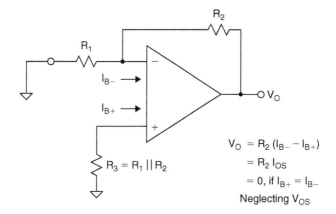

$$V_O = R_2 (I_{B-} - I_{B+})$$
$$= R_2 I_{OS}$$
$$= 0, \text{ if } I_{B+} = I_{B-}$$
Neglecting $V_{OS}$

**Figure 1-39: Bias current compensation**

Care should be used when applying this technique. It obviously will not work with a bias-compensated op amp, since the bias currents are not equal. With FET input amps, the impedance levels tend to be high and the bias currents are small, so the added effects of the Johnson noise of the high input impedances might be worse than the effects of the bias current flowing through them. Analysis needs to be performed.

## Calculating Total Output Offset Error Due to IB and $V_{OS}$

The equations shown in Figure 1-40 below are useful in referring all the offset voltage and induced offset voltage from bias current errors to the either the input (RTI) or the output (RTO) of the op amp. The choice of RTI or RTO is a matter of preference.

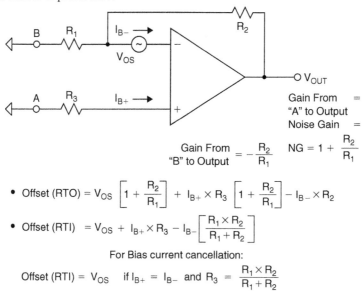

$$\text{Gain From "B" to Output} = -\frac{R_2}{R_1} \qquad NG = 1 + \frac{R_2}{R_1}$$

- Offset (RTO) $= V_{OS} \left[ 1 + \frac{R_2}{R_1} \right] + I_{B+} \times R_3 \left[ 1 + \frac{R_2}{R_1} \right] - I_{B-} \times R_2$

- Offset (RTI) $= V_{OS} + I_{B+} \times R_3 - I_{B-} \left[ \frac{R_1 \times R_2}{R_1 + R_2} \right]$

For Bias current cancellation:

$$\text{Offset (RTI)} = V_{OS} \quad \text{if } I_{B+} = I_{B-} \text{ and } R_3 = \frac{R_1 \times R_2}{R_1 + R_2}$$

**Figure 1-40: Total offset voltage calculations**

The RTI value is useful in comparing the cumulative op amp offset error to the input signal. The RTO value is more useful if the op amp drives additional circuitry, to compare the net errors with those of the next stage. In any case, the RTO value is simply obtained by multiplying the RTI value by the stage noise gain, which is $1 + R_2/R_1$.

There are some simple rules towards minimization offset voltage and bias current errors. First, keep input/feedback resistance values low, to minimize offset voltage due to bias current effects. Second, use bias compensation resistors. Bypass these resistors with fairly large values of capacitance. This gives the advantage of the resistors at DC for bias currents, but shorts out the resistances at higher frequencies to minimize noise at higher frequencies. Next, it is probably not wise to use this technique with FET input devices, since the value of the compensation resistor will likely add more noise than it will save in bias current compensation. If an op amp uses internal bias current compensation, *do not* use the compensation resistance, since the bias currents will not match. When necessary, use *external* offset trim networks, for lowest induced drift. Select an appropriate precision op amp specified for low offset and drift, as opposed to trimming.

## Input Impedance

VFB op amps normally have both differential and common-mode input impedances specified. CFB op amps normally specify the impedance to ground at each input. Different models may be used for different VFB op amps, but in the absence of other information, it is usually safe to use the model in Figure 1-41. In this model the bias currents flow into the inputs from infinite impedance current sources.

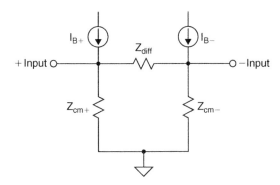

**Figure 1-41: Input impedance**

The common-mode input impedance data sheet specification ($Z_{cm+}$ and $Z_{cm-}$) is the impedance from either input to ground (NOT from both to ground). The differential input impedance ($Z_{diff}$) is the impedance between the two inputs. These impedances are usually resistive and high ($10^5$–$10^{12}\Omega$.) with some shunt capacitance (generally a few pF, sometimes 20–25 pF). In most op amp circuits, the inverting input impedance is reduced to a very low value by negative feedback, and only $Z_{cm+}$ and $Z_{diff}$ are of importance.

A CFB op amp is even more simple, as shown in Figure 1-42. Z+ is resistive, generally with some shunt capacitance, and high ($10^5$–$10^9\Omega$) while Z− is reactive (L or C, depending on the device) but has a resistive component of 10–100 $\Omega$, varying from type to type.

## Input Capacitance

In general, the input capacitance is not an issue with high speed op amps. In certain applications, such as a photodiode amp, where the source impedance is high, it could come into play. With a very large source

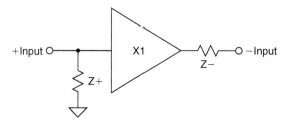

**Figure 1-42: CFB input resistance**

impedance, a relatively small capacitance could set up a zero in the transmission function. This could lead to instability. Since the noise gain of the amplifier is rising a 6 dB/octave, and the open-loop gain is falling at 6 dB/octave, the intersection will be at 12 dB/octave, which is unstable.

Another issue with FET input devices driven from a high impedance source in the non-inverting configuration is the modulation of the input capacitance by the common-mode voltage. This leads to a level dependent distortion. To compensate for this effect, balancing the impedances as seen by the inputs is used. This is similar to the balancing used for input bias current, except that the balance is not just for DC.

## Input Common-Mode Voltage Range

The input common-mode range is the allowable voltage on the input pins. It usually is not the full supply range. Classical system design used $\pm 15$ V supplies with an expected dynamic range of $\pm 10$ V, so the inputs really needed only to cover those ranges.

However, the current trend is to smaller and smaller supply voltages. This increases the need to maximize the input dynamic range. Many low voltage op amps utilize "rail-to-rail" inputs. While there is no industry standard definition for "rail-to rail," at ADI it is defined as swinging within 100 mV of either rail. Note that not all devices marketed as single supply are rail-to-rail, and not all devices that are marketed as rail-to-rail are able to swing to the rails on both input and output. You must read the data sheet carefully.

Certain inputs, such as bias-compensated and super-beta op amps, will further limit the input voltage range.

## Differential Input Voltage

Certain input structures require limiting of differential input voltage to prevent damage. These op amps will generally have back-to-back diodes across the inputs. This will not always show up in the simplified schematics of the amps. It will show up, however, as a differential input voltage specification of $\pm 700$ mV maximum.

In addition you may find a specification for the maximum input differential current. Some amps have current limiting resistors built in, but these resistors raise the noise, so for low noise op amps there are left off.

## Supply Voltage

Classical system design was $\pm 15$ V supplies with an expected signal dynamic range of $\pm 10$ V. Most early op amps were designed to operate on these voltages. The supply voltage typically had a very wide range. On the data sheet a range of allowable supply voltages was generally listed. It could be something like $\pm 4.5$ V to $\pm 18$ V, which is the specification for the AD712. In general there are some small changes in the specifications for the same op amp operated on different supplies. This usually shows up as multiple specification pages, each at a different set of conditions, which usually means different supplies.

Although the voltage specification was generally given as a symmetrical bipolar voltage, there is no reason that it had to be either symmetrical or bipolar. To the op amp a ±15V supply is the same as a +30/0V supply or a +20/−10V supply, as long as the inputs are biased in the active region (within the common-mode range).

The current trend is to lower supply voltages. For high speed amps this is partially due to process limitations. Higher speeds imply small physical structures, which, in turn, imply lower breakdown voltages. Lower breakdown voltages imply lower supply voltages. Currently most high speed op amps require ±5V or single +5V supplies. For general purpose op amps, supplies are getting as low as +1.8V. Note that the term single supply is sometimes used to indicate lower supply voltages. The two concepts are related, but, as pointed out above, single supply does not necessarily mean low voltage. Keep the concepts separate.

CMOS op amps are also generally operated with lower supplies. The trend in CMOS processes, again driven by digital circuits, emphasizes small and smaller geometries, with their attendant lower breakdown voltages.

## Quiescent Current

The quiescent current is the current internally consumed by the op amp itself (no load). In general, high speed amps tend to draw more quiescent current than general purpose amps. In addition, for general purpose op amps, some performance parameters (noise and distortion in particular) tend to improve with higher current. On the other end of the spectrum, the lowest quiescent current amps have severely limited bandwidth. Currently the lowest quiescent current device from ADI is the OP-290 at 3.5 µA.

There is a strong demand for low quiescent current op amps. One driving application is battery powered equipment. While there is no industry standard for what "low power" means, at ADI "low power" is defined as less than 1 mA of quiescent current. "Micropower" is defined as less than 100 µA quiescent current. Note that this is per amplifier, so a quad op amp will draw 4× this amount. Also note that this applies only to amplifiers. Low power can mean many things to many people. For instance a very high speed analog-to-digital converter (ADC) may dissipate over 1 W! This can still be considered low power, since competing solutions can be over 4 W.

## Output Voltage Swing (Output Voltage High/Output Voltage Low)

As pointed out above, classical system design used ±15V supplies with an expected dynamic range of ±10V. The standard output structure was an emitter follower (common collector) circuit. The base is a diode drop above the output. There must be some voltage above that for biasing the drive signal. So we need a specification on how much voltage we can expect from the output. If using reduced supply voltages, this specification for overhead will remain constant. For example, if the specification is ±12V (minimum) on a ±15V supply, we should expect to achieve ±6V on a ±9V supply.

Again, as we shrink the supply voltage, we need to maximize the output dynamic range. After all, if we lose 3V to each of the supply rails, as in the example above, and we are operating on a ±3V supply, we will have a severely compressed dynamic range. What is typically done to increase the dynamic range is to change the configuration of the output stage from an emitter follower to a common emitter. The output will then be able to swing to within the $V_{CEsat}$ of the output transistor.

Allowing the output to swing close to the rail is referred to as "rail-to-rail." As we discussed in the input voltage section, there is no industry standard rail-to-rail specification. At ADI we define it, again, as being able to swing within 100 mV of either rail, with the added constraint of driving a 10 kΩ load. The value of the load is important, since the $V_{CEsat}$ of the output transistor is dependent on output current. Remember not all "single supply" op amps are "rail-to-rail" and not all "rail-to-rail" are so on both input and output. You must read the data sheet.

## Output Current (Short Circuit Current)

Most general purpose op amps have output stages which are protected against short circuits to ground or to either supply. This is commonly referred to as "infinite" short circuit protection, since the amplifier can drive that value of current into the short circuit indefinitely. The output current that can be expected to be delivered by the op amp is the output current. Typically the limit is set so that the op amp can deliver 10 mA for general purpose op amps.

If an op amp is required to have both high precision and a large output current, it is advisable to use a separate output stage (within the feedback loop) to minimize self-heating of the precision op amp. This added amplifier is often called a buffer, since it typically will have a voltage gain = 1.

There are some op amps that are designed to give large output currents. An example is the AD8534, which is a quad device that has an output current of 250 mA for each of the four sections. A word of warning: if you try to supply 250 mA from of all four sections at the same time, you will exceed the package dissipation specification. The amp will overheat, and could destroy itself. This problem gets worse with smaller packages, which have lower dissipation.

High speed op amps typically do not have output currents limited to a low value, since it would affect their slew rate and ability to drive low impedances. Most high speed op amps will source and sink between 50–100 mA, though a few are limited to less than 30 mA. Even for high speed op amps that have short circuit protection, junction temperatures may be exceeded (because of the high short circuit current) resulting in device damage for prolonged shorts.

## *AC Specifications*

### Noise

This section discusses the noise generated within op amps, not external noise which they may pick up. External noise is important, and is discussed in detail in other texts, but in this section we are concerned solely with internal noise.

There are three noise sources in an op amp: a voltage noise, which appears differentially across the two inputs and a current noise in each input. These sources are effectively uncorrelated (independent of each other). In fact, there is a slight correlation between the two noise currents, but it is too small to need consideration in practical noise analyses. In addition to these three internal noise sources, it is necessary to consider the Johnson noise of the external resistors, which are used with the op amp in the feedback network.

### Voltage Noise

The voltage noise of different op amps may vary from under 1 to $20 \, \text{nV}/\sqrt{\text{Hz}}$, or even more. Bipolar op amps tend to have lower voltage noise than JFET amps. Voltage noise is specified on the data sheet, and it is not possible to predict it from other parameters (Figure 1-43).

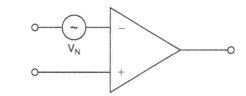

Figure 1-43: Voltage noise

Until recently, JFET input amplifiers tended to have comparatively high voltage noise (though they have very low current noise), and were thus more suitable for low noise applications in high impedance rather than low impedance circuitry. The AD645 and AD743/AD745 have very low values of both voltage and current noise. The AD645 specifications at 10 kHz are $10 \, nV/\sqrt{Hz}$ and $0.6 \, fA/\sqrt{Hz}$, and the AD743/AD745 specifications at 10 kHz are $2.9 \, nV/\sqrt{Hz}$ and $6.9 \, fA/\sqrt{Hz}$. These make possible the design of low noise amplifier circuits, which have low noise over a wide range of source impedances. The cost of the lower voltage noise is large input devices, and hence large input capacitance.

## Noise Bandwidth

When calculating the bandwidth of the noise contribution we always use a bandwidth of $1.57 \, f_c$ to calculate the noise. The reason for this is that a Gaussian (white) noise source passed through a single pole filter with a cutoff frequency of $f_c$ has the same spectral energy as the same source passed through a brick wall filter with a cutoff frequency of $1.57 \, f_c$. A brick wall filter has a flat response up to the cutoff frequency above which it has infinite attenuation. Similarly, a two pole filter has an apparent corner frequency of approximately $1.2 \, f_c$. The error correction factor is usually negligible for filters having more than two poles (Figure 1-44).

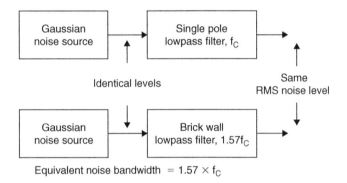

Figure 1-44: Equivalent noise bandwidth

## Noise Figure

Noise figure is rarely used with op amps. The *noise figure* of an amplifier is the amount (in dB) by which the noise of the amplifier exceeds the noise of a perfect noise-free amplifier in the same environment. The concept comes from RF and TV applications, where 50 or $75 \, \Omega$ transmission lines and terminations are ubiquitous, but is useless for an op amp, which may be used with a wide variety of impedances. Voltage noise spectral density and current noise spectral density are much more useful specifications.

## Current Noise

Current noise can vary much more widely, from around $0.1 \, fA/\sqrt{Hz}$ (in JFET electrometer op amps) to several $pA/\sqrt{Hz}$ (in high speed bipolar op amps). It is not always specified on data sheets, but may be calculated in cases (like simple BJT or JFET input devices) where all the bias current flows in the input junction, because in these cases it is simply the Schottky (or shot) noise of the bias current. It cannot be calculated for bias-compensated or CFB op amps, where the external bias current is the *difference* of two internal current sources. The shot noise spectral density is simply $\sqrt{2I_{bq}}/\sqrt{Hz}$, where $I_B$ is the bias current (in amps) and q is the charge on an electron ($1.6 \times 10^{-19}$ C) (Figure 1-45).

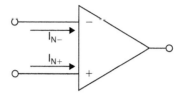

**Figure 1-45: Current noise**

The current noise for the inputs of a VFB op amp are uncorrelated and roughly equal in value. In the simple input structures, the current noise is the shot noise of the input bias current. In a Bias-Compensated op amp, the current noise cannot be calculated. Also, since the inputs of a CFB op amp are different, the current noise for the two inputs can be very different. The 1/f corners will typically not match either.

Current noise is only important when it flows in an impedance and generates a noise voltage. Therefore, the choice of a low noise op amp depends on the impedances around it. Consider an OP-27, a bias-compensated op amp with low voltage noise ($3\,\text{nV}/\sqrt{\text{Hz}}$), but quite high current noise ($1\,\text{pA}/\sqrt{\text{Hz}}$). With zero source impedance, the voltage noise will dominate. With a source resistance of $3\,\text{k}\Omega$, the current noise ($1\,\text{pA}/\sqrt{\text{Hz}}$ flowing in $3\,\text{k}\Omega$) will equal the voltage noise, but the Johnson noise of the $3\,\text{k}\Omega$ resistor is $7\,\text{nV}/\sqrt{\text{Hz}}$ and so is dominant. With a source resistance of $300\,\text{k}\Omega$, the current noise increases a hundred-fold to $300\,\text{nV}/\sqrt{\text{Hz}}$, while the voltage noise continues unchanged, and the Johnson noise (which is proportional to the *square root* of the resistance) only increases ten-fold. Here, current noise is dominant.

### Total Noise (Sum of Noise Sources)

Uncorrelated noise voltages add in a "root-sum-of-squares" manner; i.e., noise voltages $V_1$, $V_2$, $V_3$ give a summed result of $\sqrt{V_1^2 + V_2^2 + V_3^2}$ . Noise powers, of course, add normally. Thus, any noise voltage that is more than 3–5 times any of the others is dominant, and the others may generally be ignored. This simplifies noise assessment. Current noises flowing through resistance equal noise voltages (Figure 1-46).

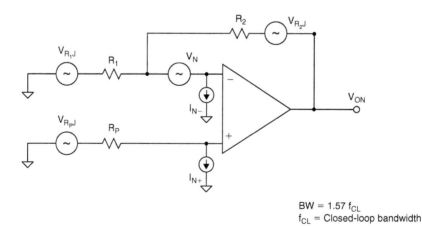

$$BW = 1.57\,f_{CL}$$
$$f_{CL} = \text{Closed-loop bandwidth}$$

$$V_{ON} = \sqrt{BW}\,\sqrt{[(I_{N-}{}^2)R_2^2]\,[NG] + [(I_{N+}{}^2)R_p^2]\,[NG] + V_N^2\,[NG] + 4kTR_2\,[NG-1] + 4kTR_1\,[NG-1] + 4kTR_p\,[NG]}$$

**Figure 1-46: Total noise calculation**

The choice of a low noise op amp depends on the source impedance of the signal, and at high impedances, current noise always dominates.

For low impedance circuitry, amplifiers with low voltage noise, such as the OP-27, will be the obvious choice, since they are inexpensive, and their comparatively large current noise will not affect the application (see Figure 1-47). At medium resistances, the Johnson noise of resistors is dominant, while at very high resistances, we must choose an op amp with the smallest possible current noise, such as the FET input devices AD549 or AD645.

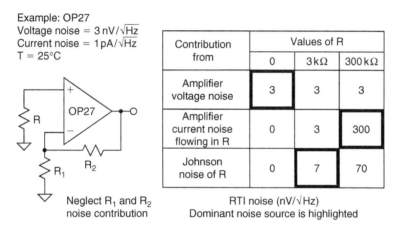

Example: OP27
Voltage noise $= 3\,nV/\sqrt{Hz}$
Current noise $= 1\,pA/\sqrt{Hz}$
$T = 25°C$

Neglect $R_1$ and $R_2$ noise contribution

| Contribution from | Values of R | | |
|---|---|---|---|
| | 0 | 3 kΩ | 300 kΩ |
| Amplifier voltage noise | 3 | 3 | 3 |
| Amplifier current noise flowing in R | 0 | 3 | 300 |
| Johnson noise of R | 0 | 7 | 70 |

RTI noise (nV/√Hz)
Dominant noise source is highlighted

**Figure 1-47: Dominant noise source determined by input impedance**

### 1/f Noise (Flicker Noise)

So far, we have assumed that noise is *white* (i.e., its spectral density does not vary with frequency). This is true over most of an op amp's frequency range, but at low frequencies the noise spectral density rises at 3 dB/octave, as shown in Figure 1-48. The power spectral density in this region is inversely proportional to frequency, and therefore the voltage noise spectral density is inversely proportional to the square root of the frequency. For this reason, this noise is commonly referred to as *1/f noise*. Note, however, that some textbooks still use the older term *flicker noise*.

The frequency at which this noise starts to rise is known as the *1/f corner frequency* ($F_C$) and is a figure of merit—the lower it is, the better. The 1/f corner frequencies are not necessarily the same for the voltage noise and the current noise of a particular amplifier, and a CFB op amp may have three 1/f corners: for its voltage noise, its inverting input current noise, and its non-inverting input current noise.

The general equation which describes the voltage or current noise spectral density in the 1/f region is:

$$e_n, i_n = k\sqrt{F_C}\,\sqrt{\frac{1}{f}} \tag{1-15}$$

where k is the level of the "white" current or voltage noise level, and $F_C$ is the 1/f corner frequency.

The best low frequency low noise amplifiers have corner frequencies in the range 1–10 Hz, while JFET devices and more general purpose op amps have values in the range 100 Hz to sometimes over 1 kHz. Very fast amplifiers, however, may make compromises in processing to achieve high speed which result in quite poor 1/f corners of several hundred Hz or even 1–2 kHz. This is generally unimportant in the wideband

41

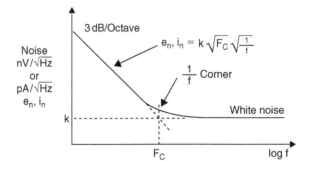

- 1/f corner frequency is a figure of merit for op amp noise performance (the lower the better)
- Typical ranges: 2 Hz to 2 kHz
- Voltage noise and current noise do not necessarily have the same 1/f corner frequency

**Figure 1-48: 1/f noise bandwidth**

applications for which they were intended, but may affect their use at audio frequencies, particularly in equalization circuits (Figure 1-49).

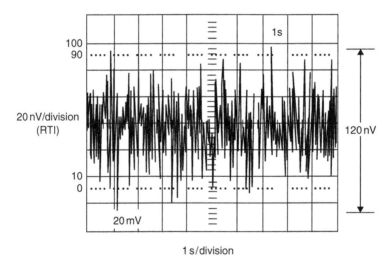

**Figure 1-49: Noise in the 0.1–10 Hz bandwidth for the OP-213**

## Popcorn Noise

*Popcorn noise* is so-called because when played through an audio system, it sounds like cooking popcorn. It consists of random step changes of offset voltage that take place at random intervals in the 10+ millisecond timeframe. Such noise results from high levels of contamination and crystal lattice dislocation at the surface of the silicon chip, which in turn results from inappropriate processing techniques or poor quality raw materials. When monolithic op amps were first introduced in the 1960s, popcorn noise was a dominant noise source. Today, however, the causes of popcorn noise are well understood, raw material purity is high, contamination is low, and production tests for it are reliable so that no op amp manufacturer should have any difficulty in shipping products that are substantially free of popcorn noise. For this reason, it is not even mentioned in most modern op amp textbooks or data sheets.

## RMS Noise Considerations

As was discussed above, noise spectral density is a function of frequency. In order to obtain the RMS noise, the noise spectral density curve must be integrated over the bandwidth of interest.

In the 1/f region, the RMS noise in the bandwidth $f_1$ to $f_2$ is given by:

$$e_{RMS} = \sqrt{\int_{f_1}^{f_2} \frac{df}{f}} = k\sqrt{\ln \frac{f_2}{f_1}} \qquad (1\text{-}16)$$

where k is the noise spectral density at 1 Hz. The total 1/f noise in a given band is a function of the ratio of the low and high band edge frequencies, since the actual frequency cancels out. It is necessary, however, that the upper band edge is still in the 1/f region for the above formula to be accurate.

It is often desirable to convert RMS noise measurements into peak-to-peak. In order to do this, one must have some understanding of the statistical nature of noise. For Gaussian noise and a given value of RMS noise, statistics tell us that the chance of a particular peak-to-peak value being exceeded decreases sharply as that value increases—but this probability never becomes zero.

Thus, for a given RMS noise, it is possible to predict the percentage of time that a given peak-to-peak value will be exceeded, but it is not possible to give a peak-to-peak value which will never be exceeded as shown in Figure 1-50.

| Nominal peak-to-peak | % of the time noise will exceed nominal peak-to peak value |
|---|---|
| 2 × RMS | 32% |
| 3 × RMS | 13% |
| 4 × RMS | 4.6% |
| 5 × RMS | 1.2% |
| 6 × RMS | 0.27% |
| 6.6 × RMS** | 0.10% |
| 7 × RMS | 0.046% |
| 8 × RMS | 0.006% |

** Most often used conversion factor is 6.6

**Figure 1-50: RMS to peak-to-peak voltage comparison chart**

Peak-to-peak noise specifications, therefore, must always be given for a specified time limit. The most common choice is for the peak-to-peak noise to be 6.6 times the RMS value, which is means the peak-to-peak level will be exceeded only 0.1% of the time.

In many cases, the low frequency noise is specified as a peak-to-peak value within the bandwidth 0.1–10 Hz. This is measured by inserting a 0.1–10 Hz bandpass filter between the op amp and the measuring device. The measurement is often presented as a scope photo with a time scale of 1 s/division as shown in Figure 1-51 for the OP-213.

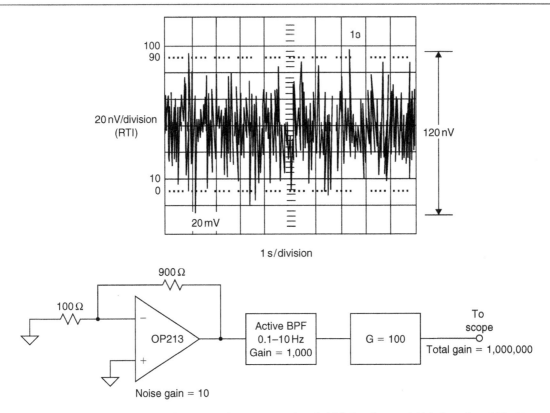

**Figure 1-51: The peak-to-peak noise in the 0.1–10 Hz bandwidth for the OP-213 is less than 120 nV**

In practice, it is virtually impossible to measure noise within specific frequency limits with no contribution from outside those limits, since practical filters have finite rolloff characteristics. Fortunately, the measurement error introduced by a single pole lowpass filter is readily computed. See the previous section on noise bandwidth.

When computing RMS noise for wide bandwidth op amps, 1/f noise becomes relatively insignificant. The dominant source of noise is *Gaussian*, or white noise. This noise has a relatively constant noise spectral density over a wide range of frequencies. The RMS noise calculation is made by multiplying the noise spectral density by the square root of the equivalent noise bandwidth.

## Total Output Noise Calculations

We have already pointed out that any noise source which produces less than one-third to one-fifth of the noise of some other source can be ignored. (Both noise voltages must be measured at the same point in the circuit.) To analyze the noise performance of an op amp circuit, we must assess the noise contributions of each part of the circuit and determine which are significant. To simplify the following calculations, we shall work with noise spectral densities, rather than actual voltages, to leave bandwidth out of the expressions (the noise spectral density, which is generally expressed in $\mu V/\sqrt{Hz}$, is equivalent to the noise in a 1 Hz bandwidth).

All resistors have a Johnson noise of $\sqrt{4\,kTBR}$ , where k is Boltzmann's Constant ($1.38 \times 10^{-23}$ J/K), T is the absolute temperature, B is the bandwidth and R is the resistance. This is intrinsic—it is not possible to obtain resistors which do not have Johnson noise (unless operated a 0°K) (Figure 1-52).

- All resistors have a voltage noise of $V_{NR} = \sqrt{(4kTBR)}$
- T = Absolute temperature = T (°C) + 273.15
- B = Bandwidth (Hz)
- k = Boltzmann's constant ($1.38 \times 10^{-23}$ J/K)
- A $1,000\,\Omega$ resistor generates $4\,nV/\sqrt{Hz}$ @ 25°C

**Figure 1-52: Resistor noise**

If we consider the circuit in Figure 1-53, which is an amplifier consisting of an op amp and three resistors ($R_p$ represents the source resistance at node A), we can find six separate noise sources: the Johnson noise of the three resistors, the op amp voltage noise, and the current noise in each input of the op amp. Each has its own contribution to the noise at the amplifier output. (Noise is generally specified RTI, or *referred to the input*, but it is often simpler to calculate the noise at the output and then divide it by the *signal* gain [not the *noise* gain] of the amplifier to obtain the RTI noise.)

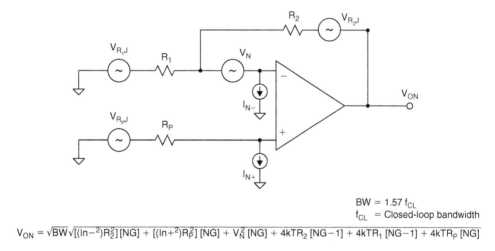

BW = 1.57 $f_{CL}$
$f_{CL}$ = Closed-loop bandwidth

$$V_{ON} = \sqrt{BW}\sqrt{[(In-^2)R_2^2][NG] + [(In+^2)R_P^2][NG] + V_N^2[NG] + 4kTR_2[NG-1] + 4kTR_1[NG-1] + 4kTR_P[NG]}$$

**Figure 1-53: Total noise calculation**

The circuit in Figure 1-54 represents a second-order system, where capacitor $C_1$ represents the source capacitance, stray capacitance on the inverting input, the input capacitance of the op amp, or any combination of these. $C_1$ causes a breakpoint in the noise gain, and $C_2$ is the capacitor which must be added to obtain stability. Because of $C_1$ and $C_2$, the noise gain is a function of frequency, and has peaking at the higher frequencies (assuming $C_2$ is selected to make the second-order system critically damped).

A DC signal applied to input A (B being grounded) sees a gain of:

$$1 + R_2/R_1 = \text{DC Noise Gain} \tag{1-17}$$

At higher frequencies, the gain from input A to the output becomes:

$$1 + C_2/C_1 = \text{AC Noise Gain} \tag{1-18}$$

45

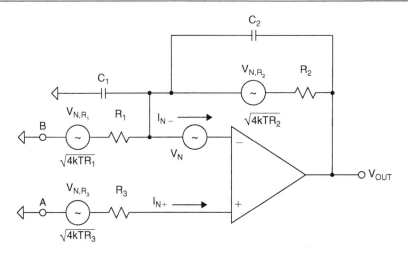

**Figure 1-54: Second-order noise model**

The closed-loop bandwidth $f_{cl}$ is the point at which the Noise Gain intersects the open-loop gain.

A DC signal applied to B (A being grounded) sees a gain of:

$$-R_2R_1 \qquad\qquad (1\text{-}19)$$

with a high frequency cutoff determined by $R_2C_2$:

$$\text{Bandwidth (B to Output)} = \frac{1}{2}\pi R_2 C_2 \qquad\qquad (1\text{-}20)$$

These are the non-inverting and inverting gains and bandwidths, respectively, of the amplifier (Figure 1-55).

The current noise of the non-inverting input, $I_{n+}$, flows in $R_p$ and gives rise to a noise voltage of $I_n + R_p$, which is amplified by Eqs (1-17 and 1-18), as are the op amp noise voltage, $V_n$, and the Johnson noise of

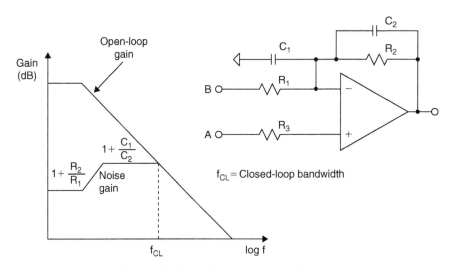

**Figure 1-55: Second-order system noise gain**

46

$R_p$, which is $\sqrt{4\,kTR_p}$ . The Johnson noise of $R_1$ is amplified by Eq. (1-19) over a bandwidth of $1/2(\pi R_2 C_2)$ Eq. (1-20), and the Johnson noise of $R_2$ is not amplified at all but is buffered directly to the output over a bandwidth of $1/2(\pi R_2 C_2)$. The current noise of the inverting input, $I_{n-}$, does not flow in $R_1$, as might be expected—negative feedback around the amplifier works to keep the potential at the inverting input unchanged, so that a current flowing from that pin is forced, by negative feedback, to flow in $R_2$ only, resulting in a voltage at the amplifier output of $I_{n-}R_2$ over a bandwidth of $1/2(\pi R_2 C_2)$ (we could equally well consider the voltage caused by $I_{n-}$ flowing in the parallel combination of $R_1$ and $R_2$ and then amplified by the noise gain of the amplifier [see below], but the results are identical—only the calculations are more involved).

If we consider these six noise contributions, we see that if $R_p$ and $R_2$ are low, then the effect of current noise and Johnson noise will be minimized, and the dominant noise will be the op amp's voltage noise. As we increase resistance, both Johnson noise and the voltage noise produced by noise currents will rise. If noise currents are low, then Johnson noise will take over from voltage noise as the dominant contributor. Johnson noise, however, rises with the square root of the resistance, while the current noise voltage rises linearly with resistance, so ultimately, as the resistance continues to rise, the voltage due to noise currents will become dominant.

These noise contributions we have analyzed are not affected by whether the input is connected to node A or node B (the other being grounded or connected to some other low impedance voltage source), which is why the non-inverting gain $(1 + R_2/R_1)$, which is seen by the voltage noise of the op amp, $V_n$, is known as the "noise gain" of the amplifier.

Calculating the total output RMS noise of the op amp requires multiplying each of the six noise voltages by the appropriate gain and integrating over the appropriate frequency. The root-sum-square of all the output contributions then represents the total RMS output noise. Fortunately, this cumbersome exercise may be greatly simplified in most cases by making the appropriate assumptions.

The noise gain for a typical second-order system is shown in Figure 1-56. It is quite easy to perform the voltage noise integration in two steps, but notice that because of peaking, the majority of the output noise

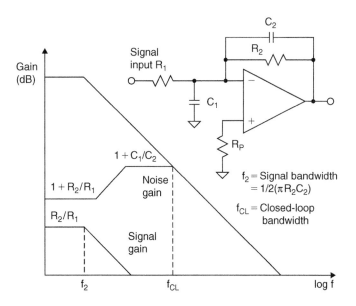

**Figure 1-56: Noise and signal gain for a second-order system**

47

due to the input voltage noise will be determined by the high frequency portion where the noise gain is $1 + C_1/C_2$. This type of response is typical of second-order systems. The noise due to the inverting input current noise, $R_1$, and $R_2$ is only integrated over the bandwidth $1/2(\pi R_2 C_2)$.

In high speed op amp applications, there are some further simplifications which can be made. The noise gain plot for a first-order system optimized for fast settling time is usually flat up to the closed-loop bandwidth frequency, with only a dB or so of gain peaking at the most. All noise sources may therefore be integrated over the closed-loop op amp bandwidth.

In high speed CFB op amp circuits, the input voltage noise and the inverting input current noise are the dominant contributors to the output noise as shown in Figure 1-57.

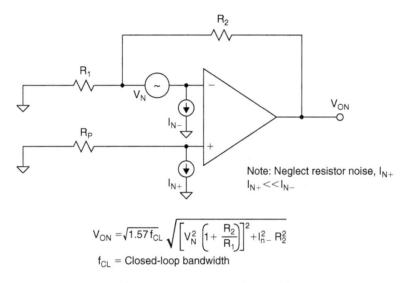

$$V_{ON} = \sqrt{1.57 f_{CL}} \sqrt{\left[ V_N^2 \left( 1 + \frac{R_2}{R_1} \right) \right]^2 + I_{n-}^2 \, R_2^2}$$

$f_{CL}$ = Closed-loop bandwidth

**Figure 1-57: CFB amp noise model**

## Distortion

Dynamic range of an op amp may be defined in several ways. One of the most common ways is to specify *harmonic distortion, total harmonic distortion* (THD), or *total harmonic distortion plus noise* (THD + N). Other related specifications include *intermodulation distortion* (IMD), *intercept points* (IP), *spurious free dynamic range* (SFDR), *multi-tone power ratio* (MTPR) among others.

### Total Harmonic Distortion

THD is the ratio of the harmonically related ($2\times$, $3\times$, $4\times$, and so on the fundamental frequency) signal components caused by amplifier nonlinearity. Only the harmonically related signals are included in the measurement. The distortion components which make up THD are usually calculated by taking the square root of the sum of the squares of the first five or six harmonics of the fundamental. In many practical situations, however, there is negligible error if only the second and third harmonics are included since the higher order terms most often are greatly reduced in amplitude.

### Total Harmonic Distortion Plus Noise

THD + N is the residual signal with only the fundamental removed. It is important to note that the THD measurement does not include noise terms, while THD + N does. The noise in the THD + N measurement must be integrated over the measurement bandwidth. In narrow-band applications, the level of the noise may be reduced by filtering, in turn lowering the THD + N which increases the signal to noise ratio (SNR). Most times when a THD specification is quoted, it is really a THD + N specification, since most measurement systems do not differentiate harmonically related signals from the other signals. The THD measurement is generally made by notching out the fundamental signal and measuring the remaining signal (the residual). The definition of THD and THD + N is shown in Figure 1-58.

- $V_s$ = Signal Amplitude (RMS Volts)
- $V_2$ = Second Harmonic Amplitude (RMS Volts)
- $V_n$ = nth Harmonic Amplitude (RMS Volts)
- $V_{noise}$ = RMS value of noise over measurement bandwidth

- THD + N = $\dfrac{\sqrt{V_2^2 + V_3^2 + V_4^2 + \ldots + V_n^2 + V_{noise}^2}}{V_s}$

- THD = $\dfrac{\sqrt{V_2^2 + V_3^2 + V_4^2 + \ldots + V_n^2}}{V_s}$

**Figure 1-58: THD and THD + N definitions**

## Intermodulation Distortion

Rather than simply examining the THD produced by a single tone sine wave input, it is often useful to look at the distortion products produced by two tones. As shown in Figure 1-59 two tones will produce intermodulation products. Intermodulation occurs when two (or more) signals are passed through a nonlinear system. And all systems are nonlinear, to some degree. Intermodulation products consist of sum and difference frequencies. The example shows the second- and third-order products produced by applying two frequencies, $f_1$ and $f_2$, to a nonlinear system. The second-order products located at $f_2 + f_1$ and $f_2 - f_1$ are located relatively far away from the two tones, and may possibly be removed by filtering, depending on the bandwidth of the system. If the system is wideband, these distortion products may still be in band. The third-order products located at $2f_1 + f_2$ and $2f_2 + f_1$ may likewise possibly be filtered. The third-order

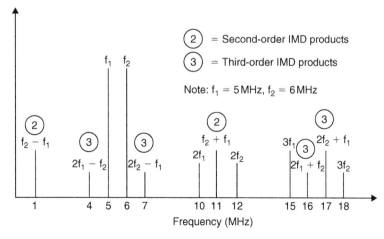

**Figure 1-59: IMD products**

49

products located at $2f_1 - f_2$ and $2f_2 - f_1$, however, are close to the original tones, and filtering them is difficult.

## Third-Order Intercept Point (IP3), Second-Order Intercept Point (IP2)

IMD products are of special interest in the RF area, and a major concern in the design of radio receivers. Third-order IMD products can mask out small signals in the presence of larger adjacent ones. Third-order IMD is often specified in terms of the *third-order intercept point* (IP3) as shown in Figure 1-60.

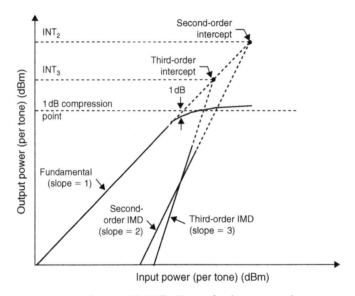

**Figure 1-60: IMD, IPs, and gain compression**

If the system nonlinearity is approximated by a power series expansion, the second-order IMD amplitudes increase 2 dB for every 1 dB of signal increase. Similarly, the third-order IMD amplitudes increase 3 dB for every 1 dB of signal increase. Once the input reaches a certain level, however, the output signal begins to soft limit, or compress due to things like power supply limits, output drive maximums, and the like. But the second- and third-order intercept lines may be extended to intersect the extension of the output signal line. These intersections are called the *second-* and *third-order intercept points*, respectively. The values are usually referenced to the output power of the device expressed in dBm. So, while the IP3 point most often will never be reached in practice, it is still used as a figure of merit in high speed systems.

To determine the IP3 point, two spectrally pure tones are applied to the system. The output signal power of a single tone (in dBm) as well as the relative amplitude of the third-order products (referenced to a single tone) is plotted as a function of input signal power. With a low level (well below clipping) two-tone input signal, and two data points, draw the second- and third-order IMD lines as are shown in Figure 1-60, because one point and a slope determine each straight line. Where they intersect will be the *second-* and *third-order intercept points*, respectively.

Figure 1-61 shows the third order intercept value as a function of frequency for a typical VFB amplifier.

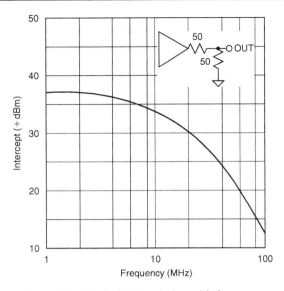

**Figure 1-61: Typical IP3 variation with frequency**

Assume the op amp output signal is 5 MHz and 2 V peak-to-peak into a 100 Ω load (50 Ω source and load termination). The voltage into the 50 Ω load is therefore 1 V peak-to peak, corresponding to +4 dBm. The value of the third-order intercept at 5 MHz is 36 dBm. The difference between +36 dBm and +4 dBm is 32 dB. This value is then multiplied by 2 to yield 64 dB (the value of the third-order intermodulation products referenced to the power in a single tone). Therefore, the intermodulation products should be −64 dBc (dB below carrier frequency), or at a level of −60 dBm. Figure 1-60 shows the graphical analysis for this example.

## 1 dB Compression Point

Another parameter, which may be of interest, is the 1 dB *compression point*. This is the point at which the output signal is compressed by 1 dB from the ideal input/output transfer function. This occurs when the dynamic range of the amplifier output is reached and the output will not increase no matter how much the input to the amplifier increases (i.e., clipping). This point is also shown in Figure 1-60.

## Signal to Noise Ratio

The SNR is the dynamic range of the system, usually expressed in dB. The reference level is the maximum signal level, and the RMS level of the noise is the floor. The bandwidth of the measurement must be specified.

## Equivalent Number of Bits

If we take the SNR of the op amp and express it in bit we have equivalent number of bits (ENOBs). The conversion formula is:

$$ENOB = \frac{SNR \text{ (in dB)} - 1.76}{6.02} \tag{1-21}$$

Although we would mainly think of bits in converter applications, they are sometimes used in the context of op amps. Again, the bandwidth of the measurement must be specified.

## Spurious Free Dynamic Range

SFDR is another measure of the dynamic range of the system. It can be measured two ways. The first is the difference between the maximum signal and the first distortion component of any type. It would be measured in dB. This would be the SFDR in dBFS. The other way to measure it is in relation to the actual signal strength. This would be the SFDR in dBc (meaning relative to the carrier). While this is again more commonly a converter specification, we sometimes see SFDR used in reference to op amps (Figure 1-62).

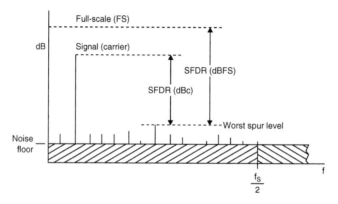

**Figure 1-62: Spurious free dynamic range**

## Slew Rate

The slew rate of an amplifier is the maximum rate of change of voltage at its output. It is expressed in V/s (or, more probably, V/μs). Op amps may have different slew rates during positive and negative going transitions, due to circuit design, but for this analysis we shall assume that good fast op amps have reasonably symmetrical slew rates.

If we consider a sine wave with a p-p amplitude of $2V_p$ and frequency f, the expression for the output voltage is:

$$v(t) = V_p \sin 2\pi ft \qquad (1-22)$$

This has a maximum slew rate:

$$\left.\frac{dv}{dt}\right|_{max} = 2\pi f V_p \qquad (1-23)$$

One note here, many high speed amplifiers will have overshoot. This means the output will go beyond the final value and will then have a damped oscillation around the final value. This is call "ringing." The amount of overshoot and ringing will be an indication of the phase margin of the amplifier. The higher the overshoot and the more ringing, the less phase margin.

The slew rate is generally measured between 10% and 90% of the final value (although 20–80% is sometimes also used) (Figure 1-63).

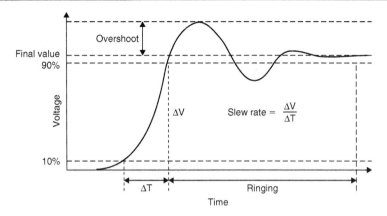

**Figure 1-63: Slew rate**

## Full Power Bandwidth

The minimum output frequency at which slew limiting *occurs* is directly proportional to slew rate and inversely proportional to the amplitude of the signal. This allows us to define the "full power bandwidth" (FPBW) of an op amp:

$$FPBW = Slew\ rate/2\pi V_p \qquad (1\text{-}24)$$

It is important to realize that both slew rate and FPBW can also depend somewhat on the power supply voltage being used and the load the amplifier is driving (particularly capacitive).

In practice, the FPBW of the op amp should be approximately 5–10 times the maximum output frequency in order to achieve acceptable distortion performance (Figure 1-64).

Slew rate = Maximum rate at which the output
voltage of an op amp can change

Ranges: A few volts/μs to several thousand volts/μs

For a sinewave, $V_{OUT} = V_p \sin 2\pi ft$

$dV/dt = 2\pi f V_p \cos 2\pi ft$

$(dv/dt)_{max} = 2\pi f V_p$

If $2 V_p$ = full output span of op amp, then

Slew rate = $(dV/dt)_{max} = 2\pi \times FPBW \times V_p$

$FPBW = Slew\ rate/2\pi V_p$

**Figure 1-64: Slew rate and FPBW**

## −3 dB Small Signal Bandwidth

The −3 dB bandwidth of an op amp will almost always be greater than the FPBW. This is because the signal does not have to swing as far. Since $V_p$ is reduced, the bandwidth is increased.

## Bandwidth for 0.1 dB Flatness

In demanding applications such as professional video, it is desirable to maintain a relatively flat bandwidth and linear phase up to some maximum specified frequency. This is because a change in gain or phase of the system will affect the color intensity or hue.

Simply specifying the 3 dB bandwidth is not enough. It has become customary to specify the *0.1 dB bandwidth*, or *0.1 dB bandwidth flatness*. This means there is no more than 0.1 dB ripple up to a specified 0.1 dB bandwidth frequency. Video buffer amplifiers generally have both the 3 dB and the 0.1 dB bandwidth specified. Figure 1-65 shows the frequency response of the AD8075 triple video buffer.

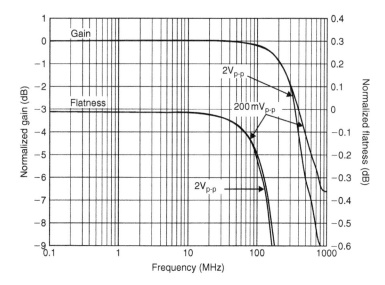

**Figure 1-65: 0.1 dB gain flatness**

Note that the 3 dB bandwidth is approximately 400 MHz. This can be determined from the response labeled "GAIN" in the graph, and the corresponding gain scale is shown on the left-hand vertical axis (at a scaling of 1 dB/division). The response scale for "FLATNESS" is on the right-hand vertical axis, at a scaling of 0.1 dB/division in this case. This allows the 0.1 dB bandwidth to be determined, which is about 65 MHz in this case. The major difference in the applicable bandwidth between 3 and 0.1 dB criteria. It requires a 400 MHz bandwidth amplifier (as conventionally measured) to provide the 65 MHz 0.1 dB flatness rating.

It should be noted that these specifications hold true when driving a 75 Ω source and load terminated cable, which represents a resistive load of 150 Ω. Any capacitive loading at the amplifier output could cause peaking in the frequency response and should be avoided.

## Gain-Bandwidth Product

For a VFB amplifier, if the gain at any particular frequency is multiplied by that frequency, the product is a constant. This is because in a first-order system a doubling of frequency causes a reduction in gain by a factor of 2. Therefore this product becomes a useful figure of merit in comparing the bandwidth of op amps (Figure 1-66).

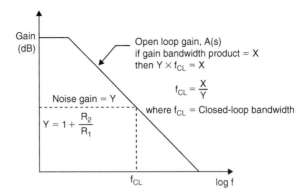

**Figure 1-66: Gain-bandwidth product**

## CFB Frequency Dependence

CFB op amps do not behave in the same way as VFB types. They are not stable with capacitive feedback, nor are they so with a short circuit from output to inverting input. With a CFB op amp, *there is an optimum feedback resistance for maximum bandwidth*. Note that the value of this resistance may vary with supply voltage. If the feedback resistance is increased, the bandwidth is reduced. Conversely, if it is reduced, bandwidth increases, and the amplifier may become unstable (Figure 1-67).

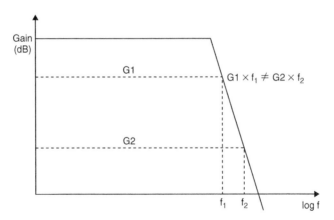

**Figure 1-67: CFB op amp open-loop gain**

In a CFB op amp, for a given value of feedback resistance, *the closed-loop bandwidth is largely unaffected by the noise gain*, as shown in Figure 1-67. Thus it is not correct to refer to gain-bandwidth product, for a CFB amplifier, because of the fact that it is not constant. Gain is manipulated in a CFB op amp application by choosing the correct feedback resistor for the device, and then selecting the input resistor to yield the desired closed-loop gain. The signal gain (as determined by the feedback network) of a CFB amplifier is identical to the case of a VFB op amp.

Typically, CFB op amp data sheets will provide a table of recommended resistor values, which provide maximum bandwidth for the device, over a range of gain, supply voltage, and package type. It simplifies the design process considerably to use these tables (see Figure 1-68).

| Component | AD8001AN (PDIP) Gain | | | | | AD8001AR (SOIC) Gain | | | | | AD8001ART (SOT-23-5) Gain | | | | |
|---|---|---|---|---|---|---|---|---|---|---|---|---|---|---|---|
| | −1 | +1 | +2 | +10 | +100 | −1 | +1 | +2 | +10 | +100 | −1 | +1 | +2 | +10 | +100 |
| $R_F$ (Ω) | 649 | 1050 | 750 | 470 | 1000 | 604 | 953 | 681 | 470 | 1000 | 845 | 1000 | 768 | 470 | 1000 |
| $R_G$ (Ω) | 649 | | 750 | 51 | 10 | 604 | | 681 | 51 | 10 | 845 | | 768 | 51 | 10 |
| $R_O$ (Nominal) (Ω) | 49.9 | 49.9 | 49.9 | 49.9 | 49.9 | 49.9 | 49.9 | 49.9 | 49.9 | 49.9 | 49.9 | 49.9 | 49.9 | 49.9 | 49.9 |
| $R_S$ (Ω) | 0 | | | | | 0 | | | | | 0 | | | | |
| $R_T$ (Nominal) (Ω) | 54.9 | 49.9 | 49.9 | 49.9 | 49.9 | 54.9 | 49.9 | 49.9 | 49.9 | 49.9 | 54.9 | 49.9 | 49.9 | 49.9 | 49.9 |
| Small signal BW (MHz) | 340 | 880 | 460 | 260 | 20 | 370 | 710 | 440 | 260 | 20 | 240 | 795 | 380 | 260 | 20 |
| 0.1 dB Flatness (MHz) | 105 | 70 | 105 | | | 130 | 100 | 120 | | | 110 | 300 | 145 | | |

**Figure 1-68: Recommended feedback resistor values for the AD8001**

## Settling Time

The settling time of an amplifier is defined as the time it takes the output to respond to a step change of input and come within *and remain within* a defined error band, as measured relative to the 50% point of the input pulse (see Figure 1-69). There is no natural error band for an op amp (a DAC naturally has an error band of 1 LSB, or perhaps ±1 LSB), so one must be chosen and defined. What is chosen will depend on the performance of the op amp, but since the value chosen will vary from device to device, comparisons are very difficult. This is true because settling is not linear, and many different time constants may be involved. Examples are early op amps using dielectrically isolated (DI) processes. These had very fast settling to 1% of full-scale, but they took almost forever to settle to 10 bits (0.1%). Similarly, some very high precision op amps have thermal effects which cause settling to 0.001% or better to take tens of ms, although they will settle to 0.025% in a few μs.

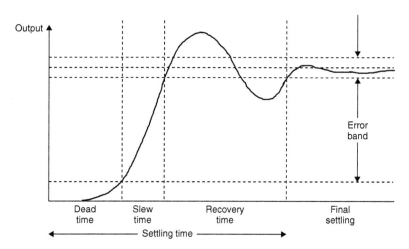

Error band is usually defined to be a percentage of the step 0.1%, 0.05%, 0.01%, etc.

Settling time is non-linear; it may take 30 times as long to settle to 0.01% as to 0.1%.

Manufacturers often choose an error band which makes the op amp look good.

**Figure 1-69: Settling time**

56

It should also be noted that thermal effects could cause significant differences between short-term settling time (generally measured in nanoseconds) and long-term settling time (generally measured in microseconds or milliseconds). In many AC applications, long-term settling time is not important; but if it is, it must be measured on a much different time scale than short-term settling time.

### Rise Time and Fall Time

For high speed op amps we might also have a specification for rise and fall times. While ideally they should be the same, there is typically some difference in practical op amps. Rise and fall times are measured by applying a square wave to the op amp and would be measured on the output waveform. This is closely related to slew rate. Also, as is done with slew rate, we generally measure between the 10% and 90% points, so that overshoot and ringing generally do not enter into the picture. The input wave generally will be full-scale, but occasionally it is specified for a smaller input signal. Overall rise and fall times are a less revealing specification than slew rate and settling time.

### Phase Margin

Phase margin is the amount of phase shift when the (VFB) amplifier's gain passes through 0 dB. It is basically a measure of how close the second pole of the system is to causing instability. Phase starts to change on the order of a decade before the corner frequency. The phase shift must be less than 180°. The phase margin is the 180°—the actual phase shift of the amplifier. Anything greater than 45° is usually acceptable. The higher the phase margin, the more stable the system. Capacitive loading will reduce the phase margin.

The graph in Figure 1-70, taken from the data sheet for the AD8054, shows that when the open-loop gain (left scale) falls below 0 dB, the phase margin is around 45° (right scale). This is a respectable value for phase margin. In general, you should avoid phase margins below 20° to 25°.

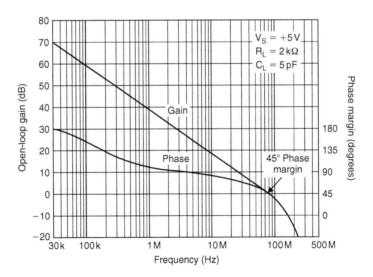

**Figure 1-70: Phase margin for the AD8054**

### Common-Mode Rejection Ratio

If a signal is applied equally to both inputs of an op amp, so that the differential input voltage is unaffected, the output should not be affected. In practice, changes in common-mode voltage will produce changes in output. The op amp *common mode rejection ratio* (CMRR) is the ratio of the common-mode gain to differential-mode gain. For example, if a differential input change of Y volts produces a change of 1 V at the output, and a common-mode change of X volts produces a similar change of 1 V, then the CMRR is X/Y. When the CMRR is expressed in dB, it is generally referred to as common-mode rejection (CMR). Typical low frequency CMR values can be between 70 and 120 dB, but at higher frequencies, CMR deteriorates. In addition to a CMRR numeric specification, many op amp data sheets show a plot of CMR versus frequency, as shown in Figure 1-71 for an OP-177 op amp.

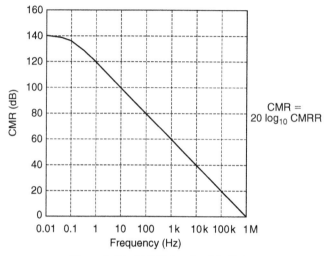

**Figure 1-71: CMRR for the OP-177**

CMRR produces a corresponding output offset voltage error in op amps configured in the non-inverting mode.

Note inverting-mode operating op amps will have less CMRR error. Since both inputs are held at a ground (or virtual ground), there is no common-mode dynamic voltage.

## Power Supply Rejection Ratio

If the supply of an op amp changes, its output should not, but it typically does. The specification of *power supply rejection ratio* or PSRR is defined similarly to the definition of CMRR. If a change of X volts in the supply produces the same output change as a differential input change of Y volts, then the PSRR on that supply is X/Y. The definition of PSRR assumes that both supplies are altered equally in opposite directions; otherwise, the change will introduce a common-mode change as well as a supply change, and the analysis becomes considerably more complex. It is this effect which causes apparent differences in PSRR between the positive and negative supplies (Figure 1-72).

Because op amp PSRR is frequency dependent, op amp power supplies must be well decoupled. At low frequencies, several devices may share a 10–50 µF capacitor on each supply, provided it is no more than 10 cm (PC track distance) from any of them.

At high frequencies, each IC should have the supply leads decoupled by a low inductance 0.1 µF (or so) capacitor with short leads and PC tracks. These capacitors must also provide a return path for high frequency currents in the op amp load. Typical decoupling circuits are shown in Figure 1-73. Further bypassing and decoupling information is found in Chapter 12.

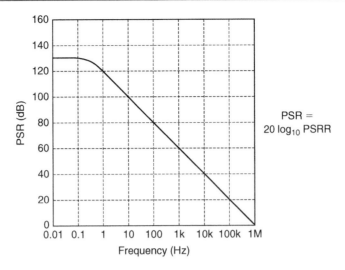

Figure 1-72: Power supply rejection ratio

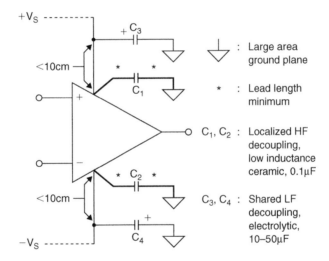

Figure 1-73: Recommended power supply decoupling

### Differential Gain

Differential gain is a specification that originated for video applications. In early video processing equipment it was found that there was sometimes a change in the gain of the amplifier with DC level. More correctly, differential gain is the change in the color saturation level (amplitude of the color modulation) for a change in low frequency luma (brightness) amplitude. This modulation is obviously a distortion, changing the intensity of the color. Professional video editing equipment commonly strives to keep the total differential gain of the system below 1%. Modern high performance video op amps have differential gain specifications of <0.01% (Figures 1-74 to 1-79).

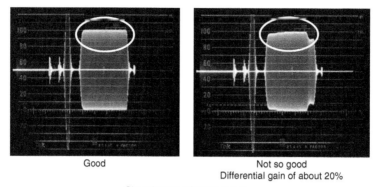

Good

Not so good
Differential gain of about 20%

Chrominance information only
Luminance information filtered out
Unfortunately, phase information is not so easily displayed

**Figure 1-74: Differential gain example**

AD829-specifications (@ $T_A$ = +25°C and $V_S$ = ±15 V dc, unless otherwise noted)

| Differential gain error[3] | $R_{Load}$ = 100Ω $C_{Comp}$ = 30pF | ±15V | 0.02 | 0.02 | % |
|---|---|---|---|---|---|
| Differential phase error[3] | $R_{Load}$ = 100Ω $C_{Comp}$ = 30 pF | ±15V | 0.04 | 0.04 | Degrees |

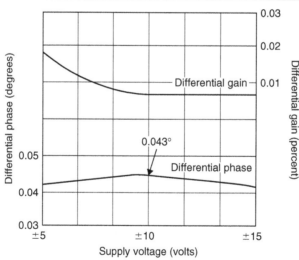

**Figure 1-75: Differential gain and differential phase specifications**

## Differential Phase

Differential phase is the change in hue (phase of the color modulation) for a change in low frequency luma (brightness) amplitude. This modulation is obviously a distortion, changing the hue of the color. Professional video editing equipment commonly strives to keep the total differential phase of the system below 1°. Modern, high performance video op amps have differential gain specifications of <0.01°.

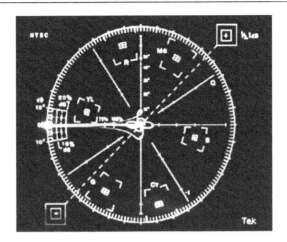

Figure 1-76: Vectorscope display of a "good" signal

Note smearing
of display line

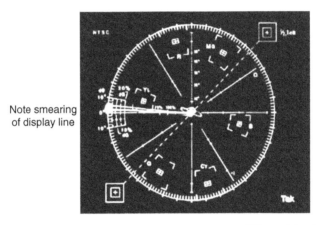

Figure 1-77: Vectorscope display showing ~15% differential gain

Note curve
on display

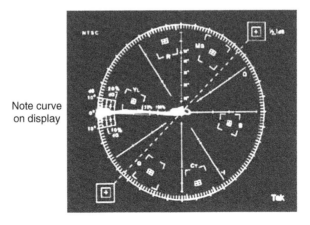

Figure 1-78: Vectorscope display showing ~5° differential phase

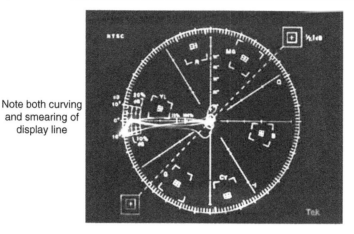

Note both curving and smearing of display line

**Figure 1-79: Vectorscope display showing ~10% differential gain and ~9° differential phase**

## Phase Reversal

Phase reversal is a problem that occurs in some op amp when the input common mode of an op amp is exceeded. The mechanism is that one of the internal stages of the op amp no longer has a bias voltage across it and subsequently turns off. The effect is that the output waveform swings to the opposite rail until the input comes back into the common-mode range (see Figure 1-80). This became a big problem with the move toward lower supply voltages and single supplies. Advances in circuit design have resulted in op amps that do not suffer from phase reversal. If the op amp is designed to avoid phase reversal it is generally noted in the bullets or "Key Features" and not necessarily in the specification table.

Input                  Output

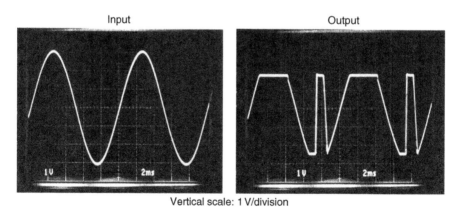

Vertical scale: 1 V/division
Horizontal scale: 2 ms/division

**Figure 1-80: Phase reversal**

## Channel Separation

Channel separation, otherwise known as crosstalk, is a signal that couples from one amplifier in a package to another amplifier in the same package. The path is typically through the power supply, which will typically be shared between the amplifiers. Careful layout of the op amp chip can minimize the crosstalk. Careful external bypassing of the power supplies can also help.

## Absolute Maximum Ratings

The absolute maximum ratings are the voltage, current, and temperature limits of the op amp. Exceeding the absolute maximums can lead to the destruction of the op amp (see Figure 1-81).

**Absolute Maximum Ratings[1]**
Supply voltage . . . . . . . . . . . . . . . . . . . . . . . . . . . . . . . . . . . . . . . .12.6 V
Internal power dissipation @ 25°C [2]
    PDIP package (N) . . . . . . . . . . . . . . . . . . . . . . . . . . . . . . . . . . . . .1.3 W
    SOIC (R) . . . . . . . . . . . . . . . . . . . . . . . . . . . . . . . . . . . . . . . . . . .0.8 W
    8-Lead CERDIP . . . . . . . . . . . . . . . . . . . . . . . . . . . . . . . . . . . . .1.1 W
    SOT-23-5 package (RT) . . . . . . . . . . . . . . . . . . . . . . . . . . . . . . .0.5 W
Input voltage (common mode) . . . . . . . . . . . . . . . . . . . . . . . . . . . . . .$\pm V_S$
Differential input voltage . . . . . . . . . . . . . . . . . . . . . . . . . . . . . . . .$\pm 1.2$ V
Output short circuit duration
    . . . . . . . . . . . . . . . . . . . . . . . . . . . . . Observe power derating curves
Storage temperature range N, R . . . . . . . . . . . . . . . . . . −65°C to +125°C
Operating temperature range (A Grade) . . . . . . . . . . . . . −40°C to +85°C
Lead temperature range (Soldering 10 sec) . . . . . . . . . . . . . . . . . .300°C

NOTES
[1] Stresses above those listed under Absolute Maximum Ratings may cause permanent damage to the device. This is a stress rating only; functional operation of the device at these or any other conditions above those indicated in the operational section of this specification is not implied. Exposure to absolute maximum rating conditions for extended periods may effect device reliability.
[2] Specification is for device in free air:
  8-Lead PDIP Package: $\theta_{JA} = 90°C/W$
  8-Lead SOIC Package: $\theta_{JA} = 155°C/W$
  8-Lead CERDIP Package: $\theta_{JA} = 110°C/W$
  5-Lead SOT-23-5 Package: $\theta_{JA} = 260°C/W$

**Figure 1-81: Typical absolute maximum ratings (from AD8001)**

Applying overvoltage to input pins is one very common way to destroy an op amp. Overvoltage conditions can be broken into two groups, overvoltage and electrostatic discharge (ESD).

ESD voltages typically run to the thousand of volts. Most of us have experienced ESD. Just shuffle your feet across a nylon carpet, especially in a dry environment, and touch a metal doorknob. Sparks will fly from your fingertips. CMOS circuits are especially prone to ESD.

Overvoltages occur when the maximum voltage allowed on the op amp is exceeded. The maximum allowable voltage is typically set by the supply voltage, although there are a few exceptions. An overvoltage on the inputs will typically cause the input devices to turn into a SCR type structure, usually through the substrate. The failure mechanism is not the overvoltage per se, but instead the current that the overvoltage causes to flow. So if the current is limited, no catastrophic damage will be done. The rule-of-thumb is to limit the current to 5 mA.

While no catastrophic damage will be done, continually overstressing the inputs can cause a change in parameters like bias current and offset voltage. So even though you will not necessarily destroy the amp, overvoltage should be avoided.

Protection for overvoltage can consist of diodes from the input pins to the supplies and current limiting resistors. The diodes are typically Schottky diodes, used because of their lower forward voltage (typically

300 mV versus 700 mV for silicon). Protection devices should be applied with caution though. Some diodes can be leaky, which causes issues similar to those of bias currents. Some can also have fairly high capacitance, which may limit frequency response. This is especially true for high speed amps. Current limiting resistors raise the noise floor. Some op amps, such as the OP-27, include protection diodes, but still require current limiting. If an op amp has protection diodes, it will typically have a specification for maximum differential input current. The protection circuit should also show up on the simplified schematic (Figure 1-82).

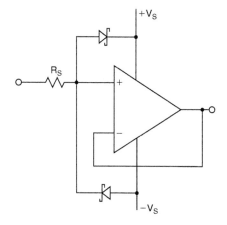

**Figure 1-82: Input protection**

Some op amps also have back to back diodes across the inputs. These are not for input overvoltage protection, but to limit the differential voltage. If these exist, there will be an absolute maximum specification of $\pm 700$ mV for the differential input voltage.

The overriding specification for temperature is the maximum junction temperature of 150°C. As this limit is approached, the life expectancy of the amp (actually any semiconductor) goes down (Figure 1-83).

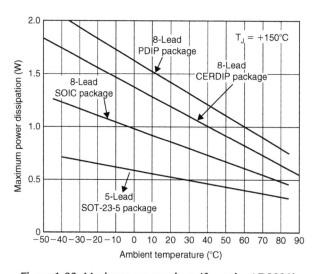

**Figure 1-83: Maximum power chart (from the AD8001)**

The temperature gradient between the junction and the case is based on the thermal resistance of the package, which is called $\theta_{JC}$. There is also a thermal resistance, $\theta_{CA}$, from the package to the ambient. These thermal resistances add up linearly, so the total thermal resistance, $\theta_{JA}$, from the junction to the ambient is $\theta_{JC} + \theta_{CA}$.

The maximum operation temperature rating has more to do with the temperature performance range of the rest of the specifications of the op amp rather than any potential damage.

### References: Op Amp Specifications

1. J.R. Ragazzini, R.H. Randall, and F.A. Russell, "Analysis of Problems in Dynamics by Electronic Circuits," **Proceedings of the IRE**, Vol. 35, No. May, 1947, pp. 444–452.

2. W. Borlase, **An Introduction to Operational Amplifiers (Parts 1–3)**, Analog Devices Seminar Notes, Analog Devices, Inc., September 1971.

3. K.D. Swartzel, Jr., "Summing Amplifier," **US Patent 2,401,779**, filed, May 1, 1941, issued June 11, 1946.

4. F.E. Terman, "Feedback Amplifier Design," **Electronics**, Vol. 10, No. 1, January 1937, pp. 12–15, 50.

5. J.M. West, "Wave Amplifying System," **US Patent 2,196,844**, filed, April 26, 1939, issued April 9, 1940.

6. H.W. Bode, "Relations Between Attenuation and Phase In Feedback Amplifier Design," **Bell System Technical Journal**, Vol. 19, No. 3, July 1940. See also: "Amplifier," **US Patent 2,173,178**, filed, June 22, 1937, issued July 12, 1938.

7. R. Stata, **"Operational Amplifiers—Parts I and II,"** Electromechanical Design, September, November 1965.

8. D. Sheingold (Ed.), **Applications Manual for Operational Amplifiers for Modeling, Measuring, Manipulating, and Much Else**, George A. Philbrick Researchers, Inc., Boston, MA, 1965. See also: **Applications Manual for Operational Amplifiers for Modeling, Measuring, Manipulating, and Much Else**, 2nd Edition, Philbrick/Nexus Research, Dedham, MA, 1966, 1984.

9. W.G. Jung, **IC Op Amp Cookbook**, **3rd Edition**, Prentice-Hall PTR, 1986, 1997, ISBN: 0-13-889601-1.

10. W. Kester (Ed.), **Linear Design Seminar**, Analog Devices, Inc., 1995, ISBN: 0-916550-15-X.

11. S. Franco, **Design with Operational Amplifiers and Analog Integrated Circuits**, 2nd Edition, McGraw-Hill, 1998, ISBN: 0-07-021857-9.

12. "Video Op Amp," **Analog Dialogue**, Vol. 24, No. 3, 1990, pp. 19.

13. G. Erdi, "Amplifier Techniques for Combining Low Noise, Precision, and High-Speed Performance," **IEEE Journal of Solid-State Circuits**, Vol. SC-16, December, 1981, pp. 653–661.

14. G. Erdi, T. Schwartz, S. Bernardi, and W. Jung, "Op Amps Tackle Noise-and for Once, Noise Loses," **Electronic Design**, December, 1980.

15. G. Erdi, "A Precision Trim Technique for Monolithic Analog Circuits", **IEEE Journal of Solid-State Circuits**, Vol. SC-10, December, 1975, pp. 412–416.

16. R. Wagner, "Laser-Trimming on the Wafer," **Analog Dialogue**, Vol. 9, No. 3, 1975, pp. 3–5.

17. D. Soderquist and G. Erdi, "The OP-07 Ultra-Low Offset Voltage Op Amp," **Precision Monolithics AN-13**, December, 1975.

18. W. Kester (Ed.), **High Speed Design Techniques**, Analog Devices, 1996, ISBN: 0-916550-17-6 (available for download at http://www.analog.com).

19. W. Kester (Ed.), **Practical Analog Design Techniques**, Analog Devices, 1995, ISBN: 0-916550-16-8 (available for download at http://www.analog.com).

20. W. Palmer and B. Hilton, "A 500 V/μs 12 Bit Transimpedance Amplifier," **ISSCC Digest**, February 1987, pp. 176–177, 386.

21. F.E. Terman, "Feedback Amplifier Design," **Electronics**, January 1937, pp. 12–15, 50.

22. E.L. Ginzton, "DC Amplifier Design Techniques," **Electronics**, March 1944, pp. 98–102.

23. S.E. Miller, "Sensitive DC Amplifier with AC Operation," **Electronics**, November 1941, pp. 27–31, 105–109.

24. J.O. Edson and H.H. Henning, "Broadband Codecs for an Experimental 224 Mb/s PCM Terminal," **Bell System Technical Journal**, Vol. 44, No. 9, November 1965, pp. 1887–1950.

25. "Op Amps Combine Superb DC Precision and Fast Settling," **Analog Dialogue**, Vol. 22, No. 2, 1988, pp. 12–15.

26. D.A. Nelson, "Settling Time Reduction in Wide-Band Direct-Coupled Transistor Amplifiers," **US Patent 4,502,020**, filed, October 26, 1983, issued February 26, 1985.

27. R.A. Gosser, "DC-Coupled Transimpedance Amplifier," **US Patent 4,970,470**, filed, October 10, 1989, issued November 13, 1990.

28. J.L. Melsa and D.G. Schultz, **Linear Control Systems**, McGraw-Hill, 1969, pp. 196–220, ISBN: 0-07-041481-5

29. L. Smith and D. Sheingold, "Noise and Operational Amplifier Circuits," **Analog Dialogue**, Vol. 3, No.1, pp. 1, 5–16. See also: **Analog Dialogue 25th Anniversary Issue**, pp. 19–31, 1991.

30. T.M. Frederiksen, **Intuitive Operational Amplifiers**, McGraw-Hill, 1988, ISBN: 0-07-021966-4.

31. J.K. Roberge, **Operational Amplifiers–Theory and Practice**, John Wiley, 1975, ISBN: 0-471-72585-4.

32. D. Stout and M. Kaufman, **Handbook of Operational Amplifier Circuit Design**, McGraw-Hill, New York, 1976, ISBN: 0-07-061797-X.

33. J. Dostal, **Operational Amplifiers**, Elsevier Scientific Publishing, New York, 1981, ISBN: 0-444-99760-1.

34. P.R. Gray and R.G. Meyer, **Analysis and Design of Analog Integrated Circuits, 3rd Edition**, John Wiley, 1993, ISBN: 0-471-57495-3.

35. R. Gosser, "Wide-Band Transconductance Generator," **US Patent 5,150,074**, filed, May 3, 1991, issued September 22, 1992.

36. R. Gosser, "DC-Coupled Transimpedance Amplifier," **US Patent 4,970,470**, filed, October 10, 1989, issued November 13, 1990.

37. Data sheet for **AD8011 300 MHz, 1 mA Current Feedback Amplifier**, http://www.analog.com.

38. W. Kester (Ed.), **Amplifier Applications Guide**, Analog Devices, 1992, ISBN: 0-916550-10-9.

39. W. Kester (Ed.), **Practical Design Techniques for Sensor Signal Conditioning**, Analog Devices, 1999, ISBN: 0-916550-20-6.

40. Data Sheet for **AD8551/AD8552/AD8554 Zero-Drift, Single-Supply, Rail-to-Rail Input/Output Operational Amplifiers**, http://www.analog.com.

41. Data Sheet for **AD8571/AD8572/AD8574 Zero-Drift, Single-Supply, Rail-to-Rail Input/Output Operational Amplifiers**, http://www.analog.com.

42. Data Sheet for **OP777/OP727/OP747 Precision Micropower Single-Supply Operational Amplifiers**, http://www.analog.com.

43. Data Sheet for **OP1177/OP2177/OP4177 Precision Low Noise, Low Input Bias Current Operational Amplifiers**, http://www.analog.com.

# How to Read a Data Sheet

While there is not an industry standard concerning the format of data sheets, what they cover, what information is included and where that information is located, for the most part data sheets from various manufacturers generally are similar in construction. In this section we will take a look at several data sheets and try to give a feel for where to find certain information and how to interpret what is found.

As a demonstration we will look at five data sheets: a precision amp (OP-1177/OP-2177/OP-4177), a single supply amp (AD8531/AD8532/AD8534), a high speed VFB amp (AD8051/AD8052/AD8054), a CFB amp (AD8001), and the AD847. The part numbers chosen are arbitrary; they were chosen only to give a range of parts.

## The Front Page

This page is designed to give you the basic information you might need to choose the part. Referring to Figure 1-84, we can break the front page up into three sections.

Section 1 is the features. These bullet points are what are considered by the manufacturer to be the more important parameters of the product for its intended application. The targeted applications are typically listed as well.

Section 2 is the product description. This typically covers some of what the manufacturer considers to be the salient features of the op amp.

The third section is the functional block diagram. For an op amp, this is typically the pin out of the various packages. For more complex parts, it will truly be a block diagram.

## The Specification Tables

There are an unlimited number of conditions possible when measuring any given specification. Obviously, it is not possible to test all possible conditions. So a representative set of conditions are chosen. The test conditions are specified (1 in Figure 1-85). Occasionally if further clarification of or modification to the conditions is required, it is handled as a footnote (2 in Figure 1-85).

In some cases, when the op amp is specified over a large range of conditions, there may be several specification pages. Each would have a different set of conditions. For instance, an op amp may be specified with a ±15V power supply, a ±5V power supply, or a +5V only supply. See the AD8051/AD8052/ AD8054 data sheet as an example (Figures 1-86 to 1-88).

On many op amps some individual specifications may have multiple entries. This is for different performance levels. It can also be for different temperature ranges (usually commercial, industrial, or military). This can be seen in Figure 1-85 (3).

Note that there are typically three possibilities for the specifications, Min, Typ, and Max (see Figure 1-85 (3)). At ADI any specification in the min (minimum) and max (maximum) columns will be guaranteed by test. This can be a direct test, or, in some instances, testing one parameter will guarantee another. A typ (typical) specification is just that, typical. Depending on the particular specification, the deviation from

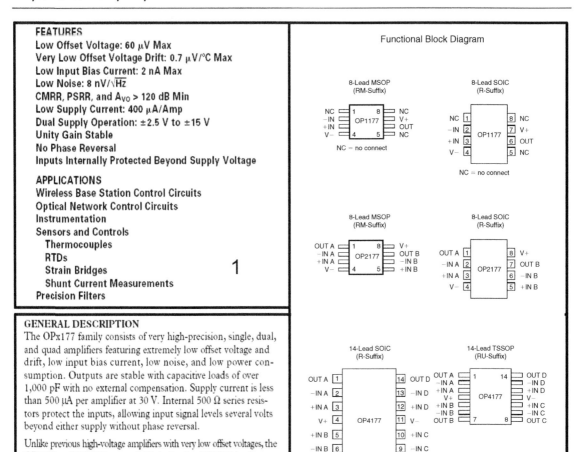

**FEATURES**
Low Offset Voltage: 60 μV Max
Very Low Offset Voltage Drift: 0.7 μV/°C Max
Low Input Bias Current: 2 nA Max
Low Noise: 8 nV/√Hz
CMRR, PSRR, and $A_{VO}$ > 120 dB Min
Low Supply Current: 400 μA/Amp
Dual Supply Operation: ±2.5 V to ±15 V
Unity Gain Stable
No Phase Reversal
Inputs Internally Protected Beyond Supply Voltage

**APPLICATIONS**
Wireless Base Station Control Circuits
Optical Network Control Circuits
Instrumentation
Sensors and Controls
    Thermocouples
    RTDs
    Strain Bridges
    Shunt Current Measurements
Precision Filters

1

**GENERAL DESCRIPTION**

The OPx177 family consists of very high-precision, single, dual, and quad amplifiers featuring extremely low offset voltage and drift, low input bias current, low noise, and low power consumption. Outputs are stable with capacitive loads of over 1,000 pF with no external compensation. Supply current is less than 500 μA per amplifier at 30 V. Internal 500 Ω series resistors protect the inputs, allowing input signal levels several volts beyond either supply without phase reversal.

Unlike previous high-voltage amplifiers with very low offset voltages, the OP1177 and OP2177 are available in the tiny MSOP 8-lead surface-mount package, while the OP4177 is available in TSSOP14. Moreover, specified performance in the MSOP/TSSOP package is identical to performance in the SOIC package.

OPx177 family offers the widest specified temperature range of any high-precision amplifier in surface-mount packaging. All versions are fully specified for operation from –40°C to +125°C for the most demanding operating environments.

Applications for these amplifiers include precision diode power measurement, voltage and current level setting, and level detection in optical and wireless transmission systems. Additional applications include line powered and portable instrumentation

2

and controls—thermocouple, RTD, strain-bridge, and other sensor signal conditioning—and precision filters.

The OP1177 (single) and the OP2177 (dual) amplifiers are available in the 8-lead MSOP and 8-lead SOIC packages. The OP4177 (quad) is available in 14-lead narrow SOIC and 14-lead TSSOP packages. MSOP and TSSOP packages are available in tape and reel only.

**Figure 1-84: Example data sheet front page**

the typical can be substantial. And you have no way of knowing what the range of variation on the typ specification is. Sometimes you will find a typ and a min (or max) for the same specification. This tells you that although the test limits are at a particular level (min or max), the typicals tend to run much better than the test limits. When designing, using typs is risky. You are much better off using mins or maxes for error budget analysis.

1

3

| Model | Conditions | Vs | AD847J | | | AD847AR | | | Units |
|---|---|---|---|---|---|---|---|---|---|
| | | | Min | Typ | Max | Min | Typ | Max | |
| Input offset voltage[1] | | ±5 V | | 0.5 | 1 | | 0.5 | 1 | MV |
| | $T_{MIN}$ to $T_{MAX}$ | | | | 3.5 | | | 4 | mV |
| Offset drift | | | | 15 | | | 15 | | µV/°C |
| Input bias current | | ±5 V. ±15 V | | 3.3 | 6.6 | | 3.3 | 6.6 | µA |
| | $T_{MIN}$ to $T_{MAX}$ | | | | 7.2 | | | 10 | µA |
| Input offset current | | ±5 V. ±15 V | | 50 | 300 | | 50 | 300 | nA |
| | $T_{MIN}$ to $T_{MAX}$ | | | | 400 | | | 500 | nA |
| Offset current draft | | | | 0.3 | | | 0.3 | | nA/°C |
| Open-loop gain | $V_{OUT} = ±2.5$ V | ±5 V | | | | | | | |
| | $R_{LOAD} = 500 \Omega$ | | 2 | 3.5 | | 2 | 3.5 | | V/mV |
| | $T_{MIN}$ to $T_{MAX}$ | | 1 | | | 1 | | | V/mV |
| | $R_{LOAD} = 150 \Omega$ | | | 1.6 | | | 1.6 | | V/mV |
| | $V_{OUT} = ±10$ V | ±15 V | | | | | | | |
| | $R_{LOAD} = 1 k\Omega$ | | 3 | 5.5 | | 3 | 5.5 | | V/mV |
| | $T_{MIN}$ to $T_{MAX}$ | | 1.5 | | | 1.5 | | | V/mV |
| Dynamic performance | | | | | | | | | |
| Unity gain bandwidth | | ±5 V | | 35 | | | 35 | | MHz |
| | | ±15 V | | 50 | | | 50 | | MHz |
| Full powder bandwidth[2] | $V_{OUT} = 5$ V p-p. | | | | | | | | |
| | $R_{LOAD} = 500 \Omega$ | ±5 V | | 12.7 | | | 12.7 | | MHz |
| | $V_{OUT} = 20$ V p-p. | | | | | | | | |
| | $R_{LOAD} = 1 k\Omega$ | ±15 V | | 4.7 | | | 4.7 | | MHz |
| Slew rate[3] | $R_{LOAD} = 1 k\Omega$ | ±5 V | | 200 | | | 200 | | V/µS |
| | | ±15 V | 225 | 300 | | 225 | 300 | | V/µS |
| Setting time | | | | | | | | | |
| to 0.1%. $R_{LOAD} = 250 \Omega$ | −2.5 V to + 2.5 V | ±5 V | | 65 | | | 65 | | ns |
| | 10 V Step, Av = −1 | ±15 V | | 65 | | | 65 | | ns |
| to 0.01%. $R_{LOAD} = 250 \Omega$ | −2.5 V to +2.5 V | ±5 V | | 140 | | | 140 | | ns |
| | 10 V Step, Av = −1 | ±15 V | | 120 | | | 120 | | ns |
| Phase margin | $C_{LOAD} = 10$ pF | ±15 V | | | | | | | |
| | $R_{LOAD} = 1 k\Omega$ | | | 50 | | | 50 | | Degree |
| Differential gain | f = 4.4 MHz., $R_{LOAD} = 1 k\Omega$ | ±15 V | | 0.04 | | | 0.04 | | % |
| Differential phase | f = 4.4 MHz., $R_{LOAD} = 1 k\Omega$ | ±15 V | | 0.19 | | | 0.19 | | Degree |
| Common-mode | $V_{CM} = ±2.5$ V | ±5 V | 78 | 95 | | 78 | 95 | | dB |
| Rejection | $V_{CM} = ±12$ V | ±15 V | 78 | 95 | | 78 | 95 | | dB |
| | $T_{MIN}$ to $T_{MAX}$ | | 75 | | | 75 | | | dB |
| Power supply | Vs = ±5V to ±15 V | | 75 | 86 | | 75 | 86 | | dB |
| rejection | $T_{MIN}$ to $T_{MAX}$ | | 72 | | | 72 | | | dB |
| Input voltage noise | f = 10 kHz | ±15 V | | 15 | | | 15 | | nV/√Hz |
| Input current noise | f = 10 kHz | ±15 V | | 1.5 | | | 1.5 | | pA/√Hz |
| Input common-mode | | | | | | | | | |
| voltage range | | ±5 V | | +4.3 | | | +4.3 | | V |
| | | | | −3.4 | | | −3.4 | | V |
| | | ±15 V | | +14.3 | | | +14.3 | | V |
| | | | | −13.4 | | | −13.4 | | V |
| Output voltage swing | $R_{LOAD} = 500 \Omega$ | ±5 V | 3.0 | 3.6 | | 3.0 | 3.6 | | ±V |
| | $R_{LOAD} = 150 \Omega$ | ±5 V | 2.5 | 3 | | 2.5 | 3 | | ±V |
| | $R_{LOAD} = 1 k\Omega$ | ±15 V | 12 | | | 12 | | | ±V |
| | $R_{LOAD} = 500 \Omega$ | ±15 V | 10 | | | 10 | | | ±V |
| Short-Circuit Current | | ±15 V | | 32 | | | 32 | | mA |
| Input resistance | | | | 300 | | | 300 | | kΩ |
| Input capacitance | | | | 1.5 | | | 1.5 | | pF |
| Output resistance | Open loop | | | 15 | | | 15 | | Ω |
| Power supply | | | | | | | | | |
| Operating range | | | ±4.5 | | ±18 | ±4.5 | | ±18 | V |
| Quiescent current | | ±5 V | | 4.8 | 6.0 | | 4.8 | 6.0 | mA |
| | $T_{MIN}$ to $T_{MAX}$ | | | | 7.3 | | | 7.3 | mA |
| | | ±15 V | | 5.3 | 6.3 | | 5.3 | 6.3 | mA |
| | $T_{MIN}$ to $T_{MAX}$ | | | | 7.6 | | | 7.6 | mA |

NOTES

2

[1] Input offset voltage specifications are guaranteed after 5 minutes at $T_A = +25°C$
[2] Full power bandwidth = slew rate/$2\pi V_{PEAK}$
[3] Slew rate is measured on rising edge
All min and max specifications are guaranteed. Specifications in boldface are 100% tested at final electrical test.
Specifications subject to change without notice.

Figure 1-85: Example specification page

## Specifications (@ $T_A$ = 25°C $V_S$ = 5 V, $R_L$ = 2 kΩ to 2.5 V. unless otherwise noted.)

| Parameter | Conditions | AD8051A/AD8052A | | | AD8054A | | | Unit |
|---|---|---|---|---|---|---|---|---|
| | | Min | Typ | Max | Min | Typ | Max | |
| **Dynamic performance** | | | | | | | | |
| −3 dB small signal | G = +1, $V_O$ = 0.2 $V_{p-p}$ | 70 | 110 | | 80 | 150 | | MHz |
| bandwidth | G = −1, +2, $V_O$ = 0.2 $V_{p-p}$ | | 50 | | | 60 | | MHz |
| Bandwidth for 0.1 dB | G = +2, $V_O$ = 0.2 $V_{p-p}$, | | | | | | | |
| flatness | $R_L$ = 150 Ω to 2.5 V, | | | | | | | |
| | $R_F$ = 806 Ω for AD8051A/ | | 20 | | | | | MHz |
| | AD8052A | | | | | | | |
| | $R_F$ = 200 Ω for AD8054A | | | | | 12 | | MHz |
| Slew rate | G = −1, $V_O$ = 2 V Step | 100 | 145 | | 140 | 170 | | V/μs |
| Full power response | G = +1, $V_O$ = 2 $V_{p-p}$ | | 35 | | | 45 | | MHz |
| Setting time to 0.1% | G = −1, $V_O$ = 2 V Step | | 50 | | | 40 | | ns |
| **Noise/distortion** | | | | | | | | |
| **performance** | | | | | | | | |
| Total harmonic distortion* | $f_C$ = 5 MHz, $V_O$ = 2 $V_{p-p}$, G = +2 | | −67 | | | −68 | | dB |
| Input voltage noise | f = 10 kHz | | 16 | | | 16 | | nV/√Hz |
| Input current none | f = 10 kHz | | 850 | | | 850 | | fA/√Hz |
| Differential gain error | G = +2, $R_L$ = 150 Ω to 2.5 V | | 0.09 | | | 0.07 | | % |
| (NTSC) | $R_L$ = 1 kΩ to 2.5 V | | 0.03 | | | 0.02 | | % |
| Differential phase error | G = +2, $R_L$ = 150 Ω to 2.5 V | | 0.19 | | | 0.26 | | Degrees |
| (NTSC) | $R_L$ = 1 kΩ to 2.5 V | | 0.03 | | | 0.02 | | Degrees |
| Crosstalk | f = 5 MHz, G = +2 | | −60 | | | −60 | | dB |
| **DC performance** | | | | | | | | |
| Input offset voltage | | | 1.7 | 10 | | 1.7 | 12 | mV |
| | $T_{MIN}$−$T_{MAX}$ | | | 25 | | | 30 | mV |
| Offset drift | | | 10 | | | 15 | | μV/°C |
| Input bias current | | | 1.4 | 2.5 | | 2 | 4.5 | μA |
| | $T_{MIN}$−$T_{MAX}$ | | | 3.25 | | | 4.5 | μA |
| Input offset current | | | 0.1 | 0.75 | | 0.2 | 1.2 | μA |
| Open-loop gain | $R_L$ = 2 kΩ to 2.5 V | 86 | 98 | | 82 | 98 | | dB |
| | $T_{MIN}$−$T_{MAX}$ | | 96 | | | 96 | | dB |
| | $R_L$ = 150 Ω to 2.5 V | 76 | 82 | | 74 | 82 | | dB |
| | $T_{MIN}$−$T_{MAX}$ | | 78 | | | 78 | | dB |
| **Input characteristics** | | | | | | | | |
| Input resistance | | | 290 | | | 300 | | kΩ |
| Input capacitance | | | 1.4 | | | 1.5 | | pF |
| Input common-mode | | | −0.2 to +4 | | | −0.2 to +4 | | V |
| Voltage range | | | | | | | | |
| Common-mode rejection | $V_{CM}$ = 0 V to 3.5 V | 72 | 88 | | 70 | 86 | | dB |
| ratio | | | | | | | | |
| **Output characteristics** | | | | | | | | |
| Output voltage swing | $R_L$ = 10 kΩ to 2.5 V | | 0.015 to 4.985 | | | 0.03 to 4.975 | | V |
| | $R_L$ = 2 kΩ to 2.5 V | 0.1 to 4.9 | 0.025 to 4.975 | | 0.125 to 4.875 | 0.05 to 4.95 | | V |
| | $R_L$ = 150 Ω to 2.5 V | 0.3 to 4.625 | 0.2 to 4.8 | | 0.55 to 4.4 | 0.25 to 4.65 | | V |
| Output current | $V_{OUT}$ = 0.5 V to 4.5 V | | 45 | | | 30 | | mA |
| | $T_{MIN}$−$T_{MAX}$ | | 45 | | | 30 | | mA |
| Short-circuit current | Sourcing | | 80 | | | 45 | | mA |
| | Sinking | | 130 | | | 85 | | mA |
| Capacitive load drive | G = +1 (AD8051/AD8052) | | 50 | | | | | pF |
| | G = +2 (AD8054) | | | | | 40 | | pF |
| **Power supply** | | | | | | | | |
| Operating range | | 3 | | 12 | 3 | | 12 | V |
| Quiescent current / | | | 4.4 | 5 | | 2.75 | 3.275 | mA |
| amplifier | | | | | | | | |
| Power supply rejection | Δ $V_s$ = ±1 V | 70 | 80 | | 68 | 80 | | dB |
| ratio | | | | | | | | |
| **Operating** | RT, RU, | | | | | | | |
| temperature range | RN-14 | −40 | | +85 | −40 | | +85 | °C |
| | RM, RN-8 | −40 | | +125 | | | | °C |

*Refer to TPC 13.
Specifications subject to change without notice.

Figure 1-86: Example specification page 2

## Specifications (@ $T_A$ = 25°C, $V_S$ = 3 V, $R_L$ = 2 kΩ to 1.5 V, unless otherwise noted.)

| Parameter | Conditions | AD8051A/AD8052A Min | Typ | Max | AD8054A Min | Typ | Max | Unit |
|---|---|---|---|---|---|---|---|---|
| **Dynamic performance** | | | | | | | | |
| −3 dB small signal bandwidth | G = +1, $V_O$ = 0.2 V p-p | 70 | 110 | | 80 | 135 | | MHz |
| | G = −1, +2, $V_O$ = 0.2 V p-p | | 50 | | | 65 | | MHz |
| Bandwidth for 0.1 dB flatness | G = +2, $V_O$ = 0.2 V p-p, $R_L$ = 150 Ω to 2.5 V, $R_F$ = 402 Ω for AD8051A/AD8052A | | 17 | | | | | MHz |
| | $R_F$ = 200 Ω for AD8054A | | | | | 10 | | MHz |
| Slew rate | G = −1, $V_O$ = 2 V Step | 90 | 135 | | 110 | 150 | | V/μs |
| Full power response | G = +1, $V_O$ = 1 V p-p | | 65 | | | 85 | | MHz |
| Settling time to 0.1% | G = −1, $V_O$ = 2 V Step | | 55 | | | 55 | | ns |
| **Noise/distortion performance** | | | | | | | | |
| Total harmonic distortion* | $f_C$ = 5 MHz, $V_O$ = 2 V p-p, G = −1, $R_L$ = 100 Ω to 1.5 V | | −47 | | | −48 | | dB |
| Input voltage noise | f = 10 kHz | | 16 | | | 16 | | nV/√Hz |
| Input current noise | f = 10 kHz | | 600 | | | 600 | | fA/√Hz |
| Differential gain error (NTSC) | G = +2, $V_{CM}$ = 1 V, $R_L$ = 150 Ω to 1.5 V | | 0.11 | | | 0.13 | | % |
| | $R_L$ = 1 kΩ to 1.5 V | | 0.09 | | | 0.09 | | % |
| Differential phase error (NTSC) | G = +2, $V_{CM}$ = 1 V, $R_L$ = 150 Ω to 1.5 V | | 0.24 | | | 0.3 | | Degrees |
| | $R_L$ = 1 kΩ to 1.5 V | | 0.10 | | | 0.1 | | Degrees |
| Crosstalk | f = 5 MHz, G = +2 | | −60 | | | −60 | | dB |
| **DC performance** | | | | | | | | |
| Input offset voltage | | | 1.6 | 10 | | 1.6 | 12 | mV |
| | $T_{MIN}$–$T_{MAX}$ | | | 25 | | | 30 | mV |
| Offset drift | | | 10 | | | 15 | | μV/°C |
| Input bias current | | | 1.3 | 2.6 | | 2 | 4.5 | μA |
| | $T_{MIN}$–$T_{MAX}$ | | | 3.25 | | | 4.5 | μA |
| Input offset current | | | 0.15 | 0.8 | | 0.2 | 1.2 | μA |
| Open-loop gain | $R_L$ = 2 kΩ | 80 | 96 | | 80 | 96 | | dB |
| | $T_{MIN}$–$T_{MAX}$ | | 94 | | | 94 | | dB |
| | $R_L$ = 150 Ω | 74 | 82 | | 72 | 80 | | dB |
| | $T_{MIN}$–$T_{MAX}$ | | 76 | | | 76 | | dB |
| **Input characteristics** | | | | | | | | |
| Input resistance | | | 290 | | | 300 | | kΩ |
| Input capacitance | | | 1.4 | | | 1.5 | | pF |
| Input common-mode voltage range | | | −0.2 to +2 | | | −0.2 to +2 | | V |
| Common-mode rejection ratio | VCM = 0 V to 1.5 V | 72 | 88 | | 70 | 86 | | dB |
| **Output characteristics** | | | | | | | | |
| Output voltage swing | $R_L$ = 10 kΩ to 1.5 V | | 0.01 to 2.99 | | | 0.025 to 2.98 | | V |
| | $R_L$ = 2 kΩ to 1.5 V | 0.075 to 2.9 | 0.02 to 2.98 | | 0.1 to 2.9 | 0.35 to 2.965 | | V |
| | $R_L$ = 150 Ω to 1.5 V | 0.2 to 2.75 | 0.125 to 2.875 | | 0.35 to 2.55 | 0.15 to 2.75 | | V |
| Output current | $V_{OUT}$ = 0.5 V to 2.5 V | | 45 | | | 25 | | mA |
| | $T_{MIN}$–$T_{MAX}$ | | 45 | | | 25 | | mA |
| Short-circuit current | Sourcing | | 60 | | | 30 | | mA |
| | Sinking | | 90 | | | 50 | | mA |
| Capacitive load drive | G = +1 (AD8051/AD8052) | | 45 | | | | | pF |
| | G = +2 (AD8054) | | | | | 35 | | pF |
| **Power supply** | | | | | | | | |
| Operating range | | 3 | | 12 | 3 | | 12 | V |
| Quiescent current/amplifier | | | 4.2 | 4.8 | | 2.625 | 3.125 | mA |
| Power supply rejection ratio | ΔVs = 0.5 V | 68 | 80 | | 68 | 80 | | dB |
| Operating temperature range | RT, RU, RN-14 | −40 | | +85 | −40 | | +85 | °C |
| | RM, RN-8 | −40 | | +125 | | | | °C |

*Refer to TPC 13.
Specifications subject to change without notice.

**Figure 1-87: Example specification page 3**

## Specifications (@ $T_A$ = 25°C, $V_S$ = ± 5 V, $R_L$ = 2 kΩ to Ground, unless otherwise noted.)

| Parameter | Conditions | AD8051A/AD8052A | | | AD8054A | | | Unit |
|---|---|---|---|---|---|---|---|---|
| | | Min | Typ | Max | Min | Typ | Max | |
| Dynamic performance | | | | | | | | |
| −3 dB small signal | G = +1, $V_O$ = 0.2 V p-p | 70 | 110 | | 85 | 160 | | MHz |
| bandwidth | G = −1, +2, $V_O$ = 0.2 V p-p | | 50 | | | 65 | | MHz |
| Bandwidth for 0.1 dB | G = +2, $V_O$ = 0.2 V p-p, | | | | | | | |
| flatness | $R_L$ = 150 Ω, | | | | | | | |
| | $R_F$ = 1.1 kΩ for | | | | | | | |
| | AD8051A/AD8052A | | 20 | | | | | MHz |
| | $R_F$ = 200 Ω for AD8054A | | | | | 15 | | MHz |
| Slew rate | G = −1, $V_O$ = 2 V Step | 105 | 170 | | 150 | 190 | | V/µs |
| Full power response | G = +1, $V_O$ = 2 V p-p | | 40 | | | 50 | | MHz |
| Settling time to 0.1% | G = −1, $V_O$ = 2 V Step | | 50 | | | 40 | | ns |
| Noise/distortion | | | | | | | | |
| performance | | | | | | | | |
| Total harmonic distortion | $f_C$ = 5 MHz, $V_O$ = 2 V p-p, | | −71 | | | −72 | | dB |
| | G = +2 | | | | | | | |
| Input voltage noise | f = 10 kHz | | 16 | | | 16 | | nV/√Hz |
| Input current noise | f = 10 kHz | | 900 | | | 900 | | fA/√Hz |
| Differential gain error | G = +2, $R_L$ = 150 Ω | | 0.02 | | | 0.06 | | % |
| (NTSC) | $R_L$ = 1 kΩ | | 0.02 | | | 0.02 | | % |
| Differential phase error | G = + 2, $R_L$ = 150 Ω | | 0.11 | | | 0.15 | | Degrees |
| (NTSC) | $R_L$ = 1 kΩ | | 0.02 | | | 0.03 | | Degrees |
| Crosstalk | f = 5 MHz, G = +2 | | −60 | | | −60 | | dB |
| DC performance | | | | | | | | |
| Input offset voltage | | | 1.8 | 11 | | 1.8 | 13 | mV |
| | $T_{MIN}$–$T_{MAX}$ | | | 27 | | | 32 | mV |
| Offset drift | | | 10 | | | 15 | | µV/°C |
| Input bias current | | | 1.4 | 2.6 | | 2 | 4.5 | µA |
| | $T_{MIN}$–$T_{MAX}$ | | | 3.5 | | | 4.5 | µA |
| Input offset current | | | 0.1 | 0.75 | | 0.2 | 1.2 | µA |
| Open-loop gain | $R_L$ = 2 kΩ | 88 | 96 | | 84 | 96 | | dB |
| | $T_{MIN}$–$T_{MAX}$ | | 96 | | | 96 | | dB |
| | $R_L$ = 150 Ω | 78 | 82 | | 76 | 82 | | dB |
| | $T_{MIN}$–$T_{MAX}$ | | 80 | | | 80 | | dB |
| Input characteristics | | | | | | | | |
| Input resistance | | | 290 | | | 300 | | kΩ |
| Input capacitance | | | 1.4 | | | 1.5 | | pF |
| Input common-mode | | | −5.2 to +4 | | | −5.2 to +4 | | V |
| voltage range | | | | | | | | |
| Common-mode rejection | $V_{CM}$ = −5 V to +3.5 V | 72 | 88 | | 70 | 86 | | dB |
| ratio | | | | | | | | |
| Output characteristics | | | | | | | | |
| Output voltage swing | $R_L$ = 10 kΩ | | −4.98 to +4.98 | | | −4.97 to +4.97 | | V |
| | $R_L$ = 2 kΩ | −4.85 to +4.85 | −4.97 to +4.97 | | −4.8 to +4.8 | −4.9 to +4.9 | | V |
| | $R_L$ = 150 Ω | −4.45 to +4.3 | −4.6 to +4.6 | | −4.0 to +3.8 | −4.5 to +4.5 | | V |
| Output current | $V_{OUT}$ = −4.5 V to +4.5 V | | 45 | | | 30 | | mA |
| | $T_{MIN}$–$T_{MAX}$ | | 45 | | | 30 | | mA |
| Short-circuit current | Sourcing | | 100 | | | 60 | | mA |
| | Sinking | | 160 | | | 100 | | mA |
| Capacitive load drive | G = +1 (AD8051/AD8052) | | 50 | | | | | pF |
| | G = +2 (AD8054) | | | | | 40 | | pF |
| Power supply | | | | | | | | |
| Operating range | | 3 | | 12 | 3 | | 12 | V |
| Quiescent current/ | | | 4.8 | 5.5 | | 2.875 | 3.4 | mA |
| amplifier | | | | | | | | |
| Power supply rejection | $\Delta V_S$ = ±1 V | 68 | 80 | | 68 | 80 | | dB |
| ratio | | | | | | | | |
| Operating temperature | RT, RU, | | | | | | | |
| range | RN-14 | −40 | | +85 | −40 | | +85 | °C |
| | RM, RN-8 | −40 | | +125 | | | | °C |

Specifications subject to change without notice.

Figure 1-88: Example specification page 4

Testing is one of the most expensive steps in the manufacturing of op amps. Therefore a more highly specified part will typically cost more than a less completely specified part. But, in your system, the higher specified part may be required to guarantee the circuit performance.

## *The Absolute Maximums*

There is always a section just after the specification tables that contains the absolute maximum ratings. These are typically voltage and temperature related.

The process used to fabricate the op amp will typically determine the maximum supply voltage. Maximum input voltages typically are limited to the supply voltages. It should be pointed out that the supply voltage is the instantaneous value, not the average or final value. So if an op amp has voltages on its input but the supply voltage is not present (which could occur during power up when one section of the system is powered but others are not) the op amp is overvoltaged, even if when the op amp power is applied, everything is within operational limits.

Looking at Figure 1-89, the maximum input voltage specification is GND to $V_S$. The differential input voltage maximum is $\pm 6V$. Note that both of these conditions must be met. So the input pins of the op amp must be between GND and $V_S$ *and* no more than 6V from each other.

The primary concern for semiconductor reliability is to keep the junction temperature below 150 °C. There will be a $\theta_{JA}$ given for the various package options. This is the thermal resistance. The units are °C/W (see Figure 1-89). To use this information first determine the power dissipation of the package. This would be the quiescent current times the supply voltage. Then take the maximum dissipation generated by the output stage (output current times the difference between the output voltage and the supply voltage). Add these two together and

**Absolute Maximum Ratings[1]**

Supply voltage . . . . . . . . . . . . . . . . . . . . . . . . . . . . . . . . . . . . . . . . . . . . .12.6 V
Internal power dissipation[2]
    Plastic DIP package (N) . . . . . . . . . . . . . . . . . . . . . . . . . . . . . . . . .1.3 W
    Small outline package (R) . . . . . . . . . . . . . . . . . . . . . . . . . . . . . . .0.9 W
    SOT-23-5 package (RT) . . . . . . . . . . . . . . . . . . . . . . . . . . . . . . . .0.5 W
Input voltage (common mode) . . . . . . . . . . . . . . . . . . . . . . . . . . . . .$\pm V_S$
Differential input voltage . . . . . . . . . . . . . . . . . . . . . . . . . . . . . . . . .$\pm 1.2$ V
Output short circuit duration
    . . . . . . . . . . . . . . . . . . . . . . . . . . . . . Observe power derating curves
Storage temperature range N, R . . . . . . . . . . . . . . . . . . . .$-65$°C to $+125$°C
Operating temperature range (A Grade) . . . . . . . . . . . . . .$-40$°C to $+85$°C
Lead temperature range (soldering 10 sec) . . . . . . . . . . . . . . . . . . .300° C

NOTES
[1] Stresses above those listed under Absolute Maximum Ratings may cause permanent damage to the device. This is a stress rating only; functional operation of the device at these or any other conditions above those indicated in the operational section of this specification is not implied. Exposure to absolute maximum rating conditions for extended periods may effect device reliability.
[2] Specification is for device in free air:
    8-Lead plastic DIP package: $\theta_{JA} = 90$°C/W
    8-Lead SOIC package: $\theta_{JA} = 155$°C/W
    8-Lead CERDIP package: $\theta_{JA} = 110$°C/W
    5-Lead SOT-23-5 package: $\theta_{JA} = 260$°C/W

**Figure 1-89: Typical absolute maximum ratings**

you will have the total package dissipation, in Watts. Multiply the thermal resistance by the dissipation and you have the temperature rise. Start with the ambient temperature (in °C), take the rise calculated above and that will give you the junction temperature. Remember that the ambient temperature should be in operation. Circuits packaged in an enclosure, which is in turn placed in a rack with other equipment, will have an internal ambient temperature that could be significantly above the air temperature where it is located. This must be considered.

As an example, let us take the AD8534. We will assume that it is being used as a line driver. The required output voltage range is 500 mV to 5 V. The maximum output current we expect from each of the four sections is 100 mA at a maximum output voltage of 5 V. This equates to a load of 50 Ω. Let us say the circuit will operate on a supply of 5.5 V. This allows for a bit of headroom for the driver. If you plot the output voltage versus output current for an amplifier with a resistive load, the maximum dissipation is approximately 55% of the maximum (see Figure 1-90). This is due to the fact that as the output voltage increases, the dissipation voltage (the difference between the output voltage and the supply voltage) decreases, even though the current keeps increasing. Remember it is the power dissipation of the package, not the load, which will rise with increasing output voltage. The quiescent current (Iq) is 1.75 mA maximum over temperature per amplifier. For the four amplifiers, then, the total quiescent dissipation is: 38.5 mW (Iq $\times$ V$_S$ $\times$ 4). The maximum output dissipation is calculated from the following equation:

$$P_D = \frac{(V_S - 0.55 \times V_{OUT}(max))^2}{R_{LOAD}} \qquad (1\text{-}25)$$

which calculates to 150 mW per amplifier or 600 mW total. The total dissipation is therefore 638.5 mW.

We chose a TSSOP package because it was the smallest available. The $\theta_{JA}$ for this package is 240°C/W. This gives a temperature rise of 154°C (240°C/W $\times$ 638.5 mW). If the ambient temperature is assumed to be 25°C (the usual value given for room temperature), the junction temperature would be 179°C!!! This is a problem. So we can see that even though we are operating the AD8534 below what would seem to

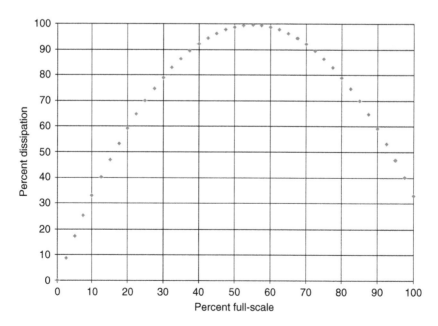

**Figure 1-90: Power dissipation versus percent full-scale**

be its maximum output current rating (which is 250 mA), the part will not be reliable since the junction temperature (150°C) will be exceeded.

$\theta_{JA}$ actually has two components, $\theta_{JC}$ (the thermal resistance from the junction to the case) and $\theta_{CA}$ (the thermal resistance from the case to the ambient). They add linearly. We cannot do anything about the $\theta_{JC}$, but by adding a heat sink we can change $\theta_{CA}$ to some degree. Most of the time with op amps, this is not an issue, but it could help for a high current output op amp in a small package as in the example above.

## The Ordering Guide

Many op amps are available in multiple packages and/or multiple temperature ranges. Each of the various combinations of package and temperature range requires a unique part number. This is spelled out in the ordering guide (see Figure 1-91.)

**Ordering guide**

| Model | Temperature range | Package description | Package option | Branding information |
|-------|-------------------|---------------------|----------------|----------------------|
| AD8531AKS* | −40°C to +85°C | 5-Lead SC70 | KS-5 | A7B |
| AD8531AR | −40°C to +85°C | 8-Lead SOIC | SO-5 | |
| AD8531ART* | −40°C to +85°C | 5-Lead SOT-23 | RT-5 | A7A |
| AD8532AR | −40°C to +85°C | 8-Lead SOIC | SO-8 | |
| AD8532ARM* | −40°C to +85°C | 8-Lead MSOP | RM-8 | ARA |
| AD8532AN | −40°C to +85°C | 8-Lead plastic DIP | N-8 | |
| AD8532ARU* | −40°C to +85°C | 8-Lead TSSOP | RU-8 | |
| AD8534AR | −40°C to +85°C | 14-Lead SOIC | SO-14 | |
| AD8534AN | −40°C to +85°C | 14-Lead plastic DIP | N-14 | |
| AD8534ARU* | −40°C to +85°C | 14-Lead TSSOP | RU-14 | |

* Available in reels only.

**Figure 1-91: Typical ordering guide**

Just as a note, in the case of op amps, the commercial (0°C to 70°C) temperature range has become much less common. The reason for this is that most circuits yield to the industrial temperature range. It is less expensive to support fewer part types. Each discrete part number requires a separate test program, separate inventorying, etc. An exception to this rule is for parts designed for a specific application which is, by definition, commercial. An example of this is consumer applications, such as audio. Wider temperature range for these parts offers no advantage.

The industrial temperature range can also mean different things. The standard industrial temperature range is −40°C to 85°C. A common variant on this is what is commonly called the automotive temperature range, −40°C to 105°C. 0°C to 100°C is also common.

The military temperature range is −55°C to 125°C.

## The Graphs

Many specifications vary over the operational range of the op amp. An example is the variation of open-loop gain with frequency (see Figure 1-92). So to completely specify the open-loop gain of a part there would be an open-loop gain specification at DC, which typically would appear in the specification table, and a graph showing variation with frequency. The information presented in the graphs is not uniform from vendor to vendor or even from part to part from the same manufacturer. Higher performance parts tend to be more completely specified. For the most part the graphs will tend to be typical values.

## The Main Body

The main body of the data sheet contains detailed information on the operation and applications of the op amp.

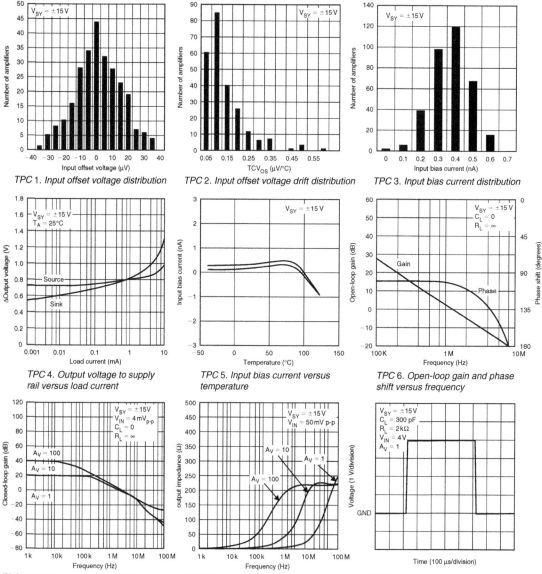

TPC 1. Input offset voltage distribution     TPC 2. Input offset voltage drift distribution     TPC 3. Input bias current distribution

TPC 4. Output voltage to supply rail versus load current     TPC 5. Input bias current versus temperature     TPC 6. Open-loop gain and phase shift versus frequency

TPC 7. Closed-loop gain versus frequency     TPC 8. Output impedance versus frequency     TPC 9. Large signal transient response

**Figure 1-92: Typical performance graphs**

The main body typically starts off with a section on the theory of operation of the part. This is usually a short description of the various specifications that are particularly to build was not the best approach. Typically simple calculations—noise, for example—are worked out as examples. The rest of the body of the data sheet contains application information. Since its founding ADI determined that just giving people an amplifier and letting them go off on their own to try to build whatever it is that they want. Therefore, ADI includes application information with the data sheet appropriate for the specific op amp. For instance a precision op amp will emphasize offset and noise, while a high speed op amp will emphasize bandwidth and speed. Much of the information in the applications section is relevant to op amps other than the one that it appears in. The last thing that is typically included in the data sheet is the package drawings (Figure 1-93).

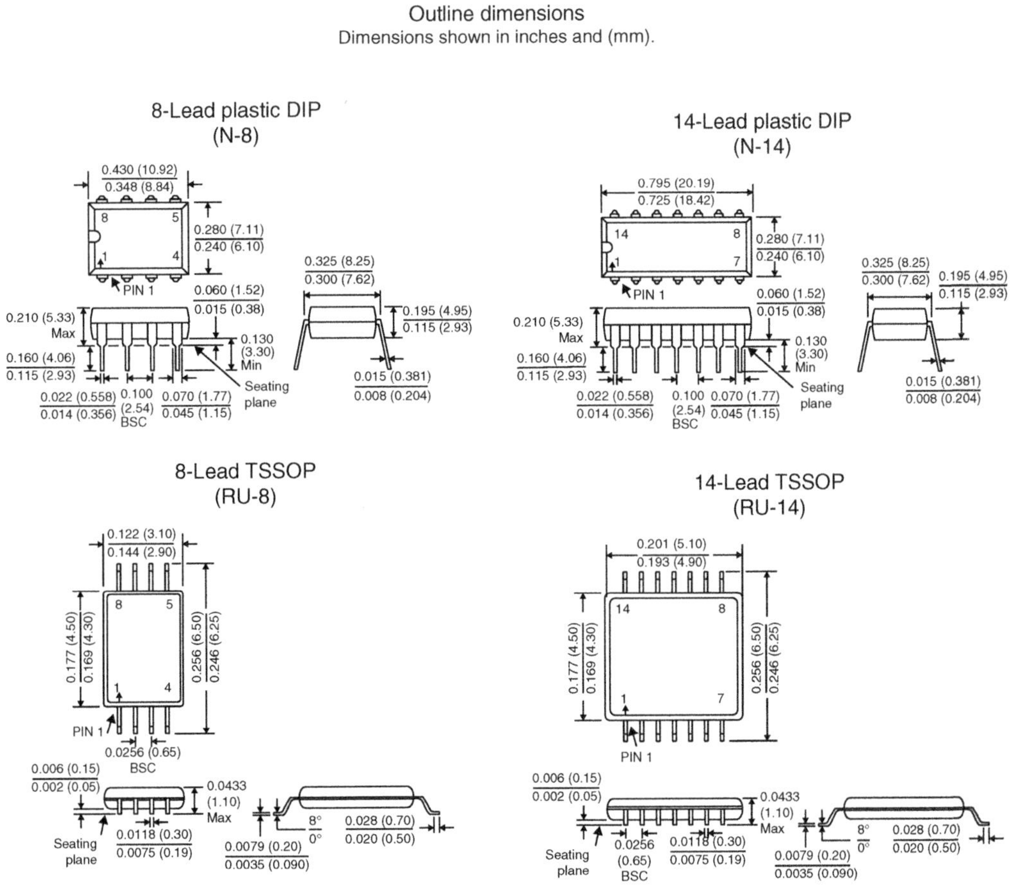

Outline dimensions
Dimensions shown in inches and (mm).

Figure 1-93: Typical package dimension drawing

# Choosing an Op Amp

As we have seen in the previous sections, an op amp can have many specifications. Now that we have gone over what those specifications mean and how to read a data sheet, we are ready to proceed to the next step. How, then, do you determine which amp best suits your needs?

## Step 1: Determine the Parameters

The first step in the process is to determine what parameters are important to your design. To do this you must have a clear idea of:

1.  The input signal
    (a)  Is it a voltage or a current?
    (b)  What are the frequency and the amplitude ranges?
    (c)  What is the impedance level of the surrounding circuit?

2   The accuracy requirements

3.  The output signal
    (a)  What are the frequency and the amplitude ranges?
    (b)  What will the circuit be driving (another op amp stage, an ADC, a cable, etc.)?

4.  The physical environment
    (a)  What is the operational temperature range?
    (b)  What is the size limitation?
    (c)  What power supplies are available?

For instance, if you are designing a single supply system that is going to be capacitively coupled, offsets probably are not a concern. If you are designing a system to interface to a low level physical sensor, then noise, DC precision, and closed-loop gain are important, but bandwidth is probably of less importance, since the bandwidth of most physical sensors is relatively low. However, you do need enough bandwidth to support the required closed-loop gain.

Part of this process is determining the values for the various parameters. In doing this you should determine both an optimum value and an acceptable range. For example, you may have a target value of $500\,\mu V$ for the offset voltage, but you may be able to live with $1\,mV$ and by relaxing this specification, a better overall fit could be made. The operating temperature range that the circuit will be required to operate in will affect this as well. The physical size of the package and the cost, as always, should be considered. It is also good practice to allow a little margin on the specifications so that aging effects, etc., do not cause the circuit to go out of specification.

## Step 2: Prioritize the Parameters

The next step is to prioritize these parameters. Typically one or two parameters are critical. A few more may be desirable but not required. Try not to overspecify the part. Remember that the more specified a part is, the harder it will be to find an exact match; and the tighter the specifications, the more expensive the part is likely to be.

## Step 3: Selecting the Part

The next step is to finally select the part. The brute force method would be to gather data sheets and randomly start to look at the specifications for each of the parts individually. This can quickly get out of hand. There are several tools that make the job much easier.

The first is to use selection guides. These appear frequently in magazine ads and promotional mailers. The problem with using these guides is that, in many instances, the lists are not all inclusive, but instead are usually focused on a specific sub group, such as new products, single supply, or the like. The narrow focus may cause you to miss some otherwise acceptable options.

An alternative is a parametric search engine. Here you enter the relevant parameters for your design. The search engine will then search the database of parts and it will come up with acceptable alternatives.

# Buffer Amplifiers

In the early days of high speed circuits, simple emitter followers were often used as high speed buffers. The term *buffer* was generally accepted to mean a unity-gain, open-loop amplifier. With the availability of matching PNP transistors, a simple emitter follower can be improved, as shown in Figure 2-1(A). This complementary circuit offers first-order cancellation of DC offset voltage, and can achieve bandwidths greater than 100 MHz. Typical offset voltages without trimming are usually less than 50 mV, even with unmatched discrete transistors.

If high input impedance is required, a dual FET can be used as an input stage ahead of a complementary emitter follower, as shown in Figure 2-1(B). This form of the buffer circuit was implemented by both National Semiconductor Corporation as the LH0033, and by Analog Devices as the ADLH0033.

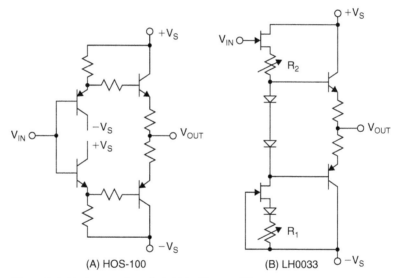

(A) HOS-100    (B) LH0033

**Figure 2-1: Early open-loop hybrid buffer amplifiers: (A) HOS-100 bipolar, (B) LH0033 FET input**

Circuits such as these achieved bandwidths of about 100 MHz at fairly respectable levels of harmonic distortion, typically better than −60 dBc. However, they suffered from DC and AC nonlinearities when driving loads less than 500 Ω.

One of the first totally monolithic implementations of these functions was the Precision Monolithics, Inc. BUF03 shown in Figure 2-2 (see Reference 1). PMI is now a division of Analog Devices. This open-loop IC buffer achieved a bandwidth of about 50 MHz for a 2 V peak-to-peak signal.

The BUF03 circuit is interesting because it demonstrates techniques that eliminated the requirement for the slow, bandwidth-limited vertical PNP transistors associated with most IC processes available at the time of the design (approximately 1979).

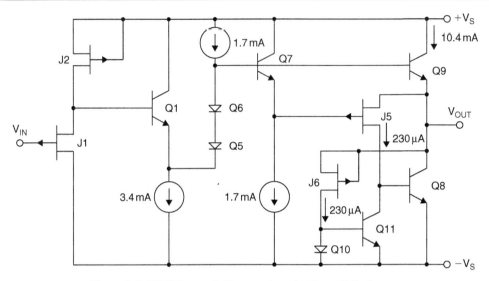

Figure 2-2: BUF03 monolithic open-loop buffer—1979 vintage

One of the problems with all the open-loop buffers discussed thus far is that although high bandwidths can be achieved, the devices discussed do not take advantage of negative feedback. Distortion and DC performance suffer considerably when open-loop buffers are loaded with typical video impedance levels of 50, 75, or 100 Ω. The solution is to use a properly compensated wide bandwidth op amp in a unity-gain follower configuration. In the early days of monolithic op amps, process limitations prevented this, so the open-loop approach provided a popular interim solution (Figure 2-3).

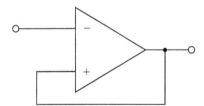

Figure 2-3: Simple unity-gain monolithic buffers

Practically all unity-gain stable voltage or current feedback op amps can be used in a simple follower configuration. Usually, however, the general-purpose op amps are compensated to operate over a wide range of gains and feedback conditions. Therefore, bandwidth suffers somewhat at low gains, especially in the unity-gain non-inverting mode, and additional external compensation is usually required, as shown in Figure 2-4.

A practical solution is to compensate the op amp for the desired closed-loop gain, while including the gain-setting resistors on-chip. Note that this form of op amp, internally configured as a buffer, may typically have no feedback pin. Also, putting the resistors and compensation on-chip serves to reduce parasitics.

There are a number of op amps optimized in this manner. Roy Gosser's AD9620 (see Reference 2) was probably the earliest monolithic implementation. The AD9620 was a 1990 product release, and achieved a bandwidth of 600 MHz using ±5 V supplies. It was optimized for unity gain, and used the voltage feedback

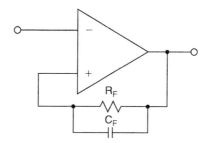

**Figure 2-4: Frequency-compensated buffer**

architecture. A newer design based on similar techniques is the AD9630, which achieves a 750 MHz bandwidth.

The BUF04 unity-gain buffer (see Reference 3) was released in 1994 and achieves a bandwidth of 120 MHz. This device was optimized for large signals and operates on supplies from ±5V to ±15V. Because of the wide supply range, the BUF04 is useful not only as a standalone unit-gain buffer, but also within a feedback loop with a standard op amp, to boost output.

Although the common definition of a buffer is unity-gain device, sometimes the term is used for a circuit with a gain of two. Closed-loop buffers with a gain of two find wide applications as transmission line drivers, as shown in Figure 2-5. The internally configured fixed gain of the amplifier compensates for the loss incurred by the source and load termination. Impedances of 50, 75, and 100 Ω are popular cable impedances. The AD8074/AD8075 500 MHz triple buffers are optimized for gains of 1 and 2, respectively. The dual AD8079A/AD8079B 260 MHz buffer is optimized for gains of 2 and 2.2, respectively.

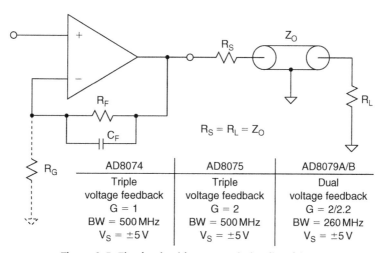

| AD8074 | AD8075 | AD8079A/B |
|---|---|---|
| Triple | Triple | Dual |
| voltage feedback | voltage feedback | voltage feedback |
| G = 1 | G = 2 | G = 2/2.2 |
| BW = 500 MHz | BW = 500 MHz | BW = 260 MHz |
| $V_S = \pm 5\,V$ | $V_S = \pm 5\,V$ | $V_S = \pm 5\,V$ |

**Figure 2-5: Fixed-gain video transmission line drivers**

In implementing a high speed unity-gain buffer with a voltage feedback op amp, there will typically be no resistor required in the feedback loop, which considerably simplifies the circuit. Note that this is not a 100% hard-and-fast rule however, so always check the device data sheet to be sure. A unity-gain buffer with a current feedback op amp will *always* require a feedback resistor, typically in the range of 500–1,000 Ω. So, be sure to use a value appropriate to not only the basic part, but also the specific power supplies in use.

# Gain Blocks

While the op amp allows gain to be set with external resistors, there are a group of circuits that are designed to operate at a fixed gain. These parts are typically RF components. They are also typically designed to be operated in a 50 Ω environment, with the inputs and outputs matched internally. Often the gain blocks are available in several gain settings.

For example, the AD8354 RF gain block is a fixed-gain amplifier with single-ended input and output ports whose impedances are nominally equal to 50 Ω over the frequency range 100 MHz to 2.7 GHz. Consequently, it can be directly inserted into a 50 Ω system with no impedance matching circuitry required. The input and output impedances are sufficiently stable versus variations in temperature and supply voltage that no impedance matching compensation is required (Figure 2-6).

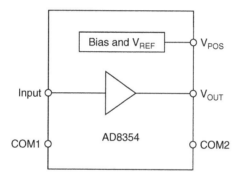

**Figure 2-6: AD8352 20 dB RF gain block**

Differential input and output gain blocks are also available. An example of a differential input, single-ended output device is the AD8129 (see Figure 2-7).

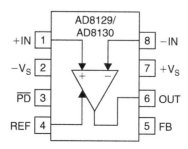

**Figure 2-7: AD8129/30 differential input, single-ended output gain block**

Fully differential input and output devices are also available, such as the AD8350 (see Figure 2-8).

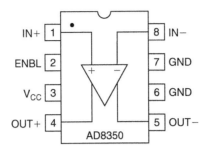

**Figure 2-8: AD8350 differential in/differential out gain block**

# *Instrumentation Amplifiers*

The instrumentation amp is primarily used to amplify small differential voltages in the presence of (typically) larger common-mode (CM) voltages.

## In Amp Definitions

An in amp is a *precision* closed-loop gain block. It has a pair of differential input terminals, and a single-ended output that works with respect to a reference or common terminal, as shown in Figure 2-9. The input impedances are balanced and high in value, typically $\geq 10^9 \Omega$. Again, unlike an op amp, an in amp uses an internal feedback resistor network, plus one (usually) gain set resistance, $R_G$. Also unlike an op amp is the fact that the internal resistance network and $R_G$ are *isolated* from the signal input terminals. In amp gain can also be preset via an internal $R_G$ by pin selection (again isolated from the signal inputs). Typical in amp gains range from 1 to 1,000.

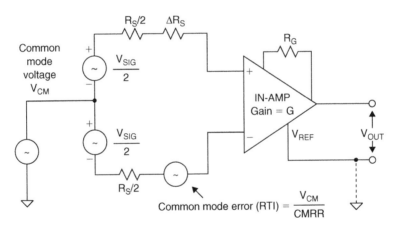

**Figure 2-9: The generic instrumentation amplifier (in amp)**

The in amp develops an output voltage which is referenced to a pin usually designated REFERENCE or $V_{REF}$. In many applications, this pin is connected to circuit ground, but it can be connected to other voltages, as long as they lie within the rated compliance range of the in amp. This feature is especially useful in single-supply applications, where the output voltage is usually referenced to mid-supply (i.e., +2.5 V in the case of a +5 V supply).

In order to be effective, an in amp needs to be able to amplify microvolt-level signals, while simultaneously rejecting volts of CM signal at its inputs. This requires that in amps have very high *common-mode rejection* (CMR). Typical values of in amp CMR are from 70 to over 100 dB (at DC), with CMR usually improving at higher gains.

It is important to note that a CMR specification for DC inputs alone is not sufficient in most practical applications. In industrial applications, the most common cause of external interference is 50/60 Hz AC power-related noise (including harmonics). In differential measurements, this type of interference tends to be induced equally onto both in amp inputs, so the interference appears as a CM input signal. Therefore, specifying CMR over frequency is just as important as specifying its DC value. Note that imbalance in the two source impedances will degrade the CMR of some in amps. Analog Devices fully specifies in amp CMR at 50/60 Hz, with a source impedance imbalance of 1 kΩ.

## Op Amp/In Amp Functionality Differences

An op amp is a general-purpose gain block—user-configurable in myriad ways using external feedback components of R, C, and (sometimes) L. The final configuration and circuit function using an op amp is truly whatever you make of it.

In contrast to this, an instrumentation amp (in amp) is a more constrained device in terms of functioning, and also the allowable range(s) of operating gain. People also often confuse in amps as to their function, calling them "op amps." But the converse is seldom (if ever) true. It should be understood that an in amp is *not* just a special type op amp; the function of the two devices is actually fundamentally different.

Perhaps a good way to differentiate the two devices is to remember that an op amp can be programmed to do almost anything, by virtue of its feedback flexibility. In contrast to this, an in amp *cannot* be programmed to do just anything. It can *only* be programmed for gain, and then over a specific range. An op amp is configured via a number of external components, while an in amp is configured by either one resistor, or by pin-selectable taps for its working gain.

## Subtractor or Difference Amplifiers

A simple subtractor or difference amplifier can be constructed with four resistors and an op amp as shown in Figure 2-10. It should be noted that this is *not* a true in amp, but it is often used in applications where

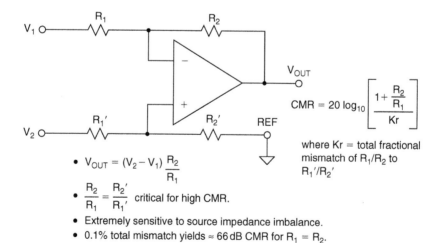

$$V_{OUT} = (V_2 - V_1)\frac{R_2}{R_1}$$

$$\frac{R_2}{R_1} = \frac{R_2'}{R_1'} \quad \text{critical for high CMR.}$$

- Extremely sensitive to source impedance imbalance.
- 0.1% total mismatch yields ≈ 66 dB CMR for $R_1 = R_2$.

$$CMR = 20\,\log_{10}\left[\frac{1+\dfrac{R_2}{R_1}}{Kr}\right]$$

where Kr = total fractional mismatch of $R_1/R_2$ to $R_1'/R_2'$

**Figure 2-10: Op amp subtractor or difference amplifier**

a simple differential to single-ended conversion is required. Because of its popularity, this circuit will be examined in more detail in order to understand its fundamental limitations before discussing true in amp architectures.

There are several fundamental problems with this simple circuit. First, the input impedance seen by $V_1$ and $V_2$ is not balanced. The input impedance seen by $V_1$ is $R_1$, but the input impedance seen by $V_2$ is $R_1' + R_2'$.

The configuration can also be quite problematic in terms of CMR, since even a small source impedance imbalance will degrade the workable CMR. This problem can be solved with well-matched open-loop buffers in series with each input (e.g., using a precision dual op amp). But this adds complexity to a simple circuit, and may introduce offset drift and nonlinearity.

The second problem with this circuit is that the *CMR is primarily determined by the resistor ratio matching*, not *the op amp*. The resistor ratios $R_1/R_2$ and $R_1'/R_2'$ must match extremely well to reject CM noise—at least as well as a typical op amp CMR of 100 dB. Note also that the *absolute* resistor values are relatively unimportant.

Picking four 1% resistors from a single batch may yield a net ratio matching of 0.1%, which will achieve a CMR of 66 dB (assuming $R_1 = R_2$). But if one resistor differs from the rest by 1%, the CMR will drop to only 46 dB. Clearly, very limited performance is possible using ordinary discrete resistors in this circuit (without resorting to hand matching). This is because the best standard off-the-shelf RNC/RNR style resistor tolerances are on the order of 0.1% (see Reference 1).

In general, the worst-case CMR for a circuit of this type is given by the following equation (see References 2 and 3):

$$\text{CMR (dB)} = 20 \log \left[ \frac{1 + R_2/R_1}{4Kr} \right] \qquad (2\text{-}1)$$

where Kr is the *individual* resistor tolerance in fractional form, for the case where four discrete resistors are used. This equation shows that the worst-case CMR for a tolerance build-up for four unselected same-nominal-value 1% resistors is no better than 34 dB.

A single resistor network with a net matching tolerance of Kr would probably be used for this circuit, in which case the expression would be as noted in the Figure, or:

$$\text{CMR (dB)} = 20 \log \left[ \frac{1 + R_2/R_1}{Kr} \right] \qquad (2\text{-}2)$$

A net matching tolerance of 0.1% in the resistor ratios therefore yields a worst-case DC CMR of 66 dB using Eq. (2-2), and assuming $R_1 = R_2$. Note that either case assumes a significantly higher amplifier CMR (i.e., >100 dB). Clearly for high CMR, such circuits need four single-substrate resistors, with very high absolute and TC matching. Such networks using thick-/thin-film technology are available from companies such as Caddock and Vishay, in ratio matches of 0.01% or better.

In implementing the simple difference amplifier, rather than incurring the higher costs and PCB real estate limitations of a precision op amp plus a separate resistor network, it is usually better to seek out a completely monolithic solution.

An interesting variation on the simple difference amplifier is found in the AD629 difference amplifier, optimized for high CM input voltages. A typical current-sensing application is shown in Figure 2-11. The AD629 is a differential to single-ended amplifier with a gain of unity. It can handle a CM voltage of ±270 V with supply voltages of ±15 V, with a small signal bandwidth of 500 kHz.

The high CM voltage range is obtained by attenuating the non-inverting input (pin 3) by a factor of 20 times, using the $R_1$–$R_2$ divider network. On the inverting input, resistor $R_5$ is chosen such that $R_5 \| R_3$ equals resistor $R_2$. The noise gain of the circuit is equal to $20 [1 + R_4/(R_3 \| R_5)]$, thereby providing unity gain for differential input voltages. Laser wafer trimming of the $R_1$–$R_5$ thin-film resistors yields a minimum CMR of 86 dB @ 500 Hz for the AD629B. Within an application, it is good practice to maintain balanced source

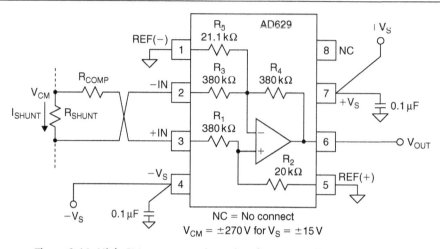

**Figure 2-11: High CM current sensing using the AD629 difference amplifier**

impedances on both inputs, so dummy resistor $R_{COMP}$ is chosen to equal to the value of the shunt-sensing resistor $R_{SHUNT}$.

## The Three Op Amp Instrumentation Amplifier Topology

For the highest precision and performance, the *three op amp* instrumentation amplifier topology is optimum for bridge and other offset transducer applications where high accuracy and low nonlinearity are required (Figure 2-12).

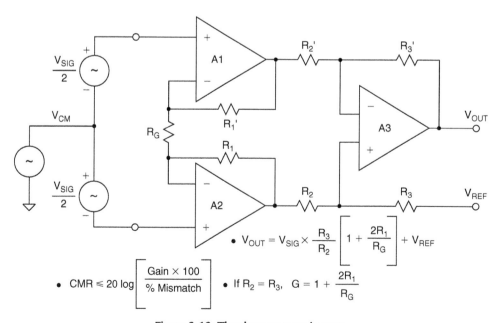

**Figure 2-12: The three op amp in amp**

Resistor $R_G$ sets the overall gain of this amplifier. It may be internal, external, or (software or pin-strap) programmable, depending on the particular in amp. In this configuration, CMR depends on the ratio matching of $R_3/R_2$ to $R_3'/R_2'$. Furthermore, CM signals are only amplified by a factor of 1 regardless of gain (no CM voltage will appear across $R_G$, hence, no CM current will flow in it because the input terminals of an op amp will have no significant potential difference between them).

As a result of the high ratio of differential to CM gain in A1–A2, CMR of this in amp theoretically increases in proportion to gain. Large CM signals (within the A1–A2 op amp headroom limits) may be handled at all gains. Finally, because of the symmetry of this configuration, CM errors in the input amplifiers, if they track, tend to be canceled out by the subtractor output stage. These features explain the popularity of this three op amp in amp configuration—it is capable of delivering the highest performance.

The classic three op amp configuration has been used in a number of monolithic IC in amps (see References 8 and 9). Besides offering excellent matching between the three internal op amps, thin-film laser-trimmed resistors provide excellent ratio matching and gain accuracy at much lower cost than using discrete precision op amps and resistor networks. The AD620 (see Reference 10) is an excellent example of monolithic IC in amp technology. A simplified device schematic is shown in Figure 2-13.

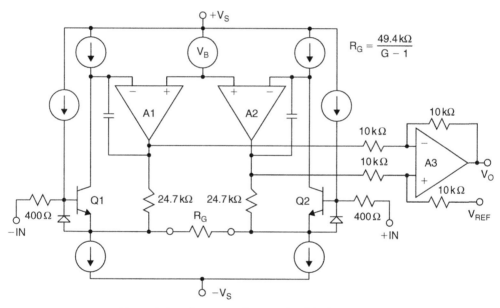

**Figure 2-13: The AD620 in amp simplified schematic**

The AD620 is a highly popular in amp and is specified for power supply voltages from $\pm2.3\,\text{V}$ to $\pm18\,\text{V}$. Input voltage noise is only $9\,\text{nV/Hz}$ @ 1 kHz. Maximum input bias current is only 1 nA, due to the use of superbeta transistors for Q1–Q2.

Overvoltage protection is provided, in part, by the internal $400\,\Omega$ thin-film current-limit resistors in conjunction with the diodes connected from the emitter-to-base of Q1 and Q2. The gain G is set with a single external $R_G$ resistor, as noted by Eq. (2-3):

$$G = \left(\frac{49.4\,\text{k}\Omega}{R_G}\right) + 1 \qquad (2\text{-}3)$$

As can be noted from this expression and Figure 2-13, the AD620 internal resistors are trimmed so that standard 1% or 0.1% resistors can be used to set gain to popular values. Single-supply operation of the three op amp in amp requires an understanding of the internal node voltages. Figure 2-14 shows a generalized diagram of the in amp operating on a single +5V supply. The maximum and minimum allowable output voltages of the individual op amps are designated $V_{OH}$ (maximum high output) and $V_{OL}$ (minimum low output), respectively.

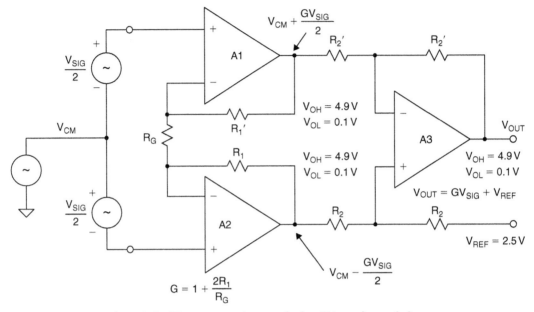

**Figure 2-14: Three op amp in amp single +5V supply restrictions**

Note that the gain from the CM voltage to the outputs of A1 and A2 is unity. It can be stated that *the sum of the CM voltage and the signal voltage at these outputs must fall within the amplifier output voltage range.* Obviously this configuration cannot handle input CM voltages of either 0V or +5V, because of saturation of A1 and A2. The output reference is positioned halfway between $V_{OH}$ and $V_{OL}$ to allow for bipolar differential input signals.

While there are a number of good single supply in amps, such as the AD627, the highest performance devices are still among those specified for traditional dual-supply operation, i.e., the just-discussed AD620. For certain applications, even devices such as the AD620, which has been designed for dual-supply operation, can be used with full precision on a single-supply power system.

## Precision Single-Supply Composite In Amp

One way to achieve both high precision and single-supply operation takes advantage of the fact that many popular sensors (e.g., strain gauges) provide an output signal which is inherently centered around an approximate midpoint of the supply voltage (and/or the reference voltage). Taking advantage of this basic point allows the inputs of a signal conditioning in amp to be biased at "mid-supply." As a consequence of this step, the inputs need not operate near ground or the positive supply voltage, and the in amp can still be used with all its precision.

Under these conditions, an AD620 dual-supply in amp referenced to the supply midpoint followed by a rail-to-rail op amp output-gain stage provides very high DC precision. Figure 2-15 illustrates one such high performance in amp, which operates on a single +5V supply.

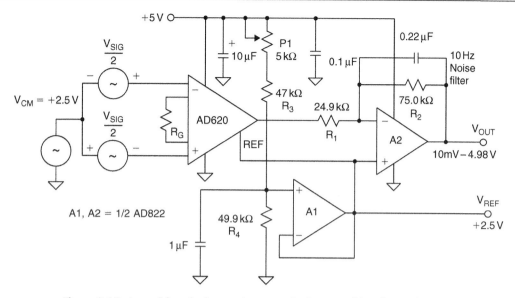

**Figure 2-15: A precision single-supply composite in amp with rail-to-rail output**

This circuit uses the AD620 as a low cost precision in amp for the input stage, along with an AD822 JFET input dual rail-to-rail output op amp for the output stage, comprising A1 and A2. The output stage operates at a fixed gain of 3, with overall gain set by $R_G$.

In this circuit, $R_3$ and $R_4$ form a voltage divider which splits the supply voltage nominally in half to +2.5 V, with fine adjustment provided by a trimming potentiometer, P1. This voltage is applied to the input of A1, an AD822 voltage follower, which buffers it and provides a low impedance source needed to drive the AD620's reference pin as well as providing the output reference voltage $V_{REF}$. *Note that this feature allows a bipolar $V_{OUT}$ to be measured with respect to this +2.5 V reference (not to GND).* This is despite the fact that the entire circuit operates from a single (unipolar) supply.

The other half of the AD822 is connected as a gain of 3 inverter, so that it can output ±2.5 V, "rail-to-rail," with only ±0.83 V required of the AD620. This output voltage level of the AD620 is well within the AD620's capability, thus ensuring high linearity for the front end.

The general-gain expression for this composite in amp is the product of the gain of the AD620 stage, and the gain of inverting amplifier:

$$\text{GAIN} = \left( \frac{49.4\,\text{k}\Omega}{R_G} + 1 \right)\left( \frac{R_2}{R_1} \right) \qquad (2\text{-}4)$$

For this example, an overall gain of 10 is realized with $R_G = 21.5\,\text{k}\Omega$ (closest standard value). The table shown in Figure 2-16 summarizes various $R_G$ gain values, and the resulting performance for gains ranging from 10 to 1,000.

In this application, the allowable input voltage on either input to the AD620 must lie between +2 V and +3.5 V in order to maintain linearity. For example, at an overall circuit gain of 10, the CM input voltage range spans from 2.25 V to 3.25 V, allowing room for the ±0.25 V full-scale differential input voltage required to drive the output ±2.5 V about $V_{REF}$.

| Circuit gain | $R_G$ ($\Omega$) | $V_{OS}$, RTI ($\mu$V) | TC $V_{OS}$, RTI ($\mu$V/$^\circ$C) | Nonlinearity (ppm)* | Bandwidth (kHz)** |
|---|---|---|---|---|---|
| 10 | 21.5 k | 1000 | 1000 | <50 | 600 |
| 30 | 5.49 k | 430 | 430 | <50 | 600 |
| 100 | 1.53 k | 215 | 215 | <50 | 300 |
| 300 | 499 | 150 | 150 | <50 | 120 |
| 1000 | 149 | 150 | 150 | <50 | 30 |

\* Nonlinearity measured over output range: 0.1 V $<$ $V_{OUT}$ $<$ 4.90 V
\*\* Without 10 Hz noise filter

**Figure 2-16: Performance summary of the +5 V single-supply AD620/AD822 composite in amp**

The inverting configuration was chosen for the output buffer to facilitate system output offset voltage adjustment by summing currents into the A2 stage buffer's feedback summing node. These offset currents can be provided by an external DAC, or from a resistor connected to a reference voltage.

To reduce the effects of unwanted noise pickup, a filter capacitor is recommended across A2's feedback resistance to limit the circuit bandwidth to the frequencies of interest. This capacitor forms a first-order lowpass filter with $R_2$. The corner frequency is 10 Hz as shown, but this may be easily modified. The capacitor should be a high quality film type, such as polypropylene.

### *The Two Op Amp Instrumentation Amplifier Topology*

The circuit shown in Figure 2-17 is referred to as the *two op amp in amp*. It is particularly applicable in single-supply systems. Dual IC op amps are used in most cases for good matching, such as the OP297 or

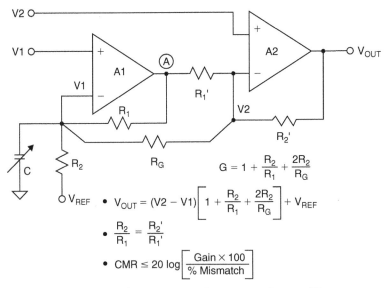

**Figure 2-17: The two op amp instrumentation amplifier**

the OP284. Most often a rail-to-rail op amp is indicated. The resistors are often a thin-film laser-trimmed array, possibly on the same chip. The in amp gain can be easily set with an external resistor, $R_G$. Without $R_G$, the gain is simply $1 + R_2/R_1$. In a practical application, the $R_2/R_1$ ratio is chosen for the desired minimum in amp gain.

The input impedance of the two op amp in amp is inherently high, permitting the impedance of the signal sources to be high and unbalanced. The DC CM rejection is limited by the matching of $R_1/R_2$ to $R_1'/R_2'$. If there is a mismatch in any of the four resistors, the DC CM rejection is limited to:

$$CMR \leq 20 \log \left( \frac{GAIN \times 100}{\%MISMATCH} \right) \qquad (2\text{-}5)$$

Notice that the net CMR of the circuit increases proportionally with the working gain of the in amp, an effective aid to high performance at higher gains.

IC in amps are particularly well suited to meeting the combined needs of ratio matching and temperature tracking of the gain-setting resistors. While thin-film resistors fabricated on silicon have an initial tolerance of up to ±20%, laser trimming during production allows the ratio error (not absolute value) between the resistors to be reduced to 0.01% (100 ppm). Furthermore, the tracking between the temperature coefficients of the thin-film resistors is inherently low and is typically less than 3 ppm/°C (0.0003%/°C).

When dual supplies are used, $V_{REF}$ is normally connected directly to ground. In single-supply applications, $V_{REF}$ is usually connected to a low impedance voltage source equal to one-half the supply voltage. The gain from $V_{REF}$ to node "A" is $R_1/R_2$, and the gain from node "A" to the output is $R_2'/R_1'$. This makes the gain from $V_{REF}$ to the output equal to unity, assuming perfect ratio matching. Note that it is critical that the source impedance seen by $V_{REF}$ be low; otherwise, CMR will be degraded.

One major disadvantage of the two op amp in amp design is that CM voltage input range must be traded off against gain. The amplifier A1 must amplify the signal at V1 by $1 + R_1/R_2$. If $R_1 \gg R_2$ (a low gain example in Figure 2-18), A1 will saturate if the V1 CM signal is too high, leaving no A1 headroom to

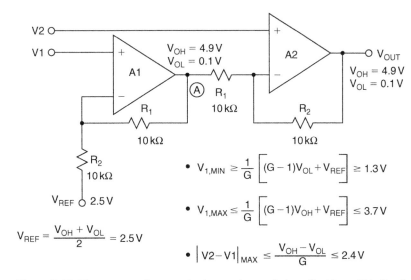

**Figure 2-18: Two op amp in amp single-supply restrictions for $V_S$ = +5 V, G = 2**

amplify the wanted differential signal. For high gains ($R_1 \ll R_2$), there is correspondingly more headroom at node "A," allowing larger CM input voltages.

The AC CM rejection of this configuration is generally poor because the signal path from V1 to $V_{OUT}$ has the additional phase shift of A1. In addition, the two amplifiers are operating at different closed-loop gains (and thus at different bandwidths). The use of a small trim capacitor "C" as shown in Figure 2-17 can improve the AC CMR somewhat.

A low gain (G = 2) single-supply two op amp in amp configuration results when $R_G$ is not used, and is shown in Figure 2-18. The input CM and differential signals must be limited to values which prevent saturation of either A1 or A2. In the example, the op amps remain linear to within 0.1 V of the supply rails, and their upper and lower output limits are designated $V_{OH}$ and $V_{OL}$, respectively. These saturation voltage limits would be typical for a single-supply, rail–rail output op amp (such as the AD822).

Using the Figure 2-18 equations, the voltage at V1 must fall between 1.3 V and 2.4 V to prevent A1 from saturating. Notice that $V_{REF}$ is connected to the average of $V_{OH}$ and $V_{OL}$ (2.5 V). This allows for bipolar differential input signals with $V_{OUT}$ referenced to +2.5 V.

A high gain (G = 100) single-supply two op amp in amp configuration is shown in Figure 2-19. Using the same equations, note that voltage at V1 can now swing between 0.124 V and 4.876 V. $V_{REF}$ is again 2.5 V, to allow for bipolar input and output signals.

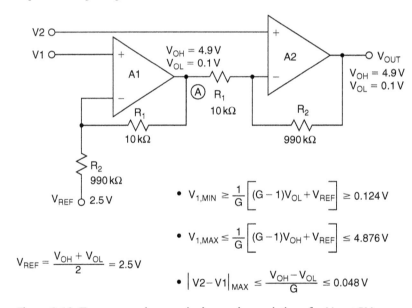

$$V_{REF} = \frac{V_{OH} + V_{OL}}{2} = 2.5\,V$$

- $V_{1,MIN} \geq \dfrac{1}{G}\left[(G-1)V_{OL} + V_{REF}\right] \geq 0.124\,V$

- $V_{1,MAX} \leq \dfrac{1}{G}\left[(G-1)V_{OH} + V_{REF}\right] \leq 4.876\,V$

- $\left|V2 - V1\right|_{MAX} \leq \dfrac{V_{OH} - V_{OL}}{G} \leq 0.048\,V$

**Figure 2-19: Two op amp in amp single-supply restrictions for $V_S$ = +5 V, G = 100**

All of these discussions show that the conventional two op amp in amp architecture is fundamentally limited, when operating from a single power supply. These limitations can be viewed in one sense as a restraint on the allowable input CM range for a given gain. Or, alternately, it can be viewed as limitation on the allowable gain range, for a given CM input voltage.

In summary, regardless of gain, the basic structure of the common two op amp in amp does not allow for CM input voltages of zero when operated on a single supply. The only route to removing these restrictions for single-supply operation is to modify the in amp architecture.

## *In Amp DC Error Sources*

The DC and noise specifications for in amps differ slightly from conventional op amps, so some discussion is required in order to fully understand the error sources.

The gain of an in amp is usually set by a single resistor. If the resistor is external to the in amp, its value is either calculated from a formula or chosen from a table on the data sheet, depending on the desired gain.

Absolute value laser wafer trimming allows the user to program gain accurately with this single resistor. The absolute accuracy and temperature coefficient of this resistor directly affects the in amp gain accuracy and drift. Since the external resistor will never exactly match the internal thin-film resistor tempcos, a low TC (<25ppm/°C) metal film resistor should be chosen, preferably with a 0.1% or better accuracy.

Often specified as having a gain range of 1 to 1,000, or 1 to 10,000, many in amps will work at higher gains, but the manufacturer will not guarantee a specific level of performance at these high gains. In practice, as the gain-setting resistor becomes smaller, any errors due to the resistance of the metal runs and bond wires become significant. These errors, along with an increase in noise and drift, may make higher single-stage gains impractical. In addition, input offset voltages can become quite sizable when reflected to output at high gains. For instance, a 0.5mV input offset voltage becomes 5V at the output for a gain of 10,000. For high gains, the best practice is to use an in amp as a preamplifier; then use a post amplifier for further amplification.

In a *pin-programmable-gain* in amp such as the AD621, the gain-set resistors are internal, well matched, and the device gain accuracy and gain drift specifications include their effects. The AD621 is otherwise generally similar to the externally gain-programmed AD620.

The *gain error* specification is the maximum deviation from the gain equation. Monolithic in amps such as the AD624C have very low factory trimmed gain errors, with its maximum error of 0.02% at G = 1 and 0.25% at G = 500 being typical for this high quality in amp. Notice that the gain error increases with increasing gain. Although externally connected gain networks allow the user to set the gain exactly, the temperature coefficients of the external resistors and the temperature differences between individual resistors within the network all contribute to the overall gain error. If the data is eventually digitized and presented to a digital processor, it may be possible to correct for gain errors by measuring a known reference voltage and then multiplying by a constant.

*Nonlinearity* is defined as the maximum deviation from a straight line on the plot of output versus input. The straight line is drawn between the end-points of the actual transfer function. Gain nonlinearity in a high quality in amp is usually 0.01% (100 ppm) or less, and is relatively insensitive to gain over the recommended gain range.

The total *input offset voltage* of an in amp consists of two components (see Figure 2-20): *input* offset voltage, $V_{OSI}$, is the input offset component that is reflected to the output of the in amp by the gain G; *output* offset voltage, $V_{OSO}$, is independent of gain.

At low gains, output offset voltage is dominant, while at high gains input offset dominates. The output offset voltage drift is normally specified as drift at G = 1 (where input effects are insignificant), while input offset voltage drift is given by a drift specification at a high gain (where output offset effects are negligible).

The total output offset error, referred to the input (RTI), is equal to $V_{OSI} + V_{OSO}/G$. In amp data sheets may specify $V_{OSI}$ and $V_{OSO}$ separately, or give the total RTI input offset voltage for different values of gain.

*Input bias currents* may also produce offset errors in in amp circuits (Figure 2-20, again). If the source resistance, $R_S$, is unbalanced by an amount, $\Delta R_S$, (often the case in bridge circuits), then there is an additional input offset voltage error due to the bias current, equal to $I_B \Delta R_S$ (assuming that $I_{B+} \approx I_{B-} = I_B$). This error is reflected to the output, scaled by the gain G.

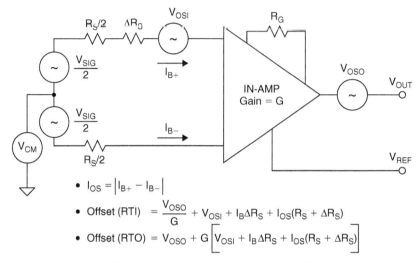

- $I_{OS} = |I_{B+} - I_{B-}|$
- Offset (RTI) $= \dfrac{V_{OSO}}{G} + V_{OSI} + I_B \Delta R_S + I_{OS}(R_S + \Delta R_S)$
- Offset (RTO) $= V_{OSO} + G\left[V_{OSI} + I_B \Delta R_S + I_{OS}(R_S + \Delta R_S)\right]$

**Figure 2-20: In amp offset voltage model**

The input offset current, $I_{OS}$, creates an input offset voltage error across the source resistance, $R_S + \Delta R_S$, equal to $I_{OS}(R_S + \Delta R_S)$, which is also reflected to the output by the gain, G.

In amp *CM error* is a function of both gain and frequency. Analog Devices specifies in amp CMR for a $1\,k\Omega$ source impedance unbalance at a frequency of 60 Hz. The RTI CM error is obtained by dividing the CM voltage, $V_{CM}$, by the common-mode rejection ratio (CMRR).

Figure 2-21 shows the CMR for the AD620 in amp as a function of frequency, with a $1\,k\Omega$ source impedance imbalance.

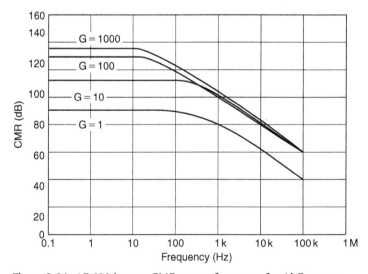

**Figure 2-21: AD620 in amp CMR versus frequency for $1\,k\Omega$ source imbalance**

*Power supply rejection* (PSR) is also a function of gain and frequency. For in amps, it is customary to specify the sensitivity to each power supply separately, as shown in Figure 2-22 for the AD620. The RTI PSR error is obtained by dividing the power supply deviation from nominal by the power supply rejection ratio (PSRR).

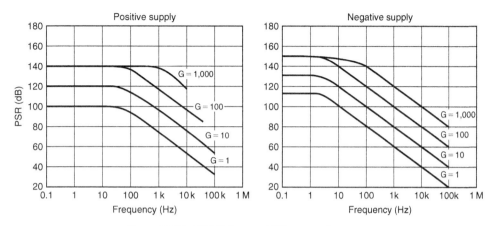

Figure 2-22: AD620 in amp PSR versus frequency

Because of the relatively poor PSR at high frequencies, decoupling capacitors are required on both power pins to an in amp. Low inductance ceramic capacitors (0.01–0.1 µF) are appropriate for high frequencies. Low ESR electrolytic capacitors should also be located at several points on the PC board for low frequency decoupling.

Note that these decoupling requirements apply to all linear devices, including op amps and data converters. Further details on power supply decoupling are found in Chapter 7.

Now that all DC error sources have been accounted for, a worst-case DC error budget can be calculated by reflecting all the sources to the in amp input, as is illustrated by the table of Figure 2-23.

| Error source | RTI value |
|---|---|
| Gain accuracy (ppm) | Gain accuracy $\times$ FS Input |
| Gain nonlinearity (ppm) | Gain nonlinearity $\times$ FS Input |
| Input offset voltage, $V_{OSI}$ | $V_{OSI}$ |
| Output offset voltage, $V_{OSO}$ | $V_{OSO} \div G$ |
| Input bias current, $I_B$, flowing in $\Delta R_S$ | $I_B \Delta R_S$ |
| Input offset current, $I_{OS}$, flowing in $R_S$ | $I_{OS}(R_S + \Delta R_S)$ |
| Common-mode input voltage, $V_{CM}$ | $V_{CM} \div CMRR$ |
| Power supply variation, $\Delta V_S$ | $\Delta V_S \div PSRR$ |

Figure 2-23: In amp DC errors RTI

It should be noted that the DC errors can be referred to the in amp output (RTO) by simply multiplying the RTI error by the in amp gain.

## In Amp Noise Sources

Since in amps are primarily used to amplify small precision signals, it is important to understand the effects of all the associated noise sources. The in amp noise model is shown in Figure 2-24.

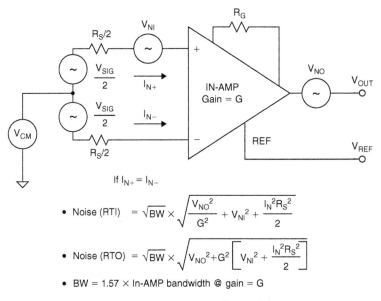

$$\text{Noise (RTI)} = \sqrt{BW} \times \sqrt{\frac{V_{NO}^2}{G^2} + V_{NI}^2 + \frac{I_N^2 R_S^2}{2}}$$

$$\text{Noise (RTO)} = \sqrt{BW} \times \sqrt{V_{NO}^2 + G^2 \left[ V_{NI}^2 + \frac{I_N^2 R_S^2}{2} \right]}$$

- BW = 1.57 × In-AMP bandwidth @ gain = G

**Figure 2-24: In amp noise model**

There are two sources of input voltage noise. The first is represented as a noise source, $V_{NI}$, in series with the input, as in a conventional op amp circuit. This noise is reflected to the output by the in amp gain, G. The second noise source is the output noise, $V_{NO}$, represented as a noise voltage in series with the in amp output. The output noise, shown here referred to $V_{OUT}$, can be RTI by dividing by the gain, G.

There are also two noise sources associated with the input noise currents $I_{N+}$ and $I_{N-}$. Even though $I_{N+}$ and $I_{N-}$ are usually equal ($I_{N+} \approx I_{N-} = I_N$), they are *uncorrelated*, and therefore, the noise they each create must be summed in a root-sum-squares (RSS) fashion. $I_{N+}$ flows through one-half of $R_S$, and $I_{N-}$ the other half. This generates two noise voltages, each having an amplitude $I_N R_S/2$. Each of these two noise sources is reflected to the output by the in amp gain, G.

The total output noise is calculated by combining all four noise sources in an RSS manner:

$$\text{NOISE (RTO)} = \sqrt{BW} \sqrt{V_{NO}^2 + G^2 \left( V_{NI}^2 + \frac{I_{N+}^2 R_S^2}{4} + \frac{I_{N-}^2 R_S^2}{4} \right)} \qquad (2\text{-}6)$$

If $I_{N+} = I_{N-} = I_N$,

$$\text{NOISE (RTO)} = \sqrt{BW} \sqrt{V_{NO}^2 + G^2 \left( V_{NI}^2 + \frac{I_N^2 R_S^2}{2} \right)} \qquad (2\text{-}7)$$

The total noise, RTI is simply the above expression divided by the in amp gain, G:

$$\text{NOISE (RTI)} = \sqrt{BW}\sqrt{\frac{V_{NO}^2}{G^2} + \left(V_{NI}^2 + \frac{I_N^2 R_S^2}{2}\right)} \tag{2-8}$$

In amp data sheets often present the total voltage noise RTI as a function of gain. This noise spectral density includes both the input ($V_{NI}$) and output ($V_{NO}$) noise contributions. The input current noise spectral density is specified separately.

As in the case of op amps, the total in amp noise RTI must be integrated over the applicable in amp closed-loop bandwidth to compute an RMS value. The bandwidth may be determined from data sheet curves that show frequency response as a function of gain.

Regarding this bandwidth, some care must be taken in computing it, as it is often *not* constant bandwidth product relationship, as is true with VFB op amps. In the case of the AD620 in amp family, e.g., the gain-bandwidth pattern is more like that of a CFB op amp. In such cases, the safest way to predict the bandwidth at a given gain is to use the curves supplied within the data sheet.

## In Amp Bridge Amplifier Error Budget Analysis

It is important to understand in amp error sources in a typical application. Figure 2-25 shows a $350\,\Omega$ load cell with a full-scale output of $100\,\text{mV}$ when excited with a $10\,\text{V}$ source. The AD620 is configured for a

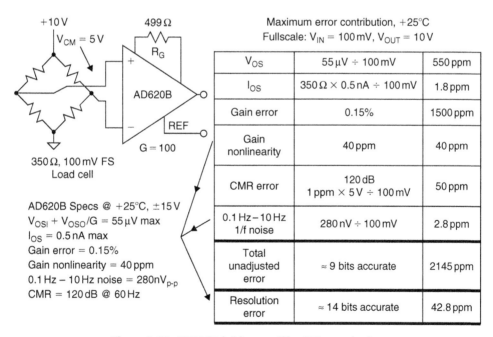

**Figure 2-25: AD620B bridge amplifier DC error budget**

gain of 100 using the external $499\,\Omega$ gain-setting resistor. The table shows how each error source contributes to a total unadjusted error of 2,145 ppm. Note however that the gain, offset, and CMR errors can all be removed with a system calibration. The remaining errors—gain nonlinearity and 0.1–10 Hz noise—cannot be removed with calibration and ultimately limit the system resolution to 42.8 ppm (approximately 14-bit accuracy).

This example is of course just an illustration, but should be useful toward the importance of addressing performance-limiting errors such as gain nonlinearity and low frequency noise.

## In Amp Input Overvoltage Protection

In their typical application as interface amplifiers for data acquisition systems, in amps are often subjected to input overloads, i.e., voltage levels in excess of the full-scale for the selected gain range. The manufacturer's "absolute maximum" input ratings for the device should be closely observed. As with op amps, many in amps have absolute maximum input voltage specifications equal to $\pm V_S$.

In some cases, external series resistors (for current limiting) and diode clamps may be used to prevent overload, if necessary (see Figure 2-26). Some in amps have built-in overload protection circuits in the form of series resistors. For example, the AD620 series have thin-film resistors, and the substrate isolation they provide allows input voltages that can exceed the supplies. Other devices use series-protection FETs, e.g. the AMP02 and the AD524, because they act as a low impedance during normal operation, and a high impedance during overvoltage fault conditions. In any instance however, there are always finite safe limits to applied overvoltage (Figure 2-26, again).

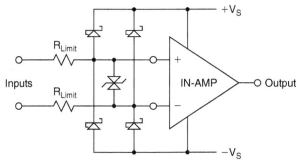

- Always observe absolute maximum data sheet specs!
- Schottky diode clamps to the supply rails will limit input to approximately $\pm V_S \pm 0.3$ V, TVSs limit differential voltage
- External resistors (or Internal thin-film resistors) can limit input current, but will increase noise
- Some in-amps have series-protection input FETs for lower noise and higher input overvoltages (up to $\pm 60$ V, depending on device)

**Figure 2-26: In amp input overvoltage considerations**

In some instances, an additional transient voltage suppressor (TVS) may be required across the input pins to limit the maximum differential input voltage. This is especially applicable to three op amp in amps operating at high gain with low values of $R_G$.

A more detailed discussion of input overvoltage and EMI/RFI protection can be found in Chapter 11 of this book.

# Differential Amplifiers

Many high performance analog-to-digital converters (ADCs) are now being designed with differential inputs. A fully differential ADC design offers the advantages of good CM rejection, reduction in second-order distortion products, and simplified DC trim algorithms. Although they can be driven single ended, a fully differential driver usually optimizes overall performance.

One of the most common ways to drive a differential input ADC is with a transformer. However, there are many applications where the ADCs cannot be driven with transformers because the frequency response must extend to DC. In these cases, differential drivers are required.

A block diagram of the AD813X family of fully differential amplifiers optimized for ADC driving is shown in Figure 2-27 (see References 3–5). Figure 2-27(A) shows the details of the internal circuit, and Figure 2-27(B) shows the equivalent circuit. The gain is set by the external $R_F$ and $R_G$ resistors, and the

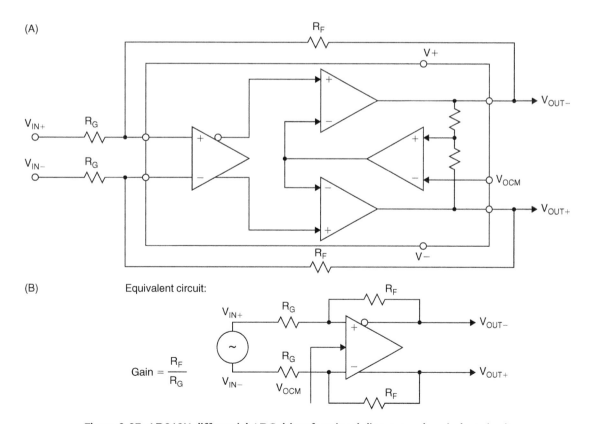

Figure 2-27: AD813X differential ADC driver functional diagram and equivalent circuit

CM voltage is set by the voltage on the $V_{OCM}$ pin. The internal CM feedback forces the $V_{OUT+}$ and $V_{OUT-}$ outputs to be balanced, i.e., the signals at the two outputs are always equal in amplitude but 180° out of phase as per the equation:

$$V_{OCM} = \frac{V_{OUT+} + V_{OUT-}}{2} \qquad (2\text{-}9)$$

The circuit can be used with either a differential or a single-ended input, and the voltage gain is equal to the ratio of $R_F$ to $R_G$.

If a buffered differential voltage output is required from a current output DAC, the AD813X-series of differential amplifiers can be used as shown in Figure 2-28.

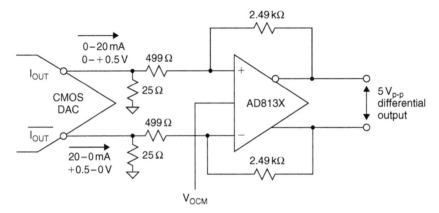

**Figure 2-28: Buffering high speed DACs using AD813X differential amplifier**

The DAC output current is first converted into a voltage that is developed across the 25 Ω resistors. The voltage is amplified by a factor of 5 using the AD813X. This technique is used in lieu of a direct I/V conversion to prevent fast slewing DAC currents from overloading the amplifier and introducing distortion. Care must be taken so that the DAC output voltage is within its compliance rating.

The $V_{OCM}$ input on the AD813X can be used to set a final output CM voltage within the range of the AD813X. If transmission lines are to be driven at the output, adding a pair of 75 Ω resistors will allow this.

Note also that these amplifiers can be used with single-ended inputs as well. Grounding one of the inputs turns these amplifiers into single ended to differential converters.

# Isolation Amplifiers

## Analog Isolation Techniques

There are many applications where it is desirable, or even essential, for a sensor to have no direct ("galvanic") electrical connection with the system to which it is supplying data. This might be in order to avoid the possibility of dangerous voltages or currents from one-half of the system doing damage in the other, or to break an intractable ground loop. Such a system is said to be "isolated," and the arrangement that passes a signal without galvanic connections is known as an *isolation barrier.*

The protection of an isolation barrier works in both directions, and may be needed in either, or even in both. The obvious application is where a sensor may encounter high voltages, such as monitoring the current in an AC induction motor, and the system it is driving must be protected. Or a sensor may need to be isolated from accidental high voltages arising downstream, in order to protect its environment: examples include the need to prevent the ignition of explosive gases by sparks at sensors and the protection from electric shock of patients whose ECG, EEG, or EMG is being monitored. The ECG case is interesting, as protection may be required in *both* directions: the patient must be protected from accidental electric shock, but if the patient's heart should stop, the ECG machine must be protected from the very high voltages ($>7.5\,\text{kV}$) applied to the patient by the defibrillator which will be used to attempt to restart it.

Just as interference, or *unwanted* information, may be coupled by electric or magnetic fields, or by electromagnetic radiation, these phenomena may be used for the transmission of *wanted* information in the design of isolated systems.

The most common isolation amplifiers use *transformers,* which exploit magnetic fields, and another common type uses small high voltage capacitors, exploiting electric fields. *Optoisolators,* which consist of an LED and a photocell, provide isolation by using light, a form of electromagnetic radiation. Different isolators have differing performance: some are sufficiently linear to pass high accuracy analog signals across an isolation barrier. With others, the signal may need to be converted to digital form before transmission for accuracy is to be maintained (note this is a common V/F converter application).

Transformers are capable of analog accuracy of 12–16 bits and bandwidths up to several hundred kHz, but their maximum voltage rating rarely exceeds $10\,\text{kV}$, and is often much lower. *Capacitively coupled* isolation amplifiers have lower accuracy, perhaps 12 bits maximum, lower bandwidth, and lower voltage ratings—but they are low cost. Optical isolators are fast and cheap, and can be made with very high voltage ratings ($4$–$7\,\text{kV}$ is one of the more common ratings), but they have poor analog domain linearity, and are not usually suitable for direct coupling of precision analog signals.

Linearity and isolation voltage are not the only issues to be considered in the choice of isolation systems. Operating power is of course, essential. Both the input and the output circuitry must be powered, and unless there is a battery on the isolated side of the isolation barrier (which is possible, but rarely convenient), some form of isolated power must be provided. Systems using transformer isolation can easily use a transformer (either the signal transformer or another one) to provide isolated power, but it is impractical to transmit useful amounts of power by capacitive or optical means. Systems using these forms of isolation must make other arrangements to obtain isolated power supplies—this is a powerful consideration in favor of choosing transformer isolated isolation amplifiers: they almost invariably include an isolated power supply.

The isolation amplifier has an input circuit that is galvanically isolated from the power supply and the output circuit. In addition, there is minimal capacitance between the input and the rest of the device. Therefore, there is no possibility for DC current flow, and minimum AC coupling. Isolation amplifiers are intended for applications requiring safe, accurate measurement of low frequency voltage or current (up to about 100 kHz) in the presence of high CM voltage (to thousands of volts) with high CMR. They are also useful for line-receiving of signals transmitted at high impedance in noisy environments, and for safety in general-purpose measurements, where DC and line-frequency leakage must be maintained at levels well below certain mandated minimums. Principal applications are in electrical environments of the kind associated with medical equipment, conventional and nuclear power plants, automatic test equipment, and industrial process control systems.

## *AD210 Three-Port Isolator*

A basic form of isolator is the *three-port isolator* (input, power, output all isolated) shown in Figure 2-29. Note that in this diagram, the input circuits, output circuits, and power source are all isolated from one another. This Figure represents the circuit architecture of a self-contained isolator, the AD210 (see References 1 and 2).

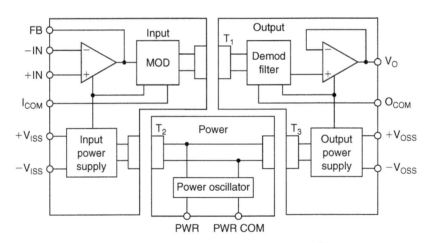

**Figure 2-29: AD210 three-port isolation amplifier**

An isolator of this type requires power from a two-terminal DC power supply (PWR, PWR COM). An internal oscillator (50 kHz) converts the DC power to AC, which is transformer-coupled to the shielded input section, then converted to DC for the input stage and the auxiliary power output. The output current capability of this output is typically limited to ±15 mA.

The AC carrier is also modulated by the input stage amplifier output, transformer-coupled to the output stage, demodulated by a phase-sensitive demodulator (using the carrier as the reference), filtered, and buffered using isolated DC power derived from the carrier.

The AD210 allows the user to select gains from 1 to 100, using external resistors with the input section op amp. Bandwidth is 20 kHz, and voltage isolation is $2,500 V_{RMS}$ (continuous) and $\pm 3,500 V_{PEAK}$ (continuous).

The AD210 is a three-port isolation amplifier, thus the power circuitry is isolated from both the input and the output stages and may therefore be connected to either (or to neither), without change in functionality. It uses transformer isolation to achieve 3,500V isolation with 12-bit accuracy.

## Motor Control Isolation Amplifier

A typical isolation amplifier application using the AD210 is shown in Figure 2-30. The AD210 is used with an AD620 instrumentation amplifier in a current-sensing system for motor control. The input of the AD210, being isolated, can be directly connected to a 110 or 230V power line without protection being necessary. The input section's isolated $\pm15$V powers the AD620, which senses the voltage drop in a small value current-sensing resistor. The AD210 input stage op amp is simply connected as a unity-gain follower, which minimizes its error contribution. The 110 or 230$V_{RMS}$ CM voltage is ignored by this isolated system.

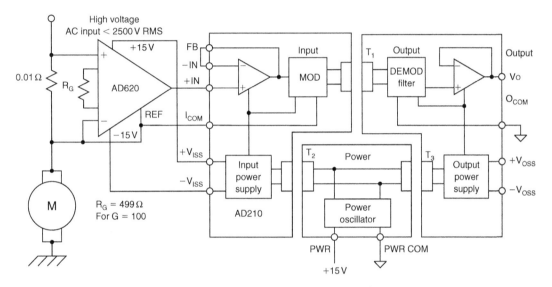

**Figure 2-30: Motor control current sensing**

Within this system the AD620 preamp is used as the system scaling control point, and will produce an output voltage proportional to motor current, as scaled by the sensing resistor value and gain as set by the AD620's $R_G$. The AD620 also improves overall system accuracy, as the AD210's $V_{OS}$ is 15mV, versus the AD620's 30$\mu$V (with less drift also). Note that if higher DC offset and drift are acceptable, the AD620 may be omitted and the AD210 connected at a gain of 100.

## Optional Noise Reduction Post Filter

Due to the nature of this type of carrier-operated isolation system, there will be certain operating situations where some residual AC carrier component will be superimposed on the recovered output DC signal. When this occurs, a low impedance passive RC filter section following the output stage may be used (if the following stage has a high input impedance, i.e., non-loading to this filter). Note that it will be the case for many high input impedance sampling ADCs, which appear essentially as small capacitors. A 150$\Omega$ resistance and 1nF capacitor will provide a corner frequency of about 1kHz. Note also that the capacitor should be a film type for low errors, such as polypropylene. As an option an active filter may be utilized. Since the output of the filter is low impedance (the output of an op amp), it may be used where the low output is required. Also note that it may be possible to include the anti-aliasing requirement of the ADC into this filter.

## Two-Port Isolator

A two-port isolator differs from a three-port isolator in that the power section is not isolated from the output section. The AD215 is an example of a high speed, two-port isolation amplifier, designed to isolate and amplify wide bandwidth analog signals (see Reference 3). The innovative circuit and transformer design of the AD215 ensures wide-band dynamic characteristics, while preserving DC performance specifications. An AD215 block diagram is shown in Figure 2-31.

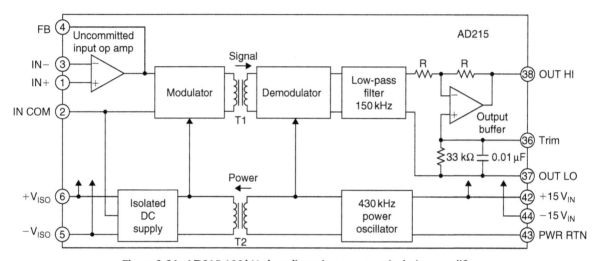

**Figure 2-31: AD215 120 kHz low distortion two-port isolation amplifier**

The AD215 provides complete galvanic isolation between the input and output of the device, which also includes the user-available front-end isolated bipolar power supply. The functionally complete design, powered by a ±15V DC supply on the output side, eliminates the need for a user supplied isolated DC/ DC converter. This permits the designer to minimize circuit overhead and reduce overall system design complexity and component costs.

The AD215 has a ±10V input/output range, a specified gain range of 1V/V to 10V/V, a buffered output with offset trim and a user-available isolated front-end power supply which produces ±15V DC at ±10mA.

# Digital Isolation Techniques

While not a linear circuit, digital isolation is closely related to isolation amplifiers, so they will be discussed here.

Analog isolation amplifiers find many applications where a high isolation is required, such as in medical instrumentation. Digital isolation techniques provide similar galvanic isolation and are a reliable method of transmitting digital signals without ground noise.

Optocouplers (also called optoisolators) are useful and available in a wide variety of styles and packages. A typical optocoupler based on an LED and a phototransistor is shown in Figure 2-32. A current of approximately 10 mA drives an LED transmitter, with light output received by a phototransistor. The light produced by the LED saturates the phototransistor. Input/output isolation of $5,000 V_{RMS}$ to $7,000 V_{RMS}$ is common. Although fine for digital signals, optocouplers are too nonlinear for most analog applications. In addition, the transfer characteristics of the optocoupler change with time. Also, since the phototransistor is often being saturated, response times can range from 10 to 20 µs in slower devices, limiting high speed applications.

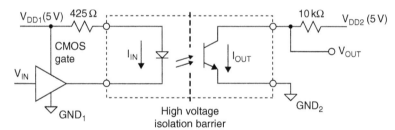

- Uses light for transmission over a high voltage barrier.
- The LED is the transmitter, and the phototransistor is the receiver.
- High voltage isolation: 5000 V to 7000 V RMS.
- Nonlinear – Best for digital or frequency information.
- Rise and fall times can be 10–20 µs in slower devices.
- Example: Siemens ILQ-1 quad (http://www.siemens.com).

**Figure 2-32: Digital isolation using LED/phototransistor optocouplers**

A much faster optocoupler architecture is shown in Figure 2-33 and is based on an LED and a photodiode. The LED is again driven with a current of approximately 10 mA. This produces a light output sufficient to generate enough current in the receiving photodiode to develop a valid high logic level at the output of the transimpedance amplifier. Speed can vary widely between optocouplers, and the fastest ones have propagation delays of 20 ns typical, and 40 ns maximum, and can handle data rates up to 25 MBd for NRZ data. This corresponds to a maximum square wave operating frequency of 12.5 MHz, and a minimum allowable passable pulse width of 40 ns.

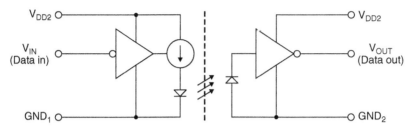

- +5 V supply voltage.
- 2500 V RMS I/O withstand voltage.
- Logic signal frequency: 12.5 MHz maximum.
- 25 MBd maximum data rate.
- 40 ns maximum propagation delay.
- 9 ns typical rise/fall time.
- Example: Agilent HCPL-7720.
- (http://www.semiconductor.agilent.com).

**Figure 2-33: Digital isolation using LED/photodiode optocouplers**

## AD260/AD261 High Speed Logic Isolators

The AD260/AD261 family of digital isolators operates on a principle of transformer-coupled isolation (see Reference 4). They provide isolation for five digital control signals to/from high speed DSPs, microcontrollers, or microprocessors. The AD260 also has a 1.5 W transformer for a $3.5\,kV_{RMS}$ isolated external AC/DC power supply circuit.

Each line of the AD260 can handle digital signals up to 20 MHz (40 MBd) with a propagation delay of only 14 ns which allows for extremely fast data transmission. Output waveform symmetry is maintained to within ±1 ns of the input so the AD260 can be used to accurately isolate time-based pulse width modulator (PWM) signals.

A simplified schematic of one channel of the AD260/AD261 is shown in Figure 2-34. The data input is passed through a Schmitt trigger circuit, through a latch, and a special transmitter circuit which

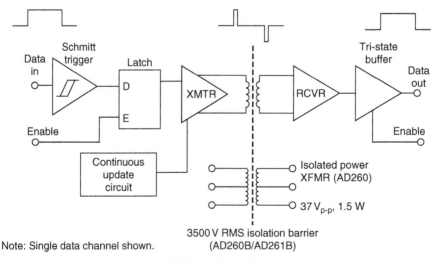

Note: Single data channel shown.

**Figure 2-34: AD260/AD261 digital isolators**

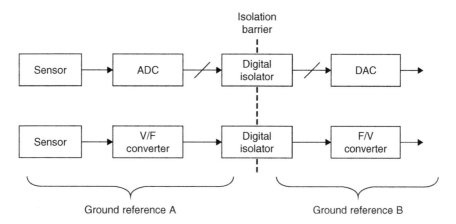

**Figure 2-35: Practical application of digital isolation in data acquisition systems**

differentiates the edges of the digital input signal and drives the primary winding of a proprietary transformer with a "set-high/set-low" signal.

The secondary of the isolation transformer drives a receiver with the same "set-high/set-low" data, which regenerates the original logic waveform. An internal circuit operating in the background interrogates all inputs about every 5 μs, and in the absence of logic transitions, sends appropriate "set-high/set-low" data across the interface. Recovery time from a fault condition or at power-up is thus between 5 μs and 10 μs.

The power transformer (available on the AD260) is designed to operate between 150 kHz and 250 kHz and will easily deliver more than 1 W of isolated power when driven push–pull (5 V) on the transmitter side. Different transformer taps, rectifier and regulator schemes will provide combinations of ±5 V, 15 V, 24 V, or even 30 V or higher.

The transformer output voltage when driven with a low voltage drop drive will be $37 V_{p-p}$ across the entire secondary with a 5 V push–pull drive. The availability of low cost digital isolators such as those previously discussed solves most system isolation problems in data acquisition systems as shown in Figure 2-35. In the upper example, digitizing the signal first and then using digital isolation eliminates the problem of analog isolation amplifiers. While digital isolation can be used with parallel output ADCs provided the bandwidth of the isolator is sufficient, it is more practical with ADCs that have *serial* outputs. This minimizes cost and component count. A three-wire interface (data, serial clock, framing clock) is all that is required in these cases.

An alternative (lower example) is to use a voltage-to-frequency converter (VFC) as a transmitter and a frequency-to-voltage converter (FVC) as a receiver. In this case, only one digital isolator is required.

## *iCoupler® Technology*

In many industrial applications, such as process control systems or data acquisition and control systems, digital signals must be transmitted from various sensors to a central controller for processing and analysis. The controller then needs to transmit commands as a result of the analysis performed, coupled with user inputs to various actuators, to achieve certain operations. To maintain safety voltage at the user interface and to prevent transients from being transmitted from the sources, galvanic isolation is required. There are three commonly known classes of isolation devices: optocouplers, capacitively coupled isolators, and transformer-based isolators. Optocouplers rely on light emitting diodes to convert the electrical signals to light signals and on photo detectors to convert the light signals back to electrical signals. The intrinsic low conversion efficiencies for electrical light conversion and slow response photo detectors lead to optocoupler limitations in terms of lifetime, speed, and power assumption. The capacitively coupled isolators have limitations in their size and ability to reject CM voltage transients, while the traditional transformer assembly based isolators are bulky and expensive. All these isolators are restricted, moreover, because of integrated circuit integration limitations and the fact that they often need hybrid packaging.

Recently *i*Coupler®, a new isolation technology, based on chip scale transformers, was developed by Analog Devices. The first product was the ADuM1100 single-channel digital isolator. *i*Coupler® technology leverages thick-film processing techniques to build microscale on-chip transformers and achieves thousands of volts of isolation on a chip.

*i*Coupler® isolated transformers can be monolithically integrated with standard silicon ICs and can be fabricated in single- or multichannel configurations. The bidirectional nature of inductive coupling further facilitates bidirectional signal transfer. The combination of high bandwidth for these on-chip transformers and fine-scale CMOS circuitry leads to isolators of unmatched performance characteristics in power, speed, timing accuracy, and ease of use.

## ADuM1100 Architecture: A Single-Channel Digital Isolator

The ADuM1100 is a single-channel 100 Mbps digital isolator. It has two ICs packaged in an eight-lead SOIC package. A cross-section view of the ADuM1100 is shown in Figure 2-36. There are two lead frame paddles inside the package, with a gap between them of about 0.4 mm. The molding compound has breakdown strength over 25 kV/mm, so the 0.4 mm gap filled with molding compound provides greater than 10 kV insulation between the substrates of two IC chips.

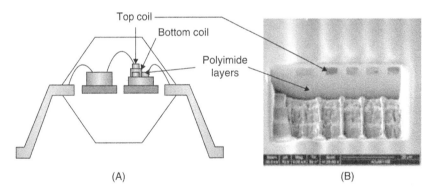

**Figure 2-36: (A) Cross-sectional view of ADuM1100 in an eight-lead SOIC package. (B) Cross-sectional view of the top coil and polyimide layers**

The driver chip sitting on the left paddle takes the input digital signal, encodes it, and drives the encoded differential signal through bond wires to the top coils of the transformers built on top of the receiver chip sitting on the right paddle. The driver die is a standard CMOS chip, and the receiver die is a CMOS chip with the additional structures of two polyimide layers and transformer primary coil fabricated on top of the passivation. The polyimide between the top and bottom coils is about 20 μm thick. The breakdown strength of the cured polyimide film is greater than 300 V/m, so 20 μm of polyimide provides greater than 6 kV of insulation between a given transformer's coils. This provides a comfortable margin over the production test voltage of $3 kV_{RMS}$.

Because of the structural quality of these wafer processed polyimide films, no partial discharge over 5 pC can be detected, even at $3\,kV_{RMS}$. The top coil is gold plated, with a 4-μm thick layer, and the coil track width and spacing between the turns are all 4 μm. The polyimide layers have good mechanical elongation and tensile strength, which also helps the adhesion between the polyimide layers or between the polyimide layer and deposited metal layer. The minimum interaction between the gold film and the polyimide film, coupled with high temperature stability of the polyimide film, results in a system that provides reliable insulation when subjected to various types of environmental stress.

In addition to the fact that thousands of volts of isolation can be achieved on-chip, the ADuM110 also makes it possible to transmit very high bandwidth signals very efficiently, accurately, and reliably. Figure 2-37 shows a simplified schematic of the ADuM1100. To guarantee input stability, the front glitch filter filters out pulses narrower than a pulse width of approximately 2 ns. Upon the receipt of a signal edge, a 1 ns pulse is sent to either Coil 1 or Coil 2. (For a leading edge signal it is sent to Coil 1, and for a falling edge signal to Coil 2.) Once the short pulses are transmitted to the secondary coils (the bottom coils in this case), they are amplified and the input signal is reconstructed through an SR flip-flop to appear as an isolated output. The wide bandwidth of these microscale transformers and high speed CMOS makes the transmission of these short nanosecond pulses possible. Since only signal edges are being used, this transmission scheme is very power efficient. With a very energetic pulse having a current ramping to 100 mA within 1 ns, the average current for a 1 Mbps input signal is only 50 μA. Some additional power is dissipated by the switching of the surrounded CMOS gates. At 5 V, an additional 50 μA/Mbps is needed if the total capacitance of the CMOS gates is 20 pF. The typical optocoupler, on the other hand, dissipates over 10 mA, even operating at 1 Mbps. This represents two orders of magnitude (100×) improvement in power dissipation provided by *i*Coupler® isolators.

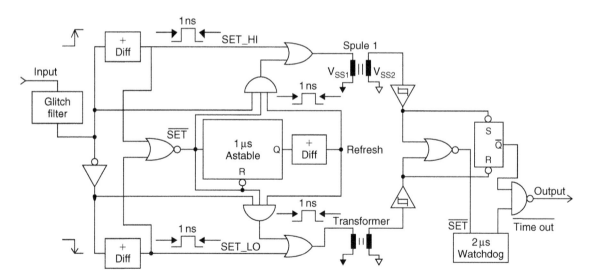

**Figure 2-37: ADuM1100 simplified schematic**

If there is no input change for a certain period of time, approximately 1 μs, the monostable generates a 1 ns pulse and sends it to Coil 1 or Coil 2, depending on the input logic level. The 1 ns refreshing pulse is sent to Coil 1 if input is high and is sent to Coil 2 if input is low. This helps maintain DC correctness for the isolator because normally pulses are transmitted only on reception of a signal edge. The receiver includes a watchdog circuit that will timeout at 2 μs if it is not reset by an incoming pulse. If a timeout happens, the receiver output will return to a default safe level (logic high in the ADuM1100). The combination of refresh and watchdog functions provides the additional advantage of detecting the failure of any field device on the system side. With other isolators, this would ordinarily require the use of an extra isolated data channel.

The bandwidth of the isolator is dependent on the input filter bandwidth within. For example, 500 Mbps can be achieved with a 2 ns input filter. For the ADuM1100, we chose a signal bandwidth of 100 MBd, still 2 times faster than the fastest optocouplers. Very tight edge symmetry between input and output logic signals is also preserved due to the instantaneous nature of the inductive coupling between these microscale on-chip coils.

The ADuM1100 has edge symmetry of better than 2 ns for 5 V operation. As the bandwidth of isolation systems continues to expand, the *i*Coupler® technology will be capable of meeting the challenge while optocoupler technology is likely to struggle.

In addition to the improvements in efficiency and bandwidth *i*Coupler® technology provides, it also offers a more robust and reliable isolation solution than competitive offerings. Because high voltage transients are present in many data acquisition and control systems, the ability of the isolator to prevent transients from affecting the logic controller is very important. High performance optocouplers have transient immunity of less than 10 kV/μs, while the ADuM1100 has a transient immunity better than 25 kV/μs. The induced error voltage at the receiver input induced by an input–output transient is given by:

$$V = C \times R \frac{dV}{dt} \tag{2-10}$$

where C is the capacitance between the input coil and the receiver coil, R, is the resistance of the bottom coil, and dV/dt is the magnitude of the transient.

In the ADuM1100, the capacitance between the top (input) coil and the bottom (receiver) coil is only 0.2 pF, while the bottom coil has a resistance of 80 Ω. Thus the error signal induced on the bottom coil by a 25 kV/μs transient on the top coil is only 0.4 V, much less than the receiver detection threshold. The transient immunity of *i*Coupler® isolators can be optimized through careful selection of the decoder detection threshold, the resistance of the receiving coil, and, of course, the capacitance between the top and bottom coils.

One recurring question about transformer-based isolators involves their magnetic immunity capability. Since *i*Couplers® use air core technology, no magnetic components are present and the problem of magnetic saturation for the core material does not exist. Therefore, *i*Couplers® have essentially infinite DC field immunity. The limitation on the ADuM1100's AC magnetic field immunity is set by the condition in which the induced error voltage in the receiving coil (the bottom coil in this case) is made sufficiently large, either to falsely set or reset the decoder. The voltage induced across the bottom coil is given by:

$$V = \left[ -\frac{d\beta}{dt} \right] \sum \pi r_n^2; \quad n = 1, 2, \ldots, N \tag{2-11}$$

where $\beta$ is the magnetic flux density (Gauss), N is the number of turns in receiving coil, and $r_n$ is the radius of nth turn in receiving coil (cm).

Because of the very small geometry of the receiving coil in the ADuM1100, even a wire carrying 1,000 A at 1 MHz and positioned only 1 cm away from the ADuM1100 would not induce an error voltage large enough to falsely trigger the decoder. Note that at combinations of strong magnetic field and high frequency, any loops formed by printed circuit board traces could induce error voltages sufficiently large to trigger the thresholds of succeeding circuitry. Typically the PC board design rather than the isolator itself is the limiting factor in the presence of such big magnetic transients. In addition to magnetic immunity, the level of electromagnetic radiation emitted from the *i*Coupler® device is a concern. Using far-field approximation,

$$P = 160\pi^2 I^2 \sum r_n^4; \quad n = 1, 2, \dots, N \tag{2-12}$$

where P is the total radiated power and I is the coil loop current.

Again, given the very small geometry of the coils, the total radiated power is still less than 50 pW, even if the part is operating at 0.5 GHz.

## ADuM130x/ADuM140x: Multichannel Products

In addition to the many performance improvements discussed previously, *i*Coupler® technology also offers tremendous advantages in terms of integration. The optical interference makes the realization of multichannel optocouplers very difficult.

Transformers based on *i*Coupler® technology can be easily integrated onto a single chip. Furthermore, one data channel can transmit signals in one direction, say from the top coil to the bottom coil, while the neighboring channel can transmit a signal in the other direction, from the bottom coil to the top coil. The bidirectional nature of inductive coupling makes this possible.

Additional products consist of five three-channel and four-channel products covering all possible channel directionality configurations. Besides providing flexible channel configurations, they support both 3 V and

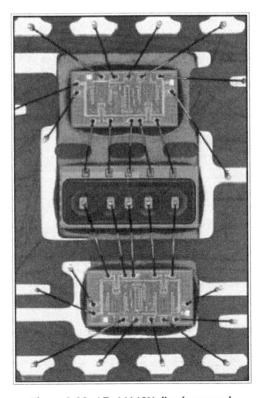

**Figure 2-38: ADuM140X die photograph**

5V operations at either side of the isolation barrier and support the use of these isolators as level translators. One side could be at 2.7V, e.g., while the other side could be at 5.5V. The edge symmetry of 2 ns is preserved over all possible supply configurations at all temperatures from −40°C to 100°C. The ability to mix bidirectional channels of isolation in a single package enables users to reduce the size and cost of their systems.

For the ADuM1100, two transformers are used to transmit a single channel of data. One is dedicated to transmit pulses representing the signal's leading edge or updating input high, and the other is dedicated to transmit pulses representing the signal's falling edge or updating input low. For the ADuM130x/ADuM140x product family, a single transformer is used for each data channel. The ADuM140x shown in Figure 2-38 has four transformers in total. The leading edge and falling edge are encoded differently, and the encoded pulses are combined in the same transformer; as a result, the receiver has responsibility for decoding the pulses to see whether they are for leading edge or falling edge. The output signal is then reconstructed correspondingly (Figure 2-39).

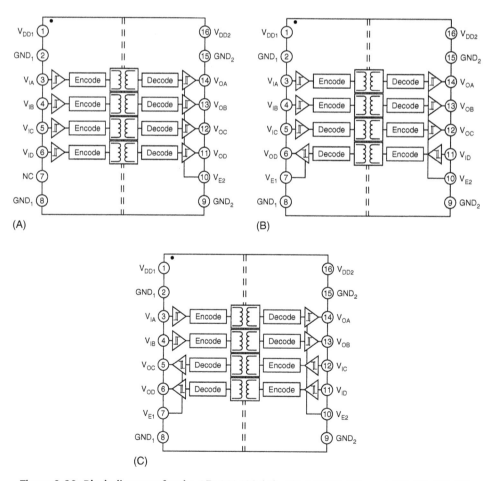

Figure 2-39: Block diagrams for the ADuM1400 (A), ADuM1401 (B), and ADuM1402 (C)

Of course, there is a penalty for using a single transformer per data channel rather than using two transformers per data channel. The propagation delay is longer for the single transformer architecture because of the additional encode and decode time needed. The penalty for bandwidth is hardly a factor, even at input speed of 100 Mbps.

In contrast to the ADuM1100, the ADuM130x/ADuM140x uses a dedicated transformer chip, separate from the receiver integrated circuit. This partitioning exemplifies the ease of integration for *i*Coupler technology. Besides standalone multichannel isolators, the *i*Coupler technology can be embedded with other data acquisition and control ICs to make the use of isolation even more transparent. Consequently, in the future, system designers will be able to devote their time to improving system functionality, rather than worrying about isolation.

# Active Feedback Amplifiers

The AD8129/AD8130 differential line receivers, along with their predecessor the AD830, utilize a novel amplifier topology called *active feedback* (see Reference 8). A simplified block diagram of these devices is shown in Figure 2-40.

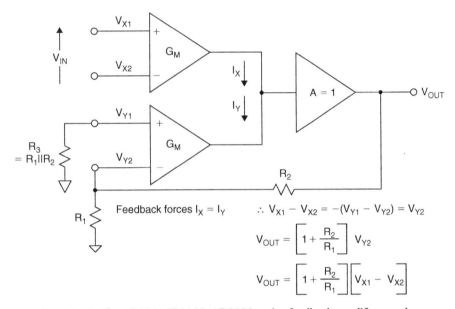

**Figure 2-40: The AD830/AD8129/AD8130 active feedback amplifier topology**

The AD830 and the AD8129/AD8130 have two sets of fully differential inputs, available at $V_{X1} - V_{X2}$ and $V_{Y1} - V_{Y2}$, respectively. Internally, the outputs of the two GM stages are summed and drive a buffer output stage.

In this device the overall feedback loop forces the internal currents $I_X$ and $I_Y$ to be equal. This condition forces the differential voltages $V_{X1} - V_{X2}$ and $V_{Y1} - V_{Y2}$ to be equal and opposite in polarity. Feedback is taken from the output back to one input differential pair, while the other pair is driven directly by an input differential input signal.

An important point of this architecture is that high CM rejection is provided by the two differential input pairs, so *CMR is not dependent on resistor bridges* and their associated matching problems. The inherently wideband balanced circuit and the quasi-floating operation of the driven input provide the high CMR, which is typically 100 dB at DC.

One way to view this topology is as a standard op amp in a non-inverting mode with a pair of differential inputs in place of the op amps standard inverting and non-inverting inputs. The general expression for the stage's gain "G" is like a non-inverting op amp, or:

$$G = \frac{V_{OUT}}{V_{IN}} = 1 + \frac{R_2}{R_1} \tag{2-13}$$

As should be noted, this expression is identical to the gain of a non-inverting op amp stage, with $R_2$ and $R_1$ in analogous positions.

The AD8129 is a low noise high gain (G = 10 or greater) version of this family, intended for applications with very long cables where signal attenuation is significant. The related AD8130 device is stable at a gain of one. It is used for those applications where lower gains are required, such as a gain of two, for driving source and load terminated cables.

The AD8129 and AD8130 have a wide power supply range, from single +5V to ±12V, allowing wide CM and differential-mode voltage ranges. The wide CM range enables the driver/receiver pair to operate without isolation transformers in many systems where the ground potential difference between driver and receiver locations is several volts. Both devices include a logic-controlled power-down function.

Both devices have high, balanced input impedances, and achieve 70 dB CMR @ 10 MHz, providing excellent rejection of high frequency CM signals. Figure 2-41 shows AD8130 CMR for various supplies. As can be noted, it can be as high as 95 dB at 1 MHz, an impressive figure considering that no trimming is required.

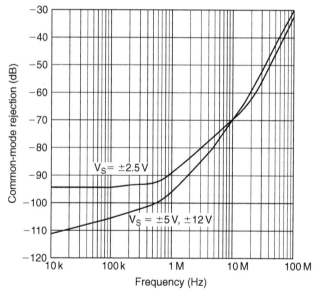

**Figure 2-41: AD8130 CMR versus frequency for ±2.5V, ±5V, and ±12V supplies**

The typical 3 dB bandwidth for the AD8129 is 200 MHz, while the 0.1 dB bandwidth is 30 MHz in the SOIC package, and 50 MHz in the μSOIC package. The conditions for these specifications are for $V_S$ = ±5V and G = 10.

The typical 3 dB bandwidth for the AD8130 is 270 MHz, and the 0.1 dB bandwidth is 45 MHz, in either package. The conditions for these specifications are for $V_S$ = ±5V and G = 1. Typical differential gain and phase specifications for the AD8130 for G = 2, $V_S$ = ±5V, and $R_L$ = 150Ω are 0.13% and 0.15°, respectively.

# Logarithmic Amplifiers

The term "logarithmic amplifier" (generally abbreviated to "log amp") is something of a misnomer, and "logarithmic converter" would be a better description. The conversion of a signal to its equivalent logarithmic value involves a nonlinear operation, the consequences of which can be confusing if not fully understood. It is important to realize that many of the familiar concepts of linear circuits are irrelevant to log amps. For example, the incremental gain of an ideal log amp approaches infinity as the input approaches zero, and a change of offset at the output of a log amp is equivalent to a change of amplitude at its input—not a change of input offset.

For the purposes of simplicity in our initial discussions, we shall assume that both the input and the output of a log amp are voltages, although there is no particular reason why logarithmic current, transimpedance, or transconductance amplifiers could not also be designed.

If we consider the equation $y = \log(x)$ we find that every time x is multiplied by a constant A, y increases by another constant A1. Thus if $\log(K) = K1$, then $\log(AK) = K1 + A1$, $\log(A^2K) = K1 + 2A1$, and $\log(K/A) = K1 - A1$. This gives a graph as shown in Figure 2-42, where y is zero when x is unity, y approaches minus infinity as x approaches zero, and where y has no values for which x is negative.

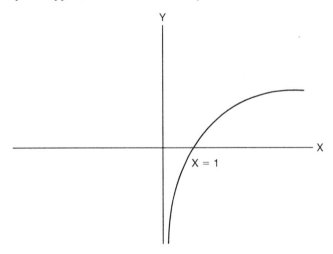

**Figure 2-42: Graph of Y = Log(X)**

On the whole, log amps do not behave in this way. Apart from the difficulties of arranging infinite negative output voltages, such a device would not, in fact, be very useful. A log amp must satisfy a transfer function of the form:

$$V_{OUT} = V_Y \log\left(\frac{V_{IN}}{V_X}\right) \qquad (2\text{-}14)$$

over some range of input values which may vary from 100:1 (40 dB) to over 1,000,000:1 (120 dB).

With inputs very close to zero, log amps cease to behave logarithmically, and most then have a linear $V_{IN}$/$V_{OUT}$ law. This behavior is often lost in device noise. Noise often limits the dynamic range of a log amp. The constant, $V_Y$, has the dimensions of voltage, because the output is a voltage. The input, $V_{IN}$, is divided by a voltage, $V_X$, because the argument of a logarithm must be a simple dimensionless ratio.

A graph of the transfer characteristic of a log amp is shown in Figure 2-43. The scale of the horizontal axis (the input) is logarithmic, and the ideal transfer characteristic is a straight line. When $V_{IN} = V_X$, the logarithm is zero (log 1 = 0). $V_X$ is therefore known as the *intercept voltage* of the log amp because the graph crosses the horizontal axis at this value of $V_{IN}$.

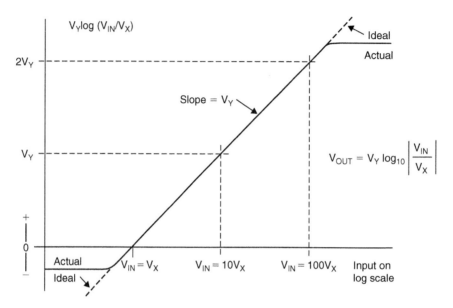

**Figure 2-43: Log amp transfer function**

The slope of the line is proportional to $V_Y$. When setting scales, logarithms to the base 10 are most often used because this simplifies the relationship to decibel values: when $V_{IN} = 10V_X$, the logarithm has the value of 1, so the output voltage is $V_Y$. When $V_{IN} = 100 V_X$, the output is $2 V_Y$, and so forth. $V_Y$ can therefore be viewed either as the "slope voltage" or as the "volts per decade factor."

The logarithm function is indeterminate for negative values of x. Log amps can respond to negative inputs in three different ways: (1) They can give a full-scale negative output as shown in Figure 2-44. (2) They can give an output which is proportional to the log of the absolute value of the input and disregards its sign as shown in Figure 2-45. This type of log amp can be considered to be a full-wave detector with a logarithmic characteristic and is often referred to as a *detecting* log amp. (3) They can give an output which is proportional to the log of the absolute value of the input and has the same sign as the input as shown in Figure 2-46. This type of log amp can be considered to be a video amp with a logarithmic characteristic, and may be known as a *logarithmic video (log video)* amplifier or, sometimes, a *true log amp* (although this type of log amp is rarely used in video-display-related applications).

There are three basic architectures which may be used to produce log amps: the *basic diode log amp*, the *successive detection log amp*, and the *true log amp* which is based on cascaded semi-limiting amplifiers. The successive detection log amp and the true log amp are discussed in the RF/IF section.

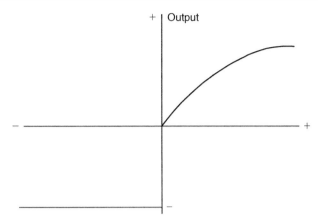

**Figure 2-44: Basic log amp (saturates with negative inputs)**

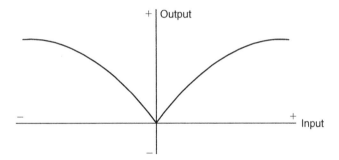

**Figure 2-45: Detecting log amp (output polarity independent of input polarity)**

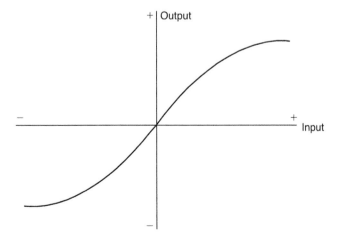

**Figure 2-46: Log video or "true log amp" (symmetrical response to positive or negative signals)**

The voltage across a silicon diode is proportional to the logarithm of the current through it. If a diode is placed in the feedback path of an inverting op amp, the output voltage will be proportional to the log of the input current as shown in Figure 2-47. In practice, the dynamic range of this configuration is limited to 40–60 dB because of non-ideal diode characteristic, but if the diode is replaced with a diode-connected transistor as shown in Figure 2-48, the dynamic range can be extended to 120 dB or more. This type of log amp has three disadvantages: (1) both the slope and intercept are temperature dependent; (2) it will only handle unipolar signals; and (3) its bandwidth is both limited and dependent on signal amplitude.

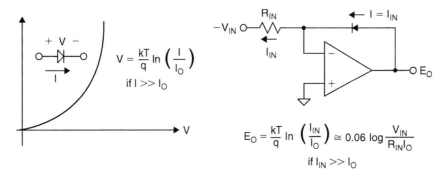

**Figure 2-47: The diode/op amp log amp**

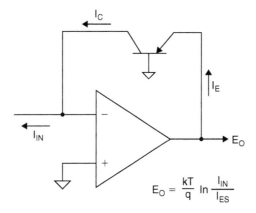

**Figure 2-48: Transistor/op amp log amp**

Where several such log amps are used on a single chip to produce an analog computer which performs both log and antilog operations, the temperature variation in the log operations is unimportant, since it is compensated by a similar variation in the antilogging. This makes possible the AD538 (Figure 2-49), a monolithic analog computer which can multiply, divide, and raise to powers. Where actual logging is required, however, the AD538 and similar circuits require temperature compensation (Reference 7). The major disadvantage of this type of log amp for high frequency applications, though, is its limited frequency response—which cannot be overcome. However carefully the amplifier is designed, there will always be a residual feedback capacitance, $C_C$ (often known as Miller capacitance), from output to input which limits the high frequency response.

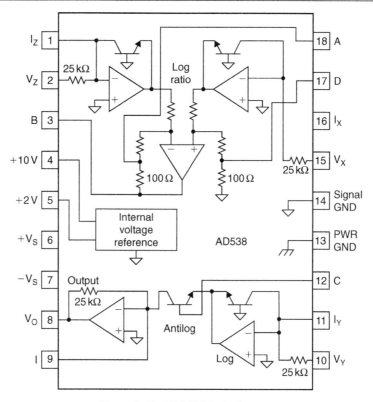

**Figure 2-49: AD538 block diagram**

What makes this Miller capacitance particularly troublesome is that the impedance of the emitter–base junction is inversely proportional to the current flowing in it—so that if the log amp has a dynamic range of 1,000,000:1, then its bandwidth will also vary by 1,000,000:1. In practice, the variation is less because other considerations limit the large signal bandwidth, but it is very difficult to make a log amp of this type with a small signal bandwidth greater than a few hundred kHz.

We also discuss high speed log amps in the RF/IF section (Section 4-4).

# *High Speed Clamping Amplifiers*

There are many situations where it is desirable to clamp the output of an op amp to prevent overdriving the circuitry which follows. Specially designed high speed, fast recovery clamping amplifiers offer an attractive alternative to designing external clamping/protection circuits. The AD8036/AD8037's low distortion, wide bandwidth clamp amplifiers represent a significant breakthrough in this technology. These devices allow the designer to specify a high ($V_H$) and low ($V_L$) clamp voltage. The output of the device clamps when the input exceeds either of these two levels. The AD8036/AD8037 offers superior clamping performance compared to competing devices that use output clamping. Recovery time from overdrive is less than 5 ns.

The key to the AD8036 and AD8037's fast, accurate clamp and amplifier performance is their proprietary input clamp architecture. This new design reduces clamp errors by more than $10\times$ over previous output clamp-based circuits, as well as substantially increasing the bandwidth, precision, and versatility of the clamp inputs.

Figure 2-50 shows an idealized block diagram of the AD8036 connected as a unity-gain voltage follower. The primary signal path comprises A1 (a 1,200 V/µs, 240 MHz high voltage gain, differential to single-ended amplifier) and A2 (a G = +1 high current gain output buffer). The AD8037 differs from the AD8036 only in that A1 is optimized for closed-loop gains of two or greater.

The input clamp section is composed of comparators $C_H$ and $C_L$, which drive switch S1 through a decoder. The unity-gain buffers in series with the $+V_{IN}$, $V_H$, and $V_L$ inputs isolate the input pins from the comparators and S1 without reducing bandwidth or precision.

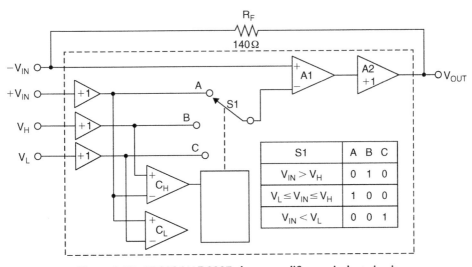

**Figure 2-50: AD8036/AD8037 clamp amplifier equivalent circuit**

The two comparators have about the same bandwidth as A1 (240 MHz), so they can keep up with signals within the useful bandwidth of the AD8036. To illustrate the operation of the input clamp circuit, consider the case where $V_H$ is referenced to +1 V, $V_L$ is open, and the AD8036 is set for a gain of +1 by connecting its output back to its inverting input through the recommended 140 Ω feedback resistor. Note that the main signal path always operates closed loop, since the clamping circuit only affects A1's non-inverting input.

If a 0 V to +2 V voltage ramp is applied to the AD8036's $+V_{IN}$ for the connection just described, $V_{OUT}$ should track $+V_{IN}$ perfectly up to +1 V, then should limit at exactly +1 V as $+V_{IN}$ continues to +2 V.

In practice, the AD8036 comes close to this ideal behavior. As the $+V_{IN}$ input voltage ramps from zero to 1 V, the output of the high limit comparator $C_H$ starts in the off state, as does the output of $C_L$. When $+V_{IN}$ just exceeds $V_H$ (practically, by about 18 mV), $C_H$ changes state, switching S1 from "A" to "B" reference level. Since the + input of A1 is now connected to $V_H$, further increases in $+V_{IN}$ have no effect on the AD8036's output voltage. The AD8036 is now operating as a unity-gain buffer for the $V_H$ input, as any variation in $V_H$, for $V_H > 1$ V, will be faithfully produced at $V_{OUT}$.

Operation of the AD8036 for negative input voltages and negative clamp levels on $V_L$ is similar, with comparator $C_L$ controlling S1. Since the comparators see the voltage on the $+V_{IN}$ pin as their common reference level, the voltage $V_H$ and $V_L$ are defined as "High" or "Low" with respect to $+V_{IN}$. For example, if $V_{IN}$ is set to 0 V, $V_H$ is open, and $V_L$ is +1 V, comparator $C_L$ will switch S1 to "C," so the AD8036 will buffer the voltage on $V_L$ and ignore $+V_{IN}$.

The performance of the AD8036/AD8037 closely matches the ideal just described. The comparator's threshold extends from 60 mV inside the clamp window defined by the voltages on $V_L$ and $V_H$ to 60 mV beyond the window's edge. Switch S1 is implemented with current steering, so that A1's + input makes a continuous transition from, say, $V_{IN}$ to $V_H$ as the input voltage traverses the comparator's input threshold from 0.9 V to 1.0 V for $V_H = 1.0$ V.

The practical effect of the non-ideal operation is to soften the transition from amplification to clamping modes, without compromising the absolute clamp limit set by the input clamping circuit. Figure 2-51 shows

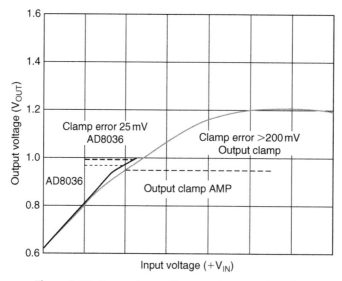

**Figure 2-51: Comparison of input and output clamping**

a graph of $V_{OUT}$ versus $V_{IN}$ for the AD8036 and a typical *output* clamp amplifier. Both amplifiers are set for $G = +1$ and $V_H = +1\,V$.

The worst-case error between $V_{OUT}$ (ideally clamped) and $V_{OUT}$ (actual) is typically $18\,mV$ times the amplifier closed-loop gain. This occurs when $V_{IN}$ equals $V_H$ (or $V_L$). As $V_{IN}$ goes above and/or below this limit, $V_{OUT}$ will stay within $5\,mV$ of the ideal value.

In contrast, the output clamp amplifier's transfer curve typically will show some compression starting at an input of $0.8\,V$, and can have an output voltage as far as $200\,mV$ over the clamp limit. In addition, since the output clamp causes the amplifier to operate open loop in the clamp mode, the amplifier's output impedance will increase, potentially causing additional errors, and the recovery time is significantly longer.

It is important that a clamped amplifier such as the AD8036/AD8037 maintain low levels of distortion when the input signals approach the clamping voltages. Figure 2-52 shows the second and third harmonic distortions for the amplifiers as the output approaches the clamp voltages. The input signal is $20\,MHz$, the output signal is $2\,V$ peak-to-peak, and the output load is $100\,\Omega$.

Recovery from step voltage which is two times over the clamping voltage is shown in Figure 2-53. The input step voltage starts at $+2\,V$ and goes to $0\,V$ (left-hand traces on scope photo). The input clamp voltage ($V_H$) is set at $+1\,V$. The right-hand trace shows the output waveform.

Figure 2-54 shows the AD9002 8-bit, 125MSPS flash converter driven by the AD8037 ($240\,MHz$ bandwidth) clamping amplifier. The clamp voltages on the AD8037 are set to $+0.55$ and $-0.55\,V$,

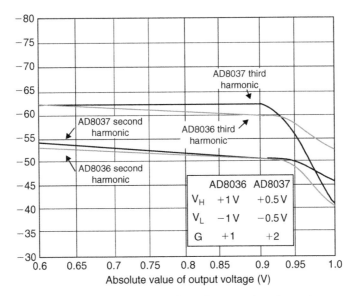

**Figure 2-52: AD8036/AD8037 distortion near clamping region, output = $2\,V_{p-p}$, load = $100\,\Omega$, f = $20\,MHz$**

referenced to the $\pm0.5\,V$ input signal, with the external resistive dividers. The AD8037 also supplies a gain of two, and an offset of $-1\,V$ (using the AD780 voltage reference), to match the 0 to $-2\,V$ input range of the AD9002 flash converter. The output signal is clamped at $+0.1\,V$ and $-2.1\,V$. This multi-function clamping circuit therefore performs several important functions as well as preventing damage to the flash converter which occurs if its input exceeds $+0.5\,V$, thereby forward biasing the substrate diode. The 1N5712 Schottky diode adds further protection during power-up.

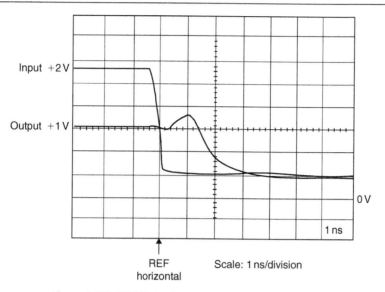

Figure 2-53: AD8036/AD8037 overdrive (2×) recovery

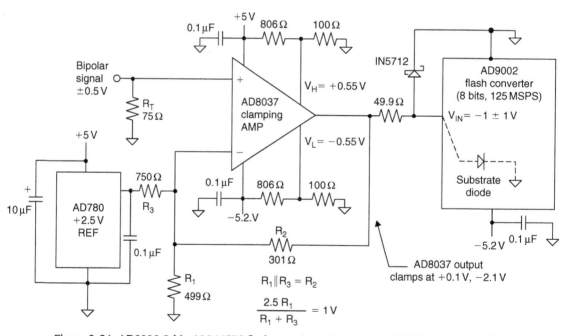

Figure 2-54: AD9002 8-bit, 125 MSPS flash converter driven by an AD8037 clamp amplifier

The feedback resistor, $R_2 = 301\,\Omega$, is selected for optimum bandwidth as per the data sheet recommendation. For a gain of two, the parallel combination of $R_1$ and $R_3$ must also equal $R_2$:

$$\frac{R_1 \times R_3}{R_1 + R_3} = R_2 = 301\,\Omega \quad \text{(nearest 1\% standard resistor value)} \qquad (2\text{-}15)$$

In addition, the Thevenin equivalent output voltage of the AD780 +2.5 V reference and the $R_3/R_1$ divider must be +1 V to provide the −1 V offset at the output of the AD8037:

$$\frac{2.5 \times R_1}{R_1 + R_3} = 1 \text{ V} \tag{2-16}$$

Solving the equations yields $R_1 = 499\,\Omega$, $R_3 = 750\,\Omega$ (using the nearest 1% standard resistor values).

Other input and output voltages ranges can be accommodated by appropriate changes in the external resistors.

Further examples of applications of these fast clamping op amps are given in Reference 9.

# *Comparators*

A comparator is similar to an op amp. It has two inputs, inverting and non-inverting, and an output. But it is specifically designed to compare the voltages between its two inputs. Therefore it operates in a nonlinear fashion. The comparator operates open loop, providing a two-state logic output voltage. These two states represent the sign of the net difference between the two inputs (including the effects of the comparator input offset voltage). Therefore, the comparator's output will be a logic "1" if the input signal on the non-inverting input exceeds the signal on the inverting input (plus the offset voltage, $V_{OS}$) and a logic "0" for the opposite case. A comparator is normally used in applications where some varying signal level is compared to a fixed level (usually a voltage reference). Since it is, in effect, a 1-bit ADC, the comparator is a basic element in all ADCs (see Figure 2-55).

Comparator DC specifications are similar to those of op amps: input offset voltage, input bias current, offset and drift, CM input range, gain, CMR, and PSR. Standard logic-related DC, timing, and interface specs are associated with the comparator outputs.

The key comparator AC specification is *propagation delay*: it is the time required for the output to reach the 50% point of a transition, after the differential input signal crosses the offset voltage—when driven by a square wave (typically 100 mV in amplitude) to a prescribed value of input overdrive (usually 5 mV or 10 mV).

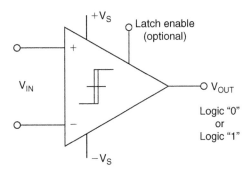

**Figure 2-55: Comparator symbol**

The propagation delay in practical comparators decreases somewhat as the input overdrive is increased. This variation in propagation delay as a function of overdrive is called *dispersion* (see Figures 2-56 and 2-57).

The addition of hysteresis, which is application of a small amount of positive feedback, to a comparator's transfer function is often useful in a noisy environment, or where it is undesirable for the comparator to toggle continuously between states when the input signal is at or near the switching threshold. This is true when a relatively slowly changing input is compared to a DC level. Noise can cause the output to toggle between the output levels many times. The transfer function for a comparator with hysteresis is shown in Figure 2-58.

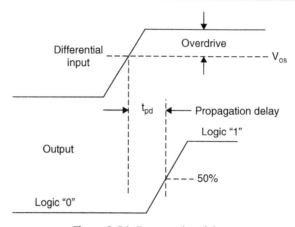

**Figure 2-56: Propagation delay**

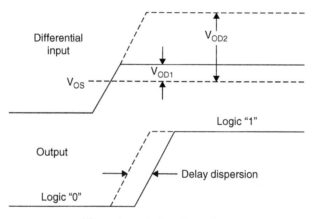

**Figure 2-57: Delay dispersion**

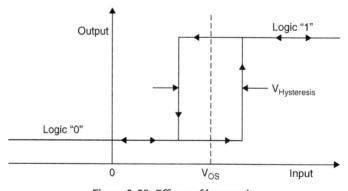

**Figure 2-58: Effects of hysteresis**

If the input voltage approaches the switching threshold ($V_{OS}$) from the negative direction, the comparator will switch from a "0" to a "1" when the input crosses $V_{OS} + V_H/2$. The "new" switching threshold now becomes $V_{OS} - V_H/2$. The comparator output will remain in a "1" state until the threshold $V_{OS} - V_H/2$ is crossed, coming from the positive direction. Input noise centered around $V_{OS}$ will not cause the comparator to switch states unless it exceeds the region bounded by $V_{OS} \pm V_H/2$.

Hysteresis can be accomplished with two resistors (see Figure 2-59); the amount of hysteresis is proportional to the resistors ratio. The signal input to the comparator may be applied to either the inverting or the non-inverting input, but if it is applied to the inverting input its source impedance must be low enough to have insignificant effect on $R_1$ (of course if the source impedance is sufficiently predictable it may be used as $R_1$).

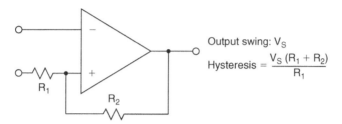

Output swing: $V_S$

$$\text{Hysteresis} = \frac{V_S (R_1 + R_2)}{R_1}$$

- Input signal may be applied to either input but its source impedance must be low if it is applied to $R_1$

**Figure 2-59: Application of hysteresis**

If the trip voltage is midway between the two comparator output voltages (as is the case with a symmetrical power supply and a ground reference), then the introduction of hysteresis will move the positive and negative thresholds equal distances from the trip point voltage, but if the trip point is nearer to one output than to the other the thresholds will be asymmetrically placed about the trip point voltage (Figure 2-60).

To calculate the hysteresis, assume the comparator output voltages are $V_p$ and $V_n$ respectively. The comparator trip point voltage is $V_{OS}$. The negative threshold is:

$$\frac{(R_1 + R_2) V_{OS} - R_1 V_n}{R_2} \tag{2-17}$$

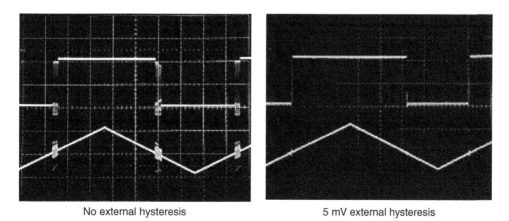

No external hysteresis                5 mV external hysteresis

**Figure 2-60: Hysteresis helps clean up comparator response**

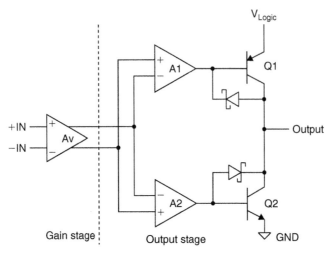

**Figure 2-61: AD790 block diagram**

And the positive threshold voltage is:

$$\frac{(R_1 + R_2)V_{OS} - R_1 V_p}{R_2} \tag{2-18}$$

A problem encountered with external hysteresis is that output voltage depends on supply voltage and loading. This means the hysteresis voltage can vary from application to application; though this affects resolution, it need not be a serious problem, since the hysteresis is usually a very small fraction of the range and can tolerate a safety margin of two or three (or more) times what one might calculate. Swapping in a few comparators can help confidence in the safety margin. Do not use wirewound resistors for feedback; their inductance can make matters worse.

Some comparators have hysteresis built in. An example of this is the AD790 (see Figure 2-61). The hysteresis voltage is nominally $500\,\mu V$. This, of course, can be overridden by applying external hysteresis.

The AD790 has an additional advantage. The supplies on the input (analog) side are not necessarily those on the output. The output swing is from $V_{LOGIC}$ to GND. The input supplies can be $\pm 15\,V$ down to $+5\,V$ and Ground.

It is quite common for the output of a comparator to be open collector (open drain). This allows interfacing to whatever logic level is appropriate to the following circuitry. Note that the maximum allowable output voltage must be observed, but this is usually not too great an issue.

A window comparator makes use of two comparators with different reference voltages and a common input voltage. The comparators are connected to logic in such a way that the final output logic level is asserted when the input signal falls between the two reference voltages (Figure 2-62).

Many comparators have an internal latch. The latch-enable signal has two states: *compare* (track) and *latch* (hold). When the latch-enable signal is in the compare state, the comparator output continuously responds

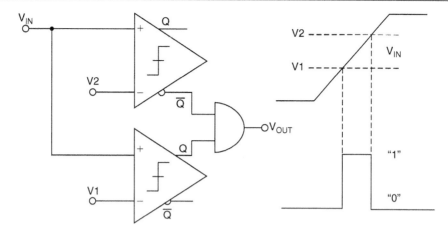

Figure 2-62: Window comparator

to the sign of the net differential input signal. When the latch-enable signal transitions to the latch state, the comparator output goes to either a logic "1" or a logic "0," depending on the sign of the differential input signal at the instant of the transition (at this point, we are neglecting the setup and hold-time, as well as the output propagation delay associated with the latch-enable function). Even though many comparators have a latch-enable function, they are often operated only in the compare mode.

The comparator internal latch-enable function is particularly useful in ADC applications because it allows the comparator decision to be recorded *at a known instant of time*. Flash converters make use of this concept and are constructed of many parallel comparators which share a common latch-enable line. Typical timing associated with the latch-enable function is shown in Figure 2-63. The delay between the assertion of latch-enable and the 50% point of the output logic swing is referred to as *latch-enable to output delay*. It may be different for positive- and negative-going outputs. The other key specification associated with the

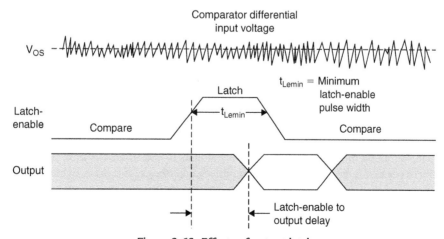

Figure 2-63: Effects of output latch

141

latch-enable function is the minimum allowable latch-enable pulse width. This specification determines the maximum frequency at which the comparator can be strobed.

Fast comparators are somewhat difficult to apply because of their high gain and bandwidth. Proper application of high speed layout, grounding, decoupling, and signal routing is mandatory when using comparators. This cannot be over-emphasized. The biggest problem is their tendency to oscillate when the input signal is very near to or equal to the switching threshold. This can also happen when a slow signal is compared to a DC reference. Hysteresis and the use of a narrow latch-enable pulse will generally help these conditions. TTL comparators are more likely to oscillate than ECL ones because of their large output swings and fast edges, often combined with power supply current spikes as the output changes state. This can feedback to the input in the form of noise.

## Using Op Amps as Comparators

Even though op amps and comparators may seem interchangeable at first glance, there are some important differences.

Comparators are designed to work open loop; they are designed to drive logic from their outputs; and they are designed to work at high speed with minimal instability. Op amps are not designed for use as comparators; they may saturate if over-driven which may cause them to recover comparatively slowly. Many have input stages which behave in unexpected ways when used with large differential voltages; in fact, in many cases, the differential input voltage range of the op amp is limited. And op amp outputs are rarely compatible with logic.

Yet many people still try to use op amps as comparators. While this may work at low speeds and low resolutions, many times the results are not satisfactory. Not all of the issues involved with using an op amp as a comparator can be resolved by reference to the op amp data sheet, since op amps are not intended for use as comparators.

The most common issues are speed (as we have already mentioned), the effects of input structures (protection diodes, phase inversion in FET amplifiers, and many others), output structures which are not intended to drive logic, hysteresis and stability, and CM effects.

### Speed

Most comparators are quite fast, but so are some op amps. Why must we expect low speed when using an op amp as a comparator?

A comparator is designed to be used with large differential input voltages, whereas op amps normally operate with their differential input voltage minimized by negative feedback. When an op amp is over-driven, sometimes by only a few millivolts, some of its stages may saturate. If this occurs the device will take a comparatively long time to come out of saturation and will therefore be much slower than if it always remained unsaturated (Figure 2-64).

The time to come out of saturation of an overdriven op amp is likely to be considerably longer than the normal group delay of the amplifier, and will often depend on the amount of overdrive. Because few op amps have this desaturation time specified for various amounts of overdrive, it will generally be necessary to determine, by experiment, the behavior of the amplifier under the conditions of overdrive to be expected in a particular application.

The results of such experiments should be regarded with suspicion and the values of propagation delay through the op amp comparator which is chosen for worst-case design calculations should be at least twice the worst value seen in any experiment.

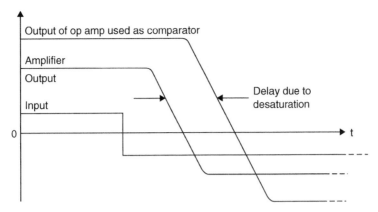

**Figure 2-64: Effects of saturation on amplifier speed when used as a comparator**

## Output Considerations

The output of a comparator will be designed to drive a particular logic family or families, while the output of an op amp is designed to swing from supply rail to supply rail.

Frequently the logic being driven by the op amp comparator will not share the op amp's supplies and the op amp rail-to-rail swing may go outside the logic supply rails—this will probably destroy the logic circuitry, and the resulting short-circuit may destroy the op amp as well.

There are three types of logic which we must consider: ECL, TTL, and CMOS.

ECL is a very fast current steering logic family. It is unlikely that an op amp would be used as a comparator in applications where ECL's highest speed is involved, for reasons given above, so we will usually be concerned only to drive ECL logic levels from an op amp's signal swing, and some additional loss of speed due to stray capacities will be unimportant. To do this we need only three resistors, as shown in Figure 2-65.

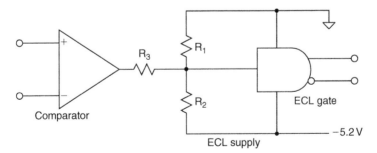

**Figure 2-65: Op amp comparator driving ECL logic**

$R_1$, $R_2$, and $R_3$ are chosen so that when the op amp output is positive the level at the gate is $-0.8\,V$, and when it is low it is $-1.6\,V$. ECL is occasionally used with positive, rather than negative, supplies (i.e., the other rail is connected to ground); the same basic interface circuit may be used but the values must be recalculated.

Although CMOS and TTL input structures, logic levels, and current flows are quite different (although some CMOS is specified to work with TTL input levels), the same interface circuitry will work perfectly well with both types of logic, since they both work for logic 0 near to 0V and logic 1 near to 5V.

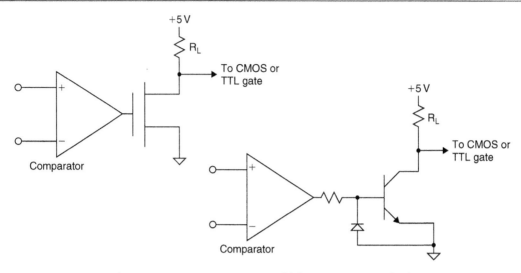

**Figure 2-66: Op amp comparator driving TTL or CMOS logic**

The simplest interface uses a single N-channel MOS transistor and a pull-up resistor, $R_L$. A similar circuit may be made with an NPN transistor, $R_L$, and an additional transistor and diode (Figure 2-66). These circuits are simple, inexpensive, and reliable, and may be connected with several transistors in parallel and a single $R_L$ to give a "wired-or" function, but the speed of the 0–1 transition depends on the value of $R_L$ and the stray capacity of the output node. The lower the value of $R_L$ the faster, but the higher, the power consumption. By using two MOS devices, one P-channel and one N-channel, it is possible to make a CMOS/TTL interface using only two components which have no quiescent power consumption in either state.

Furthermore, it may be made inverting or non-inverting by simple positioning of components. It does, however, have a large current surge during switching, when both devices are on at once, and unless MOS devices with high channel resistance are used, a current-limiting resistor may be necessary to reduce this effect. It is also important, in this application and the one in Figure 2-67, to use MOS devices with gate-source breakdown voltages, $V_{BGS}$, greater than the output voltages of the comparator *in either direction*. A value of $V_{BGS} > \pm 25V$ is common in MOS devices and is usually adequate, but many MOS devices contain gate protection diodes which reduce the value—these should not be used.

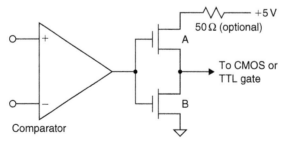

Can be inverting or non-inverting, depending on placing of VMOS devices.
Inverting: A = P-channel/B = N-channel
Non-inverting: A = N-channel/B = P-channel
($V_{BGS} > \pm 25$ V for both devices)

**Figure 2-67: Op amp comparator with CMOS drive**

## Input Circuitry

There are a number of effects which must be considered regarding the inputs of op amps used as comparators. The first-level assumption engineers make about all op amps and comparators is that they have infinite input impedance and can be regarded as open circuits (except for current feedback (transimpedance) op amps, which have a high impedance on their non-inverting input but a low impedance of a few tens of Ω on their inverting input).

But many op amps (especially bias-compensated ones such as the OP-07 and its many descendants) contain protective circuitry to prevent large voltages damaging input devices (Figure 2-68).

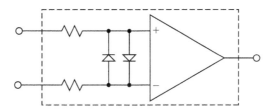

**Figure 2-68: Op amp input structure with protection**

Others contain more complex input circuitry, which only has high impedance when the differential voltage applied to it is less than a few tens of mV, or which may actually be damaged by differential voltages of more than a few volts. It is therefore necessary, when using an op amp as a comparator, to study the data sheet to determine how the input circuitry behaves when large differential voltages are applied to it. (It is always necessary to study the data sheet when using an integrated circuit to ensure that its non-ideal behavior (and every integrated circuit ever made has some non-ideal behavior) is compatible with the proposed application—it is just more important than usual in the present case.)

Of course some comparator applications never involve large differential voltages—or if they do the comparator input impedance when large differential voltages are present is comparatively unimportant. In such cases it may be appropriate to use as a comparator an op amp whose input circuitry behaves nonlinearly—but the issues involved must be considered, not just ignored.

As mentioned elsewhere, nearly all BIFET op amps exhibit anomalous behavior when their inputs are close to one of their supplies (usually the negative supply). Their inverting and non-inverting inputs may become interchanged. If this should occur when the op amp is being used as a comparator, the phase of the system involved will be inverted, which could well be inconvenient. The solution is, again, careful reading of the data sheet to determine just what CM range is acceptable.

Also, absence of negative feedback means that, unlike that of op amp circuits, the input impedance is not multiplied by the loop gain. As a result, the input current varies as the comparator switches. Therefore the driving impedance, along with parasitic feedbacks, can play a key role in affecting circuit stability. While negative feedback tends to keep amplifiers within their linear region, positive feedback forces them into saturation.

# *Analog Multipliers*

A multiplier is a device having two input ports and an output port. The signal at the output is the product of the two input signals. If both input and output signals are voltages, the transfer characteristic is the product of the two voltages divided by a scaling factor, K, which has the dimension of voltage (see Figure 2-69). From a mathematical point of view, multiplication is a "four-quadrant" operation—that is to say that both inputs may be either positive or negative and the output can be positive or negative. Some of the circuits used to produce electronic multipliers, however, are limited to signals of one polarity. If both signals must be unipolar,

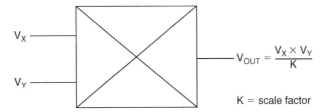

$$V_{OUT} = \frac{V_X \times V_Y}{K}$$

K = scale factor

**Figure 2-69: Multiplier block diagram**

we have a "single-quadrant" multiplier, and the output will also be unipolar. If one of the signals is unipolar, but the other may have either polarity, the multiplier is a "two-quadrant" multiplier, and the output may have either polarity (and is "bipolar"). The circuitry used to produce one- and two-quadrant multipliers may be simpler than that required for four-quadrant multipliers, and since there are many applications where full four-quadrant multiplication is not required, it is common to find accurate devices which work only in one or two quadrants. An example is the AD539, a wideband dual two-quadrant multiplier which has a single unipolar $V_Y$ input with a relatively limited bandwidth of 5 MHz, and two bipolar $V_X$ inputs, one per multiplier, with bandwidths of 60 MHz (Figure 2-70). A block diagram of the AD539 is shown in Figure 2-71.

| Type | $V_x$ | $V_y$ | $V_{out}$ |
|---|---|---|---|
| Single quadrant | Unipolar | Unipolar | Unipolar |
| Two quadrant | Bipolar | Unipolar | Bipolar |
| Four quadrant | Bipolar | Bipolar | Bipolar |

**Figure 2-70: Multiplier input/output relationships**

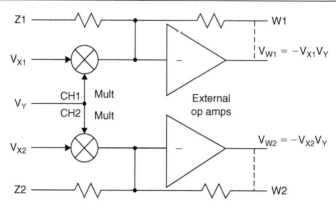

**Figure 2-71: AD539 block diagram**

The simplest electronic multipliers use logarithmic amplifiers. The computation relies on the fact that the antilog of the sum of the logs of two numbers is the product of those numbers (see Figure 2-72).

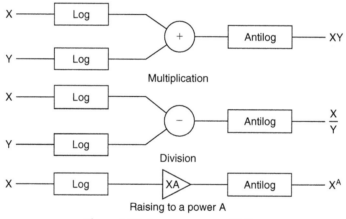

**Figure 2-72: Log amps as multiplier**

The disadvantages of this type of multiplication are the very limited bandwidth and single-quadrant operation. A far better type of multiplier uses the "Gilbert Cell." This structure was invented by Barrie Gilbert, now of Analog Devices, in the late 1960s (see References 1 and 2).

There is a linear relationship between the collector current of a silicon junction transistor and its transconductance (gain) which is given by:

$$\frac{dI_c}{dV_{be}} = \frac{qI_c}{kT} \tag{2-19}$$

where $I_c$ = the collector current, $V_{be}$ = the base–emitter voltage, q = the electron charge ($1.60219 \times 10^{-19}$), k = Boltzmann's constant ($1.38062 \times 10^{-23}$), and T = the absolute temperature.

This relationship may be exploited to construct a multiplier with a differential (long-tailed) pair of silicon transistors, as shown in Figure 2-73.

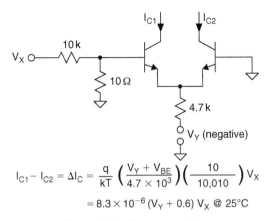

$$I_{C1} - I_{C2} = \Delta I_C = \frac{q}{kT} \left( \frac{V_Y + V_{BE}}{4.7 \times 10^3} \right) \left( \frac{10}{10{,}010} \right) V_X$$

$$= 8.3 \times 10^{-6} (V_Y + 0.6) V_X \ @ \ 25°C$$

**Figure 2-73: Simple multiplier**

This is a rather poor multiplier because (1) the Y input is offset by the $V_{BE}$—which changes nonlinearly with $V_Y$; (2) the X input is nonlinear as a result of the exponential relationship between $I_c$ and $V_{BE}$; and (3) the scale factor varies with temperature.

Gilbert realized that this circuit could be linearized and made temperature stable by working with currents, rather than voltages, and by exploiting the logarithmic $I_c/V_{BE}$ properties of transistors (see Figure 2-74). The

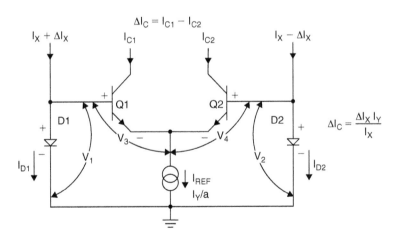

**Figure 2-74: Four-quadrant Gilbert Cell**

X input to the Gilbert Cell takes the form of a differential current, and the Y input is a unipolar current. The differential X currents flow in two diode-connected transistors, and the logarithmic voltages compensate for the exponential $V_{BE}/I_c$ relationship. Furthermore, the q/kT scale factors cancel. This gives the Gilbert Cell the linear transfer function,

$$\Delta I_c = \frac{\Delta I_x I_y}{I_x} \tag{2-20}$$

As it stands, the Gilbert Cell has three inconvenient features: (1) its X input is a differential current; (2) its output is a differential current; and (3) its Y input is a unipolar current—so the cell is only a two-quadrant multiplier.

By cross-coupling two such cells and using two voltage-to-current converters (as shown in Figure 2-75), we can convert the basic architecture to a four-quadrant device with voltage inputs, such as the AD534. At low and medium frequencies, a subtractor amplifier may be used to convert the differential current at the output to a voltage. Because of its voltage output architecture, the bandwidth of the AD534 is only about 1 MHz, although the AD734, a later version, has a bandwidth of 10 MHz.

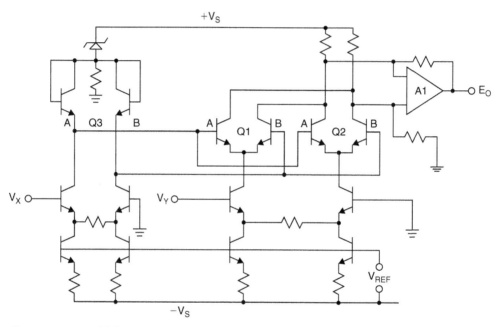

**Figure 2-75: A multiplier and an op amp configured as a divider in both inverting and non-inverting modes**

In Figure 2-75, Q1A and Q1B, and Q2A and Q2B form the two core long-tailed pairs of the two Gilbert Cells, while Q3A and Q3B are the linearizing transistors for both cells. In Figure 2-75 there is an operational amplifier acting as a differential current to single-ended voltage converter, but for higher speed applications, the cross-coupled collectors of Q1 and Q2 form a differential open collector current output (as in the AD834 500 MHz multiplier).

The translinear multiplier relies on the matching of a number of transistors and currents. This is easily accomplished on a monolithic chip. Even the best IC processes have some residual errors, however, and these show up as four DC error terms in such multipliers. In early Gilbert Cell multipliers, these errors had to be trimmed by means of resistors and potentiometers external to the chip, which was somewhat inconvenient. With modern analog processes, which permit the laser trimming of SiCr thin-film resistors on the chip itself, it is possible to trim these errors during manufacture so that the final device has very high accuracy. Internal trimming has the additional advantage that it does not reduce the high frequency performance, as may be the case with external trimpots.

Because the internal structure of the translinear multiplier is necessarily differential, the inputs are usually differential as well (after all, if a single-ended input is required it is not hard to ground one of the inputs). This is not only convenient in allowing CM signals to be rejected, it also permits more complex computations to be performed. The AD534 (shown previously in Figure 2-71) is the classic example of a four-quadrant multiplier based on the Gilbert Cell. It has an accuracy of 0.1% in the multiplier mode, fully differential inputs, and a voltage output. However, as a result of its voltage output architecture, its bandwidth is only about 1 MHz.

Multipliers can be placed in the feedback loop of op amps to form several useful functions. Figure 2-76 illustrates the basic principle of analog computation that a function generator in a negative feedback loop computes the inverse function (provided, of course, that the function is monotonic over the range of operations).

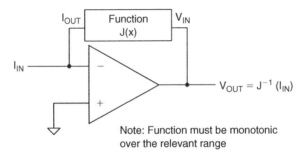

**Figure 2-76: Generating an inverse function**

High speed multipliers are also discussed in the RF/IF section (Section 4-3).

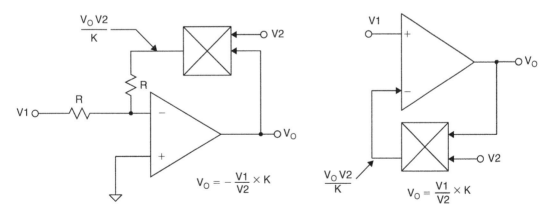

**Figure 2-77: A divider circuit**

# RMS to DC Converters

The root mean square (RMS) is a fundamental measurement of the magnitude of an AC signal. Defined practically, the RMS value assigned to the AC signal is the amount of DC required to produce an equivalent amount of heat in the same load. Defined mathematically, the RMS value of a voltage is defined as the value obtained by squaring the signal, taking the average, and then taking the square root. The averaging time must be sufficiently long to allow filtering at the lowest frequencies of operation desired. A complete discussion of RMS to DC converters can be found in Reference 13, but we will show a few examples of how efficiently analog circuits can perform this function.

The first method, called the *explicit* method, is shown in Figure 2-78. The input signal is first squared by a multiplier. The average value is then taken by using an appropriate filter, and the square root is taken using an op amp with a second squarer in the feedback loop. This circuit has limited dynamic range because the stages following the squarer must try to deal with a signal that varies enormously in amplitude. This restricts this method to inputs which have a maximum dynamic range of approximately 10:1 (20 dB). However, excellent bandwidth (greater than 100 MHz) can be achieved with high accuracy if a multiplier such as the AD834 is used as a building block (see Figure 2-79).

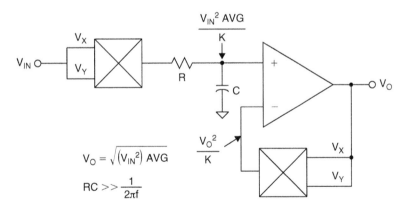

**Figure 2-78: Explicit RMS computation**

Figure 2-80 shows the circuit for computing the RMS value of a signal using the *implicit* method. Here, the output is fed back to the direct-divide input of a multiplier such as the AD734. In this circuit, the output of the multiplier varies linearly (instead of as the square) with the RMS value of the input. This considerably increases the dynamic range of the implicit circuit as compared to the explicit circuit. The disadvantage of this approach is that it generally has less bandwidth than the explicit computation.

While it is possible to construct such an RMS circuit from an AD734, it is far simpler to design a dedicated RMS circuit. The $V_{IN}^2/V_Z$ circuit may be current driven and need only be one quadrant if the input first passes through an absolute value circuit.

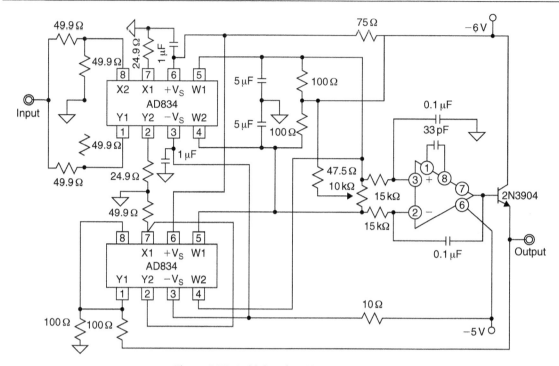

**Figure 2-79: Wideband RMS measurement**

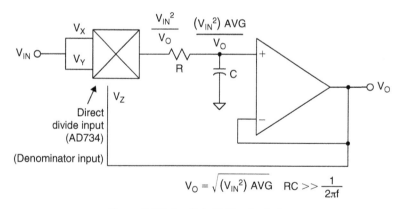

$$V_O = \sqrt{(V_{IN}^2)\ AVG} \quad RC \gg \frac{1}{2\pi f}$$

**Figure 2-80: Implicit RMS calculation**

Figure 2-81 shows a simplified diagram of a typical monolithic RMS/DC converter, the AD536A. It is subdivided into four major sections: absolute value circuit (active rectifier), squarer/divider, current mirror, and buffer amplifier. The input voltage $V_{IN}$, which can be AC or DC, is converted to a unipolar current, $I_1$, by the absolute value circuit $A_1$, $A_2$. $I_1$ drives one input of the one-quadrant squarer/divider which has the transfer function: $I_4 = I_1^2/I_3$. The output current, $I_4$, of the squarer/divider drives the current mirror through a lowpass filter formed by $R_1$ and externally connected capacitor, $C_{AV}$. If the $R_1C_{AV}$ time constant is much greater than

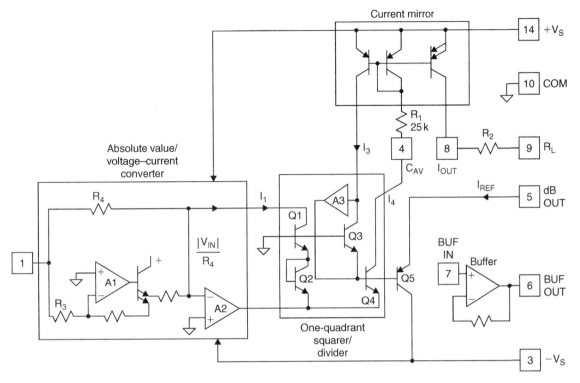

**Figure 2-81: The AD536A monolithic RMS to DC converter**

the longest period of the input signal, then $I_4$ is effectively averaged. The current mirror returns a current, $I_3$, which equals AVG[$I_4$], back to the squarer/divider to complete the implicit RMS computation. Thus:

$$I_4 = AVG\ [I_1^2/I_4] = I_1\ RMS \qquad (2\text{-}21)$$

The current mirror also produces the output current, $I_{OUT}$, which equals $2I_4$. $I_{OUT}$ can be used directly or converted to a voltage with $R_2$ and buffered by A4 to provide a low impedance voltage output. The transfer function becomes:

$$V_{OUT} = 2R_2 \times I_{RMS} = V_{IN}\ RMS \qquad (2\text{-}22)$$

The dB output is derived from the emitter of Q3, since the voltage at this point is proportional to $-\log V_{IN}$. Emitter follower, Q5, buffers and level shifts this voltage, so that the dB output voltage is zero when the externally supplied emitter current ($I_{REF}$) to Q5 approximates $I_3$. However, the gain of the dB circuit has a TC of approximately 3,300ppm/°C and must be temperature compensated.

There are a number of commercially available RMS/DC converters in monolithic form which make use of these principles. The AD536A is a true RMS/DC converter with a bandwidth of approximately 450 kHz for $V_{RMS} > 100\,mV_{RMS}$, and 2 MHz bandwidth for $V_{RMS} > 1\,V_{RMS}$. The AD636 is designed to provide 1 MHz bandwidth for low level signals up to $200\,mV_{RMS}$. The AD637 has a 600 kHz bandwidth for $100\,mV_{RMS}$ signals, and an 800 MHz bandwidth for $1\,V_{RMS}$ signals. Low cost, general-purpose RMS/DC converters such as the AD736 and AD737 (power-down option) are also available.

# Programmable Gain Amplifiers

Most systems with wide dynamic range need some method of adjusting the input signal level to the ADC. The ADC compares the input signal to a fixed voltage reference (+5V or +10V are typical values). To achieve the rated precision of the converter, the maximum input should be fairly near its full-scale voltage. However, transducers have a wide range of output voltages. High gain is needed for a small sensor voltage, but with a large transducer output, a high gain will cause the amplifier or ADC to saturate. So some type of controllable gain device is needed. Such a device has a gain that is controlled by a DC voltage or, more commonly, a digital input. This device is known as a *programmable gain amplifier* (PGA, see Figure 2-82).

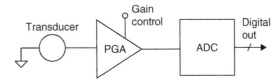

- Used to increase dynamic range of circuit.
- A PGA with a gain from 1 to 2 theoretically increases the dynamic range by 6 dB, a gain of 1 to 4 gives 12 dB increase, etc.

**Figure 2-82: Programmable gain amplifier (PGA)**

To understand the benefits of variable gain, assume an ideal PGA with two settings, gains of one and two. The dynamic range of the system is increased by 6 dB. Increasing the gain to four results in a 12 dB increase in dynamic range.

If the LSB of an ADC is equivalent to 10 mV of input voltage, the ADC cannot resolve smaller signals, but when the gain of the PGA is increased to two, input signals of 5 mV may be resolved. Thus, the processor can combine PGA gain information with the digital output of the ADC to increase its resolution by 1 bit. Essentially, this is the same as adding additional resolution to the ADC.

In practice, PGAs are not ideal, and their error sources must be studied. The most fundamental problem with PGA design is accurate gain programming. Electromechanical relays have minimal $R_{ON}$, but are otherwise unsuitable for gain switching. They are slow, large, and expensive. Silicon switches, as discussed in the section on switches and multiplexers (Chapter 7 of this book), have quite large $R_{ON}$, which is both voltage- and temperature-variable, and stray capacities, which may affect the AC parameters of a PGA using them.

To understand how $R_{ON}$ can affect the performance of a PGA, let us consider a poor PGA design (Figure 2-83). An op amp is configured in the standard non-inverting gain circuit with four different gain-setting resistors, each grounded by a switch. Most silicon switches have ON resistance in the range of 100–500 Ω. Even if the ON resistance were as low as 25 Ω, the error for a gain of 16 would be 2.4%, much

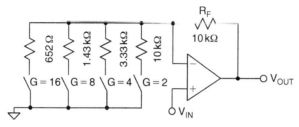

- Gain accuracy limited by switch's on resistance, R and $R_{ON}$ modulation.
- $R_{ON}$ typically 100–500 $\Omega$ for a CMOS or JFET switch.
- Even with $R_{ON} = 25 \Omega$, there is a 2.4% gain error for $A_V = 16$.
- $R_{ON}$ drift over temperature limits accuracy.
- Only solution is to use very low $R_{ON}$ switches (relays).

**Figure 2-83: How not to build a PGA**

worse than 8 bits. Furthermore, $R_{ON}$ drifts over temperature and varies from switch to switch. If the value of the feedback and gain-setting resistors were increased, noise and offset would become a problem. The only way to achieve accuracy with this circuit is to replace silicon switches with relays which have virtually no ON resistance.

It is better to use a circuit where $R_{ON}$ is unimportant. In Figure 2-84, the switch is placed in series with the inverting input of an op amp. Since the input impedance of an op amp is very large, the $R_{ON}$ of the switch is irrelevant. The gain is now determined by the external resistors. The $R_{ON}$ may add a small offset error if the op amp bias current is significant.

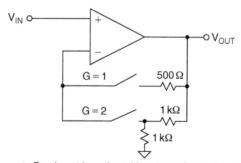

- $R_{ON}$ is not in series with gain setting resistors.
- $R_{ON}$ is very small compared to input impedance.
- Only a slight offset error occurs due to the bias current flowing through the switch.

**Figure 2-84: Alternative PGA configuration negates effect of $R_{ON}$**

The AD526 amplifier uses this method of building a PGA and integrates it onto a single chip. The AD526 has five binary gain settings from 1 to 16, and its internal JFET switches are connected to the inverting input of the amplifier. The gain resistors are laser trimmed. The maximum gain error is only 0.02%, far better than the 2.4% error in Figure 2-85. The linearity is also very good at 0.001%. The AD526 is controlled by a latched digital interface.

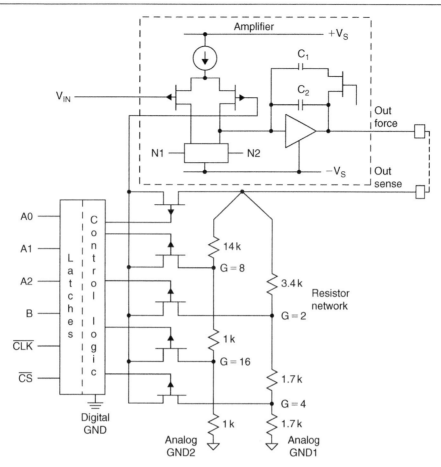

**Figure 2-85: Monolithic software programmable PGA instrumentation amplifier (AD526)**

This same design can be used to build the discrete PGA shown in Figure 2-86. It uses a single op amp, a quad switch, and precision resistors. The low noise AD797 replaces the JFET input op amp of the AD526, but almost any voltage feedback op amp could be used in this circuit. The ADG412 was picked for its low ON resistance of 35 Ω. The resistors were chosen to give gains of 1, 10, 100, and 1,000, but if other gains are required, the resistor values may easily be altered. Ideally, a trimmed resistor network should be used both for initial gain accuracy and for low drift over temperature. The 20 pF capacitor ensures stability and holds the output voltage when the gain is switched. The control signal to the switches turns one switch off a few nanoseconds before the second switch turns on. During this break, the op amp is open loop. If the capacitor was not used, the output would start slewing. Instead, the capacitor holds the output voltage during the switching. Since the time that both switches are open is very short, only 20 pF is needed. For slower switches, a larger capacitor may be necessary.

The PGA's input voltage noise spectral density is only $1.65 \, nV/\sqrt{Hz}$ at 1 kHz, only slightly higher than the noise performance of the AD797 alone. The increase is due to the noise of the ADG412, and the current noise of the AD797 flowing through the ON resistance. The noise was measured at a gain of 1,000 (worst case).

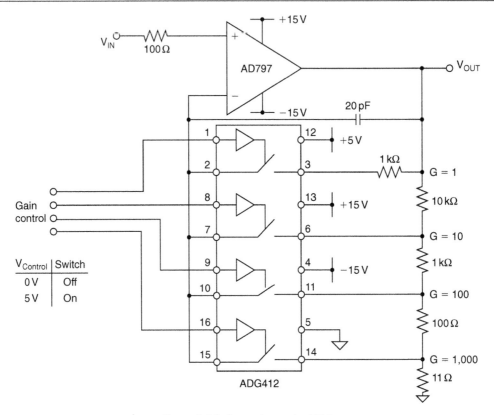

**Figure 2-86: A very low noise PGA**

The accuracy of the PGA is important in determining the overall accuracy of a system. The AD797 has a bias current of 0.9 μA, which, flowing in 35 Ω $R_{ON}$, results in an additional offset error of 31.5 μV. Combined with the AD797 offset, the total $V_{OS}$ becomes 71.5 μV (max). Offset temperature drift is affected by the change in bias current and ON resistance. Calculations show that the total temperature coefficient increases from 0.6 μV/°C to 1.6 μV/°C. These errors are small, and may not matter, but it is important to be aware of them. In practice, circuit accuracy and TC will be determined by the external resistors. Input characteristics such as CM range and input bias current are determined solely by the AD797. The circuit could be converted to single supply simply by changing the op amp. The switches do not need to be changed.

Another PGA configuration uses a DAC in the feedback loop of an op amp to adjust the gain under digital control (Figure 2-87). The digital code of the DAC controls its attenuation. Attenuating the feedback signal increases the closed-loop gain. A non-inverting PGA of this type requires a multiplying DAC with a voltage output (a multiplying DAC is a DAC with a wide reference voltage range *which includes zero*). For most applications of the PGA, the reference input must be capable of handling bipolar signals. The AD7846 is a 16-bit converter that meets these requirements. In this application, it is used in standard two-quadrant multiplying mode. The OP-213 is a low drift, low noise amplifier, but the choice of amplifier is flexible, and depends on the application. The input voltage range depends on the output swing of the AD7846, which is 3 V less than the positive supply and 4 V above the negative supply. A 1,000 pF capacitor is used in the feedback loop for stability.

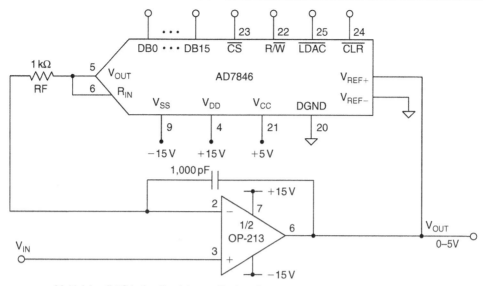

- Multiplying DAC in feedback loop adjusts gain.

- $G = \dfrac{2^{16}}{\text{Decimal value of digital code}}$.

**Figure 2-87: Using a multiplying DAC in a feedback loop to create a divider**

The gain of the circuit is set by adjusting the digital inputs of the DAC, according to the equation given in Figure 2-87. $D_{0-15}$ represents the decimal value of the digital code. For example, if all the bits were set high, the gain would be $65{,}536/65{,}535 = 1.000015$. If the eight least significant bits are set high and the rest low, the gain would be $65{,}536/255 = 257$.

Figure 2-88 shows the small signal response at a gain of 1 with a 100 mV square wave input. The bandwidth is a fairly high 4 MHz. However, this does reduce with gain, and for a gain of 256, the bandwidth

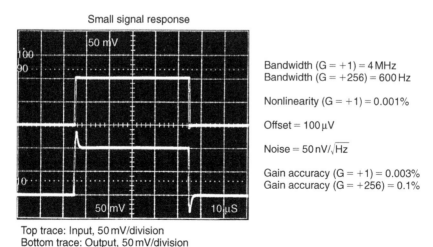

Bandwidth (G = +1) = 4 MHz
Bandwidth (G = +256) = 600 Hz

Nonlinearity (G = +1) = 0.001%

Offset = 100 μV

Noise = 50 nV/√Hz

Gain accuracy (G = +1) = 0.003%
Gain accuracy (G = +256) = 0.1%

Top trace: Input, 50 mV/division
Bottom trace: Output, 50 mV/division
Horizontal scale: 10 μs/division

**Figure 2-88: Performance of the circuit in Figure 2-87**

is only 600 Hz. If the gain-bandwidth product were constant, the bandwidth in a gain of 256 should be 15.6 kHz; but the internal capacitance of the DAC reduces the bandwidth to 600 Hz.

The gain accuracy of the circuit is determined by the resolution of the DAC and the gain setting. At a gain of 1, all bits are on, and the accuracy is determined by the DNL specification of the DAC, which is ±1 LSB maximum. Thus, the gain accuracy is equivalent to 1 LSB in a 16-bit system, or 0.003%. However, as the gain is increased, fewer of the bits are on. For a gain of 256, only bit 8 is turned on. The gain accuracy is still dependent on the ±1 LSB of DNL, but now that is compared to only the lowest 8 bits. Thus, the gain accuracy is reduced to 1 LSB in an 8-bit system, or 0.4%. If the gain is increased above 256, the gain accuracy is reduced further. The designer must determine an acceptable level of accuracy. In this particular circuit, the gain was limited to 256.

Non-inverting PGA circuits using an op amp are easily adaptable to single-supply operation, but the instrumentation amplifier topology does not lend itself to single-supply applications. However, the AMP-04 can be used with an external switch to produce the single-supply instrumentation PGA shown in Figure 2-89. This circuit has selectable gains of 1, 10, 100, and 500, which are controlled by an ADG511. The ADG511 was chosen as a single-supply switch with a low $R_{ON}$ of 45 Ω. The gain of this circuit is dependent on the $R_{ON}$

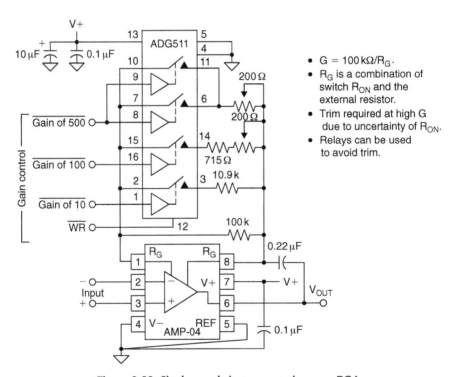

**Figure 2-89: Single-supply instrumentation amp PGA**

of the switches. Trimming is required at the higher gains to achieve accuracy. At a gain of 500, two switches are used in parallel, but their resistance causes a 10% gain error in the absence of adjustment.

Certain ΣΔ ADCs (e.g., AD7710, AD7711, AD7712, and AD7713) have built-in PGAs. Circuit design is much easier because an external PGA and its control logic are not needed. Furthermore, all the errors of the PGA are included in the specifications of the ADC, making error calculations simple. The PGA gain is controlled over the same serial interface as the ADC, and the gain setting is factored into the conversion, saving additional calculations to determine input voltage. This combination of ADC and PGA is very powerful and enables the realization of a highly accurate system, with a minimum of circuit design. The PGA function in this case is not a separate block requiring matching of resistors for accuracy in line with the expectation of the $\Sigma\Delta$ ADC. It is accomplished by modulating the duty cycle of the switched capacitors in the modulator, thus changing the gain (Figure 2-90).

High speed VGAs are discussed in the RF/IF section (Section 4.6).

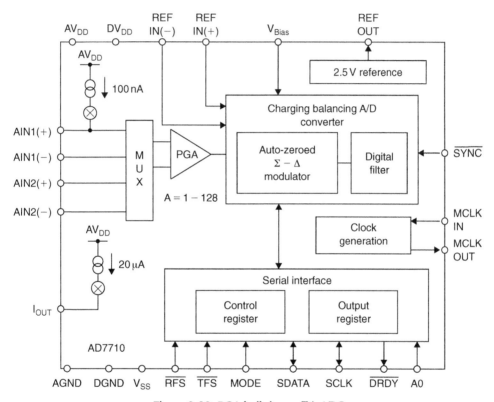

**Figure 2-90: PGA built into a $\Sigma\Delta$ ADC**

# Audio Amplifiers

## Amplifiers

There are no specific audio specifications that apply to amplifiers. Obviously the amplifier needs to be of appropriate bandwidth and be low distortion. Several types have found application in the audio field. These include the AD797, OP275, and the AD711/12/13.

One application specific audio IC is the SSM2019 microphone preamplifier (see Figure 2-91). For use as a microphone preamplifier in high fidelity applications, a primary concern is that the circuit has to be low noise. The specification for the SSM2019 is $1\,\text{nV}/\sqrt{\text{Hz}}$. The input to the SSM2019 is fully differential to interface with balanced microphones.

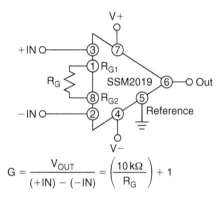

$$G = \frac{V_{OUT}}{(+IN) - (-IN)} = \left(\frac{10\,\text{k}\Omega}{R_G}\right) + 1$$

**Figure 2-91: SSM2019 microphone amplifier**

There is another application for microphone preamplifiers. Here the emphasis is on voice intelligibility rather than low noise. The target application is in communication systems and public address systems. The SSM2165/66/67 family is complete and flexible solutions for conditioning microphone inputs. The block diagram of the SSM2165 is shown in Figure 2-92. A low noise voltage controlled amplifier (VCA) provides a gain that is dynamically adjusted by a control loop to maintain a set compression characteristic. The compression ratio is set by a single resistor and can be varied from 1:1 to over 15:1 relative to the fixed rotation point. Signals above the rotation point are limited to prevent overload and to eliminate "popping." A downward expander (noise gate) prevents amplification of noise or hum. This results in optimized signal levels prior to digitization, thereby eliminating the need for additional gain or attenuation in the digital domain that could add noise or impair accuracy of speech recognition algorithms (see also Figure 2-93).

Speaker driver power amplifiers are another application specific audio area. The main application challenge here is delivering enough audio power in a limited supply voltage environment such as is typically found in computers and games, while keeping the package power dissipation down to safe levels. As an example, the SSM2211 (Figure 2-94) is a high performance audio amplifier that delivers $1\,\text{W}_{RMS}$ of low distortion audio power into a bridge-connected $8\,\Omega$ speaker load (or $1.5\,\text{W}_{RMS}$ into $4\,\Omega$ load) (Figure 2-95). The SSM2211

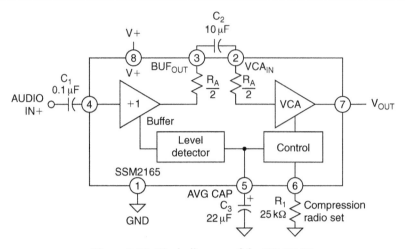

Figure 2-92: Block diagram of the SSM2165

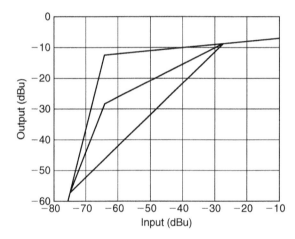

Figure 2-93: Typical transfer characteristics for the
SSM2165

is available in SO-8 and LFCSP (lead frame chip scale package) surface mount packages. The SO-8 features the patented Thermal Coastline lead frame. The Thermal Coastline package is further discussed in the power section.

## Voltage-Controlled Amplifiers

Audio signal levels are often controlled by using low distortion VCAs (voltage-controlled amplifiers) in the signal path. By using controlled rate of change drive to the VCAs, the "clicking" associated with switched resistive networks is eliminated. For example, the SSM2018T is a low noise, low distortion VCA applicable in high performance audio systems (Figures 2-96 and 2-97). The "T" suffix indicates a version that is factory trimmed for distortion and requires no subsequent user adjustments.

Despite the sonic advantages of using analog control of the signal level, it is sometimes useful to have the control voltage under digital control. In this case a DAC can be added to the VCA. An example of this

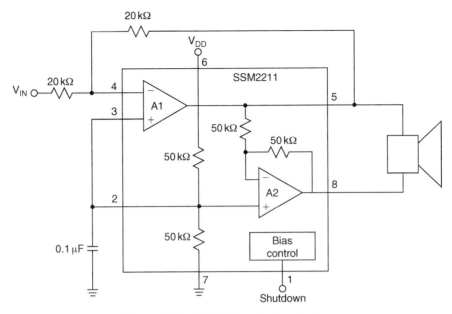

**Figure 2-94: SSM2211 typical application**

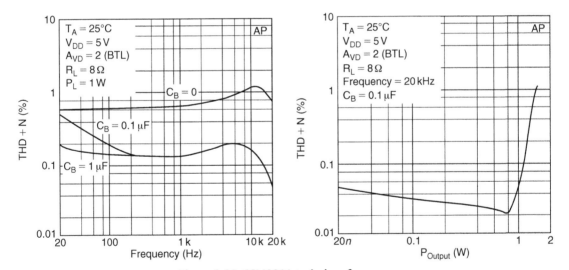

**Figure 2-95: SSM2211 typical performance**

configuration is the SSM2160 (Figure 2-98) which allows digital control of volume of six audio channels, with a master level control and individual channel controls. Low distortion VCAs are used in the signal path. Each channel is controlled by a dedicated 5-bit DAC providing 32 levels of gain. A master 7-bit DAC feeds every control port giving 128 levels of attenuation. Step sizes are nominally 1 dB and can be changed by external resistors. Channel balance is maintained over the entire master control range. Upon power-up, all outputs are automatically muted. A three-wire or four-wire serial data bus enables interfacing with most popular microcontrollers.

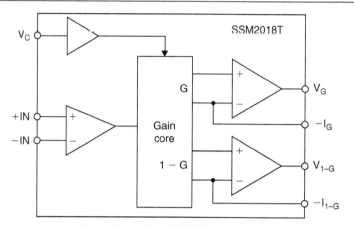

**Figure 2-96: SSM2018 block diagram**

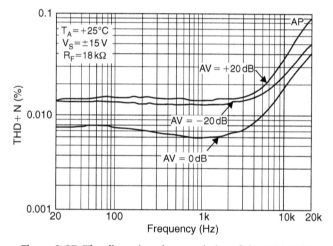

**Figure 2-97: The distortion characteristics of the SSM2018**

## Line Drivers and Receivers

The function of sending/receiving audio signals between various system components has traditionally involved tradeoffs of one form or another. Fully differential or *balanced* transmission systems are best at rejecting low frequency and RF noise, so they are used for highest performance and are discussed in some detail in the following.

A typical audio system block diagram using differential or balanced transmission is shown in Figure 2-99. In concept, a balanced transmission system like this could use several input/output coupling schemes within the driver and receiver. Some major points distinguishing coupling method details are discussed briefly below, before addressing actual circuits.

A point worth noting is that the ± voltage drive to the line need *not* be exactly balanced to reap the benefits of balanced transmission. In fact the drive can be asymmetrical to some degree, and the signal will still be received at $V_{OUT}$ with correct amplitude and with good noise rejection. What *does* need to be provided is

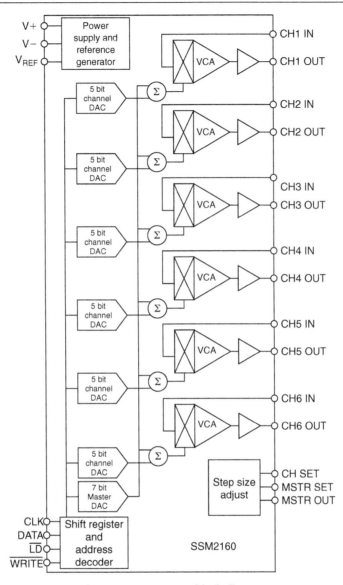

**Figure 2-98: SSM2160 block diagram**

two well-balanced line-driving impedances, $R_{O1}$ and $R_{O2}$. Also, in conjunction with these balanced drive impedances, the associated $(+)$ and $(-)$ receiver input impedances should also be equal. The technical reasons for this will be apparent shortly.

## Audio Line Receivers

An audio line receiver is simply a subtractor amplifier (Figure 2-100). From a DC and AC trim/balance perspective, the Figure 2-100 topology is most effective with resistors and amplifier made simultaneously in a single monolithic IC.

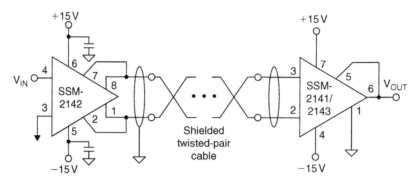

**Figure 2-99: An audio balanced transmission system**

In applying circuits of the Figure 2-100 type (or other topologies which resistively load the source), a designer must bear in mind that all *external* resistances added to the four resistances can potentially degrade CMR, unless kept to proportional value increases. To place this in perspective, a $2.5\,\Omega$ or $0.01\%$ mismatch can easily occur with wiring, and if not balanced out, this mismatch will degrade the CMR of otherwise perfectly matched $25\,k\Omega$ resistors to $86\,dB$. These circuits are therefore best fed from balanced, low impedance drive sources, preferably $25\,\Omega$ or less.

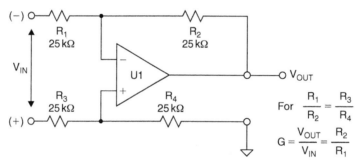

$$\text{For}\quad \frac{R_1}{R_2}=\frac{R_3}{R_4}$$

$$G=\frac{V_{OUT}}{V_{IN}}=\frac{R_2}{R_1}$$

**Figure 2-100: A simple line receiver using a four-resistor differential amplifier**

The SSM2141 and SSM2143 are monolithic IC line receivers which work very much like the circuit of Figure 2-100 differing only in their individual gains. The SSM2141 operates as a unity-gain device, while the SSM2143 operates either at a nominal gain of 0.5 ($-6\,dB$), or it can optionally be strapped with the input/output of the resistor pairs reversed, to operate at a gain of 2 ($6\,dB$).

Both devices operate from supplies up to $\pm18\,V$, can drive $600\,\Omega$ loads, and have low distortion and excellent CMR characteristics. For reference, the op amp used in these receivers is similar to one-half of an OP271. The output appears at pin 6 and is uncommitted; with conventional use it gets tied to $R_4$ (pin 5) for feedback. However, if desired, an external in-loop buffer can optionally be added. This step will allow either line receiver device to drive even lower Z loads if desired.

Perhaps the most outstanding attribute of these devices is their CMR performance, shown in Figure 2-101(A) (these data are for the SSM2141, but for the SSM2143 they are similar). For the SSM2141 the DC to 1 kHz CMR is typically $100\,dB$, and even at $10\,kHz$ it is still about $80\,dB$. The SSM2143 (not shown), using lower resistor values, has a somewhat lower typical CMR of $90\,dB$, but maintains this to about $10\,kHz$.

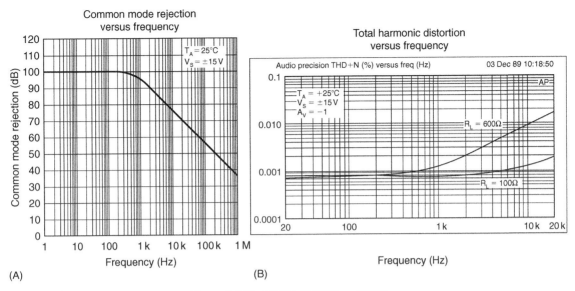

**Figure 2-101: SSM2143 CMR and THD**

The SSM2141 THD + N performance also shown in Figure 2-101(B) is also very good for both 600 Ω and 100 kΩ loads.

With a companion differential line driver (next section), these two line receivers allow convenient as well as flexible interfacing between points in audio systems, as well as other instrumentation up to 100 kHz. However, they both are also more generally useful as flexible gain blocks within a system, not necessarily requiring the full CM performance aspects. For example, they are useful as either precise inverting or non-inverting gains blocks, due to the very accurate internal resistor ratios. With the SSM2141 typical gain accuracy of 0.001%, very precise, single chip unity-gain inverters and summers can be built at low overall cost.

## *Audio Line Drivers*

Unlike the case for the differential line receiver, a standard circuit topology for differential line drivers is not quite as clear-cut. Two circuit types are discussed in this section, with their contrasts in performance and complexity.

On the other hand, the inherent features of laser-trimmed monolithic technology can make a complex circuit such as the balanced line driver thoroughly practical. Like the SSM2141 and SSM2143 line receivers, applying these concepts to a driver circuit results in an efficient and useful IC. This product, the SSM2142 balanced line driver, is shown in functional form in Figure 2-102.

The SSM2142 is designed for a single ended to differential gain of 2 times, and in use can be simply strapped with the respective FORCE/SENSE pins tied together. In a system application, the SSM2142 is used with either an SSM2143 or an SSM2141 line receiver, with the differential-mode signal being transmitted via shielded twisted pair cable. This hookup comprises a complete single ended to differential and back to single-ended transmission system, with noise isolation in the process.

With the use of the SSM2143 gain of 0.5, the SSM2142 gain of 2 is complemented, and the overall system gain is unity (Figure 2-103). If the SSM2141 is used as the receiver, the gain is 2 overall, implying the use of a terminating resistor at the receiver. The THD + N performance of the unity-gain SSM2142/SSM2143 system is shown in Figure 2-104, for the conditions of a $5V_{RMS}$ input/output signal, both with/without a 500′ cable.

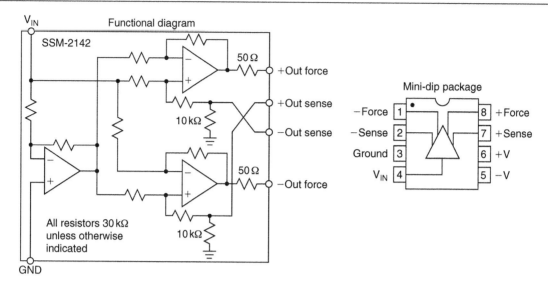

**Figure 2-102: SSM2142 block diagram**

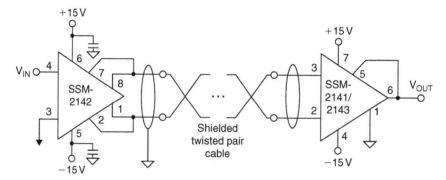

**Figure 2-103: Balanced audio transmission system**

As should be obvious, these drivers do *not* offer galvanic isolation, which means that in all applications there must be a DC current path between the grounds of the driver and the final receiver. In practice, however, this is not necessarily a problem.

## Class D Audio Power Amplifiers

### Theory of Operation

A Class D audio amplifier is basically a switch-mode or PWM amplifier and is one of a number of different classes of amplifiers. Following is a look at the definitions for the main classifications:

Class A—In a Class A amplifier, the output device(s) are continuously conducting for the entire cycle, or in other words there is always bias current flowing in the output devices. This topology has the least distortion and is the most linear, but at the same time is the least efficient, at around 20%. Therefore, the quiescent dissipation is high. In fact the dissipation is constant, regardless of how much power is delivered to the load. A Class A amplifier output is typically not complementary, with a high and low side output device(s).

*172*

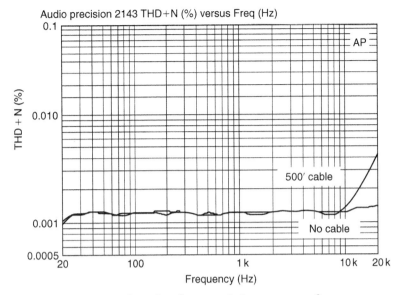

**Figure 2-104: Balanced audio transmission system performance**

Class B—In a Class B amplifier the output device(s) only conduct for half the sinusoidal cycle (one conducts for the positive half cycle, and one conducts for the negative half cycle). If there is no signal, then there is no current flow in the output devices. This class of amplifier is obviously more efficient than Class A, at about 50%, but has some distortion at the crossover point due to the time it takes to turn one device off and turn the other device on. This is referred to as crossover notch distortion. Since it occurs at the point of minimum signal (the zero crossing), its effect is very obvious.

Class AB—This type of amplifier is a combination of the above two types, and is probably the most common type of power amplifier in existence. Here both devices are allowed to conduct at the same time, but just by a small amount near the crossover point. Hence each device is conducting for more than half a cycle but less than the whole cycle, so the inherent nonlinearity (crossover distortion) of Class B designs is overcome, without the inefficiencies of a Class A design. Efficiencies for Class AB amplifiers can also be about 50%. There are variations of Class AB, depending on how much of the cycle both of the output devices conduct. Obviously, the more they both conduct, the lower the efficiency, but also the higher the linearity (Figure 2-105).

Class D—This class of amplifier is a switching amplifier as mentioned above. In this type of amplifier, the switches are either fully on or fully off, significantly reducing the power losses in the output devices. This is very similar to the difference between linear power regulators and switch-mode regulators.

Efficiencies of 90–95% are possible. Audio Class D amplifiers use a modulator to convert the input audio signal into a switching waveform used to control the output switches. PWM is the most commonly used modulation scheme. In PWM, the audio signal is used to modulate a PWM carrier signal which drives the output devices. The output devices then drive a lowpass filter to remove the high frequency PWM carrier frequency, while retaining the desired audio content. The speaker is one of the elements in this filter, and is situated at the filter output.

Class D amplifiers take on many different forms; some can have digital inputs and some can have analog inputs.

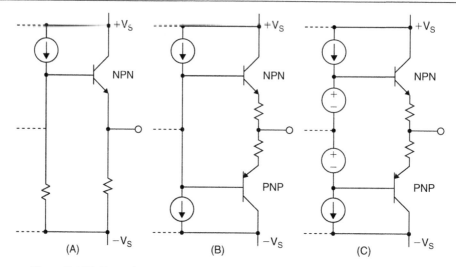

**Figure 2-105: Example output stages (A) Class A, (B) Class B, and (C) Class AB**

However, audio quality in PWM amplifiers can be limited: THD is typically 0.1% or worse, and PSRR is poor (see Reference 1). PSRR can be improved by sensing power supply variations and adjusting the modulator's behavior to compensate, as proposed in Reference 1. However, this alone will not suppress the THD produced by inherent PWM nonlinearity or power stage nonlinearity.

These THD and power supply noises can both be suppressed with feedback from the power stage outputs (see Reference 2), which incorporates the feedback around an analog PWM modulator.

PWM is attractive because it allows >100 dB audio-band SNR at low clock rates near 400 kHz, limiting switching losses. Also, many PWM modulators are stable to near 100% modulation, allowing high output power before overloading. However, PWM has several problems. First, the PWM process inherently adds distortion in many modulation schemes (see Reference 3) and second, harmonics of the PWM switching frequency produce EMI in the AM band.

$\Sigma\Delta$ modulation does not share these problems, but nonetheless has not traditionally been used for Class D (see Reference 3), because conventional 1-bit $\Sigma\Delta$ modulators are only stable to 50% modulation, and power efficiency is limited because typical output data rates are >1 MHz, when $\geq 64\times$ oversampling rate is needed to achieve sufficient audio-band SNR. However, Analog Devices has enhanced the traditional 1-bit $\Sigma\Delta$ architecture to overcome these problems, and created $\Sigma\Delta$-based Class D amplifier chips which have performance advantages over competitor PWM-based products.

## Device Architecture

The AD1990/2/4/6 family of chips are two-channel bridge tied load (BTL) switching audio power amplifiers with integrated $\Sigma\Delta$ modulator. Hereafter, "AD199x" will be used to refer to this product family.

The AD199x modulator accepts a low power analog input signal (of $5V_{p-p}$ maximum amplitude), and generates a switching waveform to drive speakers directly. One of the two modulators can control both output stages thereby providing twice the current for single-channel applications. A digital, microcontroller-compatible interface provides control of reset, mute, and PGA gain as well as output signals for thermal and over-current error conditions. The output stage can operate from supply voltages ranging from 8 V to 20 V. The analog modulator and digital logic operate from a 5 V supply.

The power stage of the AD199x is arranged internally as four transistor pairs, which are used as two H-bridge outputs to provide stereo amplification. The transistor pairs are driven by the output of the $\Sigma\Delta$ modulator. A user selectable non-overlap time is provided between the switching of the high side transistor and low side transistor to ensure that both transistors are never on at the same time. The AD199x implements turn-on pop suppression to eliminate any pops or clicks following a reset or un-mute.

## Analog Input Section

The analog input section uses an internal amplifier to bias the input signal to the reference level. A DC blocking capacitor should be connected to remove any external DC bias contained in the input signal.

## The Sigma–Delta Modulator

The modulator uses a 1-bit, seventh-order feedforward architecture. The quantizer output drives the switching power stage, whose pulses are fed back to a continuous-time (CT) first integrator. This allows fullest possible integration of the pulse waveform, maximizing error correction. If the first integrator were discrete-time (DT), its sampling process would often miss important information about errors in pulse edge timing and shape, which would reduce the error-correcting effectiveness of the feedback loop.

The CT integrator bandwidth of 100 kHz gives anti-alias filtering for the subsequent DT, switched-capacitor (SC) integrators. The SC integrators and quantizer are clocked at 6 MHz, corresponding to 128× oversampling.

For the modulator, seventh order is more than enough to achieve 100 dB SNR with traditional aggressive noise shaping (see Reference 5). However, this gives instability for modulation >50%, limiting the maximum output power with stable operation to just 25% of theoretical full power. To overcome this limitation, we use less aggressive noise shaping to maintain stability to 90% modulation. This gives good sound quality at higher power, but requires high modulator order to get acceptable SNR.

Fortunately, all integrators after the first can be SC. The high first integrator gain relaxes noise requirements for the SC ones, allowing small sampling caps (50 fF), and low power single-stage op amps. Resonator feedbacks to integrators 2, 4, and 6 reduce low frequency noise, by placing NTF zeroes at 12 kHz, 22 kHz, and 40 kHz. When PVDD = 12 V, integrated audio-band quantization noise is $25\,\mu V_{RMS}$ and additional thermal noise yields total integrated audio noise of $50\,\mu V_{RMS}$. Maximum output is $7.8\,V_{RMS}$, giving 104 dB dynamic range.

The modulator described to this point would become unstable for large inputs >90% full-scale. Output transients resulting from the instability would bear little relationship to the desired signal, and would sound bad. To solve this problem, the modulator input is monitored, and when large signals that could cause instability are detected, integrators 3–7 of the modulator are reset. This effectively converts the modulator to a second order, unconditionally stable configuration. The loop gain is reduced relative to the "normal" seventh-order configuration, so that noise shaping is less effective and more quantization noise reaches the output. However, this elevated noise is superimposed on a large output signal that is now closer to the desired waveform, and the composite sound is better than when the modulator is allowed to become unstable.

## Driving the H-Bridge

Each channel of the switching amplifier is controlled by a four-transistor H-bridge to give a differential output stage. The outputs of the H-bridges, OUTR+, OUTR−, OUTL+, and OUTL− will switch between PVDD and PGND as determined by the sigma delta ($\Sigma\Delta$) modulator. The power supply that is used to drive the power stage of the AD199x should be in the range of +8 V to + 20 V and be capable of supplying enough current to drive the load. This power supply is connected across the PVDD and PGND pins. The feedback pins, NFR+, NFR−, NFL+, and NFL− are used to supply negative feedback to the modulator. The pins are connected to the outputs of the H-bridge using a resistor divider network as shown in Figure 2-106.

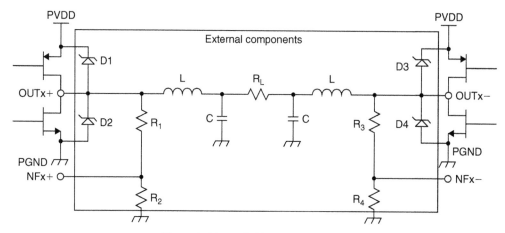

**Figure 2-106: H-bridge configuration**

External Schottky diodes can be used to reduce power loss during the non-overlap time when neither of the high side or low side transistors is on. During this time neither transistor is driving the OUTx pin. The purpose of the inductors is to keep current flowing.

For example, the OUTx pin may approach and pass the PGND level to achieve this. When the voltage at the OUTx pin is 0.7V below PGND, the parasitic diode associated with the low side transistor will become forward biased and turn on. When the high side transistor turns on, the voltage at OUTx will rise to PVDD and will reverse bias the parasitic diode. However, by its nature the parasitic diode has a long reverse recovery time and current will continue to flow through it to PGND thus causing the entire circuit to draw more current than necessary. The addition of the Schottky diodes prevents this from happening. When the OUTx pin goes more than 0.3V below PGND the Schottky diode becomes forward biased. When the high side transistor turns on, the Schottky diode becomes reverse biased. The reverse recovery time of the Schottky diode is significantly faster than the parasitic diode so far less current is wasted. A similar effect happens when the inductor induces a current which drives the OUTx pin above PVDD. Figure 2-106 shows how the external components of a system are connected to the pins of the AD199x to form the H-bridge configuration.

## Amplifier Gain

### Selecting the Modulator Gain

The AD199x modulator can be thought of as a switching analog amplifier with a voltage gain controlled by two external resistors forming a resistor divider between the OUTxx pins and PGND. The center of the resistor divider is connected to the appropriate feedback pin NFx. Selecting the gain along with the PVDD voltage will determine how much power can be delivered to a load for a fixed input signal. The gain of the modulator is controlled by the values of $R_1$ and $R_2$ (see Figure 2-107) according to the equation:

$$\text{Gain} = \left( \frac{R_2 + R_1}{R_2} \right)$$

(2-23)

If the voltage at the NFx pins exceeds 5 V, ESD protection circuitry will turn on, to protect low voltage circuitry inside the chip that is connected to NFx. When the protection circuit is active, it introduces nonlinear behavior into the modulator feedback loop, which degrades audio quality. To avoid this, $R_1$, $R_2$, and the gain should be selected in a manner that limits max voltage at NFx to <5 V. For optimal

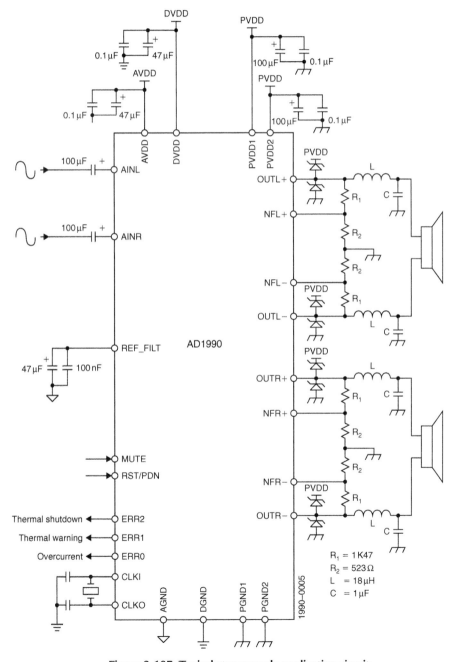

**Figure 2-107: Typical stereo mode application circuit**

*177*

modulator stability and audio quality, use the formula:

$$\text{Gain} = \left( \frac{R_1 + R_2}{R_2} \right) = \frac{\text{PVDD}}{3.635} \tag{2-24}$$

The ratio of the resistances sets the gain rather than the absolute values. However, the dividers provide a path from the high voltage supply to ground, so the values should be large enough to produce negligible loss due to quiescent current.

The chip contains a calibration circuit to minimize voltage offsets at the speaker, which helps to minimize clicks and pops when muting or un-muting. Optimal performance is achieved for the offset calibration circuit when the feedback divider resistors sum to $6\,k\Omega$ (meaning that $(R_1 + R_2) = 6\,k\Omega$).

### Power-Up Considerations

Careful power-up of the AD199x is necessary to ensure correct operation and avoid possible latch-up issues. The AD199x should be powered up with RST/PDN and MUTE held low until all the power supplies have stabilized. Once the supplies have stabilized, the AD199x can be brought out of reset by bringing RST/PDN high and then MUTE can be brought high as required.

### On/Off/Mute Pop Noise Suppression

The AD199x features pop suppression which is activated when the part is reset or taken out of mute. The pop suppression is achieved by pulsing the power outputs to bring the outputs of the LC filter from 0V to mid-scale in a controlled fashion. This feature eliminates unwanted transients on both the outputs and the high voltage power supply.

### Thermal Protection

The AD199x features thermal protection. When the die temperature exceeds approximately 135°C, the thermal warning error output (ERR1) is asserted. If the die temperature exceeds approximately 150°C, the thermal shutdown error output (ERR2) is asserted. If this occurs, the part shuts down to prevent damage. When the die temperature drops below approximately 120°C, both error outputs are negated and the part returns to normal operation.

### Over-Current Protection

The AD199x features over-current or short-circuit protection. If the current through any power transistors exceeds 4A, the part goes into mute and the over-current error output (ERR0) is asserted. This is a latched error and does not clear automatically. To clear the error condition and restore normal operation, the part must be either reset, or MUTE must be asserted and negated.

Good board layout and decoupling are vital for correct operation of the AD199x. Due to the fact that the part switches high currents, there is the potential for large PVDD bounce each time a transistor switches. This can cause unpredictable operation of the part. To avoid this potential problem, close chip decoupling is essential. It is also recommended that the decoupling capacitors are placed on the same side of the board as the AD199x, and connected directly to the PVDD and PGND pins. By placing the decoupling capacitors on the other side of the board and decoupling through vias, the effectiveness of the decoupling is reduced.

This is because vias have inductive properties and therefore prevent very fast discharge of the decoupling capacitors. Best operation is achieved with at least one decoupling capacitor on each side of the AD199x, or (optionally) two capacitors per side can be used to further reduce the series resistance of the capacitor. If these decoupling recommendations cannot be followed and decoupling through vias is the only option, the vias should be made as large as possible to increase surface area, thereby reducing inductance and resistance (Figure 2-108).

## Application Considerations

### Audio Fidelity and EMI Reduction

The AD199x amplifiers deliver audiophile sound quality (THD < 0.003%; SNR > 103 dB; PSRR > 65 dB) with 50% lower heat dissipation than traditional linear amplifiers. The THD performance is 40 dB better than typical "open-loop" competitors and 10–20 dB better than most closed-loop ones. This breakthrough performance is achieved through Analog Devices' closed-loop, mixed-signal integration of seventh-order Sigma–Delta ($\Sigma\Delta$) modulator technology with high power output drive circuitry and bridge circuitry. Radiated and conducted out-of-band RF emissions are minimized with Analog Devices' advanced modulation techniques and closed-loop, Sigma–Delta ($\Sigma\Delta$) architecture to enable a significant reduction in EMI.

Power levels range from Stereo 5 W (Mono 10 W) to Stereo 40 W (Mono 80 W). The AD1994 can be configured in a modulator-only mode. This, coupled with external high power FETs, enables very high power amplification, limited only by the power stage design. The parts also incorporate critical peripheral functions, including pop/click suppression circuitry as well as short-circuit, overload, and temperature protection.

### THD + N for 1 kHz Sinewave

Figures 2-109 and 2-110 show FFTs measured with signals of 1 kHz at 1 μW and 1 W output power levels. The 1 μW FFT (Figure 2-109) demonstrates that the noise floor is tonefree for low power inputs.

The 1 W power level in Figure 2-110 is intended to represent a realistic listening level. Harmonic distortion is evident, but the 0.00121% THD is an unprecedented level for this signal condition in a single chip Class D amplifier.

Figure 2-111 shows how THD varies with frequency, when the signal condition is a sinewave of 1 W output power.

Higher modulator loop gain at low frequencies enables better correction of modulator and power stage errors, giving better THD than at higher frequencies. THD is actually 0.001% ($-100$ dB) or better up to hundreds of Hz. The apparent THD improvement at audio frequencies above 6 kHz is a misleading artifact of the measurement setup, which was unable to detect harmonics beyond its 20 kHz bandwidth. For 20 kHz fundamentals, actual THD is near 0.01% ($-80$ dB), at ultrasonic, inaudible frequencies.

### THD + N Versus Output Power, 1 kHz Sine

Figure 2-112 shows how THD + N varies with output power, for 1 kHz sinewaves. There are two curves in the plot. The first (o) is for a low power application where PVDD = 12 V and the load is 6 $\Omega$ (the default measurement setup). The second (x) is for a higher power application where PVDD = 20 V and the load is 4 $\Omega$.

In these curves, there are three distinct regions of performance. The first is at the lowest output power levels, where the modulator is seventh order, and THD + N is best. The second performance region is at significant output power, when the modulator order is lowered from seventh to second to prevent instability.

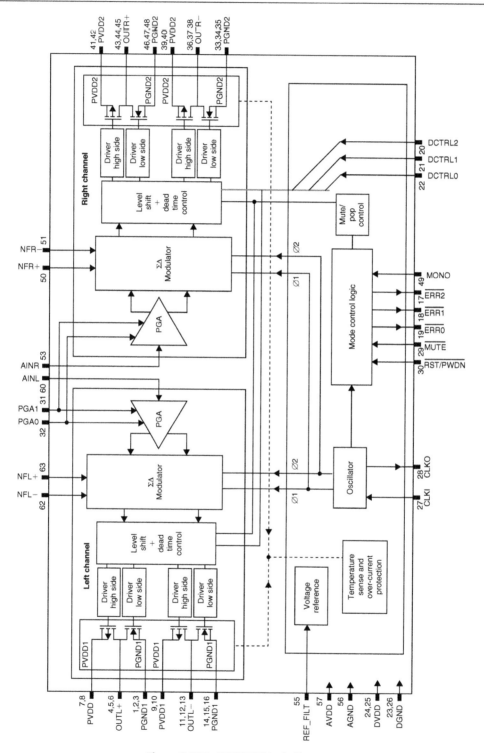

**Figure 2-108: AD199X block diagram**

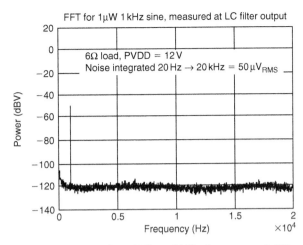

Figure 2-109: THD + N for a 1 kHz sinewave at 1 μW

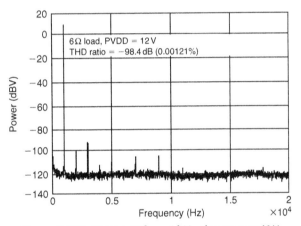

Figure 2-110: THD + N for a 1 kHz sinewave at 1 W

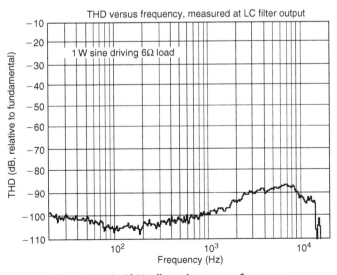

Figure 2-111: 1 kHz distortion versus frequency

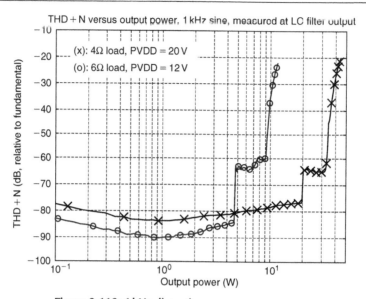

**Figure 2-112: 1 kHz distortion versus output power**

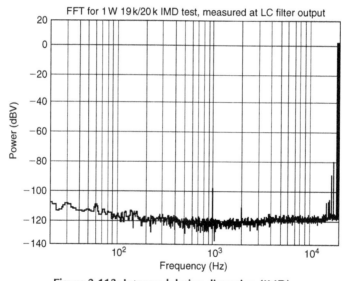

**Figure 2-113: Intermodulation distortion (IMD)**

The second-order configuration only allows 65 dB THD + N, because quantization noise is now elevated due to the lower modulator order. However, this higher noise is difficult to hear above the loud, energetic output. The third performance region is at highest output powers, where clipping occurs, and distortion associated with clipping causes THD to degrade rapidly.

## IMD

Figure 2-113 shows the intermodulation distortion (IMD) resulting from a 1 W, 19 kHz, and 20 kHz twin-tone stimulus. The 1 kHz second-order product is approximately 98 dB below the tones.

## Crosstalk

Crosstalk between channels is a concern in chips with multiple audio channels. To investigate crosstalk, we drove one channel of the chip with a 1 kHz, 1 W (+7.8 dBV) sinewave, while leaving the other channel idle (0 input). We then measured the idle channel: results are shown in Figure 2-114. The −89 dBV 1 kHz tone in the idle channel is 97 dB below the driven channel's signal.

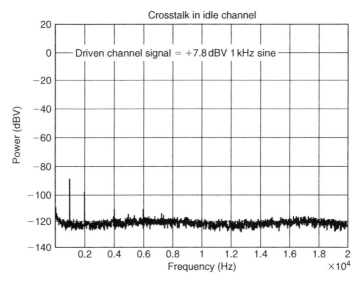

**Figure 2-114: Crosstalk**

## Power Efficiency

Figure 2-115 shows power efficiency up to 5 W output power. The 50 mW/channel modulator power consumption and power stage consumption are both included in this calculation. (If we included only the power stage consumption but excluded the modulator, as is sometimes done, the efficiency number would improve.)

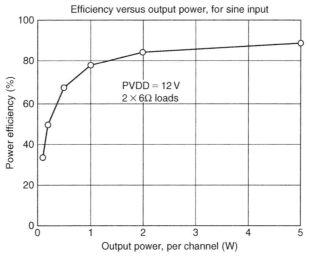

**Figure 2-115: Efficiency versus output power**

# Auto-Zero Amplifiers

## Chopper Amplifiers

Chopper type amplifier topologies have existed for decades. Initial chopper designs actually involved switching the AC coupled input signal and synchronous demodulation of the AC signal to re-establish the DC signal (see Figure 2-116). While these amplifiers achieved very low offset, low offset drift, and very high gain, they had limited bandwidth (it is a sampled system after all) and required filtering to remove the large ripple voltages generated by the chopping action. In the earliest implementations the chopping switches were actually relays, commonly switching on the order of 400 Hz.

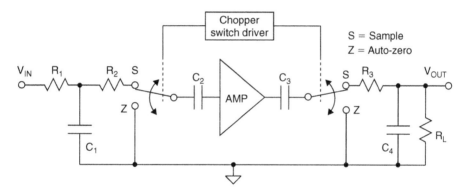

**Figure 2-116: Classic chopper amplifier simplified schematic**

Virtually all modern IC chopper amplifiers actually use an auto-zero approach utilizing a two (or more) stage composite amplifier structure similar to the chopper-stabilized scheme (see Figure 2-117). One stage provides nulling action, while the other provides wideband response. Together, the two stages provide very high voltage gain as they are connected in series.

Chopper-stabilized amplifiers solved the bandwidth limitations of the classic implementation by combining the chopper amplifier (used as a stabilizing amplifier) with a conventional wideband amplifier that remained in the signal path. Since the main signal path is not sampled, the bandwidth of the system is determined by the bandwidth of the signal amplifier. It can exceed the chopping frequency.

These chopper-stabilized designs are capable of inverting operation only since the stabilizing amplifier is connected to the non-inverting input of the wideband amplifier.

In this approach, the inputs of the nulling stage are shorted together during the first phase of the operational cycle. During this nulling phase, amplified feedback is used to virtually eliminate the offset of the nulling stage. The feedback voltage is impressed on a storage capacitor so that during the second, or "output," phase the offset remains nulled while the inputs are now connected to the signal of interest.

In the output phase, the nulled input stage and the wideband stage in series amplify the signal. The output of the nulled stage is impressed on a storage capacitor so that when the cycle returns to the nulling phase

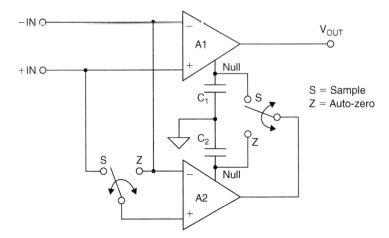

Figure 2-117: Auto-zero amplifier simplified schematic

(inputs shorted together), the output continues to reflect the last input voltage value. Higher frequency signals bypass the nulling stage through feedforward techniques, making wide bandwidth operation possible.

While this technique provides DC accuracy and better frequency response, along with the flexibility of inverting and non-inverting configurations, it is prone to high levels of digital switching noise that may limit the usefulness of the wider bandwidth.

## Auto-Zero Amplifiers Improve on Choppers

ADI's auto-zero amplifiers use a similar architecture with some major improvements. Dual nulling loops, special switching logic, and innovative compensation techniques result in dynamic performance improvements while minimizing total die area. The result: amplifiers retain the high gain and DC precision of the auto-zero approach while minimizing the negative effects of digital switching on the analog signal— at half the cost. Typical offset voltage is under $1 \mu V$ and the offset drift is *less than* $10 nV/°C$. Voltage gain is more than 10 million, while PSRR and CMRR are well above $120 dB$. Input voltage noise is only $1 \mu V_{p-p}$ from DC to $10 Hz$.

Many auto-zero amplifiers are plagued by long overload recovery times due to the complicated settling behavior of the internal nulling loops after saturation of the outputs. Analog Devices' auto-zero amplifiers have been designed so that internal settling occurs within one or two clock cycles after output saturation occurs. The result is that the overload recovery time is more than an order of magnitude shorter than previous designs and is comparable to conventional amplifiers.

The careful design and layout of the AD855x amplifiers reduce digital clock noise and aliasing effects by as much as $40 dB$ versus older designs.

In many cases the bandwidth required by the applications is such that the small amount of digital feedthrough can be eliminated by filtering. Output filtering is also useful in limiting the broadband noise of the signal amplifier.

The AD857x reduces the effects of digital switching on the analog signal by using a patented digital spread-spectrum technique. As can be seen from Figures 2-118 and 2-119, the AD857x virtually eliminates the energy spike seen in other auto-zero amplifiers at the switching frequency. It also reduces aliasing products between the chopping clock and the input signal to the noise floor. The only penalty for this breakthrough

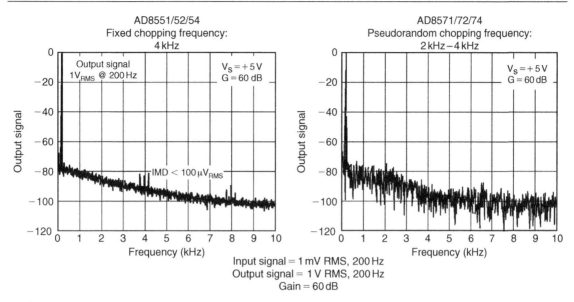

Figure 2-118: Output spectrum of auto-zero amplifiers with fixed frequency and spread spectrum chopping

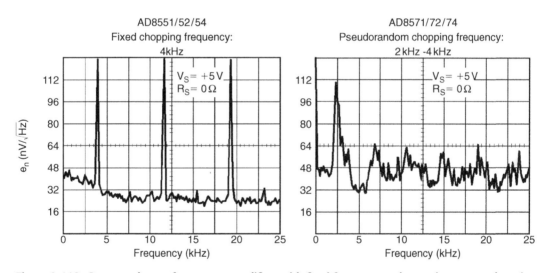

Figure 2-119: Output voltage of auto-zero amplifiers with fixed frequency and spread spectrum chopping

performance is a slight increase in voltage noise from the industry-best $1\,\mu V_{p-p}$ from DC to 10 Hz of the AD855x design.

## Implementation

The actual circuit implementation of an IC auto-zero amplifier is much more complicated than the simplified version described above. Multiple nulling loops are combined with innovative compensation, and signal paths are fully differential. Internal voltages are controlled carefully to prevent saturation of the nulling circuitry. In addition, special logic designs are utilized, and careful layout is required to minimize parasitic effects. These techniques result in stable, reliable operation and minimize unwanted digital interaction with the analog signals.

The frequency response of the nulling and wideband amplifiers is carefully tailored so that low frequency errors (DC circuit offsets and low frequency noise) are nulled while high frequency signals are amplified as in a conventional op amp. This nulling of low frequency errors has an important consequence for voltage noise. The very low frequency 1/f noise behavior seen in conventional amplifiers is not present in auto-zero amplifiers. For applications with long measurement times on slowly varying signals, the noise performance is better than the best low noise conventional amplifier designs (Figure 2-120).

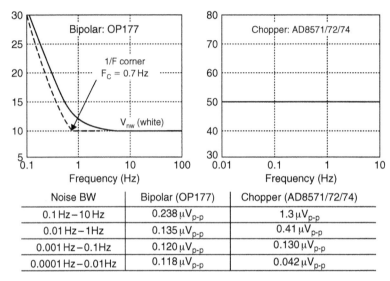

| Noise BW | Bipolar (OP177) | Chopper (AD8571/72/74) |
|---|---|---|
| 0.1 Hz – 10 Hz | $0.238\,\mu V_{p\text{-}p}$ | $1.3\,\mu V_{p\text{-}p}$ |
| 0.01 Hz – 1Hz | $0.135\,\mu V_{p\text{-}p}$ | $0.41\,\mu V_{p\text{-}p}$ |
| 0.001 Hz – 0.1Hz | $0.120\,\mu V_{p\text{-}p}$ | $0.130\,\mu V_{p\text{-}p}$ |
| 0.0001 Hz – 0.01Hz | $0.118\,\mu V_{p\text{-}p}$ | $0.042\,\mu V_{p\text{-}p}$ |

**Figure 2-120: Noise comparison between conventional precision amplifiers and chopper-stabilized op amps**

In this IC implementation, the size of the on-chip storage capacitors is limited to achieve a cost-effective die size. The small storage capacitors require careful attention to the switch design and layout so that charge injection effects do not create large offset errors. Switch leakage must also be minimized to maintain circuit accuracy, especially at high temperatures. In the AD855x and AD857x amplifiers, the switches have been optimized for accurate operation up to +125°C.

## Operation Description

The simplified circuit consists of the nulling amplifier ($A_A$), the wideband amplifier ($A_B$), storage capacitors ($C_{M1}$ and $C_{M2}$), and switches for the inputs and storage capacitors. There are two phases (A and B) per clock cycle.

In Phase A, the auto-zero phase (Figure 2-121), the nulling amplifier auto-zeros itself while the wideband amplifier amplifies the input signal directly. The inputs of the nulling amp are shorted together and to the inverting input terminal (CM input voltage). The nulling amplifier nulls its inherent offset voltage through its nulling terminal gain ($-B_A$). The nulling voltage is also impressed on $C_{M1}$. The signal at the input terminals is amplified directly by the wideband amplifier.

In Phase B, the output phase (Figure 2-122), both amplifiers amplify the input signal. The inputs of the nulling amplifier are connected to the input terminals. The nulling voltage of the nulling amplifier is now

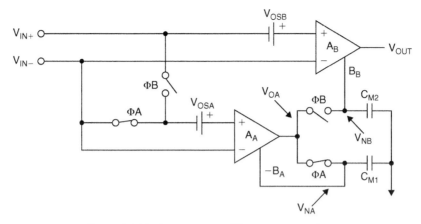

**Figure 2-121: Auto-zero amplifier, auto-zero phase**

stored on capacitor $C_{M1}$ and continues to minimize its output offset voltage. The instantaneous input signal is amplified by the nulling amplifier into the wideband amplifier through the wideband amplifier nulling terminal gain ($B_B$). The output voltage of the nulling amplifier is also impressed on storage capacitor $C_{M2}$. The total amplifier gain is approximately equal to the product of the nulling amplifier gain and the wideband amplifier gain. The total offset voltage is approximately equal to the sum of the nulling amplifier and wideband amplifier offset voltages divided by the gain of the wideband amplifier nulling terminal. By making this gain very large, the total amplifier effective offset voltage becomes very small.

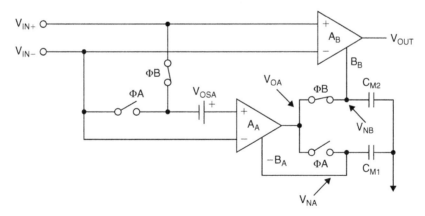

**Figure 2-122: Auto-zero amplifier, output phase**

Both $V_{OSA}$ and $V_{OSB}$ are highpass filtered "corner frequency" of highpass filter set by chopping frequency.

As the cycle returns to the nulling phase, the stored voltage on $C_{M2}$ continues to effectively correct the DC offset of the composite amplifier. The cycle from nulling to output phase is repeated continuously at a rate set by the internal clock and logic circuits. This model circuit, while simplified from the actual design, accurately depicts the essentials of the auto-zero technique.

A more rigorous analysis is available in the data sheets for the AD855X.

## References: Auto-Zero Amplifiers

1. D.H. Sheingold, (Ed.), **Transducer Interfacing Handbook**, Analog Devices, Inc., Boston, MA, 1981.

2. C. Kitchin and L. Counts, **Instrumentation Amplifier Applications Guide**, Analog, Devices, Inc., Boston, MA, 1991.

3. **Amplifier Applications Guide**, Analog Devices, Inc., Boston, MA, 2002.

4. **System Applications Guide**, Analog Devices, Inc., Boston, MA, 1993.

5. J. Sylvan, "High-speed comparators provide many useful circuit functions when used correctly." **Ask The Applications Engineer—5**.

6. R. Moghimi, "Curing Comparator Instability with Hysteresis," **Analog Dialogue**, Vol. 34, No. 7, 2000.

7. G. Erdi, "A 300V/μs Monolithic Voltage Follower," **IEEE Journal of Solid State Circuits**, Vol. SC-14, No. 6, December, 1979, pp. 1059–1065.

8. R.A. Gosser, "Wideband Transconductance Generator," **US Patent 5,150,074**, filed, May 3, 1991, issued September 22, 1992.

9. D.F. Bowers, "A 6.8mA Closed-Loop Monolithic Buffer with 120MHz Bandwidth, 4000V/μs Slew Rate, and ±12V Signal Compatibility," **1994 Bipolar/BiCMOS Circuits and Technology Meeting 1.3**, pp. 23–26.

10. B. Gilbert, **ISSCC Digest of Technical Papers 1968**, February 16, 1968, pp. 114–115.

11. B. Gilbert, **Journal of Solid State Circuits**, Vol. SC-3, December 1968, pp. 353–372.

12. C.L. Ruthroff, "Some Broadband Transformers," **Proc. I.R.E.**, Vol. 47, August, 1959, pp. 1337–1342.

13. J.M. Bryant, *Mixers for High Performance Radio*, **Wescon 1981: Session 24** (Published by Electronic Conventions, Inc., Sepulveda Blvd., El Segundo, CA).

14. P.E. Chadwick, *High Performance IC Mixers*, **IERE Conference on Radio Receivers and Associated Systems**, Leeds, 1981, IERE Conference Publication No. 50.

15. P.E. Chadwick, *Phase Noise, Intermodulation, and Dynamic Range*, **RF Expo**, Anaheim, CA, January, 1986.

16. H. Daniel and R. Sheingold, **Nonlinear Circuits Handbook**, 3rd Edition, Analog Devices, Inc., 1974.

17. R.S. Hughes, **Logarithmic Amplifiers**, Artech House, Inc., Dedham, MA., 1986.

18. W.L. Barber and E.R. Brown, "A True Logarithmic Amplifier for Radar IF Applications," **IEEE Journal of Solid State Circuits**, Vol. SC-15, No. 3, June, 1980, pp. 291–295.

19. **Broadband Amplifier Applications**, Plessey Co. Publication P.S. 1938, September, 1984.

20. M.S. Gay, **SL521 Application Note**, Plessey Co., Norwood, Ma, 1966.

21. **Amplifier Applications Guide**, Analog Devices, Inc., Boston, MA, 1992, Section 9.

22. C. Kitchen and L. Counts, **RMS-to-DC Conversion Application Guide, Second Edition**, Analog Devices, Inc., New York, 1986.

23. B. Gilbert, *A Low Noise Wideband Variable-Gain Amplifier Using an Interpolated Ladder Attenuator*, **IEEE ISSCC Technical Digest**, 1991, pp. 280, 281, 330.

24. B. Gilbert, "A Monolithic Microsystem for Analog Synthesis of Trigonometric Functions and their Inverses," **IEEE Journal of Solid State Circuits**, Vol. SC-17, No. 6, December.

25. L. Zhang, et al., "Real-time Power Supply Compensation for Noise-shaped Class D Amplifier," Presented at 117th AES Convention, San Francisco CA, USA, October 28–31, 2004.

26. M. Berkhout, "Integrated 200-W class-D audio amplifier," **Journal of Solid State Circuits**, Vol. 38, July 2003, pp. 1198–1206.

27. K. Nielsen, "A review and comparison of pulse width modulation (PWM) methods for analog and digital input switching power amplifiers," Presented at 102nd AES Convention, Munich, Germany, March 22–25, 1997.

28. P. Morrow, et al., "A 20 Watt Stereo Class-D Audio Output Power Stage in 0.6μm BCDMOS Technology," **Journal of Solid State Circuits**, Vol. 39, November, 2004.

29. S. R Norsworthy, R. Schreier, G. C Temes, (Ed.), **Delta–Sigma Data Converters**, IEEE press, 1997, pp. 153–155.

30. E. Gaalaas, B. Y. Liu, N. Nishimura, R. Adams, K. Sweetland, and R. Morajkar, "Integrated Stereo ΣΔ Class D Amplifier," Presented at the 118th Convention, Barcelona, Spain, May 28–31, 2005.

31. E. Gaalaas, B. Y. Liu, N. Nishimura, R. Adams, and K. Sweetland, "Integrated Stereo ΣΔ Class D Amplifier," **IEEE Journal of Solid-State Circuits**, Vol. 40, No. 12, 2005, pp. 2388–2397.

# CHAPTER 3

## *Sensors*

# *Positional Sensors*

## *Linear Variable Differential Transformers*

The linear variable differential transformer (LVDT) is an accurate and reliable method for measuring linear distance. LVDTs find uses in modern machine-tool, robotics, avionics, and computerized manufacturing.

The LVDT (see Figure 3-1) is a position-to-electrical sensor whose output is proportional to the position of a movable magnetic core. The core moves linearly inside a transformer consisting of a center primary coil and two outer secondary coils wound on a cylindrical form. The primary winding is excited with an AC voltage source (typically several kHz), inducing secondary voltages which vary with the position of the magnetic core within the assembly. The core is usually threaded in order to facilitate attachment to a non-ferromagnetic rod which in turn is attached to the object whose movement or displacement is being measured.

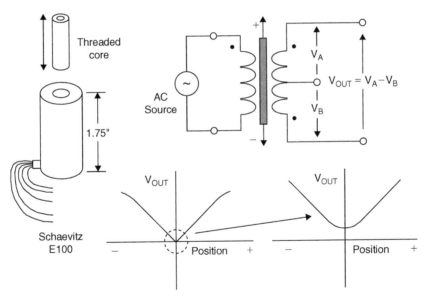

**Figure 3-1: Linear variable differential transformer (LVDT)**

The secondary windings are wound out of phase with each other, and when the core is centered the voltages in the two secondary windings oppose each other, and the net output voltage is zero. When the core is moved off center, the voltage in the secondary toward which the core is moved increases, while the opposite voltage decreases. The result is a differential voltage output which varies linearly with the core's position. Linearity is excellent over the design range of movement, typically 0.5% or better. The LVDT offers good accuracy, linearity, sensitivity, infinite resolution, as well as frictionless operation and ruggedness.

A wide variety of measurement ranges are available in different LVDTs, typically from $\pm 100\,\mu m$ to $\pm 25\,cm$. Typical excitation voltages range from $1\,V$ to $24\,V_{RMS}$, with frequencies from $50\,Hz$ to $20\,kHz$.

Note that a true null does not occur when the core is in center position because of mismatches between the two secondary windings and leakage inductance. Also, simply measuring the output voltage $V_{OUT}$ will not tell on which side of the null position the core resides.

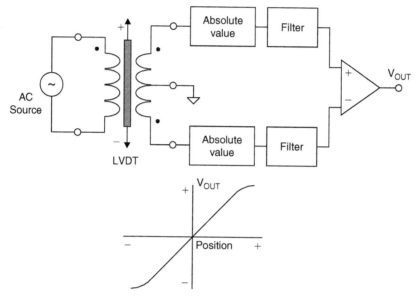

**Figure 3-2: Improved LVDT output signal processing**

A signal conditioning circuit which removes these difficulties is shown in Figure 3-2 where the absolute values of the two output voltages are subtracted. Using this technique, both positive and negative variations about the center position can be measured. While a diode/capacitor-type rectifier could be used as the absolute value circuit, the precision rectifier shown in Figure 3-3 is more accurate and linear. The input is applied to a V/I converter which in turn drives an analog multiplier. The sign of the differential input is detected by the comparator whose output switches the sign of the V/I output via the analog multiplier. The final output is a precision replica of the absolute value of the input. These circuits are well-understood by integrated circuit (IC) designers and are easy to implement on modern bipolar processes.

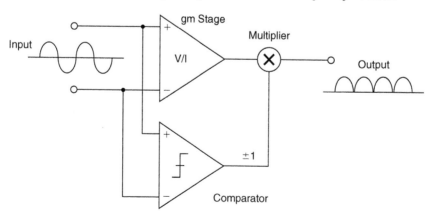

**Figure 3-3: Precision absolute value circuit (full wave rectifier)**

*196*

The industry-standard AD598 LVDT signal conditioner shown in Figure 3-4 (simplified form) performs all required LVDT signal processing. The on-chip excitation frequency oscillator can be set from 20 Hz to 20 kHz with a single external capacitor. Two absolute value circuits followed by two filters are used to detect the amplitude of the A and B channel inputs. Analog circuits are then used to generate the ratiometric function $(A - B)/(A + B)$. Note that this function is independent of the amplitude of the primary winding excitation voltage, assuming the sum of the LVDT output voltage amplitudes remains constant over the operating range. This is usually the case for most LVDTs, but the user should always check with the manufacturer if it is not specified on the LVDT data sheet. Note also that this approach requires the use of a five-wire LVDT.

A single external resistor sets the AD598 excitation voltage from approximately 1 to $24 V_{RMS}$. Drive capability is $30 mA_{RMS}$. The AD598 can drive an LVDT at the end of 300 feet of cable, since the circuit is not affected by phase shifts or absolute signal magnitudes. The position output range of $V_{OUT}$ is $\pm 11$ V for a 6 mA load, and it can drive up to 1,000 feet of cable. The $V_A$ and $V_B$ inputs can be as low as $100 mV_{RMS}$.

The AD698 LVDT signal conditioner (see Figure 3-5) has similar specifications as the AD598 but processes the signals slightly differently and uses synchronous demodulation. The A and B signal processors each consist of an absolute value function and a filter. The A output is then divided by the B output to produce a final output which is ratiometric and independent of the excitation voltage amplitude. Note that the sum of the LVDT secondary voltages does not have to remain constant in the AD698.

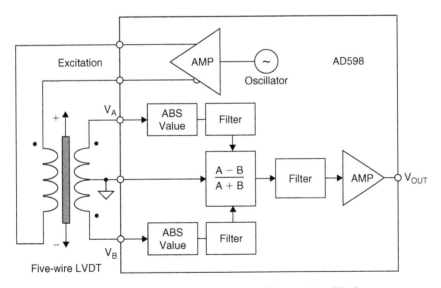

**Figure 3-4: AD598 LVDT signal conditioner (simplified)**

The AD698 can also be used with a half-bridge (similar to an auto-transformer) LVDT as shown in Figure 3-6. In this arrangement, the entire secondary voltage is applied to the B processor, while the center-tap voltage is applied to the A processor. The half-bridge LVDT does not produce a null voltage, and the A/B ratio represents the range-of-travel of the core.

It should be noted that the LVDT concept can be implemented in rotary form, in which case the device is called a *rotary variable differential transformer* (RVDT). The shaft is equivalent to the core in an LVDT, and the transformer windings are wound on the stationary part of the assembly. However, the RVDT is linear over a relatively narrow range of rotation and is not capable of measuring a full 360° rotation. Although capable of continuous rotation, typical RVDTs are linear over a range of about ±40° about the

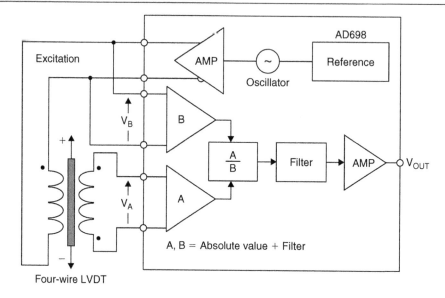

**Figure 3-5: AD698 LVDT signal conditioner (simplified)**

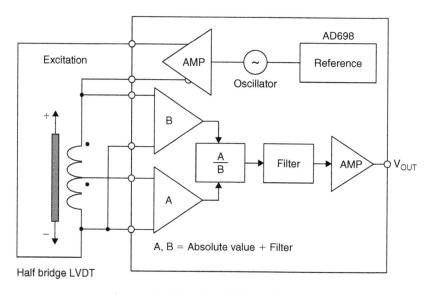

**Figure 3-6: Half-bridge LVDT configuration**

null position (0°). Typical sensitivity is 2–3 mV/V/degree of rotation, with input voltages in the range of $3 V_{RMS}$ at frequencies between 400 Hz and 20 kHz. The 0° position is marked on the shaft and the body.

## Hall Effect Magnetic Sensors

If a current flows in a conductor (or semiconductor) and there is a magnetic field present which is perpendicular to the current flow, then the combination of current and magnetic field will generate a voltage perpendicular to both (see Figure 3-7). This phenomenon is called the *Hall Effect*, was discovered by E. H. Hall in 1879.

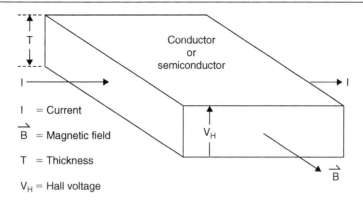

**Figure 3-7: Hall Effect sensor**

The voltage, $V_H$, is known as the *Hall Voltage*. $V_H$ is a function of the current density, the magnetic field, and the charge density and carrier mobility of the conductor.

The Hall Effect may be used to measure magnetic fields (and hence in contact-free current measurement), but its commonest application is in motion sensors where a fixed Hall sensor and a small magnet attached to a moving part can replace a cam and contacts with a great improvement in reliability. (Cams wear and contacts arc become fouled, but magnets and Hall sensors can contact free and do neither.) Since $V_H$ is proportional to magnetic field and not to the rate of change of magnetic field like an inductive sensor, the Hall Effect provides a more reliable low speed sensor than an inductive pickup.

Although several materials can be used for Hall Effect sensors, silicon has the advantage that signal conditioning circuits can be integrated on the same chip as the sensor. Complementary-MOS (CMOS) processes are common for this application. A simple rotational speed detector can be made with a Hall sensor, a gain stage, and a comparator as shown in Figure 3-8. The circuit is designed to detect rotation speed as in automotive applications. It responds to small changes in field, and the comparator has built-in hysteresis to prevent oscillation. Several companies manufacture such Hall switches, and their usage is widespread.

There are many other applications, particularly in automotive throttle, pedal, suspension, and valve position sensing, where a linear representation of the magnetic field is desired. The AD22151 is a linear magnetic field sensor whose output voltage is proportional to a magnetic field applied perpendicularly to the package

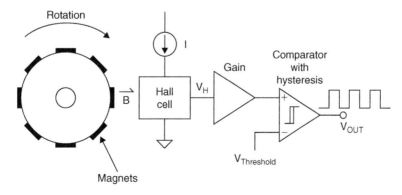

**Figure 3-8: Hall Effect sensor used as a rotational sensor**

top surface (see Figure 3-9). The AD22151 combines integrated bulk Hall cell technology and conditioning circuitry to minimize temperature-related drifts associated with silicon Hall cell characteristics.

The architecture maximizes the advantages of a monolithic implementation while allowing sufficient versatility to meet varied application requirements with a minimum number of external components. Principal features include dynamic offset drift cancellation using a chopper-type op amp and a built-in temperature sensor. Designed for single +5V supply operation, low offset, and gain drift allows operation over a −40°C to +150°C range. Temperature compensation (set externally with a resistor $R_1$) can accommodate a number of magnetic materials commonly utilized in position sensors. Output voltage range and gain can be easily set with external resistors. Typical gain range is usually set from 2mV/G to 6mV/G. Output voltage can be adjusted from fully bipolar (reversible) field operation to fully unipolar field sensing. The voltage output achieves near rail-to-rail dynamic range (+0.5V to +4.5V), capable of supplying 1 mA into large capacitive loads. The output signal is ratiometric to the positive supply rail in all configurations.

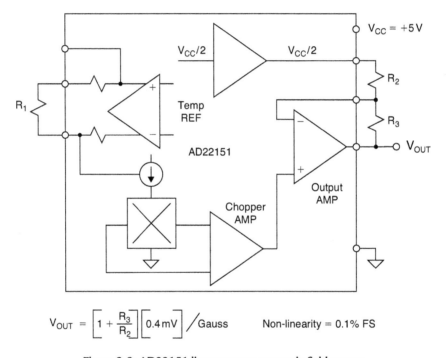

$$V_{OUT} = \left[ 1 + \frac{R_3}{R_2} \right] \left[ 0.4\,mV \right] / Gauss \qquad \text{Non-linearity} = 0.1\% \text{ FS}$$

**Figure 3-9: AD22151 linear output magnetic field sensor**

## Resolvers and Synchros

Machine-tool and robotics manufacturers have increasingly turned to resolvers and synchros to provide accurate angular and rotational information. These devices excel in demanding factory applications requiring small size, long-term reliability, absolute position measurement, high accuracy, and low noise operation.

A diagram of a typical synchro and resolver is shown in Figure 3-10. Both synchros and resolvers employ single-winding rotors that revolve inside fixed stators. In the case of a simple synchro, the stator has three windings oriented 120° apart and electrically connected in a Y-connection. Resolvers differ from synchros in that their stators have only two windings oriented at 90°.

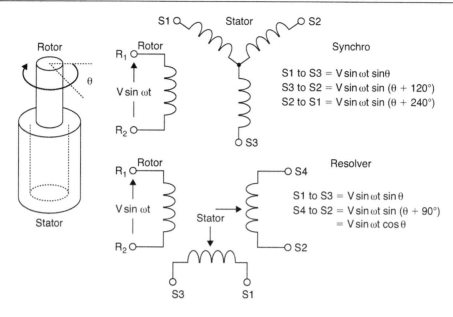

**Figure 3-10: Synchros and resolvers**

Because synchros have three stator coils in a 120° orientation, they are more difficult to manufacture than resolvers and are therefore more costly. Today, synchros find decreasing use, except in certain military and avionic retrofit applications.

Modern resolvers, in contrast, are available in a brushless form that employ a transformer to couple the rotor signals from the stator to the rotor. The primary winding of this transformer resides on the stator, and the secondary on the rotor. Other resolvers use more traditional brushes or slip rings to couple the signal into the rotor winding. Brushless resolvers are more rugged than synchros because there are no brushes to break or dislodge, and the life of a brushless resolver is limited only by its bearings. Most resolvers are specified to work over $2\,V - 40\,V_{RMS}$ and at frequencies from 400 Hz to 10 kHz. Angular accuracies range from 5 arc-minutes to 0.5 arc-minutes. (There are 60 arc-minutes in 1°, and 60 arc-seconds in 1 arc-minute. Hence, 1 arc-minute is equal to 0.0167°.)

In operation, synchros and resolvers resemble rotating transformers. The rotor winding is excited by an AC reference voltage, at frequencies up to a few kHz. The magnitude of the voltage induced in any stator winding is proportional to the sine of the angle ($\theta$) between the rotor coil axis and the stator coil axis. In the case of a synchro, the voltage induced across any pair of stator terminals will be the vector sum of the voltages across the two connected coils.

For example, if the rotor of a synchro is excited with a reference voltage, $V \sin \omega t$, across its terminals $R_1$ and $R_2$, then the stator's terminal will see voltages in the form:

$$\text{S1 to S3} = V \sin \omega t \sin \theta \tag{3-1}$$

$$\text{S3 to S2} = V \sin \omega t \sin(\theta + 120°) \tag{3-2}$$

$$\text{S2 to S1} = V \sin \omega t \sin(\theta + 240°) \tag{3-3}$$

where $\theta$ is the shaft angle.

In the case of a resolver, with a rotor AC reference voltage of $V \sin \omega t$, the stator's terminal voltages will be:

$$S1 \text{ to } S3 = V \sin \omega t \sin \theta \qquad (3\text{-}4)$$

$$S4 \text{ to } S2 = V \sin \omega t \sin(\theta + 90°) = V \sin \omega t \cos \theta \qquad (3\text{-}5)$$

It should be noted that the three-wire synchro output can be easily converted into the resolver-equivalent format using a Scott-T transformer. Therefore, the following signal processing example describes only the resolver configuration.

A typical resolver-to-digital converter (RDC) is shown functionally in Figure 3-11. The two outputs of the resolver are applied to cosine and sine multipliers. These multipliers incorporate sine and cosine lookup tables and function as multiplying digital-to-analog converters (DACs). Begin by assuming that the current state of the up/down counter is a digital number representing a trial angle, $\varphi$. The converter seeks to adjust the digital angle, $\varphi$, continuously to become equal to, and to track $\theta$, the analog angle being measured. The resolver's stator output voltages are written as:

$$V1 = V \sin \omega t \sin \theta \qquad (3\text{-}6)$$

$$V2 = V \sin \omega t \cos \theta \qquad (3\text{-}7)$$

where $\theta$ is the angle of the resolver's rotor. The digital angle $\varphi$ is applied to the cosine multiplier, and its cosine is multiplied by V1 to produce the term:

$$V \sin \omega t \sin \theta \cos \varphi \qquad (3\text{-}8)$$

The digital angle $\varphi$ is also applied to the sine multiplier and multiplied by V2 to produce the term:

$$V \sin \omega t \cos \theta \sin \varphi \qquad (3\text{-}9)$$

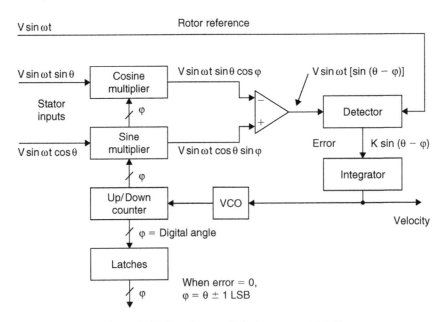

**Figure 3-11: Resolver-to-digital converter (RDC)**

These two signals are subtracted from each other by the error amplifier to yield an AC error signal of the form:

$$V \sin \omega t(\sin \theta \cos \varphi - \cos \theta \sin \varphi) \tag{3-10}$$

Using a simple trigonometric identity, this reduces to:

$$V \sin \omega t(\sin(\theta - \varphi)) \tag{3-11}$$

The detector synchronously demodulates this AC error signal, using the resolver's rotor voltage as a reference. This results in a DC error signal proportional to $\sin(\theta - \varphi)$.

The DC error signal feeds an integrator, the output of which drives a voltage-controlled-oscillator (VCO). The VCO, in turn, causes the up/down counter to count in the proper direction to cause:

$$\sin(\theta - \varphi) \to 0 \tag{3-12}$$

When this is achieved,

$$\theta - \varphi \to 0 \tag{3-13}$$

and therefore

$$\varphi = \theta \tag{3-14}$$

to within one count. Hence, the counter's digital output, $\varphi$, represents the angle $\theta$. The latches enable this data to be transferred externally without interrupting the loop's tracking.

## Inductosyns

Synchros and resolvers inherently measure rotary position, but they can make linear position measurements when used with lead screws. An alternative, the Inductosyn™ (registered trademark of Farrand Controls, Inc.) measures linear position directly. In addition, Inductosyns are accurate and rugged, well-suited to severe industrial environments, and do not require ohmic contact.

The linear Inductosyn consists of two magnetically coupled parts; it resembles a multipole resolver in its operation (see Figure 3-12). One part, the scale, is fixed (e.g., with epoxy) to one axis, such as a machine-tool

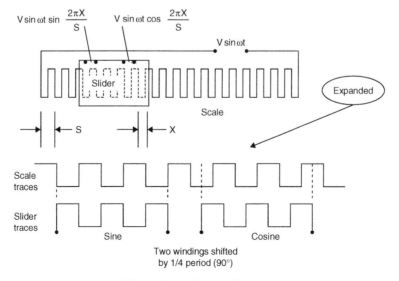

**Figure 3-12: Linear Inductosyn**

bed. The other part, the slider, moves along the scale in conjunction with the device to be positioned (e.g., the machine-tool carrier).

The scale is constructed of a base material such as steel, stainless steel, aluminum, or a tape of spring steel, covered by an insulating layer. Bonded to this is a printed circuit trace, in the form of a continuous rectangular waveform pattern. The pattern typically has a cyclic pitch of 0.1 inch, 0.2 inch, or 2 mm. The slider, about 4 inches long, has two separate but identical printed circuit traces bonded to the surface that faces the scale. These two traces have a waveform pattern with exactly the same cyclic pitch as the waveform on the scale, but one trace is shifted one-quarter of a cycle relative to the other. The slider and the scale remain separated by a small air gap of about 0.007 inch.

Inductosyn operation resembles that of a resolver. When the scale is energized with a sinewave, this voltage couples to the two slider windings, inducing voltages proportional to the sine and cosine of the slider's spacing within the cyclic pitch of the scale. If S is the distance between pitches, and X is the slider displacement within a pitch, and the scale is energized with a voltage $V \sin \omega t$, then the slider windings will see terminal voltages of:

$$V \text{ (sine output)} = V \sin \omega t \sin(2\pi X/S) \tag{3-15}$$

$$V \text{ (cosine output)} = V \sin \omega t \cos(2\pi X/S) \tag{3-16}$$

As the slider moves the distance of the scale pitch, the voltages produced by the two slider windings are similar to those produced by a resolver rotating through 360°. The absolute orientation of the Inductosyn is determined by counting successive pitches in either direction from an established starting point. Because the Inductosyn consists of a large number of cycles, some form of coarse control is necessary in order to avoid ambiguity. The usual method of providing this is to use a resolver or synchro operated through a rack and pinion or a lead screw.

In contrast to a resolver's highly efficient transformation of 1:1 or 2:1, typical Inductosyns operate with transformation ratios of 100:1. This results in a pair of sinusoidal output signals in the millivolt range which generally require amplification.

Since the slider output signals are derived from an average of several spatial cycles, small errors in conductor spacing have minimal effects. This is an important reason for the Inductosyn's very high accuracy. In combination with 12-bit RDCs, linear Inductosyns readily achieve 25 μ inch resolutions.

Rotary Inductosyns can be created by printing the scale on a circular rotor and the slider's track pattern on a circular stator. Such rotary devices can achieve very high resolutions. For instance, a typical rotary Inductosyn may have 360 cyclic pitches per rotation, and might use a 12-bit RDC. The converter effectively divides each pitch into 4,096 sectors. Multiplying by 360 pitches, the rotary Inductosyn divides the circle into a total of 1,474,560 sectors. This corresponds to an angular resolution of less than 0.9 arc-seconds. As in the case of the linear Inductosyn, a means must be provided for counting the individual pitches as the shaft rotates. This may be done with an additional resolver acting as the coarse measurement.

## Accelerometers

Accelerometers are widely used to measure tilt, inertial forces, shock, and vibration. They find wide usage in automotive, medical, industrial control, and other applications. Modern micromachining techniques allow these accelerometers to be manufactured on CMOS processes at low cost with high reliability. Analog Devices iMEMS® (integrated micro electro mechanical systems) accelerometers represent a breakthrough in this technology. A significant advantage of this type of accelerometer over piezoelectric-type charge-output accelerometers is that DC acceleration can be measured (e.g., they can be used in tilt measurements where the acceleration is a constant 1 g).

The basic unit cell sensor building block for these accelerometers is shown in Figure 3-13. The surface-micromachined sensor element is made by depositing polysilicon on a sacrificial oxide layer that is then etched away leaving the suspended sensor element. The actual sensor has tens of unit cells for sensing acceleration, but the diagram shows only one cell for clarity. The electrical basis of the sensor is the differential capacitor (CS1 and CS2) which is formed by a center plate which is part of the moving beam and two fixed outer plates. The two capacitors are equal at rest (no applied acceleration). When acceleration is applied, the mass of the beam causes it to move closer to one of the fixed plates while moving further from the other. This change in differential capacitance forms the electrical basis for the conditioning electronics as shown in Figure 3-14.

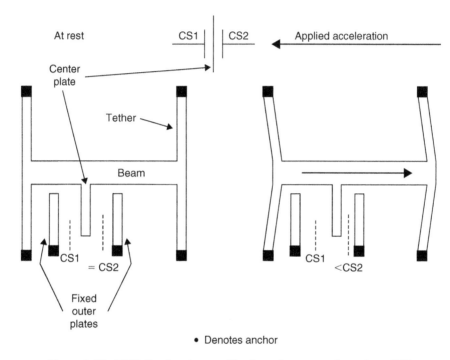

**Figure 3-13: ADXL family micromachined accelerometers (top view of IC)**

The sensor's fixed capacitor plates are driven differentially by a 1 MHz square wave: the two square wave amplitudes are equal but are 180° out of phase. When at rest, the values of the two capacitors are the same, and therefore the voltage output at their electrical center (i.e., at the center plate attached to the movable beam) is zero. When the beam begins to move, a mismatch in the capacitance produces an output signal at the center plate. The output amplitude will increase with the acceleration experienced by the sensor. The center plate is buffered by A1 and applied to a synchronous demodulator. The direction of beam motion affects the phase of the signal, and synchronous demodulation is therefore used to extract the amplitude information. The synchronous demodulator output is amplified by A2 which supplies the acceleration output voltage, $V_{OUT}$.

An interesting application of low-g accelerometers is measuring tilt. Figure 3-15 shows the response of an accelerometer to tilt. The accelerometer output on the diagram has been normalized to 1 g full-scale. The accelerometer output is proportional to the sine of the tilt angle with respect to the horizon. Note that

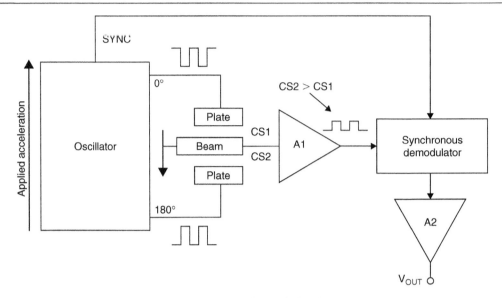

**Figure 3-14: Accelerometer internal signal conditioning**

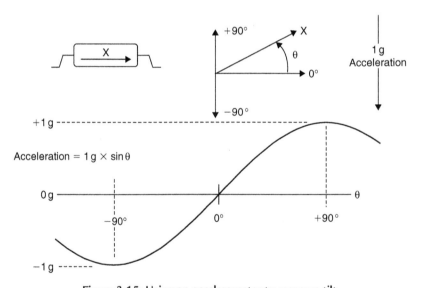

**Figure 3-15: Using an accelerometer to measure tilt**

maximum sensitivity occurs when the accelerometer axis is perpendicular to the acceleration. This scheme allows tilt angles from $-90°$ to $+90°$ ($180°$ of rotation) to be measured. However, in order to measure a full $360°$ rotation, a dual axis accelerometer must be used.

Figure 3-16 shows a simplified block diagram of the ADXL202 dual axis $\pm 2\,g$ accelerometer. The output is a pulse whose duty cycle contains the acceleration information. This type of output is extremely useful because of its high noise immunity, and the data are transmitted over a single wire. Standard low cost

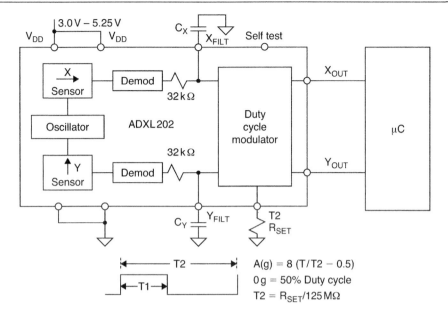

**Figure 3-16: ADXL202 ±2 g dual axis accelerometer**

microcontrollers have timers which can be easily used to measure the T1 and T2 intervals. The acceleration in g is then calculated using the formula:

$$A(g) = 8(T1/T2 - 0.5) \tag{3-17}$$

Note that a duty cycle of 50% (T1 = T2) yields a 0 g output. T2 does not have to be measured for every measurement cycle. It need only be updated to account for changes due to temperature. Since the T2 time period is shared by both X and Y channels, it is necessary to measure it on only one channel. The T2 period can be set from 0.5 ms to 10 ms with an external resistor.

Analog voltages representing acceleration can be obtained by buffering the signal from the $X_{FILT}$ and $Y_{FILT}$ outputs or by passing the duty cycle signal through an RC filter to reconstruct its DC value.

A single accelerometer cannot work in all applications. Specifically, there is a need for both low-g and high-g accelerometers. Low-g devices are useful in such applications as tilt measurements, but higher-g accelerometers are needed in applications such as airbag crash sensors.

## iMEMS® Angular-Rate-Sensing Gyroscope

The ADXRS150 and ADXRS300 gyros, with full-scale ranges of 150°/seconds and 300°/seconds, represent a quantum jump in gyro technology. The first commercially available surface-micromachined angular-rate sensors with integrated electronics, are smaller—with lower power consumption, and better immunity to shock and vibration—than any gyros having comparable functionality.

### Gyroscope Description

Gyroscopes are used to measure angular rate—how quickly an object turns. The rotation is typically measured in reference to one of three axes: yaw, pitch, or roll. Figure 3-17 shows a diagram representing each axis of sensitivity relative to a package mounted to a flat surface. Depending on how a gyro normally sits, its primary axis of sensitivity can be one of the three axes of motion: yaw, pitch, or roll. The ADXRS150

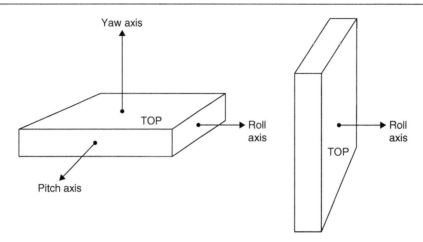

**Figure 3-17: Gyro axes of rotational sensitivity**

and ADXRS300 are yaw-axis gyros, but they can measure rotation about other axes by appropriate mounting orientation. For example, at the right of Figure 3-17 a yaw-axis device is positioned to measure roll.

A gyroscope with one axis of sensitivity can also be used to measure other axes by mounting the gyro differently, as shown in the right-hand diagram. Here, a yaw-axis gyro, such as the ADXRS150 or ADXRS300, is mounted on its side so that the yaw-axis becomes the roll axis.

As an example of how a gyro could be used, a yaw-axis gyro mounted on a turntable rotating at 33 1/3 rpm (revolutions per minute) would measure a constant rotation of 360° times 33 1/3 rpm divided by 60 seconds, or 200°/seconds. The gyro would output a voltage proportional to the angular rate, as determined by its sensitivity, measured in millivolts per degree per second (mV/°/s). The full-scale voltage determines how much angular rate can be measured, so in the example of the turntable, a gyro would need to have a full-scale voltage corresponding to at least 200°/seconds. Full-scale is limited by the available voltage swing divided by the sensitivity. The ADXRS300, for example, with 1.5 V full-scale and a sensitivity of 5 mV/°/ seconds, handles a full-scale of 300°/seconds. The ADXRS150, has a more limited full-scale of 150°/ seconds but a greater sensitivity of 12.5 mV/°/seconds.

One practical application is to measure how quickly a car turns by mounting a gyro inside the vehicle; if the gyro senses that the car is spinning out of control, differential braking engages to bring it back into control. The angular rate can also be integrated over time to determine angular position—particularly useful for maintaining continuity of GPS-based navigation when the satellite signal is lost for short periods of time.

## Coriolis Acceleration

Analog Devices' (ADI) ADXRS gyros measure angular rate by means of Coriolis acceleration. The Coriolis effect can be explained as follows, starting with Figure 3-18. Consider yourself standing on a rotating platform, near the center. Your speed relative to the ground is shown as the arrow lengths in Figure 3-18. If you were to move to a point near the outer edge of the platform, your speed would increase relative to the ground, as indicated by the longer arrow. The rate of increase of your tangential speed, caused by your radial velocity, is the *Coriolis* acceleration (after Gaspard G. de Coriolis, 1792–1843—a French mathematician).

If $\Omega$ is the angular rate and r the radius, the tangential velocity is $\Omega r$. So, if r changes at speed, v, there will be a tangential acceleration $\Omega v$. This is half of the Coriolis acceleration. There is another half from

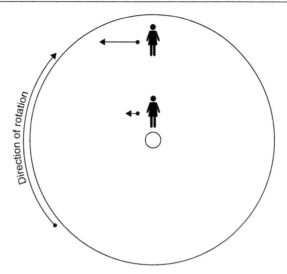

**Figure 3-18: Coriolis acceleration example**

changing the direction of the radial velocity giving a total of $2\Omega v$. If you have mass, M, the platform must apply a force, $2M\Omega v$, to cause that acceleration, and the mass experiences a corresponding reaction force.

## Motion in two dimensions

Consider the position coordinate, $z = r\varepsilon^{j\theta}$, in the complex plane. Differentiating with respect to time, t, the velocity is:

$$\frac{dz}{dt} = \frac{dr}{dt}\,\varepsilon^{j\theta} + ir\frac{d\theta}{dt}\,\varepsilon^{j\theta} \qquad (3\text{-}18)$$

The two terms are the respective radial and tangential components, the latter arising from the angular rate. Differentiating again, the acceleration is:

$$\frac{d^2z}{dt^2} = \left[\frac{d^2r}{dt^2}\,\varepsilon^{j\theta} + i\frac{dr}{dt}\frac{d\theta}{dt}\,\varepsilon^{j\theta}\right] + \left[i\frac{dr}{dt}\frac{d\theta}{dt}\,\varepsilon^{j\theta} + ir\frac{d^2\theta}{dt^2}\,\varepsilon^{j\theta} - r\left(\frac{d\theta}{dt}\right)^2\varepsilon^{j\theta}\right] \qquad (3\text{-}19)$$

The first term is the radial linear acceleration and the fourth term is the tangential component arising from angular acceleration. The last term is the familiar centripetal acceleration needed to constrain r. The second and third terms are tangential and are the Coriolis acceleration components. They are equal, respectively arising from the changing direction of the radial velocity and from the changing magnitude of the tangential velocity. If the angular rate and radial velocities are constant,

$$\frac{d\theta}{dt} = \Omega \qquad (3\text{-}20)$$

and

$$\frac{dr}{dt} = v \qquad (3\text{-}21)$$

then

$$\frac{d^2z}{dt^2} = i2\,\Omega\,v\,\varepsilon^{i\theta} - \Omega^2 r\varepsilon^{i\theta} \qquad (3\text{-}22)$$

where the angular component, $i\varepsilon^{i\theta}$, indicates a tangential direction in the sense of positive $\theta$ for the Coriolis acceleration, $2\,\Omega v$, and $-\varepsilon^{i\theta}$ indicates toward the center (i.e., *centripetal*) for the $\Omega^2 r$ component.

The ADXRS gyros take advantage of this effect by using a resonating mass analogous to the person moving out and in on a rotating platform. The mass is micromachined from polysilicon and is tethered to a polysilicon frame so that it can resonate along only one direction.

Figure 3-19 shows that when the resonating mass moves toward the outer edge of the rotation, it is accelerated to the right and exerts on the frame a reaction force to the left. When it moves toward the center of the rotation, it exerts a force to the right, as indicated by the arrows.

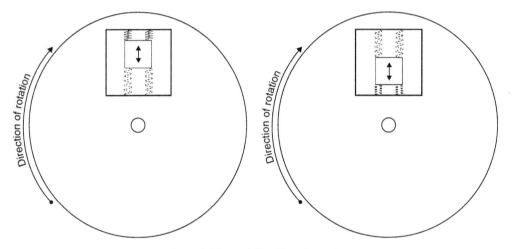

Figure 3-19: Coriolis effect demo 1

To measure the Coriolis acceleration, the frame containing the resonating mass is tethered to the substrate by springs at 90° relative to the resonating motion, as shown in Figure 3-20. This figure also shows the Coriolis sense fingers that are used to capacitively sense displacement of the frame in response to the force exerted by the mass, as described in Figure 3-19, a demonstration of the Coriolis effect in response to a

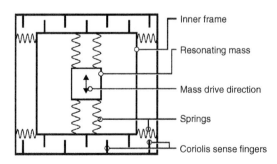

Figure 3-20: Schematic of the gyro's mechanical structure

resonating silicon mass suspended inside a frame. The arrows indicate the force applied to the structure, based on the status of the resonating mass.

In Figure 3-21 the frame and resonating mass are displaced laterally in response to the Coriolis effect. The displacement is determined from the change in capacitance between the Coriolis sense fingers on the frame and those attached to the substrate.

If the springs have a stiffness, K, then the displacement resulting from the reaction force will be 2 $\Omega$vM/K.

Figure 3-21, which shows the complete structure, demonstrates that as the resonating mass moves, and as the surface to which the gyro is mounted rotates, the mass and its frame experience the Coriolis acceleration and are translated 90° from the vibratory movement. As the rate of rotation increases, so does the displacement of the mass and the signal derived from the corresponding capacitance change.

It should be noted that the gyro may be placed anywhere on the rotating object and at any angle, as long as its sensing axis is parallel to the axis of rotation. The above explanation is intended to give an intuitive sense of the function and has been simplified by the placement of the gyro.

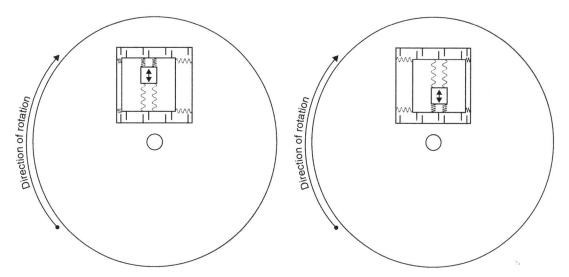

**Figure 3-21: Displacement due to the Coriolis effect**

## Capacitive Sensing

ADXRS gyros measure the displacement of the resonating mass and its frame due to the Coriolis effect through capacitive sensing elements attached to the resonator, as shown in Figures 3-19, 20, and 21. These elements are silicon beams inter-digitated with two sets of stationary silicon beams attached to the substrate, thus forming two nominally equal capacitors. Displacement due to angular rate induces a differential capacitance in this system. If the total capacitance is C and the spacing of the beams is g, then the differential capacitance is 2$\Omega$vMC/gK, and is directly proportional to the angular rate. The fidelity of this relationship is excellent in practice, with nonlinearity less than 0.1%.

The ADXRS gyro electronics can resolve capacitance changes as small as $12 \times 10^{-21}$ F (12 zF) from beam deflections as small as 0.00016 Å (16 fm). The only way this can be utilized in a practical device is by situating the electronics, including amplifiers and filters, on the same die as the mechanical sensor. The differential signal alternates at the resonator frequency and can be extracted from the noise by correlation.

These sub-atomic displacements are meaningful as the *average* positions of the surfaces of the beams, even though the individual atoms on the surface are moving randomly by much more. There are about 1,012 atoms on the surfaces of the capacitors, so the statistical averaging of their individual motions reduces the uncertainty by a factor of 106. So why can't we do 100 times better? The answer is that the impact of the air *molecules* causes the structure to move—although similarly averaged, their effect is far greater! So why not remove the air? The device is not operated in a vacuum because it is a very fine, thin film weighing only 4 µg; its flexures, only 1.7 µ wide, are suspended over the silicon substrate. Air cushions the structure, preventing it from being destroyed by violent shocks—*even those experienced during firing of a guided shell from a howitzer* (as demonstrated recently).

Figure 3-22 shows that the ADXRS gyros include two structures to enable differential sensing in order to reject environmental shock and vibration.

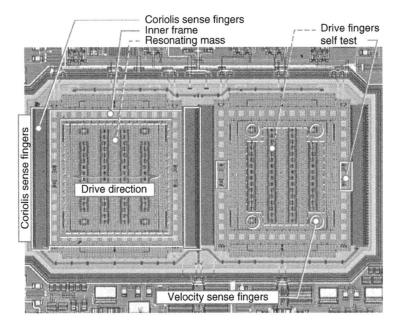

Figure 3-22: Photograph of mechanical sensor

Integration of electronics and mechanical elements is a key feature of products such as the ADXRS150 and ADXRS300, because it makes possible the smallest size and cost for a given performance level. Figure 3-23 is a photograph of the ADXRS die, highlighting the integration of the mechanical rate sensor and the signal conditioning electronics.

The ADXRS150 and ADXRS300 are housed in an industry-standard package that simplifies users' product development and production. The ceramic package—a 32-pin ball grid array (BGA)—measures 7 mm wide by 7 mm deep by 3 mm tall. It is at least 100 times smaller than any other gyro having similar performance. Besides their small size, these gyros consume 30 mW, far less power than similar gyros. The combination of

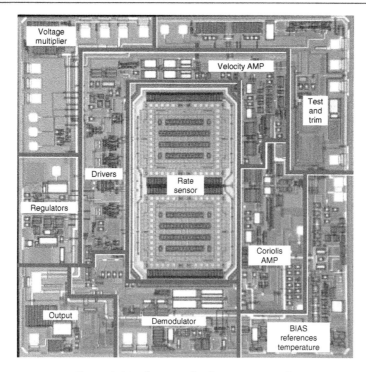

**Figure 3-23: Photograph of ADXRS gyro die**

small size and low power make these products ideally suited for consumer applications such as toy robots, scooters, and navigation devices.

## Immunity to Shock and Vibration

One of the most important concerns for a gyro user is the device's ability to reliably provide an accurate angular-rate output signal—even in the presence of environmental shock and vibration. One example of such an application is automotive rollover detection, in which a gyro is used to detect whether *or not* a car (or SUV) is rolling over. Some rollover events are triggered by an impact with another object, such as a curb, that results in a shock to the vehicle. If the shock saturates the gyro sensor, and the gyro cannot filter it out, then the airbags may not deploy. Similarly, if a bump in the road results in a shock or vibration that translates into a rotational signal, the airbags might deploy when not needed—a considerable safety hazard!

As can be seen, the ADXRS gyros employ a novel approach to angular-rate sensing that makes it possible to reject shocks of up to 1,000 g—they use two resonators to differentially sense signals and reject common-mode external accelerations that are unrelated to angular motion. This approach is, in part, the reason for the excellent immunity of the ADXRS gyros to shock and vibration. The two resonators in Figure 3-22 are mechanically independent, and they operate antiphase. As a result, they measure the same magnitude of rotation, but give outputs in opposite directions. Therefore, the difference between the two sensor signals is used to measure angular rate. This cancels non-rotational signals that affect both sensors. The signals are combined in the internal hard-wiring ahead of the very sensitive preamplifiers. Thus, extreme acceleration overloads are largely prevented from reaching the electronics—thereby allowing the signal conditioning to preserve the angular-rate output during large shocks. This scheme requires that the two sensors be well-matched, precisely fabricated copies of each other.

## References: Positional Sensors

1. H. Schaevitz, "The Linear Variable Differential Transformer," **Proceedings of the SASE**, Vol. 4, No. 2, 1946.

2. Dr. E.D.D. Schmidt, "Linear Displacement—Linear Variable Differential Transformers – LVDTs," Schaevitz Sensors, http://www.schaevitz.com.

3. E-Series LVDT Data Sheet, Schaevitz Sensors, http://www.schaevitz.com. Schaevitz Sensors is now a division of Lucas Control Systems, 1000 Lucas Way, Hampton, VA 23666.

4. R. Pallas-Areny and J.G. Webster, **Sensors and Signal Conditioning**, John Wiley, New York, 1991.

5. H.L. Trietley, **Transducers in Mechanical and Electronic Design**, Marcel Dekker, Inc., 1986.

6. AD598 and AD698 Data Sheet, Analog Devices, Inc., http://www.analog.com.

7. B. Travis, "Hall-Effect Sensor ICs Sport Magnetic Personalities," **EDN,** Vol. XX, April 9, 1998, pp. 81–91.

8. AD22151 Data Sheet, **Analog Devices**, Inc., http://www.analog.com.

9. D. Sheingold, **Analog-Digital Conversion Handbook**, **3rd Edition**, Prentice-Hall, Norwood, MA, 1986.

10. F.P. Flett, "Vector Control Using a Single Vector Rotation Semiconductor for Induction and Permanent Magnet Motors," **PCIM Conference, Intelligent Motion, September 1992 Proceedings**, Available from Analog Devices.

11. F.P. Flett, "Silicon Control Algorithms for Brushless Permanent Magnet Synchronous Machines," **PCIM Conference, Intelligent Motion, June 1991 Proceedings**, Available from Analog Devices.

12. P.J.M. Coussens, et al., "Three Phase Measurements with Vector Rotation Blocks in Mains and Motion Control," **PCIM Conference, Intelligent Motion, April 1992 Proceedings**, Available from Analog Devices.

13. Dennis Fu, "Digital to Synchro and Resolver Conversion with the AC Vector Processor AD2S100," Available from Analog Devices.

14. Dennis Fu, "Circuit Applications of the AD2S90 Resolver-to-Digital Converter, AN-230," Analog Devices.

15. A. Murray and P. Kettle, "Towards a Single Chip DSP Based Motor Control Solution," **Proceedings PCIM—Intelligent Motion**, May 1996, Nurnberg, Germany, pp. 315–326. Also available at http://www.analog.com.

16. D.J. Lucey, P.J. Roche, M.B. Harrington, and J.R. Scannell, "Comparison of Various Space Vector Modulation Strategies," **Proceedings Irish DSP and Control Colloquium**, July 1994, Dublin, Ireland, pp. 169–175.

17. N. Lyne, "ADCs Lend Flexibility to Vector Motor Control Application," **Electronic Design**, May 1, 1998, pp. 93–100.

18. F. Goodenough, "Airbags Boom when IC Accelerometer Sees 50 g," **Electronic Design**, 1991.

# Temperature Sensors

## Introduction

Measurement of temperature is critical in modern electronic devices, especially expensive laptop computers and other portable devices with densely packed circuits which dissipate considerable power in the form of heat. Knowledge of system temperature can also be used to control battery charging as well as prevent damage to expensive microprocessors.

Compact high power portable equipment often has fan cooling to maintain junction temperatures at proper levels. In order to conserve battery life, the fan should only operate when necessary. Accurate control of the fan requires a knowledge of critical temperatures from the appropriate temperature sensor.

Accurate temperature measurements are required in many other measurement systems such as process control and instrumentation applications. In most cases, because of low level nonlinear outputs, the sensor output must be properly conditioned and amplified before further processing can occur.

Except for IC sensors, all temperature sensors have nonlinear transfer functions. In the past, complex analog conditioning circuits were designed to correct for the sensor nonlinearity. These circuits often required manual calibration and precision resistors to achieve the desired accuracy. Today, however, sensor outputs may be digitized directly by high resolution analog-to-digital converters (ADCs). Linearization and calibration is then performed digitally, thereby reducing cost and complexity.

Resistance Temperature Detectors (RTDs) are accurate, but require excitation current and are generally used in bridge circuits. Thermistors have the most sensitivity but are the most nonlinear. However, they are popular in portable applications such as measurement of battery temperature and other critical temperatures in a system.

Modern semiconductor temperature sensors offer high accuracy and high linearity over an operating range of about −55°C to +150°C. Internal amplifiers can scale the output to convenient values, such as 10 mV/°C. They are also useful in cold-junction compensation circuits for wide temperature range thermocouples. Semiconductor temperature sensors can be integrated into multi-function ICs which perform a number of other hardware monitoring functions.

Figure 3-24 lists the most popular types of temperature transducers and their characteristics.

## Semiconductor Temperature Sensors

Modern semiconductor temperature sensors offer high accuracy and high linearity over an operating range of about −55°C to +150°C. Internal amplifiers can scale the output to convenient values, such as 10 mV/°C. They are also useful in cold-junction compensation circuits for wide temperature range thermocouples.

All semiconductor temperature sensors make use of the relationship between a bipolar junction transistor's (BJT) base-emitter voltage to its collector current:

$$V_{BE} = \frac{kT}{q} \ln\left(\frac{I_c}{I_s}\right) \qquad (3\text{-}23)$$

| Thermocouple | RTD | Thermistor | Semiconductor |
|---|---|---|---|
| Widest range: −184°C to +2300°C | Range: −200°C to +850°C | Range: 0°C to +100°C | Range: −55°C to +150°C |
| High accuracy and repeatability | Fair linearity | Poor linearity | Linearity: 1°C Accuracy: 1°C |
| Needs cold junction compensation | Requires excitation | Requires excitation | Requires excitation |
| Low-voltage output | Low cost | High sensitivity | 10 mV/K, 20 mV/K, or 1 μA/K typical output |

**Figure 3-24: Types of temperature sensors**

where k is Boltzmann's constant, T is the absolute temperature, q is the charge of an electron, and $I_s$ is a current related to the geometry and the temperature of the transistors. (The equation assumes a voltage of at least a few hundred mV on the collector, and ignores early effects.)

If we take N transistors identical to the first (see Figure 3-25) and allow the total current $I_c$ to be shared equally among them, we find that the new base-emitter voltage is given by the equation

$$V_N = \frac{kT}{q} \ln\left(\frac{I_c}{N \times I_s}\right) \tag{3-24}$$

Neither of these circuits is of much use by itself because of the strongly temperature dependent current $I_s$, but if we have equal currents in one BJT and N similar BJTs then the expression for the *difference* between the two base-emitter voltages is proportional to absolute temperature (PTAT) and does not contain $I_s$.

$$\Delta V_{BE} = V_{BE} - V_N = \frac{kT}{q} \ln\left(\frac{I_c}{I_s}\right) - \frac{kT}{q} \ln\left(\frac{I_c}{N \times I_s}\right) \tag{3-25}$$

$$\Delta V_{BE} = V_{BE} - V_N = \frac{kT}{q} \left[ \ln\left(\frac{I_c}{I_s}\right) - \ln\left(\frac{I_c}{N \times I_s}\right) \right] \tag{3-26}$$

$$\Delta V_{BE} = V_{BE} - V_N = \frac{kT}{q} \ln\left[ \left(\frac{I_c}{I_s}\right) \middle/ \left(\frac{I_c}{N \times I_s}\right) \right] = \frac{kT}{q} \ln(N) \tag{3-27}$$

The circuit shown in Figure 3-26 implements the above equation and is known as the "Brokaw Cell" (see Reference 10). The voltage $\Delta V_{BE} = V_{BE} - V_N$ appears across resistor $R_2$. The emitter current in Q2 is therefore $\Delta V_{BE}/R_2$. The op amp's servo loop and the resistors, R, force the same current to flow through Q1. The Q1 and Q2 currents are equal and are summed and flow into resistor $R_1$. The corresponding voltage developed across $R_1$ is PTAT and given by:

$$V_{PTAT} = \frac{2R_1(V_{BE} - V_N)}{R_2} = 2 \frac{R_1}{R_2} \frac{kT}{q} \ln(N) \tag{3-28}$$

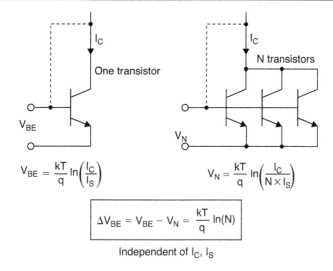

$$V_{BE} = \frac{kT}{q} \ln\left(\frac{I_C}{I_S}\right)$$

$$V_N = \frac{kT}{q} \ln\left(\frac{I_C}{N \times I_S}\right)$$

$$\Delta V_{BE} = V_{BE} - V_N = \frac{kT}{q} \ln(N)$$

Independent of $I_C$, $I_S$

**Figure 3-25: Basic relationships for semiconductor temperature sensors**

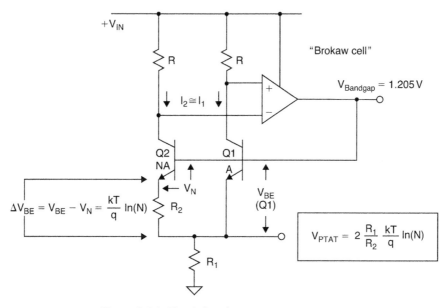

$$\Delta V_{BE} = V_{BE} - V_N = \frac{kT}{q} \ln(N)$$

$$V_{PTAT} = 2 \frac{R_1}{R_2} \frac{kT}{q} \ln(N)$$

**Figure 3-26: Classic bandgap temperature sensor**

The bandgap cell reference voltage, $V_{BANDGAP}$, appears at the base of Q1 and is the sum of $V_{BE}(Q1)$ and $V_{PTAT}$. $V_{BE(Q1)}$ is complementary to absolute temperature (CTAT), and summing it with $V_{PTAT}$ causes the bandgap voltage to be constant with respect to temperature (assuming proper choice of $R_1/R_2$ ratio and N to make the bandgap voltage equal to 1.205 V). This circuit is the basic bandgap temperature sensor and is widely used in semiconductor temperature sensors.

*217*

## Current Out Temperature Sensors

This type of temperature sensor produces a current output PTAT. For supply voltages between 4 V and 30 V the device acts as a high impedance constant current regulator with an output PTAT with a typical transfer function of 1 µA/°K. This means that at 25°C there will be 298 µA flowing in the loop.

A current output temperature sensor such as the AD590 is particularly useful in remote sensing applications. These devices are insensitive to voltage drops over long lines due to their high impedance current outputs. The output characteristics also make this type of device easy to multiplex: the current can be switched by a simple logic gate as shown in Figure 3-27.

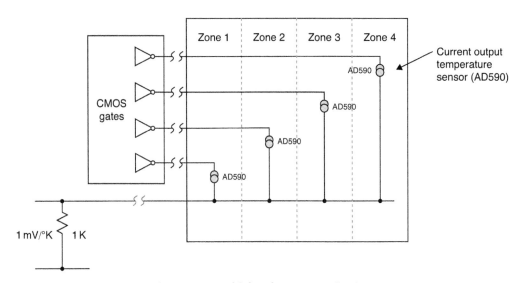

**Figure 3-27: Multiplexed AD590 application**

## Current and Voltage Output Temperature Sensors

The concepts used in the bandgap temperature sensor discussion above can be used as the basis for a variety of IC temperature sensors to generate either current or voltage outputs.

In some cases, it is desirable for the output of a temperature sensor to be ratiometric with its supply voltage. The AD22100 (see Figure 3-28) has an output that is ratiometric with its supply voltage (nominally 5 V) according to the equation:

$$V_{OUT} = \frac{V_S}{5} * \left[ 1.375\,V \frac{22 \times m}{C} * T_A \right] \qquad (3\text{-}29)$$

The circuit shown in Figure 3-28 uses the AD22100 power supply as the reference to the ADC, thereby eliminating the need for a precision voltage reference.

The thermal time constant of a temperature sensor is defined to be the time required for the sensor to reach 63.2% of the final value for a step change in the temperature. Figure 3-29 shows the thermal time constant of the ADT45/ADT50 series of sensors with the SOT-23-3 package soldered to 0.338″ × 0.307″ copper PC board as a function of air flow velocity. Note the rapid drop from 32 seconds to 12 seconds as the air velocity increases from 0 (still air) to 100 LFPM. As a point of reference, the thermal time constant of the

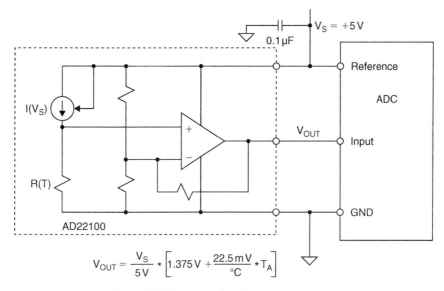

$$V_{OUT} = \frac{V_S}{5V} * \left[1.375\,V + \frac{22.5\,mV}{°C} * T_A\right]$$

Figure 3-28: Ratiometric voltage output sensor

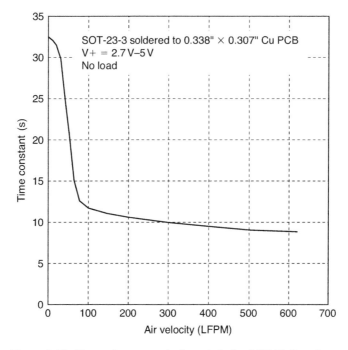

Figure 3-29: Thermal response in forced air for SOT-23-2 package

ADT45/ADT50 series in a stirred oil bath is less than 1 second, which verifies that the major part of the thermal time constant is determined by the case.

The power supply pin of these sensors should be bypassed to ground with a 0.1 μF ceramic capacitor having very short leads (preferably surface mount) and located as close to the power supply pin as possible.

Since these temperature sensors operate on very little supply current and could be exposed to very hostile electrical environments, it is important to minimize the effects of electromagnetic interference (EMI)/radio frequency interference (RFI) on these devices. The effect of RFI on these temperature sensors is manifested as abnormal DC shifts in the output voltage due to rectification of the high frequency noise by the internal IC junctions. In those cases where the devices are operated in the presence of high frequency radiated or conducted noise, a large value tantalum electrolytic capacitor ($>2.2\,\mu F$) placed across the $0.1\,\mu F$ ceramic may offer additional noise immunity.

## Thermocouple Principles and Cold-Junction Compensation

Thermocouples are small, rugged, relatively inexpensive, and operate over the widest range of all temperature sensors. They are especially useful for making measurements at extremely high temperatures (up to $+2300°C$) in hostile environments. They produce only millivolts of output, however, and require precision amplification for further processing. They also require cold-junction compensation (CJC) techniques which will be discussed shortly. They are more linear than many other sensors, and their nonlinearity has been well-characterized. Some common thermocouples are shown in Figure 3-30. The most common metals used are Iron, Platinum, Rhodium, Rhenium, Tungsten, Copper, Alumel (composed of Nickel and Aluminum), Chromel (composed of Nickel and Chromium), and Constantan (composed of Copper and Nickel).

| Junction materials | Typical useful range (°C) | Nominal sensitivity (μV/°C) | ANSI designation |
|---|---|---|---|
| Platinum (6%)/Rhodium–Platinum (30%)/Rhodium | 38–1,800 | 7.7 | B |
| Tungsten (5%)/Rhenium–Tungsten (26%)/Rhenium | 0–2,300 | 16 | C |
| Chromel–Constantan | 0–982 | 76 | E |
| Iron–Constantan | 0–760 | 55 | J |
| Chromel–Alumel | −184–1,260 | 39 | K |
| Platinum (13%)/Rhodium–Platinum | 0–1,593 | 11.7 | R |
| Platinum (10%)/Rhodium–Platinum | 0–1,538 | 10.4 | S |
| Copper–Constantan | −184–400 | 45 | T |

**Figure 3-30: Common thermocouples**

Figure 3-31 shows the voltage-temperature curves of three commonly used thermocouples, referred to a $0°C$ fixed-temperature reference junction. Of the thermocouples shown, Type J thermocouples are the most sensitive, producing the largest output voltage for a given temperature change. On the other hand, Type S thermocouples are the least sensitive. These characteristics are very important to consider when designing signal conditioning circuitry in that the thermocouples' relatively low output signals require low noise, low-drift, high gain amplifiers.

To understand thermocouple behavior, it is necessary to consider the nonlinearities in their response to temperature differences. Figure 3-31 shows the relationships between sensing junction temperature and voltage output for a number of thermocouple types (in all cases, the reference *cold* junction is maintained at $0°C$). It is evident that the responses are not quite linear, but the nature of the nonlinearity is not so obvious.

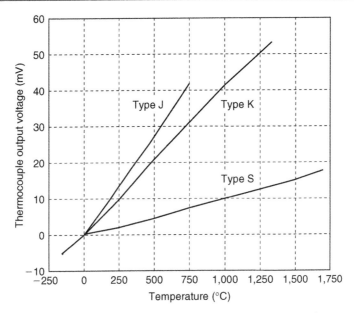

**Figure 3-31: Thermocouple output voltages for type J, K, and S thermocouples**

Figure 3-32 shows how the Seebeck coefficient (the *change* of output voltage with *change* of sensor junction temperature—i.e., the first derivative of output with respect to temperature) varies with sensor junction temperature (we are still considering the case where the reference junction is maintained at 0°C).

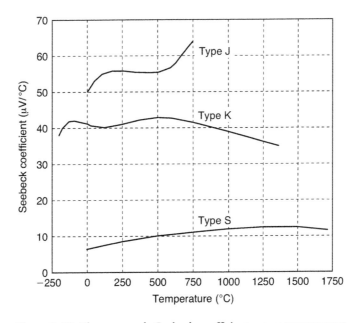

**Figure 3-32: Thermocouple Seebeck coefficient versus temperature**

When selecting a thermocouple for making measurements over a particular range of temperature, we should choose a thermocouple whose Seebeck coefficient varies as little as possible over that range.

For example, a Type J thermocouple has a Seebeck coefficient which varies by less than $1\,\mu V/°C$ between 200 and 500°C, which makes it ideal for measurements in this range.

Presenting these data on thermocouples serves two purposes. First, Figure 3-30 illustrates the range and sensitivity of the three thermocouple types so that the system designer can, at a glance, determine that a Type S thermocouple has the widest useful temperature range, but a Type J thermocouple is more sensitive. Second, the Seebeck coefficients provide a quick guide to a thermocouple's linearity. Using Figure 3-31 the system designer can choose a Type K thermocouple for its linear Seebeck coefficient over the range of 400°C–800°C or a Type S over the range of 900°C–1,700°C. The behavior of a thermocouple's Seebeck coefficient is important in applications where variations of temperature rather than absolute magnitude are important. These data also indicate what performance is required of the associated signal conditioning circuitry.

To use thermocouples successfully we must understand their basic principles. Consider the diagrams in Figure 3-33.

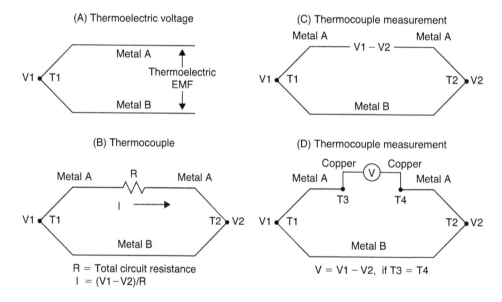

**Figure 3-33: Thermocouple basics**

If we join two dissimilar metals at any temperature above absolute zero, there will be a potential difference between them (their "thermoelectric e.m.f." or "contact potential") which is a function of the temperature of the junction (Figure 3-33(A)). If we join the two wires at two places, two junctions are formed (Figure 3-33(B)). If the two junctions are at different temperatures, there will be a net e.m.f. in the circuit, and a current will flow determined by the e.m.f. and the total resistance in the circuit (Figure 3-33(B)). If we break one of the wires, the voltage across the break will be equal to the net thermoelectric e.m.f. of the circuit, and if we measure this voltage, we can use it to calculate the temperature difference between the two junctions (Figure 3-33(C)). *We must always remember that a thermocouple measures the temperature difference between two junctions, not the absolute temperature at one junction.* We can only measure the temperature at the measuring junction if we know the temperature of the other junction (often called the "reference" junction or the "cold" junction).

But it is not so easy to measure the voltage generated by a thermocouple. Suppose that we attach a voltmeter to the circuit in Figure 3-33(C) (Figure 3-33(D)). The wires attached to the voltmeter will form further thermojunctions where they are attached. If both these additional junctions are at the same temperature (it does not matter what temperature), then the "Law of Intermediate Metals" states that they will make no net contribution to the total e.m.f. of the system. If they are at different temperatures, they will introduce errors. Since *every pair of dissimilar metals in contact generates a thermoelectric e.m.f.* (including copper/solder, kovar/copper (kovar is the alloy used for IC leadframes) and aluminum/kovar (at the bond inside the IC)), it is obvious that in practical circuits the problem is even more complex, and it is necessary to take extreme care to ensure that all the junction pairs in the circuitry around a thermocouple, except the measurement and reference junctions themselves, are at the same temperature.

Thermocouples generate a voltage, albeit a very small one, and do not require excitation. As shown in Figure 3-33(D), however, two junctions (T1, the measurement junction and T2, the reference junction) are involved. If T2 = T1, then V2 = V1, and the output voltage V = 0. Thermocouple output voltages are often defined with a reference junction temperature of 0°C (hence the term *cold* or *ice-point* junction), so the thermocouple provides an output voltage of 0V at 0°C. To maintain system accuracy, the reference junction must therefore be at a well-defined temperature (but not necessarily 0°C). A conceptually simple approach to this need is shown in Figure 3-34. Although an ice/water bath is relatively easy to define, it is quite inconvenient to maintain.

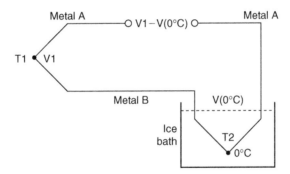

**Figure 3-34: Classic cold-junction compensation using an ice-point (0°C) reference junction**

Today an ice-point reference, and its inconvenient ice/water bath, is generally replaced by electronics. A temperature sensor of another sort (often a semiconductor sensor, sometimes a thermistor) measures the temperature of the cold junction and is used to inject a voltage into the thermocouple circuit which compensates for the difference between the actual cold-junction temperature and its ideal value (usually 0°C) as shown in Figure 3-35. Ideally, the compensation voltage should be an exact match for the difference voltage required, which is why the diagram gives the voltage as f(T2) (*a function* of T2) rather than KT2, where K is a simple constant. In practice, since the cold junction is rarely more than a few tens of degrees from 0°C, and generally varies by little more than ±10°C, a linear approximation (V = KT2) to the more complex reality is sufficiently accurate and is what is often used. (The expression for the output voltage of a thermocouple with its measuring junction at T°C and its reference at 0°C is a polynomial of the form $V = K_1T + K_2T^2 + K_3T^3 + \cdots$, but the values of the coefficients $K_2$, $K_3$, etc. are very small for most common types of thermocouple. References 8 and 9 give the values of these coefficients for a wide range of thermocouples.)

When electronic cold-junction compensation is used, it is common practice to eliminate the additional thermocouple wire and terminate the thermocouple leads in the isothermal block in the arrangement shown in Figure 3-36. The metal A (copper) and the metal B (copper) junctions, if at the same temperature, are equivalent to the metal A–metal B thermocouple junctions in Figure 3-35.

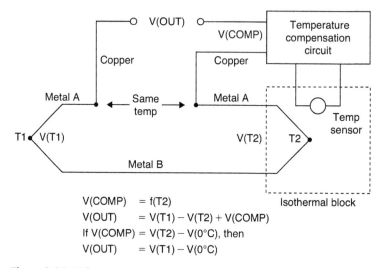

$$V(COMP) = f(T2)$$
$$V(OUT) = V(T1) - V(T2) + V(COMP)$$
If $V(COMP) = V(T2) - V(0°C)$, then
$$V(OUT) = V(T1) - V(0°C)$$

**Figure 3-35: Using a temperature sensor for cold-junction compensations**

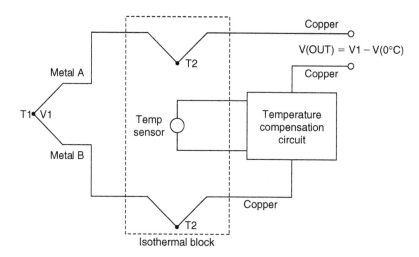

**Figure 3-36: Terminating thermocouple leads directly to an isothermal block**

The circuit in Figure 3-37 conditions the output of a Type K thermocouple, while providing cold-junction compensation, for temperatures between 0°C and 250°C. The circuit operates from single +3.3 V to +12 V supplies and has been designed to produce an output voltage transfer characteristic of 10 mV/°C.

A Type K thermocouple exhibits a Seebeck coefficient of approximately 41 μV/°C; therefore, at the cold junction, the TMP35 voltage output sensor with a temperature coefficient (TC) of 10 mV/°C is used with $R_1$ and $R_2$ to introduce an opposing cold-junction TC of −41 μV/°C. This prevents the isothermal, cold-junction connection between the circuit's printed circuit board traces and the thermocouple's wires from introducing an error in the measured temperature. This compensation works extremely well for circuit ambient temperatures

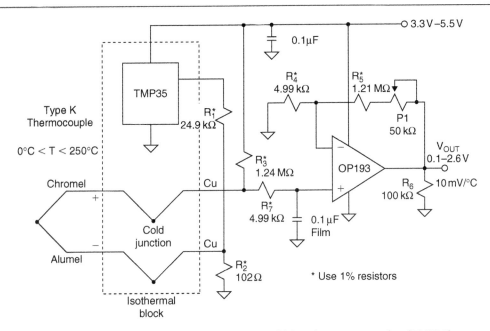

**Figure 3-37: Using a temperature sensor for cold-junction compensation (TMP35)**

in the range of 20–50°C. Over a 250°C measurement temperature range, the thermocouple produces an output voltage change of 10.151 mV. Since the required circuit's output full-scale voltage change is 2.5 V, the gain of the circuit is set to 246.3. Choosing $R_4$ equal to 4.99 kΩ sets $R_5$ equal to 1.22 MΩ. Since the closest 1% value for $R_5$ is 1.21 MΩ, a 50 kΩ potentiometer is used with $R_5$ for fine trim of the full-scale output voltage. Although the OP193 is a single-supply op amp, its output stage is not rail-to-rail, and will only go down to about 0.1 V above ground. For this reason, $R_3$ is added to the circuit to supply an output offset voltage of about 0.1 V for a nominal supply voltage of 5 V. This offset (10°C) must be subtracted when making measurements referenced to the OP193 output. $R_3$ also provides an open thermocouple detection, forcing the output voltage to greater than 3 V should the thermocouple open. Resistor $R_7$ balances the DC input impedance of the OP193, and the 0.1 μF film capacitor reduces noise coupling into its non-inverting input.

The AD594/AD595 is a complete instrumentation amplifier and thermocouple cold-junction compensator on a monolithic chip (see Figure 3-38). It combines an ice-point reference with a precalibrated amplifier to provide a high level (10 mV/°C) output directly from the thermocouple signal. Pin-strapping options allow it to be used as a linear amplifier–compensator or as a switched output setpoint controller using either fixed or remote setpoint control. It can be used to amplify its compensation voltage directly, thereby becoming a stand-alone Celsius transducer with 10 mV/°C output. In such applications it is very important that the IC chip is at the same temperature as the cold junction of the thermocouple, which is usually achieved by keeping the two in close proximity and isolated from any heat sources.

The AD594/AD595 includes a thermocouple failure alarm that indicates if one or both thermocouple leads open. The alarm output has a flexible format which includes TTL drive capability. The device can be powered from a single-ended supply (which may be as low as +5 V), but by including a negative supply, temperatures below 0°C can be measured. To minimize self-heating, an unloaded AD594/AD595 will operate with a supply current of 160 μA, but is also capable of delivering ±5 mA to a load.

The AD594 is precalibrated by laser wafer trimming to match the characteristics of type J (iron/constantan) thermocouples, and the AD595 is laser trimmed for type K (chromel/alumel). The temperature transducer

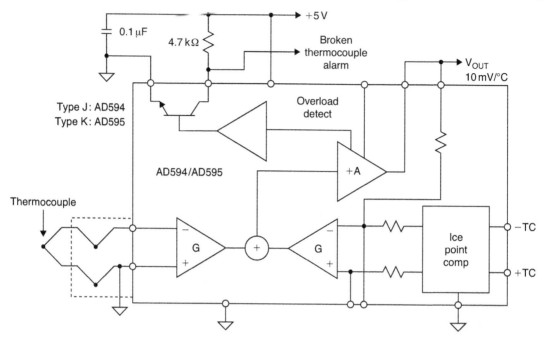

**Figure 3-38: AD594/AD595 monolithic thermocouple amplifier with cold-junction compensation**

voltages and gain control resistors are available at the package pins so that the circuit can be recalibrated for other thermocouple types by the addition of resistors. These terminals also allow more precise calibration for both thermocouple and thermometer applications. The AD594/AD595 is available in two performance grades. The C and the A versions have calibration accuracies of ±1°C and ±3°C, respectively. Both are designed to be used with cold junctions between 0°C and +50°C. The circuit shown in Figure 3-38 will provide a direct output from a type J thermocouple (AD594) or a type K thermocouple (AD595) capable of measuring 0 to +300°C.

The AD596/AD597 are monolithic setpoint controllers which have been optimized for use at elevated temperatures as are found in oven control applications. The device cold-junction compensates and amplifies a type J/K thermocouple to derive an internal signal proportional to temperature. They can be configured to provide a voltage output (10mV/°C) directly from type J/K thermocouple signals. The device is packaged in a 10-pin metal can and is trimmed to operate over an ambient range from +25°C to +100°C. The AD596 will amplify thermocouple signals covering the entire −200°C to +760°C temperature range recommended for type J thermocouples while the AD597 can accommodate −200°C to +1,250°C type K inputs. They have a calibration accuracy of ±4°C at an ambient temperature of 60°C and an ambient temperature stability specification of 0.05°C/°C from +25°C to +100°C.

None of the thermocouple amplifiers previously described compensates for thermocouple nonlinearity; they only provide conditioning and voltage gain. High resolution ADCs such as the AD77XX family can be used to digitize the thermocouple output directly, allowing a microcontroller to perform the transfer function linearization as shown in Figure 3-39. The two multiplexed inputs to the ADC are used to digitize the thermocouple voltage and the cold-junction temperature sensor outputs directly. The input programmable gain amplifier (PGA) gain is programmable from 1 to 128, and the ADC resolution is between 16 and 22 bits (depending on the particular ADC selected). The microcontroller performs both the cold-junction compensation and the linearization arithmetic.

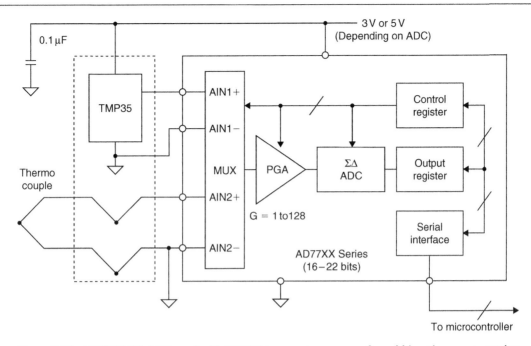

**Figure 3-39: AD77XX ΣΔ ADC used with TMP35 temperature sensor for cold-junction compensation**

## Auto-Zero Amplifier for Thermocouple Measurements

In addition to the devices mentioned above, ADI has released an auto-zero instrumentation amplifier, the AD8230, designed to amplify thermocouple and bridge outputs. Through the use of auto-zeroing, this product has an offset voltage drift of less than 50 nV/°C which is 1,000 times less than the signal produced by a typical thermocouple. This allows very accurate measurement of the thermocouple signal. In addition, the instrumentation amplifier architecture rejects common-mode voltages that often appear when using thermocouples for temperature measurement. This product is typically used in applications involving a bank of thermocouples with one temperature reference point which is compensated for in the system microcontroller. Other applications include highly accurate bridge transducer measurements.

Auto-zeroing is a dynamic offset and drift cancellation technique that reduces input referred voltage offset to the μV level and voltage offset drift to the nV/°C level. A further advantage of dynamic offset cancellation is the reduction of low frequency noise, in particular the 1/f component.

The AD8230 is an instrumentation amplifier that uses an auto-zeroing topology and combines it with high common-mode signal rejection. The internal signal path consists of an active differential sample-and-hold stage (preamp) followed by a differential amplifier (gain amp). Both amplifiers implement auto-zeroing to minimize offset and drift. A fully differential topology increases the immunity of the signals to parasitic noise and temperature effects. Amplifier gain is set by two external resistors for convenient TC matching.

The signal sampling rate is controlled by an on-chip, 6 kHz oscillator and logic to derive the required non-overlapping clock phases. For simplification of the functional description, two sequential clock phases, A and B, are used to distinguish the order of internal operation as depicted in the first figure, respectively.

During Phase A (Figure 3-40), the sampling capacitors are connected to the inputs. The input signal's difference voltage, $V_{DIFF}$, is stored across the sampling capacitors, $C_{SAMPLE}$. Since the sampling capacitors

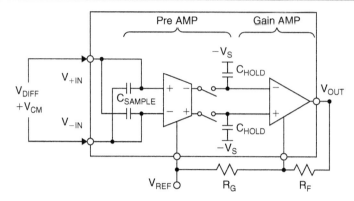

**Figure 3-40: Phase A of the sampling phase**

only retain the difference voltage, the common-mode voltage is rejected. During this period, the gain amplifier is not connected to the preamplifier so its output remains at the level set by the previously sampled input signal held on $C_{HOLD}$, as shown in Figure 3-41.

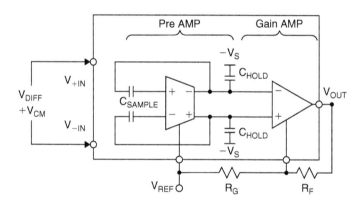

**Figure 3-41: Phase B of the sampling phase**

In Phase B, the differential signal is transferred to the hold capacitors refreshing the value stored on $C_{HOLD}$. The output of the preamplifier is held at a common-mode voltage determined by the reference potential, $V_{REF}$. In this manner, the AD8230 is able to condition the difference signal and set the output voltage level. The gain amplifier conditions the updated signal stored on the hold capacitors, $C_{HOLD}$.

## Resistance Temperature Detectors

The resistance temperature detector, or the RTD, is a sensor whose resistance changes with temperature. Typically built of a platinum (Pt) wire wrapped around a ceramic bobbin, the RTD exhibits behavior which is more accurate and more linear over wide temperature ranges than that of a thermocouple. Figure 3-42 illustrates the TC of a $100\Omega$ RTD and the Seebeck coefficient of a Type S thermocouple. Over the entire range (approximately $-200°C$ to $+850°C$), the RTD is a more linear device. Hence, linearizing an RTD is less complex.

- Platinum (Pt) the most common

- 100 Ω, 1,000 Ω standard values

- Typical TC = 0.385%/°C,
  0.385 Ω/°C for 100 Ω Pt RTD

- Good linearity—better than thermocouple, easily compensated

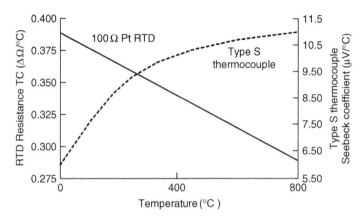

**Figure 3-42: Resistance temperature detectors (RTD)**

Unlike a thermocouple, however, an RTD is a passive sensor and requires current excitation to produce an output voltage. The RTD's low TC of 0.385%/°C requires similar high performance signal conditioning circuitry to that used by a thermocouple; however, the voltage drop across an RTD is much larger than a thermocouple output voltage. A system designer may opt for large value RTDs with higher output, but large-valued RTDs exhibit slow response times. Furthermore, although the cost of RTDs is higher than that of thermocouples, they use copper leads, and thermoelectric effects from terminating junctions do not affect their accuracy. And finally, because their resistance is a function of the absolute temperature, RTDs require no cold-junction compensation.

Caution must be exercised using current excitation because the current through the RTD causes heating. This self-heating changes the temperature of the RTD and appears as a measurement error. Hence, careful attention must be paid to the design of the signal conditioning circuitry so that self-heating is kept below 0.5°C. Manufacturers specify self-heating errors for various RTD values and sizes in still and in moving air. To reduce the error due to self-heating, the minimum current should be used for the required system resolution, and the largest RTD value chosen that results in acceptable response time.

Another effect that can produce measurement error is voltage drop in RTD lead wires. This is especially critical with low value two-wire RTDs because the TC and the absolute value of the RTD resistance are both small. If the RTD is located at a long distance from the signal conditioning circuitry, then the lead resistance can be ohms or tens of ohms, and a small amount of lead resistance can contribute a significant error to the temperature measurement. To illustrate this point, let us assume that a 100 Ω platinum RTD with 30-gauge copper leads is located about 100 feet from a controller's display console. The resistance of 30-gauge copper wire is 0.105 Ω/feet, and the two leads of the RTD will contribute a total 21 Ω to the network which is shown in Figure 3-43. This additional resistance will produce a 55°C error in the measurement! The leads' TC can contribute an additional, and possibly significant, error to the measurement. To eliminate the effect of the lead resistance, a four-wire technique is used.

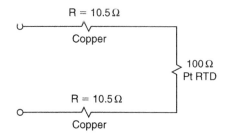

Resistance TC of Copper = 0.40%/°C @ 20°C
Resistance TC of Pt RTD = 0.385%/°C @ 20°C

**Figure 3-43: A 100 Ω Pt RTD with 100 feet of 30-gauge lead wires**

In Figure 3-44, a four-wire, or Kelvin, connection is made to the RTD. A constant current is applied though the FORCE leads of the RTD, and the voltage across the RTD itself is measured remotely via the SENSE leads. The measuring device can be a digital voltmeter (DVM) or an instrumentation amplifier, and high accuracy can be achieved provided that the measuring device exhibits high input impedance and/or low input bias current. Since the SENSE leads do not carry appreciable current, this technique is insensitive to lead wire length. Sources of errors are the stability of the constant current source and the input impedance and/or bias currents in the amplifier or DVM.

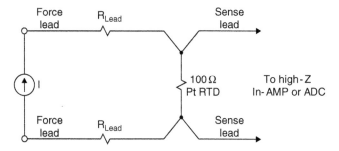

**Figure 3-44: Four-wire or Kelvin connection to Pt RTD for accurate measurements**

RTDs are generally configured in a four-resistor bridge circuit. The bridge output is amplified by an instrumentation amplifier for further processing. However, high resolution measurement ADCs such as the AD77XX series allow the RTD output to be digitized directly. In this manner, linearization can be performed digitally, thereby easing the analog circuit requirements.

Figure 3-45 shows a 100 Ω Pt RTD driven with a 400 μA excitation current source. The output is digitized by one of the AD77XX series ADCs. Note that the RTD excitation current source also generates the 2.5 V reference voltage for the ADC via the 6.25 kΩ resistor. Variations in the excitation current do not affect the circuit accuracy, since both the input voltage and the reference voltage vary ratiometrically with the excitation current. However, the 6.25 kΩ resistor must have a low TC to avoid errors in the measurement. The high resolution of the ADC and the input PGA (gain of 1–128) eliminates the need for additional conditioning circuits.

The ADT70 is a complete Pt RTD signal conditioner which provides an output voltage of 5 mV/°C when using a 1 kΩ RTD (see Figure 3-46). The Pt RTD and the 1 kΩ reference resistor are both excited with 1 mA matched current sources. This allows temperature measurements to be made over a range of approximately −50°C to +800°C.

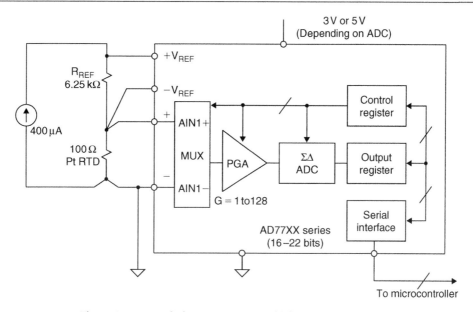

Figure 3-45: Interfacing a Pt RTD to a high resolution ΣΔ ADC

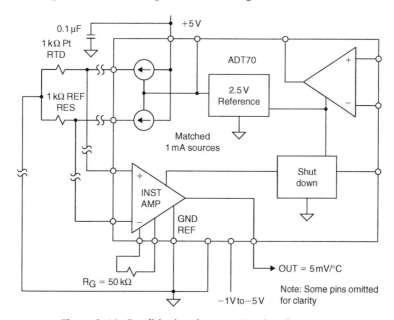

Figure 3-46: Conditioning the Pt RTD using the ADT70

The ADT70 contains the two matched current sources, a precision rail-to-rail output instrumentation amplifier, a 2.5V reference, and an uncommitted rail-to-rail output op amp. The ADT71 is the same as the ADT70 except the internal voltage reference is omitted. A shutdown function is included for battery powered equipment that reduces the quiescent current from 3 mA to 10 µA. The gain or full-scale range for the Pt RTD and ADT701 system is set by a precision external resistor connected to the instrumentation amplifier. The uncommitted op amp may be used for scaling the internal voltage reference, providing a "Pt RTD open" signal or "over temperature" warning, providing a heater switching signal, or other external conditioning determined by the user. The ADT70 is specified for operation from $-40°C$ to $+125°C$ and is available in 20-pin DIP and SOIC packages.

## Thermistors

Similar in function to the RTD, thermistors are low cost temperature-sensitive resistors and are constructed of solid semiconductor materials which exhibit a positive or negative temperature coefficient (NTC). Although positive TC devices are available, the most commonly used thermistors are those with an NTC. Figure 3-47 shows the resistance–temperature characteristic of a commonly used NTC thermistor. The thermistor is highly nonlinear and, of the three temperature sensors discussed, is the most sensitive.

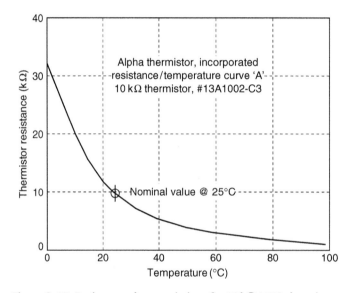

Figure 3-47: Resistance characteristics of a 10 kΩ NTC thermistor

The thermistor's high sensitivity (typically, $-44,000$ ppm/°C at 25°C, as shown in Figure 3-48), allows it to detect minute variations in temperature which could not be observed with an RTD or thermocouple. This high sensitivity is a distinct advantage over the RTD in that four-wire Kelvin connections to the thermistor are not needed to compensate for lead wire errors. To illustrate this point, suppose a 10 kΩ NTC thermistor, with a typical 25°C TC of $-44,000$ ppm/°C, were substituted for the 100 Ω Pt RTD in the example given earlier; then a total lead wire resistance of 21 Ω would generate less than 0.05°C error in the measurement. This is roughly a factor of 500 improvements in error over an RTD.

However, the thermistor's high sensitivity to temperature does not come without a price. As was shown in Figure 3-48, the TC of thermistors does not decrease linearly with increasing temperature as it does with RTDs; therefore, linearization is required for all but the narrowest of temperature ranges. Thermistor applications are limited to a few hundred degrees at best because they are more susceptible to damage at high temperatures. Compared to thermocouples and RTDs, thermistors are fragile in construction and require careful mounting procedures to prevent crushing or bond separation. Although a thermistor's response time is short due to its small size, its small thermal mass makes it very sensitive to self-heating errors.

Thermistors are very inexpensive, highly sensitive temperature sensors. However, we have shown that a thermistor's TC varies from $-44,000$ ppm/°C at 25°C to $-29,000$ ppm/°C at 100°C. Not only is this

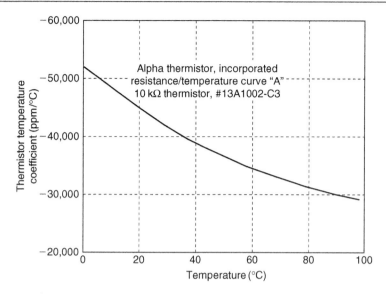

**Figure 3-48: Temperature coefficient of a 10 kΩ NTC thermistor**

nonlinearity the largest source of error in a temperature measurement, but also it limits useful applications to very narrow temperature ranges if linearization techniques are not used.

It is possible to use a thermistor over a wide temperature range only if the system designer can tolerate a lower sensitivity to achieve improved linearity. One approach to linearizing a thermistor is simply shunting it with a fixed resistor. Paralleling the thermistor with a fixed resistor increases the linearity significantly. As shown in Figure 3-49, the parallel combination exhibits a more linear variation with temperature compared to the thermistor itself. Also, the sensitivity of the combination still is high compared to a thermocouple or RTD. The primary disadvantage to this technique is that linearization can only be achieved within a narrow range.

The value of the fixed resistor can be calculated from the following equation:

$$R = \frac{RT2 \times (RT1 + RT3) - 2 \times RT1 \times RT3}{RT1 + RT3 - 2 \times RT2} \qquad (3\text{-}30)$$

where RT1 is the thermistor resistance at T1, the lowest temperature in the measurement range, RT3 is the thermistor resistance at T3, the highest temperature in the range, and RT2 is the thermistor resistance at T2, the midpoint, T2 = (T1 + T3)/2.

For a typical 10 kΩ NTC thermistor, RT1 = 32,650 Ω at 0°C, RT2 = 6,532 Ω at 35°C, and RT3 = 1,752 Ω at 70°C. This results in a value of 5.17 kΩ for R. The accuracy needed in the signal conditioning circuitry depends on the linearity of the network. For the example given above, the network shows a nonlinearity of −2.3°C /+2.0°C.

The output of the network can be applied to an ADC to perform further linearization as shown in Figure 3-50. Note that the output of the thermistor network has a slope of approximately −10 mV/°C, which implies a 12-bit ADC has more than sufficient resolution.

## Digital Output Temperature Sensors

Temperature sensors which have digital outputs have a number of advantages over those with analog outputs, especially in remote applications. Opto-isolators can also be used to provide galvanic isolation

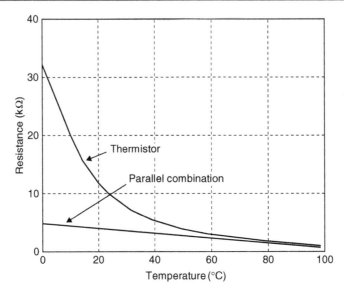

**Figure 3-49: Linearization of NTC thermistor using a 5.17 kΩ shunt resistor**

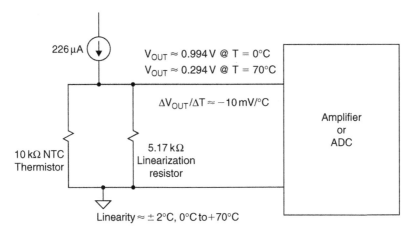

**Figure 3-50: Linearized thermistor amplifier**

between the remote sensor and the measurement system. A voltage-to-frequency converter driven by a voltage output temperature sensor accomplishes this function; however, more sophisticated ICs are now available which are more efficient and offer several performance advantages.

The TMP03/TMP04 digital output sensor family includes a voltage reference, $V_{PTAT}$ generator, sigma–delta ADC, and a clock source (see Figure 3-51).The sensor output is digitized by a first-order sigma–delta modulator, also known as the "charge balance" type ADC. This converter utilizes time-domain oversampling and a high accuracy comparator to deliver 12 bits of effective accuracy in an extremely compact circuit.

The output of the sigma–delta modulator is encoded using a proprietary technique which results in a serial digital output signal with a mark-space ratio format (see Figure 3-52) that is easily decoded by any microprocessor into either degrees centigrade or degrees Fahrenheit, and readily transmitted over a single wire. Most importantly, this encoding method avoids major error sources common to other modulation

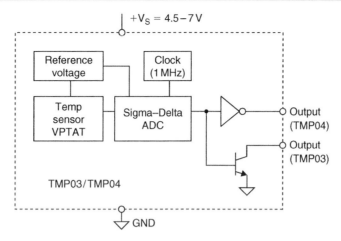

**Figure 3-51: Digital output temperature sensor: TMP03/04**

$$\text{Temperature (°C)} = 235 - \left(\frac{400 \times T1}{T2}\right)$$

$$\text{Temperature (°F)} = 455 - \left(\frac{720 \times T1}{T2}\right)$$

- T1 Nominal pulse width = 10 ms
- ±1.5°C Error over temp, ±0.5°C Nonlinearity (typical)
- Specified −40°C to +100°C
- Nominal T1/T2 @ 0°C = 60%
- Nominal frequency @ +25°C = 35 Hz
- 6.5 mW Power consumption @ 5 V
- TO-92, SO-8, or TSSOP packages

**Figure 3-52: TMP03/TMP04 output format**

techniques, as it is clock-independent. The nominal output frequency is 35 Hz at +25°C, and the device operates with a fixed high level pulse width (T1) of 10 ms.

The TMP03/TMP04 output is a stream of digital pulses, and the temperature information is contained in the mark-space ratio as per the equations:

$$\text{Temperature (°C)} = 235 - \left(\frac{400 \times T1}{T2}\right) \tag{3-31}$$

$$\text{Temperature (°F)} = 455 - \left(\frac{720 \times T1}{T2}\right) \tag{3-32}$$

Popular microcontrollers, such as the 80C51 and 68HC11, have on-chip timers which can easily decode the mark-space ratio of the TMP03/TMP04. A typical interface to the 80C51 is shown in Figure 3-53. Two timers, labeled *Timer 0* and *Timer 1*, are 16 bits in length. The 80C51's system clock, divided by 12,

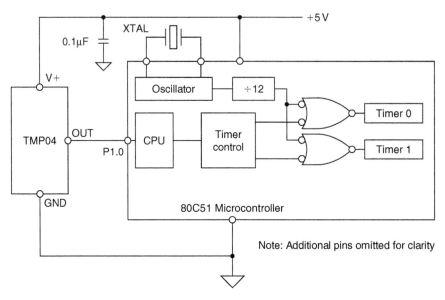

**Figure 3-53: Interfacing a TMP04 to a microcontroller**

provides the source for the timers. The system clock is normally derived from a crystal oscillator, so timing measurements are quite accurate. Since the sensor's output is ratiometric, the actual clock frequency is not important. This feature is important because the microcontroller's clock frequency is often defined by some external timing constraint, such as the serial baud rate.

Software for the sensor interface is straightforward. The microcontroller simply monitors I/O port P1.0, and starts *Timer 0* on the rising edge of the sensor output. The microcontroller continues to monitor P1.0, stopping *Timer 0* and starting *Timer 1* when the sensor output goes low. When the output returns high, the sensor's T1 and T2 times are contained in registers *Timer 0* and *Timer 1*, respectively. Further software routines can then apply the conversion factor shown in the equations above and calculate the temperature.

## Thermostatic Switches and Setpoint Controllers

Temperature sensors used in conjunction with comparators can act as thermostatic switches. ICs such as the AD22105 accomplish this function at low cost and allow a single external resistor to program the setpoint to 2°C accuracy over a range of −40°C to +150°C (see Figure 3-54). The device asserts an open collector output when the ambient temperature exceeds the user-programmed setpoint temperature. The ADT05 has approximately 4°C of hysteresis which prevents rapid thermal on/off cycling. The ADT05 is designed to operate on a single-supply voltage from +2.7 to +7.0V facilitating operation in battery powered applications as well as industrial control systems. Because of low power dissipation ($200\,\mu W @ 3.3\,V$), self-heating errors are minimized, and battery life is maximized. An optional internal $200\,k\Omega$ pull-up resistor is included to facilitate driving light loads such as CMOS inputs.

The setpoint resistor is determined by the equation:

$$R_{SET} = \frac{39M\Omega°C}{T_{SET}(°C) + 281.6°C} - 90.3\,k\Omega \tag{3-33}$$

The setpoint resistor should be connected directly between the $R_{SET}$ pin (Pin 4) and the GND pin (Pin 5). If a ground plane is used, the resistor may be connected directly to this plane at the closest available point.

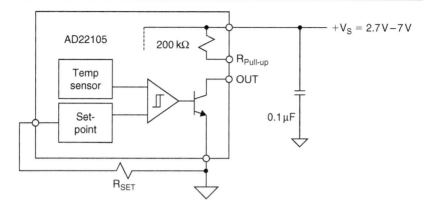

**Figure 3-54: AD22105 thermostatic switch**

The setpoint resistor can be of nearly any resistor type, but its initial tolerance and thermal drift will affect the accuracy of the programmed switching temperature. For most applications, a 1% metal-film resistor will provide the best tradeoff between cost and accuracy. Once $R_{SET}$ has been calculated, it may be found that the calculated value does not agree with readily available standard resistors of the chosen tolerance. In order to achieve a value as close as possible to the calculated value, a compound resistor can be constructed by connecting two resistors in series or parallel.

The TMP01 is a dual setpoint temperature controller which also generates a PTAT output voltage (see Figure 3-55). It also generates a control signal from one of two outputs when the device is either above or below a specific temperature range. Both the high/low temperature trip points and hysteresis band are determined by user-selected external resistors.

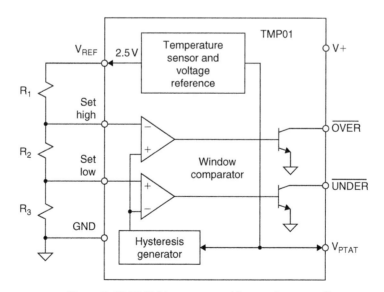

**Figure 3-55: TMP01 programmable setpoint controller**

The TMP01 consists of a bandgap voltage reference combined with a pair of matched comparators. The reference provides both a constant 2.5V output and a PTAT output voltage which has a precise TC of 5 mV/K and is 1.49V (nominal) at +25°C. The comparators compare $V_{PTAT}$ with the externally set temperature trip points and generate an open collector output signal when one of their respective thresholds has been exceeded.

Hysteresis is also programmed by the external resistor chain and is determined by the total current drawn out of the 2.5V reference. This current is mirrored and used to generate a hysteresis offset voltage of the appropriate polarity after a comparator has been tripped. The comparators are connected in parallel, which guarantees that there is no hysteresis overlap and eliminates erratic transitions between adjacent trip zones.

## Microprocessor Temperature Monitoring

Today's computers require that hardware as well as software operate properly, in spite of the many things that can cause a system crash or lockup. The purpose of hardware monitoring is to monitor the critical items in a computing system and take corrective action so that problems do not occur.

Microprocessor supply voltage and temperature are two critical parameters. If the supply voltage drops below a specified minimum level, further operations should be halted until the voltage returns to acceptable levels. In some cases, it is desirable to reset the microprocessor under "brownout" conditions. It is also common practice to reset the microprocessor on power-up or power-down. Switching to a battery backup may be required if the supply voltage is low.

Under low voltage conditions it is mandatory to inhibit the microprocessor from writing to external CMOS memory by inhibiting the Chip Enable signal to the external memory.

Many microprocessors can be programmed to periodically output a "watchdog" signal. Monitoring this signal gives an indication that the processor and its software are functioning properly and that the processor is not stuck in an endless loop.

The need for hardware monitoring has resulted in a number of ICs, traditionally called "microprocessor supervisory products," which perform some or all of the above functions. These devices range from simple manual reset generators (with debouncing) to complete microcontroller-based monitoring sub-systems with on-chip temperature sensors and ADCs. ADI' ADM family of products is specifically to perform the various microprocessor supervisory functions required in different systems.

CPU temperature is critically important in the Pentium microprocessors. For this reason, all new Pentium devices have an on-chip substrate PNP transistor which is designed to monitor the actual chip temperature. The collector of the substrate PNP is connected to the substrate, and the base and emitter are brought out on two separate pins of the Pentium II.

The ADM1021 Microprocessor Temperature Monitor is specifically designed to process these outputs and convert the voltage into a digital word representing the chip temperature. The simplified analog signal processing portion of the ADM1021 is shown in Figure 3-56.

The technique used to measure the temperature is identical to the "$\Delta V_{BE}$" principle previously discussed. Two different currents (I and N · I) are applied to the sensing transistor, and the voltage measured for each. In the ADM1021, the nominal currents are I = 6 µA, (N = 17), N · I = 102 µA.

The change in the base-emitter voltage, $\Delta V_{BE}$, is a PTAT voltage and given by the equation:

$$\Delta V_{BE} = \frac{kT}{q} \ln(N) \qquad (3\text{-}34)$$

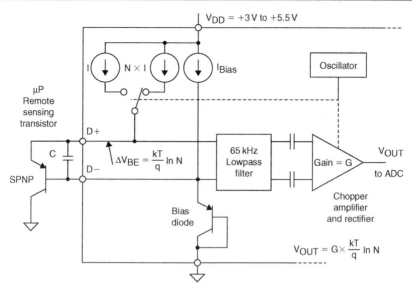

**Figure 3-56: ADM1021 microprocessor temperature monitor input signal conditioning circuits**

Figure 3-56 shows the external sensor as a substrate transistor, provided for temperature monitoring in the microprocessor, but it could equally well be a discrete transistor. If a discrete transistor is used, the collector should be connected to the base and not grounded. To prevent ground noise interfering with the measurement, the more negative terminal of the sensor is not referenced to ground, but is biased above ground by an internal diode. If the sensor is operating in a noisy environment, C may be optionally added as a noise filter. Its value is typically 2200 pF, but should be no more than 3000 Pf.

To measure $\Delta V_{BE}$, the sensing transistor is switched between operating currents of I and N · I. The resulting waveform is passed through a 65 kHz lowpass filter to remove noise, then to a chopper-stabilized amplifier which performs the function of amplification and synchronous rectification. The resulting DC voltage is proportional to $\Delta V_{BE}$ and is digitized by an 8-bit ADC. To further reduce the effects of noise, digital filtering is performed by averaging the results of 16 measurement cycles.

In addition, the ADM1021 contains an on-chip temperature sensor, and its signal conditioning and measurement are performed in the same manner.

One least significant bit (LSB) of the ADC corresponds to 1°C, so the ADC can theoretically measure from −128°C to +127°C, although the practical lowest value is limited to −65°C due to device maximum ratings. The results of the local and remote temperature measurements are stored in the local and remote temperature value registers, and are compared with limits programmed into the local and remote high and low limit registers as shown in Figure 3-57. An $\overline{\text{ALERT}}$ output signals when the on-chip or remote temperature is out of range. This output can be used as an interrupt, or as a system management bus (SMBus) alert.

The limit registers can be programmed, and the device controlled and configured, via the serial SMBus. The contents of any register can also be read back by the SMBus. Control and configuration functions consist of: switching the device between normal operation and standby mode, masking or enabling the $\overline{\text{ALERT}}$ output, and selecting the conversion rate which can be set from 0.0625 to 8 Hz.

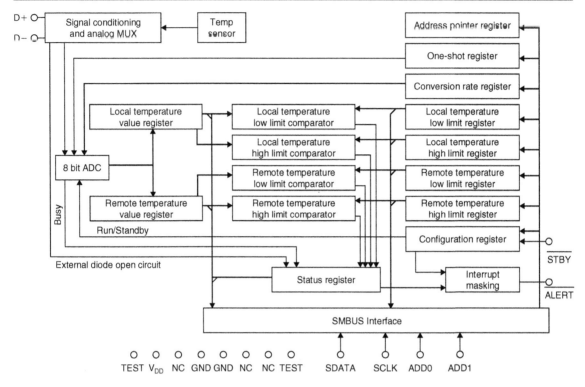

**Figure 3-57: ADM1021 simplified block diagram**

## References: Temperature Sensors

1.  R. Pallas-Areny and J.G. Webster, **Sensors and Signal Conditioning**, John Wiley, New York, 1991.

2.  D. Sheingold, (Ed.), **Transducer Interfacing Handbook**, Analog Devices, Inc., 1980, Norwood, Ma.

3.  W. Kester, (Ed.), **1992 Amplifier Applications Guide**, Section 2, 3, Analog Devices, Inc., 1992, Norwood, Ma.

4.  W. Kester, (Ed.), **System Applications Guide**, Section 1, 6, Analog Devices, Inc., 1993, Norwood, Ma.

5.  D. Sheingold, **Nonlinear Circuits Handbook**, Analog Devices, Inc., 1980, Norwood, Ma.

6.  J. Wong, "Temperature Measurements Gain from Advances in High-precision Op Amps," **Electronic Design**, May 15, 1986.

7.  OMEGA Temperature Measurement Handbook, Omega Instruments, Inc.

8.  **Handbook of Chemistry and Physics**, CRC.

9.  P. Brokaw, "A Simple Three-Terminal IC Bandgap Voltage Reference," **IEEE Journal of Solid State Circuits**, No. SC-9, 1974.

# *Charge Coupled Devices*

Charge coupled devices (CCDs) contain a large number of small photocells called photosites or pixels which are arranged either in a single row (linear arrays) or in a matrix (area arrays). CCD area arrays are commonly used in video applications, while linear arrays are used in facsimile machines, graphics scanners, and pattern recognition equipment.

The linear CCD array consists of a row of image sensor elements (photosites, or pixels) which are illuminated by light from the object or document. During one exposure period each photosite acquires an

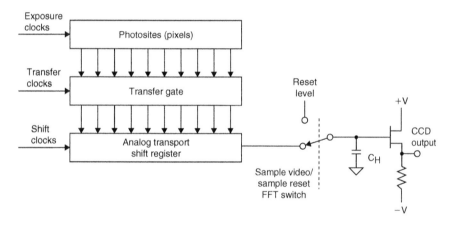

**Figure 3-58: Linear CCD array**

amount of charge which is proportional to its illumination. These photosite charge packets are subsequently switched simultaneously via transfer gates to an analog shift register. The charge packets on this shift register are clocked serially to a charge detector (storage capacitor) and buffer amplifier (source follower), which convert them into a string of photo-dependent output voltage levels (see Figure 3-58). While the charge packets from one exposure are being clocked out to the charge detector, another exposure is underway. The analog shift register typically operates at frequencies between 1 and 10 MHz.

The charge detector readout cycle begins with a reset pulse which causes a FET switch to set the output storage capacitor to a known voltage. The switching FET's capacitive feedthrough causes a reset glitch at the output as shown in Figure 3-59. The switch is then opened, isolating the capacitor, and the charge from the last pixel is dumped onto the capacitor causing a voltage change. The difference between the reset voltage and the final voltage (video level), shown in Figure 3.59, represents the amount of charge in the pixel. CCD charges may be as low as 10 electrons, and a typical CCD output sensitivity is 0.6 µV/electron. Most CCDs have a saturation output voltage of about 1 V (see Reference 16).

Since CCDs are generally fabricated on MOS processes, they have limited capability to perform on-chip signal conditioning. Therefore the CCD output is generally processed by external conditioning circuits.

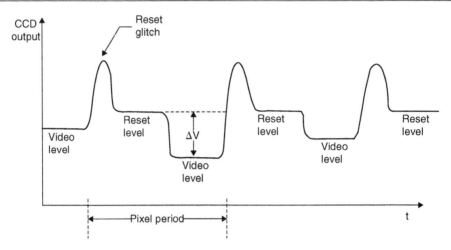

**Figure 3-59: CCD output waveform**

CCD output voltages are small and quite often buried in noise. The largest source of noise is the thermal noise in the resistance of the FET reset switch. This noise may have a typical value of 100–300 electrons root mean square (RMS) (approximately $60–180\,mV_{RMS}$). This noise occurs as a *sample-to-sample* variation in the CCD output level and is common to both the reset level and the video level for a given pixel period. A technique called *correlated double sampling* (CDS) is often used to reduce the effect of this noise. Figure 3.60 shows two circuit implementations of the CDS scheme. In the top circuit, the CCD output drives both sample-and-hold amplifiers (SHAs). At the end of the reset interval, SHA1 holds the reset voltage level. At the end of the video interval, SHA2 holds the video level. The SHA outputs are applied to a difference amplifier which subtracts one from the other. In this scheme, there is only a short interval during which both SHA outputs are stable, and their difference represents $\Delta V$, so the difference amplifier must settle quickly.

Another arrangement is shown in the bottom half of Figure 3-60, which uses three SHAs and allows for either faster operation or more time for the difference amplifier to settle. In this circuit, SHA1 holds the reset level so that it occurs simultaneously with the video level at the input to SHA2 and SHA3. When the video clock is applied simultaneously to SHA2 and SHA3, the input to SHA2 is the reset level, and the input

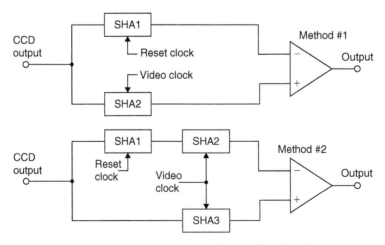

**Figure 3-60: Correlated double sampling (CDS)**

to SHA3 the video level. This arrangement allows the entire pixel period (less the acquisition time of SHA2 and SHA3) for the difference amplifier to settle.

## References: Charge Coupled Devices

1. W. Kester, (Ed.), **1992 Amplifier Applications Guide**, Section 2, 3, Analog Devices, Inc., 1992, Norwood, Ma.

2. W. Kester, (Ed.), **System Applications Guide**, Section 1, 6, Analog Devices, Inc., 1993, Norwood, Ma.

3. **Optoelectronics Data Book**, EG&G Vactec, St. Louis, MO, 1990.

4. *Silicon Detector Corporation*, Camarillo, CA, Part Number SD-020-12-001 Data Sheet.

5. **Photodiode 1991 Catalog**, Hamamatsu Photonics, Bridgewater, NJ.

6. R. Morrison, **Grounding and Shielding Techniques in Instrumentation**, 3rd Edition, John Wiley, Inc., 1986.

7. H. W. Ott, **Noise Reduction Techniques in Electronic Systems**, 2nd Edition, John Wiley, Inc., 1988.

8. *An Introduction to the Imaging CCD Array*, Technical Note 82W-4022, Tektronix, Inc., Beaverton, OR., 1987.

9. **Handbook of Chemistry and Physics**, CRC.

# CHAPTER 4

# *RF/IF Circuits*

## Chapter Introduction

From cellular phones to two-way pagers to wireless Internet access, the world is becoming more connected, even though wirelessly. No matter the technology, these devices are basically simple radio transceivers (transmitters and receivers). In the vast majority of cases the receivers and transmitters are a variation on the superheterodyne radio shown in Figure 4-1 for the receiver and Figure 4-2 for the transmitter.

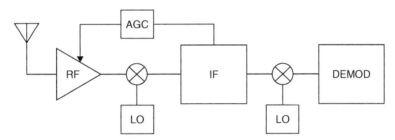

**Figure 4-1: Basic superheterodyne radio receiver**

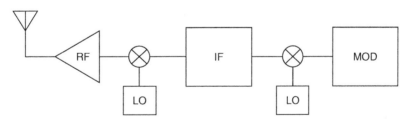

**Figure 4-2: Basic superheterodyne radio transmitter**

The basic concept of operation is as follows. For the receiver, the signal from the antenna is amplified in the radio frequency (RF) stage. The output of the RF stage is one input of a mixer. A local oscillator (LO) is the other input. The output of the mixer is at the intermediate frequency (IF). The concept here is that it is much easier to build a high gain amplifier string at a narrow frequency band than it is to build a wideband, high gain amplifier. Also, the modulation bandwidth is typically very much smaller than the carrier frequency. A second mixer stage converts the signal to the baseband. The signal is then demodulated (demod). The modulation technique is independent from the receiver technology. The modulation scheme could be amplitude modulation (AM), frequency modulation (FM), phase modulation, or some form of quadrature amplitude modulation (QAM), which is a combination of amplitude and phase modulation.

To put some numbers around it, let us consider a broadcast FM signal. The carrier frequency is in the range of 98–108 MHz. The IF frequency is almost always 10.7 MHz. The baseband is 0 Hz–15 kHz. This is the sum of the right and left audio frequencies. There is also a modulation band centered at 38 kHz that is the difference of the left and right audio signals. This difference signal is demodulated and summed with the sum signal to generate the separate left and right audio signals.

On the transmit side the mixers convert the frequencies up instead of down.

These simplified block diagrams neglect some of the refinements that may be incorporated into these designs, such as power monitoring and control of the transmitter power amplifier as achieved with the "Tru-Power" circuits.

As technology has improved, we have seen the proliferation of IF sampling. Analog-to-digital converters (ADCs) of sufficient performance have been developed which allow the sampling of the signal at the IF frequency range, with demodulation occurring in the digital domain. This allows for system simplification by eliminating a mixer stage.

In addition to the basic building blocks that are the subject of this chapter, these circuit blocks often appear as building blocks in larger application specific integrated circuits (ASIC).

## SECTION 4-1

# *Mixers*

## The Ideal Mixer

An idealized mixer is shown in Figure 4-3. An RF (or IF) mixer (not to be confused with video and audio mixers) is an active or passive device that converts a signal from one frequency to another. It can either modulate or demodulate a signal. It has three signal connections, which are called *ports* in the language of radio engineers. These three ports are the RF input, the LO input, and the IF output.

A mixer takes an RF input signal at a frequency $f_{RF}$, mixes it with a LO signal at a frequency $f_{LO}$, and produces an IF output signal that consists of the sum and difference frequencies, $f_{RF} \pm f_{LO}$. The user provides a bandpass filter that follows the mixer and selects the sum ($f_{RF} + f_{LO}$) or difference ($f_{RF} - f_{LO}$) frequency.

Some points to note about mixers and their terminology:

*   When the sum frequency is used as the IF, the mixer is called an *upconverter*; when the difference is used, the mixer is called a *downconverter*. The former is typically used in a transmit channel; the latter in a receive channel.

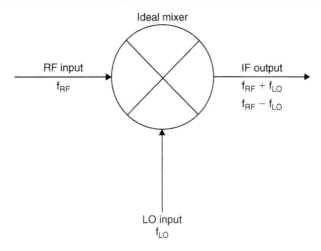

**Figure 4-3: The mixing process**

- In a receiver, when the LO frequency is below the RF, it is called *low side injection* and the mixer *a low side downconverter*; when the LO is above the RF, it is called *high side injection*, and the mixer *a high side downconverter*.

- Each of the outputs is only half the amplitude (one-quarter the power) of the individual inputs; thus, there is a loss of 6 dB in this ideal linear mixer. (In a practical multiplier, the conversion loss may be greater than 6 dB, depending on the scaling parameters of the device. Here, we assume a *mathematical* multiplier, having no dimensional attributes).

A mixer can be implemented in several ways, using active or passive techniques.

Ideally, to meet the low noise, high linearity objectives of a mixer we need some circuit that implements a polarity-switching function in response to the LO input. Thus, the mixer can be reduced to Figure 4-4, which shows the RF signal being split into in-phase (0°) and antiphase (180°) components; a changeover switch, driven by the LO signal, alternately selects the in-phase and antiphase signals. Thus reduced to essentials, the ideal mixer can be modeled as a sign-switcher.

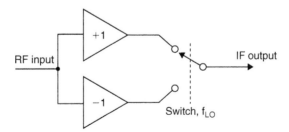

**Figure 4-4: An ideal switching mixer**

In a perfect embodiment, this mixer would have no noise (the switch would have zero resistance), no limit to the maximum signal amplitude, and would develop no intermodulation between the various RF signals. Although simple in concept, the waveform at the IF output can be very complex for even a small number of signals in the input spectrum. Figure 4-6 shows the result of *mixing* just a single input at 11 MHz with an LO of 10 MHz.

The *wanted* IF at the difference frequency of 1 MHz is still visible in this waveform, and the 21 MHz sum is also apparent. How are we to analyze this?

We still have a product, but now it is that of a sinusoid (the RF input) at $\omega_{RF}$ and a variable that can only have the values $+1$ or $-1$, that is, a unit square wave at $\omega_{LO}$. The latter can be expressed as a Fourier series.

$$S_{LO} = 4/\pi \{\sin \omega_{LO}t - 1/3 \sin 3\omega_{LO}t + 1/5 \sin 5\omega_{LO}t - ....\} \tag{4-1}$$

Thus, the output of the switching mixer is its RF input, which we can simplify as $\sin \omega_{RF}t$, multiplied by the above expansion for the square wave, producing:

$$S_{IF} = 4/\pi \{\sin \omega_{RF}t \sin \omega_{LO}t - 1/3 \sin \omega_{RF}t \sin 3\omega_{LO}t \\ + 1/5 \sin 5\omega_{RF}t \sin 5\omega_{LO}t - ....\} \tag{4-2}$$

Now expanding each of the products, we obtain:

$$S_{IF} = 2/\pi \{\sin(\omega_{RF} + \omega_{LO})t + \sin(\omega_{RF} - \omega_{LO})t - 1/3 \sin(\omega_{RF} + 3\omega_{LO})t \\ - 1/3 \sin(\omega_{RF} - 3\omega_{LO})t \\ + 1/5 \sin(\omega_{RF} + 5\omega_{LO})t + 1/5 \sin(\omega_{RF} - 5\omega_{LO})t - ...\} \tag{4-3}$$

or simply

$$S_{IF} = 2/\pi \{\sin(\omega_{RF} + \omega_{LO})t + \sin(\omega_{RF} - \omega_{LO})t + harmonics\} \tag{4-4}$$

The most important of these harmonic components are sketched in Figure 4-5 for the particular case used to generate the waveform shown in Figure 4-6, that is, $f_{RF} = 11$ MHz and $f_{LO} = 10$ MHz. Because of the $2/\pi$ term, a mixer has a minimum 3.92 dB insertion loss (and noise figure) in the absence of any gain.

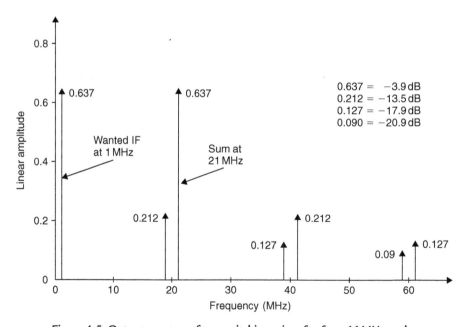

Figure 4-5: Output spectrum for a switching mixer for $f_{RF} = 11$ MHz and $f_{LO} = 10$ MHz

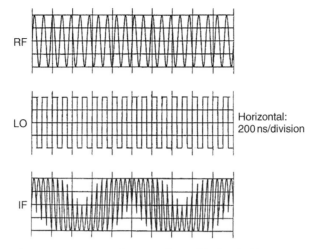

**Figure 4-6: Inputs and output for an ideal switching mixer for $f_{RF} = 11\,MHz$, $f_{LO} = 10\,MHz$**

Note that the ideal (switching) mixer has exactly the same problem of image response to $\omega_{LO} - \omega_{RF}$ as the linear multiplying mixer. The image response is somewhat subtle, as it does not immediately show up in the output spectrum: it is a latent response, awaiting the occurrence of the "wrong" frequency in the input spectrum.

## Diode-Ring Mixer

For many years, the most common mixer topology for high performance applications has been the diode-ring mixer, one form of which is shown in Figure 4-7. The diodes, which may be silicon junction, silicon Schottky-barrier, or gallium–arsenide (GaAs) types, provide the essential switching action. We do not need to analyze this circuit in great detail, but note in passing that the LO drive needs to be quite high—often a substantial fraction of 1 W—in order to ensure that the diode conduction is strong enough to achieve low noise and to allow large signals to be converted without excessive spurious nonlinearity.

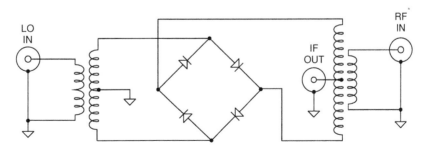

**Figure 4-7: Diode-ring mixer**

Because of the highly nonlinear nature of the diodes, the impedances at the three ports are poorly controlled, making matching difficult. Furthermore, there is considerable coupling between the three ports; this, and the high power needed at the LO port, make it very likely that there will be some component of the (highly distorted) LO signal coupled back toward the antenna. Finally, it will be apparent that a passive mixer such as this cannot provide conversion gain; in the idealized scenario, there will be a conversion loss of $2/\pi$ (as Eq. 4-4 shows), or 3.92 dB. A practical mixer will have higher losses, due to the resistances of the diodes and the losses in the transformers.

Users of this type of mixer are accustomed to judging the signal-handling capabilities by a "Level" rating. Thus, a Level-17 mixer needs +17 dBm (50 mW) of LO drive and can handle an RF input as high as +10 dBm (±1 V). A typical mixer in this class would be the Mini-Circuits LRMS-1H, covering 2–500 MHz, having a nominal insertion loss of 6.25 dB (8.5 dB maximum), a worst-case LO–RF isolation of 20 dB, and a worst-case LO–IF isolation of 22 dB (these figures for an LO frequency of 250–500 MHz). The price of this component is approximately $10.00 in small quantities. Even the most expensive diode-ring mixers have similar drive power requirements, high losses, and high coupling from the LO port.

The diode-ring mixer not only has certain performance limitations, but also is not amenable to fabrication using integrated circuit (IC) technologies, at least in the form shown in Figure 4-7. In the mid-1960s it was realized that the four diodes could be replaced by four transistors to perform essentially the same switching function. This formed the basis of the now-classical bipolar circuit shown in Figure 4-8, which is a minimal configuration for the fully balanced version. Millions of such mixers have been made, including variants in complementary-MOS (CMOS) and GaAs. We will limit our discussion to the bipolar junction transistor (BJT) form, an example of which is the Motorola MC1496, which, although quite rudimentary in structure, has been a mainstay in semi-discrete receiver designs for about 25 years.

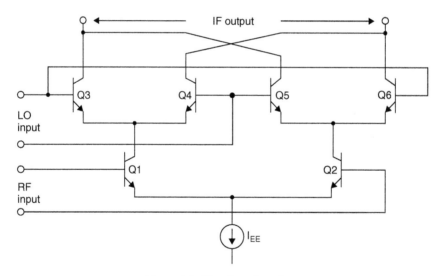

**Figure 4-8: Classic active mixer**

The *active mixer* is attractive for the following reasons:

- It can be monolithically integrated with other signal processing circuitry.

- It can provide conversion gain, whereas a diode-ring mixer always has an insertion loss. (Note: active mixers may have gain. The Analog Devices' AD831 active mixer, example, amplifies the result in Eq. 4-4 by $\pi/2$ to provide unity gain from RF to IF.)

- It requires much less power to drive the LO port.

- It provides excellent isolation between the signal ports.

- Is far less sensitive to load matching, requiring neither diplexer nor broadband termination.

Using appropriate design techniques it can provide tradeoffs between third-order intercept (3OI or IP3) and the 1 dB gain-compression point ($P_{1dB}$), on the one hand, and total power consumption ($P_D$) on the other. (That is, including the LO power, which in a passive mixer is "hidden" in the drive circuitry.)

## Basic Operation of the Active Mixer

Unlike the diode-ring mixer, which performs the polarity-reversing switching function in the voltage domain, the active mixer performs the switching function in the current domain. Thus the active mixer core (transistors Q3–Q6 in Figure 4-8) must be driven by current-mode signals. The voltage-to-current converter formed by Q1 and Q2 receives the voltage-mode RF signal at their base terminals and transforms it into a differential pair of currents at their collectors.

A second point of difference between the active mixer and diode-ring mixer, therefore, is that the active mixer responds only to magnitude of the input voltage, not to the input power; that is, the active mixer is not matched to the source. (The concept of matching is that both the current and the voltage at some port are used by the circuitry which forms that port.) By altering the bias current, $I_{EE}$, the transconductance of the input pair Q1–Q2 can be set over a wide range. Using this capability, an active mixer can provide variable gain.

A third point of difference is that the output (at the collectors of Q3–Q6) is in the form of a current, and can be converted back to a voltage at some other impedance level to that used at the input; hence, it can provide further gain. By combining both output currents (typically, using a transformer) this voltage gain can be doubled. Finally, it will be apparent that the isolation between the various ports, in particular, from the LO port to the RF port, is inherently much lower than can be achieved in the diode-ring mixer, due to the reversed-biased junctions that exist between the ports.

Briefly stated, though, the operation is as follows. In the absence of any voltage difference between the bases of Q1 and Q2, the collector currents of these two transistors are essentially equal. Thus, a voltage applied to the LO input results in no change of output current. Should a small DC offset voltage be present at the RF input (due typically to mismatch in the emitter areas of Q1 and Q2), this will only result in a small feedthrough of the LO signal to the IF output, which will be blocked by the first IF filter.

Conversely, if an RF signal is applied to the RF port, but no voltage difference is applied to the LO input, the output currents will again be balanced. A small offset voltage (due now to emitter mismatches in Q3–Q6) may cause some RF signal feedthrough to the IF output; as before, this will be rejected by the IF filters. It is only when a signal is applied to both the RF and LO ports that a signal appears at the output; hence, the term doubly balanced mixer.

Active mixers can realize their gain in one other way: the matching networks used to transform a $50\,\Omega$ source to the (usually) high input impedance of the mixer provide an impedance transformation and thus voltage gain due to the impedance step up. Thus, an active mixer that has loss when the input is terminated in a broadband $50\,\Omega$ termination can have "gain" when an input matching network is used.

## References: Mixers

1. B. Gilbert, **ISSCC Digest of Technical Papers 1968**, February, 16, 1968, pp. 114–115.

2. B. Gilbert, *Journal of Solid State Circuits*, Vol. SC-3, December, 1968, pp. 353–372.

3.  C.L. Ruthroff, "Some Broadband Transformers," **Proceedings of the I.R.E.**, Vol. 47, August, 1959, pp. 1337–1342.

4.  J.M. Bryant, "Mixers for High Performance Radio," **Wescon 1981: Session 24**, Electronic Conventions, Inc., Sepulveda Blvd., El Segundo, CA.

5.  P.E. Chadwick, "High Performance IC Mixers," **IERE Conference on Radio Receivers and Associated Systems**, Leeds, 1981, IERE Conference Publication No. 50.

6.  P.E. Chadwick, **Phase Noise, Intermodulation, and Dynamic Range**, RF Expo, Anaheim, CA, January, 1986.

7.  AD831 Data Sheet, *Rev. B*, **Analog Devices**.

# Modulators

*Modulators* (sometimes called *balanced modulators, doubly balanced modulators,* or even on occasions *high level mixers*) can be viewed as *sign-changers*. The two inputs, X and Y, generate an output W, which is simply one of these inputs (say, Y) multiplied by just the sign of the other (say, X), that is $W = Y \times \text{sign}(X)$. Therefore, no reference voltage is required. A good modulator exhibits very high linearity in its signal path, with precisely equal gain for positive and negative values of Y, and precisely equal gain for positive and negative values of X. Ideally, the amplitude of the X input needed to fully switch the output sign is very small; that is, the X-input exhibits a comparator-like behavior. In some cases, where this input may be a logic signal, a simpler X-channel can be used.

As an example, the AD8345 is a silicon RFIC quadrature modulator, designed for use from 250 to 1,000 MHz. Its excellent phase accuracy and amplitude balance enable the high performance direct modulation of an IF carrier (Figure 4-9).

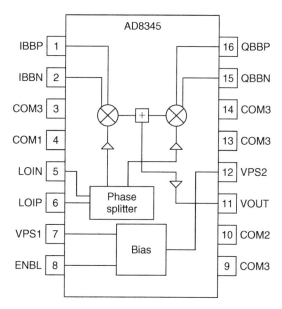

**Figure 4-9: AD8345 block diagram**

The AD8345 accurately splits the external LO signal into two quadrature components through the polyphase phase-splitter network. The two I and Q LO components are mixed with the baseband I and Q differential input signals. Finally, the outputs of the two mixers are combined in the output stage to provide a single-ended 50 Ω drive at $V_{OUT}$.

# Analog Multipliers

A multiplier is a device having two input ports and an output port. The signal at the output is the product of the two input signals. If both input and output signals are voltages, the transfer characteristic is the product of the two voltages divided by a scaling factor, K, which has the dimension of voltage (see Figure 4-10). From a mathematical point of view, multiplication is a "four-quadrant" operation—that is to say that both inputs may be either positive or negative and the output can be positive or negative (Figure 4-11). Some of the circuits used to produce electronic multipliers, however, are limited to signals of one polarity. If both signals must be unipolar, we have a "single-quadrant" multiplier, and the output will also be unipolar. If one of the signals is unipolar, but the other may have either polarity, the multiplier is a "two-quadrant" multiplier, and the output may have either polarity (and is "bipolar"). The circuitry used to produce one- and two-quadrant multipliers may be simpler than that required for four-quadrant multipliers, and since there are many applications where full four-quadrant multiplication is not required, it is common to find accurate devices which work only in one or two quadrants. An example is the AD539, a wideband dual

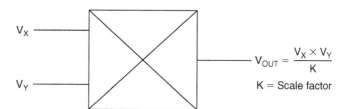

$$V_{OUT} = \frac{V_X \times V_Y}{K}$$

K = Scale factor

Figure 4-10: An analog multiplier block diagram

| Type | $V_X$ | $V_Y$ | $V_{OUT}$ |
|---|---|---|---|
| Single quadrant | Unipolar | Unipolar | Unipolar |
| Two quadrant | Bipolar | Unipolar | Bipolar |
| Four quadrant | Bipolar | Bipolar | Bipolar |

Figure 4-11: Definition of multiplier quadrants

two-quadrant multiplier which has a single unipolar $V_Y$ input with a relatively limited bandwidth of 5 MHz, and two bipolar $V_X$ inputs, one per multiplier, with bandwidths of 60 MHz. A block diagram of the AD539 is shown in Figure 4-12.

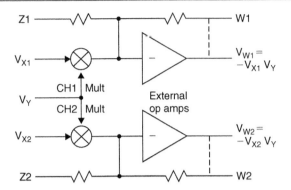

**Figure 4-12: AD539 block diagram**

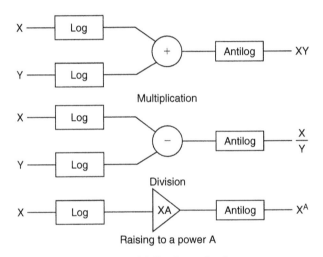

**Figure 4-13: Multiplication using log amps**

The simplest electronic multipliers use logarithmic amplifiers. The computation relies on the fact that the antilog of the sum of the logs of two numbers is the product of those numbers (see Figure 4-13).

The disadvantages of this type of multiplication are the very limited bandwidth and single-quadrant operation. A far better type of multiplier uses the "Gilbert Cell." This structure was invented by Barrie Gilbert, now of Analog Devices, in the late 1960s (see References 1 and 2).

There is a linear relationship between the collector current of a silicon junction transistor and its transconductance (gain) which is given by:

$$dI_C/dV_{BE} = qI_C/kT \tag{4-5}$$

where $I_C$ is the collector current, $V_{BE}$ is the base–emitter voltage, q is the electron charge ($1.60219 \times 10^{-19}$), k is Boltzmann's constant ($1.38062 \times 10^{-23}$), and T is the absolute temperature.

This relationship may be exploited to construct a multiplier with a differential (long-tailed) pair of silicon transistors, as shown in Figure 4-14.

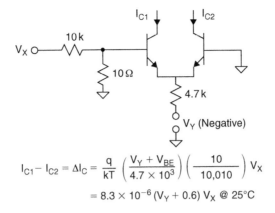

$$I_{C1} - I_{C2} = \Delta I_C = \frac{q}{kT} \left( \frac{V_Y + V_{BE}}{4.7 \times 10^3} \right) \left( \frac{10}{10,010} \right) V_X$$

$$= 8.3 \times 10^{-6} (V_Y + 0.6) V_X \text{ @ } 25°C$$

**Figure 4-14: Basic transconductance multiplier**

This is a rather poor multiplier because (1) the Y input is offset by the $V_{BE}$ which changes nonlinearly with $V_Y$; (2) the X input is nonlinear as a result of the exponential relationship between $I_C$ and $V_{BE}$; and (3) the scale factor varies with temperature.

Gilbert realized that this circuit could be linearized and made temperature stable by working with currents, rather than voltages, and by exploiting the logarithmic $I_C/V_{BE}$ properties of transistors (see Figure 4-15). The X input to the Gilbert Cell takes the form of a differential current, and the Y input is a unipolar current. The differential X currents flow in two diode-connected transistors, and the logarithmic voltages compensate for the exponential $V_{BE}/I_C$ relationship. Furthermore, the q/kT scale factors cancel. This gives the Gilbert Cell the linear transfer function.

$$\Delta I_C = \frac{\Delta I_X I_Y}{I_X} \tag{4-6}$$

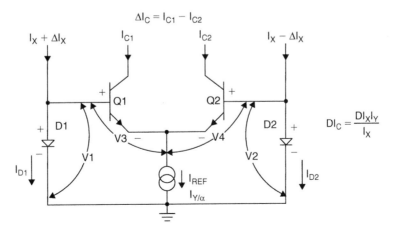

**Figure 4-15: Gilbert Cell**

As it stands, the Gilbert Cell has three inconvenient features: (1) its X input is a differential current; (2) its output is a differential current; and (3) its Y input is a unipolar current — so the cell is only a two-quadrant multiplier.

By cross-coupling two such cells and using two voltage-to-current converters (as shown in Figure 4-16), we can convert the basic architecture to a four-quadrant device with voltage inputs, such as the AD534. At low and medium frequencies, a subtractor amplifier may be used to convert the differential current at the output to a voltage. Because of its voltage output architecture, the bandwidth of the AD534 is only about 1 MHz, although the AD734, a later version, has a bandwidth of 10 MHz.

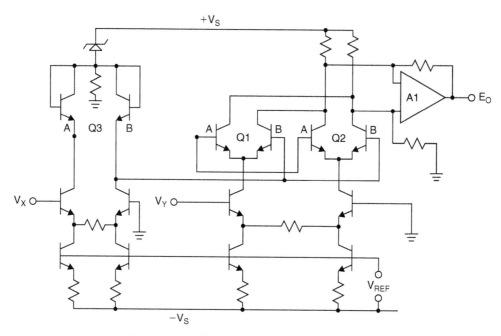

**Figure 4-16: A four-quadrant translinear multiplier**

In Figure 4-16, Q1A and Q1B, and Q2A and Q2B form the two core long-tailed pairs of the two Gilbert Cells, while Q3A and Q3B are the linearizing transistors for both cells. In Figure 3-35 there is an operational amplifier acting as a differential current to single-ended voltage converter, but for higher speed applications, the cross-coupled collectors of Q1 and Q2 form a differential open collector current output (as in the AD834 500 MHz multiplier).

The translinear multiplier relies on the matching of a number of transistors and currents. This is easily accomplished on a monolithic chip. Even the best IC processes have some residual errors, however, and these show up as four DC error terms in such multipliers. Offset voltage on the X input shows up as feedthrough of the Y input. Conversely, offset voltage on the Y input shows up as feedthrough of the X input. Offset voltage on the Z input causes offset of the output signal, and resistor mismatch causes gain error. In early Gilbert Cell multipliers, these errors had to be trimmed by means of resistors and potentiometers external to the chip, which was somewhat inconvenient. With modern analog processes, which permit the laser trimming of SiCr thin film resistors on the chip itself, it is possible to trim these errors during manufacture so that the final device has very high accuracy. Internal trimming has the additional advantage that it does not reduce the high frequency performance, as may be the case with external trimpots.

Because the internal structure of the translinear multiplier is necessarily differential, the inputs are usually differential as well (after all, if a single-ended input is required it is not hard to ground one of the inputs). This is not only convenient in allowing common-mode signals to be rejected, it also permits more complex computations to be performed. The AD534 (shown previously in Figure 4-16) is the classic example of a four-quadrant multiplier based on the Gilbert Cell. It has an accuracy of 0.1% in the multiplier mode, fully differential inputs, and a voltage output. However, as a result of its voltage output architecture, its bandwidth is only about 1 MHz.

For wideband applications, the basic multiplier with open collector current outputs is used. The AD834 is an 8-pin device with differential X inputs, differential Y inputs, differential open collector current outputs, and a bandwidth of over 500 MHz. A block diagram is shown in Figure 4-17.

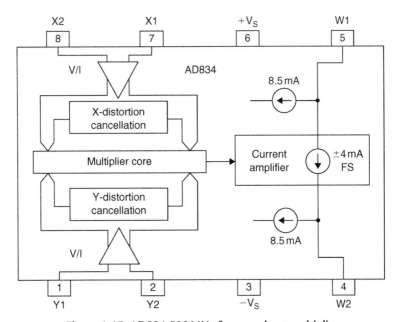

**Figure 4-17: AD834 500 MHz four-quadrant multiplier**

The AD834 is a true linear multiplier with a transfer function of:

$$I_{OUT} = \frac{V_x \cdot V_y}{1\,V \cdot 250\,\Omega} \tag{4-7}$$

Its X and Y offsets are trimmed to $500\,\mu V$ ($3\,mV$ maximum), and it may be used in a wide variety of applications including multipliers (broadband and narrowband), squarers, frequency doublers, and high frequency power measurement circuits. A consideration when using the AD834 is that, because of its very wide bandwidth, its input bias currents, approximately $50\,\mu A$/input, must be considered in the design of input circuitry lest, flowing in source resistances, they give rise to unplanned offset voltages.

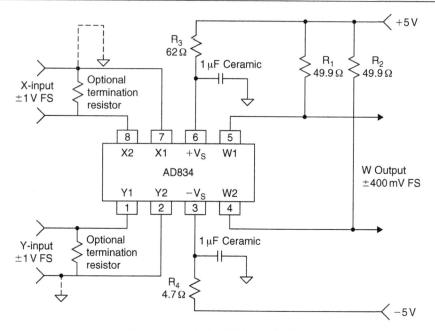

**Figure 4-18: Basic AD834 multiplier**

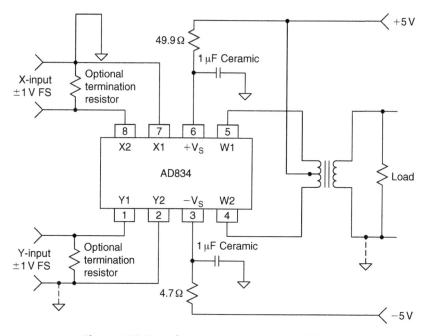

**Figure 4-19: Transformer coupled AD834 multiplier**

A basic wideband multiplier using the AD834 is shown in Figure 4-18. The differential output current flows in equal load resistors, R1 and R2, to give a differential voltage output. This is the simplest application circuit for the device. Where only the high frequency outputs are required, transformer coupling may be used, with either simple transformers (see Figure 4-19), or for better wideband performance, transmission line or "Ruthroff" transformers.

Low speed multipliers are also discussed in Chapter 2 (Section 2-11).

## References: Analog Multipliers

1.  D.H. Sheingold, (Ed.), *Nonlinear Circuits Handbook*, Analog Devices, Inc., 1974.

2.  AN-309: Build Fast VCAs and VCFs with Analog Multipliers.

# *Logarithmic Amplifiers*

In Chapter 2 (Section 2-8) we discussed low frequency logarithmic (log) amps. In this section we discuss high frequency applications.

The classic diode/op amp (or transistor/op amp) log amp suffers from limited frequency response, especially at low levels. For high frequency applications, *detecting* and *true log* architectures are used. Although these differ in detail, the general principle behind their design is common to both: instead of one amplifier having a logarithmic characteristic, these designs use a number of similar cascaded linear stages having well-defined large signal behavior.

Consider N cascaded limiting amplifiers, the output of each driving a summing circuit as well as the next stage (Figure 4-20). If each amplifier has a gain of A dB, the small signal gain of the strip is NA dB.

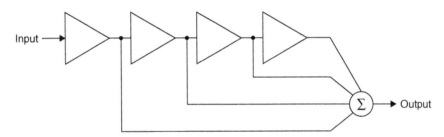

Input → Output

Σ

**Figure 4-20: Basic multistage log amp architecture**

If the input signal is small enough for the last stage not to limit, the output of the summing amplifier will be dominated by the output of the last stage.

As the input signal increases, the last stage will limit, and so will not add any more gain. Therefore it will now make a fixed contribution to the output of the summing amplifier, but the incremental gain to the summing amplifier will drop to $(N - 1)$A dB. As the input continues to increase, this stage in turn will limit and make a fixed contribution to the output, and the incremental gain will drop to $(N - 2)$A dB, and so forth—until the first stage limits, and the output ceases to change with increasing signal input.

The response curve is thus a set of straight lines as shown in Figure 4-21. The total of these lines, though, is a very good approximation to a logarithmic curve, and in practical cases, is an even better one, because few limiting amplifiers, especially high frequency ones, limit quite as abruptly as this model assumes.

The choice of gain, A, will also affect the log linearity. If the gain is too high, the log approximation will be poor. If it is too low, too many stages will be required to achieve the desired dynamic range. Generally, gains of 10–12 dB (3 times to 4 times) are chosen.

This is, of course, an ideal and very general model—it demonstrates the principle, but its practical implementation at very high frequencies is difficult. Assume that there is a delay in each limiting amplifier of t ns (this delay may also change when the amplifier limits but let's consider first-order effects!).

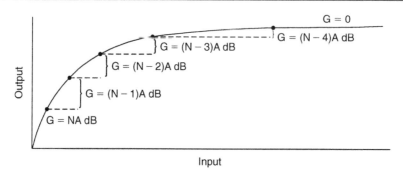

**Figure 4-21: Basic multistage log amp response (uniploar case)**

The signal which passes through all N stages will undergo delay of Nt ns, while the signal which only passes one stage will be delayed only t ns. This means that a small signal is delayed by Nt ns, while a large one is "smeared," and arrives spread over Nt ns. A nanosecond equals a foot at the speed of light, so such an effect represents a spread in position of Nt feet in the resolution of a radar system which may be unacceptable in some systems (for most log amp applications this is not a problem).

A solution is to insert delays in the signal paths to the summing amplifier, but this can become complex. Another solution is to alter the architecture slightly so that instead of limiting gain stages, we have stages with small signal gain of A and large signal (incremental) gain of unity (0 dB). We can model such stages as two parallel amplifiers, a limiting one with gain, and a unity gain buffer, which together feed a summing amplifier as shown in Figure 4-22.

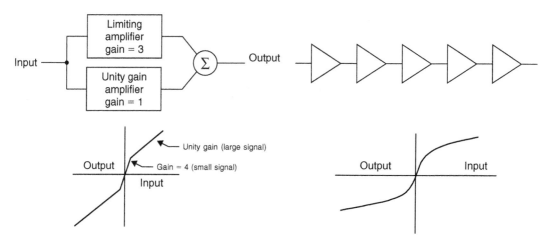

**Figure 4-22: Structure and performance of "true" log amp element and of a log amp formed by several such elements**

The *successive detection* log amp consists of cascaded limiting stages as described above, but instead of summing their outputs directly, these outputs are applied to detectors, and the detector outputs are summed as shown in Figure 4-23. If the detectors have current outputs, the summing process may involve no more than connecting all the detector outputs together.

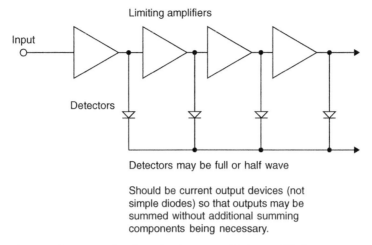

**Figure 4-23: Successive detection log amp with log and limiter outputs**

Log amps using this architecture have two outputs: the log output and a limiting output. In many applications, the limiting output is not used, but in some (e.g., FM receivers with "S"-meters), both are necessary. The limited output is especially useful in extracting the phase information from the input signal in polar demodulation techniques.

The log output of a successive detection log amplifier generally contains amplitude information, and the phase and frequency information is lost. This is not necessarily the case, however, if a half-wave detector is used, and attention is paid to equalizing the delays from the successive detectors—but the design of such log amps is demanding.

The specifications of log amps will include *noise, dynamic range, frequency response* (some of the amplifiers used as successive detection log amp stages have low frequency as well as high frequency cutoff), the *slope of the transfer characteristic* (which is expressed as V/dB or mA/dB depending on whether we are considering a voltage- or current-output device), the *intercept point* (the input level at which the output voltage or current is zero), and the *log linearity* (see Figures 4-24).

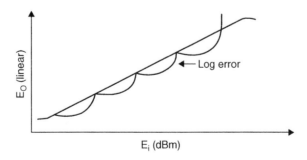

**Figure 4-24: Successive detection log linearity**

In the past, it has been necessary to construct high performance, high frequency successive detection log amps (called log strips) using a number of individual monolithic limiting amplifiers such as the Plessey SL-1521-series. Recent advances in IC processes, however, have allowed the complete log strip function to be integrated into a single chip, thereby eliminating the need for costly hybrid log strips.

The AD641 log amp contains five limiting stages (10dB/stage) and five full-wave detectors in a single IC package, and its logarithmic performance extends from DC to 250 MHz. Furthermore, its amplifier and full-wave detector stages are balanced so that, with proper layout, instability from feedback via supply rails is unlikely. A block diagram of the AD641 is shown in Figure 4-25. Unlike many previous IC log amps, the AD641 is laser trimmed to high absolute accuracy of both slope and intercept, and is fully temperature compensated. The transfer function for the AD641 as well as the log linearity is shown in Figure 4-26.

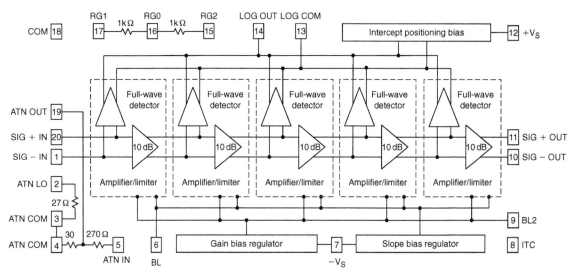

**Figure 4-25: Block diagram of the AD641 monolithic log amp**

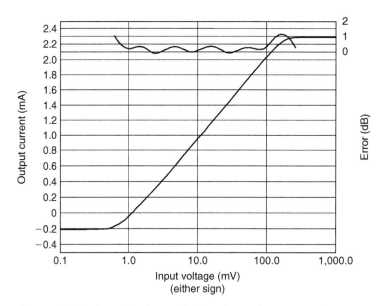

**Figure 4-26: DC logarithmic transfer function and error curve for a single AD641**

Because of its high accuracy, the actual waveform driving the AD641 must be considered when calculating responses. When a waveform passes through a log function generator, the mean value of the resultant waveform changes. This does not affect the slope of the response, but the apparent intercept is modified.

The AD641 is calibrated and laser trimmed to give its defined response to a DC level or a symmetrical 2 kHz square wave. It is also specified to have an intercept of 2 mV for a sinewave input (that is to say a 2 kHz sinewave of amplitude 2 mV peak (not peak-to-peak) gives the same mean output signal as a DC or square wave signal of 1 mV).

The waveform also affects the ripple or nonlinearity of the log response. This ripple is greatest for DC or square wave inputs because every value of the input voltage maps to a single location on the transfer function, and thus traces out the full nonlinearities of the log response. By contrast, a general time-varying signal has a continuum of values within each cycle of its waveform. The averaged output is thereby "smoothed" because the periodic deviations away from the ideal response, as the waveform "sweeps over" the transfer function, tend to cancel. As is clear in Figure 4-27, this smoothing effect is greatest for a triwave.

| Input waveform | Peak or RMS | Intercept factor | Error (relative to a DC input) |
|---|---|---|---|
| Square wave | Either | 1 | 0.00 dB |
| Sine wave | Peak | 2 | −6.02 dB |
| Sine wave | RMS | 1.414 ($\sqrt{2}$) | −3.01 dB |
| Triwave | Peak | 2.718 (e) | −8.68 dB |
| Triwave | RMS | 1.569 (e/$\sqrt{3}$) | −3.91 dB |
| Gaussian noise | RMS | 1.887 | −5.52 dB |

**Figure 4-27: The effects of waveform on intercept point**

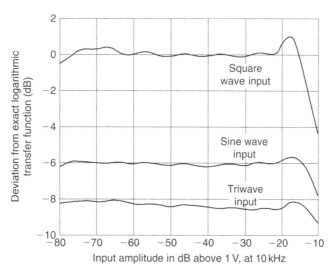

**Figure 4-28: The effect of the waveform on AD641 log linearity**

Each of the five stages in the AD641 has a gain of 10 dB and a full-wave detected output. The transfer function for the device was shown in Figure 4-26 along with the error curve. Note the excellent log linearity over an input range of 1–100 mV (40 dB)(Figure 4-28). Although well suited to RF applications, the AD641 is DC-coupled throughout. This allows it to be used in low frequency and very low frequency systems, including audio measurements, sonar, and other instrumentation applications requiring operation to low frequencies or even DC.

The limiter output of the AD641 has better than 1.6 dB gain flatness ($-44$ dBm–0 dBm @ 10.7 MHz) and less than $2°$ phase variation, allowing it to be used as a polar demodulator.

### References: Logarithmic Amplifiers

1.  D.H. Sheingold, (Ed.), **Nonlinear Circuits Handbook**, Analog Devices, Inc., Norwood, MA., 1974.

2.  R.S. Hughes, **Logarithmic Amplifiers**, Artech House, Inc., Dedham, MA., 1986.

3.  W.L. Barber and E.R. Brown, "A True Logarithmic Amplifier for Radar IF Applications," **IEEE Journal of Solid State Circuits**, Vol. SC-15, No. 3, June, 1980, pp. 291–295.

4.  **Broadband Amplifier Applications**, Plessey Co. Publication P.S. Norwood, MA., 1938, September, 1984.

5.  M.S. Gay, **SL521 Application Note**, Plessey Co., 1966.

6.  **Amplifier Applications Guide**, Analog Devices, Inc., Norwood, MA., 1992, Section 9.

7.  "Ask the Applications Engineer – 28 Logarithmic Amplifiers-Explained," **Analog Dialogue**, Vol. 33, No. 3, March, 1999.

8.  "Detecting Fast RF Bursts Using Log Amps," **Analog Dialogue**, Vol. 36, No. 5, September–October, 2002.

9.  R. Moghimi, "Log-Ratio Amplifier has Six-decade Dynamic Range," **EDN**, November, 2003.

# Tru-Power Detectors

In many systems, cellular phones as an example, monitoring of the transmit signal amplitude is required. The AD8362 is a true root mean square (RMS)-responding power detector that has a 60 dB measurement range (Figures 4-29 and 4-30). It is intended for use in a variety of high frequency communication systems and in instrumentation requiring an accurate response to signal power. It can operate from arbitrarily low frequencies to over 2.7 GHz and can accept inputs that have RMS values from 1 mV to at least $1 V_{RMS}$, with peak crest factors of up to 6, exceeding the requirements for accurate measurement of CDMA signals. Unlike earlier RMS-to-DC converters, the response bandwidth is completely independent of the signal magnitude. The $-3 dB$ point occurs at about 3.5 GHz.

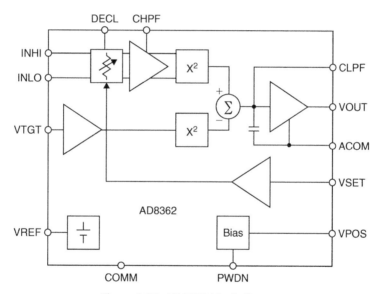

**Figure 4-29: AD8362 block diagram**

The input signal is applied to a resistive ladder attenuator that comprises the input stage of a variable gain amplifier (VGA). The 12-tap points are smoothly interpolated using a proprietary technique to provide a continuously variable attenuator, which is controlled by a voltage applied to the VSET pin. The resulting signal is applied to a high performance broadband amplifier. Its output is measured by an accurate square-law detector cell. The fluctuating output is then filtered and compared with the output of an identical squarer, whose input is a fixed DC voltage applied to the VTGT pin, usually the accurate reference of 1.25 V provided at the VREF pin.

The difference in the outputs of these squaring cells is integrated in a high gain error amplifier, generating a voltage at the $V_{OUT}$ pin with rail-to-rail capabilities. In a controller mode, this low noise output can be used to vary the gain of a host system's RF amplifier, thus balancing the setpoint against the input power.

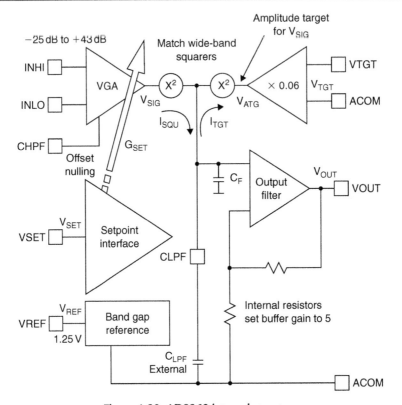

**Figure 4-30: AD8362 internal structure**

Optionally, the voltage at VSET may be a replica of the RF signal's AM, in which case the overall effect is to remove the modulation component prior to detection and lowpass filtering. The corner frequency of the averaging filter may be lowered without limit by adding an external capacitor at the CLPF pin.

The AD8362 can be used to determine the true power of a high frequency signal having a complex low FM envelope (or simply as a low frequency RMS voltmeter). The high pass corner generated by its offset-nulling loop can be lowered by a capacitor added on the CHPF pin (Figure 4-31).

Used as a power measurement device, $V_{OUT}$ is strapped to VSET, and the output is then proportional to the logarithm of the RMS value of the input; that is, the reading is presented directly in decibels, and is conveniently scaled 1 V/decade, that is, 50 mV/dB; other slopes are easily arranged. In controller modes, the voltage applied to VSET determines the power level required at the input to null the deviation from the setpoint. The output buffer can provide high load currents.

The AD8362 can be powered down by a logic high applied to the PWDN pin (i.e., the consumption is reduced to about 1.3 mW). It powers up within about 20 μs to its nominal operating current of 20 mA at 25°C.

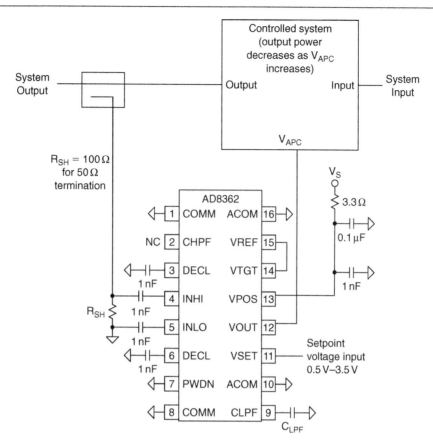

Figure 4-31: AD8362 typical application

# VGAs

## Voltage Controlled Amplifiers

Many monolithic VGAs use techniques that share common principles that are broadly classified as translinear, a term referring to circuit cells whose functions depend directly on the very predictable properties of BJTs, notably the linear dependence of their transconductance on collector current. Since the discovery of these cells in 1967, and their commercial exploitation in products developed during the early 1970s, accurate wide bandwidth analog multipliers, dividers, and VGAs have invariably employed translinear principles.

While these techniques are well understood, the realization of a high performance VGA requires special technologies and attention to many subtle details in its design. As an example, the AD8330 is fabricated on a proprietary silicon-on-insulator, complementary bipolar IC process and draws on decades of experience in developing many leading-edge products using translinear principles to provide an unprecedented level of versatility. Figure 4-32 shows a basic representative cell comprising just four transistors. This, or a very closely related form, is at the heart of most translinear multipliers, dividers, and VGAs. The key concepts are as follows: First, the ratio of the currents in the left-hand and right-hand pairs of transistors are identical; this is represented by the modulation factor, x, which may have values between $-1$ and $+1$. Second, the input signal is arranged to modulate the fixed tail current $I_D$ to cause the variable value of x introduced in the left-hand pair to be replicated in the right-hand pair, and thus generate the output by

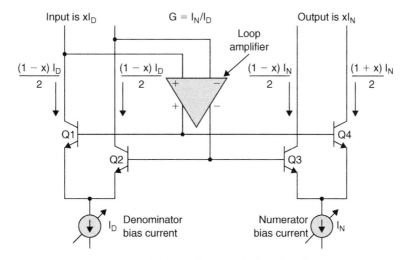

**Figure 4-32: Translinear variable gain cell**

modulating its nominally fixed tail current $I_N$. Third, the current gain of this cell is very exactly $G = I_N/I_D$ over many decades of variable bias current.

In practice, the realization of the full potential of this circuit involves many other factors, but these three elementary ideas remain essential. By varying $I_N$, the overall function is that of a two-quadrant analog multiplier, exhibiting a linear relationship to both the signal modulation factor x and this numerator current. On the other hand, by varying $I_D$, a two-quadrant analog divider is realized, having a hyperbolic gain function with respect to the input factor x, controlled by this denominator current. The AD8330 exploits both modes of operation. However, since a hyperbolic gain function is generally of less value than one in which the decibel gain is a linear function of a control input, a special interface is included to provide either increasing or decreasing exponential control of $I_D$.

The VGA core of the AD8330 (Figure 4-33) contains a much elaborated version of the cell shown in Figure 4-32. The current called $I_D$ is controlled exponentially (linear in decibels) through the decibel gain interface at the pin $V_{DBS}$ and its local common $C_{MGN}$. The gain span (i.e., the decibel difference between maximum and minimum values) provided by this control function is slightly more than 50 dB. The absolute gain from input to output is a function of source and load impedance and also depends on the voltage on a second gain—control pin, $V_{MAG}$.

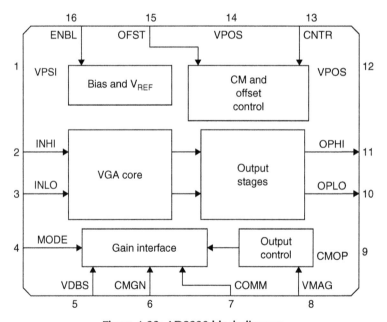

**Figure 4-33: AD8330 block diagram**

# X-AMP®

Most voltage controlled amplifiers (VCAs) made with analog multipliers have gain which is *linear in volts* with respect to the control voltage; moreover they tend to be noisy. There is a demand, however, for a VCA which combines a wide gain range with constant bandwidth and phase, low noise with large signal-handling capabilities, and low distortion with low power consumption, while providing accurate, stable, *linear-in-dB* gain. The X-AMP® family achieves these demanding and conflicting objectives with a unique and elegant solution (for *exponential amplifier*). The concept is simple: a fixed-gain amplifier follows a passive,

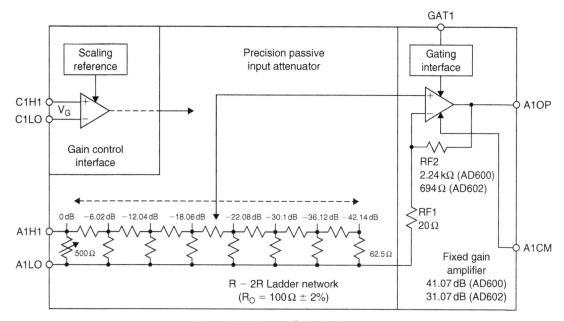

**Figure 4-34: X-AMP® block diagram**

broadband attenuator equipped with special means to alter its attenuation under the control of a voltage (see Figure 4-34). The amplifier is optimized for low input noise, and negative feedback is used to accurately define its moderately high gain (about 30–40 dB) and minimize distortion. Because this amplifier's gain is fixed, so also are its AC and transient response characteristics, including distortion and group delay; Because its gain is high, its input is never driven beyond a few millivolts. Therefore, it is always operating within its small signal response range.

The attenuator is a 7-section (8-tap) R–2R ladder network. The voltage ratio between all adjacent taps is exactly 2, or 6.02 dB. This provides the basis for the precise linear-in-dB behavior. The overall attenuation is 42.1 4 dB. As will be shown, the amplifier's input can be connected to any one of these taps, or even *interpolated* between them, with only a small deviation error of about ±0.2 dB. The overall gain can be varied all the way from the fixed (maximum) gain to a value 42.14 dB less. For example, in the AD600, the fixed gain is 41.07 dB (a voltage gain of 113); using this choice, the full gain range is −1.07 dB to +41.07 dB. The gain is related to the control voltage by the relationship $G_{dB} = 32V_G + 20$ where $V_G$ is in volts.

The gain at $V_G = 0$ is laser trimmed to an absolute accuracy of ±0.2 dB. The gain scaling is determined by an on-chip bandgap reference (shared by both channels), laser trimmed for high accuracy and low temperature coefficient. Figure 4-35 shows the gain versus the differential control voltage for both the AD600 and the AD602.

In order to understand the operation of the X-AMP® family, consider the simplified diagram shown in Figure 4-36. Notice that each of the eight taps is connected to an input of one of eight bipolar differential pairs, used as current controlled transconductance ($g_m$) stages; the other input of all these $g_m$ stages is connected to the amplifier's gain-determining feedback network, $R_{F1}/R_{F2}$. When the emitter bias current, $I_E$, is directed to one of the eight transistor pairs (by means not shown here), it becomes the input stage for the complete amplifier.

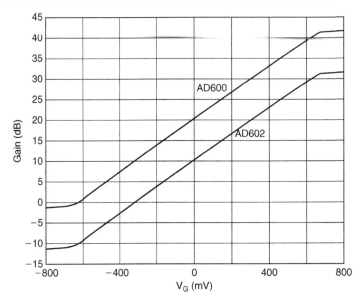

Figure 4-35: X-AMP® transfer function

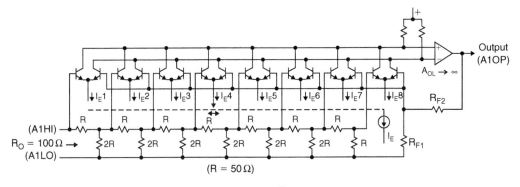

Figure 4-36: X-AMP® schematic

When $I_E$ is connected to the pair on the left-hand side, the signal input is connected directly to the amplifier, giving the maximum gain. The distortion is very low, even at high frequencies, due to the careful open-loop design, aided by the negative feedback. If $I_E$ were now to be abruptly switched to the second pair, the overall gain would drop by exactly 6.02 dB, and the distortion would remain low, because only one $g_m$ stage remains active.

In reality, the bias current is *gradually* transferred from the first pair to the second. When $I_E$ is equally divided between two $g_m$ stages, both are active, and the situation arises where we have an op amp with two input stages fighting for control of the loop, one getting the full signal and the other getting a signal exactly half as large.

Analysis shows that the effective gain is reduced, not by 3 dB, as one might first expect, but rather by 20 log 1.5, or 3.52 dB. This error, when divided equally over the whole range, would amount to a gain ripple of ±0.25 dB; however, the interpolation circuit actually generates a Gaussian distribution of bias currents, and a significant fraction of $I_E$ always flows in adjacent stages. This smoothes the gain function and actually lowers the ripple. As $I_E$ moves further to the right, the overall gain progressively drops.

The total input-referred noise of the X-AMP® is $1.4\,nV/\sqrt{Hz}$; only slightly more than the thermal noise of a $100\,\Omega$ resistor, which is $1.29\,nV/\sqrt{Hz}$ at $25°C$. The input-referred noise is constant regardless of the attenuator setting; therefore, the output noise is always constant and independent of gain.

The AD8367 is a high performance 45 dB VGA with linear-in-dB gain control for use from low frequencies up to several hundred megahertz (Figure 4-37). It includes an onboard detector which is used to build an automatic gain-controlled amplifier. The range, flatness, and accuracy of the gain response are achieved using Analog Devices' X-AMP® architecture, the most recent in a series of powerful proprietary concepts for variable gain applications, which far surpasses what can be achieved using competing techniques.

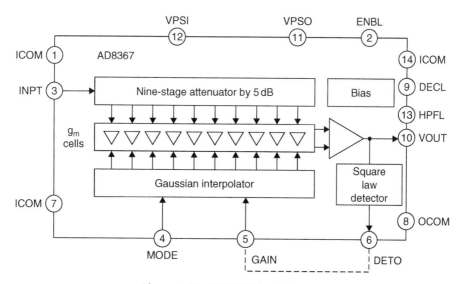

**Figure 4-37: AD8367 block diagram**

The input is applied to a $200\,\Omega$ resistive ladder network, having nine sections each of 5 dB loss, for a total attenuation of 45 dB. At maximum gain, the first tap is selected; at progressively lower gains, the tap moves smoothly and continuously toward higher attenuation values. The attenuator is followed by a 42.5 dB fixed-gain feedback amplifier—essentially an operational amplifier with a gain bandwidth product of 100 GHz—and is very linear, even at high frequencies. The output third-order intercept is +20 dBV at 100 MHz (+27 dBm re200 $\Omega$), measured at an output level of 1 Vp–p with $V_S = 5\,V$. The analog gain-control interface is very simple to use. It is scaled at 20 mV/dB, and the control voltage, $V_{GAIN}$, runs from 50 mV at $-2.5$ dB to 950 mV at +42.5 dB. In the inverse-gain mode of operation, selected by a simple pin-strap, the gain decreases from +42.5 dB at $V_{GAIN} = 50\,mV$ to $-2.5$ dB at $V_{GAIN} = 950\,mV$. This inverse mode is needed in AGC applications, which are supported by the integrated square-law detector, whose setpoint is chosen to level the output to $354\,mV_{RMS}$, regardless of the waveshape. A single external capacitor sets up the loop averaging time.

## Digitally Controlled VGAs

In some cases it may be advantageous to have the control of the signal level under digital control. The AD8370 is a low cost, digitally controlled VGA that provides precision gain control, high IP3, and low noise figure (Figure 4-38). The AD8370 has excellent distortion performance and wide bandwidth. For wide input, dynamic range applications, the AD8370 provides two input ranges: high gain mode and low gain mode. A Vernier 7-bit transconductance (Gm) stage provides 28 dB of gain range at better than 2 dB resolution, and

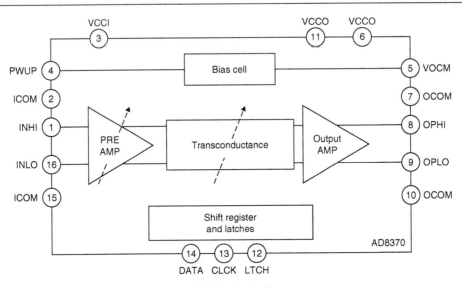

**Figure 4-38: AD8370 block diagram**

22 dB of gain range at better than 1 dB resolution. A second gain range, 17 dB higher than the first, can be selected to provide improved noise performance. The AD8370 is powered on by applying the appropriate logic level to the PWUP pin. When powered down, the AD8370 consumes less than 4 mA and offers excellent input to output isolation. The gain setting is preserved when operating in a power-down mode.

Gain control of the AD8370 is through a serial 8-bit gain-control word. The most significant bit (MSB) selects between the two gain ranges, and the remaining 7 bits adjust the overall gain in precise linear gain steps.

VGAs are also discussed in Chapter 2 (Sections 2-3 and 2-14).

## References: VGAs

1. B. Gilbert, "A Low Noise Wideband Variable-Gain Amplifier Using an Interpolated Ladder Attenuator," **IEEE ISSCC Technical Digest**, 1991, pp. 280, 281, 330.

2. B. Gilbert, ""A Monolithic Microsystem for Analog Synthesis of Trigonometric Functions and Their Inverses"," *IEEE Journal of Solid State Circuits*, Vol. SC-17, No. 6, December, 1982, pp. 1179–1191.

3. **Linear Design Seminar**, Analog Devices, 1995, Section 3.

4. E. Newman, "X-amp, A New 45-dB, 500-MHz Variable-Gain Amplifier (VGA) Simplifies Adaptive Receiver Designs," **Analog Dialogue**, Vol. 36, No. 1, January–February, 2002.

5. B. Gilbert and E. Nash, A 10.7 MHz, 120 dB Logarithmic Amp ... An extract from "Demodulating Logamps Bolster Wide-Dynamic-Range Measurements" **Microwaves and RF**, March, 1998.

6. S. Bonadio and E. Newman, "Variable Gain Amplifiers Enable Cost Effective IF Sampling Receiver Designs," **Microwave Product Digest**, October, 2003.

7. E. Newman and S. Lee, "Linear-in-dB Variable Gain Amplifier Provides True RMS Power nts," **Wireless Design 2004**.

8. P. Halford and E. Nash, "Integrated VGA Aids Precise Gain Control," **Microwaves & RF**, March, 2002.

# Direct Digital Synthesis

A frequency synthesizer generates multiple frequencies from one or more frequency references. These devices have been used for decades, especially in communications systems. Many are based on switching and mixing frequency outputs from a bank of crystal oscillators. Others have been based on well understood techniques utilizing phase-locked loops (PLLs). These will be discussed in the following section.

## DDS (Direct Digital Synthesis)

With the widespread use of digital techniques in instrumentation and communications systems, a digitally controlled method of generating multiple frequencies from a reference frequency source has evolved called direct digital synthesis (DDS). The basic architecture is shown in Figure 4-39. In this simplified model, a stable clock drives a programmable-read-only memory (PROM) which stores one or more integral number of cycles of a sinewave (or other arbitrary waveform, for that matter). As the address counter steps through each memory location, the corresponding digital amplitude of the signal at each location drives a digital-to-analog converter (DAC) which in turn generates the analog output signal. The spectral purity of the final analog output signal is determined primarily by the DAC. The phase noise is basically that of the reference clock.

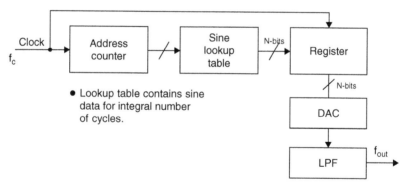

**Figure 4-39: Fundamental direct digital synthesis system**

Because a DDS system is a sampled data system, all the issues involved in sampling must be considered: quantization noise, aliasing, filtering, etc. For instance, the higher order harmonics of the DAC output frequencies fold back into the Nyquist bandwidth, making them unfilterable; whereas, the higher order harmonics of the output of PLL-based synthesizers can be filtered. There are other considerations which will be discussed shortly.

A fundamental problem with this simple DDS system is that the final output frequency can be changed only by changing the reference clock frequency or by reprogramming the PROM, making it rather inflexible. A practical DDS system implements this basic function in a much more flexible and efficient manner using digital hardware called a numerically controlled oscillator (NCO). A block diagram of such a system is shown in Figure 4-40.

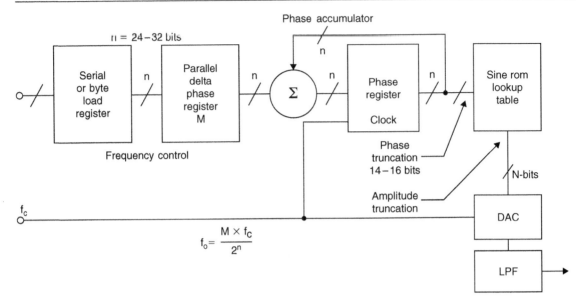

Figure 4-40: A flexible DDS system

The heart of the system is the *phase accumulator* whose content is updated once each clock cycle (Figure 4-41). Each time the phase accumulator is updated, the digital number, M, stored in the *delta phase register* is added to the number in the phase accumulator register. Assume that the number in the delta phase register is 00...01 and that the initial content of the phase accumulator is 00...00. The phase accumulator is updated by 00...01 on each clock cycle. If the accumulator is 32-bits wide, $2^{32}$ clock cycles (over 4 billion) are required before the phase accumulator returns to 00...00, and the cycle repeats.

The truncated output of the phase accumulator serves as the address to a sine (or cosine) lookup table. Each address in the lookup table corresponds to a phase point on the sinewave from 0° to 360°. The lookup table

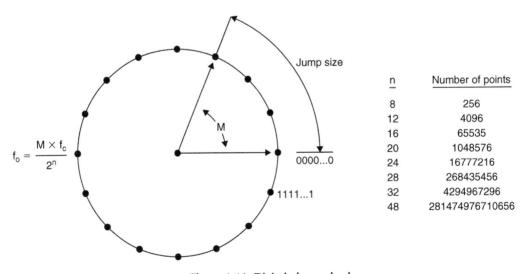

| n | Number of points |
|---|---|
| 8 | 256 |
| 12 | 4096 |
| 16 | 65535 |
| 20 | 1048576 |
| 24 | 16777216 |
| 28 | 268435456 |
| 32 | 4294967296 |
| 48 | 281474976710656 |

Figure 4-41: Digital phase wheel

282

contains the corresponding digital amplitude information for one complete cycle of a sinewave. (Actually, only data for 90° is required because the quadrature data are contained in the two MSBs.) The lookup table therefore maps the phase information from the phase accumulator into a digital amplitude word, which in turn drives the DAC.

Consider the case for n = 32, and M = 1. The phase accumulator steps through each of $2^{32}$ possible outputs before it overflows and restarts. The corresponding output sinewave frequency is equal to the input clock frequency divided by $2^{32}$. If M = 2, then the phase accumulator register "rolls over" twice as fast, and the output frequency is doubled. This can be generalized as follows.

For an n-bit phase accumulator (n generally ranges from 24 to 32 in most DDS systems), there are $2^n$ possible phase points. The digital word in the delta phase register, M, represents the amount the phase accumulator is incremented each clock cycle. If $f_c$ is the clock frequency, then the frequency of the output sinewave is equal to:

$$f_o = \frac{M \cdot f_c}{2^n} \tag{4-8}$$

This equation is known as the DDS "tuning equation." Note that the frequency resolution of the system is equal to $f_c/2^n$. For n = 32, the resolution is greater than one part in four billion! In a practical DDS system, all the bits out of the phase accumulator are not passed on to the lookup table but are truncated, leaving only the first 13–15 MSBs. This reduces the size of the lookup table and does not affect the frequency resolution. The phase truncation only adds a small but acceptable amount of phase noise to the final output.

The resolution of the DAC is typically 2–4 bits less than the width of the lookup table. Even a perfect N-bit DAC will add quantization noise to the output. Figure 4-42 shows the calculated output spectrum for a 32-bit phase accumulator, 15-bit phase truncation, and an ideal 12-bit DAC. The value of M was chosen so that the output frequency was slightly offset from 0.25 times the clock frequency. Note that the spurs caused by the phase truncation and the finite DAC resolution are all at least 90 dB below the full-scale output. This performance far exceeds that of any commercially available 12-bit DAC and is adequate for most applications.

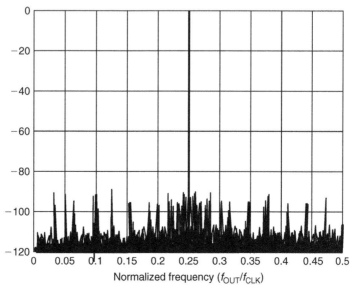

**Figure 4-42: Calculated output spectrum shows 90 dB SFDR for a 15-bit phase truncation and an ideal 12-bit DAC**

The basic DDS system described above is extremely flexible and has high resolution. The frequency can be changed instantaneously with no phase discontinuity by simply changing the contents of the M-register. However, practical DDS systems first require the execution of a serial, or byte-loading sequence to get the new frequency word into an internal buffer register which precedes the parallel-output M-register. This is done to minimize package pin count. After the new word is loaded into the buffer register, the parallel-output delta phase register is clocked, thereby changing all the bits simultaneously. The number of clock cycles required to load the delta phase buffer register determines the maximum rate at which the output frequency can be changed.

## Aliasing in DDS Systems

There is one important limitation to the range of output frequencies that can be generated from the simple DDS system. The Nyquist criteria state that the clock frequency (sample rate) must be at least twice the output frequency. Practical limitations restrict the actual highest output frequency to about 1/3 the clock frequency. Figure 4-43 shows the output of a DAC in a DDS system where the output frequency is 30 MHz and the clock frequency is 100 MHz. An antialiasing filter must follow the reconstruction DAC to remove the lower image frequency ($100 - 30 = 70$ MHz) as shown in Figure 4-43.

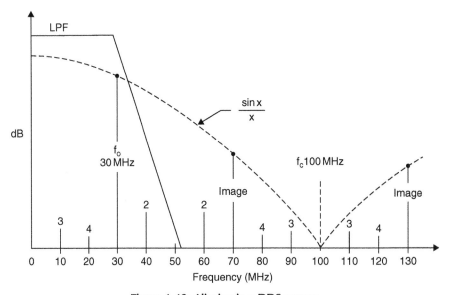

**Figure 4-43: Aliasing in a DDS system**

Note that the amplitude response of the DAC output (before filtering) follows a $\sin(x)/x$ response with zeros at the clock frequency and multiples thereof. The exact equation for the normalized output amplitude, $A(f_o)$, is given by:

$$A(f_o) = \frac{\sin\left(\dfrac{\pi f_o}{f_c}\right)}{\dfrac{\pi f_o}{f_c}} \tag{4-9}$$

where $f_o$ is the output frequency and $f_c$ is the clock frequency.

This rolloff is because the DAC output is not a series of zero-width impulses (as in a perfect re-sampler), but a series of rectangular pulses whose width is equal to the reciprocal of the update rate. The amplitude of the sin(x)/x response is down 3.92 dB at the Nyquist frequency (1/2 the DAC update rate). In practice, the transfer function of the reconstruction (antialiasing) filter can be designed to compensate for the sin(x)/x rolloff so that the overall frequency response is relatively flat up to the maximum output DAC frequency (generally 1/3 the update rate).

Another important consideration is that, unlike a PLL-based system, the higher order harmonics of the fundamental output frequency in a DDS system will fold back into the baseband because of aliasing. These harmonics cannot be removed by the antialiasing filter. For instance, if the clock frequency is 100 MHz, and the output frequency is 30 MHz, the second harmonic of the 30 MHz output signal appears at 60 MHz (out of band), but also at $100-60 = 40$ MHz (the aliased component). Similarly, the third harmonic (which would occur at 90 MHz) appears in band at $100-90 = 10$ MHz, and the fourth harmonic at $120-100$ MHz $= 20$ MHz. Higher order harmonics also fall within the Nyquist bandwidth (DC to $f_c/2$). The location of the first four harmonics is shown in Figure 4-43.

## DDS Systems as ADC Clock Drivers

DDS systems such as the AD9850 provide an excellent method of generating the sampling clock to the ADC, especially when the ADC sampling frequency must be under software control and locked to the system clock (see Figure 4-44). The *true* DAC output current $I_{out}$, drives a 200 $\Omega$, 42 MHz lowpass filter which is source and load terminated, thereby making the equivalent load 100 $\Omega$. The filter removes spurious frequency components above 42 MHz. The filtered output drives one input of the AD9850 internal comparator. The *complementary* DAC output current drives a 100 $\Omega$ load. The output of the 100 k$\Omega$ resistor divider placed between the two outputs is decoupled and generates the reference voltage for the internal comparator.

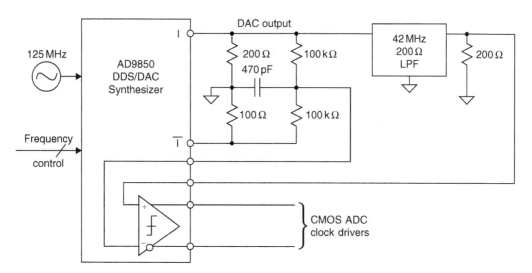

**Figure 4-44: Using a DDS system as an ADC clock driver**

The comparator output has a 2 ns rise and fall time and generates a TTL/CMOS-compatible square wave. The jitter of the comparator output edges is less than 20 ps RMS. True and complementary outputs are available if required.

In the circuit shown (Figure 4-44), the total output RMS jitter for a 40 MSPS ADC clock is 50 ps RMS, and the resulting degradation in SNR must be considered in wide dynamic range applications

## AM in a DDS System

AM in a DDS system can be accomplished by placing a digital multiplier between the lookup table and the DAC input as shown in Figure 4-45. Another method to modulate the DAC output amplitude is to vary the reference voltage to the DAC. In the case of the AD9850, the bandwidth of the internal reference control amplifier is approximately 1 MHz. This method is useful for relatively small output amplitude changes as long as the output signal does not exceed the $+1$ V compliance specification.

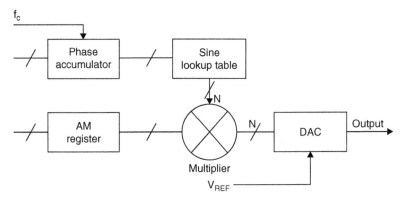

Figure 4-45: AM in a DDS System

## Spurious Free Dynamic Range Considerations in DDS Systems

In many DDS applications, the spectral purity of the DAC output is of primary concern. Unfortunately, the measurement, prediction, and analysis of this performance is complicated by a number of interacting factors.

Even an ideal N-bit DAC will produce harmonics in a DDS system. The amplitude of these harmonics is highly dependent on the ratio of the output frequency to the clock frequency. This is because the spectral content of the DAC quantization noise varies as this ratio varies, even though its theoretical RMS value remains equal to q/12 (where q is the weight of the least significant bit (LSB)). The assumption that the quantization noise appears as white noise and is spread uniformly over the Nyquist bandwidth is simply not true in a DDS system (it is more apt to be a true assumption in an ADC-based system, because the ADC adds a certain amount of noise to the signal which tends to "dither" or randomize the quantization error. However, a certain amount of correlation still exists). For instance, if the DAC output frequency is set to an exact submultiple of the clock frequency, then the quantization noise will be concentrated at multiples of the output frequency (i.e., it is highly signal dependent). If the output frequency is slightly offset, however, the quantization noise will become more random, thereby giving an improvement in the effective spurious free dynamic range (SFDR).

This is illustrated in Figure 4-46, where a 4096 (4 k) point Fourier transform (FFT) is calculated based on digitally generated data from an ideal 12-bit DAC. In the left-hand diagram, the ratio between the clock frequency and the output frequency was chosen to be exactly 32 (128 cycles of the sinewave in the FFT record length), yielding an SFDR of about 78 dBc. In the right-hand diagram, the ratio was changed to

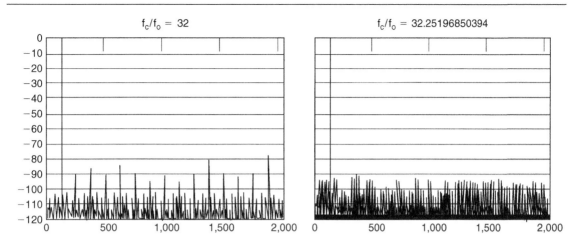

**Figure 4-46: Effect of ratio of clock to output frequency on theoretical 12-bit DAC SFDR using 4096-point FFT**

32.25196850394 (127 cycles of the sinewave within the FFT record length), and the effective SFDR is now increased to 92 dBc. In this ideal case, we observed a change in SFDR of 14 dB just by slightly changing the frequency ratio.

Best SFDR can therefore be obtained by the careful selection of the clock and output frequencies. However, in some applications, this may not be possible. In ADC-based systems, adding a small amount of random noise to the input tends to randomize the quantization errors and reduce this effect. The same thing can be done in a DDS system as shown in Figure 4-47 (Reference 5). The pseudo-random digital noise generator output is added to the DDS sine amplitude word before being loaded into the DAC. The amplitude of the digital noise is set to about 1/2 LSB. This accomplishes the randomization process at the expense of a slight increase in the overall output noise floor. In most DDS applications, however, there is enough flexibility in selecting the various frequency ratios so that dithering is not required.

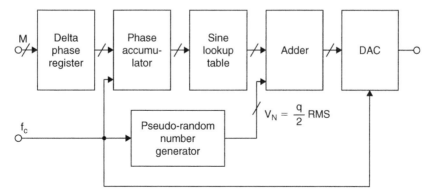

**Figure 4-47: Injection of digital dither in a DDS system to randomize quantization noise and increase SFDR**

## References: Direct Digital Synthesis

1.  "Ask the Application Engineer—33: All About Direct Digital Synthesis," **Analog Dialogue**, Vol. 38, August, 2004.

2.  "Innovative Mixed-Signal Chipset Targets Hybrid-Fiber Coaxial Cable Modems," **Analog Dialogue**, Vol. 31, No. 3, 1997.

3.  "Single-Chip Direct Digital Synthesis vs. the Analog PLL," **Analog Dialogue**, Vol. 30, No. 3, 1996.

4.  V. Kroupa (Ed.), **Direct Digital Frequency Synthesizers** Wiley-IEEE Press, 1998.

5.  D. Brandon, **DDS Design**, Analog Devices, Inc.

6.  **Jitter Reduction in DDS Clock Generator Systems** Copyright ©. Analog Devices, Inc.

7.  **A Technical Tutorial on Digital Signal Synthesis** Copyright © 1999 Analog Devices, Inc.

# PLLs

A PLL is a feedback system combining a voltage-controlled oscillator (VCO) and a phase comparator so connected that the oscillator maintains a constant phase angle relative to a reference signal. PLLs can be used, for example, to generate stable output frequency signals from a fixed low frequency signal. The PLL can be analyzed in general as a negative feedback system with a forward gain term and a feedback term. A simple block diagram of a voltage-based negative feedback system is shown in Figure 4-48.

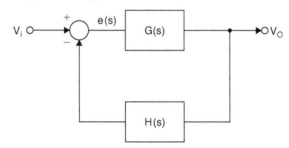

**Figure 4-48: Standard negative feedback control system model**

In a PLL, the error signal from the phase comparator is proportional to the relative phase of the input and feedback signals. The average output of the phase detector will be constant when the input and feedback signals are the same frequency. The usual equations for a negative feedback system apply.

$$\text{Forward Gain} = G(s) \tag{4-10}$$

$$s = j\omega = j2\pi f \tag{4-11}$$

$$\text{Closed Loop Gain} = \frac{G(s)}{1 + G(s)H(s)} \tag{4-12}$$

$$\text{Loop Gain} = G(s) \times H(s) \tag{4-13}$$

Because of the integration in the loop, at low frequencies the steady state gain, $G(s)$, is high and,

$$\frac{V_o}{V_b} \text{ Closed Loop Gain} = \frac{1}{H} \tag{4-14}$$

The components of a PLL that contribute to the loop gain include (Figure 4-49):

1. The phase detector (PD) and charge pump (CP).
2. The loop filter, with a transfer function of $Z(s)$.
3. The VCO, with a sensitivity of $KV/s$.
4. The feedback divider, $1/N$.

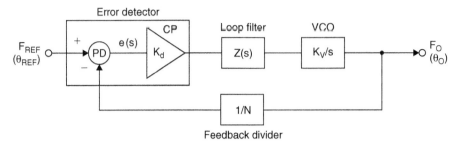

Figure 4-49: Basic phase-locked loop model

If a linear element like a four-quadrant multiplier is used as the phase detector, and the loop filter and VCO are also analog elements, this is called an analog, or *linear PLL* (LPLL). If a *digital* phase detector (Exclusive-or (EXOR) gate or J–K flip-flop) is used, and everything else stays the same, the system is called *a digital PLL* (DPLL). If the PLL is built exclusively from digital blocks, without any passive components or linear elements, it becomes an *all-digital PLL* (ADPLL).

In commercial PLLs, the phase detector and CP together form the error detector block. When $F_o \times N F_{REF}$, the error detector will output source/sink current pulses to the lowpass loop filter. This smoothes the current pulses into a voltage which in turn drives the VCO. The VCO frequency will then increase or decrease as necessary, by $K_v \times \Delta V$, where $K_v$ is the VCO sensitivity in MHz/V and $\Delta V$ is the change in VCO input voltage. This will continue until e(s) is zero and the loop is locked. The CP and VCO thus serves as an integrator, seeking to increase or decrease its output frequency to the value required so as to restore its input (from the phase detector) to zero.

The overall transfer function (CLG or closed-loop gain) of the PLL can be expressed simply by using the CLG expression for a negative feedback system as given above.

$$\frac{F_o}{F_{ref}} = \frac{\text{Forward Gain}}{1 + \text{Loop Gain}} \qquad (4\text{-}15)$$

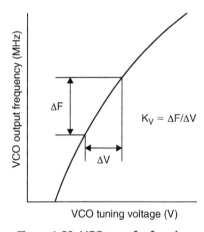

Figure 4-50: VCO transfer function

When GH is much greater than 1, we can say that the closed-loop transfer function for the PLL system is N and so:

$$F_{out} = N \times F_{REF} \qquad (4\text{-}16)$$

The loop filter is a lowpass type, typically with one pole and one zero. The transient response of the loop depends on:

1.  The magnitude of the pole/zero.

2.  The CP magnitude.

3.  The VCO sensitivity.

4.  The feedback factor, N.

All of the above must be taken into account when designing the loop filter. In addition, the filter must be designed to be stable (usually a phase margin of 90° is recommended). The 3-dB cutoff frequency of the response is usually called the loop bandwidth, BW. Large loop bandwidths result in very fast transient response. However, this is not always advantageous, since there is a tradeoff between fast transient response and reference spur attenuation.

## PLL Synthesizer Basic Building Blocks

A PLL synthesizer can be considered in terms of several basic building blocks. Already touched on, they will now be dealt with in greater detail:

Phase-Frequency Detector (PFD).

Reference Counter (R).

Feedback Counter (N).

The heart of a synthesizer is the phase detector—or PFD. This is where the reference frequency signal is compared with the signal fed back from the VCO output, and the resulting error signal is used to drive the loop filter and VCO. In a DPLL the phase detector or PFD is a logical element.

The three most common implementations are:

EXOR gate.

J–K Flip-Flop.

Digital Phase-Frequency Detector.

Here we will consider only the PFD, the element used in the ADF411X and ADF421X synthesizer families, because—unlike the EXOR gate and the J–K flip-flop—its output is a function of both the frequency difference and the phase difference between the two inputs when it is in the unlocked state. One implementation of a PFD, basically consisting of two D-type flip-flops is shown in Figure 4-51. One Q output enables a positive current source, and the other Q output enables a negative current source. Assuming that, in this design, the D-type flip-flop is positive-edge triggered, the states are these (Q1, Q2):

**11**—both outputs high, is disabled by the AND gate (U3) back to the CLR pins on the flip-flops.

**00**—both P1 and N1 are turned off and the output, OUT, is essentially in a high impedance state.

**10**—P1 is turned on, N1 is turned off, and the output is at V+.

**01**—P1 is turned off, N1 is turned on, and the output is at V−.

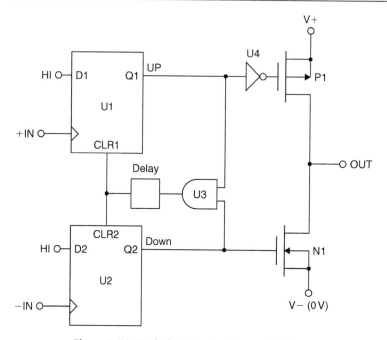

**Figure 4-51: Typical PFD using D-type flip-flops**

Consider now how the circuit behaves if the system is out of lock and the frequency at +IN is much higher than the frequency at −IN, as exemplified in Figure 4-52.

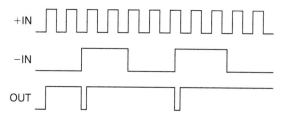

**Figure 4-52: PFD waveforms, out of frequency and phase lock**

Since the frequency at +IN is much higher than that at −IN, the output spends most of its time in the high state. The first rising edge on +IN forces the output high and this state is maintained until the first rising edge occurs on −IN. In a practical system this means that the output, and thus the input to the VCO, is driven higher, resulting in an increase in frequency at −IN. This is exactly what is desired. If the frequency on +IN were much lower than on −IN, the opposite effect would occur. The output at OUT would spend most of its time in the low condition. This would have the effect of driving the VCO in the negative direction and again bring the frequency at −IN much closer to that at +IN, to approach the locked condition. Figure 4-53 shows the waveforms when the inputs are frequency-locked and close to phase lock.

Since +IN is leading −IN, the output is a series of positive current pulses. These pulses will tend to drive the VCO so that the −IN signal become phase-aligned with that on +IN.

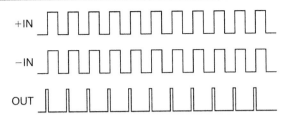

**Figure 4-53: PFD waveforms, in frequency lock but out of phase lock**

When this occurs, if there were no delay element between U3 and the CLR inputs of U1 and U2, it would be possible for the output to be in high impedance state, producing neither positive nor negative current pulses. This would not be a good situation. The VCO would drift until a significant phase error developed and started producing either positive or negative current pulses once again. Over a relatively long period of time, the effect of this cycling would be for the output of the CP to be modulated by a signal that is a subharmonic of the PFD input reference frequency. Since this could be a low frequency signal, it would not be attenuated by the loop filter and would result in very significant spurs in the VCO output spectrum, a phenomenon known as the *backlash* effect. The delay element between the output of U3 and the CLR inputs of U1 and U2 ensures that it does not happen. With the delay element, even when the $+IN$ and $-IN$ are perfectly phase-aligned, there will still be a current pulse generated at the CP output. The duration of this delay is equal to the delay inserted at the output of U3 and is known as the *antibacklash pulse width*.

## The Reference Counter

In the classical integer-N synthesizer, the resolution of the output frequency is determined by the reference frequency applied to the phase detector. So, for example, if 200 kHz spacing is required (as in GSM phones), then the reference frequency must be 200 kHz. However, getting a stable 200 kHz frequency source is not easy. A sensible approach is to take a good crystal-based high frequency source and divide it down. For example, the desired frequency spacing could be achieved by starting with a 10 MHz frequency reference and dividing it down by 50. This approach is shown in the diagram in Figure 4-54.

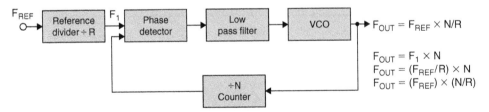

**Figure 4-54: Using a reference counter in a PLL synthesizer**

## The Feedback Counter, N

The N counter, also known as the N divider, is the programmable element that sets the relationship between the input and output frequencies in the PLL. The complexity of the N counter has grown over the years. In addition to a straightforward N counter, it has evolved to include a prescaler, which can have a dual modulus.

This structure has emerged as a solution to the problems inherent in using the basic divide-by-N structure to feed back to the phase detector when very high frequency outputs are required. For example, let's assume that a 900 MHz output is required with 10 kHz spacing. A 10 MHz reference frequency might be used, with the R divider set at 1,000. Then, the N-value in the feedback would need to be of the order of 90,000. This would mean at least a 17-bit counter capable of operating at an input frequency of 900 MHz.

To handle this range, it makes sense to precede the programmable counter with a fixed counter element to bring the very high input frequency down to a range at which standard CMOS counters will operate. This counter, called a *prescaler*, is shown in Figure 4-55.

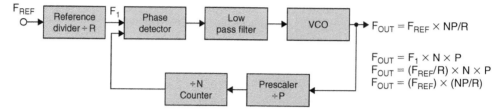

**Figure 4-55: Basic prescaler**

However, using a standard prescaler introduces other complications. The system resolution is now degraded ($F_1 \times P$). This issue can be addressed by using a dual-modulus prescaler (Figure 4-56). It has the advantages of the standard prescaler but without any loss in system resolution. A dual-modulus prescaler is a counter whose division ratio can be switched from one value to another by an external control signal. By using the dual-modulus prescaler with an A and B counter, one can still maintain output resolution of $F_1$.

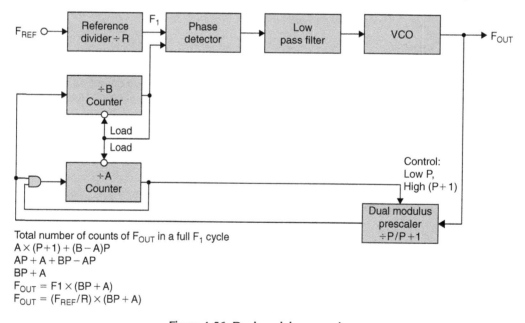

**Figure 4-56: Dual-modulus prescaler**

However, the following conditions must be met:

1.  The output signals of both counters are high if the counters have not timed out.

2.  When the B counter times out, its output goes low, and it immediately loads both counters to their preset values.

3.  The value loaded to the B counter must always be greater than that loaded to the A counter.

Assume that the B counter has just timed out and both counters have been reloaded with the values A and B. Let's find the number of VCO cycles necessary to get to the same state again.

As long as the A counter has not timed out, the prescaler is dividing down by P + 1. So, both the A and B counters will count down by 1 every time the prescaler counts (P+1) VCO cycles. This means the A counter will time out after ((P + 1) × A) VCO cycles.

At this point the prescaler is switched to divide-by-P. It is also possible to say that at this time the B counter still has (B − A) cycles to go before it times out. How long will it take to do this: ((B − A) × P). The system is now back to the initial condition where we started.

The total number of VCO cycles needed for this to happen is:

$$N = (A \times (P + 1)) + ((B - A) \times P) \tag{4-17}$$

$$= AP + A + BP - AP \tag{4-18}$$

$$= A + BP \tag{4-19}$$

When using a dual-modulus prescaler, it is important to consider the lowest and highest values of N. What we really want here is the range over which it is possible to change N in discrete integer steps. Consider the expression $N = A + BP$. To ensure a continuous integer spacing for N, A must be in the range 0 to (P − 1). Then, every time B is incremented there is enough resolution to fill in all the integer values between BP and (B + 1)P. As was already noted for the dual-modulus prescaler, B must be greater than or equal to A for the dual-modulus prescaler to work. From these we can say that the smallest division ratio possible while being able to increment in discrete integer steps is:

$$N_{MIN} = (B_{MIN} \times P) + A_{MIN} \tag{4-20}$$

$$= ((P - 1) \times P) + 0 \tag{4-21}$$

$$= P2 - P \tag{4-22}$$

The highest value of N is given by:

$$N_{MAX} = (B_{MAX} \times P) + A_{MAX} \tag{4-23}$$

In this case $A_{MAX}$ and $B_{MAX}$ are simply determined by the size of the A and B counters.

Now for a practical example with the ADF4111, let's assume that the prescaler is programmed to 32/33. The A counter is 6-bits wide, which means A can be $2^6 - 1 = 63$. The B counter is 13-bits wide, which means B can be $2^{13} - 1 = 8191$.

$$N_{MIN} = P2 - P = 992 \tag{4-24}$$

$$N_{MAX} = (B_{MAX} \times P) + A_{MAX} \tag{4-25}$$

$$\begin{aligned} &= (8191 \times 32) + 63 \\ &= 262175 \end{aligned} \tag{4-26}$$

## *Fractional-N Synthesizers*

Many of the emerging wireless communication systems have a need for faster switching and lower phase noise in the LO. Integer-N synthesizers require a reference frequency that is equal to the channel spacing. This can be quite low and thus necessitates a high N. This high N produces a phase noise that is proportionally high. The low reference frequency limits the PLL lock time. Fractional-N synthesis is a means of achieving both low phase noise and fast lock time in PLLs. The technique was originally developed in the early 1970s. This early work was done mainly by Hewlett Packard and Racal. The technique originally went by the name of "digiphase," but it later became popularly named fractional-N. In the standard synthesizer, it is possible to divide the RF signal by an integer only. This necessitates the use of a relatively low reference frequency (determined by the system channel spacing) and results in a high value of N in the feedback. Both of these facts have a major influence on the system settling time and the system phase noise. The low reference frequency means a long settling time, and the high value of N means larger phase noise.

If division by a fraction could occur in the feedback, it would be possible to use a higher reference frequency and still achieve the desired channel spacing. This lower fractional number would also mean lower phase noise.

In fact it is possible to implement division by a fraction over a long period of time by alternately dividing by two integers (divide by 2.5 can be achieved by dividing successively by 2 and 3). So, how does one divide by X or (X + 1) (assuming that the fractional number is between these two values)?

Well, the fractional part of the number can be allowed to accumulate at the reference frequency rate. Then every time the accumulator overflows, this signal can be used to change the N divide ratio. This is done in Figure 4-57 by removing one pulse being fed to the N counter. This effectively increases the divide ratio by one every time the accumulator overflows. Also, the bigger the number in the F-register, the more often the accumulator overflows and the more often division by the larger number occurs. This is exactly what is desired from the circuit. There are some added complications, however. The signal being fed to the phase detector from the divide-by-N circuit is not a uniform stream of regularly spaced pulses. Instead the pulses are being modulated at a rate determined by the reference frequency and the programmed fraction. This

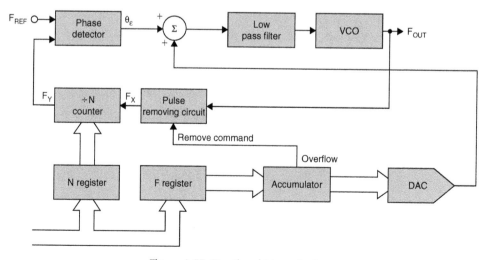

**Figure 4-57: Fractional-N synthesizer**

in turn modulates the phase detector output and drives the VCO input. The end result is a high spurious content at the output of the VCO. Major efforts are currently under way to minimize these spurs. Up to now, monolithic fractional-N synthesizers have failed to live up to expectations but the eventual benefits that may be realized mean that development is continuing at a rapid pace.

## Noise in Oscillator Systems

In any oscillator design, frequency stability is of critical importance. We are interested in both long-term and short-term stability. *Long-term* frequency stability is concerned with how the output signal varies over a long period of time (hours, days, or months). It is usually specified as the ratio, $\Delta f/f$ for a given period of time, expressed as a percentage or in dB. *Short-term* stability, on the other hand, is concerned with variations that occur over a period of seconds or less. These variations can be random or periodic. A spectrum analyzer can be used to examine the short-term stability of a signal. Figure 4-58 shows a typical spectrum, with random and discrete frequency components causing both a broad skirt and spurious peaks.

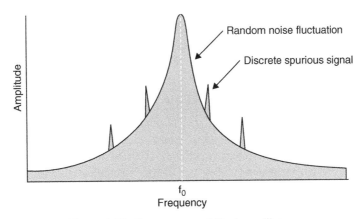

**Figure 4-58: Short-term stability in oscillators**

The discrete spurious components could be caused by known clock frequencies in the signal source, power line interference, and mixer products. The broadening caused by random noise fluctuation is due to *phase noise*. It can be the result of thermal noise, shot noise, and/or flicker noise in active and passive devices.

## Phase Noise in VCOs

Before we look at phase noise in a PLL system, it is worth considering the phase noise in a VCO. An ideal VCO would have no phase noise. Its output as seen on a spectrum analyzer would be a single spectral line. In practice, of course, this is not the case. There will be jitter on the output, and a spectrum analyzer would show phase noise. To help understand phase noise, consider a phasor representation, such as that shown in Figure 4-59.

A signal of angular velocity $\omega_O$ and peak amplitude $V_{SPK}$ is shown. Superimposed on this is an error signal of angular velocity $\omega_m$. $\Delta\theta_{RMS}$ represents the RMS value of the phase fluctuations and is expressed in RMS degrees.

In many radio systems, an overall integrated phase error specification must be met. This overall phase error is made up of the PLL phase error, the modulator phase error, and the phase error due to base band components. In GSM, for example, the total allowed is $5°$ RMS.

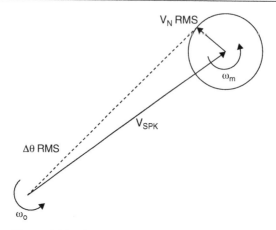

**Figure 4-59: Phasor representation of phase noise**

## Leeson's Equation

Leeson (see Reference 6) developed an equation to describe the different noise components in a VCO.

$$L_{PM} \sim 10 \log \left[ \frac{F\,kT}{A}\, \frac{1}{8Q_L}\left(\frac{f_o}{f_m}\right)^2 \right]$$
(4-27)

where $L_{PM}$ is single-sideband phase noise density (dBc/Hz), F is the device noise factor at operating power level A (linear), k is Boltzmann's constant, $1.38 \times 10-23\,\text{J/K}$, T is temperature (K), A is oscillator output power (W), $Q_L$ is loaded Q (dimensionless), $f_O$ is the oscillator carrier frequency, and $f_m$ is the frequency offset from the carrier.

For Leeson's equation to be valid, the following must be true:

- fm, the offset frequency from the carrier, is greater than the 1/f;
- flicker corner frequency;
- the noise factor at the operating power level is known;
- the device operation is linear;

Q includes the effects of component losses, device loading and buffer loading;

- a single resonator is used in the oscillator.

Leeson's equation only applies in the knee region between the break ($f_1$) to the transition from the "1/f" (more generally 1/fg) flicker noise frequency to a frequency beyond which amplified white noise dominates ($f_2$). This is shown in Figure 4-60 (g = 3). $f_1$ should be as low as possible; typically, it is less than 1 kHz, while $f_2$ is in the region of a few MHz. High performance oscillators require devices specially selected for low 1/f transition frequency.

Some guidelines to minimizing the phase noise in VCOs are:

1.  Keep the tuning voltage of the varactor sufficiently high (typically between 3 and 3.8 V).
2.  Use filtering on the DC voltage supply.

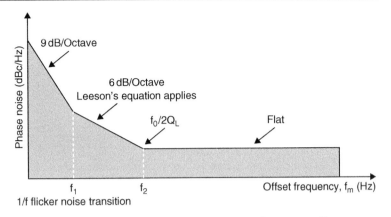

**Figure 4-60: Phase noise in a VCO versus frequency offset**

3. Keep the inductor Q as high as possible. Typical off-the-shelf coils provide a Q of between 50 and 60.

4. Choose an active device that has minimal noise figure as well as low flicker frequency. The flicker noise can be reduced by the use of feedback elements.

5. Most active device exhibit a broad U-shaped noise figure versus bias-current curve. Use this information to choose the optimal operating bias current for the device.

6. Maximize the average power at the tank circuit output.

7. When buffering the VCO, use devices with the lowest possible noise figure.

## Closing the Loop

Having looked at phase noise in a free-running VCO and how it can be minimized, we will now consider the effect of closing the loop on phase noise.

$$\text{Closed Loop Gain} = \frac{G}{1 + GH} \tag{4-28}$$

Figure 4-61 shows the main phase noise contributors in a PLL. The system transfer function may be described by the following equations.

$$G = \frac{K_d \times K_v \times Z(s)}{s} \tag{4-29}$$

$$H = \frac{1}{N} \tag{4-30}$$

For the discussion that follows, we will define $S_{REF}$ as the noise that appears on the reference input to the phase detector. It is dependent on the reference divider circuitry and the spectral purity of the main reference signal. $S_N$ is the noise due to the feedback divider appearing at the frequency input to the phase detector. $S_{CP}$ is the noise due to the phase detector (depending on its implementation). And $S_{VCO}$ is the phase noise of the VCO as described by equations developed earlier.

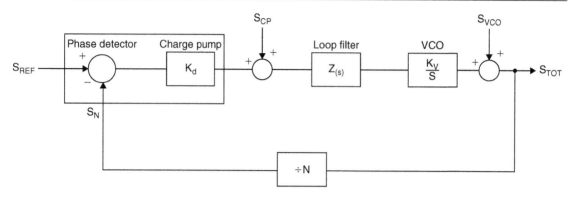

**Figure 4-61: PLL phase noise contributors**

$$\text{Closed Loop Gain} = \frac{\dfrac{K_d \times K_v \times Z(s)}{s}}{\dfrac{K_d \times K_v \times Z(s)}{N \times s}} \tag{4-31}$$

The overall phase noise performance at the output depends on the terms described above. All the effects at the output are added in an RMS fashion to give the total noise of the system. Thus:

$$S_{TOT}^2 = X^2 + Y^2 + Z^2 \tag{4-32}$$

where $S_{TOT}^2$ is the total phase noise power at the output, $X^2$ is the noise power at the output due to $S_N$ and $S_{REF}$, $Y^2$ is the noise power at the output due to $S_{CP}$, and $Z^2$ is the noise power at the output due to $S_{VCO}$.

The noise terms at the $P_D$ inputs, $S_{REF}$ and $S_N$, will be operated on in the same fashion as $F_{REF}$ and will be multiplied by the CLG of the system.

$$X^2 = (S_{REF}^2 + S_N^2) \times \left(\frac{G}{1 + GH}\right)^2 \tag{4-33}$$

At low frequencies, inside the loop bandwidth,

$$GH \gg 1 \tag{4-34}$$

and

$$X^2 = (S_{REF}^2 + S_N^2) \times N^2 \tag{4-35}$$

At high frequencies, outside the loop bandwidth,

$$GH \gg 1 \tag{4-36}$$

and

$$X^2 \rightarrow 0 \tag{4-37}$$

The overall output noise contribution due to the phase detector noise, $S_{CP}$, can be calculated by referencing $S_{CP}$ back to the input of the PFD. The equivalent noise at the PD input is $S_{CP}/K_d$. This is then multiplied by the CLG:

$$Y^2 = S_{CP}{}^2 \left(\frac{1}{K_d}\right)^2 \left(\frac{G}{1+GH}\right)^2 \tag{4-38}$$

Finally, the contribution of the VCO noise, $S_{VCO}$, to the output phase noise is calculated in a similar manner. The forward gain this time is simply 1. Therefore its contribution to the output noise is:

$$Z^2 = S_{TCO}{}^2 \left(\frac{1}{1+GH}\right)^2 \tag{4-39}$$

G, the forward loop gain of the closed-loop response, is usually a lowpass function; it is very large at low frequencies and small at high frequencies. H is a constant, 1/N. The denominator of the above expression is therefore lowpass, so $S_{VCO}$ is actually highpass filtered by the closed loop. A similar description of the noise contributors in a PLL/VCO can be found in Reference 1. Recall that the closed-loop response is a lowpass filter with a 3 dB cutoff frequency, BW, denoted the *loop bandwidth*. For frequency offsets at the output less than BW, the dominant terms in the output phase noise response are X and Y, the noise terms due to reference noise, N (counter noise), and CP noise.

Keeping $S_N$ and $S_{REF}$ to a minimum, keeping $K_d$ large and keeping N small will thus minimize the phase noise inside the loop bandwidth, BW. Because N programs the output frequency, it is not generally available as a factor in noise reduction. For frequency offsets much greater than BW, the dominant noise term is that due to the VCO, $S_{VCO}$. This is due to the highpass filtering of the VCO phase noise by the loop. A small value of BW would be desirable as it would minimize the total integrated output noise (phase error). However a small BW results in a slow transient response and increased contribution from the VCO phase noise inside the loop bandwidth. The loop bandwidth calculation therefore must tradeoff transient response and total output integrated phase noise.

To show the effect of closing the loop on a PLL, Figure 4-62 shows an overlay of the output of a free-running VCO and the output of a VCO as part of a PLL. Note that the in-band noise of the PLL has been attenuated compared to that of the free-running VCO.

## Phase Noise Measurement

One of the most common ways of measuring phase noise is with a high frequency spectrum analyzer (Figure 4-64). Figure 4-65 is a typical example of what would be seen.

With the spectrum analyzer we can measure the spectral density of phase fluctuations per unit bandwidth. VCO phase noise is best described in the frequency domain where the spectral density is characterized by measuring the noise sidebands on either side of the output signal center frequency. Phase noise power is specified in decibels relative to the carrier (dBc/Hz) at a given frequency offset from the carrier. The following equation describes this single side band (SSB) phase noise (dBc/Hz).

$$S_C(f) = 10 \log \frac{P_S}{P_{SSB}} \tag{4-40}$$

The 10 MHz, 0 dBm reference oscillator, available on the spectrum analyzer's rear-panel connector, has excellent phase noise performance. The R divider, N divider, and the phase detector are part of ADF4112

301

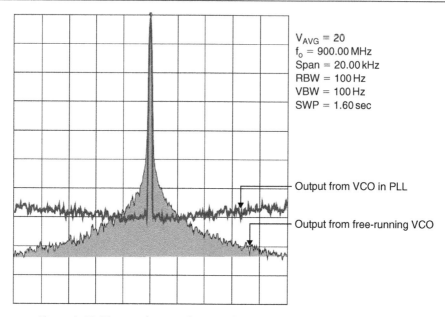

$V_{AVG} = 20$
$f_0 = 900.00\,MHz$
Span = 20.00 kHz
RBW = 100 Hz
VBW = 100 Hz
SWP = 1.60 sec

— Output from VCO in PLL

— Output from free-running VCO

Figure 4-62: Phase noise on a free-running VCO and a PLL connected VCO

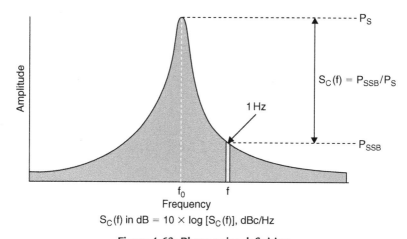

$P_S$

$S_C(f) = P_{SSB}/P_S$

1 Hz

$P_{SSB}$

Amplitude

$f_0$  f
Frequency

$S_C(f)$ in dB $= 10 \times \log [S_C(f)]$, dBc/Hz

Figure 4-63: Phase noise definition

frequency synthesizer. These dividers are programmed serially under the control of a PC. The frequency and phase noise performance are observed on the spectrum analyzer.

Figure 4-65 illustrates a typical phase noise plot of a PLL synthesizer using an ADF4112 PLL with a Murata VCO, MQE520-1880. The frequency and phase noise were measured in a 5 kHz span. The reference frequency used was $f_{REF} = 200\,kHz$ (R = 50) and the output frequency was 1880 MHz (N = 9400). If this were an ideal world PLL synthesizer, a single discrete tone would be displayed rising up above the spectrum analyzer's noise floor. What is displayed here is the tone, with the phase noise due to the loop components. The loop filter values were chosen to give a loop bandwidth of approximately 20 kHz. The flat part of the phase noise for frequency offsets less than the loop bandwidth is actually the

Test setup

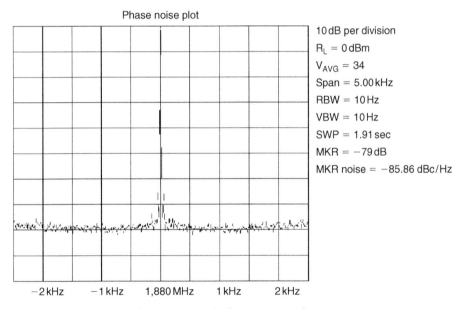

**Figure 4-64: Measuring phase noise with a spectrum analyzer**

Phase noise plot

10 dB per division

$R_L = 0$ dBm

$V_{AVG} = 34$

Span = 5.00 kHz

RBW = 10 Hz

VBW = 10 Hz

SWP = 1.91 sec

MKR = −79 dB

MKR noise = −85.86 dBc/Hz

−2 kHz    −1 kHz    1,880 MHz    1 kHz    2 kHz

**Figure 4-65: Typical spectrum-analyzer output**

phase noise as described by X2 and Y2 in the section "closing the loop" for cases where f is inside the loop bandwidth. It is specified at a 1 kHz offset. The value measured, the phase noise power in a 1 Hz bandwidth, was 85.86 dBc/Hz. It is made up of the following:

1.  Relative power in dBc between the carrier and the sideband noise at 1 kHz offset.

2.  The spectrum analyzer displays the power for a certain resolution bandwidth (RBW). In the plot, a 10 Hz RBW is used. To represent this power in a 1 Hz bandwidth, 10 log(RBW) must be subtracted from the value obtained from (1).

3.  A correction factor, which takes into account the implementation of the RBW, the log display mode and detector characteristic, must be added to the result obtained in (2).

4.    Phase noise measurement with the HP 8561E can be made quickly by using the marker noise function, MKR NOISE. This function takes into account the above three factors and displays the phase noise in dBc/Hz.

The phase noise measurement above is the total output phase noise at the VCO output. If we want to estimate the contribution of the PLL device (noise due to phase detector, R and N dividers, and the phase detector gain constant), the result must be divided by N2 (or $20 \times \log N$ be subtracted from the above result). This gives a phase noise floor of $(-85.86 - 20 \times \log(9400)) = -165.3\,\text{dBc/Hz}$.

## Reference Spurs

In an integer-N PLL (where the output frequency is an integer multiple of the reference input), reference spurs are caused by the fact that the CP output is being continuously updated at the reference frequency rate. Consider again the basic model for the PLL. This is shown again in Figure 4-66.

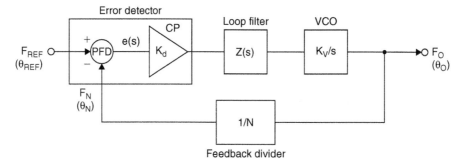

**Figure 4-66: Basic PLL model**

When the PLL is in lock, the phase and frequency inputs to the PFD ($f_{REF}$ and $f_N$) are essentially equal, and, in theory, one would expect that there to be no output from the PFD. However, this can create problems so the PFD is designed such that, in the locked condition, the current pulses from the CP will typically be as shown in Figure 4-67.

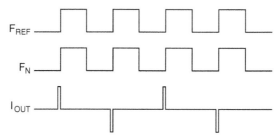

**Figure 4-67: Output current pulses from the PFD CP**

Although these pulses have a very narrow width, the fact that they exist means that the DC voltage driving the VCO is modulated by a signal of frequency $f_{REF}$. This produces *reference spurs* in the RF output occurring at offset frequencies that are integer multiples of $f_{REF}$. A spectrum analyzer can be used to detect reference spurs. Simply increase the span to greater than twice the reference frequency. A typical plot is shown in Figure 4-68.

In this case the reference frequency is 200 kHz and the diagram clearly shows reference spurs at ±200 kHz from the RF output of 1880 MHz. The level of these spurs is $-90\,\text{dB}$. If the span were increased to more than 4 times the reference frequency, we would also see the spurs at $(2 \times f_{REF})$.

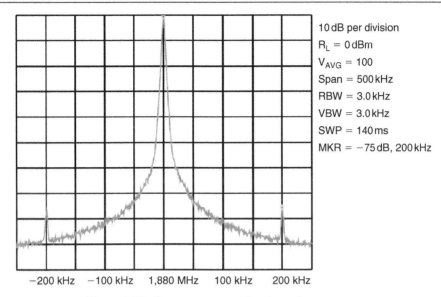

Figure 4-68: Output spectrum showing reference spurs

## CP Leakage Current

When the CP output from the synthesizer is programmed to the high impedance state, there should, in theory, be no leakage current flowing. In practice, in some applications the level of leakage current will have an impact on overall system performance. For example, consider an application where a PLL is used in open-loop mode for FM—a simple and inexpensive way of implementing FM that also allows higher data rates than modulating in closed-loop mode. For FM, a closed-loop method works fine but the data rate is limited by the loop bandwidth.

A system that uses open-loop modulation is the European cordless telephone system, DECT. The output carrier frequencies are in a range of 1.77–1.90 GHz and the data rate is high, 1.152 Mbps.

A block diagram of open-loop modulation is shown in Figure 4-69. The principle of operation is as follows: The loop is initially closed to lock the RF output, $f_{out} = N\, f_{REF}$. The modulating signal is turned on and at first the modulation signal is simply the DC mean of the modulation. The loop is then opened, by putting

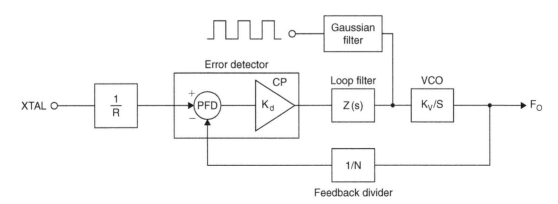

Figure 4-69: Block diagram of open-loop modulation

305

the CP output of the synthesizer into high impedance mode, and the modulation data are fed to the Gaussian filter. The modulating voltage then appears at the VCO where it is multiplied by $K_V$. When the data burst finishes, the loop is returned to the closed-loop mode of operation.

As the VCO usually has a high sensitivity (typical figures are between 20 and 80 MHz/V), any small voltage drift before the VCO will cause the output carrier frequency to drift. This voltage drift, and hence the system frequency drift, is directly dependent on the leakage current of the CP, when in the high impedance state. This leakage will cause the loop capacitor to charge or discharge depending on the polarity of the leakage current. For example, a leakage current of 1 nA would cause the voltage on the loop capacitor (1,000 pF for example) to charge or discharge by dV/dt = I/C (1 V/s in this case). This, in turn, would cause the VCO to drift. So, if the loop is open for 1 ms and the $K_V$ of the VCO is 50 MHz/V, the frequency drift caused by 1 nA leakage into a 1,000 pF loop capacitor would be 50 kHz. In fact, the DECT bursts are generally shorter (0.5 ms), so the drift will be even less in practice for the loop capacitance and leakage current used in the example. However, it does serve to illustrate the importance of charge pump leakage in this type of application.

### References: PLLs

1.  R.E. Best, *Phase-Locked Loops*, McGraw-Hill, New York, 1984.

2.  F.M. Gardner, *Phaselock Techniques,* 2nd Edition, John Wiley, New York, 1979.

3.  **Phase-Locked Loop Design Fundamentals**, *Applications Note AN-535*, Motorola, Inc.

4.  **The ARRL Handbook for Radio Amateurs**, American Radio Relay League, Newington, CT, 1992.

5.  R.J. Kerr and L.A. Weaver, "Pseudorandom Dither for Frequency Synthesis Noise," **United States Patent Number 4,901,265**, February 13, 1990.

6.  H.T. Nicholas, III and H. Samueli, "An Analysis of the Output Spectrum of Direct Digital Frequency Synthesizers in the Presence of Phase-Accumulator Truncation," **IEEE 41st Annual Frequency Control Symposium Digest of Papers**, 1987, pp. 495–502, IEEE Publication No. CH2427-3/87/0000-495.

7.  H.T. Nicholas, III and H. Samueli, "The Optimization of Direct Digital Frequency Synthesizer Performance in the Presence of Finite Word Length Effects," **IEEE 42nd Annual Frequency Control Symposium Digest of Papers**, 1988, pp. 357–363, IEEE Publication No. CH2588-2/88/0000-357.

8.  M. Curtin and P. O'Brien, "Phase-Locked Loops for High-Frequency Receivers and Transmitters–Part 1," **Analog Dialogue**, Vol. 33, No. 3, 1999.

9.  M. Curtin and P. O'Brien, "Phase-Locked Loops for High-Frequency Receivers and Transmitters–Part 2," **Analog Dialogue**, Vol. 33, No. 5, 1999.

10. M. Curtin and P. O'Brien, "Phase Locked Loops for High-Frequency Receivers and Transmitters—Part 3", **Analog Dialogue**, Vol. 33, No. 7, 1999.

11. **VCO Designers' Handbook**, Mini-Circuits Corporation, 1996.

12. L.W. Couch, **Digital and Analog Communications Systems**, Macmillan Publishing Company, New York, 1990.

13. P. Vizmuller, **RF Design Guide**, Artech House, 1995.

14. R.L. Best, **Phase Locked Loops: Design, Simulation and Applications**, 3rd Edition, McGraw-Hill, New York, 1997.

15. D.E. Fague, "Open Loop Modulation of VCOs for Cordless Telecommunications," **RF Design**, 1994.

16. D.B. Leeson, "A Simplified Model of Feedback Oscillator Noise Spectrum," **Proceedings of the IEEE**, Vol. 42, February, 1965, pp. 329–330.

## CHAPTER 5

# Fundamentals of Sampled Data Systems

## Chapter Introduction

To fully understand the specifications for converters it is beneficial to cover the fundamentals of sampling theory.

# Coding and Quantizing

Analog-to-digital converters (ADCs) translate analog measurements, which are characteristic of most phenomena in the "real world," to digital language, used in information processing, computing, data transmission, and control systems. Digital-to-analog converters (DACs) are used in transforming transmitted or stored data, or the results of digital processing, back to "real-world" variables for control, information display, or further analog processing. The relationships between inputs and outputs of ADCs and DACs are shown in Figure 5-1.

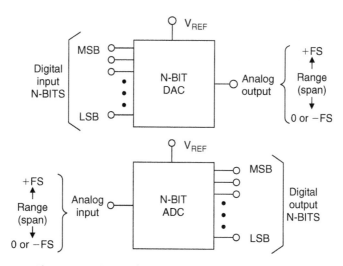

**Figure 5-1: ADC and DAC input and output definitions**

Analog input variables, whatever their origin, are most frequently converted by transducers into voltages or currents. These electrical quantities may appear as fast or slow "DC" continuous direct measurements of a phenomenon in the time domain, as modulated AC waveforms (using a wide variety of modulation techniques), or in some combination, with a spatial configuration of related variables to represent shaft angles. Examples of the first are outputs of thermocouples, potentiometers on DC references, and analog computing circuitry; of the second, "chopped" optical measurements, AC strain gauge or bridge outputs, and digital signals buried in noise; and of the third, synchros and resolvers.

The analog variables to be dealt with in this chapter are those involving voltages or currents representing the actual analog phenomena. They may be either wideband or narrowband. They may be either scaled from the direct measurement or subjected to some form of analog pre-processing, such as linearization, combination, demodulation, filtering, sample–hold, etc.

As part of the process, the voltages and currents are "normalized" to ranges compatible with assigned ADC input ranges. Analog output voltages or currents from DACs are direct and in normalized form, but they may be subsequently post-processed (e.g., scaled, filtered, amplified, etc.).

Information in digital form is normally represented by arbitrarily fixed voltage levels referred to "ground," either occurring at the outputs of logic gates or applied to their inputs. The digital numbers used are all basically binary; i.e., each "bit," or unit of information, has one of two possible states. These states are "off," "false," or "0," and "on," "true," or "1." It is also possible to represent the two logic states by two different levels of current; however, this is much less popular than using voltages. There is also no particular reason why the voltages need be referenced to ground—as in the case of emitter-coupled logic (ECL), positive-emitter-coupled logic (PECL), or low-voltage-differential-signaling logic (LVDS), for example.

*Words* are groups of levels representing digital numbers; the levels may appear simultaneously in *parallel*, on a bus or groups of gate inputs or outputs, *serially* (or in a time sequence) on a single line, or as a sequence of parallel bytes (i.e., "byte-serial") or nibbles (small bytes). For example, a 16-bit word may occupy the 16 bits of a 16-bit bus, or it may be divided into two sequential bytes for an 8-bit bus, or four 4-bit nibbles for a 4-bit bus.

A unique parallel or serial grouping of digital levels, or a *number*, or *code*, is assigned to each analog level which is quantized (i.e., represents a unique portion of the analog range). A typical digital code would be this array:

$$a7\ a6\ a5\ a4\ a3\ a2\ a1\ a0\ =\ 1\ 0\ 1\ 1\ 1\ 0\ 0\ 1$$

It is composed of 8 bits. The "1" at the extreme left is called the "most significant bit" (MSB, or Bit 1), and the one at the right is called the "least significant bit" (LSB, or Bit N: 8 in this case). The meaning of the code, as either a number, a character, or a representation of an analog variable, is unknown until the *code* and the *conversion relationship* have been defined. It is important not to confuse the designation of a particular bit (i.e., Bit 1, Bit 2, etc.) with the subscripts associated with the "a" array. The subscripts correspond to power of 2 associated with the weight of a particular bit in the sequence.

The best-known code is *natural or straight binary* (base 2). Binary codes are most familiar in representing integers; i.e., in a natural binary integer code having N bits, the LSB has a weight of $2^0$ (i.e., 1), the next bit has a weight of $2^1$ (i.e., 2), and so on up to the MSB, which has a weight of $2^{N-1}$ (i.e., $2^N/2$). The value of a binary number is obtained by adding up the weights of all non-zero bits. When the weighted bits are added up, they form a unique number having any value from 0 to $2^{N-1}$.

Often, for convenience, a binary number is expressing in *hexadecimal* (base 16). This reduces the length of the word and makes it easier to read. Figure 5-2 shows the relationship between binary and hexadecimal (commonly referred to as "hex").

| Binary | Hex | Binary | Hex |
|--------|-----|--------|-----|
| 0000 | 0 | 1000 | 8 |
| 0001 | 1 | 1001 | 9 |
| 0010 | 2 | 1010 | A |
| 0011 | 3 | 1011 | B |
| 0100 | 4 | 1100 | C |
| 0101 | 5 | 1101 | D |
| 0110 | 6 | 1110 | E |
| 0111 | 7 | 1111 | F |

**Figure 5-2: The relationship between binary and hexadecimal**

In converter technology, full-scale (abbreviated *FS*) is independent of the number of bits of resolution, N. A more useful coding is *fractional* binary which is always normalized to full-scale. Integer binary can be interpreted as fractional binary if all integer values are divided by $2^N$. For example, the MSB has a weight of 1/2 (i.e., $2^{(N-1)}/2^N = 2^{-1}$), the next bit has a weight of 1/4 (i.e., $2^{-2}$), and so forth down to the LSB, which has a weight of $1/2^N$ (i.e., $2^{-N}$). When the weighted bits are added up, they form a number with any of $2^N$ values, from 0 to $(1-2^{-N})$ of full-scale. Additional bits simply provide more fine structure without affecting full-scale range. The relationship between base 10 numbers and binary numbers (base 2) are shown in Figure 5-3 along with examples of each.

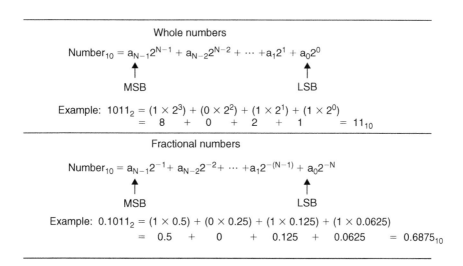

**Figure 5-3: Representing a base 10 number with a binary number (base 2)**

## Unipolar Codes

In data conversion systems, the coding method must be related to the analog input range (or span) of an ADC or the analog output range (or span) of a DAC. The simplest case is when the input to the ADC or the output of the DAC is always a unipolar positive voltage (current outputs are very popular for DAC outputs, much less for ADC inputs). The most popular code for this type of signal is *straight binary* and is shown in Figure 5-4 for a 4-bit converter. Notice that there are 16 distinct possible levels, ranging from the all-zeros code 0000, to the all-ones code 1111. It is important to note that the analog value represented by the all-ones code is not full-scale (abbreviated FS), but FS − 1 LSB. This is a common convention in data conversion notation and applies to both ADCs and DACs. Figure 5-4 gives the base 10 equivalent number, the value of the base 2 binary code relative to full-scale (FS), and also the corresponding voltage level for each code (assuming a +10V full-scale converter).

Figure 5-5 shows the transfer function for an ideal 3-bit DAC with straight binary input coding. Notice that the analog output is zero for the all-zeros input code. As the digital input code increases, the analog output increases 1 LSB (1/8 scale in this example) per code. The most positive output voltage is 7/8 FS, corresponding to a value equal to FS − 1 LSB. The mid-scale output of 1/2 FS is generated when the digital input code is 100.

The transfer function of an ideal 3-bit ADC is shown in Figure 5-6. There is a range of analog input voltage over which the ADC will produce a given output code, and this range is the *quantization uncertainty* and is equal to 1 LSB. Note that the width of the transition regions between adjacent codes is zero for an ideal ADC. In practice, however, there is always transition noise associated with these levels, and therefore the width is non-zero. It is customary to define the analog input corresponding to a given code by the *code center* which lies halfway between two adjacent transition regions (illustrated by the black dots in the diagram). This requires that the first transition region occur at 1/2 LSB. The full-scale analog input voltage is defined by 7/8 FS (FS − 1 LSB).

| Base 10 number | Scale | +10 V FS | Binary |
|---|---|---|---|
| +15 | +FS − 1 LSB = 15/16 FS | 9.375 | 1111 |
| +14 | +7/8 FS | 8.750 | 1110 |
| +13 | +13/16 FS | 8.125 | 1101 |
| +12 | +3/4 FS | 7.500 | 1100 |
| +11 | +11/16 FS | 6.875 | 1011 |
| +10 | +5/16 FS | 6.250 | 1010 |
| +9 | +9/16 FS | 5.625 | 1001 |
| +8 | +1/2 FS | 5.000 | 1000 |
| +7 | +7/16 FS | 4.375 | 0111 |
| +6 | +3/8 FS | 3.750 | 0110 |
| +5 | +5/16 FS | 3.125 | 0101 |
| +4 | +1/4 FS | 2.500 | 0100 |
| +3 | +3/16 FS | 1.875 | 0011 |
| +2 | +1/8 FS | 1.250 | 0010 |
| +1 | 1 LSB = +1/16 FS | 0.625 | 0001 |
| 0 | 0 | 0.000 | 0000 |

Figure 5-4: Unipolar binary code, 4-bit converter

## Bipolar Codes

In many systems, it is desirable to represent both positive and negative analog quantities with binary codes. Either *offset binary*, *twos complement*, *ones complement*, or *sign-magnitude* codes will accomplish this, but

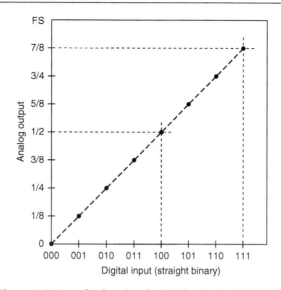

**Figure 5-5: Transfer function for ideal unipolar 3-bit DAC**

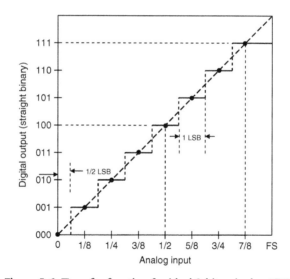

**Figure 5-6: Transfer function for ideal 3-bit unipolar ADC**

offset binary and twos complement are by far the most popular. The relationships between these codes for a 4-bit system are shown in Figure 5-7. Note that the values are scaled for a ±5V full-scale input/output voltage range.

For *offset binary*, the zero signal value is assigned the code 1000. The sequence of codes is identical to that of straight binary. The only difference between a straight and offset binary system is the half-scale offset

*313*

| Base 10 number | Scale | | ±5 V FS | Offset binary | Twos complement | Ones complement | Sign Magnitude |
|---|---|---|---|---|---|---|---|
| +7 | +FS − 1LSB = | +7/8 FS | +4.375 | 1 1 1 1 | 0 1 1 1 | 0 1 1 1 | 0 1 1 1 |
| +6 | | +3/4 FS | +3.750 | 1 1 1 0 | 0 1 1 0 | 0 1 1 0 | 0 1 1 0 |
| +5 | | +5/8 FS | +3.125 | 1 1 0 1 | 0 1 0 1 | 0 1 0 1 | 0 1 0 1 |
| +4 | | +1/2 FS | +2.500 | 1 1 0 0 | 0 1 0 0 | 0 1 0 0 | 0 1 0 0 |
| +3 | | +3/8 FS | +1.875 | 1 0 1 1 | 0 0 1 1 | 0 0 1 1 | 0 0 1 1 |
| +2 | | +1/4 FS | +1.250 | 1 0 1 0 | 0 0 1 0 | 0 0 1 0 | 0 0 1 0 |
| +1 | | +1/8 FS | +0.625 | 1 0 0 1 | 0 0 0 1 | 0 0 0 1 | 0 0 0 1 |
| 0 | | 0 | 0.000 | 1 0 0 0 | 0 0 0 0 | *0 0 0 0 | *1 0 0 0 |
| −1 | | −1/8 FS | −0.625 | 0 1 1 1 | 1 1 1 1 | 1 1 1 0 | 1 0 0 1 |
| −2 | | −1/4 FS | −1.250 | 0 1 1 0 | 1 1 1 0 | 1 1 0 1 | 1 0 1 0 |
| −3 | | −3/8 FS | −1.875 | 0 1 0 1 | 1 1 0 1 | 1 1 0 0 | 1 0 1 1 |
| −4 | | −1/2 FS | −2.500 | 0 1 0 0 | 1 1 0 0 | 1 0 1 1 | 1 1 0 0 |
| −5 | | −5/8 FS | −3.125 | 0 0 1 1 | 1 0 1 1 | 1 0 1 0 | 1 1 0 1 |
| −6 | | −3/4 FS | −3.750 | 0 0 1 0 | 1 0 1 0 | 1 0 0 1 | 1 1 1 0 |
| −7 | −FS + 1LSB = | −7/8 FS | −4.375 | 0 0 0 1 | 1 0 0 1 | 1 0 0 0 | 1 1 1 1 |
| −8 | | −FS | −5.000 | 0 0 0 0 | 1 0 0 0 | | |

|  |  | Ones complement | Sign Magnitude |
|---|---|---|---|
| * | 0+ | 0 0 0 0 | 0 0 0 0 |
|  | 0− | 1 1 1 1 | 1 0 0 0 |

Codes not normally used in computations (see text)

**Figure 5-7: Bipolar codes, 4-bit converter**

associated with analog signal. The most negative value (−FS + 1 LSB) is assigned the code 0001, and the most positive value (+FS − 1 LSB) is assigned the code 1111. Note that in order to maintain perfect symmetry about mid-scale, the all-zeros code (0000) representing negative full-scale (−FS) is not normally used in computation. It can be used to represent a negative off-range condition or simply assigned the value of the 0001 (−FS + 1 LSB).

The relationship between the offset binary code and the analog output range of a bipolar 3-bit DAC is shown in Figure 5-8. The analog output of the DAC is zero for the zero-value input code 100. The most negative output voltage is generally defined by the 001 code (−FS + 1 LSB), and the most positive by 111 (+FS − 1 LSB). The output voltage for the 000 input code is available for use if desired but makes the output non-symmetrical about zero and complicates the mathematics.

The offset binary output code a bipolar 3-bit ADC as a function of its analog input is shown in Figure 5-9. Note that zero analog input defines the center of the mid-scale code 100. As in the case of bipolar DACs, the most negative input voltage is generally defined by the 001 code (−FS + 1 LSB), and the most positive by 111 (+FS − 1 LSB). As discussed above, the 000 output code is available for use if desired but makes the output non-symmetrical about zero and complicates the mathematics.

*Twos complement* is identical to offset binary with the MSB complemented (inverted). This is obviously very easy to accomplish in a data converter, using a simple inverter or taking the complementary output of a "D" flip-flop. The popularity of twos complement coding lies in the ease with which mathematical operations can be performed in computers. Twos complement, for conversion purposes, consists of a binary code for positive magnitudes (0 sign bit), and the twos complement of each positive number to represent its

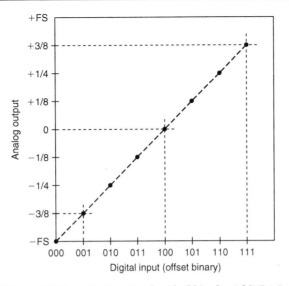

**Figure 5-8: Transfer function for ideal bipolar 3-bit DAC**

negative. The twos complement is formed arithmetically by complementing the number and adding 1 LSB. For example, −3/8 FS is obtained by taking the twos complement of +3/8 FS. This is done by first complementing +3/8 FS, 0011 obtaining 1100. Adding 1 LSB, we obtain 1101.

Twos complement makes subtraction easy. For example, to subtract 3/8 FS from 4/8 FS, add 4/8 to −3/8, or 0100 to 1101. The result is 0001, disregarding the extra carry, or 1/8.

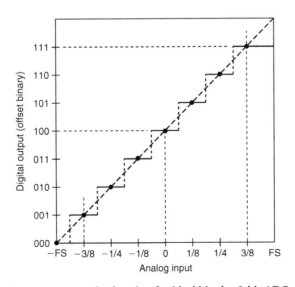

**Figure 5-9: Transfer function for ideal bipolar 3-bit ADC**

*Ones complement* can also be used to represent negative numbers, although it is much less popular than twos complement and rarely used today. The ones complement is obtained by simply complementing all of a positive numbers digits. For instance, the ones complement of 3/8 FS (0011) is 1100. A ones complemented code can be formed by complementing each positive value to obtain its corresponding negative value. This includes zero, which is then represented by either of two codes, 0000 (referred to as 0+) and 1111 (referred to as 0−). This ambiguity must be dealt with mathematically, and presents obvious problems relating to ADCs and DACs for which there is a single code which represents zero.

*Sign magnitude* would appear to be the most straightforward way of expressing signed analog quantities digitally. Simply determine the code appropriate for the magnitude and add a polarity bit. Sign-magnitude binary-coded decimal (BCD) is popular in bipolar digital voltmeters (DVMs), but has the problem of two allowable codes for zero. It is therefore unpopular for most applications involving ADCs or DACs.

Figure 5-10 summarizes the relationships between the various bipolar codes: offset binary, twos complement, ones complement, and sign magnitude and shows how to convert between them.

The last code to be considered in this section is *BCD*, where each base 10 digit (0 to 9) in a decimal number is represented as the corresponding 4-bit straight binary word as shown in Figure 5-11. The minimum digit 0 is represented as 0000, and the digit 9 by 1001. This code is relatively inefficient, since only 10 of the 16 code states for each decade are used. It is, however, a very useful code for interfacing to decimal displays such as in DVMs.

| To Convert From To↓ | Sign Magnitude | Twos Complement | Offset Binary | Ones Complement |
|---|---|---|---|---|
| Sign Magnitude | No change | If MSB = 1, complement other bits, add 00...01 | Complement MSB If new MSB = 1, complement other bits, add 00...01 | If MSB = 1, complement other bits |
| Twos Complement | If MSB = 1, complement other bits, add 00...01 | No change | Complement MSB | If MSB = 1, add 00...01 |
| Offset Binary | Complement MSB If new MSB = 0, complement other bits, add 00...01 | Complement MSB | No change | Complement MSB If new MSB = 0, add 00...01 |
| Ones Complement | If MSB = 1, complement other bits | If MSB = 1, add 11...11 | Complement MSB If new MSB = 1, add 11...11 | No change |

Figure 5-10: Relationships among bipolar codes

## Complementary Codes

Some forms of data converters (e.g., early DACs using monolithic NPN quad current switches) require standard codes such as natural binary or BCD, but with all bits represented by their complements. Such

| Base 10 number | Scale | +10 V FS | Binary |
|---:|---:|---:|---|
| +15 | +FS – 1 LSB = 15/16 FS | 9.375 | 1111 |
| +14 | +7/8 FS | 8.750 | 1110 |
| +13 | +13/16 FS | 8.125 | 1101 |
| +12 | +3/4 FS | 7.500 | 1100 |
| +11 | +11/16 FS | 6.875 | 1011 |
| +10 | +5/16 FS | 6.250 | 1010 |
| +9 | +9/16 FS | 5.625 | 1001 |
| +8 | +1/2 FS | 5.000 | 1000 |
| +7 | +7/16 FS | 4.375 | 0111 |
| +6 | +3/8 FS | 3.750 | 0110 |
| +5 | +5/16 FS | 3.125 | 0101 |
| +4 | +1/4 FS | 2.500 | 0100 |
| +3 | +3/16 FS | 1.875 | 0011 |
| +2 | +1/8 FS | 1.250 | 0010 |
| +1 | 1 LSB = +1/16 FS | 0.625 | 0001 |
| 0 | 0 | 0.000 | 0000 |

**Figure 5-11: BCD code**

codes are called *complementary codes*. All the codes discussed thus far have complementary codes which can be obtained by this method.

In a 4-bit complementary-binary converter, 0 is represented by 1111, half-scale by 0111, and FS − 1 LSB by 0000. In practice, the complementary code can usually be obtained by using the complementary output of a register rather than the true output, since both are available.

Sometimes the complementary code is useful in inverting the analog output of a DAC. Today many DACs provide differential outputs which allow the polarity inversion to be accomplished without modifying the input code. Similarly, many ADCs provide differential logic inputs which can be used to accomplish the polarity inversion.

## DAC and ADC Static Transfer Functions and DC Errors

The most important thing to remember about both DACs and ADCs is that either the input or output is digital, and therefore the signal is quantized. That is, an N-bit word represents one of $2^N$ possible states, and therefore an N-bit DAC (with a fixed reference) can have only $2^N$ possible analog outputs, and an N-bit ADC can have only $2^N$ possible digital outputs. As previously discussed, the analog signals will generally be voltages or currents.

The resolution of data converters may be expressed in several different ways: the weight of the least significant bit (LSB), parts per million of full-scale (ppm FS), millivolts (mV), etc. Different devices (even from the same manufacturer) will be specified differently, so converter users must learn to translate between the different types of specifications if they are to compare devices successfully. The size of the LSB for various resolutions is shown in Figure 5-12.

Before we can consider the various architectures used in data converters, it is necessary to consider the performance to be expected, and the specifications which are important. The following sections will consider the definition of errors and specifications used for data converters. This is important in understanding the strengths and weaknesses of different ADC/DAC architectures.

Figure 5-13 shows the ideal transfer characteristics for a 3-bit unipolar DAC and a 3-bit unipolar ADC. In a DAC, both the input and the output are quantized, and the graph consists of eight points—while it is

| Resolution N | $2^N$ | Voltage (10 V FS) | ppm FS | % FS | dB FS |
|---|---|---|---|---|---|
| 2-bit | 4 | 2.5 V | 250,000 | 25 | −12 |
| 4-bit | 16 | 625 mV | 62,500 | 6.25 | −24 |
| 6-bit | 64 | 156 mV | 15,625 | 1.56 | −36 |
| 8-bit | 256 | 39.1 mV | 3,906 | 0.39 | −48 |
| 10-bit | 1,024 | 9.77 mV (10 mV) | 977 | 0.098 | −60 |
| 12-bit | 4,096 | 2.44 mV | 244 | 0.024 | −72 |
| 14-bit | 16,384 | 610 µV | 61 | 0.0061 | −84 |
| 16-bit | 65,536 | 153 µV | 15 | 0.0015 | −96 |
| 18-bit | 262,144 | 38 µV | 4 | 0.0004 | −108 |
| 20-bit | 1,048,576 | 9.54 µV (10 µV) | 1 | 0.0001 | −120 |
| 22-bit | 4,194,304 | 2.38 µV | 0.24 | 0.000024 | −132 |
| 24-bit | 16,777,216 | 596 nV* | 0.06 | 0.000006 | −144 |

*600 nV is the Johnson Noise in a 10 kHz BW of a 2.2 kΩ Resistor @ 25°C

Remember: 10-bits and 10 V FS yields an LSB of 10 mV, 1000 ppm, or 0.1%.
All other values may be calculated by powers of 2.

**Figure 5-12: Quantization: the size of an LSB**

reasonable to discuss the line through these points, it is very important to remember that the actual transfer characteristic is *not* a line, but a number of discrete points.

The input to an ADC is analog and is not quantized, but its output is quantized. The transfer characteristic therefore consists of eight horizontal steps. When considering the offset, gain, and linearity of an ADC we consider the line joining the midpoints of these steps—often referred to as the *code centers*.

For both DACs and ADCs, digital full-scale (all "1"s) corresponds to 1 LSB below the analog full-scale (FS). The (ideal) ADC transitions take place at 1/2 LSB above zero, and thereafter every LSB, until 1 1/2 LSB below analog full-scale. Since the analog input to an ADC can take any value, but the digital output is quantized, there may be a difference of up to 1/2 LSB between the actual analog input and the exact value of the digital output. This is known as the *quantization error* or *quantization uncertainty* as shown in Figure 5-15. In AC (sampling) applications this quantization error gives rise to *quantization noise* which will be discussed in Section 5-2 of this chapter.

As previously discussed, there are many possible digital coding schemes for data converters: *straight binary, offset binary, ones complement, twos complement, sign-magnitude, gray code, BCD*, and others. This section, being devoted mainly to the *analog* issues surrounding data converters, will use simple *binary* and *offset binary* in its examples and will not consider the merits and disadvantages of these, or any other forms of digital code.

The examples in Figure 5-13 use *unipolar* converters, whose analog port has only a single polarity. These are the simplest type, but *bipolar* converters are generally more useful in real-world applications.

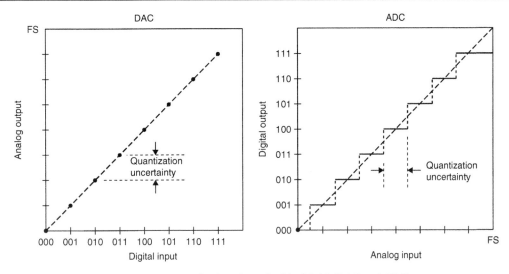

**Figure 5-13: Transfer functions for ideal 3-bit DAC and ADC**

There are two types of bipolar converters: the simpler is merely a unipolar converter with an accurate 1 MSB of negative offset (and many converters are arranged so that this offset may be switched in and out so that they can be used as either unipolar or bipolar converters at will), but the other, known as a *sign-magnitude* converter, is more complex, and has N bits of magnitude information and an additional bit which corresponds to the sign of the analog signal. Sign-magnitude DACs are quite rare, and sign-magnitude ADCs are found mostly in DVMs. The unipolar, offset binary, and sign-magnitude representations are shown in Figure 5-14.

The four DC errors in a data converter are *offset error*, *gain error*, and two types of *linearity error* (*differential and integral*). Offset and gain errors are analogous to offset and gain errors in amplifiers as

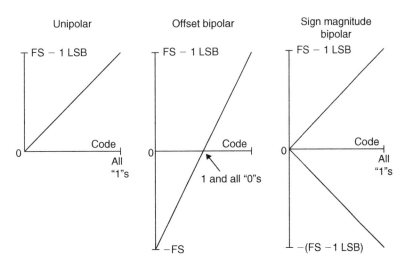

**Figure 5-14: Unipolar and bipolar converters**

shown in Figure 5-15 for a bipolar input range (though offset error and zero error, which are identical in amplifiers and unipolar data converters, are not identical in bipolar converters and should be carefully distinguished). The transfer characteristics of both DACs and ADCs may be expressed as $D = K + GA$, where D is the digital code, A is the analog signal, and K and G are constants. In a unipolar converter, K is

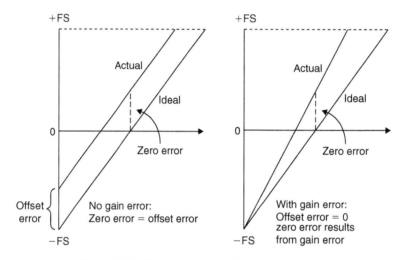

**Figure 5-15: Data converter offset and gain error**

zero, and in an offset bipolar converter, it is $-1$ MSB. The offset error is the amount by which the actual value of K differs from its ideal value.

The gain error is the amount by which G differs from its ideal value, and is generally expressed as the percentage difference between the two, although it may be defined as the gain error contribution (in mV or LSB) to the total error at full-scale. These errors can usually be trimmed by the data converter user. Note, however, that amplifier offset is trimmed at zero input, and then the gain is trimmed near to full-scale. The trim algorithm for a bipolar data converter is not so straightforward.

The integral linearity error of a converter is also analogous to the linearity error of an amplifier, is defined as the maximum deviation of the actual transfer characteristic of the converter from a straight line, and is generally expressed as a percentage of full-scale (but may be given in LSBs). For an ADC, the most popular convention is to draw the straight line through the midpoints of the codes, or the code centers. There are two common ways of choosing the straight line: *end point* and *best straight line* as shown in Figure 5-16.

In the *end point* system, the deviation is measured from the straight line through the origin and the full-scale point (after gain adjustment). This is the most useful integral linearity measurement for measurement and control applications of data converters (since error budgets depend on deviation from the ideal transfer characteristic, not from some arbitrary "best fit"), and is the one normally adopted by Analog Devices, Inc.

The *best straight line*, however, does give a better prediction of distortion in AC applications, and also gives a lower value of "linearity error" on a data sheet. The best fit straight line is drawn through the transfer characteristic of the device using standard curve fitting techniques, and the maximum deviation is measured from this line. In general, the integral linearity error measured in this way is only 50% of the value measured by end point methods. This makes the method good for producing impressive data sheets, but it is less useful

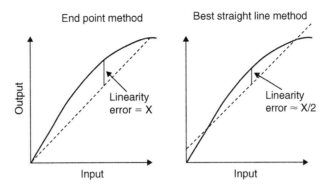

**Figure 5-16: Method of measuring integral linearity errors (same converter on both graphs)**

for error budget analysis. For AC applications, it is even better to specify distortion than DC linearity, so it is rarely necessary to use the best straight line method to define converter linearity.

The other type of converter nonlinearity is *differential nonlinearity* (DNL). This relates to the linearity of the code transitions of the converter. In the ideal case, a change of 1 LSB in digital code corresponds to a change of exactly 1 LSB of analog signal. In a DAC, a change of 1 LSB in digital code produces exactly 1 LSB change of analog output, while in an ADC there should be exactly 1 LSB change of analog input to move from one digital transition to the next. Differential linearity error is defined as the maximum amount of deviation of any quantum (or LSB change) in the entire transfer function from its ideal size of 1 LSB.

Where the change in analog signal corresponding to 1 LSB digital change is more or less than 1 LSB, there is said to be a DNL error. The DNL error of a converter is normally defined as the maximum value of DNL to be found at any transition across the range of the converter. Figure 5-17 shows the non-ideal transfer functions for a DAC and an ADC and shows the effects of the DNL error.

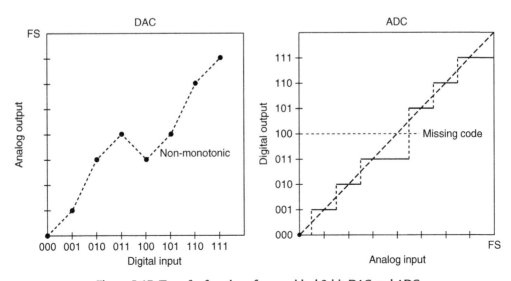

**Figure 5-17: Transfer functions for non-ideal 3-bit DAC and ADC**

The DNL of a DAC is examined more closely in Figure 5-18. If the DNL of a DAC is less than −1 LSB at any transition, the DAC is *non-monotonic*, i.e., its transfer characteristic contains one or more localized maxima or minima. A DNL greater than +1 LSB does not cause non-monotonicity, but is still undesirable. In many DAC applications (especially closed-loop systems where non-monotonicity can change negative feedback to positive feedback), it is critically important that DACs are monotonic. DAC monotonicity is often explicitly specified on data sheets, although if the DNL is guaranteed to be less than 1 LSB (i.e., |DNL| ≤ 1 LSB) then the device must be monotonic, even without an explicit guarantee.

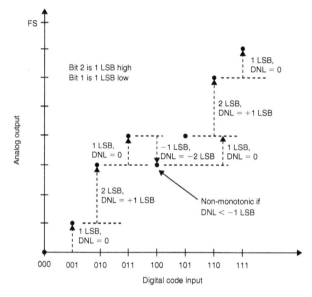

**Figure 5-18: Details of DAC differential nonlinearity**

In Figure 5-19, the DNL of an ADC is examined more closely on an expanded scale. ADCs can be non-monotonic, but a more common result of excess DNL in ADCs is *missing codes*. Missing codes in an ADC are as objectionable as non-monotonicity in a DAC. Again, they result from DNL <−1 LSB.

Not only can ADCs have missing codes, they can also be non-monotonic as shown in Figure 5-20. As in the case of DACs, this can present major problems—especially in servo applications.

In a DAC, there can be no missing codes—each digital input word will produce a corresponding analog output. However, DACs can be non-monotonic as previously discussed. In a straight binary DAC, the most likely place a non-monotonic condition can develop is at mid-scale between the two codes: 011…11 and 100…00. If a non-monotonic condition occurs here, it is generally because the DAC is not properly calibrated or trimmed. A successive approximation ADC with an internal non-monotonic DAC will generally produce missing codes but remain monotonic. However it is possible for an ADC to be non-monotonic—again depending on the particular conversion architecture. Figure 5-20 shows the transfer function of an ADC which is non-monotonic and has a missing code.

ADCs which use the *subranging* architecture divide the input range into a number of coarse segments, and each coarse segment is further divided into smaller segments—and ultimately the final code is derived. This process is described in more detail in Chapter 6 of this book. An improperly trimmed subranging ADC may exhibit non-monotonicity, wide codes, or missing codes at the subranging points as shown in Figure 5-21(A)–(C),

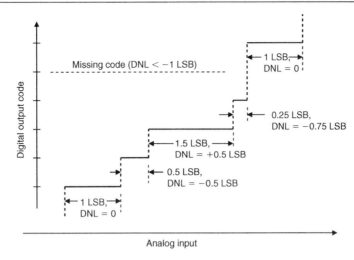

**Figure 5-19: Details of ADC differential nonlinearity**

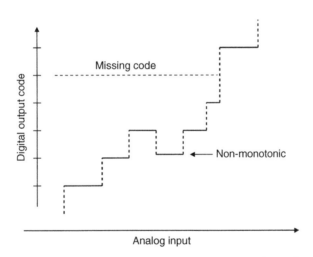

**Figure 5-20: Non-monotonic ADC with missing code**

respectively. This type of ADC should be trimmed so that drift due to aging or temperature produces wide codes at the sensitive points rather than non-monotonic or missing codes.

Defining missing codes is more difficult than defining non-monotonicity. All ADCs suffer from some inherent transition noise as shown in Figure 5-22 (think of it as the flicker between adjacent values of the last digit of a DVM). As resolutions and bandwidths become higher, the range of input over which transition noise occurs may approach, or even exceed, 1 LSB. High resolution wideband ADCs generally have internal noise sources which can be reflected to the input as effective input noise summed with the signal. The effect of this noise, especially if combined with a negative DNL error, may be that there are some (or even all) codes where transition noise is present for the whole range of inputs. There are therefore some codes for which there is *no* input which will *guarantee* that code as an output, although there may be a range of inputs which will *sometimes* produce that code.

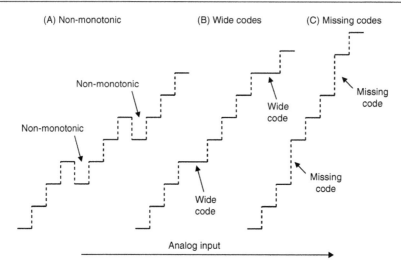

**Figure 5-21: Errors associated with improperly trimmed subranging ADC**

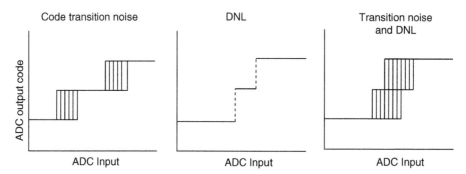

**Figure 5-22: Combined effects of code transition noise and DNL**

For low resolution ADCs, it may be reasonable to define *no missing codes* as a combination of transition noise and DNL which guarantees some level (perhaps 0.2 LSB) of noise-free code for all codes. However, this is impossible to achieve at the very high resolutions achieved by modern sigma–delta ADCs, or even at lower resolutions in wide bandwidth sampling ADCs. In these cases, the manufacturer must define noise levels and resolution in some other way. Which method is used is less important, but the data sheet should contain a clear definition of the method used and the performance to be expected.

The discussion thus far has not dealt with the most important DC specifications associated with data converters. Other less important specifications require only a definition. There are also AC specifications. Converter specifications are covered in Chapter 6.

## References: Coding and Quantizing

1.  K.W. Cattermole, **Principles of Pulse Code Modulation**, American Elsevier Publishing Company, Inc., New York, 1969, ISBN: 444-19747-8 *(an excellent tutorial and historical discussion of data conversion theory and practice, oriented towards PCM, but covers practically all aspects. This one is a must for anyone serious about data conversion! Try Internet secondhand bookshops such as* http://www.abebooks.com *for starters).*

2.  Frank Gray, "Pulse Code Communication," **US Patent 2,632,058**, filed, November 13, 1947, issued March 17, 1953 *(detailed patent on the Gray code and its application to electron beam coders).*

3.  R.W. Sears, "Electron Beam Deflection Tube for Pulse Code Modulation," **Bell System Technical Journal**, Vol. 27, January 1948, pp. 44–57 *(describes an electron-beam deflection tube 7-bit,100kSPS flash converter for early experimental PCM work).*

4.  J.O. Edson and H. H. Henning, "Broadband Codecs for an Experimental 224Mb/s PCM Terminal," **Bell System Technical Journal**, Vol. 44, November 1965, pp. 1887–1940 *(summarizes experiments on ADCs based on the electron tube coder as well as a bit-per-stage Gray code 9-bit solid state ADC. The electron beam coder was 9-bits at 12MSPS, and represented the fastest of its type).*

5.  Dan Sheingold, **Analog–Digital Conversion Handbook**, **3rd Ed**., Analog Devices and Prentice-Hall, 1986, ISBN: 0-13-032848-0 *(the defining and classic book on data conversion).*

# Sampling Theory

This section discusses the basics of sampling theory. A block diagram of a typical real-time sampled data system is shown in Figure 5-23. Prior to the actual analog-to-digital conversion, the analog signal usually passes through some sort of signal conditioning circuitry which performs such functions as amplification, attenuation, and filtering. The lowpass/bandpass filter is required to remove unwanted signals outside the bandwidth of interest and prevent aliasing.

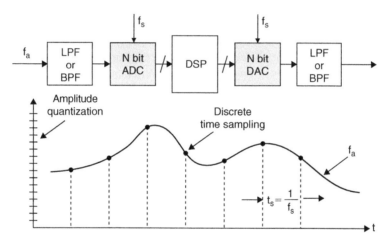

Figure 5-23: Sampled data system

The system shown in Figure 5-23 is a real-time system, i.e., the signal to the ADC is continuously sampled at a rate equal to $f_s$, and the ADC presents a new sample to the DSP at this rate. In order to maintain real-time operation, the DSP must perform all its required computation within the sampling interval, $1/f_s$, and present an output sample to the DAC before arrival of the next sample from the ADC. An example of a typical DSP function would be a digital filter.

In the case of FFT analysis, a block of data is first transferred to the DSP memory. The FFT is calculated at the same time a new block of data is transferred into the memory, in order to maintain real-time operation. The DSP must calculate the FFT during the data transfer interval so it will be ready to process the next block of data.

Note that the DAC is required only if the DSP data must be converted back into an analog signal (as would be the case in a voiceband or audio application, for example). There are many applications where the signal remains entirely in digital format after the initial A/D conversion. Similarly, there are applications where the DSP is solely responsible for generating the signal to the DAC, such as in CD player electronics. If a DAC is used, it must be followed by an analog anti-imaging filter to remove the image frequencies. Finally, there are slower speed industrial process control systems where sampling rates are much lower—regardless of the system, the fundamentals of sampling theory still apply.

There are two key concepts involved in the actual analog-to-digital and digital-to-analog conversion process: *discrete time sampling* and *finite amplitude resolution due to quantization*. An understanding of these concepts is vital to data converter applications.

## The Need for a Sample-and-Hold Amplifier Function

The generalized block diagram of a sampled data system shown in Figure 5-23 assumes some type of AC signal at the input. It should be noted that this does not necessarily have to be so, as in the case of modern DVMs or ADCs optimized for DC measurements, but for this discussion assume that the input signal has some upper frequency limit $f_a$.

Most ADCs today have a built-in sample-and-hold amplifier (SHA) function, thereby allowing them to process AC signals. This type of ADC is referred to as a *sampling ADC*. However many early ADCs, such as the Analog Devices' industry-standard AD574, were not of the sampling type, but simply *encoders*. If the input signal to a SAR ADC (assuming no SHA function) changes by more than 1 LSB during the conversion time (8 μs in the example), the output data can have large errors, depending on the location of the code as shown in Figure 5-24. Most ADC architectures are subject to this type of error—some more, some less—with the possible exception of very well flash converters having well-matched comparators.

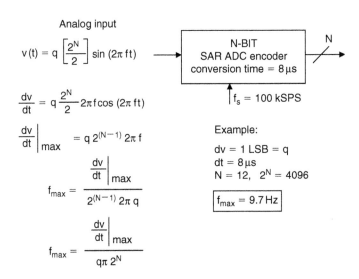

**Figure 5-24: Input frequency limitations of non-sampling ADC (encoder)**

Assume that the input signal to the encoder is a sinewave with a full-scale amplitude ($q2^N/2$), where q is the weight of 1 LSB:

$$v(t) = q(2^N/2)\sin(2\pi\, ft) \tag{5-1}$$

Taking the derivative:

$$dv/dt = q2f\,(2^N/2)\cos(2\,ft) \tag{5-2}$$

The maximum rate of change is therefore:

$$dv/dt \mid_{max} = q2\ f\ (2^N/2) \tag{5-3}$$

Solving for f:

$$f = (dv/dt \mid_{max})/(q\pi\ 2^N) \tag{5-4}$$

If N = 12, and 1 LSB change (dv = q) is allowed during the conversion time (dt = 8 μs), then the equation can be solved for $f_{max}$, the maximum full-scale signal frequency that can be processed without error:

$$f_{max} = 9.7\,Hz$$

This implies any input frequency greater than 9.7 Hz is subject to conversion errors, even though a sampling frequency of 100 kSPS is possible with the 8 μs ADC (this allows an extra 2 μs interval for an external SHA to re-acquire the signal after coming out of the hold mode).

To process AC signals, an SHA function is added. The ideal SHA is simply a switch driving a hold capacitor followed by a high input impedance buffer. The input impedance of the buffer must be high enough so that the capacitor is discharged by less than 1 LSB during the hold time. The SHA samples the signal in the *sample* mode, and holds the signal constant during the *hold* mode. The timing is adjusted so that the encoder performs the conversion during the hold time. A sampling ADC can therefore process fast signals—the upper frequency limitation is determined by the SHA aperture jitter, bandwidth, distortion, etc., not the encoder. In the example shown, a good sample-and-hold could acquire the signal in 2 μs, allowing a sampling frequency of 100 kSPS, and the capability of processing input frequencies up to 50 kSPS. A complete discussion of the SHA function including these specifications follows later in this chapter.

## *The Nyquist Criteria*

A continuous analog signal is sampled at discrete intervals, $t_s = 1/f_s$, which must be carefully chosen to ensure an accurate representation of the original analog signal. It is clear that the more samples taken (faster sampling rates), the more accurate the digital representation, but if fewer samples are taken (lower sampling rates), a point is reached where critical information about the signal is actually lost. The mathematical basic of sampling was set forth by Harry Nyquist of Bell Telephone Laboratories in two classic papers published in 1924 and 1928, respectively (see References 1 and 2). Nyquist's original work was shortly supplemented by R.V.L. Hartley (Reference 3). These papers formed the basis for the PCM work to follow in the 1940s, and in 1948 Claude Shannon wrote his classic paper on communication theory (Reference 4).

Simply stated, the Nyquist criteria requires that the sampling frequency be at least twice the highest frequency contained in the signal, or information about the signal will be lost. If the sampling frequency is less than twice the maximum analog signal frequency, a phenomenon known as aliasing will occur (Figure 5-25).

---

- A signal with a *maximum BANDWIDTH* $f_a$ must be sampled at a rate $f_a > 2f_a$ or information about the signal will be lost because of aliasing
- Aliasing occurs whenever $f_a < 2f_a$
- The concept of aliasing is widely used in communications applications such as direct IF-to-digital conversion
- A signal which has frequency components between $f_a$ and $f_b$ must be sampled at least at a rate $f_a > 2(f_b - f_a)$ to prevent alias components from overlapping the signal frequencies.

---

**Figure 5-25: Nyquist's criteria**

In order to understand the implications of *aliasing* in both the time and frequency domain, first consider the case of a time domain representation of a single tone sinewave sampled as shown in Figure 5-26. In this example, the sampling frequency $f_s$ is not at least $2f_a$, but only slightly more than the analog input frequency $f_a$—the Nyquist criteria is violated. Notice that the pattern of the actual samples produces an *aliased* sinewave at a lower frequency equal to $f_s - f_a$.

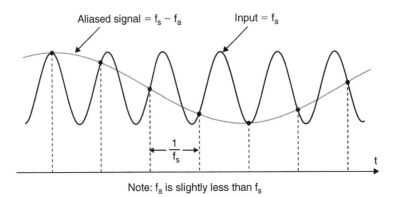

Aliased signal = $f_s - f_a$    Input = $f_a$

$\frac{1}{f_s}$

t

Note: $f_a$ is slightly less than $f_s$

**Figure 5-26: Aliasing in the time domain**

The corresponding frequency domain representation of this scenario is shown in Figure 5-27(B). Now consider the case of a single frequency sinewave of frequency $f_a$ sampled at a frequency $f_s$ by an ideal impulse sampler (see Figure 5-27(A)). Also assume that $f_s > 2f_a$ as shown. The frequency domain output of the sampler shows *aliases* or *images* of the original signal around every multiple of $f_s$, i.e., at frequencies equal to $|\pm Kf_s \pm f_a|$, $K = 1, 2, 3, 4, \ldots$.

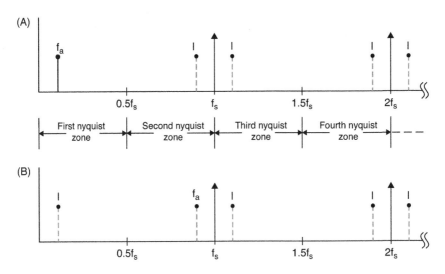

**Figure 5-27: Analog signal $f_a$ sampled at $f_s$ using ideal sampler has images (aliases) at $1 \pm Kf_s \pm f_{a1}$, $K = 1, 2, 3, \ldots$**

The *Nyquist* bandwidth is defined to be the frequency spectrum from DC to $f_s/2$. The frequency spectrum is divided into an infinite number of *Nyquist zones*, each having a width equal to $0.5f_s$ as shown. In practice, the ideal sampler is replaced by an ADC followed by an FFT processor. The FFT processor only provides an output from DC to $f_s/2$, i.e., the signals or aliases which appear in the first Nyquist zone.

Now consider the case of a signal which is outside the first Nyquist zone (Figure 5-27(B)). The signal frequency is only slightly less than the sampling frequency, corresponding to the condition shown in the time domain representation in Figure 5-26. Notice that even though the signal is outside the first Nyquist zone, its image (or *alias*), $f_s - f_a$, falls inside. Returning to Figure 5-27(A), it is clear that if an unwanted signal appears at any of the image frequencies of $f_a$, it will also occur at $f_a$, thereby producing a spurious frequency component in the first Nyquist zone.

This is similar to the analog mixing process and implies that some filtering ahead of the sampler (or ADC) is required to remove frequency components which are outside the Nyquist bandwidth, but whose aliased components fall inside it. The filter performance will depend on how close the out-of-band signal is to $f_s/2$ and the amount of attenuation required.

## Baseband Antialiasing Filters

Baseband sampling implies that the signal to be sampled lies in the first Nyquist zone. It is important to note that with no input filtering at the input of the ideal sampler, *any frequency component (either signal or noise) that falls outside the Nyquist bandwidth in any Nyquist zone will be aliased back into the first Nyquist zone*. For this reason, an antialiasing filter is used in almost all sampling ADC applications to remove these unwanted signals.

Properly specifying the antialiasing filter is important. The first step is to know the characteristics of the signal being sampled. Assume that the highest frequency of interest is $f_a$. The antialiasing filter passes signals from DC to $f_a$ while attenuating signals above $f_a$.

Assume that the corner frequency of the filter is chosen to be equal to $f_a$. The effect of the finite transition from minimum to maximum attenuation on system dynamic range is illustrated in Figure 5-28(A).

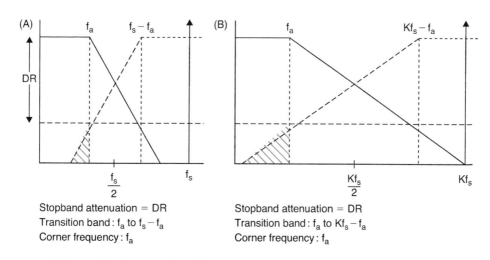

Figure 5-28: Oversampling relaxes requirements on baseband antialiasing filter

Assume that the input signal has full-scale components well above the maximum frequency of interest, $f_a$. The diagram shows how full-scale frequency components above $f_s - f_a$ are aliased back into the bandwidth DC to $f_a$. These aliased components are indistinguishable from actual signals and therefore limit the dynamic range to the value on the diagram which is shown as *DR*.

Some texts recommend specifying the antialiasing filter with respect to the Nyquist frequency, $f_s/2$, but this assumes that the signal bandwidth of interest extends from DC to $f_s/2$ which is rarely the case. In the example shown in Figure 5-28, the aliased components between $f_a$ and $f_s/2$ are not of interest and do not limit the dynamic range.

The antialiasing filter transition band is therefore determined by the corner frequency $f_a$, the stopband frequency $f_s - f_a$, and the desired stopband attenuation, DR. The required system dynamic range is chosen based on the requirement for signal fidelity.

Filters become more complex as the transition band becomes sharper, all other things being equal. For instance, a Butterworth filter gives 6 dB attenuation per octave for each filter pole. Achieving 60 dB attenuation in a transition region between 1 and 2 MHz (1 octave) requires a minimum of 10 poles—not a trivial filter, and definitely a design challenge.

Therefore, other filter types are generally more suited to high speed applications where the requirement is for a sharp transition band and in-band flatness coupled with linear phase response. Elliptic filters meet these criteria and are a popular choice. There are a number of companies which specialize in supplying custom analog filters. TTE is an example of such a company (Reference 5).

From this discussion, we can see how the sharpness of the antialiasing transition band can be traded off against the ADC sampling frequency. Choosing a higher sampling rate (oversampling) reduces the requirement on transition band sharpness (hence, the filter complexity) at the expense of using a faster ADC and processing data at a faster rate. This is illustrated in Figure 5-28(B) which shows the effects of increasing the sampling frequency by a factor of K, while maintaining the same analog corner frequency, $f_a$, and the same dynamic range, DR, requirement. The wider transition band ($f_a$ to $Kf_s - f_a$) makes this filter easier to design than for the case of Figure 5-28(A).

The antialiasing filter design process is started by choosing an initial sampling rate of 2.5 to 4 times $f_a$. Determine the filter specifications based on the required dynamic range and see if such a filter is realizable within the constraints of the system cost and performance. If not, consider a higher sampling rate which may require using a faster ADC. It should be mentioned that sigma–delta ADCs are inherently oversampling converters, and the resulting relaxation in the analog antialiasing filter requirements is therefore an added benefit of this architecture.

The antialiasing filter requirements can also be relaxed somewhat if it is certain that there will never be a full-scale signal at the stopband frequency $f_s - f_a$. In many applications, it is improbable that full-scale signals will occur at this frequency. If the maximum signal at the frequency $f_s - f_a$ will never exceed X dB below full-scale, then the filter stopband attenuation requirement is reduced by that same amount. The new requirement for stopband attenuation at $f_s - f_a$ based on this knowledge of the signal is now only DR – X dB. When making this type of assumption, be careful to treat any noise signals which may occur above the maximum signal frequency $f_a$ as unwanted signals which will also alias back into the signal bandwidth.

## Undersampling

Thus far we have considered the case of baseband sampling, i.e., all the signals of interest lie within the first Nyquist zone. Figure 5-29(A) shows such a case, where the band of sampled signals is limited to the first Nyquist zone, and images of the original band of frequencies appear in each of the other Nyquist zones.

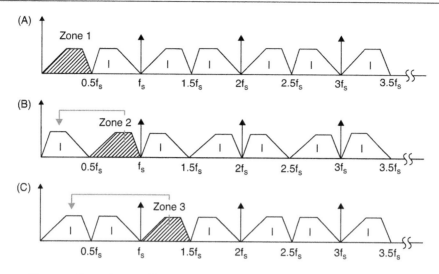

**Figure 5-29: Undersampling and frequency translation between Nyquist zones**

Consider the case shown in Figure 5-29(B), where the sampled signal band lies entirely within the second Nyquist zone. The process of sampling a signal outside the first Nyquist zone is often referred to as *undersampling*, or *harmonic sampling (also referred to as bandpass sampling, IF sampling, direct IF to digital conversion)*. Note that the first Nyquist zone image contains all the information in the original signal, with the exception of its original location (the order of the frequency components within the spectrum is reversed, but this is easily corrected by re-ordering the output of the FFT).

Figure 5-29(C) shows the sampled signal restricted to the third Nyquist zone. Note that the first Nyquist zone image has no frequency reversal. In fact, the sampled signal frequencies may lie in *any* unique Nyquist zone, and the first Nyquist zone image is still an accurate representation (with the exception of the frequency reversal which occurs when the signals are located in even Nyquist zones). At this point we can clearly restate the Nyquist criteria:

> A signal must be sampled at a rate equal to or greater than twice its **bandwidth** in order to preserve all the signal information.

Notice that there is no mention of the precise *location* of the band of sampled signals within the frequency spectrum relative to the sampling frequency. The only constraint is that the band of sampled signals be restricted to a *single* Nyquist zone, i.e., the signals must not overlap any multiple of $f_s/2$ (this, in fact, is the primary function of the antialiasing filter).

Sampling signals above the first Nyquist zone has become popular in communications because the process is equivalent to analog demodulation. It is becoming common practice to sample IF signals directly and then use digital techniques to process the signal, thereby eliminating the need for the IF demodulator and filters. Clearly, however, as the IF frequencies become higher, the dynamic performance requirements on the ADC become more critical. The ADC input bandwidth and distortion performance must be adequate at the IF frequency, rather than only baseband. This presents a problem for most ADCs designed to process signals in the first Nyquist zone; therefore, an ADC suitable for undersampling applications must maintain dynamic performance into the higher order Nyquist zones.

## Antialiasing Filters in Undersampling Applications

Figure 5-30 shows a signal in the second Nyquist zone centered around a carrier frequency, $f_c$, whose lower and upper frequencies are $f_1$ and $f_2$, respectively. The antialiasing filter is a bandpass filter. The desired dynamic range is DR, which defines the filter stopband attenuation. The upper transition band is $f_2$ to $2f_s - f_2$, and the lower is $f_1$ to $f_s - f_1$. As in the case of baseband sampling, the antialiasing filter requirements can be relaxed by proportionally increasing the sampling frequency, but $f_c$ must also be increased so that it is always centered in the second Nyquist zone.

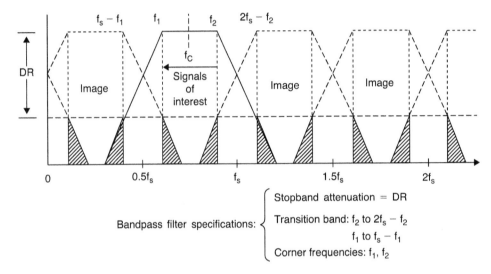

Figure 5-30: Antialiasing filter for undersampling

Two key equations can be used to select the sampling frequency, $f_s$, given the carrier frequency, $f_c$, and the bandwidth of its signal, $\Delta f$. The first is the Nyquist criteria:

$$f_s > 2\Delta f \qquad (5\text{-}5)$$

The second equation ensures that $f_c$ is placed in the center of a Nyquist zone:

$$f_s = \frac{4f_c}{2NZ - 1} \qquad (5\text{-}6)$$

where NZ = 1, 2, 3, 4, ... and NZ corresponds to the Nyquist zone in which the carrier and its signal fall (see Figure 5-31).

NZ is normally chosen to be as large as possible while still maintaining $f_s$ lt; $2\Delta f$. This results in the minimum required sampling rate. If NZ is chosen to be odd, then $f_c$ and its signal will fall in an odd Nyquist zone, and the image frequencies in the first Nyquist zone will not be reversed. Tradeoffs can be made between the sampling frequency and the complexity of the antialiasing filter by choosing smaller values of NZ (hence a higher sampling frequency).

334

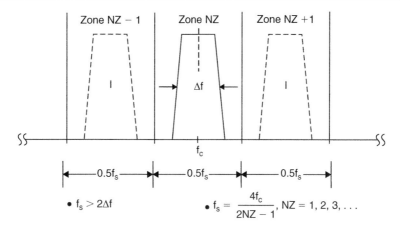

**Figure 5-31: Centering an undersampled signal within a Nyquist zone**

As an example, consider a 4 MHz wide signal centered around a carrier frequency of 71 MHz. The minimum required sampling frequency is therefore 8 MSPS. Solving Eq. (5.6) for NZ using $f_c = 71$ MHz and $f_s = 8$ MSPS yields NZ = 18.25. However, NZ must be an integer, so we round 18.25 to the next lowest integer, 18. Solving Eq. (5.6) again for $f_s$ yields $f_s = 8.1143$ MSPS. The final values are therefore $f_s = 8.1143$ MSPS, $f_c = 71$ MHz, and NZ = 18.

Now assume that we desire more margin for the antialiasing filter, and we select $f_s$ to be 10 MSPS. Solving Eq. (5.6) for NZ, using $f_c = 71$ MHz and $f_s = 10$ MSPS yields NZ = 14.7. We round 14.7 to the next lowest integer, giving NZ = 14. Solving Eq. (5.6) again for $f_s$ yields $f_s = 10.519$ MSPS. The final values are therefore $f_s = 10.519$ MSPS, $f_c = 71$ MHz, and NZ = 14.

The above iterative process can also be carried out starting with $f_s$ and adjusting the carrier frequency to yield an integer number for NZ.

## References: Sampling Theory

1.  H. Nyquist, "Certain Factors Affecting Telegraph Speed," **Bell System Technical Journal**, Vol. 3, April 1924, pp. 324–346.

2.  H. Nyquist, "Certain Topics in Telegraph Transmission Theory," **A.I.E.E. Transactions**, Vol. 47, April 1928, pp. 617–644.

3.  R.V. L. Hartley, "Transmission of Information," **Bell System Technical Journal**, Vol. 7, July 1928, pp. 535–563.

4.  C.E. Shannon, "A Mathematical Theory of Communication," **Bell System Technical Journal**, Vol. 27, July 1948, pp. 379–423 and October 1948, pp. 623–656.

5.  TTE, Inc., 11652 Olympic Blvd., Los Angeles, CA 90064, http://www.tte.com.

# CHAPTER 6
# *Converters*

## Chapter Introduction

There are two basic types of converters, digital-to-analog (DACs or D/As) and analog-to-digital (ADCs or A/Ds). Their purpose is fairly straightforward. In the case of DACs, they output an analog voltage that is a proportion of a reference voltage, the proportion based on the digital word applied. In the case of ADCs, a digital representation of the analog voltage that is applied to the ADCs input is outputted, the representation proportional to a reference voltage.

In both cases the digital word is almost always based on a binarily weighted proportion. The digital input or output is arranged in words of varying widths, referred to as bits, typically anywhere from 6 bits to 24 bits. In a binarily weighted system each bit is worth half of the bit to its left and twice the bit to its right. The greater the number of bits in the digital word, the finer the resolution. These bits are typically arranged in groups of 4, called bytes, for convenience.

For a better understanding of the relationship between the digital domain and the analog domain please refer to the section on sampling theory.

As stated earlier, we shall look at the operation of converters primarily from a "black box" view. We will concern ourselves less with the internal construction of the converter and more with its operation. We cannot, however, completely ignore the internal architecture because in many cases it is relevant to operational advantages or limitations. There are a number of works that cover the internal workings of the converters in much more detail (see references).

Another point that should be kept in mind is the difference between accuracy and resolution. The resolution of a converter is the number of bits in its digital word. The accuracy is the number of those bits that meet the specifications. For instance, a DAC might have 16 bits of resolution, but might only be monotonic to 14 bits. This means that the assured accuracy of the DAC will be no better than 14 bits. Also, an audio ADC might have a digital word width of 16 bits, but the signal-to-noise ratio (SNR) may be only 70dB. This means that the accuracy will only be at the 12-bit level. This is not to say that the other bits are irrelevant. With further processing, typically filtering, often the accuracy can be improved. While these terms are similar and sometimes used interchangeably, the distinction between the two should be remembered.

We shall examine the DAC first.

# DAC Architectures

## DACs or D/As Introduction

What we commonly refer to as a DAC today is typically quite a bit more. The DAC will typically have the converter itself and a collection of support circuitry built into the chip (Figure 6-1).

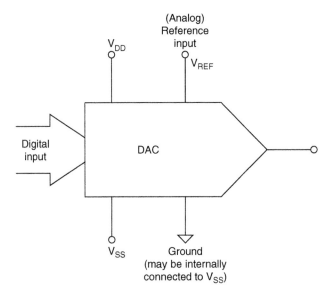

**Figure 6-1: The basic DAC**

The first DACs were board level designs, built from discrete components, including vacuum tubes as the switching elements. Monolithic DACs began to appear in the early 1970s. These early examples were actually sub-blocks of the DAC. An example of this would be the AD550, which was a 4 bit binarily weighted current source. This current source block would be mated to a separate part, such as the AD850, which contained a resistor array and complementary-MOS (CMOS) switches. Together these would form the basic DAC. As we moved on in time these functions were integrated on the same die, additional digital circuitry, specifically latches to store the digital input, were added. Then a second rank of latches were often added. The purpose of the second rank was to allow the microprocessor or microcontroller to write many DACs in a system and update them all at the same time. The input rank of latches could also be a shift register, which would allow a serial interface.

On the back end, since the output of the DAC is often a current, an op amp is often added to perform the current-to-voltage (I/V) conversion. On the front end a voltage reference is often added.

Process limitations did not allow the integration of all these sub-blocks to occur at once. Initially, the processes used to make the various sub-blocks were not compatible. The process that made the best switches was typically not the best for the amplifier and the reference. As the processes became more advanced these limitations became less. Today CMOS can make acceptable amplifiers and processes combining bipolar and CMOS together exist.

There are several advantages to including all this additional circuitry in one package. The first is the obvious advantage of reducing the chip count. This reduces the size of the circuitry and increases the reliability. Probably more important is that the circuit designer now does not have to concern himself with the accuracy of several parts in a system. The system is now one part and tested by the manufacturer as a unit.

Next we will look at the various DAC architectures. When we refer to DACs here we are referring to the basic converter rather than the complete system.

## Kelvin Divider (String DAC)

The simplest structure of all is the Kelvin divider or string DAC as shown in Figure 6-2. An N-bit version of this DAC simply consists of $2^N$ equal resistors in series and $2^N$ switches (usually CMOS), one between each node of the chain and the output. The output is taken from the appropriate tap by closing just one of the switches (there is some slight digital complexity involved in decoding to 1 of $2^N$ switches from N-bit data).

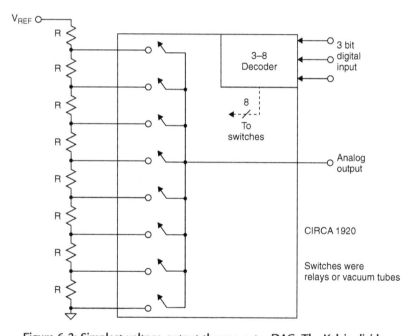

**Figure 6-2: Simplest voltage-output thermometer DAC: The Kelvin divider**

This architecture is simple, has a voltage output, and is inherently monotonic—even if a resistor is accidentally short-circuited, output n cannot exceed output n + 1. It is linear if all the resistors are equal, but may be made deliberately nonlinear if a nonlinear DAC is required. The output is a voltage, but it has the disadvantage of having a relatively large output impedance. This output impedance is also code dependent (the impedance changes with changes to the digital input). In many cases it will be beneficial to follow the output of the DAC with an op amp to buffer this output impedance and present a low impedance source to the following circuitry.

Since only two switches operate during a transition it is a low glitch architecture (the concept of glitch will be examined in a following section). Also, the switching glitch is not code-dependent, making it ideal for low distortion applications. Because the glitch is constant regardless of the code transition, the frequency content of the glitch is at the DAC update rate and its harmonics—not at the harmonics of the DAC output signal frequency. The major drawback of the Kelvin DAC is the large number of resistors and switches required for high resolution. There are $2^N$ resistors required, so a 10-bit DAC would require 1,024 switches and resistors, and as a result it was not commonly used as a simple DAC architecture until the recent advent of very small integrated circuit (IC) feature sizes made it very practical for low and medium resolution (typically up to 10 bits) DACs.

As we mentioned in the section on sampling theory, the output of a DAC for an all 1s code is 1 least significant bit (LSB) below the reference, so a Kelvin divider DAC intended for use as a general purpose DAC has a resistor between the reference terminal and the first switch as shown in Figure 6-2.

## Segmented String DACs

A variation of the Kelvin divider is the segmented string DAC. Here we reduce the number of resistors required by segmenting. Figure 6-3 shows two varieties of segmented voltage-output DAC. The architecture in Figure 6-3(A) is sometimes called a Kelvin–Varley divider. Since there are buffers between the first and second stages, the second string DAC does not load the first, and the resistors in the second string do not need to have the same value as the resistors in the first. All the resistors in each string, however, do need to be equal to each other or the DAC will not be linear. The examples shown have 3-bit first and second stages but for the sake of generality, let us refer to the first (most significant bit (MSB)) stage resolution as M bits and the second (LSB) as K bits for a total of $N = M + K$ bits. The MSB DAC has a string of $2^M$ equal resistors, and a string of $2^K$ equal resistors in the LSB DAC. As an example if we make a 10-bit string DAC out of two 5-bit sections, each segment would have $2^5$ or 32 resistors, for a total of 64, as opposed to the 1,024 required for a standard Kelvin divider. This is an obvious advantage.

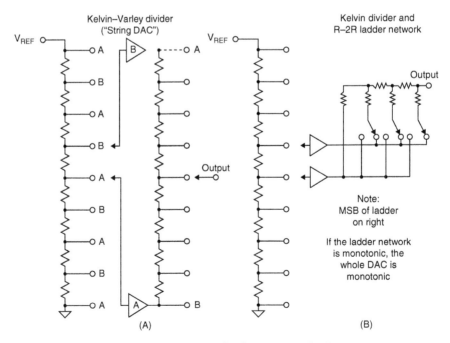

Figure 6-3: Segmented voltage-output DACs

Buffer amplifiers can have offset, of course, and this can cause non-monotonicity in a buffered segmented string DAC.

In the simpler configuration of a buffered Kelvin–Varley divider buffer (Figure 6-3(A)), buffer A is always "below" (at a lower potential than) buffer B, and the extra tap labeled "A" on the LSB string DAC is not necessary. The data decoding is just two priority encoders.

But if the decoding of the MSB string DAC is made more complex so that buffer A can only be connected to the taps labeled "A" in the MSB string DAC, and buffer B to the taps labeled "B," then it is not possible for buffer offsets to cause non-monotonicity. Of course, the LSB string DAC decoding must change direction each time one buffer "leapfrogs" the other, and taps A and B on the LSB string DAC are alternately not used—but this involves a fairly trivial increase in logic complexity and is justified by the increased performance.

Rather than using a second string of resistors, a binary R–2R DAC can be used to generate the three LSBs as shown in Figure 6-3(B). This voltage-output DAC (Figure 6-3(B)) consists of a 3-bit string DAC followed by a 3-bit buffered voltage-mode ladder network. Again the number of resistors required for the DAC is reduced.

An unbuffered version of the segmented string DAC is shown in Figure 6-4. This version is more clever in concept. Here, the resistors in the two strings must be equal, except that the top resistor in the MSB string must be smaller—$1/(2^K)$ of the value of the others—and the LSB string has $2^K - 1$ resistors rather than $2^K$. Because there are no buffers, the LSB string appears in parallel with the resistor in the MSB string that is switched across and loads it. This drops the voltage across that MSB resistor by 1 LSB of the LSB DAC—which is exactly what is required. The output impedance of this DAC, being unbuffered, varies with changing digital code. This circuit is intrinsically monotonic since it is unbuffered (and, of course, can be manufactured on CMOS processes which make resistors and switches but not high precision amplifiers, so it may be cheaper as well).

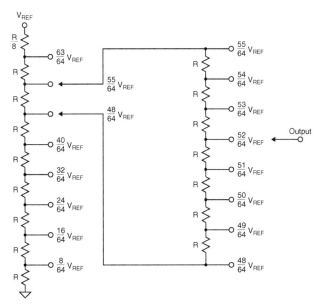

*Adapted from*: Dennis Dempsey and Christopher "Digital-to-Analog Converter," US Patent 5,969,657, filed, July 27,1997, issued October 19, 1999.

**Figure 6-4: Segmented unbuffered string DACs use patented architecture**

In order to understand this clever concept better, the actual voltages at each of the taps has been worked out and labeled for the 6-bit segmented DAC composed of two 3-bit string DACs shown in Figure 6-4. The reader is urged to go through this simple analysis with the second string DAC connected across any other resistor in the first string DAC and verify the numbers. A detailed mathematical analysis of the unbuffered segmented string DAC can be found in the relevant patent filed by Dennis Dempsey and Christopher Gorman of Analog Devices in 1997 (Reference 14).

## Digital Pots

Another variation of the string DAC is the digital potentiometer. A simple digital potentiometer is shown in Figure 6-5.

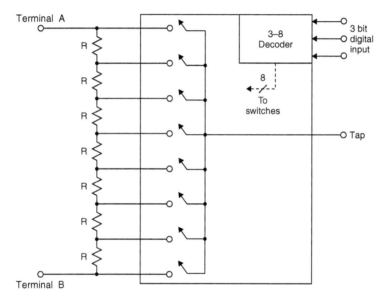

**Figure 6-5: A slight modification to a Kelvin DAC yields a "digital potentiometer"**

The major difference is that the lower arm of the pot (terminal B) is not connected to ground, but is instead left floating. The absolute values of the resistors in a Kelvin DAC typically are not critical. They are limited by the available material. They must, of course, be the same as each other. In a digital pot the end-to-end resistance is specified. The accuracy of the end-to-end resistance is on the order of a mechanical pot.

Digital pots are typically available in end-to-end resistance values from $10\,k\Omega$ to $1\,M\Omega$. Lower values of end-to-end resistance are difficult since the on-resistance of the CMOS switches is on the order of the resistor segment, so the linearity of the pot suffers at the low end.

The advantages to digital pots are many. Even the lowest resolution digital pots have better setability than their mechanical counterparts. Also they are immune to mechanical vibration and oxidation of the wiper contact. Obviously, adjustments can be made without human intervention.

In most digital pots the voltage on the input pins cannot exceed the supplies (typically $3\,V$ or $5\,V$) due to the CMOS switches used in their construction, but certain models are designed for $\pm15\,V$ operation.

Another design feature on many of the digital pots is that on power-up (sometimes from an internal timer, sometimes controlled by an external pin) the wiper is shorted to one of the terminals. This is useful since output on power-up is undefined until it is written to. Since it might take a while (relatively) for the microcontroller to initialize itself and then get around to initializing the rest of the system, having the digital pot in a known state can be useful. Some digital potentiometers incorporate non-volatile logic so that their settings are retained when they are turned off.

One time programmable (OTP) versions of digital pots have become available. Here the digital code is locked into the pot once the setting had been determined. The technology used is fusible links. A variation on this theme is the two times programmable (TTP) digital pot. This allows the non-volatile settings to be modified one time. The block diagram of a TTP digital pot is shown in Figure 6-6.

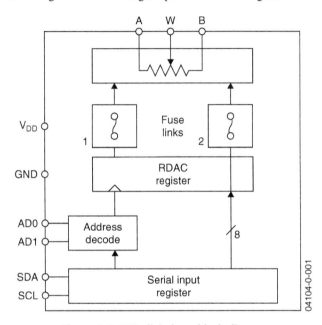

**Figure 6-6: TTP digital pot block diagram**

## *Thermometer (Fully Decoded) DACs*

There is a current-output DAC architecture analogous to a string DAC which consists of $2^N - 1$ switchable current sources (which may be resistors and a voltage reference or may be active current sources) connected to an output terminal. This output must be at, or close to, ground. Figure 6-7 shows a thermometer DAC which uses resistors connected to a reference voltage to generate the currents.

If active current sources are used as shown in Figure 6-8, the output may have more compliance (the allowable voltage on the output pin which still guarantees performance), and a resistive load is typically used to develop an output voltage. The load resistor must be chosen so that at maximum output current the output terminal remains within its rated compliance voltage.

Once a current in a thermometer DAC is switched into the circuit by increasing the digital code, any further increases do not switch it out again. The structure is thus inherently monotonic, irrespective of inaccuracies in the currents. Again, like the Kelvin divider, only the advent of high density IC processes has made this architecture practical for general purpose medium resolution DACs, although a slightly more complex version—shown in the next diagram—is quite widely used in high speed applications. Unlike the Kelvin divider, this type of current-mode DAC does not have a unique name, although both types may be referred to as *fully decoded* DACs or *thermometer* DACs.

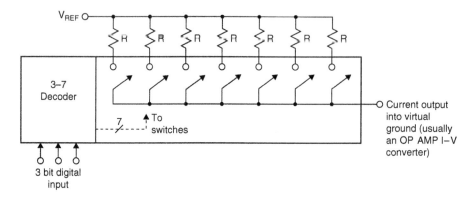

**Figure 6-7: The simplest current-output thermometer (fully decoded) DAC**

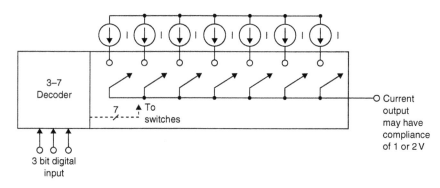

**Figure 6-8: Current sources improve the basic current-output thermometer DAC**

A DAC where the currents are switched between two output lines—one of which is often grounded, but may, in the more general case, be used as the inverted output—is more suitable for high speed applications because switching a current between two outputs is far less disruptive, and so causes a far lower glitch than simply switching a current on and off. This architecture is shown in Figure 6-9.

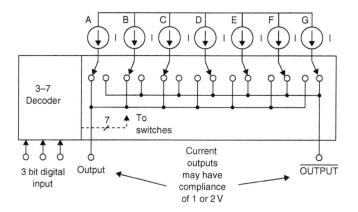

**Figure 6-9: High speed thermometer DAC with complementary current outputs**

But the settling time of this DAC still varies with initial and final code, giving rise to *intersymbol distortion* (ISI). This can be addressed with even more complex switching where the output current is returned to zero before going to its next value. Note that although the current in the output is returned to zero it is not "turned off"—the current is dumped to ground when it is not being used, rather than being switched on and off. The techniques involved are too complex to discuss in detail here but can be found in the references.

In the normal (linear) version of this DAC, all the currents are nominally equal. Where it is used for high speed reconstruction, its linearity can also be improved by dynamically changing the order in which the currents are switched by ascending code. Instead of code 001 always turning on current A; code 010 always turning on currents A and B; code 011 always turning on currents A, B, and C; etc. the order of turn-on relative to ascending code changes for each new data point. This can be done quite easily with a little extra logic in the decoder. The simplest way of achieving it is with a counter which increments with each clock cycle so that the order advances: ABCDEFG, BCDEFGA, CDEFGAB, etc. but this algorithm may give rise to spurious tones in the DAC output. A better approach is to set a new pseudo-random order on each clock cycle—this requires a little more logic, but even complex logic is now very cheap and easily implemented on CMOS processes. There are other, even more complex, techniques which involve using the data itself to select bits and thus turn current mismatch into shaped noise. Again they are too complex for a book of this sort (see references for a more detailed discussion).

## Binary-Weighted Current Source

The voltage-mode binary-weighted resistor DAC shown in Figure 6-10 is usually the simplest textbook example of a DAC. However, this DAC is not inherently monotonic and is actually quite hard to manufacture successfully at high resolutions due to the large spread in component (resistor) values. In addition, the output impedance of the voltage-mode binary DAC changes with the input code.

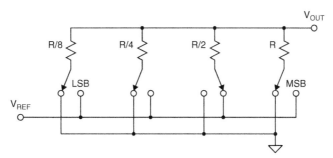

*Adapted from:* B. D. Smith, "Coding by Feedback Methods," *Proceedings of the IRE,* Vol. 41, August 1953, pp. 1053–1058.

**Figure 6-10: Voltage-mode binary-weighted resistor DAC**

Current-mode binary-weighted DACs are shown in Figure 6-11(A) (resistor-based), and Figure 6-11(B) (current-source based). An N-bit DAC of this type consists of N-weighted current sources (which may simply be resistors and a voltage reference) in the ratio $1:2:4:8:...:2^{N-1}$. The LSB switches the $2^{N-1}$ current, the MSB the 1 current, etc. The theory is simple but the practical problems of manufacturing an IC of an economical size with current or resistor ratios of even 128:1 for an 8-bit DAC are enormous, especially as they must have matched temperature coefficients. This architecture is virtually never used on its own in IC DACs, although, again, 3- or 4-bit versions have been used as components in more complex structures. For example, the AD550 mentioned at the beginning of this section is an example of a binary-weighted DAC.

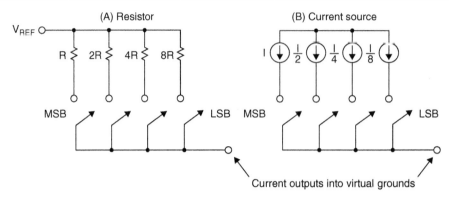

Difficult to fabricate in IC form due to large resistor or current ratios for high resolutions.

**Figure 6-11: Current-mode binary-weighted DACs**

If the MSB current is slightly low in value, it will be less than the sum of the other bit currents, and the DAC will not be monotonic (the differential nonlinearity (DNL) of most types of DAC is worst at major bit transitions).

However, there is another binary-weighted DAC structure which has recently become widely used. This uses binary-weighted capacitors as shown in Figure 6-12. The problem with a DAC using capacitors is that leakage causes it to lose its accuracy within a few milliseconds of being set. This may make capacitive DACs unsuitable for general purpose DAC applications, but it is not a problem in successive approximation ADCs, since the conversion is complete in a few microseconds or less—long before leakage has any appreciable effect.

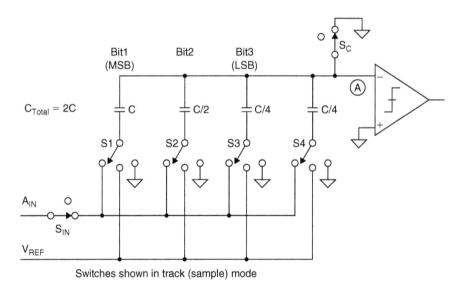

Switches shown in track (sample) mode

**Figure 6-12: Capacitive binary-weighted DAC in successive approximation ADC**

The use of capacitive charge redistribution DACs offers another advantage as well—the DAC itself behaves as a sample-and-hold amplifier (SHA) circuit, so not only is an external SHA unnecessary with these ADCs, there is no need to allocate separate chip area for a separate integral SHA.

## R–2R Ladder

One of the most common DAC building-block structures is the R–2R resistor ladder network shown in Figure 6-13. It uses resistors of only two different values, and their ratio is 2:1. An N-bit DAC requires 2N resistors, and they are quite easily trimmed. There are also relatively few resistors to trim.

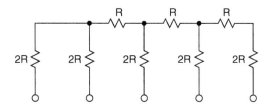

**Figure 6-13: 4-Bit R–2R ladder network**

This structure is the basis of a large family of DACs. Figure 6-14 is the block diagram of the AD7524, which is typical of a basic current-output CMOS DAC. The diagram shows the structure of the DAC.

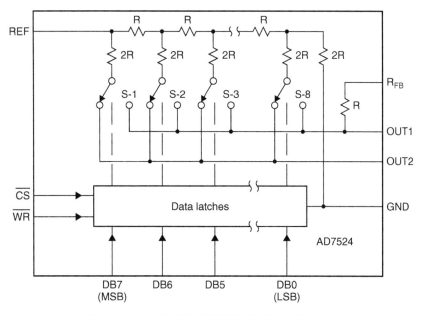

**Figure 6-14: AD7524 CMOS DAC block diagram**

The input impedance (basically the value of the resistors) is not a closely specified parameter. The specified range is 4:1 ($5 \, k\Omega$ minimum, $20 \, k\Omega$ maximum, although it is typically closer than that). It is the relative accuracy, not the absolute accuracy of the resistors that is of interest. In most applications the absolute value is not important. Certain applications exist where the value does matter. In these instances, the parts must be selected at test.

Note the extra resistor added at the $R_{FEEDBACK}$ pin. This is designed to be the feedback resistor for the I/V op amp. This resistor is trimmed along with the rest of the resistors so it tracks. Also, since it is made of the same material as the rest of the resistors, therefore having the same temperature coefficient, and is on the same substrate, hence at the same temperature, it will track over temperature.

Figure 6-15 shows a more modern example of a CMOS DAC, the AD7394. Several trends are obvious here. First off all, the output is voltage, not current. Advancements in process technology have allowed reasonable quality CMOS op amps to be created. Also note the two ranks of latches. The purpose of these latches is to allow the microcontroller to write to all converters in a system and then update them all at the same time. This will be covered in more detail in a later section. Note also the power on reset circuit. Since the wake up state of a CMOS DAC is undefined and not repeatable, many modern DACs include a circuit to force the output to either half-scale of minimum scale, depending on whether the intended application is unipolar or bipolar. Probably the most obvious difference is that this is a multiple DAC package. Shrinking device geometries have allowed more circuitry to be included, even with the smaller packages in use today.

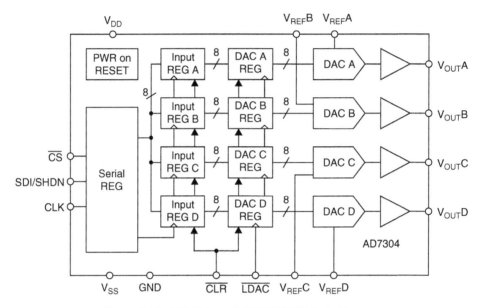

**Figure 6-15: AD7394 quad CMOS DAC block diagram**

The previous examples were CMOS devices, that is to say that the switches were implemented with CMOS switches. The switches could also be implemented with bipolar junction transistors (BJT). An example of this is the classic DAC-08. Its block diagram is shown in Figure 6-16. One major difference in the BJT implementation is that the switch allows current in one direction, versus the CMOS switch, which can allow bidirectional current. This limits the BJT DAC to two-quadrant operation while the CMOS version can be four-quadrant. Supplies tend to be different as well.

There are two ways in which the R–2R ladder network may be used as a DAC—known respectively as the *voltage mode* and the *current mode* (they are sometimes called "normal" mode and "inverted" mode, but as there is no consensus on whether the voltage mode or the current mode is the "normal" mode for a ladder network this nomenclature can be misleading, although in most cases the current mode would be considered the "normal" mode). Each mode has its advantages and disadvantages.

In the current-mode R–2R ladder DAC shown in Figure 6-17, the gain of the DAC may be adjusted with a series resistor at the $V_{REF}$ terminal, since in the current mode, the end of the ladder, with its code-independent impedance, is used as the $V_{REF}$ terminal; and the ends of the arms are switched between ground and an output line which must be held at ground potential. The normal connection of a current-mode ladder network output

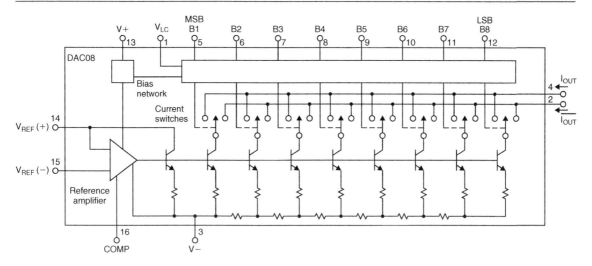

Figure 6-16: DAC-08 block diagram

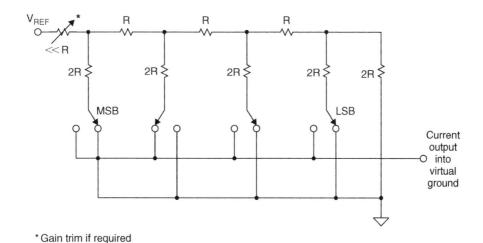

* Gain trim if required

Figure 6-17: Current-mode R–2R ladder network DAC

is to an op amp's inverting input (virtual ground), but stabilization of this op amp is complicated by the DAC output impedance variation with digital code.

Current-mode operation has a larger switching glitch than voltage mode since the switches connect directly to the output line(s). However, since the switches of a current-mode ladder network are always at ground potential, their design is less demanding and, in particular, their voltage rating does not affect the reference voltage rating. If switches capable of carrying current in either direction (such as CMOS devices) are used, the reference voltage may have either polarity, or may even be AC. Such a structure is one of the most common types used as a multiplying DAC (MDAC) which will be discussed later in this section.

Since the switches are always at, or very close to, ground potential, the maximum reference voltage may greatly exceed the logic voltage, provided the switches are make-before-break—which they are in this type of DAC. It is not unknown for a CMOS MDAC to accept a $\pm30\,V$ reference (or even a 60V peak-to-peak AC reference) while working from a single 5V supply.

In the voltage-mode R–2R ladder DAC shown in Figure 6-18, the "rungs" or arms of the ladder are switched between $V_{REF}$ and ground, and the output is taken from the end of the ladder. The output may be taken as a voltage, but the output impedance is independent of code, so it may equally well be taken as a current into a virtual ground.

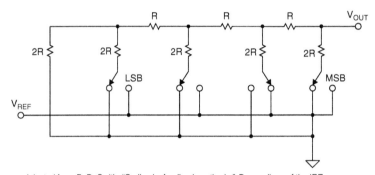

Adapted from: B. D. Smith, "Coding by feedback methods," *Proceedings of the IRE*, Vol. 41, August 1953, pp.1053–1058.

**Figure 6-18: Voltage-mode R–2R ladder network DAC**

The voltage output is an advantage of this mode, as is the constant output impedance, which eases the stabilization of any amplifier connected to the output node. Additionally, the switches switch the arms of the ladder between a low impedance $V_{REF}$ connection and ground, which is also, of course, low impedance, so capacitive glitch currents tend not to flow in the load. On the other hand, the switches must operate over a wide voltage range ($V_{REF}$ to ground), which is difficult from a design and manufacturing viewpoint, and the reference input impedance varies widely with code, so that the reference input must be driven from a very low impedance. In addition, the gain of the DAC cannot be adjusted by means of a resistor in series with the $V_{REF}$ terminal.

Probably the most important advantage to the voltage mode is that it allows single-supply operation. This is because the op amp that is commonly used as I/V converter in the current-mode converter is in the inverting configuration so would require a negative output for a positive input, assuming ground reference. Of course you could bias everything up to a rail-splitter ground, but that introduces other issues into the system.

## Multiplying DACs

In most cases the reference to a DAC is a highly stable DC voltage. In some instances, however, it is useful to have a variable reference. The R–2R ladder structure using CMOS switches can easily handle a bipolar signal on its input. Having the ability to have bipolar (positive and negative) signals on the input allows construction of two-quadrant and four-quadrant MDACs. Figure 6-19 shows the schematic and the table in this figure outlines the operation of a two-quadrant MDAC, and Figure 6-20 shows the schematic and the table in this figure outlines the operation of a four-quadrant MDAC for an 8-bit DAC.

DACs utilizing BJTs as switches, such as the DAC-08 above, cannot accommodate bipolar signals on the reference. Therefore they can only implement two-quadrant MDACs. In addition, the reference voltage cannot go all the way to 0V. The maximum allowable range is typically from 10% to 100% of the allowable reference voltage range.

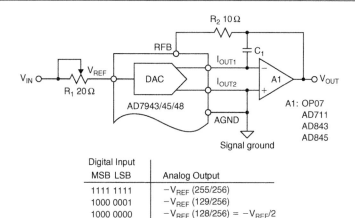

| Digital Input MSB LSB | Analog Output |
|---|---|
| 1111 1111 | $-V_{REF}$ (255/256) |
| 1000 0001 | $-V_{REF}$ (129/256) |
| 1000 0000 | $-V_{REF}$ (128/256) = $-V_{REF}/2$ |
| 0111 1111 | $-V_{REF}$ (127/256) |
| 0000 0001 | $-V_{REF}$ (1/256) |
| 0000 0000 | $-V_{REF}$ (0/256) = 0 |

Note: 1 LSB = $(2^{-8})$ $(V_{REF})$ = 1/256 $(V_{REF})$

**Figure 6-19: Two-quadrant MDAC**

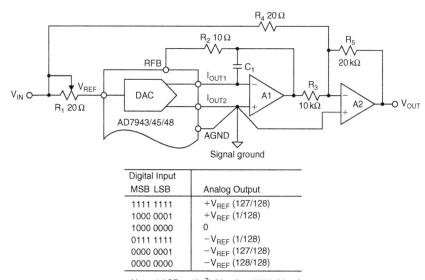

| Digital Input MSB LSB | Analog Output |
|---|---|
| 1111 1111 | $+V_{REF}$ (127/128) |
| 1000 0001 | $+V_{REF}$ (1/128) |
| 1000 0000 | 0 |
| 0111 1111 | $-V_{REF}$ (1/128) |
| 0000 0001 | $-V_{REF}$ (127/128) |
| 0000 0000 | $-V_{REF}$ (128/128) |

Note: 1 LSB = $(2^{-7})$ $(V_{REF})$ = 1/128 $(V_{REF})$

**Figure 6-20: Four-quadrant MDAC**

One of the main applications of the MDAC is as a variable gain amplifier, where the gain is controlled by the digital word applied to the MDAC.

The frequency response of the MDAC is limited by the parasitic capacitance across the switches in the off condition. As the frequency goes up the impedance of the capacitors goes down, effectively bypassing the switch. This reduces the off isolation at higher frequencies. Typically the frequency response of an MDAC will be on the order of 1 MHz.

## Segmented DACs

So far we have considered mostly basic DAC architectures. When we are required to design a DAC with a specific performance, it may well be that no single architecture is ideal. In such cases, two or more DACs may be combined in a single higher resolution DAC to give the required performance. These DACs may be of the same type or of different types and need not each have the same resolution. For example, the segmented string DAC is a segmented DAC where 2 Kelvin DACs are cascaded.

Typically, one DAC handles the MSBs, another handles the LSBs, and their outputs are added in some way. The process is known as "segmentation," and these more complex structures are called "segmented DACs." There are many different types of segmented DACs and some, but by no means all, of them will be illustrated in the next few diagrams. It is sometimes not obvious from looking at the data sheet that a particular DAC is segmented.

Very high speed DACs for video, communications, and other high frequency reconstruction applications are often built with arrays of fully decoded current sources. The 2 or 3 LSBs may use binary-weighted current sources. It is extremely important that such DACs have low distortion at high frequency, and there are several important issues to be considered in their design.

Two examples of segmented current-output DAC structures are shown in Figure 6-21. Figure 6-21(A) shows a resistor-based approach for the 7-bit DAC where the 3 MSBs are fully decoded, and the 4 LSBs are

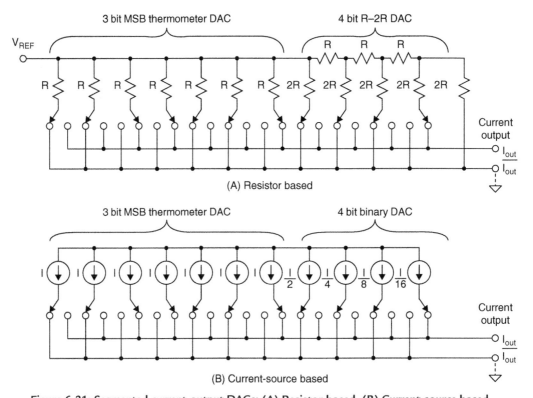

Figure 6-21: Segmented current-output DACs: (A) Resistor-based, (B) Current-source based

derived from an R–2R network. Figure 6-21(B) shows a similar implementation using current sources. The current source implementation is by far the most popular for today's high speed reconstruction DACs.

It is also often desirable to utilize more than one fully decoded thermometer section to make up the total DAC. Figure 6-22 shows a 6-bit DAC constructed from two fully decoded 3-bit DACs. As previously discussed, these current switches must be driven simultaneously from parallel latches in order to minimize the output glitch.

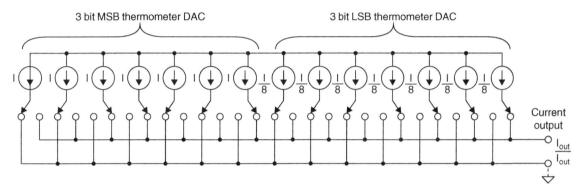

Figure 6-22: 6-Bit current-output segmented DAC based on two 3-bit thermometer DACs

The AD9775 14-bit, 160 MSPS (input)/400 MSPS (output) TxDAC™ uses three sections of segmentation as shown in Figure 6-23. Other members of the AD977x-family and the AD985x-family also use this same basic core.

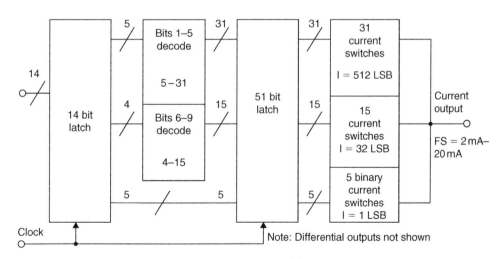

Figure 6-23: AD9775 TxDAC™ 14-bit CMOS DAC core

The first 5 bits (MSBs) are fully decoded and drive 31 equally weighted current switches, each supplying 512 LSBs of current. The next 4 bits are decoded into 15 lines which drive 15 current switches, each supplying 32 LSBs of current. The 5 LSBs are latched and drive a traditional binary-weighted DAC which supplies 1 LSB per output level. A total of 51 current switches and latches are required to implement this ultra low glitch architecture.

Decoding must be done before the new data is applied to the DAC so that all the data is ready and can be applied simultaneously to all the switches in the DAC. This is generally implemented by using a separate parallel latch for the individual switches in fully decoded array. If all switches were to change state instantaneously and simultaneously there would be no skew glitch—by very careful design of propagation delays around the chip and time constants of switch resistance and stray capacitance the update synchronization can be made very good, and hence the glitch-related distortion is very small.

## Sigma–Delta DACs

Sigma–delta DACs will be discussed in detail in the sigma–delta section.

## I/V Converters

Modern IC DACs provide either voltage or current outputs. Figure 6-24 below shows three fundamental configurations, all with the objective of using an op amp for a buffered output voltage.

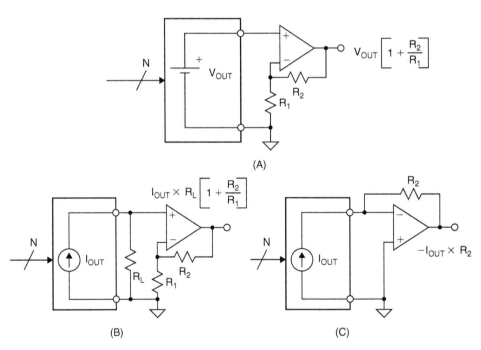

**Figure 6-24: Buffering DAC outputs with op amps**

Figure 6-24(A) shows a buffered voltage-output DAC. In many cases, the DAC output can be used directly, without additional buffering. If an additional op amp is needed, it is usually configured in a non-inverting mode, with gain determined by $R_1$ and $R_2$.

There are two basic methods for dealing with a current-output DAC.

A direct method to convert the output current into a voltage is shown in Figure 6-24(C). This circuit is usually called a current-to-voltage converter, or I/V. In this circuit, the DAC output drives the inverting input of an op amp, with the output voltage developed across the $R_2$ feedback resistor. In this approach the DAC output always operates at virtual ground (which may give a linearity improvement vis-à-vis Figure. 6-24(B)).

In Figure 6-24(B) a voltage is simply developed across external load resistor, RL. This is typically done with high speed op amps. An external op amp can be used to buffer and/or amplify this voltage if required. The output current is dumped into a resistor instead of into an op amp directly since the fast edges may exceed the slew rate of the amplifier and cause distortion. Many DACs supply full-scale currents of 20 mA or more, thereby allowing reasonable voltages to be developed across fairly low value load resistors. For instance, fast settling video DACs typically supply nearly 30 mA full-scale current, allowing 1 V to be developed across a source and load terminated 75 Ω coaxial cable (representing a DC load of 37.5 Ω to the DAC output).

The general selection process for an op amp used as a DAC buffer is that the performance of the op amp should not compromise the performance of the DAC. The basic specifications of interest are DC accuracy, noise, settling time, bandwidth, distortion, etc.

## *Differential to Single-Ended Conversion Techniques*

A general model of a modern current-output DAC is shown in Figure 6-25. This model is typical of the AD976X and AD977X TxDAC™ series (see Reference 1).

Current output is more popular than voltage output, especially at audio frequencies and above. If the DAC is fabricated on a bipolar or BiCMOS process, it is likely that the output will sink current, and that the output impedance will be less than 500 Ω (due to the internal R–2R resistive ladder network). On the other hand, a CMOS DAC is more likely to source output current and have a high output impedance, typically greater than 100 kΩ.

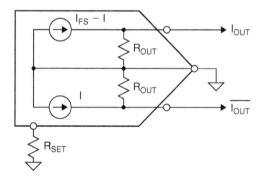

**Figure 6-25: Model of high speed DAC output**

Another consideration is the output *compliance voltage*—the maximum voltage swing allowed at the output in order for the DAC to maintain its linearity. This voltage is typically 1–1.5 V, but will vary depending on the DAC. Best DAC linearity is generally achieved when driving a virtual ground, such as an op amp I/V converter. Modern current-output DACs usually have differential outputs, to achieve high CM rejection and reduce the even-order distortion products. Full-scale output currents in the range of 2–20 mA are common.

In most applications, it is desirable to convert the differential output of the DAC into a single-ended signal, suitable for driving a coax line. This can be readily achieved with a radio frequency (RF) transformer, provided low frequency response is not required. Figure 6-26 shows a typical example of this approach. The high impedance current-output of the DAC is terminated differentially with 50 Ω, which defines the source impedance to the transformer as 50 Ω.

The resulting differential voltage drives the primary of a 1:1 RF transformer, to develop a single-ended voltage at the output of the secondary winding. The output of the 50 Ω LC filter is matched with the 50 Ω load resistor RL, and a final output voltage of 1 $V_{p-p}$ is developed.

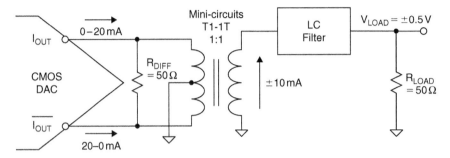

**Figure 6-26: Differential transformer coupling**

The transformer not only serves to convert the differential output into a single-ended signal, but it also isolates the output of the DAC from the reactive load presented by the LC filter, thereby improving overall distortion performance.

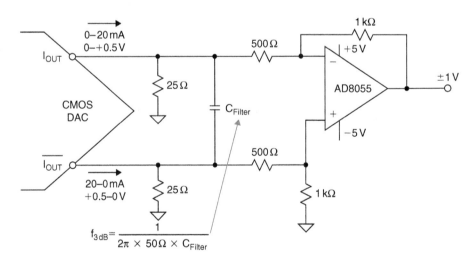

**Figure 6-27: Differential DC coupled output using a dual-supply op amp**

An op amp connected as a differential to single-ended converter can be used to obtain a single-ended output when frequency response to DC is required. In Figure 6-28 the AD8055 op amp is used to achieve high bandwidth and low distortion (see Reference 2). The current-output DAC drives balanced 25 Ω resistive loads, thereby developing an out of phase voltage of 0V to +0.5V at each output. The AD8055 is configured for a gain of 8, to develop a final single-ended ground-referenced output voltage of 2$V_{p-p}$. Note that because the output signal swings above and below ground, a dual-supply op amp is required (Figure 6-27).

The $C_{FILTER}$ capacitor forms a differential filter with the equivalent $50\,\Omega$ differential output impedance. This filter reduces any slew induced distortion of the op amp, and the optimum cutoff frequency of the filter is determined empirically to give the best overall distortion performance.

A modified form of Figure 6-26 circuit can also be operated on a single supply, provided the CM voltage of the op amp is set to mid-supply ($+2.5\,V$). This is shown in Figure 6-28. The output voltage is $2\,V_{p-p}$ centered around a CM voltage of $+2.5\,V$. This CM voltage can be either developed from the $+5\,V$ supply using a resistor divider, or directly from a $+2.5\,V$ voltage reference. If the $+5\,V$ supply is used as the CM voltage, it must be heavily decoupled to prevent supply noise from being amplified.

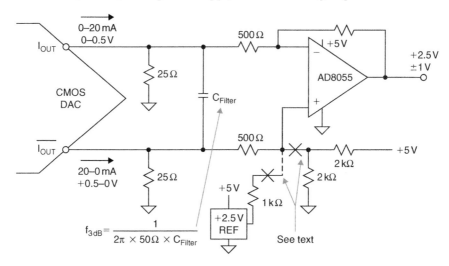

**Figure 6-28: Differential DC coupled output using a single-supply op amp**

## Single-Ended Current-to-Voltage Conversion

Single-ended current-to-voltage conversion is easily performed using a single op amp as an I/V converter, as shown in Figure 6-29. The $10\,mA$ full-scale DAC current from the AD768 (see Reference 3) develops a $0\,V$ to $+2\,V$ output voltage across the $200\,\Omega$ $R_F$.

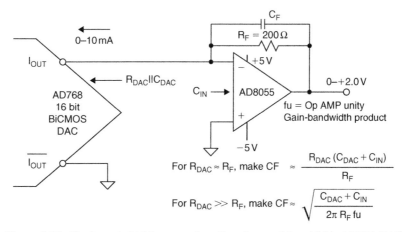

**Figure 6-29: Single-ended I/V op amp interface for precision 16-bit AD768 DAC**

Driving the virtual ground of the AD8055 op amp minimizes any distortion due to nonlinearity in the DAC output impedance. In fact, most high resolution DACs of this type are factory trimmed using an I/V converter.

It should be recalled, however, that using the single-ended output of the DAC in this manner will cause degradation in the CM rejection and increased second-order distortion products, compared to a differential operating mode.

The $C_F$ feedback capacitor should be optimized for best pulse response in the circuit. The equations given in the diagram should only be used as guidelines. A more detailed analysis of this circuit is given in the References.

## Differential Current-to-Differential Voltage Conversion

If a buffered differential voltage output is required from a current-output DAC, the AD813X-series of differential amplifiers can be used as shown in Figure 6-30.

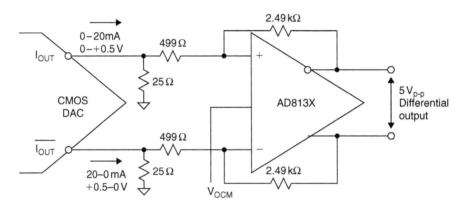

**Figure 6-30: Buffering high speed DACs using AD813X differential amplifier**

The DAC output current is first converted into a voltage that is developed across the $25\,\Omega$ resistors. The voltage is amplified by a factor of five using the AD813X. This technique is used in lieu of a direct I/V conversion to prevent fast slewing DAC currents from overloading the amplifier and introducing distortion. Care must be taken so that the DAC output voltage is within its compliance rating.

The $V_{OCM}$ input on the AD813X can be used to set a final output CM voltage within the range of the AD813X. If transmission lines are to be driven at the output, adding a pair of $75\,\Omega$ resistors will allow this.

## Digital Interfaces

The earliest monolithic DACs contained little, if any, logic circuitry, and parallel data had to be maintained on the digital input to maintain the digital output. Today almost all DACs are latched and data need only be written to them, not maintained. Some even have nonvolatile latches and remember settings while turned off.

There are innumerable variations of DAC digital input structure, which will not be discussed here, but nearly all are described as "double-buffered." A double-buffered DAC has two sets of latches. Data is initially latched in the first rank and subsequently transferred to the second as shown in Figure 6-31. There are two reasons why this arrangement is useful.

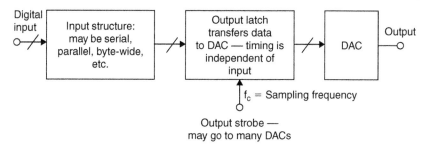

**Figure 6-31: Double-buffered DAC permits complex input structures and simultaneous update**

The first is that it allows data to enter the DAC in many different ways. A DAC without a latch, or with a single latch, must be loaded in parallel with all bits at once, since otherwise its output during loading may be totally different from what it was or what it is to become. A double-buffered DAC, on the other hand, may be loaded with parallel data, or with serial data, or with 4-bit or 8-bit words, or whatever, and the output will be unaffected until the new data is completely loaded and the DAC receives its update instruction.

The other convenience of the double-buffered structure is that many DACs may be updated simultaneously: data is loaded into the first rank of each DAC in turn, and when all is ready, the output buffers of all the DACs are updated at once. There are many DAC applications where the output of a number of DACs must change simultaneously, and the double-buffered structure allows this to be done very easily.

Most early monolithic high resolution DACs had parallel or byte-wide data ports and tended to be connected to parallel data buses and address decoders and addressed by microprocessors as if they were very small write-only memories. (Some parallel DACs are not write-only, but can have their contents read as well—this is convenient for some applications, but is not very common.) A DAC connected to a data bus is vulnerable to capacitive coupling of logic noise from the bus to the analog output. Serial interfaces are less vulnerable to such noise (since fewer noisy pins are involved), use fewer pins and therefore take less board space, and are frequently more convenient for use with modern microprocessors, most of which have serial data ports. Some, but not all, of such serial DACs have data outputs as well as data inputs so that several DACs may be connected in series and data clocked to them all from a single serial port. This arrangement is often referred to as "daisy-chaining."

Of course, serial DACs cannot be used where high update rates are involved, since the clock rate of the serial data would be too high. Some very high speed DACs actually have two parallel data ports and use them alternately in a multiplexed fashion (sometimes this is called a "ping-pong" input) to reduce the data rate on each port as shown in Figure 6-32. The alternate loading (ping-pong) DAC in the diagram loads from port A and port B alternately on the rising and falling edges of the clock, which must have a mark-space ratio close to 50:50. The internal clock multiplier ensures that the DAC itself is updated with data A and data B alternately at exactly 50:50 time ratio, even if the external clock is not so precise.

Historically IC logic circuitry (with the exception of emitter coupled logic or ECL) operated from 5V supplies and had compatible logic levels—with a few exceptions 5V logic would interface with other 5V logic. Today, with the advent of low voltage logic operating with supplies of 3.3V, 2.7V or even less, it is important to ensure that logic interfaces are compatible. There are several issues which must be considered—absolute maximum ratings, worst case logic levels, and timing. The logic inputs of ICs generally have absolute maximum ratings, as do most other inputs, of 300mV outside the power supply.

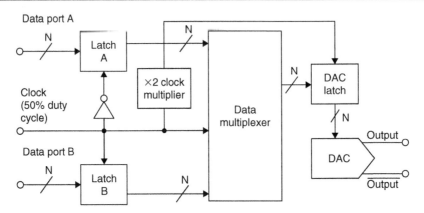

**Figure 6-32: Alternate loading (ping-pong) high speed DAC**

Note that these are instantaneous ratings. If an IC has such a rating and is currently operating from a $+5\,\text{V}$ supply then the logic inputs may be between $-0.3$ and $+5.3\,\text{V}$—but if the supply is not present then that input must be between $+0.3\,\text{V}$ and $-0.3\,\text{V}$ not the $-0.3\,\text{V}$ to $+5.3\,\text{V}$ which are the limits once the power is applied—ICs cannot predict the future.

The reason for the rating of $0.3\,\text{V}$ is to ensure that no parasitic diode inside the IC is ever turned on by a voltage outside the IC's absolute maximum rating. It is quite common to protect an input from such overvoltage with a Schottky diode clamp. At low temperatures the clamp voltage of a Schottky diode may be a little more than $0.3\,\text{V}$, and so the IC may see voltages just outside its absolute maximum rating. Although, strictly speaking, this subjects the IC to stresses outside its absolute maximum ratings and so is forbidden, this is an acceptable exception to the general rule provided the Schottky diode is at a similar temperature to the IC that it is protecting (say within $\pm 10°\text{C}$).

Some low voltage devices, however, have inputs with absolute maximum ratings which are substantially greater than their supply voltage. This allows such circuits to be driven by higher voltage logic without additional interface or clamp circuitry. But it is important to read the data sheets and ensure that both logic levels and absolute maximum voltages are compatible for all combinations of high and low supplies.

This is the general rule when interfacing different low voltage logic circuitry—it is always necessary to check both that at the lowest value of its power supply the logic 1 output from the driving circuit applied to its worst case load is greater than the specified minimum logic 1 input for the receiving circuit, and that, again with its lowest value of power supply and with its output sinking maximum allowed current, the logic 0 output is less than the specified logic 0 input of the receiver. If the logic specifications of your chosen devices do not meet these criteria it will be necessary to select different devices, use different power supplies, or use additional interface circuitry to ensure that the required levels are available. Note that additional interface circuitry will introduce extra delays in timing.

It is not sufficient to build an experimental set-up and test it. In general, logic thresholds are generously specified and usually logic circuits will work correctly well outside their specified limits—but it is not possible to rely on this in a production design. At some point a batch of devices near the limit on low output swing will be required to drive some devices needing slightly more drive than usual—and will be unable to do so.

One of the latest developments in high speed logic interface is low voltage differential signaling (LVDS). LVDS presents a solution to the high speed converter interface problem by mitigating the effects of CMOS single-ended interfaces and accommodating higher data rates. The LVDS standard specifies a p-p voltage swing of 350 mV around a common-mode voltage (CMV) of 1.2 V, which facilitates transmission of high speed differential digital signals with balanced current, thereby reducing the slew rate requirement. Reducing the slew rate eliminates the gradients that result in noise from ground bounce that are present in conventional CMOS drivers. Ground-bounce noise can couple back into sensitive analog circuits and degrade the converter's dynamic range. Parallel LVDS interfaces enable much higher data rates and optimum dynamic performance, in high speed data converters.

LVDS also offers some benefit in reduced electromagnetic interference (EMI). The EMI fields generated by the opposing currents will tend to cancel each other (for matched edge rates). Trace length, skew, and discontinuities will reduce this benefit and should be avoided.

LVDS also offers simpler timing constraints compared to a demuxed CMOS solution at similar data rates. A demuxed databus requires a synchronization signal that is not required in LVDS. In demuxed CMOS buses, a clock equal to one-half the ADC sample rate is needed, adding cost and complexity, that is not required in LVDS. In general, the LVDS is more forgiving and can lead to a simpler, cleaner design.

**Figure 6-33: LVDS output levels**

The LVDS specification (IEEE Standard 1596.3) was developed as an extension to the 1992 SCI protocol (IEEE Standard 1596-1992). The original SCI protocol was suitable for high speed packet transmissions in high end computing and used ECL levels. However, for low end and power-sensitive applications, a new standard was needed. LVDS signals were chosen because the voltage swing is smaller than that of ECL outputs, allowing for lower power supplies in power-sensitive designs.

Unlike CMOS, which is typically a voltage output, LVDS is a current-output technology. LVDS outputs for high performance converters should be treated differently than standard LVDS outputs used in digital logic (Figure 6-33). While standard LVDS can drive 1–10 m in high speed digital applications (dependent on data rate), it is not recommended to let high performance converters drive that distance. It is recommended to keep the output trace lengths short (<2 inches), minimizing the opportunity for any noise coupling onto the outputs from the adjacent circuitry, which may get back to the analog outputs.

The differential output traces should be routed close together, maximizing common-mode rejection (CMR) with the 100 Ω termination resistor close to the receiver. Users should pay attention to printed circuit board (PCB) trace lengths to minimize any delay skew.

A typical differential microstrip PCB trace cross section is shown in Figure 6-34.

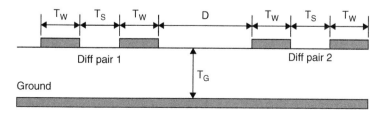

**Figure 6-34: PCB trace spacing**

**Layout Guidelines**

Keep $T_W$, $T_S$, and D constant over the trace length

Keep $T_S \sim <2\, T_W$

Avoid use of vias where possible

Keep $D > 2\, T_S$

Avoid 90° bends if possible

Design $T_W$ and $T_G$ for $\sim 50\,\Omega$

Power supply decoupling is very important with these fast (<0.5 ns) edge rates. A low inductance, surface mount capacitor should be placed at every power supply and ground pin as close to the converter as possible. Placing the decoupling caps on the other side of the PCB is not recommended, since the via inductance will reduce the effective decoupling. The differential $Z_O$ will tend to be slightly lower than twice the single-ended $Z_O$ of each conductor due to proximity effects—the $Z_O$ of each line should be designed to be slightly higher than $50\,\Omega$. Simulation can be used in critical applications to verify impedance matching. In short runs, this should not be critical.

## Data Converter Logic: Timing and Other Issues

It is not the purpose of this brief section to discuss logic architectures, so we shall not define the many different data converter logic interface operations and their timing specifications except to note that data converter logic interfaces may be more complex than you expect—do not expect that because there is a pin with the same name on memory and interface chips it will behave in exactly the same way in a data converter. Unfortunately, there is not a standard nomenclature for pin functionality, even for the same manufacturer. The data sheet should always be consulted to determine the operation of all control pins. Also some data converters reset to a known state on power-up but many more do not.

But it is very necessary to consider general timing issues. The new low voltage processes which are used for many modern data converters have a number of desirable features. One which is often overlooked by users (but not by converter designers!) is their higher logic speed. DACs built on older processes frequently had logic which was orders of magnitude slower than the microprocessors that they interfaced with and it was sometimes necessary to use separate buffers, or multiple WAIT instructions, to make the two compatible. Today it is much more common for the write times of DACs to be compatible with those of the fast logic with which they interface.

Nevertheless not all DACs are speed compatible with all logic interfaces and it is still important to ensure that minimum data set-up times and write pulse widths are observed. Again, experiments will often show that devices work with faster signals than their specification requires—but at the limits of temperature or supply voltage some may not and interfaces should be designed on the basis of specified rather than measured timing.

## Interpolating DACs (Interpolating TxDACs)

The concept of oversampling, to be discussed in another section (on sampling theory), can be applied on high speed DACs typically used in communications applications. Oversampling relaxes the requirements on the output filter as well as increasing the SNR due to process gain.

Assume a traditional DAC is driven at an input word rate of 30 MSPS (see Figure 6-35(A)). Assume the DAC output frequency is 10 MHz. The image frequency component at $30 - 10 = 20$ MHz must be attenuated by the analog reconstruction filter, and the transition band of the filter is therefore 10–20 MHz. Assume that the image frequency must be attenuated by 60 dB. The filter must therefore go from a passband of 10 MHz to 60 dB stopband attenuation over the transition band lying between 10 and 20 MHz (one octave). Filter gives 6 dB attenuation per octave for each pole. Therefore, a minimum of 10 poles is required to provide the desired attenuation. This is a fairly aggressive filter and would involve high Q sections which would be difficult to align and manufacture. Filters become even more complex as the transition band becomes narrower.

Assume that we increase the DAC update rate to 60 MSPS and insert a "zero" between each original data sample. The parallel data stream is now 60 MSPS, but we must now determine the value of the zero-value data points. This is done by passing the 60 MSPS data stream with the added zeros through a digital interpolation filter which computes the additional data points. The response of the digital filter relative to the 2-times oversampling frequency is shown in Figure 6-35(B). The analog antialiasing filter transition zone is now 10–50 MHz (the first image occurs at $2f_c - f_o = 60 - 10 = 50$ MHz). This transition zone is a little greater than 2 octaves, implying that a 5- or 6-pole Butterworth filter is sufficient.

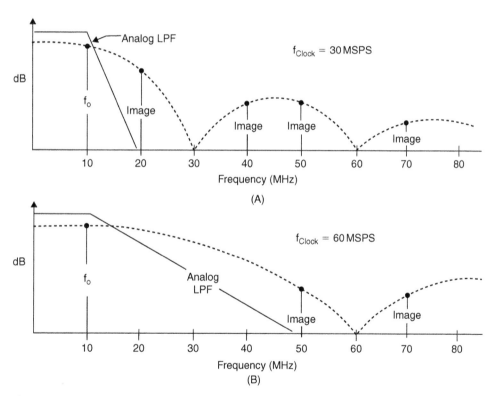

**Figure 6-35: Analog filter requirements for $f_o = 10$ MHz: (A) $f_c = 30$ MSPS, and (B) $f_c = 60$ MSPS**

The AD9773/AD9775/AD9777 (12/14/16-bit) series of Transmit DACs (TxDAC™) are selectable 2 times, 4 times, or 8 times oversampling interpolating dual DACs, and a simplified block diagram is shown in Figure 6-36. These devices are designed to handle 12/14/16-bit input word rates up to 160 MSPS. The output word rate is 400 MSPS maximum. For an output frequency of 50 MHz, an input update rate of 160 MHz, and an oversampling ratio of 2 times, the image frequency occurs at 320 MHz $-50$ MHz $= 270$ MHz. The transition band for the analog filter is therefore 50–270 MHz. Without 2 times oversampling, the image frequency occurs at 160 MHz − 50 MHz = 110 MHz, and the filter transition band is 60–110 MHz.

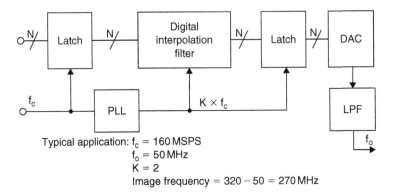

Figure 6-36: Oversampling interpolating TxDAC™ simplified block diagram

## Reconstruction Filters

The output of a DAC is not a continuously varying waveform, but instead a series of DC levels. This output must be passed through a filter to remove the high frequency components and smooth waveform into a more truly analog waveform.

The concept of filtering is discussed in more detail in Chapter 8.

In general, to preserve spectral purity, the images of the DAC output must be attenuated below the resolution of DAC. To use the example sited above, we assume that the DAC output passband is 10 MHz. The sample rate is 30 MHz. Therefore the image of the passband that must be attenuated is 30 MHz − 10 MHz = 20 MHz. This is the sample rate minus the passband frequency. The DAC in this example is a 10-bit device, which would indicate a distortion level of −60 dB. So a reconstruction filter should reduce the image by 60 dB while not attenuating the fundamental at all. Since a filter attenuates at 6 dB/pole, this would indicate that a tenth-order filter would be required.

There are several other considerations that must be taken into account.

First is that most filter cutoffs are measured at the −3 dB point. Therefore, if we do not want the fundamental attenuated, some margin in the filter is required. The graphs in the filter section will help illustrate this point. This will cause the transition band to become narrower and thus the order of the filter to increase.

Secondly, there is a phenomenon called "sinc."

## Sin(x)/(x)(Sinc)

The output of a DAC is not a continually varying waveform but instead a series of DC levels. The DAC puts out a DC level until it is told to put out a new level. This is illustrated in Figure 6-37.

The width of the pulses is $1/F_S$. The spectrum of each pulse is the sin(x)/x curve. This is also known as the sinc curve. This response is added to the response of the reconstruction filter to provide the overall response of the converter. This will cause an amplitude error as the output frequency approaches the Nyquist frequency ($F_S/2$). The value of the sinc function is shown in Figure 6-38. Some high speed DACs incorporate an inverse filter (in the digital domain) to compensate for this rolloff.

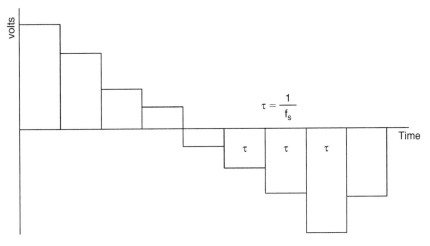

**Figure 6-37: Output of a DAC**

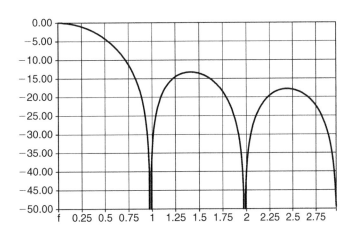

**Figure 6-38: Sinc (sin x/x) curve (normalized to $F_S$)**

## *Intentionally Nonlinear DACs*

Thus far, we have emphasized the importance of maintaining good differential and integral linearity. However, there are situations where ADCs and DACs which have been made intentionally nonlinear (but maintaining good differential linearity) are useful, especially when processing signals having a wide dynamic range. One of the earliest uses of nonlinear data converters was in the digitization of voiceband signals for pulse code modulation (PCM) systems. Major contributions were made at Bell Labs during the development of the T1 carrier system. The motive for the nonlinear ADCs and DACs was to reduce the total number of bits (and therefore the serial transmission rate) required to digitize voice channels. Straight linear encoding of a voice channel required 11 or 12 bits at an 8 kSPS per channel sampling rate. In the 1960s Bell Labs determined that 7-bit nonlinear encoding was sufficient, and later in the 1970s went to 8-bit nonlinear encoding for better performance.

The nonlinear transfer function allocates more quantization levels out of the total range for small signals and fewer for large amplitude signals. In effect, this reduces the quantization noise associated with small signals (where it is most noticeable) and increases the quantization noise for larger signals (where it is less noticeable). The term *companding* is generally used to describe this form of encoding.

The logarithmic transfer function chosen is referred to as the "Bell μ-255" standard, or simply "μ-law." A similar standard developed in Europe is referred to as "A-law." The Bell μ-law allows a dynamic range of about 4,000:1 using 8 bits, whereas an 8-bit linear data converter provides a range of only 256:1.

The first generation channel bank (D1) used temperature-controlled resistor-diode networks for "compressors" ahead of a 7-bit linear ADC in the transmitter to generate the logarithmic transfer function. Corresponding resistor-diode "expandors" having an inverse transfer function followed the 7-bit linear DAC in the receiver. The next generation D2 channel banks used nonlinear ADCs and DACs to accomplish the compression/expansion functions in a much more reliable and cost-effective manner and eliminated the need for the temperature-controlled diode networks.

In his 1953 classic paper, B.D. Smith proposed that the transfer function of a successive approximation ADC utilizing a nonlinear internal DAC in the feedback path would be the inverse transfer function of the DAC (Reference 8). The same basic DAC could therefore be used in the ADC and also for the reconstruction DAC. Later in the 1960s and early 1970s, nonlinear ADC and DAC technology using piecewise linear approximations of the desired transfer function allowed low cost high volume implementations (References 18–23). These nonlinear 8-bit, 8-kSPS data converters became popular telecommunications building blocks.

The nonlinear transfer function of the 8-bit DAC is first divided into 16 segments (chords) of different slopes—the slopes are determined by the desired nonlinear transfer function. The 4 MSBs determine the segment containing the desired data point, and the individual segment is further subdivided into 16 equal quantization levels by the 4 LSBs of the 8-bit word. This is shown in Figure 6-39 for a 6-bit DAC, where the first 3 bits identify one of the 8 possible chords, and each chord is further subdivided into 8 equal levels defined by the 3 LSBs. The 3 MSBs are generated using a nonlinear string DAC, and the 3 LSBs are generated using a 3-bit binary R–2R DAC.

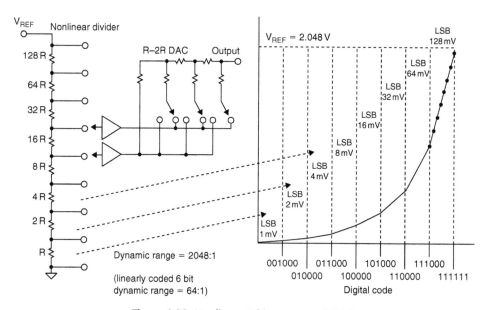

**Figure 6-39: Nonlinear 6-bit segmented DAC**

In 1982, Analog Devices introduced the LOGDAC™ AD7111 monolithic MDAC featuring wide dynamic range using a logarithmic transfer function. The basic DAC in the LOGDAC is a linear 17-bit voltage-mode R–2R DAC preceded by an 8-bit input decoder (a functional diagram of the LOGDAC is shown in Figure 6-40). The LOGDAC can attenuate an analog input signal, $V_{IN}$, over the range 0–88.5 dB in 0.375 dB steps. The degree of attenuation across the DAC is determined by a nonlinear-coded 8-bit word applied to the onboard decode logic. This 8-bit word is mapped into the appropriate 17-bit word which is then applied to a 17-bit, R–2R ladder. In addition to providing the logarithmic transfer function, the LOGDAC also acts as a full four-quadrant MDAC.

With the introduction of high resolution linear ADCs and DACs, the method used in the LOGDAC™ is widely used today to implement various nonlinear transfer functions such as the μ-law and A-law companding functions required for telecommunications and other applications. Figure 6-41 shows a general block diagram of the modern approach. The μ-law or A-law companded input data is mapped into data points on the transfer function of a high resolution DAC. This mapping can be easily accomplished by a simple lookup table in either hardware, software, or firmware. A similar nonlinear ADC can be constructed by digitizing the analog input signal using a high resolution ADC and mapping the data points into a shorter word using the appropriate transfer function. A big advantage of this method is that the transfer curve does not have to be approximated with straight line segments as in the earlier method, thereby providing more accuracy.

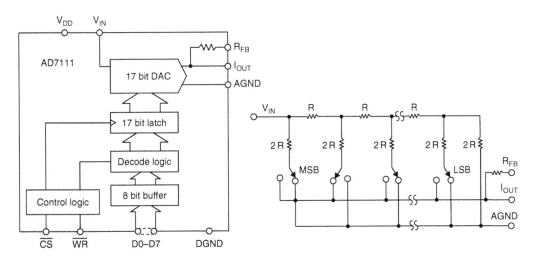

**Figure 6-40: AD7111 LOGDAC™ (released 1982)**

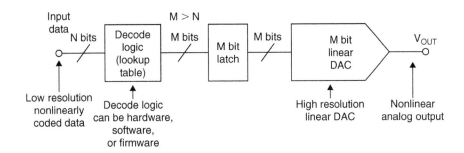

**Figure 6-41: General nonlinear DAC**

369

# ADC Architectures

The basic ADC function is shown in Figure 6-42. This could also be referred to as a quantizer. Most ADC chips also include some of the support circuitry, such as clock oscillator for the sampling clock, reference (REF), the SHA function, and output data latches. In addition to these basic functions, some ADCs have additional circuitry built in. These functions could include multiplexers, sequencers, auto-calibration circuits, programmable gain amplifiers (PGAs), etc.

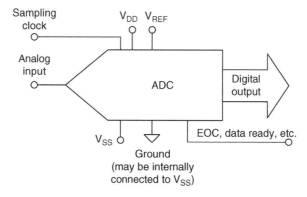

**Figure 6-42: Basic ADC function**

Similar to DACs, some ADCs use external references and have a reference input terminal, while others have an output from an internal reference. In some instances the ADC may have an internal reference that is pinned out through a resistor. This connection allows the reference to be filtered (using the internal R and an external C) or by allowing the internal reference to be overdriven by an external reference. The AD789X family of parts are examples of ADC that use this type of connection. The simplest ADCs, of course, have neither—the reference is on the ADC chip and has no external connections.

If an ADC has an internal reference, its overall accuracy is specified when using that reference. If such an ADC is used with a perfectly accurate external reference, its absolute accuracy may actually be worse than when it is operated with its own internal reference. This is because it is trimmed for absolute accuracy when working with its own actual reference voltage, not with the nominal value. Twenty years ago it was common for converter references to have accuracies as poor as $\pm5\%$ since these references were trimmed for low temperature coefficient rather than absolute accuracy, and the inaccuracy of the reference was compensated in the gain trim of the ADC itself. Today the problem is much less severe, but it is still important to check for possible loss of absolute accuracy when using an external reference with an ADC which has a built-in one.

ADCs which have reference terminals must, of course, specify their behavior and parameters. If there is a reference input the first specification will be the reference input voltage—and of course this has two values, the absolute maximum rating, and the range of voltages over which the ADC performs correctly.

Most ADCs require that their reference voltage is within quite a narrow range whose maximum value is less than or equal to the ADC's $V_{DD}$.

The reference input terminal of an ADC may be buffered as shown in Figure 6-43, in which case it has input impedance (usually high) and bias current (usually low) specifications, or it may connect directly to the ADC. In either case, the transient currents developed on the reference input due to the internal conversion process need good decoupling with external low inductance capacitors. Good ADC data sheets recommend appropriate decoupling networks.

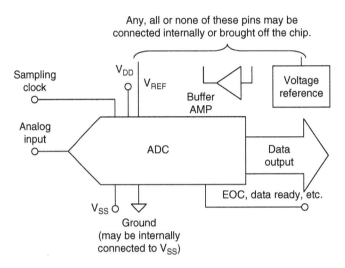

**Figure 6-43: ADC with reference and buffer**

The reference output may be buffered or unbuffered. If it is buffered, the maximum output current will probably be specified. In general such a buffer will have a unidirectional output stage which sources current but does not allow current to flow into the output terminal. If the buffer does have a push–pull output stage (not as common), the output current will probably be defined as ± (some value) mA. If the reference output is unbuffered, the output impedance may be specified, or the data sheet may simply advise the use of a high input impedance external buffer.

There are some instances where the power supply is the reference. In these cases it is imperative to make sure the power supply is clean.

The *sampling clock* input is a critical function in an ADC and a source of some confusion. It could truly be the sampling clock. This frequency would typically be several times higher than the sampling rate of the converter. It could also be a convert start (or encode) command which would happen once per conversion. Pipeline architecture devices and sigma–delta ($\Sigma\Delta$) converters are continuously converting and have no convert start command.

Regardless of the ADC, it is extremely important to read the data sheet and determine exactly what the external clock requirements are, because they can vary widely from one ADC to another.

At some point after the assertion of the sampling clock, the output data is valid. This data may be in parallel or serial format depending on the ADC. Early successive approximation ADCs such as the AD574 simply provided a STATUS output (STS) which went high during the conversion, and returned to the low state when the output data was valid. In other ADCs, this line is variously called *busy, end-of-conversion (EOC), data ready* (DRDY), etc. Regardless of the ADC, there must be some method of knowing when the output data is valid—and again, the data sheet is where this information can always be found.

Another detail which can cause trouble is the difference between EOC and DRDY. EOC indicates that conversion has finished, DRDY that data is available at the output. In some ADCs, EOC functions as DRDY—in others, data is not valid until several tenths of nanoseconds *after* the EOC has become valid, and if EOC is used as a data strobe, the results will be unreliable.

There are one or two other practical points which are worth remembering about the logic of ADCs. On power-up, many ADCs do not have logic reset circuitry and may enter an anomalous logical state. Several conversions may be necessary to restore their logic to proper operation so: (a) the first few conversions after power-up should never be trusted, and (b) control outputs (EOC, DRDY, etc.) may behave in unexpected ways at this time (and not necessarily in the same way at each power-up), and (c) care should be taken to ensure that such anomalous behavior cannot cause system latch-up. For example, EOC should not be used to initiate conversion if there is any possibility that EOC will not occur until the first conversion has taken place, as otherwise initiation will never occur.

Some low power ADCs now have power-saving modes of operation variously called *standby*, *power-down*, *sleep*, etc. When an ADC comes out of one of these low power modes, there is a certain recovery time required before the ADC can operate at its full specified performance. The data sheet should therefore be carefully studied when using these modes of operation.

As a final example, some ADCs use CS (chip select) edges to reset internal logic, and it may not be possible to perform another conversion without asserting or reasserting CS (or it may not be possible to read the same data twice, or both).

For more detail, it is important to read the whole data sheet before using an ADC since there are innumerable small logic variations from type to type. Unfortunately, many data sheets are not as clear as one might wish, so it is also important to understand the general principles of ADCs in order to interpret data sheets correctly. That is one of the purposes of this section.

There are a couple of general trend in ADCs that should be addressed. The first is the general trend toward lower supply voltages. This is partially due to the processes, particularly CMOS, which are used to manufacture the chips. Increasing demand for speed has driven the feature size of the processes down. This typically results in lower breakdown voltages for the transistors. This, in turn, requires lower supply voltages. Very few new parts are developed with the legacy $\pm 15\,$V supplies and $\pm 10\,$V input range.

Since the input signal range of the ADCs is shrinking, there is also a trend toward differential inputs. This helps improve the dynamic range of a converter, typically by 6 dB. There could be even further improvement since the common-mode ground-referenced noise is rejected. In many cases the differential input can be driven single endedly (with the resultant reduction of SNR). Occasionally the REF input might also be differential.

## The Comparator: A 1-Bit ADC

A comparator is a 1-bit ADC (see Figure 6-44). If the input is above a threshold, the output has one logic value, below it has another. There is no ADC architecture which does not use at least one comparator of some sort. So while a 1-bit ADC is of very limited usefulness it is a building block for other architectures.

Comparators used as building blocks in ADCs need good resolution which implies high gain. This can lead to uncontrolled oscillation when the differential input approaches zero. In order to prevent this, *hysteresis* is often added to comparators using a small amount of positive feedback. Figure 6-44 shows the effects of hysteresis on the overall transfer function. Many comparators have a millivolt or two of hysteresis to encourage "snap" action and to prevent local feedback from causing instability in the transition region. Note that the resolution of the comparator can be no less than the hysteresis, so large values of hysteresis are generally not useful.

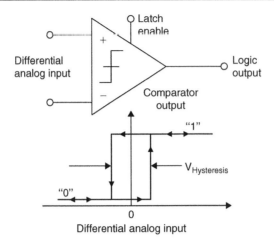

Figure 6-44: The comparator: A 1-bit ADC

## Successive Approximation ADCs

The successive approximation ADC has been the mainstay of data acquisition for many years. Recent design improvements have extended the sampling frequency of these ADCs into the megahertz region.

The basic successive approximation ADC is shown in Figure 6-45. It performs conversions on command. On the assertion of the CONVERT START command, the SHA is placed in the *hold* mode, and all the bits of the successive approximation register (SAR) are reset to "0" except the MSB which is set to "1." The SAR output drives the internal DAC. If the DAC output is greater than the analog input, this bit in the SAR is reset, otherwise it is left set. The next MSB is then set to "1." If the DAC output is greater than the analog input, this bit in the SAR is reset, otherwise it is left set. The process is repeated with each bit in turn. When all the bits have been set, tested, and reset or not as appropriate, the contents of the SAR correspond to the value of the analog input, and the conversion is complete. These bit "tests" can form the basis of a serial output version SAR-based ADC.

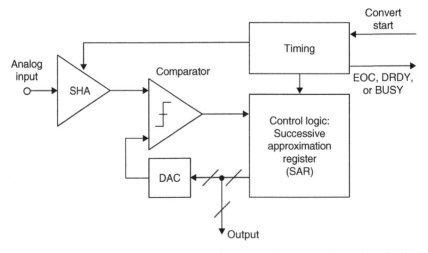

Figure 6-45: Basic successive approximation ADC (feedback subtraction ADC)

The fundamental timing diagram for a typical SAR ADC is shown in Figure 6-46. The EOC is generally indicated by an EOC, DRDY, or a busy signal (actually, *not*-BUSY indicates EOC). The polarities and name of this signal may be different for different SAR ADCs, but the fundamental concept is the same. At the beginning of the conversion interval, the signal goes high (or low) and remains in that state until the conversion is completed, at which time it goes low (or high). The trailing edge is generally an indication of valid output data, but the data sheet should be carefully studied—in some ADCs additional delay is required before the output data is valid.

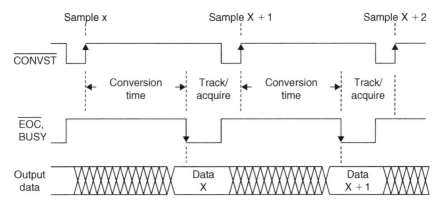

**Figure 6-46: Typical SAR ADC timing**

An N-bit conversion takes N steps. It would seem on superficial examination that a 16-bit converter would have twice the conversion time of an 8-bit one, but this is not the case. In an 8-bit converter, the DAC must settle to 8-bit accuracy before the bit decision is made, whereas in a 16-bit converter, it must settle to 16-bit accuracy, which takes a lot longer. In practice, 8-bit successive approximation ADCs can convert in a few hundred nanoseconds, while 16-bit ones will generally take several microseconds.

While there are some variations, the fundamental timing of most SAR ADCs is similar and relatively straightforward. The conversion process is initiated by asserting a CONVERT START signal. This signal is typically named something like $\overline{\text{CONVST}}$ or CS. This signal is usually a negative-going pulse whose positive-going edge actually initiates the conversion. The internal SHA is placed in the hold mode on this edge, and the various bits are determined using the SAR algorithm. The negative-going edge of the $\overline{\text{CONVST}}$ pulse causes a signal typically called $\overline{\text{EOC}}$ or BUSY to go high. When the conversion is complete, the BUSY line goes low (or $\overline{\text{EOC}}$ goes high), indicating the completion of the conversion process. In most cases the trailing edge of the BUSY line can be used as an indication that the output data is valid and can be used to strobe the output data into an external register.

There may also be other control lines. And sometimes control lines have dual function. This is primarily done when the chip is pin limited. Because of the many variations in terminology and design, the individual data sheet should always be consulted when using a specific ADC.

It should also be noted that some SAR ADCs require an external high frequency clock in addition to the CONVERT START command. In most cases, there is no need to synchronize the two. The frequency of the external clock, if required, generally falls in the range of 1–30 MHz depending on the conversion time and resolution of the ADC. Other SAR ADCs have an internal oscillator which is used to perform the conversions and only requires the CONVERT START command. Because of their architecture, SAR

ADCs generally allow single-shot conversion at any repetition rate from DC to the converter's maximum conversion rate.

Notice that the overall accuracy and linearity of the SAR ADC is determined primarily by the internal DAC. Until recently, most precision SAR ADCs used laser-trimmed thin-film DACs to achieve the desired accuracy and linearity. The thin-film resistor trimming process adds cost, and the thin-film resistor values may be affected when subjected to the mechanical stresses of packaging.

For these reasons, switched capacitor (or charge redistribution) DACs have become popular in newer SAR ADCs. The advantage of the switched capacitor DAC is that the accuracy and linearity are primarily determined by photolithography, which in turn controls the capacitor plate area and the capacitance as well as matching. In addition, small capacitors can be placed in parallel with the main capacitors which can be switched in and out under control of autocalibration routines to achieve high accuracy and linearity without the need for thin-film laser trimming. Temperature tracking between the switched capacitors can be better than 1 ppm/°C, thereby offering a high degree of temperature stability.

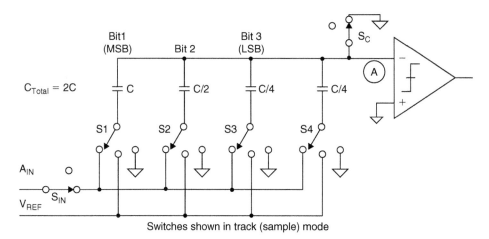

**Figure 6-47: 3-bit switched capacitor DAC**

A simple 3-bit capacitor DAC is shown in Figure 6-47. The switches are shown in the *track*, or *sample* mode where the analog input voltage, $A_{IN}$, is constantly charging and discharging the parallel combination of all the capacitors. The *hold* mode is initiated by opening $S_{IN}$, leaving the sampled analog input voltage on the capacitor array. Switch $S_C$ is then opened allowing the voltage at node A to move as the bit switches are manipulated. If S1, S2, S3, and S4 are all connected to ground, a voltage equal to $-A_{IN}$ appears at node A. Connecting S1 to $V_{REF}$ adds a voltage equal to $V_{REF}/2$ to $-A_{IN}$. The comparator then makes the MSB bit decision, and the SAR either leaves S1 connected to $V_{REF}$ or connects it to ground depending on the comparator output (which is high or low depending on whether the voltage at node A is negative or positive, respectively). A similar process is followed for the remaining two bits. At the EOC interval, S1, S2, S3, S4, and $S_{IN}$ are connected to $A_{IN}$, $S_C$ is connected to ground, and the converter is ready for another cycle.

Note that the extra LSB capacitor (C/4 in the case of the 3-bit DAC) is required to make the total value of the capacitor array equal to 2C so that binary division is accomplished when the individual bit capacitors are manipulated.

The operation of the capacitor DAC (cap DAC) is similar to an R–2R resistive DAC. When a particular bit capacitor is switched to $V_{REF}$, the voltage divider created by the bit capacitor and the total array capacitance

(2C) adds a voltage to node A equal to the weight of that bit. When the bit capacitor is switched to ground, the same voltage is subtracted from node A.

An example of charge redistribution successive approximation ADCs is Analog Devices' PulSAR™ series. The AD7677 is a 16-bit, 1 MSPS, PulSAR, fully differential ADC that operates from a single 5 V power supply (see Figure 6-48). The part contains a high speed 16-bit sampling ADC, an internal conversion clock, error correction circuits, and both serial and parallel system interface ports. The AD7677 is hardware factory calibrated and comprehensively tested to ensure such AC parameters as SNR and total harmonic distortion (THD), in addition to the more traditional DC parameters of gain, offset, and linearity. It features a very high sampling rate mode (Warp) and, for asynchronous conversion rate applications, a fast mode (Normal) and, for low power applications, a reduced power mode (Impulse) where the power is scaled with the throughput.

The operation of a successive approximation ADC is as follows. Using Figure 6-49 as an example, one side of the balance is loaded with half-scale (in this case 32 lbs.). Call this the proof mass. The test mass is then put on the other side of the balance. If the test mass is greater, as it is in this case, the proof mass is retained, otherwise it is discarded. Next a proof mass equal to 1/4 scale is added. Again, if the test mass is still greater the proof mass is retained, otherwise it is rejected. In the example it is rejected. This process is continued, each time cutting the proof mass in half, until the desired resolution is reached. The proof masses are added up. This will equal the mass of the test mass, to the resolution of the test.

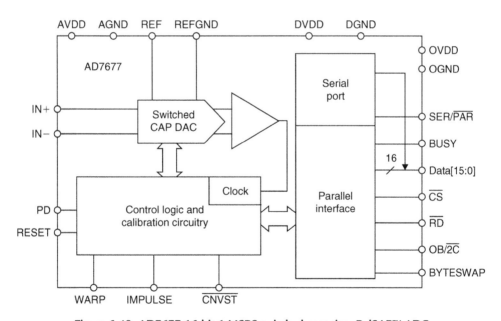

**Figure 6-48: AD7677 16-bit 1 MSPS switched capacitor PulSAR™ ADC**

In an SAR ADC the proof mass is a voltage provided by the DAC. It is compared to the input, corresponding to the test mass, by the comparator. Keeping track of output of each test and setting the DAC is accomplished by the SAR.

The digital output is basically serial in nature, but SAR ADCs are generally available in both serial and parallel output formats.

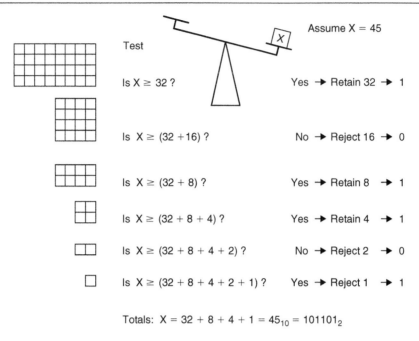

Assume X = 45

Test

Is X ≥ 32 ?  →  Yes → Retain 32 → 1

Is X ≥ (32 +16) ?  →  No → Reject 16 → 0

Is X ≥ (32 + 8) ?  →  Yes → Retain 8 → 1

Is X ≥ (32 + 8 + 4) ?  →  Yes → Retain 4 → 1

Is X ≥ (32 + 8 + 4 + 2) ?  →  No → Reject 2 → 0

Is X ≥ (32 + 8 + 4 + 2 + 1) ?  →  Yes → Reject 1 → 1

Totals: $X = 32 + 8 + 4 + 1 = 45_{10} = 101101_2$

**Figure 6-49: Successive approximation ADC algorithm**

## Flash Converters

Flash ADCs (sometimes called *parallel* ADCs) are the fastest type of ADC and use large numbers of comparators. An N-bit flash ADC consists of $2^N$ resistors and $2^N-1$ comparators arranged as in Figure 6-50. Each comparator has a reference voltage which is 1 LSB higher than that of the one below it in the chain. For a given input voltage, all the comparators below a certain point will have their input voltage larger than their reference voltage and a "1" logic output, and all the comparators above that point will have a reference voltage larger than the input voltage and a "0" logic output. The $2^N-1$ comparator outputs therefore behave in a way analogous to a mercury thermometer, and the output code at this point is sometimes called a *thermometer* code. Since $2^N-1$ data outputs are not really practical, they are processed by a decoder to generate an N-bit binary output.

The input signal is applied to all the comparators at once, so the thermometer output is delayed by only one comparator delay from the input, and the encoder N-bit output by only a few gate delays on top of that, so the process is very fast. However, the architecture uses large numbers of resistors and comparators and is limited to low resolutions, and if it is to be fast, each comparator must run at relatively high power levels. Hence, the problems of flash ADCs include limited resolution, high power dissipation because of the large number of high speed comparators (especially at sampling rates greater than 50 MSPS), and relatively large (and therefore expensive) chip sizes. In addition, the resistance of the reference resistor chain must be kept low to supply adequate bias current to the fast comparators, so the voltage reference has to source quite large currents (typically >10 mA).

Each comparator has a voltage-variable junction capacitance, and this signal-dependent capacitance results in most flash ADCs having reduced effective number of bits (ENOB) and higher distortion at high input frequencies. For this reason, most flash converters must be driven with a wideband op amp which is tolerant to the capacitive load presented by the converter as well as high speed transients developed on the input.

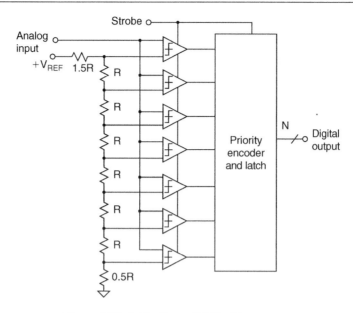

**Figure 6-50: 3-bit all-parallel (flash) converter**

Power dissipation is always a big consideration in flash converters, especially at resolutions above 8 bits. A clever technique was used for AD9410 10-bit, 210 MSPS ADC called *interpolation* to minimize the number of preamplifiers in the flash converter comparators and also reduce the power (2.1 W). The method is shown in Figure 6-51 (see reference).

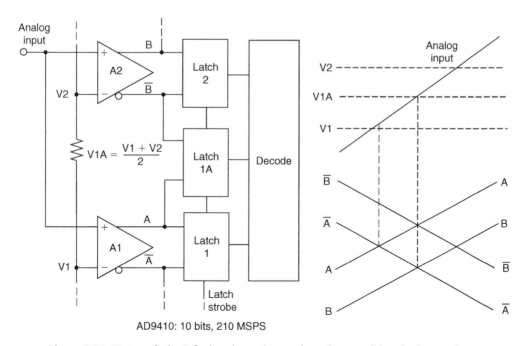

AD9410: 10 bits, 210 MSPS

**Figure 6-51: "Interpolating" flash reduces the number of preamplifiers by factor of two**

The preamplifiers (labeled "A1," "A2," etc.) are low gain $g_m$ stages whose bandwidth is proportional to the tail currents of the differential pairs. Consider the case for a positive-going ramp input which is initially below the reference to AMP A1, V1. As the input signal approaches V1, the differential output of A1 approaches zero (i.e., $A = \overline{A}$), and the decision point is reached. The output of A1 drives the differential input of LATCH 1. As the input signal continues to go positive, A continues to go positive, and $\overline{B}$ begins to go negative. The interpolated decision point is determined when $A = \overline{B}$. As the input continues positive, the third decision point is reached when $B = \overline{B}$. This novel architecture reduces the ADC input capacitance and thereby minimizes its change with signal level and the associated distortion. The AD9410 also uses an input SHA circuit for improved AC linearity.

## Subranging, Error Corrected, and Pipelined ADCs

A basic two-stage N bit subranging ADC is shown in Figure 6-52. The ADC is based on two separate conversions—a coarse conversion (N1 bits) in the MSB sub-ADC (SADC) followed by a fine conversion (N2 bits) in the LSB SADC. Early subranging ADCs nearly always used flash converters as building blocks, but a number of recent ADCs utilize other architectures for the individual ADCs.

The conversion process begins placing the SHA in the hold mode followed by a coarse N1-bit SADC conversion of the MSBs. The digital outputs of the MSB converter drive an N1-bit sub-DAC (SDAC) which generates a coarsely quantized version of the analog input signal. The N1-bit SDAC output is subtracted from the held analog signal, amplified, and applied to the N2-bit LSB SADC. The amplifier provides gain, G, sufficient to make the "residue" signal exactly fill the input range of the N2 SADC. The output data from the N1 SADC and the N2 SADC are latched into the output registers yielding the N-bit digital output code, where N = N1 + N2.

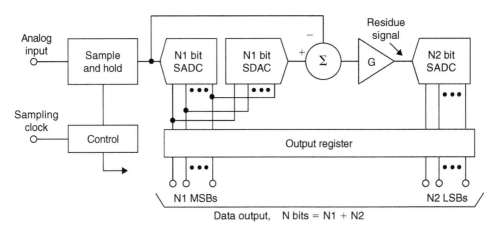

*Adapted from:* R. Staffin and R. Lohman, "Signal Amplitude Quantizer," US Patent 2,869,079, filed, December 19, 1956, issued January 13, 1959.

**Figure 6-52: N-bit two-stage subranging ADC**

In order for this simple subranging architecture to work satisfactorily, both the N1 SADC and SDAC (although they only have N1 bits of resolution) must be better than N bits accurate. The residue signal offset and gain must be adjusted such that it precisely fills the range of the N2 SADC as shown in Figure 3-66(A). If the residue signal drifts by more than 1 LSB (referenced to the N2 SADC), then there will be missing codes as shown in Figure 6-53(A) where the residue signal enters the out-of-range regions labeled "X" and

"Y." Any nonlinearity or drift in the N1 SADC will also cause missing codes if it exceeds 1 LSB referenced to N bits. In practice, an 8 bit subranging ADC with N1 = 4 bits and N2 = 4 bits represents a realistic limit to this architecture in order to maintain no missing codes over a reasonable operating temperature range.

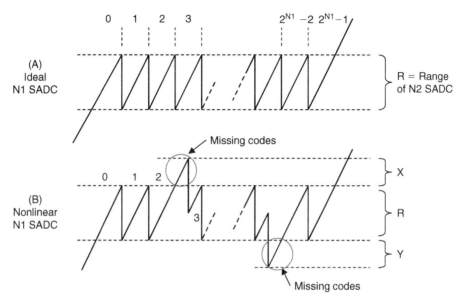

**Figure 6-53: Residue waveforms at input of N2 SADC**

When the interstage alignment is not correct, missing codes will appear in the overall ADC transfer function as shown in Figure 6-54. If the residue signal goes into positive overrange (the "X" region), the output first "sticks" on a code and then "jumps" over a region leaving missing codes. The reverse occurs if the residue signal is negative overrange.

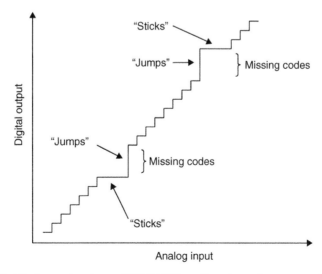

**Figure 6-54: Missing codes due to MSB SADC nonlinearity or interstage misalignment**

In order to reliably achieve higher than 8-bit resolution using the subranging approach, a technique generally referred to as *digital corrected subranging, digital error correction, overlap bits, redundant bits,* etc. is utilized.

Figure 6 55 shows two methods that can be used to design a pipeline stage in a subranging ADC. Figure 6-55 shows two pipelined stages which use an interstage T/H in order to provide interstage gain and give each stage the maximum possible amount of time to process the signal at its input. In Figure 6-55(B) an MDAC is used to provide the appropriate amount of interstage gain as well as the subtraction function.

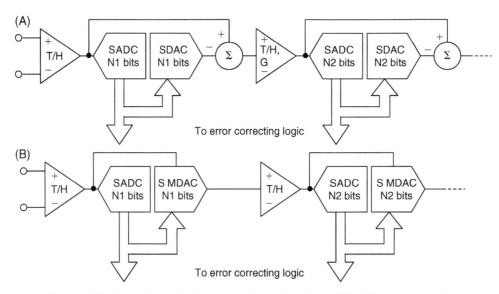

Figure 6-55: Generalized pipeline stages in a subranging ADC with error correction

The term "pipelined" architecture refers to the ability of one stage to process data from the previous stage during any given clock cycle. At the end of each phase of a particular clock cycle, the output of a given stage is passed on to the next stage using the T/H functions and new data is shifted into the stage. Of course this means that the digital outputs of all but the last stage in the "pipeline" must be stored in the appropriate number of shift registers so that the digital data arriving at the correction logic corresponds to the same sample.

Figure 6-56 shows a timing diagram of a typical pipelined subranging ADC. Notice that the phases of the clocks to the T/H amplifiers are alternated from stage to stage such that when a particular T/H in the ADC enters the hold mode it holds the sample from the preceding T/H, and the preceding T/H returns to the track mode. The held analog signal is passed along from stage to stage until it reaches the final stage in the pipelined ADC—in this case, a flash converter. When operating at high sampling rates, it is critical that the differential sampling clock be kept at a 50% duty cycle for optimum performance. Duty cycles other than 50% affect all the T/H amplifiers in the chain—some will have longer than optimum track times and shorter than optimum hold times; while others suffer exactly the reverse condition. Several newer pipelined ADCs including the 12-bit, 65 MSPS AD9235 and the 12-bit, 210 MSPS AD9430 have on-chip clock conditioning circuits to control the internal duty cycle while allowing some variation in the external clock duty cycle.

The effects of the "pipeline" delay (sometimes called latency) in the output data are shown in Figure 6-57 for the AD9235 12-bit 65 MSPS ADC where there is a seven-clock cycle pipeline delay.

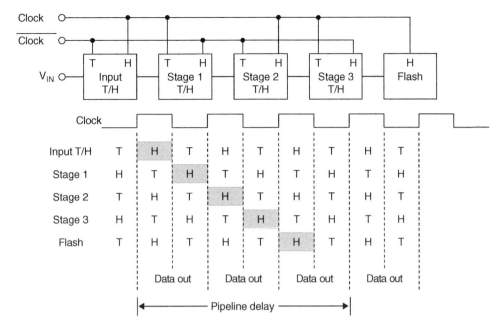

Figure 6-56: Clock issues in pipelined ADCs

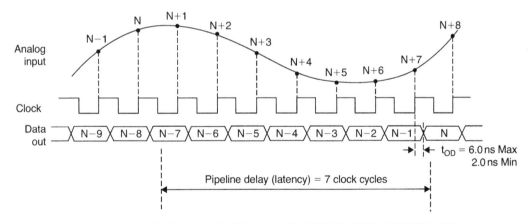

Figure 6-57: Typical pipelined ADC timing for AD9235 12-bit, 65 MSPS ADC

Note that the pipeline delay is a function of the number of stages and the particular architecture of the ADC under consideration—the data sheet should always be consulted for the exact details of the relationship between the sampling clock and the output data timing. In many applications the pipeline delay will not be a problem, but if the ADC is inside a feedback loop the pipeline delay may cause instability. The pipeline delay can also be troublesome in multiplexed applications or when operating the ADC in a "single-shot" mode. Other ADC architectures—such as successive approximation—may be better suited to these types of applications.

The pipelined error correcting ADC has become very popular in modern ADCs requiring wide dynamic range and low levels of distortion. There are many possible ways to design a pipelined ADC, and we will now look at just a few of the tradeoffs. Figure 6-58(A) shows a pipelined ADC designed with identical stages of k-bits each. This architecture uses the same core hardware in each stage, offers a few other advantages, but does necessarily optimize the ADC for best possible performance. Figure 6-58(B) shows the simplest form of this architecture where k = 1.

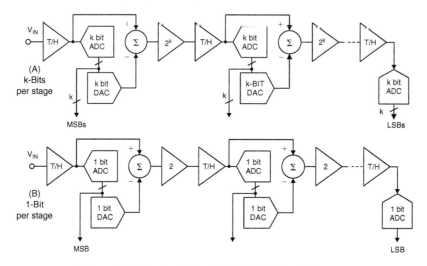

**Figure 6-58: Basic pipelined ADC with identical stages**

In order to optimize performance at the 12-bit level, e.g., 1-bit-per-stage pipeline is more commonly used with a multibit front-end and back-end ADC as shown in Figure 6-59.

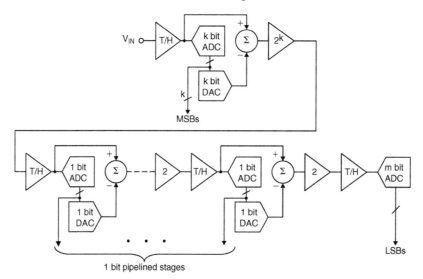

**Figure 6-59: Multibit and 1-bit pipelined core combined**

Another less popular type of error corrected subranging architecture is the *recirculating* subranging ADC (Figure 6-60). The concept is similar to the error corrected subranging architecture previously discussed, but in this architecture, the residue signal is recirculated through a single ADC and DAC stage using switches and a PGA. The major problem with this technique is the PGA. Its gain-bandwidth product will limit the frequency response at higher gains. Also matching of the various gains could be problematic.

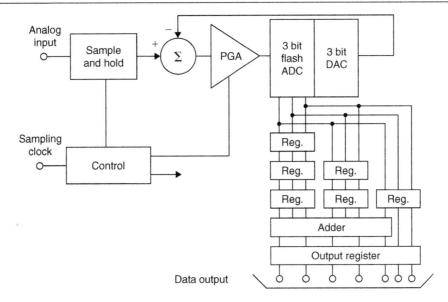

Adapted from: D. J. Kinnimet, D. Aspinall, and D. B. G. Edwards High Speed Analogue–Digital Converter, *IEE Proceedings*, Vol. 113, pp. 2.61–2.69, December 1966.

**Figure 6-60: Kinniment, et al., 1966 Pipelined 7-bit, 9 MSPS recirculating ADC architecture**

## Serial Bit-per-Stage Binary and Gray Coded (Folding) ADCs

Various architectures exist for performing A/D conversion using one stage per bit. Figure 6-61 shows the overall concept. In fact, a multistage subranging ADC with 1 bit-per-stage and no error correction is one form as previously discussed. In this approach, the input signal must be held constant during the entire conversion cycle. There are N stages, each of which have a bit output and a *residue* output. The residue output of one stage is the input to the next. The last bit is detected with a single comparator as shown.

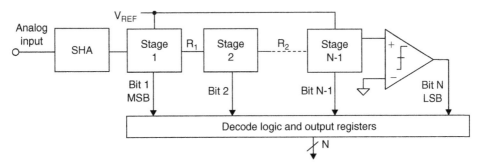

Adapted from: B. D. Smith, "An Unusual Electronic Analog–Digital Conversion Method" *IRE Transactions on Instrumentation*, June 1956, pp. 155–160.

**Figure 6-61: Generalized bit-per-stage ADC architecture**

The basic stage for performing a single binary bit conversion is shown in Figure 6-62. It consists of a gain-of-two amplifier, a comparator, and a 1-bit DAC. Assume that this is the first stage of the ADC. The MSB is simply the polarity of the input, and that is detected with the comparator which also controls the 1-bit DAC. The 1-bit DAC output is summed with the output of the gain-of-two amplifier. The resulting residue output

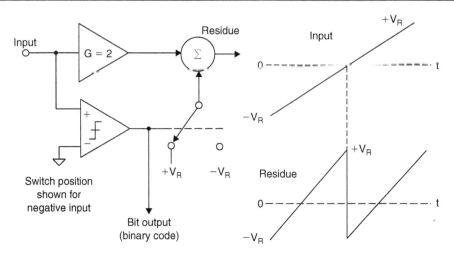

**Figure 6-62: Single-stage transfer function for binary ADC**

is then applied to the next stage. In order to better understand how the circuit works, the diagram shows the residue output for the case of a linear ramp input voltage which traverses the entire ADC range, $-V_R$ to $+V_R$. Notice that the polarity of the residue output determines the binary bit output of the next stage.

A simplified 3-bit serial-binary ADC is shown in Figure 6-63, and the residue outputs are shown in Figure 6-64. Again, the case is shown for a linear ramp input voltage whose range is between $-V_R$ and $+V_R$. Each residue output signal has discontinuities which correspond to the point where the comparator changes state and causes the DAC to switch. The fundamental problem with this architecture is the discontinuity in the residue output waveforms. Adequate settling time must be allowed for these transients to propagate through all the stages and settle at the final comparator input. As presented here, the prospects of making this architecture operate at high speed are dismal. However using the 1.5 bit-per-stage pipelined architecture previously discussed in this section makes it much more attractive at high speeds.

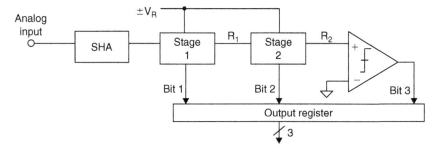

**Figure 6-63: 3-bit serial ADC with binary output**

Although the binary method is discussed in his paper, B. D. Smith also describes a much preferred bit-per-stage architecture based on absolute value amplifiers (magnitude amplifiers, or simply *MagAMPs*™). This scheme has often been referred to as *serial-Gray* (since the output coding is in Gray code), or *folding* converter because of the shape of the transfer function. Performing the conversion using a transfer function that produces an initial Gray code output has the advantage of minimizing discontinuities in the residue output waveforms and offers the potential of operating at much higher speeds than the binary approach.

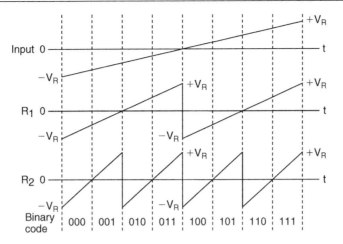

**Figure 6-64: Input and residue waveforms of 3-bit binary ripple ADC**

The basic folding stage is shown functionally in Figure 6-65 along with its transfer function. The input to the stage is assumed to be a linear ramp voltage whose range is between $-V_R$ and $+V_R$. The comparator detects the polarity of the input signal and provides the Gray bit output for the stage. It also determines whether the overall stage gain is $+2$ or $-2$. The reference voltage $V_R$ is summed with the switch output to generate the residue signal which is applied to the next stage. The polarity of the residue signal determines the Gray bit for the next stage. The transfer function for the folding stage is also shown in Figure 6-65.

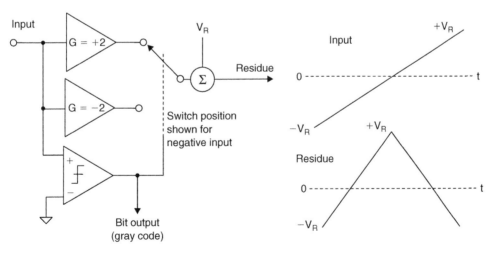

**Figure 6-65: Folding stage functional equivalent circuit**

A 3-bit MagAMP folding ADC is shown in Figure 6-66, and the corresponding residue waveforms in Figure 6-67. As in the case of the binary ripple ADC, the polarity of the residue output signal of a stage determines the value of the Gray bit for the next stage. The polarity of the input to the first stage determines the Gray MSB; the polarity of R1 output determines the Gray bit-2; and the polarity of $R_2$ output determines the Gray bit-3. Notice that unlike the binary ripple ADC, there is no abrupt transition in any of the folding stage residue output waveforms. This makes operation at high speeds quite feasible.

387

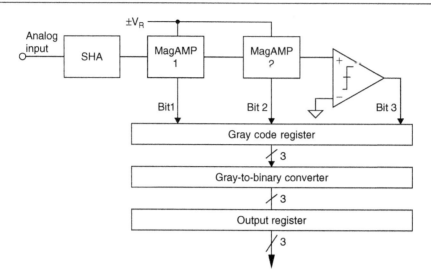

Figure 6-66: 3-bit folding ADC block diagram

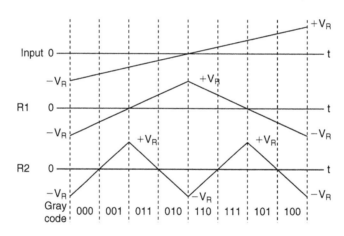

Figure 6-67: Input and residue waveforms for 3-bit folding ADC

Modern IC circuit designs implement the transfer function using current-steering open-loop gain techniques which can be made to operate much faster. Fully differential stages (including the SHA) also provide speed, lower distortion, and yield 8 bit accurate folding stages with no requirement for thin-film resistor laser trimming.

An example of a fully differential gain-of-two MagAMP folding stage is shown in Figure 6-68. The differential input signal is applied to the degenerated-emitter differential pair Q1, Q2 and the comparator. The differential input voltage is converted into a differential current which flows in the collectors of Q1, Q2. If +IN is greater than −IN, cascode-connected transistors Q3, Q6 are on, and Q4, Q6 are off. The differential signal currents therefore flow through the collectors of Q3, Q6 into level-shifting transistors Q7, Q8 and into the output load resistors, developing the differential output voltage between +OUT and −OUT. The overall differential voltage gain of the circuit is two.

If +IN is less than −IN (negative differential input voltage), the comparator changes stage and turns Q4, Q5 on and Q3, Q6 off. The differential signal currents flow from Q5 to Q7 and from Q4 to Q8, thereby maintaining the same relative polarity at the differential output as for a positive differential input voltage. The required offset voltage is developed by adding a current $I_{OFF}$ to the emitter current of Q7 and subtracting it from the emitter current of Q8.

The differential residue output voltage of the stage drives the next stage input, and the comparator output represents the Gray code output for the stage.

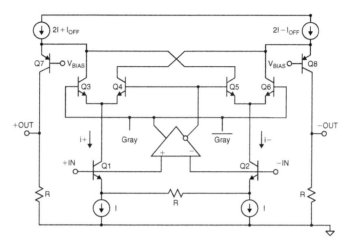

**Figure 6-68: A modern current-steering MagAMP™ stage**

The MagAMP architecture offers lower power and can be extended to sampling rates previously dominated by flash converters. For example, the AD9054A 8-bit, 200 MSPS ADC is shown in Figure 6-69. The first 5 bits (Gray code) are derived from five differential MagAMP stages. The differential residue output of the fifth MagAMP stage drives a 3-bit flash converter, rather than a single comparator.

The Gray-code output of the five MagAMPs and the binary-code output of the 3-bit flash are latched, all converted into binary, and latched again in the output data register. Because of the high data rate, a demultiplexed output option is provided.

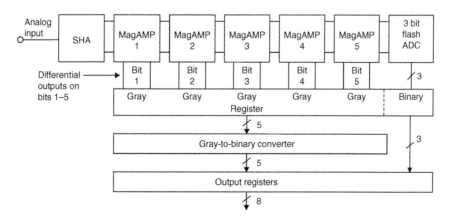

**Figure 6-69: AD9054A 8-bit, 200 MSPS ADC functional diagram**

## Counting and Integrating ADC Architectures

Although counting-based ADCs are not well suited for high speed applications, they are ideal for high resolution low frequency applications, especially when combined with integrating techniques.

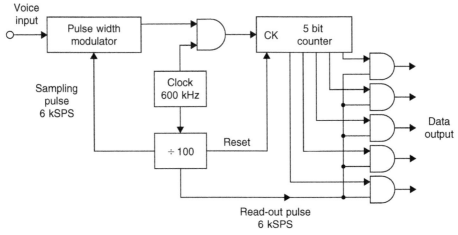

*Adapted from*: A. H. Reeves "Electric signaling system," US Patent 2,272,070, filed, November 22, 1939, issued February 3, 1942.

**Figure 6-70: A. H. Reeves' 5-bit counting ADC**

The counting ADC technique (see Figure 6-70) basically uses a sampling pulse to take a sample of the analog signal, set an R/S flip-flop, and simultaneously start a controlled ramp voltage. The ramp voltage is compared with the input, and when they are equal, a pulse is generated which resets the R/S flip-flop. The output of the flip-flop is a pulse whose width is proportional to the analog signal at the sampling instant. This pulse width modulated (PWM) pulse controls a gated oscillator, and the number of pulses out of the gated oscillator represents the quantized value of the analog signal. This pulse train can be easily converted to a binary word by driving a counter. In Reeves' system, a master clock of 600 kHz was used, and a 100:1 divider generated the 6 kHz sampling pulses. The system uses a 5-bit counter, and 31 counts (out of the 100 counts between sampling pulses) therefore represent a full-scale signal. The technique can obviously be extended to higher resolutions.

## Charge Run-Down ADCs

The charge run-down ADC architecture shown in Figure 6-71 first samples the analog input and stores the voltage on a fixed capacitor. The capacitor is then discharged with a constant current source, and the time required for complete discharge is measured using a counter. Notice that in this approach, the overall accuracy is dependent on the magnitude of the capacitor, the magnitude of the current source, as well as the accuracy of the timebase.

## Ramp Run-Up ADCs

In the ramp run-up architecture shown in Figure 6-72, a ramp generator is started at the beginning of the conversion cycle. The counter then measures the time required for the ramp voltage to equal the analog input voltage. The counter output is therefore proportional to the value of the analog input. In an alternate version (shown dotted in Figure 6-72), the ramp voltage generator is replaced by a DAC which is driven by

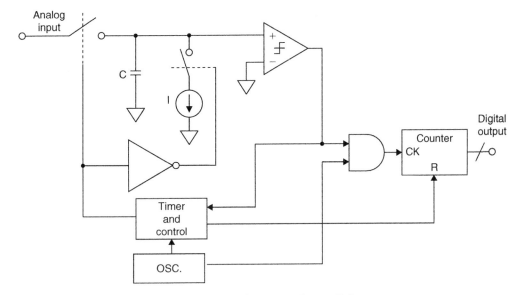

**Figure 6-71: Charge run-down ADC**

the counter output. The advantage of using the ramp is that the ADC is always monotonic, whereas overall monotonicity is determined by the DAC when it is used as a substitute.

The accuracy of the ramp run-up ADC depends on the accuracy of the ramp generator (or the DAC) as well as the oscillator.

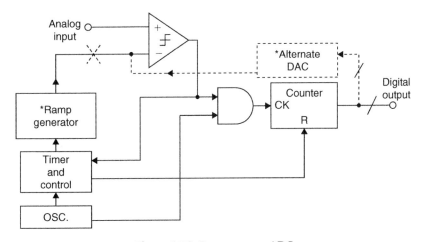

**Figure 6-72: Ramp run-up ADC**

## Tracking ADCs

The tracking ADC architecture shown in Figure 6-73 continually compares the input signal with a reconstructed representation of the input signal. The up/down counter is controlled by the comparator output. If the analog input exceeds the DAC output, the counter counts up until they are equal. If the DAC output exceeds the analog input, the counter counts down until they are equal. It is evident that if the analog input changes slowly, the counter will follow, and the digital output will remain close to its correct value. If the analog input suddenly undergoes a large step change, it will be many hundreds or thousands of clock cycles before the output is again valid. The tracking ADC therefore responds quickly to slowly changing signals, but slowly to a quickly changing one.

The simple analysis above ignores the behavior of the ADC when the analog input and DAC output are nearly equal. This will depend on the exact nature of the comparator and counter. If the comparator is a simple one, the DAC output will cycle by 1 LSB from just above the analog input to just below it, and the digital output will, of course, do the same—there will be 1 LSB of flicker. Note that the output in such a case steps every clock cycle, irrespective of the exact value of analog input, and hence always has unity mark-space ratio. In other words, there is no possibility of taking a mean value of the digital output and increasing resolution by oversampling.

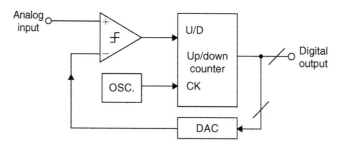

**Figure 6-73: Tracking ADC**

A more satisfactory, but more complex arrangement would be to use a window comparator with a window 1–2 LSB wide. When the DAC output is high or low the system behaves as in the previous description, but if the DAC output is within the window, the counter stops. This arrangement eliminates the flicker, provided that the DAC DNL never allows the DAC output to step across the window for 1 LSB change in code.

Tracking ADCs are not very common. Their slow step response makes them unsuitable for many applications, but they do have one asset: their output is *continuously* available. Most ADCs perform conversions: i.e., on receipt of a "start convert" command (which may be internally generated), they perform a conversion and, after a delay, a result becomes available. Providing that the analog input changes slowly, the output of a tracking ADC is always available. This is valuable in synchro-to-digital and resolver-to-digital converters (SDCs and RDCs), and this is the application where tracking ADCs are most often used. Another valuable characteristic of tracking ADCs is that a fast transient on the analog input causes the output to change only one count. This is very useful in noisy environments. Notice the similarity between a tracking ADC and a successive approximation ADC. Replacing the up/down counter with SAR logic yields the architecture for a successive approximation ADC.

## Voltage-to-Frequency Converters

A voltage-to-frequency converter (VFC) is an oscillator whose frequency is linearly proportional to a control voltage (a high accuracy voltage-controlled oscillator (VCO)). The VFC/counter ADC is monotonic and free of missing codes, integrates noise, and can consume very little power. It is also very useful for telemetry applications, since the VFC, which is small, cheap and low powered, can be mounted on the experimental subject (patient, wild animal, artillery shell, etc.) and communicate with the counter by a telemetry link as shown in Figure 6-74.

There are two common VFC architectures: the *current-steering multivibrator VFC* and the *charge-balanced VFC*. The charge-balanced VFC may be made in *asynchronous* or *synchronous* (clocked) forms. There are many more VFO (variable frequency oscillator) architectures, including the ubiquitous 555 timer, but the key feature of VFCs is linearity—few VFOs are very linear.

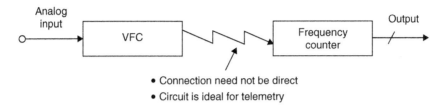

- Connection need not be direct
- Circuit is ideal for telemetry

**Figure 6-74: VFC and frequency counter make a low cost, versatile, high resolution ADC**

The current-steering multivibrator VFC is actually a current-to-frequency converter rather than a VFC, but, as shown in Figure 6-75, practical circuits invariably contain a voltage-to-current converter at the input. The principle of operation is evident: the current discharges the capacitor until a threshold is reached, and when the capacitor terminals are reversed, the half-cycle repeats itself. The waveform across the capacitor is a linear tri-wave, but the waveform on either terminal with respect to ground is the more complex waveform shown.

Practical VFCs of this type have linearities around 14 bits, and comparable stability, although they may be used in ADCs with higher resolutions without missing codes. The performance limits are set by comparator threshold noise, threshold temperature coefficient, and the stability and dielectric absorption (DA) of the capacitor, which is generally a discrete component. The comparator/voltage reference structure shown in the diagram is more of a representation of the function performed than the actual circuit used, which is much more integrated with the switching, and correspondingly harder to analyze.

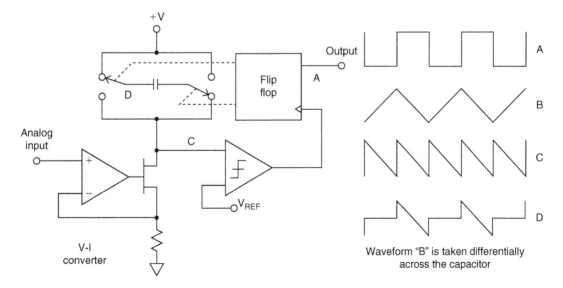

Waveform "B" is taken differentially across the capacitor

**Figure 6-75: A current-steering VFC**

This type of VFC is simple, inexpensive, and low powered, and most run from a wide range of supply voltages. They are ideally suited for low cost medium accuracy (12 bit) ADC and data telemetry applications.

The charge-balance VFC shown in Figure 6-76 is more complex, more demanding in its supply voltage and current requirements, and more accurate. It is capable of 16–18-bit linearity.

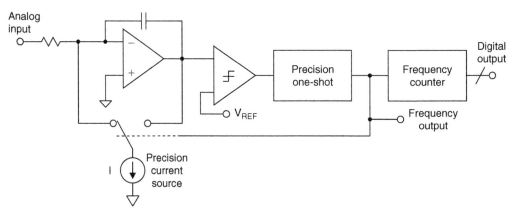

**Figure 6-76: Charge-balance VFC**

The integrator capacitor charges from the signal as shown in Figure 6-76. When it passes the comparator threshold, a fixed charge is removed from the capacitor, but the input current continues to flow during the discharge, so no input charge is lost. The fixed charge is defined by the precision current source and the pulse width of the precision monostable. The output pulse rate is thus accurately proportional to the rate at which the integrator charges from the input.

At low frequencies, the limits on the performance of this VFC are set by the stability of the current source and the monostable timing (which depends on the monostable capacitor, among other things). The absolute value and temperature stability of the integration capacitor do not affect the accuracy, although its leakage and dielectric absorption (DA) do. At high frequencies, second-order effects, such as switching transients in the integrator and the precision of the monostable when it is retriggered very soon after the end of a pulse, take their toll on accuracy and linearity.

The changeover switch in the current source addresses the integrator transient problem. By using a changeover switch instead of the on/off switch more common on older VFC designs: (a) there are no on/off transients in the precision current source and (b) the output stage of the integrator sees a constant load— most of the time the current from the source flows directly in the output stage; during charge balance, it still flows in the output stage, but through the integration capacitor.

The stability and transient behavior of the precision monostable present more problems, but the issue may be avoided by replacing the monostable with a clocked bistable multivibrator. This arrangement is known as a *synchronous* VFC or SVFC and is shown in Figure 6-77.

The difference from the previous circuit is quite small, but the charge-balance pulse length is now defined by two successive edges of the external clock. If this clock has low jitter, the charge will be very accurately defined. The output pulse will also be synchronous with the clock. SVFCs of this type are capable of up to 18-bit linearity and excellent temperature stability.

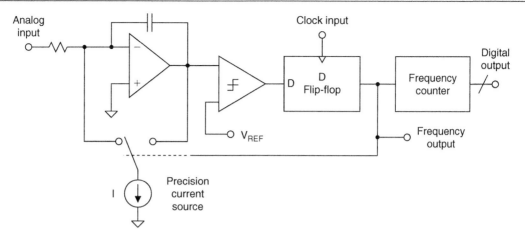

**Figure 6-77: Synchronous VFC (SVFC)**

This synchronous behavior is convenient in many applications, since synchronous data transfer is often easier to handle than asynchronous. It does mean, however, that the output of an SVFC is not a pure tone (plus harmonics, of course) like a conventional VFC, but contains components harmonically related to the clock frequency. The display of an SVFC output on an oscilloscope is especially misleading and is a common cause of confusion—a change of input to a VFC produces a smooth change in the output frequency, but a change to an SVFC produces a change in probability density of output pulses N and N + 1 clock cycles after the previous output pulse, which is often misinterpreted as severe jitter and a sign of a faulty device (see Figure 6-78).

Another problem with SVFCs is nonlinearity at output frequencies related to the clock frequency. If we study the transfer characteristic of an SVFC, we find nonlinearities close to sub-harmonics of the clock frequency $F_C$ as shown in Figure 6-79. They can be found at $F_C/3$, $F_C/4$, and $F_C/6$. This is due to stray capacitance on the chip (and in the circuit layout!) and coupling the clock signal into the SVFC comparator which causes the device to behave as an injection-locked phase-locked loop (PLL). This problem is intrinsic to SVFCs, but is not often serious: if the circuit card is well laid out, and clock amplitude and dv/dts kept as low as practical, the effect is a discontinuity in the transfer characteristic of less than 8 LSBs (at 18-bit resolution) at $F_C/3$ and $F_C/4$, and less at other sub-harmonics. This is frequently tolerable, since the frequencies where it occurs are known. Of course, if the circuit layout or decoupling is poor, the effect may be much larger, but this is the fault of poor design and not the SVFC itself.

It is evident that the SVFC is quantized, while the basic VFC is not. It does *not* follow from this that the counter/VFC ADC has higher resolution (neglecting nonlinearities) than the counter/SVFC ADC, because the clock in the counter also sets a limit to the resolution.

When a VFC has a large input, it runs quickly and (counting for a short time) gives good resolution, but it is hard to get good resolution in a reasonable sample time with a slow-running VFC. In such a case, it may be more practical to measure the period of the VFC output (this does not work for an SVFC), but of course the resolution of this system deteriorates as the input (and the frequency) increases. However, if the counter/timer arrangement is made "smart," it is possible to measure the approximate VFC frequency and the exact period of not one, but N cycles (where the value of N is determined by the approximate frequency), and maintain high resolution over a wide range of inputs. The AD1170 modular ADC released in 1986 is an example of this architecture.

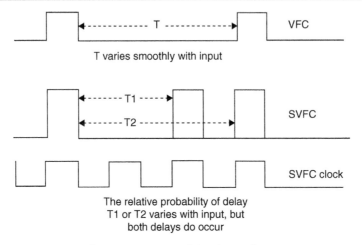

Figure 6-78: VFC and SVFC waveforms

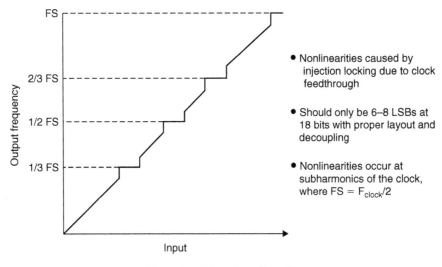

- Nonlinearities caused by injection locking due to clock feedthrough

- Should only be 6–8 LSBs at 18 bits with proper layout and decoupling

- Nonlinearities occur at subharmonics of the clock, where $FS = F_{clock}/2$

Figure 6-79: SVFC nonlinearity

VFCs have more applications than as a component in ADCs. Since their output is a pulse stream, it may easily be sent over a wide range of transmission media (PSN, radio, optical, IR, ultrasonic, etc.). It need not be received by a counter, but by another VFC configured as a frequency-to-voltage converter (FVC). This gives an analog output, and a VFC–FVC combination is a very useful way of sending a precision analog signal across an isolation barrier.

## Dual-Slope/Multi-Slope ADCs

The dual-slope ADC architecture was truly a breakthrough in ADCs for high resolution applications such as digital voltmeters (DVMs), etc. A simplified diagram is shown in Figure 6-80, and the integrator output waveforms are shown in Figure 6-81.

The input signal is applied to an integrator; at the same time a counter is started, counting clock pulses. After a predetermined amount of time (T), a reference voltage having opposite polarity is applied to the

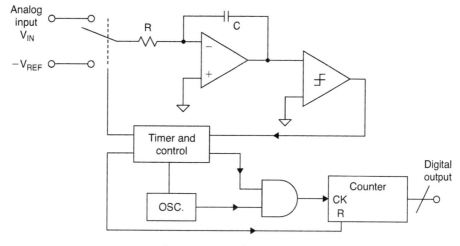

**Figure 6-80: Dual-slope ADC**

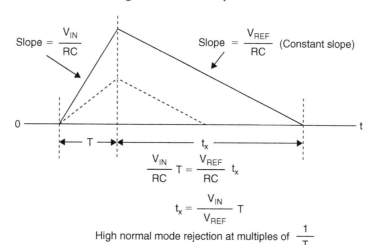

**Figure 6-81: Dual-slope ADC integrator output waveforms**

integrator. At that instant, the accumulated charge on the integrating capacitor is proportional to the average value of the input over the interval T. The integral of the reference is an opposite-going ramp having a slope of $V_{REF}/RC$. At the same time, the counter is again counting from zero. When the integrator output reaches zero, the count is stopped, and the analog circuitry is reset. Since the charge gained is proportional to $V_{IN} \times T$, and the equal amount of charge lost is proportional to $V_{REF} \times t_x$, then the number of counts relative to the full-scale count is proportional to $t_x/T$, or $V_{IN}/V_{REF}$. If the output of the counter is a binary number, it will therefore be a binary representation of the input voltage.

Dual-slope integration has many advantages. Conversion accuracy is independent of both the capacitance and the clock frequency, because they affect both the up-slope and the down-slope by the same ratio.

The fixed input signal integration period results in rejection of noise frequencies on the analog input that have periods that are equal to or a sub-multiple of the integration time T. Proper choice of T can therefore result in excellent rejection of 50 Hz or 60 Hz line ripple as shown in Figure 6-82.

Errors caused by bias currents and the offset voltages of the integrating amplifier and the comparator as well as gain errors can be canceled by using additional charge/discharge cycles to measure "zero" and "full-scale" and using the results to digitally correct the initial measurement, as in the quad-slope architecture.

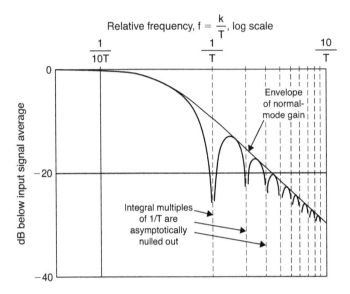

**Figure 6-82: Frequency response of integrating ADC**

The triple-slope architecture retains the advantages of the dual-slope, but greatly increases the conversion speed at the cost of added complexity. The increase in conversion speed is achieved by accomplishing the reference integration (ramp-down) at two distinct rates: a high speed rate, and a "vernier" lower speed rate. The counter is likewise divided into two sections, one for the MSBs and one for the LSBs. In a properly designed triple-slope converter, a significant increase in speed can be achieved while retaining the inherent linearity, differential linearity, and stability characteristics associated with dual-slope ADCs.

## RDCs and Synchros

Machine-tool and robotics manufacturers have increasingly turned to resolvers and synchros to provide accurate angular and rotational information. These devices excel in demanding factory applications requiring small size, long-term reliability, absolute position measurement, high accuracy, and low noise operation.

A diagram of a typical synchro and resolver is shown in Figure 6-83. Both synchros and resolvers employ single-winding rotors that revolve inside fixed stators. In the case of a simple synchro, the stator has three windings oriented 120° apart and electrically connected in a Y-connection. Resolvers differ from synchros in that their stators have only two windings oriented at 90°.

Because synchros have three stator coils in a 120° orientation, they are more difficult than resolvers to manufacture and are therefore more costly. Today, synchros find decreasing use, except in certain military and avionic retrofit applications.

Modern resolvers, in contrast, are available in a brushless form that employ a transformer to couple the rotor signals from the stator to the rotor. The primary winding of this transformer resides on the stator, and the secondary on the rotor. Other resolvers use more traditional brushes or slip rings to couple the signal

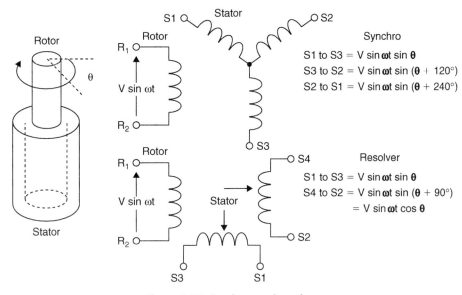

**Figure 6-83: Synchros and resolvers**

into the rotor winding. Brushless resolvers are more rugged than synchros because there are no brushes to break or dislodge, and the life of a brushless resolver is limited only by its bearings. Most resolvers are specified to work over 2 V to 40 V$_{RMS}$ and at frequencies from 400 Hz to 10 kHz. Angular accuracies range from 5 arc-minutes to 0.5 arc-minutes. (There are 60 arc-minutes in one degree, and 60 arc-seconds in 1 arc-minute. Hence, 1 arc-minute is equal to 0.0167°.)

In operation, synchros and resolvers resemble rotating transformers. The rotor winding is excited by an AC reference voltage, at frequencies up to a few kHz. The magnitude of the voltage induced in any stator winding is proportional to the sine of the angle θ, between the rotor coil axis and the stator coil axis. In the case of a synchro, the voltage induced across any pair of stator terminals will be the vector sum of the voltages across the two connected coils.

For example, if the rotor of a synchro is excited with a reference voltage, V sin ωt, across its terminals R$_1$ and R$_2$, then the stator's terminal will see voltages in the form:

$$\text{Sl to S3} = V \sin \omega t \sin \theta \tag{6-1}$$

$$\text{S3 to S2} = V \sin \omega t \sin (\theta + 120°) \tag{6-2}$$

$$\text{S2 to Sl} = V \sin \omega t \sin (\theta + 240°) \tag{6-3}$$

where θ is the shaft angle.

In the case of a resolver, with a rotor AC reference voltage of V sin ωt, the stator's terminal voltages will be:

$$\text{Sl to S3} = V \sin \omega t \sin \theta \tag{6-4}$$

$$\text{S4 to S2} = V \sin \omega t \sin(\theta + 90°) = V \sin \omega t \cos \theta \tag{6-5}$$

It should be noted that the three-wire synchro output can be easily converted into the resolver-equivalent format using a Scott-T transformer. Therefore, the following signal processing example describes only the resolver configuration.

A typical RDC is shown functionally in Figure 6-84. The two outputs of the resolver are applied to cosine and sine multipliers. These multipliers incorporate sine and cosine lookup tables and function as multiplying DACs. Begin by assuming that the current state of the up/down counter is a digital number representing a trial angle, φ. The converter seeks to adjust the digital angle, φ, continuously to become equal to, and to track θ, the analog angle being measured. The resolver's stator output voltages are written as:

$$V_1 = V \sin \omega t \sin \theta \qquad (6\text{-}6)$$

$$V_2 = V \sin \omega t \cos \theta \qquad (6\text{-}7)$$

where θ is the angle of the resolver's rotor. The digital angle φ is applied to the cosine multiplier, and its cosine is multiplied by $V_1$ to produce the term:

$$V \sin \omega t \sin \theta \cos \varphi \qquad (6\text{-}8)$$

The digital angle φ is also applied to the sine multiplier and multiplied by $V_2$ to produce the term:

$$V \sin \omega t \cos \theta \sin \varphi \qquad (6\text{-}9)$$

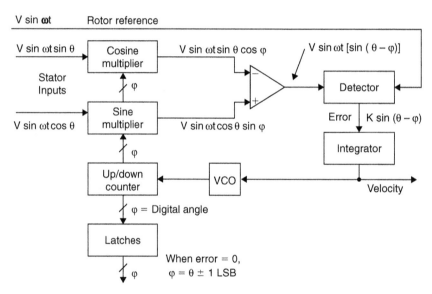

**Figure 6-84: Resolver-to-digital converter (RDC)**

These two signals are subtracted from each other by the error amplifier to yield an AC error signal of the form:

$$V \sin \omega t [\sin \theta \cos \varphi - \cos \theta \sin \varphi] \qquad (6\text{-}10)$$

Using a simple trigonometric identity, this reduces to:

$$V \sin \omega t [\sin (\theta - \varphi)] \qquad (6\text{-}11)$$

The detector synchronously demodulates this AC error signal, using the resolver's rotor voltage as a reference. This results in a DC error signal proportional to sin(θ – φ).

The DC error signal feeds an integrator, the output of which drives a VCO. The VCO, in turn, causes the up/down counter to count in the proper direction to cause:

$$\sin(\theta - \varphi) \rightarrow 0 \qquad (6\text{-}12)$$

When this is achieved,

$$\theta - \varphi \to 0 \qquad (6\text{-}13)$$

and therefore

$$\varphi = \theta \qquad (6\text{-}14)$$

to within one count. Hence, the counter's digital output $\varphi$, represents the angle $\theta$. The latches enable this data to be transferred externally without interrupting the loop's tracking.

This circuit is equivalent to a so-called type-2 servo loop, because it has, in effect, two integrators. One is the counter, which accumulates pulses; the other is the integrator at the output of the detector. In a type-2 servo loop with a constant rotational velocity input, the output digital word continuously follows, or tracks the input, without needing externally derived convert commands, and with no steady state phase lag between the digital output word and actual shaft angle. An error signal appears only during periods of acceleration or deceleration.

As an added bonus, the tracking RDC provides an analog DC output voltage directly proportional to the shaft's rotational velocity. This is a useful feature if velocity is to be measured or used as a stabilization term in a servo system, and it makes tachometers unnecessary.

Since the operation of an RDC depends only on the ratio between input signal amplitudes, attenuation in the lines connecting them to resolvers does not substantially affect performance. For similar reasons, these converters are not greatly susceptible to waveform distortion. In fact, they can operate with as much as 10% harmonic distortion on the input signals; some applications actually use square-wave references with little additional error.

Tracking ADCs are therefore ideally suited to RDCs. While other ADC architectures, such as successive approximation, could be used, the tracking converter is the most accurate and efficient for this application.

Because the tracking converter doubly integrates its error signal, the device offers a high degree of noise immunity (12 dB/octave rolloff). The net area under any given noise spike produces an error. However, typical inductively coupled noise spikes have equal positive and negative-going waveforms. When integrated, this results in a zero net error signal. The resulting noise immunity, combined with the converter's insensitivity to voltage drops, lets the user locate the converter at a considerable distance from the resolver. Noise rejection is further enhanced by the detector's rejection of any signal not at the reference frequency, such as wideband noise.

The AD2S90 is one of a number of integrated RDCs offered by Analog Devices. The general architecture is similar to that of Figure 6-83. Further details on synchro and RDCs can be found in the references.

Syncros and resolvers are also discussed in Chapter 3 (Section 3-1).

## References: ADC Architectures

1.  W. Kester, Editor, **Amplifier Applications Guide**, Analog Devices, 1992, ISBN-0-916550-10-9, Chapter 10. (*An excellent discussion by James Bryant on the use of op amps as comparators.*)

2.  J.N. Giles, "High Speed Transistor Difference Amplifier," **US Patent 3,843,934**, filed, January 31, 1973, issued October 22, 1974. (*Describes one of the first high-speed ECL comparators, the AM685.*)

3.  C.W. Mangelsdorf, "A 400-MHz Input Flash Converter with Error Correction," **IEEE Journal of Solid-State Circuits**, Vol. 25, No. 1, February, 1990, pp. 184–191. (*A discussion of the AD770, an*

*8-bit 200 MSPS flash ADC. The paper describes the comparator metastable state problem and how to optimize the ADC design to minimize its effects.*)

4.  C.E. Woodward, "A Monolithic Voltage-Comparator Array for A/D Converters," **IEEE Journal of Solid State Circuits**, Vol. SC-10, No. 6, December, 1975, pp. 392–399. (*An early paper on a 3-bit flash converter optimized to minimize metastable state errors.*)

5.  P.M. Rainey, "Facsimile Telegraph System," **US Patent 1,608,527**, filed, July 20, 1921, issued November 30, 1926. (*Although A.H. Reeves is generally credited with the invention of PCM, this patent discloses an electro-mechanical PCM system complete with A/D and D/A converters. The 5-bit electro-mechanical ADC described is probably the first documented flash converter. The patent was largely ignored and forgotten until many years after the various Reeves' patents were issued in 1939–1942.*)

6.  R.W. Sears, "Electron Beam Deflection Tube for Pulse Code Modulation," **Bell System Technical Journal**, Vol. 27, January, 1948, pp. 44–57. (*Describes an electron-beam deflection tube 7-bit, 100-kSPS flash converter for early experimental PCM work.*)

7.  F. Gray, "Pulse Code Communication," **US Patent 2,632,058**, filed, November 13, 1947, issued March 17, 1953. (*Detailed patent on the Gray code and its application to electron beam coders.*)

8.  J.O. Edson and H.H. Henning, "Broadband Codecs for an Experimental 224 Mb/s PCM Terminal," **Bell System Technical Journal**, Vol. 44, November, 1965, pp. 1887–1940. (*Summarizes experiments on ADCs based on the electron tube coder as well as a bit-per-stage Gray code 9-bit solid state ADC. The electron beam coder was 9-bits at 12 MSPS, and represented the fastest of its type at the time.*)

9.  R. Staffin and R.D. Lohman, "Signal Amplitude Quantizer," **US Patent 2,869,079**, filed, December 19, 1956, issued January 13, 1959. (*Describes flash and subranging conversion using tubes and transistors.*)

10. E. Goto et al., "Esaki Diode High-Speed Logical Circuits, " **IRE Transactions on Electronic Computers**, Vol. EC-9, March, 1960, pp. 25–29. (*Describes how to use tunnel diodes as logic elements.*)

11. T. Kiyomo, K. Ikeda, and H. Ichiki, "Analog-to-Digital Converter Using an Esaki Diode Stack," **IRE Transactions on Electronic Computers**, Vol. EC-11, December, 1962, pp. 791–792. (*Description of a low resolution 3-bit flash ADC using a stack of tunnel diodes.*)

12. H.R. Schindler, "Using the Latest Semiconductor Circuits in a UHF Digital Converter," **Electronics**, August, 1963, pp. 37–40. (*Describes a 6-bit 50-MSPS subranging ADC using three 2-bit tunnel diode flash converters.*)

13. J.B. Earnshaw, "Design for a Tunnel Diode-Transistor Store with Nondestructive Read-out of Information," **IEEE Transactions on Electronic Computers**, Vol. EC-13, 1964, pp. 710–722. (*Use of tunnel diodes as memory elements.*)

14. W.K. Bucklen, "A Monolithic Video A/D Converter," **Digital Video, Vol. 2**, Society of Motion Picture and Television Engineers, March, 1979, pp. 34–42. (*Describes the revolutionary TDC1007J 8-bit 20MSPS video flash converter. Originally introduced at the February 3, 1979, SMPTE Winter Conference in San Francisco, Bucklen accepted an Emmy award for this product in 1988 and was responsible for the initial marketing and applications support for the device.*)

15. J. Peterson, "A Monolithic Video A/D Converter," **IEEE Journal of Solid-State Circuits**, Vol. SC-14, No. 6, December, 1979, pp. 932–937. (*Another detailed description of the TRW TDC1007J 8-bit, 20-MSPS flash converter.*)

16. Y. Akazawa et. al., *A 400MSPS 8 Bit Flash A/D Converter*, **1987 ISSCC Digest of Technical Papers**, pp. 98–99. (*Describes a monolithic flash converter using Gray decoding.*)

17. Matsuzawa *et al., An 8b 600 MHz Flash A/D Converter with Multi-stage Duplex-gray Coding*, **Symposium VLSI Circuits, Digest of Technical Papers**, May, 1991, pp. 113–114. (*Describes a monolithic flash converter using Gray decoding.*)

18. Chuck Lane, *A 10-bit 60MSPS Flash ADC*, **Proceedings of the 1989 Bipolar Circuits and Technology Meeting**, IEEE Catalog No. 89CH2771-4, September, 1989, pp. 44–47. (*Describes an interpolating method for reducing the number of preamps required in a flash converter.*)

19. W.W. Rouse Ball and H.S.M. Coxeter, **Mathematical Recreations and Essays**, 13th Edition, Dover Publications, 1987, pp. 50, 51. (*Describes a mathematical puzzle for measuring unknown weights using the minimum number of weighing operations. The solution proposed in the 500's is the same basic successive approximation algorithm used today.*)

20. A.H. Reeves, "Electric Signaling System," **US Patent 2,272,070**, filed, November 22, 1939, issued February 3, 1942. Also **French Patent 852,183** issued 1938, and **British Patent 538,860** issued 1939. (*The ground-breaking patent on PCM. Interestingly enough, the ADC and DAC proposed by Reeves are counting types, and not successive approximation.*)

21. J.C. Schelleng, "Code Modulation Communication System," **US Patent 2,453,461**, filed, June 19, 1946, issued November 9, 1948. (*An interesting description of a rather cumbersome successive approximation ADC based on vacuum tube technology. This converter was not very practical, but did illustrate the concept. Also in the patent is a description of a corresponding binary DAC.*)

22. W.M. Goodall, "Telephony by Pulse Code Modulation," **Bell System Technical Journal**, Vol. 26, July, 1947, pp. 395–409. (*Describes an experimental PCM system using a 5-bit, 8KSPS successive approximation ADC based on the subtraction of binary weighted charges from a capacitor to implement the internal subtraction/DAC function. It required 5 internal reference voltages.*)

23. H.R. Kaiser, et al., "High-Speed Electronic Analogue-to-Digital Converter System," **US Patent 2,784,396**, filed, April 2, 1953, issued March 5, 1957. (*One of the first SAR ADCs to use an actual binary-weighted DAC internally.*)

24. B.D. Smith, "Coding by Feedback Methods," **Proceedings of the IRE**, Vol. 41, August, 1953, pp. 1053–1058. (*Smith uses an internal DAC and also points out that a non-linear transfer function can be achieved by using a DAC with non-uniform bit weights, a technique which is widely used in today's voiceband ADCs with built-in companding.*)

25. L.A. Meacham and E. Peterson, "An Experimental Multichannel Pulse Code Modulation System of Toll Quality," **Bell System Technical Journal**, Vol. 27, No. 1, January, 1948, pp. 1–43. (*Describes non-linear diode-based compressors and expanders for generating a non-linear ADC/DAC transfer function.*)

26. B.M. Gordon and R.P. Talambiras, "Signal Conversion Apparatus," **US Patent 3,108,266**, filed, July 22, 1955, issued October 22, 1963. (*Classic patent describing Gordon's 11-bit, 20kSPS vacuum tube successive approximation ADC done at Epsco. The internal DAC represents the first known use of equal currents switched into an R/2R ladder network.*)

27. B.M. Gordon and E.T. Colton, "Signal Conversion Apparatus," **US Patent 2,997,704**, filed, February 24, 1958, issued August 22, 1961. (*Classic patent describes the logic to perform the successive approximation algorithm in an SAR ADC.*)

28. J.R. Gray and S.C. Kitsopoulos, "A Precision Sample-and-Hold Circuit with Subnanosecond Switching," **IEEE Transactions on Circuit Theory**, Vol. CT11, September, 1964, pp. 389–396. (*One of the first papers on the detailed analysis of a sample-and-hold circuit.*)

29. T.C. Verster, "A Method to Increase the Accuracy of Fast Serial-Parallel Analog-to-Digital Converters," **IEEE Transactions on Electronic Computers**, Vol. EC-13, 1964, pp. 471–473. (*One of the first references to the use of error correction in a subranging ADC.*)

30. G.G. Gorbatenko, "High-Performance Parallel-Serial Analog to Digital Converter with Error Correction," **IEEE National Convention Record**, New York, March, 1966. (*Another early reference to the use of error correction in a subranging ADC.*)

31. D.J. Kinniment, D. Aspinall, and D.B.G. Edwards, "High-Speed Analogue-Digital Converter," **IEE Proceedings**, Vol. 113, December, 1966, pp. 2061–2069. (*A 7-bit 9MSPS three-stage pipelined error corrected converter is described based on recirculating through a 3-bit stage three times. Tunnel (Esaki) diodes are used for the individual comparators. The article also shows a proposed faster pipelined 7-bit architecture using three individual 3-bit stages with error correction. The article also describes a fast bootstrapped diode-bridge sample-and-hold circuit.*)

32. O.A. Horna, "A 150Mbps A/D and D/A Conversion System," **Comsat Technical Review**, Vol. 2, No. 1, 1972, pp. 52–57. (*A detailed description and analysis of a subranging ADC with error correction.*)

33. J.L. Fraschilla, R.D. Caveney, and R.M. Harrison, "High Speed Analog-to-Digital Converter," **US Patent 3,597,761**, filed, November 14, 1969, issued August 13, 1971. (*Describes an 8-bit, 5-MSPS subranging ADC with switched references to second comparator bank.*)

34. S.H. Lewis, S. Fetterman, G.F. Gross Jr., R. Ramachandran, and T.R. Viswanathan, "A 10-b 20-Msample/s Analog-Digital Converter," **IEEE Journal of Solid-State Circuits**, Vol. 27, No. 3, March, 1992, pp. 351–358. (*A detailed description and analysis of an error corrected subranging ADC using 1.5-bit pipelined stages.*)

35. R. Gosser and F. Murden, "A 12-bit 50MSPS Two-Stage A/D Converter," **1995 ISSCC Digest of Technical Papers**, p. 278. (*A description of the AD9042 error corrected subranging ADC using MagAMP stages for the internal ADCs.*)

36. B.D. Smith, "An Unusual Electronic Analog-Digital Conversion Method," **IRE Transactions on Instrumentation**, June, 1956, pp. 155–160. (*Possibly the first published description of the binary-coded and Gray-coded bit-per-stage ADC architectures. Smith mentions similar work partially covered in R. P. Sallen's 1949 thesis at M.I.T.*)

37. N.E. Chasek, "Pulse Code Modulation Encoder," **US Patent 3,035,258**, filed, November 14, 1960, issued May 15, 1962. (*An early patent showing a diode-based circuit for realizing the Gray code folding transfer function.*)

38. F.D. Waldhauer, "Analog-to-Digital Converter," **US Patent 3,187,325**, filed, July 2, 1962, issued June 1, 1965. (*A classic patent using op amps with diode switches in the feedback loops to implement the Gray code folding transfer function.*)

39. J.O. Edson and H.H. Henning, "Broadband Codecs for an Experimental 224Mb/s PCM Terminal," **Bell System Technical Journal**, Vol. 44, November, 1965, pp. 1887–1940. (*A further description of a 9-bit ADC based on Waldhauer's folding stage.*)

40. U. Fiedler and D. Seitzer, "A High-Speed 8 Bit A/D Converter Based on a Gray-Code Multiple Folding Circuit," **IEEE Journal of Solid-State Circuits**, Vol. SC-14, No. 3, June, 1979, pp. 547–551. (An early monolithic folding ADC.)

41. R.J. van de Plassche and R.E.J. van der Grift, "A High-Speed 7 Bit A/D Converter," **IEEE Journal of Solid-State Circuits**, Vol. SC-14, No. 6, December, 1979, pp. 938–943. (*A monolithic folding ADC.*)

42. R.E.J. van de Grift and R.J. van der Plassche, "A Monolithic 8-bit Video A/D Converter," **IEEE Journal of Solid State Circuits**, Vol. SC-19, No. 3, June, 1984, pp. 374–378. (*A monolithic folding ADC.*)

43. R.E.J. van der Grift, I.W.J.M. Rutten, and M. van der Veen, "An 8-bit Video ADC Incorporating Folding and Interpolation Techniques," **IEEE Journal of Solid State Circuits**, Vol. SC-22, No. 6, December, 1987, pp. 944–953. (Another monolithic folding ADC.)

44. R. van de Plassche, **Integrated Analog-to-Digital and Digital-to-Analog Converters**, Kluwer Academic Publishers,, Norwell, Ma., 1994, pp. 148–187. (*A good textbook on ADCs and DACs with a section on folding ADCs indicated by the referenced page numbers.*)

45. C. Moreland, "An 8-bit 150 MSPS Serial ADC," **1995 ISSCC Digest of Technical Papers**, Vol. 38, p. 272. (*A description of an 8-bit ADC with 5 folding stages followed by a 3-bit flash converter.*)

46. C. Moreland, **An Analog-to-Digital Converter Using Serial-Ripple Architecture**, Masters' Thesis, Florida State University College of Engineering, Department of Electrical Engineering, 1995. (Moreland's early work on folding ADCs.)

47. F. Murden, "Analog to Digital Converter Using Complementary Differential Emitter Pairs," **US Patent 5,550,492**, filed, December 1, 1994, issued August 27, 1996. (A description of an ADC based on the MagAMP folding stage.)

48. C.W. Moreland, "Analog to Digital Converter Having a Magnitude Amplifier with an Improved Differential Input Amplifier," **US Patent 5,554,943**, filed, December 1, 1994, issued September 10, 1996. (*A description of an 8-bit ADC with 5 folding stages followed by a 3-bit flash converter.*)

49. F. Murden and C.W. Moreland, "N-bit Analog-to-Digital Converter with N-1 Magnitude Amplifiers and N Comparators," **US Patent 5,684,419**, filed, December 1, 1994, issued November 4, 1997. (*Another patent on the MagAMP folding architecture applied to an ADC.*)

50. C. Moreland, F. Murden, M. Elliott, J. Young, M. Hensley, and R. Stop, "A 14-bit 100-Msample/s Subranging ADC," **IEEE Journal of Solid State Circuits**, Vol. 35, No. 12, December, 2000, pp. 1791–1798. (*Describes the architecture used in the 14-bit AD6645 ADC.*)

51. F. Murden and M.R. Elliott, "Linearizing Structures and Methods for Adjustable-Gain Folding Amplifiers," **US Patent 6,172,636B1**, filed, July 13, 1999, issued January 9, 2001. (*Describes methods for trimming the folding amplifiers in an ADC.*)

52. B.M. Oliver and C.E. Shannon, "Communication System Employing Pulse Code Modulation," **US Patent 2,801,281**, filed, February 21, 1946, issued July 30, 1957. (*Charge run-down ADC and Shannon-Rack DAC.*)

53. A.H. Dickinson, "Device to Manifest an Unknown Voltage as a Numerical Quantity," **US. Patent 2,872,670**, filed, May 26, 1951, issued February 3, 1959. (*Ramp run-up ADC.*)

54. K. Howard Barney, "Binary Quantizer," **US Patent 2,715,678**, filed, May 26, 1950, issued August 16, 1955. (*Tracking ADC.*)

55. B.M. Gordon and R.P. Talambiras, "Information Translating Apparatus and Method," **US Patent 2,989,741**, filed, July 22, 1955, issued June 20, 1961. (*Tracking ADC.*)

56. J.L. Lindesmith, "Voltage-to-Digital Measuring Circuit," **US Patent 2,835,868**, filed, September 16, 1952, issued May 20, 1958. (*Voltage-to-frequency ADC.*)

57. P. Klonowski, "Analog-to-Digital Conversion Using Voltage-to-Frequency Converters," **Application Note AN-276**, Analog Devices, Inc. (*A good application note on VFCs.*) Norwood, Ma.

58. J.M. Bryant, "Voltage-to-Frequency Converters," **Application Note AN-361**, Analog Devices, Inc. (*A good overview of VFCs.*) Norwood, Ma.

59. W. Jung, "Operation and Applications of the AD654 IC V-F Converter," **Application Note AN-278**, Analog Devices, Inc., Norwood, Ma.

60. S. Martin, "Using the AD650 Voltage-to-Frequency Converter as a Frequency-to-Voltage Converter," **Application Note AN-279**, Analog Devices, Inc. (*A description of a frequency-to-voltage converter using the popular AD650 VFC.*) Norwood, Ma.

61. R.N. Anderson and H.A. Dorey, "Digital Voltmeters," **US Patent 3,267,458**, filed, August 20, 1962, issued August 16, 1966. (*Charge balance dual slope voltmeter ADC.*)

62. R. Olshausen, "Analog-to-Digital Converter," **US Patent 3,281,827**, filed, June 27, 1963, issued October 25, 1966. (*Charge balance dual slope ADC.*)

63. R.W. Gilbert, "Analog-to-Digital Converter," **US Patent 3,051,939**, filed, May 8, 1957, issued August 28, 1962. (*Dual-slope ADC.*)

64. S.K. Ammann, "Integrating Analog-to-Digital Converter," **US Patent 3,316,547**, filed, July 15, 1964, issued April 25, 1967. (*Dual-slope ADC.*)

65. I. Wold, "Integrating Analog-to-Digital Converter Having Digitally Derived Offset Error Compensation and Bipolar Operation without Zero Discontinuity," **US Patent 3,872,466**, filed, July 19, 1973, issued March 18, 1975. (*Quad-slope ADC.*)

66. H.B. Aasnaes, "Triple Integrating Ramp Analog-to-Digital Converter," **US Patent 3,577,140**, filed, June 27, 1967, issued May 4, 1971. (*Triple-slope ADC.*)

67. F. Bondzeit, L.J. Neelands, "Multiple Slope Analog-to-Digital Converter," **US Patent 3,564,538**, filed, January 29, 1968, issued February 16, 1971. (*Triple-slope ADC.*)

68. D. Wheable, "Triple-Slope Analog-to-Digital Converters," **US Patent 3,678,506**, filed, October 2, 1968, issued July 18, 1972. (*Triple-slope ADC.*)

69. D. Sheingold, **Analog–Digital Conversion Handbook**, Prentice-Hall, Norwood, Ma., 1986, ISBN-0-13-032848-0, pp. 441-471. (*This chapter contains an excellent tutorial on optical, synchro, and resolver-to-digital conversion.*)

70. Dennis Fu, "Circuit Applications of the AD2S90 Resolver-to-Digital Converter," **Application Note AN-230**, Analog Devices. (*Applications of the AD2S90 RTD.*) Norwood, Ma.

# Sigma–Delta Converters

## Historical Perspective

The sigma–delta (Σ–Δ) ADC architecture had its origins in the early development phases of PCM systems—specifically, those related to transmission techniques called *delta modulation* and *differential PCM*. (An excellent discussion of both the history and concepts of the sigma–delta ADC can be found by Max Hauser in Reference 1).

The driving force behind delta modulation and differential PCM was to achieve higher transmission efficiency by transmitting the *changes* (delta) in value between consecutive samples rather than the actual samples themselves.

In *delta modulation*, the analog signal is quantized by a 1-bit ADC (a comparator) as shown in Figure 6-85(A). The comparator output is converted back to an analog signal with a 1-bit DAC, and subtracted from the input after passing through an integrator. The shape of the analog signal is transmitted as follows: a "1" indicates that a positive excursion has occurred since the last sample, and a "0" indicates that a negative excursion has occurred since the last sample.

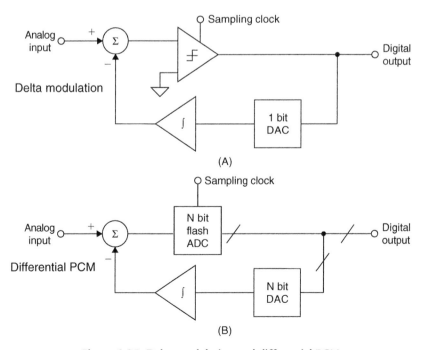

(A)

(B)

**Figure 6-85: Delta modulation and differential PCM**

If the analog signal remains at a fixed DC level for a period of time, a pattern alternating of "0s" and "1s" is obtained. It should be noted that *differential PCM* (see Figure 6-85(B)) uses exactly the same concept except a multibit ADC is used rather than a comparator to derive the transmitted information.

Since there is no limit to the number of pulses of the same sign that may occur, delta modulation systems are capable of tracking signals of any amplitude. In theory, there is no peak clipping. However, the theoretical limitation of delta modulation is that the analog signal must not change too rapidly. The problem of slope clipping is shown in Figure 6-86. Here, although each sampling instant indicates a positive excursion, the analog signal is rising too quickly, and the quantizer is unable to keep pace.

Slope clipping can be reduced by increasing the quantum step size or increasing the sampling rate. Differential PCM uses a multibit quantizer to effectively increase the quantum step sizes at the increase of complexity. Tests have shown that in order to obtain the same quality as classical PCM, delta modulation requires very high sampling rates, typically 20 times the highest frequency of interest, as opposed to Nyquist rate of 2 times.

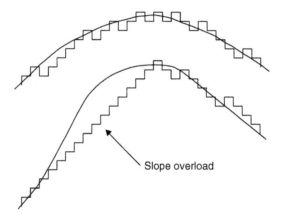

**Figure 6-86: Quantization using delta modulation**

For these reasons, delta modulation and differential PCM have never achieved any significant degree of popularity, however a slight modification of the delta modulator leads to the basic sigma–delta architecture, one of the most popular high resolution ADC architectures in use today.

The basic single and multibit first-order sigma–delta ADC architecture is shown in Figure 6-87(A) and 6-87(B), respectively. Note that the integrator operates on the error signal, whereas in a delta modulator, the integrator is in the feedback loop. The basic oversampling sigma–delta modulator increases the overall SNR at low frequencies by shaping the quantization noise such that most of it occurs outside the bandwidth of interest. The digital filter then removes the noise outside the bandwidth of interest, and the decimator reduces the output data rate back to the Nyquist rate.

The IC sigma–delta ADC offers several advantages over the other architectures, especially for high resolution, low frequency applications. First and foremost, the single-bit sigma–delta ADC is inherently monotonic and requires no laser trimming. The sigma–delta ADC also lends itself to low cost foundry CMOS processes because of the digitally intensive nature of the architecture. Examples of early monolithic sigma–delta ADCs are given in References 13–21. Since that time there have been a constant stream of process and design improvements in the fundamental architecture proposed in the early works cited above.

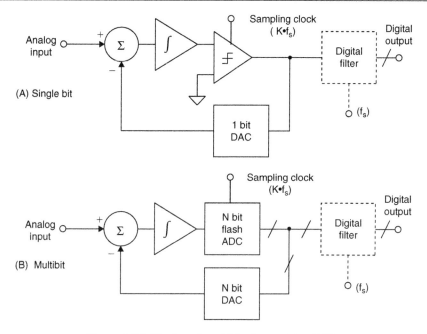

**Figure 6-87: Single and multibit sigma–delta ADCs**

## Sigma–Delta ($\Sigma$–$\Delta$) or Delta–Sigma ($\Delta$–$\Sigma$)? Editor's Notes from *Analog Dialogue* Vol. 24-2, 1990, by Dan Sheingold

This is not the most earth-shaking of controversies, and many readers may wonder what the fuss is all about—if they wonder at all. The issue is important to both editor and readers because of the need for consistency; we would like to use the same name for the same thing whenever it appears. But *which* name? In the case of the modulation technique that led to a new oversampling A/D conversion mechanism, we chose *sigma–delta*. Here is why.

Ordinarily, when a new concept is named by its creators, the name sticks; it should not be changed unless it is erroneous or flies in the face of precedent. The seminal paper on this subject was published in 1962 (References 9, 10), and its authors chose the name "delta–sigma modulation," since it was based on *delta* modulation but included an integration (summation, hence $\Sigma$).

*Delta–sigma* was apparently unchallenged until the 1970s, when engineers at AT&T were publishing papers using the term *sigma–delta*. Why? According to Hauser (Reference 1), the precedent had been to name variants of delta modulation with adjectives preceding the word "delta." Since the form of modulation in question is a variant of delta modulation, the sigma, used as an adjective—so the argument went—should precede the delta.

Many engineers who came upon the scene subsequently used whatever term caught their fancy, often without knowing why. It was even possible to find *both* terms used interchangeably in the same paper. As matters stand today, sigma–delta is in widespread use, probably for the majority of citations. Would its adoption be an injustice to the inventors of the technique?

We think not. Like others, we believe that the name delta–sigma is a departure from precedent. Not just in the sense of grammar, but also in relation to the hierarchy of operations. Consider a block diagram for embodying an analog root mean square (RMS) (finding the square root of the mean of a squared signal) computer. First the signal is squared, then it is integrated, and finally it is rooted (see Figure 6-88).

If we were to name the overall function after the causal order of operations, it would have to be called a "square mean root" function. But naming in order of the *hierarchy* of its mathematical operations gives us the familiar—and undisputed—name, *root mean square*. Consider now a block diagram for taking a difference (delta), and then integrating it (sigma).

Its causal order would give *delta–sigma*, but in functional hierarchy it is *sigma–delta*, since it computes the integral of a difference. We believe that the latter term is correct and follows precedent; and we have adopted it as our standard.

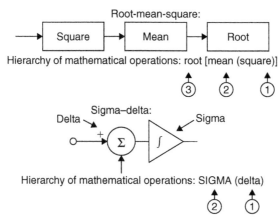

**Figure 6-88: Sigma–delta (Σ–Δ) or delta–sigma (Δ–Σ)?**

## *Basics of Sigma–Delta ADCs*

Sigma–delta analog-to-digital converters (Σ–Δ ADCs) have been known for over 30 years, but only recently has the technology (high density digital VLSI) existed to manufacture them as inexpensive monolithic ICs. They are now used in many applications where a low cost, low bandwidth, low power, high resolution ADC is required.

There have been innumerable descriptions of the architecture and theory of Σ–Δ ADCs, but most commence with a maze of integrals and deteriorate from there. (Some engineers who do not understand the theory of operation of Σ–Δ ADCs and are convinced, from study of a typical published article, which it is too complex to comprehend easily.)

There is nothing particularly difficult to understand about Σ–Δ ADCs, as long as you avoid the detailed mathematics, and this section has been written in an attempt to clarify the subject. A Σ–Δ ADC contains very simple analog electronics (a comparator, voltage reference, a switch, and one or more integrators and analog summing circuits), and quite complex digital computational circuitry. This circuitry consists of a digital signal processor (DSP) which acts as a filter (generally, but not invariably, a lowpass filter (LPF)). It is not necessary to know precisely how the filter works to appreciate what it does. To understand how a Σ–Δ ADC works, familiarity with the concepts of *oversampling, quantization noise shaping, digital filtering,* and *decimation* is required.

Let us consider the technique of oversampling with an analysis in the frequency domain. Where a DC conversion has a *quantization error* of up to ½ LSB, a sampled data system has *quantization noise*.

A perfect classical N-bit sampling ADC has an RMS quantization noise of $q/\sqrt{12}$ uniformly distributed within the Nyquist band of DC to $f_s/2$ (where q is the value of an LSB and $f_s$ is the sampling rate) as shown in Figure 6-89(A). Therefore, its SNR with a full-scale sinewave input will be $(6.02N + 1.76)\,\text{dB}$. If the ADC is less than perfect, and its noise is greater than its theoretical minimum quantization noise, then its *effective* resolution will be less than N-bits. Its actual resolution (often known as its effective number of bits or ENOB) will be defined by:

$$\text{ENOB} = \frac{\text{SNR} - 1.76\,\text{dB}}{6.02\,\text{dB}} \qquad (6\text{-}15)$$

If we choose a much higher sampling rate, $Kf_s$ (see Figure 6-89(B)), the RMS quantization noise remains $q/\sqrt{12}$, but the noise is now distributed over a wider bandwidth DC to $Kf_s/2$. If we then apply a digital LPF to the output, we remove much of the quantization noise, but do not affect the wanted signal—so the ENOB is improved. We have accomplished a high resolution A/D conversion with a low resolution ADC. The factor K is generally referred to as the *oversampling ratio*. It should be noted at this point that oversampling has an added benefit in that it relaxes the requirements on the analog antialiasing filter.

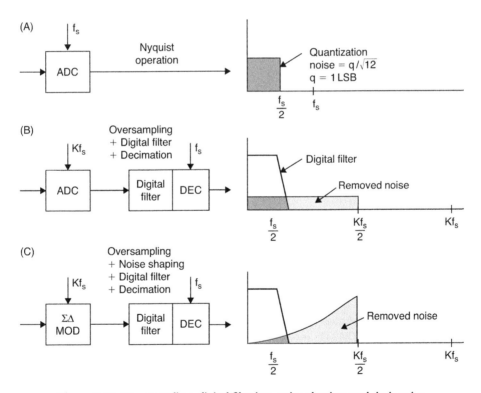

**Figure 6-89: Oversampling, digital filtering, noise shaping, and decimation**

Since the bandwidth is reduced by the digital output filter, the output data rate may be lower than the original sampling rate ($Kf_s$) and still satisfy the Nyquist criterion. This may be achieved by passing every Mth result to the output and discarding the remainder. The process is known as "decimation" by a factor of M. Despite the origins of the term (*decem* is Latin for 10), M can have any integer value, provided that the output data rate is more than twice the signal bandwidth. Decimation does not cause any loss of information (see Figure 6-89(B)).

If we simply use oversampling to improve resolution, we must oversample by a factor of $2^{2N}$ to obtain an N-bit increase in resolution. The Σ–Δ converter does not need such a high oversampling ratio because it not only limits the signal passband, but also shapes the quantization noise so that most of it falls outside this passband as shown in Figure 6-89(C).

If we take a 1-bit ADC (generally known as a comparator), drive it with the output of an integrator, and feed the integrator with an input signal summed with the output of a 1-bit DAC fed from the ADC output, we have a first-order Σ–Δ modulator as shown in Figure 6-90. Add a digital LPF and decimator at the digital output, and we have a Σ–Δ ADC—the Σ–Δ modulator shapes the quantization noise so that it lies above the passband of the digital output filter, and the ENOB is therefore much larger than would otherwise be expected from the oversampling ratio.

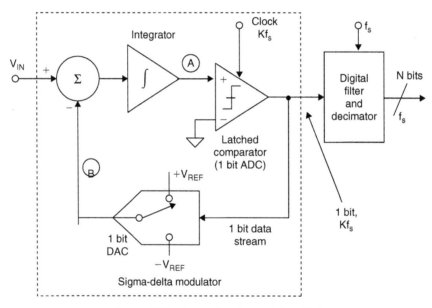

**Figure 6-90: First-order sigma–delta ADC**

Intuitively, a Σ–Δ ADC operates as follows. Assume a DC input at $V_{IN}$. The integrator is constantly ramping up or down at node A. The output of the comparator is fed back through a 1-bit DAC to the summing input at node B. The negative feedback loop from the comparator output through the 1-bit DAC back to the summing point will force the average DC voltage at node B to be equal to $V_{IN}$. This implies that the average DAC output voltage must be equal to the input voltage $V_{IN}$. The average DAC output voltage is controlled by the *ones-density* in the 1-bit data stream from the comparator output. As the input signal increases toward $+V_{REF}$, the number of "ones" in the serial bit stream increases, and the number of "zeros" decreases. Similarly, as the signal goes negative toward $-V_{REF}$, the number of "ones" in the serial bit stream decreases, and the number of "zeros" increases. From a very simplistic standpoint, this analysis shows that the average value of the input voltage is contained in the serial bit stream out of the comparator. The digital filter and decimator process the serial bit stream and produce the final output data.

For any given input value in a single sampling interval, the data from the 1-bit ADC is virtually meaningless. Only when a large number of samples are averaged will a meaningful value result. The sigma–delta modulator is very difficult to analyze in the time domain because of this apparent randomness of the single-bit data output. If the input signal is near positive full-scale, it is clear that there will be more 1 second than 0 second in the bit stream. Likewise, for signals near negative full-scale, there will be more 0 second than 1 second in the bit stream. For signals near mid-scale, there will be approximately an equal number of 1 seconds and 0 seconds. Figure 6-91 shows the output of the integrator for two input conditions. The first is for an input of zero (mid-scale). To decode the output, pass the output samples through a simple digital LPF that averages every four samples. The output of the filter is 2/4. This value represents bipolar zero. If more samples are averaged, more dynamic range is achieved. For example, averaging 4 samples gives 2 bits of resolution, while averaging 8 samples yields 4/8, or 3 bits of resolution. In the bottom waveform of Figure 6-91, the average obtained for 4 samples is 3/4, and the average for 8 samples is 6/8.

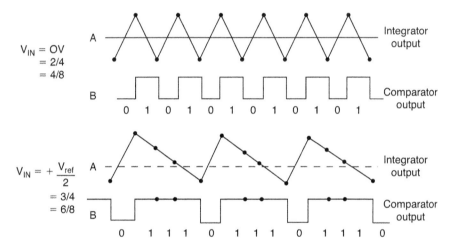

**Figure 6-91: Sigma–delta modulator waveforms**

Further time-domain analysis is not productive, and the concept of noise shaping is best explained in the frequency domain by considering the simple $\Sigma$–$\Delta$ modulator model in Figure 6-92.

The integrator in the modulator is represented as an analog LPF with a transfer function equal to $H(f) = 1/f$. This transfer function has an amplitude response which is inversely proportional to the input frequency. The 1-bit quantizer generates quantization noise, Q, which is injected into the output summing block. If we let the input signal be X, and the output Y, the signal coming out of the input summer must be $X - Y$. This is multiplied by the filter transfer function, $1/f$, and the result goes to one input to the output summer. By inspection, we can then write the expression for the output voltage Y as:

$$Y = \frac{1}{f}(X - Y) + Q \qquad (6\text{-}16)$$

This expression can easily be rearranged and solved for Y in terms of X, f, and Q:

$$Y = \frac{X}{f+1} + \frac{Q \times f}{f+1} \qquad (6\text{-}17)$$

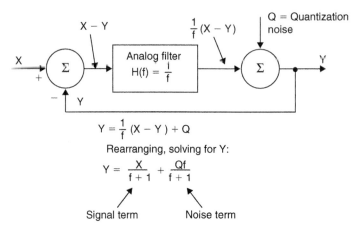

**Figure 6-92: Simplified frequency domain linearized model** *of a sigma–delta modulator*

Note that as the frequency f approaches zero, the output voltage Y approaches X with no noise component. At higher frequencies, the amplitude of the signal component approaches zero, and the noise component approaches Q. At high frequency, the output consists primarily of quantization noise. In essence, the analog filter has a lowpass effect on the signal, and a highpass effect on the quantization noise. Thus the analog filter performs the noise shaping function in the $\Sigma$–$\Delta$ modulator model.

For a given input frequency, higher order analog filters offer more attenuation. The same is true of $\Sigma$–$\Delta$ modulators, provided certain precautions are taken.

By using more than one integration and summing stage in the $\Sigma$–$\Delta$ modulator, we can achieve higher orders of quantization noise shaping and even better ENOB for a given oversampling ratio as is shown in Figure 6-93 for both a first- and second-order $\Sigma$–$\Delta$ modulator.

The block diagram for the second-order $\Sigma$–$\Delta$ modulator is shown in Figure 6-94. Third, and higher, order $\Sigma$–$\Delta$ ADCs were once thought to be potentially unstable at some values of input—recent analyses using *finite* rather than infinite gains in the comparator have shown that this is not necessarily so, but even if instability does start to occur, it is not important, since the DSP in the digital filter and decimator can be made to recognize incipient instability and react to prevent it.

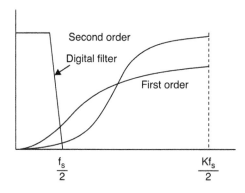

**Figure 6-93: Sigma–delta modulators shape quantization noise**

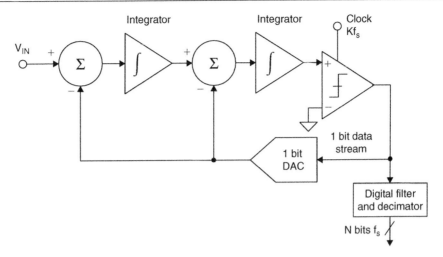

**Figure 6-94: Second-order sigma–delta ADC**

Figure 6-95 shows the relationship between the order of the $\Sigma$–$\Delta$ modulator and the amount of oversampling necessary to achieve a particular SNR. For instance, if the oversampling ratio is 64, an ideal second-order system is capable of providing an SNR of about 80 dB. This implies approximately 13 ENOB. Although the filtering done by the digital filter and decimator can be done to any degree of precision desirable, it would be pointless to carry more than 13 binary bits to the outside world. Additional bits would carry no useful signal information, and would be buried in the quantization noise unless post-filtering techniques were employed. Additional resolution can be obtained by increasing the oversampling ratio and/ or by using a higher order modulator.

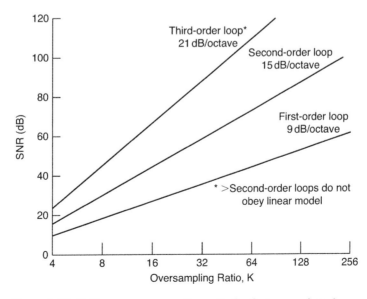

**Figure 6-95: SNR versus oversampling ratio for first, second, and third-order loops**

## Idle-Tone Considerations

In our discussion of sigma–delta ADCs up to this point, we have made the assumption that the quantization noise produced by the sigma–delta modulator is random and uncorrelated with the input signal. Unfortunately, this is not entirely the case, especially for the first-order modulator. Consider the case where we are averaging 16 samples of the modulator output in a 4-bit sigma–delta ADC.

Figure 6-96 shows the bit pattern for two input signal conditions: an input signal having the value 8/16, and an input signal having the value 9/16. In the case of the 9/16 signal, the modulator output bit pattern has an extra "1" every 16th output. This will produce energy at $f_s/16$, which translates into an unwanted tone. If the oversampling ratio is less than 16, this tone will fall into the passband. In audio applications these tones are referred to as "idle tones."

Figure 6-97 shows the correlated idling pattern behavior for a first-order sigma–delta modulator, and Figure 6-98 shows the relatively uncorrelated pattern for a second-order modulator. For this reason, virtually all sigma–delta ADCs contain at least a second-order modulator loop.

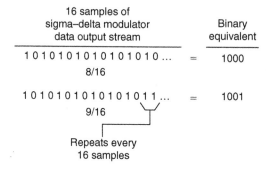

**Figure 6-96: Repetitive bit pattern in sigma–delta modulator output**

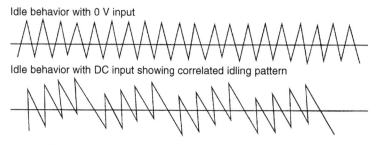

**Figure 6-97: Idling patterns for first-order sigma–delta modulator (integrator output)**

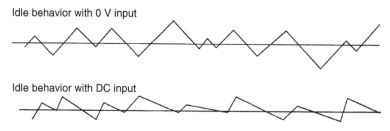

**Figure 6-98: Idling patterns for second-order sigma–delta modulator (*integrator output*)**

## *Higher Order Loop Considerations*

In order to achieve wide dynamic range, sigma–delta modulator loops greater than second-order are necessary, but present real design challenges. First of all, the simple linear models previously discussed are no longer fully accurate. Loops of order greater than two are generally not guaranteed to be stable under all input conditions. The instability arises because the comparator is a nonlinear element whose effective "gain" varies inversely with the input level. This mechanism for instability causes the following behavior: if the loop is operating normally, and a large signal is applied to the input that overloads the loop, the average gain of the comparator is reduced. The reduction in comparator gain in the linear model causes loop instability. This causes instability even when the signal that caused it is removed. In actual practice, such a circuit would normally oscillate on power-up due to initial conditions caused by turn-on transients. As an example, the AD1879 dual audio ADC released in 1994 by Analog Devices used a fifth-order loop. Extensive nonlinear stabilization techniques were required in this and similar higher order loop designs (References 22–26).

## *MultiBit Sigma–Delta Converters*

So far we have considered only sigma–delta converters which contain a single-bit ADC (comparator) and a single-bit DAC (switch). The block diagram of Figure 6-99 shows a multibit sigma–delta ADC which uses an n-bit flash ADC and an n-bit DAC. Obviously, this architecture will give a higher dynamic range for a given oversampling ratio and order of loop filter. Stabilization is easier, since second-order loops can generally be used. Idling patterns tend to be more random thereby minimizing tonal effects.

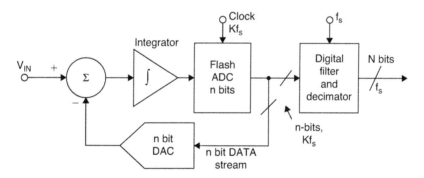

**Figure 6-99: Multibit sigma–delta ADC**

The real disadvantage of this technique is that the linearity depends on the DAC linearity, and thin-film laser trimming is required to approach 16-bit performance levels. This makes the multibit architecture extremely impractical to implement on mixed-signal ICs using traditional binary DAC techniques.

However, fully decoded thermometer DACs coupled with proprietary data scrambling techniques as used in a number of Analog Devices' audio ADCs and DACs, including the 24-bit stereo AD1871 (see References 27 and 28) can achieve high SNR and low distortion using the multibit architecture. A simplified block diagram of the AD1871 ADC is shown in Figure 6-100.

The AD1871's analog Σ–Δ modulator section comprises a second-order multibit implementation using Analog Device's proprietary technology for best performance. As shown in Figure 6-101, the two analog integrator blocks are followed by a flash ADC section that generates the multibit samples.

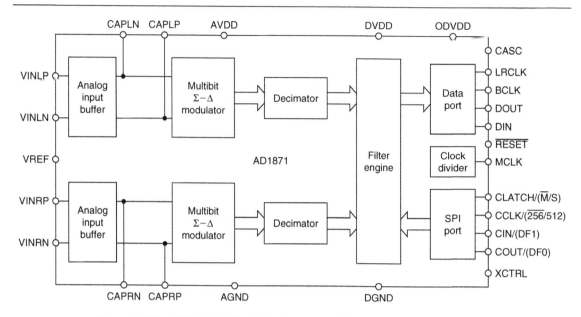

**Figure 6-100: AD1871 24-bit 96-kSPS stereo audio multibit sigma–delta ADC**

The output of the flash ADC, which is thermometer encoded, is decoded to binary for output to the filter sections and is scrambled for feedback to the two integrator stages. The modulator is optimized for operation at a sampling rate of 6.144 MHz (which is $128 \times f_s$ at 48 kHz sampling and $64 \times f_s$ at 96 kHz sampling). The A-weighted dynamic range of the AD1871 is typically 105 dB.

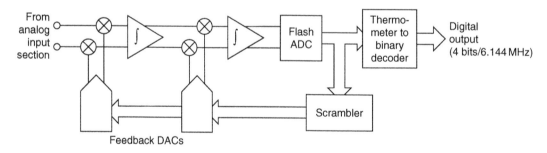

**Figure 6-101: Details of the AD1871 second-order modulator and data scrambler**

## Digital Filter Implications

The digital filter is an integral part of all sigma–delta ADCs—there is no way to remove it. The settling time of this filter affects certain applications especially when using sigma–delta ADCs in multiplexed applications. The output of a multiplexer can present a step function input to an ADC if there are different input voltages on adjacent channels. In fact, the multiplexer output can represent a full-scale step voltage to the sigma–delta ADC when channels are switched. Adequate filter settling time must be allowed, therefore, in such applications. This does not mean that sigma–delta ADCs should not be used in multiplexed applications, just that the settling time of the digital filter must be considered. Some newer sigma–delta ADCs are actually optimized for use in multiplexed applications.

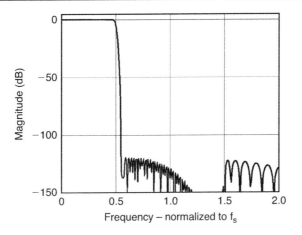

**Figure 6-102: 24-Bit, 96-kSPS stereo sigma–delta
ADC digital filter characteristics**

For example, the group delay through the AD1871 digital filter is 910 μs (sampling at 48 kSPS) and 460 μs (sampling at 96 kSPS)—this represents the time it takes for a step function input to propagate through one-half the number of taps in the digital filter. The total settling time is therefore approximately twice the group delay time. The input oversampling frequency is 6.144 MSPS for both conditions. The frequency response of the digital filter in the AD1871 ADC is shown in Figure 6-102.

In other applications, such as low frequency, high resolution 24-bit measurement sigma–delta ADCs (such as the AD77xx-series), other types of digital filters may be used. For instance, the SINC$^3$ response is popular because it has zeros at multiples of the throughput rate. For instance a 10 Hz throughput rate produces zeros at 50 Hz and 60 Hz which aid in AC power line rejection. The frequency response of a typical ΣΔ ADC, the AD7730, is shown in Figure 6-103.

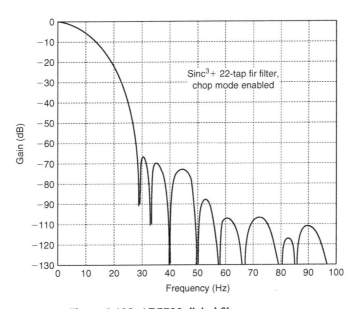

**Figure 6-103: AD7730 digital filter response**

419

## High Resolution Measurement Sigma–Delta ADCs

In order to better understand the capability of sigma–delta measurement ADCs and the power of the technique, a modern example, the AD7730, will be examined in detail. The AD7730 is a member of the AD77XX family and is shown in Figure 6-104. This ADC was specifically designed to interface directly to bridge outputs in weigh scale applications. The device accepts low level signals directly from a bridge and outputs a serial digital word. There are two buffered differential inputs which are multiplexed, buffered, and drive a PGA. The PGA can be programmed for four differential unipolar analog input ranges: 0V to +10mV, 0V to +20mV, 0V to +40mV, and 0V to +80mV, and four differential bipolar input ranges: ±10mV, ±20mV, ±40mV, and ±80mV.

The maximum peak-to-peak, or noise-free resolution achievable is 1 in 230,000 counts, or approximately 18 bits. It should be noted that the noise-free resolution is a function of input voltage range, filter cutoff, and output word rate. Noise is greater using the smaller input ranges where the PGA gain must be increased. Higher output word rates and associated higher filter cutoff frequencies will also increase the noise.

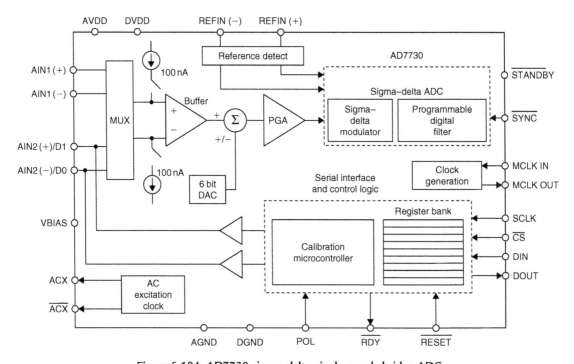

**Figure 6-104: AD7730 sigma–delta single-supply bridge ADC**

The analog inputs are buffered on-chip allowing relatively high source impedances. Both analog channels are differential, with a CMV range that comes within 1.2V of analog ground (AGND) and 0.95V of $AV_{DD}$. The reference input is also differential, and the common-mode range is from AGND to $AV_{DD}$.

The 6-bit DAC is controlled by on-chip registers and can remove TARE (pan weight) values of up to ±80mV from the analog input signal range. The resolution of the TARE function is 1.25mV with a +2.5V reference and 2.5mV with a +5V reference.

The output of the PGA is applied to the $\Sigma$–$\Delta$ modulator and programmable digital filter. The serial interface can be configured for three-wire operation and is compatible with microcontrollers and DSPs. The AD7730 contains self-calibration and system-calibration options and has an offset drift of less than 5 nV/°C and a gain drift of less than 2 ppm/°C. This low offset drift is obtained using a *chop* mode which operates similarly to a chopper-stabilized amplifier.

The oversampling frequency of the AD7730 is 4.9152 MHz, and the output data rate can be set from 50 Hz to 1,200 Hz. The accuracy of the output of the ADC is dependent on the output data rate as shown in Tables I and II of Figure 6-105. These are taken from the AD7730. Note that the accuracy is also dependent on the PGA gain as well.

This is easy to understand. The quantization is performed at the master clock rate (4.9152 MHz). If the data rate is increased, there is less time for filtering, so the measured result is noisier. Also as gain is increased, noise is increased as well.

While the output data word is 24-bits wide, there will not be a constant 24-bit data output, even with the input grounded. As seen in Table I, the maximum accuracy is on the order of 18 bits peak-to-peak. This gives rise to a new way of specifying accuracy. This is noise-free counts. For the AD7730 this is 230,000.

The clock source can be provided via an external clock or by connecting a crystal oscillator across the MCLK IN and MCLK OUT pins.

The AD7730 can accept input signals from a DC-excited bridge. It can also handle input signals from an AC-excited bridge by using the AC excitation clock signals (ACX and $\overline{ACX}$). These are non-overlapping clock signals used to synchronize the external switches which drive the bridge. The ACX clocks are demodulated on the AD7730 input.

The AD7730 contains two 100 nA constant current generators, one source current from $AV_{DD}$ to $A_{IN}(+)$ and one sink current from $A_{IN}(-)$ to AGND. The currents are switched to the selected analog input pair under the control of a bit in the mode register. These currents can be used in checking that a sensor is still operational before attempting to take measurements on that channel. If the currents are turned on and a full-scale reading is obtained, then the sensor has gone open circuit. If the measurement is 0 V, the sensor has gone short circuit. In normal operation, the burnout currents are turned off by setting the proper bit in the mode register to 0.

The AD7730 contains an internal programmable digital filter. The filter consists of two sections: a first stage filter, and a second stage filter. The first stage is a sinc³ LPF. The cutoff frequency and output rate of this first stage filter is programmable. The second stage filter has three modes of operation. In its normal mode, it is a 22-tap FIR filter that processes the output of the first stage filter. When a step change is detected on the analog input, the second stage filter enters a second mode (FASTStep™) where it performs a variable number of averages for some time after the step change, and then the second stage filter switches back to the FIR filter mode. The third option for the second stage filter (SKIP mode) is that it is completely bypassed so the only filtering provided on the AD7730 is the first stage. Both the FASTStep mode and SKIP mode can be enabled or disabled via bits in the control register. Again, there will be an affect on accuracy.

Figure 6-106 shows the full frequency response of the AD7730 when the second stage filter is set for normal FIR operation. This response is with the chop mode enabled and an output word rate of 200 Hz and a clock frequency of 4.9152 MHz. The response is shown from DC to 100 Hz. The rejection at 50 Hz ±1 Hz and 60 Hz ±1 Hz is better than 88 dB.

Figure 6-107 shows the step response of the AD7730 with and without the FASTStep mode enabled. The vertical axis shows the code value and indicates the settling of the output to the input step change. The horizontal axis shows the number of output words required for that settling to occur. The positive input step change occurs at the 5th output.

Table I. Output noise vs. input range and update rate (CHP = 1)
typical output RMS noise in nV

| Output Data Rate | −3 dB Frequency | SF Word | Settling Time Normal Mode | Settling Time Fast Mode | Input Range = ±80 mV | Input Range = ±40 mV | Input Range = ±20 mV | Input Range = ±10 mV |
|---|---|---|---|---|---|---|---|---|
| 50 Hz | 1.97 Hz | 2048 | 460 ms | 60 ms | 115 | 75 | 55 | 40 |
| 100 Hz | 3.95 Hz | 1024 | 230 ms | 30 ms | 155 | 105 | 75 | 60 |
| 150 Hz | 5.92 Hz | 683 | 153 ms | 20 ms | 200 | 135 | 95 | 70 |
| 200 Hz | 7.9 Hz | 512 | 115 ms | 15 ms | 225 | 145 | 100 | 80 |
| 400 Hz | 15.8 Hz | 256 | 57.5 ms | 7.5 ms | 335 | 225 | 160 | 110 |

*Power-On Default

Table II. Peak-to-peak resolution vs. input range and update rate (CHP = 1)
peak-to-peak resolution in counts (Bits)

| Output Data Rate | −3 dB Frequency | SF Word | Settling Time Normal Mode | Settling Time Fast Mode | Input Range = ±80 mV | Input Range = ±40 mV | Input Range = ±20 mV | Input Range = ±10 mV |
|---|---|---|---|---|---|---|---|---|
| 50 Hz | 1.97 Hz | 2048 | 460 ms | 60 ms | 230k (18) | 175k (17.5) | 120k (17) | 80k (16.5) |
| 100 Hz | 3.95 Hz | 1024 | 230 ms | 30 ms | 170k (17.5) | 125k (17) | 90k (16.5) | 55k (16) |
| 150 Hz | 5.92 Hz | 683 | 153 ms | 20 ms | 130k (17) | 100k (16.5) | 70k (16) | 45k (15.5) |
| 200 Hz* | 7.9 Hz | 512 | 115 ms | 15 ms | 120k (17) | 90k (16.5) | 65k (16) | 40k (15.5) |
| 400 Hz | 15.8 Hz | 256 | 57.5 ms | 7.5 ms | 80k (16.5) | 55k (16) | 40k (15.5) | 30k (15) |

*Power-On Default

**Figure 6-105: Resolution versus output data rate and gain for the AD7730**

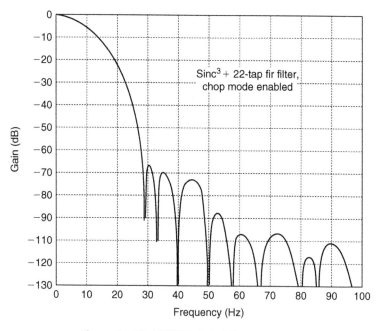

**Figure 6-106: AD7730 digital filter response**

In the normal mode (FASTStep disabled), the output has not reached its final value until the 23rd output word. In FASTStep mode with chopping enabled, the output has settled to the final value by the 7th output word. Between the 7th and the 23rd output, the FASTStep mode produces a settled result, but with additional noise compared to the specified noise level for normal operating conditions. It starts at a noise

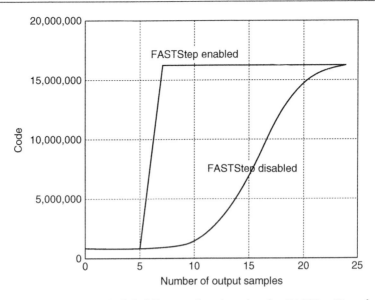

Figure 6-107: AD7730 digital filter settling time showing FASTStep™ mode

level comparable to the SKIP mode, and as the averaging increases ends up at the specified noise level. The complete settling time required for the part to return to the specified noise level is the same for FASTStep mode and normal mode. The FASTStep mode gives a much earlier indication of where the output channel is going and its new value. This feature is very useful in weigh scale applications to give a much earlier indication of the weight, or in an application scanning multiple channels where the user does not have to wait the full settling time to see if a channel has changed.

Note, however, that the FASTStep mode is not particularly suitable for multiplexed applications because of the excess noise associated with the settling time. For multiplexed applications, the full 23-cycle output

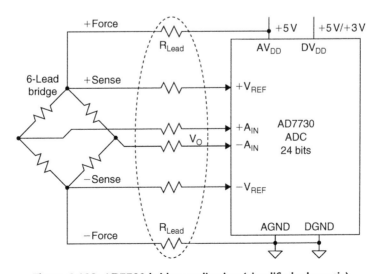

Figure 6-108: AD7730 bridge application (simplified schematic)

word interval should be allowed for settling to a new channel. This points out the fundamental issue of using $\Sigma-\Delta$ ADCs in multiplexed applications. There is no reason why they won't work, provided the internal digital filter is allowed to settle fully after switching channels.

The AD7730 gives the user access to the on-chip calibration registers allowing an external microprocessor to read the device's calibration coefficients and also to write its own calibration coefficients to the part from prestored values in external $E^2$PROM. This gives the microprocessor much greater control over the AD7730's calibration procedure. It also means that the user can verify that the device has performed its calibration correctly by comparing the coefficients after calibration with prestored values in $E^2$PROM. Since the calibration coefficients are derived by performing a conversion on the input voltage provided, the accuracy of the calibration can only be as good as the noise level the part provides in the normal mode. To optimize calibration accuracy, it is recommended to calibrate the part at its lowest output rate where the noise level is lowest. The coefficients generated at any output rate will be valid for all selected output update rates. This scheme of calibrating at the lowest output data rate does mean that the duration of the calibration interval is longer.

The AD7730 requires an external voltage reference, however, the power supply may be used as the reference in the ratiometric bridge application shown in Figure 6-108. In this configuration, the bridge output voltage is directly proportional to the bridge drive voltage which is also used to establish the reference voltages to the AD7730. Variations in the supply voltage will not affect the accuracy. The SENSE outputs of the bridge are used for the AD7730 reference voltages in order to eliminate errors caused by voltage drops in the lead resistances.

## Bandpass Sigma–Delta Converters

The $\Sigma-\Delta$ ADCs that we have described so far contain integrators, which are LPFs, whose passband extends from DC. Thus, their quantization noise is pushed up in frequency. At present, most commercially available $\Sigma-\Delta$ ADCs are of this type (although some which are intended for use in audio or telecommunications applications contains bandpass rather than lowpass digital filters to eliminate any system DC offsets). But there is no particular reason why the filters of the $\Sigma-\Delta$ modulator should be LPFs, except that traditionally ADCs have been thought of as being baseband devices, and that integrators are somewhat easier to construct than bandpass filters (BPFs). If we replace the integrators in a $\Sigma-\Delta$ ADC with BPFs as shown in Figure 6-109, the quantization noise is moved up and down in frequency to leave a virtually noise-free region in the passband (see References 31–33). If the digital filter is then programmed to have its passband in this region, we have a $\Sigma-\Delta$ ADC with a bandpass, rather than a lowpass characteristic. Such devices would appear to be useful in direct IF-to-digital conversion, digital radios, ultrasound, and other undersampling applications. However, the modulator and the digital BPF must be designed for the specific set of frequencies required by the system application, thereby somewhat limiting the flexibility of this approach.

In an undersampling application of a bandpass $\Sigma-\Delta$ ADC, the minimum sampling frequency must be at least twice the signal bandwidth, BW. The signal is centered around a carrier frequency, $f_c$. A typical digital radio application using a 455 kHz center frequency and a signal bandwidth of 10 kHz is described in Reference 102. An oversampling frequency $Kf_S = 2$ MSPS and an output rate $f_S = 20$ kSPS yielded a dynamic range of 70 dB within the signal bandwidth.

Another example of a bandpass is the AD9870 IF Digitizing Subsystem having a nominal oversampling frequency of 18 MSPS, a center frequency of 2.25 MHz, and a bandwidth of 10–150 kHz (see details in Reference 32).

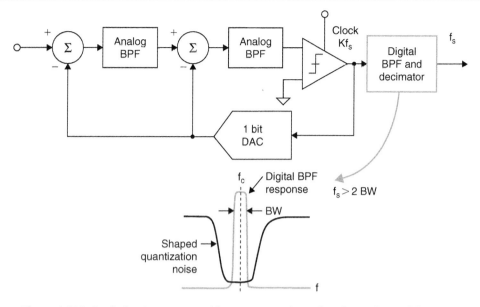

**Figure 6-109: Replacing integrators with resonators gives a bandpass sigma–delta ADC**

## Sigma–Delta DACs

Sigma–delta DACs operate very similarly to sigma–delta ADCs, however in a sigma–delta DAC, the noise shaping function is accomplished with a digital modulator rather than an analog one.

A Σ–Δ DAC, unlike the Σ–Δ ADC, is mostly digital (see Figure 6-110(A)). It consists of an "interpolation filter" (a digital circuit which accepts data at a low rate, inserts zeros at a high rate, and then applies a

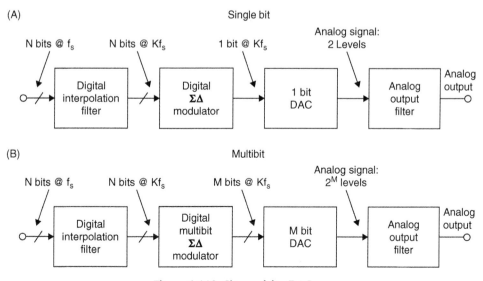

**Figure 6-110: Sigma–delta DACs**

digital filter algorithm and outputs data at a high rate), a $\Sigma-\Delta$ modulator (which effectively acts as an LPF to the signal but as a highpass filter to the quantization noise, and converts the resulting data to a high speed bit stream), and a 1-bit DAC whose output switches between equal positive and negative reference voltages. The output is filtered in an external analog LPF. Because of the high oversampling frequency, the complexity of the LPF is much less than the case of traditional Nyquist operation.

It is possible to use more than 1 bit in the $\Sigma-\Delta$ DAC, and this leads to the *multibit* architecture shown in Figure 6-110B. The concept is similar to that of interpolating DACs previously discussed in this Chapter, with the addition of the digital sigma–delta modulator. In the past, multibit DACs have been difficult to design because of the accuracy requirement on the n-bit internal DAC (this DAC, although only n-bits,

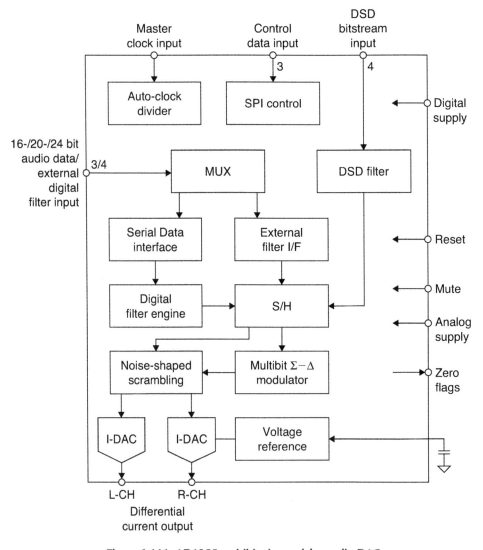

**Figure 6-111: AD1955 multibit sigma–delta audio DAC**

must have the linearity of the final number of bits, N). The AD185x-series of audio DACs, however, use a proprietary *data scrambling* technique (*called data directed scrambling*) which overcomes this problem and produces excellent performance with respect to all audio specifications (see References 27 and 28). For instance, the AD1853 dual 24-bit, 92-kSPS DAC has greater than 104 dB total harmonic distortion plus noise (THD + N) at a 48-kSPS sampling rate.

One of the newest members of this family is the AD1955 multibit sigma–delta audio DAC shown in Figure 6-111. The AD1955 also uses data directed scrambling, supports a multitude of DVD audio formats and has an extremely flexible serial port. THD + N is typically 110 dB.

## Summary

Sigma–delta ADCs and DACs have proliferated into many modern applications including measurement, voiceband, audio, etc. The technique takes full advantage of low cost CMOS processes and therefore makes integration with highly digital functions such as DSPs practical. Resolutions up to 24-bits are currently available, and the requirements on analog antialiasing/antiimaging filters are greatly relaxed due to oversampling. Modern techniques such as the multibit data scrambled architecture minimize problems with idle tones which plagued early sigma–delta products.

Many sigma–delta converters offer a high level of user programmability with respect to output data rate, digital filter characteristics, and self-calibration modes. Multi-channel sigma–delta ADCs are now available for data acquisition systems, and most users are well-educated with respect to the settling time requirements of the internal digital filter in these applications.

### References: Sigma–Delta Converters

1. M.W. Hauser, "Principles of Oversampling A/D Conversion," **Journal Audio Engineering Society**, Vol. 39, No. 1/2, January/December, 1991, pp. 3–26. (*One of the best tutorials and practical discussions of the sigma–delta ADC architecture and its history.*)

2. E.M. Deloraine, S. Van Mierlo, and B. Derjavitch, "Methode et systéme de transmission par impulsions," **French Patent 932,140**, issued August, 1946. Also **British Patent 627,262**, issued 1949.

3. E.M. Deloraine, S. Van Mierlo, and B. Derjavitch, "Communication System Utilizing Constant Amplitude Pulses of Opposite Polarities," **US Patent 2,629,857**, filed, October 8, 1947, issued February 24, 1953.

4. F. de Jager, "Delta Modulation: A Method of PCM Transmission Using the One Unit Code," **Phillips Research Reports**, Vol. 7, 1952, pp. 542–546. (*Additional work done on delta modulation during the same time period.*)

5. H. Van de Weg, "Quantizing Noise of a Single Integration Delta Modulation System with an N-Digit Code," **Phillips Research Reports**, Vol. 8, 1953, pp. 367–385. (*Additional work done on delta modulation during the same time period.*)

6. C.C. Cutler, "Differential Quantization of Communication Signals," **US Patent 2,605,361**, filed, June 29, 1950, issued July 29, 1952. (*Recognized as the first patent on differential PCM or delta modulation, although actually first invented in the Paris labs of the International Telephone and Telegraph Corporation by E. M. Deloraine, S. Mierlo and B. Derjavitch a few years earlier.*)

7. C.C. Cutler, "Transmission Systems Employing Quantization," **US Patent 2,927,962**, filed, April 26, 1954, issued March 8, 1960. (*A ground-breaking patent describing oversampling and noise shaping*

*using first and second-order loops to increase effective resolution. The goal was transmission of oversampled noise shaped PCM data without decimation, not a Nyquist-type ADC.)*

8.  C.B. Brahm, "Feedback Integrating System," **US Patent 3,192,371**, filed, September 14, 1961, issued June 29, 1965. (*Describes a second-order multibit oversampling noise shaping ADC.*)

9.  H. Inose, Y. Yasuda, and J. Murakami, "A Telemetering System by Code Modulation: $\Delta - \Sigma$ Modulation," **IRE Transactions on Space Electronics Telemetry**, Vol. SET-8, September, 1962, pp. 204–209, Reprinted in N. S. Jayant, Waveform Quantization and Coding, IEEE Press and John Wiley, 1976, ISBN 0-471-01970-4. (*An elaboration on the 1-bit form of Cutler's noise-shaping oversampling concept. This work coined the description of the architecture as "delta–sigma modulation."*)

10. H. Inose and Y. Yasuda, "A Unity Bit Coding Method by Negative Feedback," **IEEE Proceedings**, Vol. 51, November, 1963, pp. 1524–1535. (*Further discussions on their 1-bit "delta–sigma" concept.*)

11. D.J. Goodman, "The Application of Delta Modulation of Analog-to-PCM Encoding," **Bell System Technical Journal**, Vol. 48, February, 1969, pp. 321–343, Reprinted in N. S. Jayant, **Waveform Quantization and Coding**, IEEE Press and John Wiley, 1976, ISBN 0-471-01970-4. (*The first description of using oversampling and noise shaping techniques followed by digital filtering and decimation to produce a true Nyquist-rate ADC.*)

12. J.C. Candy, "A Use of Limit Cycle Oscillations to Obtain Robust Analog-to-Digital Converters," **IEEE Transactions on Communications**, Vol. COM-22, December, 1974, pp. 298–305. (*Describes a multibit oversampling noise shaping ADC with output digital filtering and decimation to interpolate between the quantization levels.*)

13. R.J. van de Plassche, "A Sigma–Delta Modulator as an A/D Converter," **IEEE Transactions on Circuits and Systems**, Vol. CAS-25, July, 1978, pp. 510–514.

14. B.A. Wooley and J.L. Henry, "An Integrated Per-Channel PCM Encoder Based on Interpolation," **IEEE Journal of Solid State Circuits**, Vol. SC-14, February, 1979, pp. 14–20. (*One of the first all-integrated CMOS sigma–delta ADCs.*)

15. B.A. Wooley et al., "An Integrated Interpolative PCM Decoder," **IEEE Journal of Solid State Circuits**, Vol. SC-14, February, 1979, pp. 20–25.

16. J.C. Candy, B.A. Wooley, and O.J. Benjamin, "A Voiceband Codec with Digital Filtering," **IEEE Transactions on Communications**, Vol. COM-29, June, 1981, pp. 815–830.

17. J.C. Candy and G.C. Temes, **Oversampling Delta–Sigma Data Converters**, IEEE Press, Piscataway, NJ, 1992, ISBN 0-87942-258-8.

18. R. Koch, B. Heise, F. Eckbauer, E. Engelhardt, J. Fisher, and F. Parzefall, "A 12-bit Sigma–Delta Analog-to-Digital Converter with a 15 MHz Clock Rate," **IEEE Journal of Solid-State Circuits**, Vol. SC-21, No. 6, December, 1986.

19. D.R. Welland, B.P. Del Signore, and E.J. Swanson, "A Stereo 16-Bit Delta–Sigma A/D Converter for Digital Audio," **Journal of Audio Engineering Society**, Vol. 37, No. 6, June, 1989, pp. 476–485.

20. B. Boser and B. Wooley, "The Design of Sigma–Delta Modulation Analog-to-Digital Converters," **IEEE Journal of Solid-State Circuits**, Vol. 23, No. 6, December, 1988, pp. 1298–1308.

21. J. Dattorro, A. Charpentier, D. Andreas, *The Implementation of a One-Stage Multirate 64:1 FIR Decimator for use in One-Bit Sigma–Delta A/D Applications*, **AES 7th International Conference**, May, 1989.

22. W.L. Lee and C.G. Sodini, "A Topology for Higher-Order Interpolative Coders," **ISCAS PROC.**, 1987.

23. P.F. Ferguson Jr., A. Ganesan, and R.W. Adams, "One-Bit Higher Order Sigma–Delta A/D Converters," **ISCAS PROC**, Vol. 2, 1990, pp. 890–893.

24. W.L. Lee, **A Novel Higher Order Interpolative Modulator Topology for High Resolution Oversampling A/D Converters**, MIT Masters Thesis, June, 1987.

25. R.W. Adams, "Design and Implementation of an Audio 18-Bit Analog-to-Digital Converter Using Oversampling Techniques," **Journal of Audio Engineering Society**, Vol. 34, March, 1986, pp. 153–166.

26. P. Ferguson, Jr., A. Ganesan, R. Adams, et. al., "An 18-Bit 20-kHz Dual Sigma–Delta A/D Converter," **ISSCC Digest of Technical Papers**, February, 1991.

27. R. Adams, K. Nguyen, and K. Sweetland, "A 113 dB SNR Oversampling DAC with Segmented Noise-Shaped Scrambling, " **ISSCC Digest of Technical Papers**, Vol. 41, 1998, pp. 62, 63, 413. (*Describes a segmented audio DAC with data scrambling.*)

28. R.W. Adams and T.W. Kwan, "Data-directed Scrambler for Multi-bit Noise-shaping D/A Converters," **US Patent 5,404,142**, filed, August 5, 1993, issued April 4, 1995. (*Describes a segmented audio DAC with data scrambling.*)

29. Y. Matsuya et al., "A 16-Bit Oversampling A/D Conversion Technology Using Triple-Integration Noise Shaping," **IEEE Journal of Solid-State Circuits**, Vol. SC-22, No. 6, December, 1987, pp. 921–929.

30. Y. Matsuya et al., "A 17-Bit Oversampling D/A Conversion Technology Using Multistage Noise Shaping," **IEEE Journal of Solid-State Circuits**, Vol. 24, No. 4, August, 1989, pp. 969–975.

31. P.H. Gailus, W.J. Turney, and F.R. Yester, Jr., "Method and Arrangement for a Sigma–Delta Converter for Bandpass Signals," **US Patent 4,857,928**, filed, January 28, 1988, issued August 15, 1989.

32. S.A. Jantzi, M. Snelgrove, and P.F. Ferguson Jr., "A 4th-Order Bandpass Sigma–Delta Modulator," **IEEE Journal of Solid State Circuits**, Vol. 38, No. 3, March, 1993, pp. 282–291.

33. P. Hendriks, R. Schreier, and J. DiPilato, "High Performance Narrowband Receiver Design Simplified by IF Digitizing Subsystem in LQFP," **Analog Dialogue**, Vol. 35, No. 3, June–July, 2001, available at http://www.analog.com (*describes an IF subsystem with a bandpass sigma–delta ADC having a nominal oversampling frequency of 18MSPS, a center frequency of 2.25 MHz, and a bandwidth of 10 kHz–150 kHz*).

# *Defining the Specifications*

In specifying a converter there are basically two sets of specifications used for data converters. They can simplistically be divided into DC and AC specifications. Which you are more concerned with is based primarily on the application. DC specifications are more common with lower frequency applications. Here we are measuring against a reference. In AC specifications we are less concerned with absolute accuracy and more concerned with relative accuracy. This is not to say the reference is unimportant. It is just that we are typically interested in a relative number rather than an absolute. Distortion is always relative to the fundamental, for instance. While there is no direct conversion between the two, it is possible to infer that you need good linearity to get good distortion. It is rare to have a converter specified both ways.

Another point that should be made here is the difference between resolution and accuracy. While the two terms sometimes tend to be used interchangeably, they are not the same thing.

Resolution can be defined as the number of bits in the data word of the converter. Accuracy is the number of those bits that meets the specifications. As an example, an audio converter may have a data bus width of 24 bits, but only a signal-to-noise range of 120 dB that roughly corresponds to an accuracy of 20 bits. While 120 dB is not poor performance, it is not 24-bit performance.

You should keep in mind the magnitude in volts as well as in bits of the resolution of the converter that you are considering. This is shown in Figure 6-112 for a full-scale level of 2 V (note that there is some rounding

| Resolution N | $2^N$ | Voltage (2 V FS) | ppm FS | % FS | dB FS |
|---|---|---|---|---|---|
| 2-bit | 4 | 500 mV | 250,000 | 25 | −12 |
| 4-bit | 16 | 125 mV | 62,500 | 6.25 | −24 |
| 6-bit | 64 | 31.2 mV | 15,625 | 1.56 | −36 |
| 8-bit | 256 | 7.81 mV | 3,906 | 0.39 | −48 |
| 10-bit | 1,024 | 1.95 mV | 977 | 0.098 | −60 |
| 12-bit | 4,096 | 488 µV | 244 | 0.024 | −72 |
| 14-bit | 16,384 | 122 µV | 61 | 0.0061 | −84 |
| 16-bit | 65,536 | 30.5 µV | 15 | 0.0015 | −96 |
| 18-bit | 262,144 | 7.62 µV | 4 | 0.0004 | −108 |
| 20-bit | 1,048,576 | 1.9 µV | 1 | 0.0001 | −120 |
| 22-bit | 4,194,304 | 476 nV | 0.24 | 0.000024 | −132 |
| 24-bit | 16,777,216 | 119 nV | 0.06 | 0.000006 | −144 |

Figure 6-112: LSB size for a 2 V full-scale input

off of the voltages in this table). This level is not uncommon for modern systems and is the typical standard for line level audio measurement. Remember that Gaussian noise of the system will probably set the lower limit of the accuracy spec. For example, 600 nV is the Johnson noise in a 10 kHz BW of a 2.2 kΩ resistor @ 25°C. This corresponds to approximately 21.5 bits.

And some systems use even smaller full-scales. Notably the AD7730 ΣΔ ADC system is designed to operate with full-scale inputs down to 10 mV. With a 24-bit resolution this means an LSB weighting of 596 μV.

# DAC and ADC Static Transfer Functions and DC Errors

The four primary DC errors in a data converter are *offset error*, *gain error*, and two types of *linearity error (differential and integral)*. Offset and gain errors are analogous to offset and gain errors in amplifiers as shown in Figure 6-113 for a bipolar input range. (Though offset error and zero error, which are identical in amplifiers and unipolar data converters, are not identical in bipolar converters and should be carefully distinguished.)

The transfer characteristics of both DACs and ADCs may be expressed as $D = K + GA$, where D is the digital code, A is the analog signal, and K and G are constants. In a unipolar converter, K is zero, and in an offset bipolar converter, it is $-1$ MSB. The offset error is the amount by which the actual value of K differs from its ideal value.

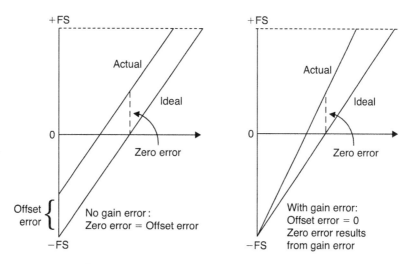

**Figure 6-113: Data converter offset and gain error**

The gain error is the amount by which G differs from its ideal value, and is generally expressed as the percentage difference between the two, although it may be defined as the gain error contribution (in millivolt or LSB) to the total error at full-scale. These errors can usually be trimmed by the data converter user. Note, however, that amplifier offset is trimmed at zero input, and then the gain is trimmed near to full-scale. The trim algorithm for a bipolar data converter is not so straightforward.

The integral linearity error of a converter is also analogous to the linearity error of an amplifier, and is defined as the maximum deviation of the actual transfer characteristic of the converter from a straight line, and is generally expressed as a percentage of full-scale (but may be given in LSBs). For an ADC, the most popular convention is to draw the straight line through the midpoints of the codes, or the code-centers. There are two common ways of choosing the straight line: *end point* and *best straight line* as shown in Figure 6-114.

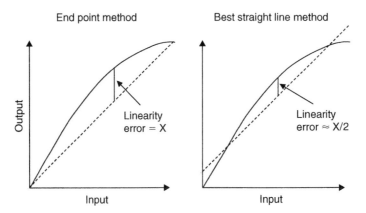

**Figure 6-114: Method of measuring integral linearity errors (same converter on both graphs)**

In the *end point* system, the deviation is measured from the straight line through the origin and the full-scale point (after gain adjustment). This is the most useful integral linearity measurement for measurement and control applications of data converters (since error budgets depend on deviation from the ideal transfer characteristic, not from some arbitrary "best fit"), and is the one normally adopted by Analog Devices, Inc.

The *best straight line*, however, does give a better prediction of distortion in AC applications, and also gives a lower value of "linearity error" on a data sheet. The best fit straight line is drawn through the transfer characteristic of the device using standard curve fitting techniques, and the maximum deviation is measured from this line. In general, the integral linearity error measured in this way is only 50% of the value measured by end point methods. This makes the method good for producing impressive data sheets, but it is less useful for error budget analysis. For AC applications, it is even better to specify distortion than DC linearity, so it is rarely necessary to use the best straight line method to define converter linearity.

The other type of converter nonlinearity is *DNL*. This relates to the linearity of the code transitions of the converter. In the ideal case, a change of 1 LSB in digital code corresponds to a change of exactly 1 LSB of analog signal. In a DAC, a change of 1 LSB in digital code produces exactly 1 LSB change of analog output, while in an ADC there should be exactly 1 LSB change of analog input to move from one digital transition to the next. Differential linearity error is defined as the maximum amount of deviation of any quantum (or LSB change) in the entire transfer function from its ideal size of 1 LSB.

Where the change in analog signal corresponding to 1 LSB digital change is more or less than 1 LSB, there is said to be a DNL error. The DNL error of a converter is normally defined as the maximum value of DNL to be found at any transition across the range of the converter. Figure 6-115 shows the non-ideal transfer functions for a DAC and an ADC and shows the effects of the DNL error.

The DNL of a DAC is examined more closely in Figure 6-116. If the DNL of a DAC is less than $-1$ LSB at any transition, the DAC is *non-monotonic*, i.e., its transfer characteristic contains one or more localized maxima or minima. A DNL greater than $+1$ LSB does not cause non-monotonicity, but is still undesirable. In many DAC applications (especially closed-loop systems where non-monotonicity can change negative feedback to positive feedback), it is critically important that DACs are monotonic. DAC monotonicity is

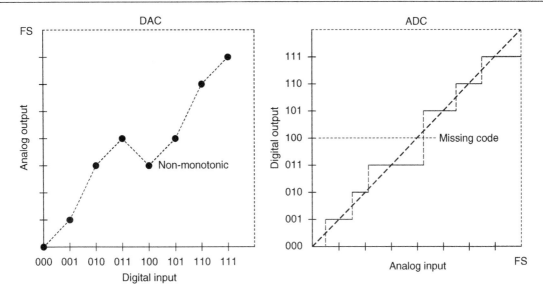

**Figure 6-115: Transfer functions for non-ideal 3-bit DAC and ADC**

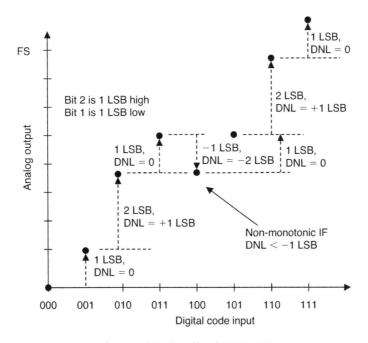

**Figure 6-116: Details of DAC DNL**

often explicitly specified on data sheets, although if the DNL is guaranteed to be less than 1 LSB (i.e., |DNL| ≤1 LSB) then the device must be monotonic, even without an explicit guarantee.

In Figure 6-117, the DNL of an ADC is examined more closely on an expanded scale. ADCs can be non-monotonic, but a more common result of excess DNL in ADCs is *missing codes*. Missing codes in an ADC are as objectionable as non-monotonicity in a DAC. Again, they result from DNL < −1 LSB.

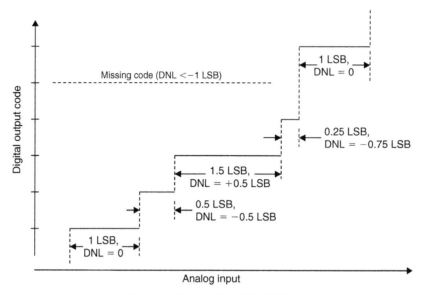

**Figure 6-117: Details of ADC DNL**

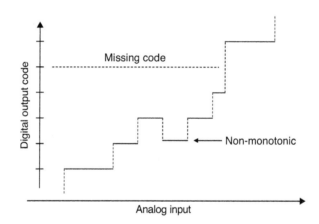

**Figure 6-118: Non-monotonic ADC with missing code**

Not only can ADCs have missing codes, they can also be non-monotonic as shown in Figure 6-118. As in the case of DACs, this can present major problems—especially in servo applications.

In a DAC, there can be no missing codes—each digital input word will produce a corresponding analog output. However, DACs can be non-monotonic as previously discussed. In a straight binary DAC, the most likely place a non-monotonic condition can develop is at mid-scale between the two codes: 011 ... 11 and 100 ... 00. If a non-monotonic condition occurs here, it is generally because the DAC is not properly calibrated or trimmed. A successive approximation ADC with an internal non-monotonic DAC will generally produce missing codes but remain monotonic. However it is possible for an ADC to be non-monotonic—again depending on the particular conversion architecture. Figure 6-118 shows the transfer function of an ADC which is non-monotonic and has a missing code.

ADCs which use the *subranging* architecture divide the input range into a number of coarse segments, and each coarse segment is further divided into smaller segments—and ultimately the final code is derived. This process is described in more detail in Chapter 4 of this book. An improperly trimmed subranging ADC may exhibit non-monotonicity, wide codes, or missing codes at the subranging points as shown in Figure 6-119 (A–C), respectively. This type of ADC should be trimmed so that drift due to aging or temperature produces wide codes at the sensitive points rather than non-monotonic or missing codes.

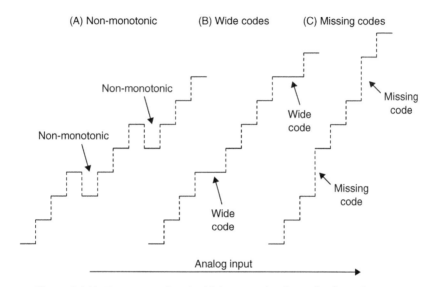

**Figure 6-119: Errors associated with improperly trimmed subranging ADC**

Defining missing codes is more difficult than defining non-monotonicity. All ADCs suffer from some inherent transition noise as shown in Figure 6-120 (think of it as the flicker between adjacent values of the last digit of a DVM). As resolutions and bandwidths become higher, the range of input over which transition noise occurs may approach, or even exceed, 1 LSB. High resolution wideband ADCs generally have internal noise sources which can be reflected to the input as effective input noise summed with the signal. The effect of this noise, especially if combined with a negative DNL error, may be that there are some (or even all) codes where transition noise is present for the whole range of inputs. There are therefore

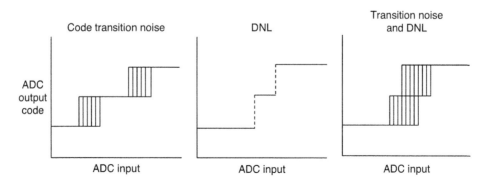

**Figure 6-120: Combined effects of code transition noise and DNL**

some codes for which there is *no* input which will *guarantee* that code as an output, although there may be a range of inputs which will *sometimes* produce that code.

For low resolution ADCs, it may be reasonable to define *no missing codes* as a combination of transition noise and DNL which guarantees some level (perhaps 0.2 LSB) of noise-free code for all codes. However, this is impossible to achieve at the very high resolutions achieved by modern sigma–delta ADCs, or even at lower resolutions in wide bandwidth sampling ADCs. In these cases, the manufacturer must define noise levels and resolution in some other way. Which method is used is less important, but the data sheet should contain a clear definition of the method used and the performance to be expected. A complete discussion of effective input noise follows later in this chapter.

The discussion thus far has not dealt with the most important DC specifications associated with data converters. Other less important specifications require only a definition.

*Accuracy, Absolute*: Absolute accuracy error of a *DAC* is the difference between actual analog output and the output that is expected when a given digital code is applied to the converter. Error is usually commensurate with resolution, i.e., less than ½ LSB of full-scale, for example. However, accuracy may be much better than resolution in some applications, e.g., a 4-bit reference supply having only 16 discrete digitally chosen levels would have a resolution of 1/16, but it might have an accuracy to within 0.01% of each ideal value.

Absolute accuracy error of an *ADC* at a given output code is the difference between the actual and the theoretical analog input voltages required to produce that code. Since the code can be produced by any analog voltage in a finite band (see *Quantizing Uncertainty*), the "input required to produce that code" is defined as the midpoint of the band of inputs that will produce that code. For example, if 5V, $\pm 1.2$ mV, will theoretically produce a 12-bit half-scale code of 100000000000, then a converter for which any voltage from 4.997 to 4.999V will produce that code will have absolute error of $(\frac{1}{2})(4.997 + 4.999)$ $-5V = +2$ mV.

Sources of error include gain (calibration) error, zero error, linearity errors, and noise. Absolute accuracy measurements should be made under a set of standard conditions with sources and meters traceable to an internationally accepted standard.

*Accuracy, Logarithmic DACs*: The difference (measured in dB) between the actual transfer function and the ideal transfer function, as measured after calibration of gain error at 0 dB.

*Accuracy, Relative*: Relative accuracy error, expressed in %, ppm, or fractions of 1 LSB, is the deviation of the analog value at any code (relative to the full analog range of the device transfer characteristic) from its theoretical value (relative to the same range), after the full-scale range (FSR) has been calibrated (see *FSR*).

Since the discrete analog values that correspond to the digital values ideally lie on a straight line, the specified worst case relative accuracy error of a linear ADC or DAC can be interpreted as a measure of end point nonlinearity (see *Linearity*).

The "discrete points" of a DAC transfer characteristic are measured by the actual analog outputs. The "discrete points" of an ADC transfer characteristic are the midpoints of the quantization bands at each code (see *Accuracy, Absolute*).

*Temperature Coefficient*: In general, temperature instabilities are expressed as %/°C, ppm/°C, fractions of 1 LSB/°C, or as a change in a parameter over a specified temperature range. Measurements are usually made at room temperature and at the extremes of the specified range, and the temperature coefficient (tempco, TC) is defined as the change in the parameter, divided by the corresponding temperature change. Parameters of interest include gain, linearity, offset (bipolar), and zero.

(a) *Gain TC*: Two factors principally affect converter gain stability with temperature. In fixed-reference converters, the reference voltage will vary with temperature. The reference circuitry and switches (and comparator in aid converters) will also contribute to the overall gain TC.

(b) *Linearity TC*: Sensitivity of linearity (integral and/or differential linearity) to temperature, in % FSR/°C or ppm FSR/°C, over the specified range. Monotonic behavior in DACs is achieved if the DNL is less than 1 LSB at any temperature in the range of interest. The *DNL temperature coefficient* may be expressed as a ratio, as a maximum change over a temperature range, and/or implied by a statement that the device is monotonic over the specified temperature range. To avoid missing codes in noiseless ADCs, it is sufficient that the DNL error magnitude be greater than −1 LSB at any temperature in the range of interest. The DNL temperature coefficient is often implied by the statement that there are no missed codes when operating within a specified temperature range. In DACs, the DNL TC is often implied by the statement that the DAC is monotonic over a specified temperature range.

(c) *Zero TC (unipolar converters)*: The temperature stability of a unipolar fixed-reference DAC, measured in % FSR/°C or ppm FSR/°C, is principally affected by current leakage (current-output DAC), and offset voltage and bias current of the output op amp (voltage-output DAC). The zero stability of an ADC is dependent on the zero stability of the DAC or integrator and/or the input buffer and the comparator. It is typically expressed in μV/°C or in % or ppm of FSR/°C.

(d) *Offset TC*: The temperature coefficient of the all-DAC-switches-off (minus full-scale) point of a bipolar converter (in % FSR/°C or ppm FSR/°C) depends on three major factors—the TC of the reference source, the voltage zero stability of the output amplifier, and the tracking capability of the bipolar-offset resistors and the gain resistors. In an ADC, the corresponding TC of the negative full-scale point depends on similar quantities—the TC of the reference source, the voltage stability of the input buffer and the SHA, and the tracking capabilities of the bipolar-offset resistors and the gain-setting resistors.

*Common-Mode Range*: CMR usually varies with the magnitude of the range through which the input signal can swing, determined by the sum of the common-mode and the differential voltage. *Common-mode range* is that range of *total* input voltage over which specified CMR is maintained. For example, if the common-mode signal is ±5V and the differential signal is ±5V, the common-mode range is ±10V.

*Common-Mode Rejection*: CMR is a measure of the change in output voltage when both inputs are changed by equal amounts of AC and/or DC voltage. CMR is usually expressed either as a ratio (e.g., CMRR = 1,000,000:1) or in decibels: CMR = $20\log_{10}$CMRR; if CMRR = $10^6$, CMR = 120 dB. A CMRR of $10^6$ means that 1 V of common mode is processed by the device as though it were a differential signal of 1 μV at the input.

CMR is usually specified for a full-range CMV change, at a given frequency, and a specified imbalance of source impedance (e.g., 1 kΩ source unbalance, at 60 Hz). In amplifiers, the common-mode rejection ratio (CMRR) is defined as the ratio of the signal gain, G, to the common-mode gain (the ratio of common-mode signal appearing at the output to the CMV at the input).

*Common-Mode Voltage (CMV)*: A voltage that appears in common at both input terminals of a device, with respect to its output reference (usually "ground"). For inputs, $V_1$ and $V_2$, with respect to ground, CMV = ½($V_1 + V_2$). An ideal differential input device would ignore CMV. *Common-mode error* (CME) is any error at the output due to the common-mode input voltage. The errors due to supply-voltage variation, an internal common-mode effect, are specified separately.

*Compliance-Voltage Range*: For a current source (e.g., a current-output DAC), the maximum range of (output) terminal voltage for which the device will maintain the specified current-output characteristics.

*Differential Analog Input Resistance, Differential Analog Input Capacitance, and Differential Analog Input Impedance*: The real and complex impedances measured at each analog input port of an ADC. The resistance is measured statically and the capacitance and differential input impedances are measured with a network analyzer.

*Differential Analog Input Voltage Range*: The peak-to-peak differential voltage that must be applied to the converter to generate a full-scale response. Peak differential voltage is computed by observing the voltage on a single pin and subtracting the voltage from the other pin, which is 180° out of phase. Peak-to-peak differential is computed by rotating the inputs phase 180° and taking the peak measurement again. Then the difference is computed between both peak measurements.

*Full-Scale Range (FSR)*: The magnitude of voltage, current or, in the case of an MDAC, gain that is the complete input dynamic range of an ADC or output dynamic range of a DAC. All bit voltage ratios are measured against this value. FSR is independent of resolution; the value of the LSB (voltage, current, or gain) is $2^{-N}$ FSR. There are several other terms, with differing meanings, that are often used in the context of discussions or operations involving FSR. They are the following.

(a) *Full-scale*: similar to FSR, but pertaining to a single polarity. Thus, full-scale for a unipolar device is twice the prescribed value of the MSB and has the same polarity. For a bipolar device, *positive or negative full-scale* is that positive or negative value, of which the next bit after the polarity bit is tested to be one-half.

(b) *Span*: the scalar voltage or current range corresponding to FSR.

(c) *All-1's*: *All bits on*, the condition used, in conjunction with *all-zeros*, for gain adjustment of an ADC or DAC, in accordance with the manufacturer's instructions. Its magnitude, for a binary device, is $(1 - 2^{-N})$ FSR. *All-1's* is a *positive-true* definition of a specific magnitude relationship; for complementary coding the "all-1's" code will actually be all-zeros. To avoid confusion, all-1's should never be called "full-scale": FSR and FS are independent of the number of bits, all-1's isn't.

(d) *All-0's*: *All-bits-off*, the condition used in offset (and gain) adjustment of a DAC or ADC, according to the manufacturer's instructions. All-0's corresponds to zero output in a unipolar DAC and negative full-scale in an offset bipolar DAC with positive output reference. In a sign-magnitude device, all-0's refers to all bits after the sign bit. Analogous to "all-1's," "all-0's" is a *positive-true* definition of the *all-bits-off* condition; in a complementary-coded device, it is expressed by all ones. To avoid confusion, all-0's should not be called "zero" unless it accurately corresponds to true analog zero output from a DAC.

The best way of defining the critical points for an actual device is a brief table of critical codes and the ideal voltages, currents, or gains to which they correspond, with the conditions for measurement defined.

*Gain*: The "gain" of a converter is that analog scale factor setting that establishes the nominal conversion relationship, e.g., 10V full-scale. In an MDAC or ratiometric ADC, it is indeed a gain. In a device with fixed internal reference, it is expressed as the full-scale magnitude of the output parameter (e.g., 10V or 2 mA). In a fixed-reference converter, where the use of the internal reference is optional, the converter gain and the reference may be specified separately. Gain and zero adjustment are discussed under *zero*.

*Impedance, Input*: The dynamic load of an ADC presented to its input source. In unbuffered CMOS switched capacitor ADCs, the presence of current transients at the converter's clock frequency mandates that the converter be driven from a low impedance (at the frequencies contained in the transients) in order to accurately convert. For buffered-input ADCs, the input impedance is generally represented by a resistive and capacitive component.

*Input-Referred Noise (Effective Input Noise)*: Input-referred noise can be viewed as the net effect of all internal ADC noise sources referred to the input. It is generally expressed in *LSBs RMS*, but can also be expressed as a voltage. It can be converted to a peak-to-peak value by multiplying by the factor 6.6. The peak-to-peak input-referred noise can then be used to calculate the *noise-free code resolution* (see *Noise-Free Code Resolution*).

*Leakage Current, Output*: Current which appears at the output terminal of a DAC with all bits "off." For a converter with two complementary outputs (e.g., many fast CMOS DACs), output leakage current is the current measured at OUT 1, with all digital inputs *low—and* the current measured at OUT 2, with all digital inputs *high*.

*Output Propagation Delay*: For an ADC having a single-ended sampling (or ENCODE) clock input, the delay between the 50% point of the sampling clock and the time when all output data bits are within valid logic levels. For an ADC having differential sampling clock inputs, the delay is measured with respect to the zero-crossing of the differential sampling clock signal.

*Output Voltage Tolerance*: For a reference, the maximum deviation from the normal output voltage at 25°C and specified input voltage, as measured by a device traceable to a recognized fundamental voltage standard.

*Overload*: An input voltage exceeding the ADC's full-scale input range producing an overload condition.

*Overvoltage Recovery Time*: Overvoltage recovery time is defined as the amount of time required for an ADC to achieve a specified accuracy after an overvoltage (usually 50% greater than FSR), measured from the time the overvoltage signal re-enters the converter's range. The ADC should act as an ideal limiter for out-of-range signals, producing a positive or negative full-scale code during the overvoltage condition. Some ADCs provide over- and under-range flags to allow gain-adjustment circuits to be activated.

*Overrange, Overvoltage*: An input signal that exceeds the input range of an ADC, but is less than an overload.

*Power Supply Rejection Ratio (PSRR)*: The ratio of a change in DC power supply voltage to the resulting change in the specified device error, expressed in percentage, parts per million, or fractions of 1 LSB. It may also be expressed logarithmically, in dB ($PSR = 20\log_{10}(PSRR)$).

*Power Supply Sensitivity*: The sensitivity of a converter to changes in the power supply voltages is normally expressed in terms of percent-of-full-scale change in analog value or fractions of 1 LSB—(DAC output, ADC output code-center) for a 1% DC change in the power supply, e.g., $0.05\%/\%$ $\Delta V_S$. Power supply sensitivity may also be expressed in relation to a specified maximum DC shift of power supply voltage. A converter might be considered "good" if the change in reading at full-scale does not exceed $\pm\frac{1}{2}$ LSB for a 3% change in power supply voltage. Even better specs are necessary for converters designed for direct battery operation.

*Ratiometric*: The output of an ADC is a digital number proportional to the *ratio* of (some measure of) the input to a reference voltage. Most requirements for conversions call for an absolute measurement, i.e., against a fixed reference; but this presumes that the signal applied to the converter is either reference-independent or in some way derived from another fixed reference. However, real references are not truly fixed; the references for both the converter and the signal source vary with time, temperature, loading, etc. Therefore, if the converter is used with signal sources that also rely on references (e.g., strain-gage bridges, resistance temperature devices (RTDs), thermistors), it makes sense to replace this multiplicity of references by a single system reference; reference-caused errors will tend to cancel out. This can be done by using the converter's internal reference (if it has one) as the system reference. Another way is to use a separate external system reference, which also becomes the reference for a *ratiometric* converter.

Over limited ranges, ratiometric conversion can also serve as a substitute for analog or digital signal division (where the denominator changes by less than $\frac{1}{2}$ LSB during the conversion). The signal input is the numerator; the reference input is the denominator.

*Total Unadjusted Error*: A comprehensive specification on some devices which includes full-scale error, relative accuracy, and zero-code errors, under a specified set of conditions.

# Data Converter AC Errors

This section examines the AC errors associated with data converters. Many of the errors and specifications apply equally to ADCs and DACs, while some are more specific to one or the other. All possible specifications are not discussed here, only the most popular ones.

In most applications, the input to the ADC is a band of frequencies (usually summed with some noise), so the quantization noise tends to be random. In spectral analysis applications (or in performing FFTs on ADCs using spectrally pure sinewaves—see Figure 6-121), however, the correlation between the quantization noise and the signal depends on the ratio of the sampling frequency to the input signal. This is demonstrated in Figure 6-122, where an ideal 12-bit ADCs output is analyzed using a 4,096-point FFT. In the left-hand FFT plot, the ratio of the sampling frequency to the input frequency was chosen to be exactly 32, and the worst harmonic is about 76 dB below the fundamental. The right-hand diagram shows the effects of slightly offsetting the ratio to 4,096/127 = 32.25196850394, showing a relatively random noise spectrum, where the spurious free dynamic range (SFDR) is now about 92 dBc. In both cases, the RMS value of all the noise components is approximately $q/\sqrt{12}$, but in the first case, the noise is concentrated at harmonics of the fundamental.

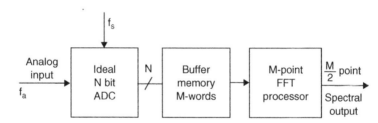

Figure 6-121: Dynamic performance analysis of an ideal N-bit ADC

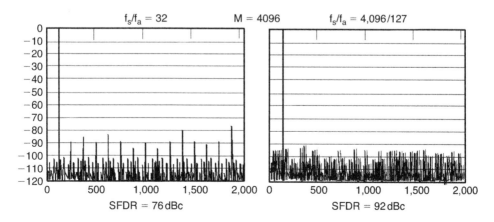

Figure 6-122: Effect of ratio of sampling clock to input frequency on SFDR for ideal 12-bit ADC

Note that this variation in the apparent harmonic distortion of the ADC is an artifact of the sampling process and the correlation of the quantization error with the input frequency. In a practical ADC application, the quantization error generally appears as random noise because of the random nature of the wideband input signal and the additional fact that there is usually a small amount of system noise which acts as a *dither* signal to further randomize the quantization error spectrum.

It is important to understand the above point, because single-tone sinewave FFT testing of ADCs is a universally accepted method of performance evaluation. In order to accurately measure the harmonic distortion of an ADC, steps must be taken to ensure that the test setup truly measures the ADC distortion, not the artifacts due to quantization noise correlation. This is done by properly choosing the frequency ratio and sometimes by injecting a small amount of noise (dither) with the input signal. The exact same precautions apply to measuring DAC distortion with an analog spectrum analyzer.

Figure 6-123 shows the FFT output for an ideal 12-bit ADC. Note that the average value of the noise floor of the FFT is approximately 100 dB below full-scale, but the theoretical SNR of a 12-bit ADC is 74 dB. The FFT noise floor is *not* the SNR of the ADC, because the FFT acts like an analog spectrum analyzer with a bandwidth of $f_s/M$, where M is the number of points in the FFT. The theoretical FFT noise floor is therefore $10 \log_{10}$ (M/2) dB below the quantization noise floor due to the *processing gain* of the FFT. In the case of an ideal 12-bit ADC with an SNR of 74 dB, a 4,096-point FFT would result in a processing gain of $10 \log_{10}(4096/2) = 33$ dB, thereby resulting in an overall FFT noise floor of $74 + 33 = 107$ dBc. In fact, the FFT noise floor can be reduced even further by going to larger and larger FFTs; just as an analog spectrum analyzer's noise floor can be reduced by narrowing the bandwidth. When testing ADCs using FFTs, it is important to ensure that the FFT size is large enough so that the distortion products can be distinguished from the FFT noise floor itself.

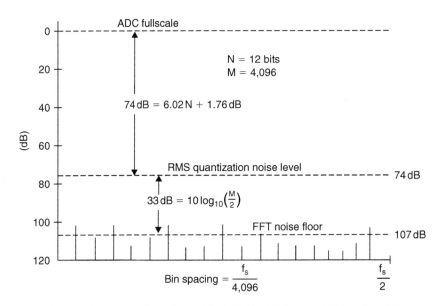

**Figure 6-123: Noise floor for an ideal 12-bit ADC using 4096-point FFT**

## Noise in Practical ADCs

A practical sampling ADC (one that has an integral SHA), regardless of architecture, has a number of noise and distortion sources as shown in Figure 6-124. The wideband analog front-end buffer has wideband noise, nonlinearity, and also finite bandwidth. The SHA introduces further nonlinearity, band-limiting, and aperture jitter. The actual quantizer portion of the ADC introduces quantization noise, and both integral and DNL. In this discussion, assume that sequential outputs of the ADC are loaded into a buffer memory of length M and that the FFT processor provides the spectral output. Also assume that the FFT arithmetic operations themselves introduce no significant errors relative to the ADC. However, when examining the output noise floor, the FFT processing gain (dependent on M) must be considered.

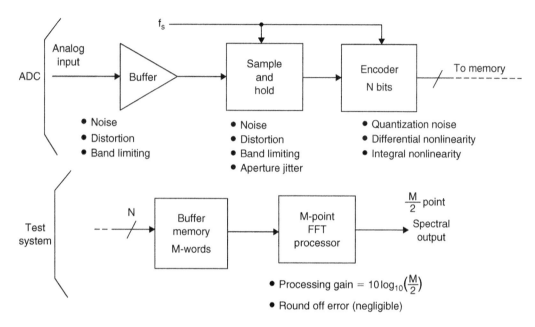

**Figure 6-124: ADC model showing noise and distortion sources**

## Equivalent Input-Referred Noise

Wideband ADC internal circuits produce a certain amount of RMS noise due to resistor and kT/C noise. This noise is present even for DC-input signals, and accounts for the fact that the output of most wideband (or high resolution) ADCs is a distribution of codes, centered around the nominal value of a DC input (see Figure 6-125). To measure its value, the input of the ADC is either grounded or connected to a heavily decoupled voltage source, and a large number of output samples are collected and plotted as a histogram (sometimes referred to as a *grounded-input* histogram). Since the noise is approximately Gaussian, the standard deviation of the histogram is easily calculated (see Reference 6), corresponding to the effective input RMS noise. It is a common practice to express this RMS noise in terms of LSBs RMS, although it can be expressed as an RMS voltage referenced to the ADC full-scale input range.

## Noise-Free (Flicker-Free) Code Resolution

The *noise-free code resolution* of an ADC is the number of bits beyond which it is impossible to distinctly resolve individual codes. The cause is the effective input noise (or input-referred noise) associated with all ADCs and described above. This noise can be expressed as an RMS quantity, usually having the units of *LSBs RMS*. Multiplying by a factor of 6.6 converts the RMS noise into peak-to-peak noise (expressed in

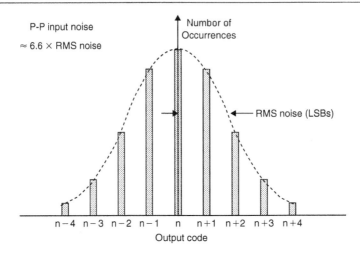

**Figure 6-125: Effect of input-referred noise on ADC "grounded-input" histogram**

*LSBs peak-to-peak*). The total range of an N-bit ADC is $2^N$ LSBs. The noise-free (or flicker-free) resolution can be calculated using the equation:

$$\text{Noise-Free Code Resolution} = \log_2(2^N/\text{Peak-to-Peak Noise}) \qquad (6\text{-}18)$$

The specification is generally associated with high resolution sigma–delta measurement ADCs, but is applicable to all ADCs.

The ratio of the FS range to the *RMS* input noise is sometimes used to calculate resolution. In this case, the term *effective resolution* is used. Note that under identical conditions, effective resolution is larger than noise-free code resolution by $\log_2(6.6)$, or approximately 2.7 bits.

$$\text{Effective Resolution} = \log_2(2^N/\text{RMS Input Noise}) \qquad (6\text{-}19)$$

$$\text{Effective Resolution} = \text{Noise-Free Code Resolution} + 2.7 \text{ bits} \qquad (6\text{-}20)$$

The calculations are summarized in Figure 6-126.

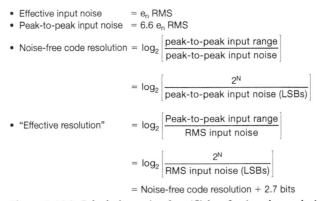

**Figure 6-126: Calculating noise-free (flicker-free) code resolution from input-referred noise**

## *Dynamic Performance of Data Converters*

There are various ways to characterize the AC performance of ADCs. In the early years of ADC technology (over 30 years ago) there was little standardization with respect to AC specifications, and measurement equipment and techniques were not well understood or available. Over nearly a 30-year period, manufacturers and customers have learned more about measuring the dynamic performance of converters, and the specifications shown in Figure 6-127 represent the most popular ones used today. Practically all the specifications represent the converter's performance in the frequency domain. The FFT is the heart of practically all these measurements and is discussed in more detail in a later section.

- Harmonic Distortion
- Worst Harmonic
- Total Harmonic Distortion (THD)
- Total Harmonic Distortion Plus Noise (THD + N)
- Signal-to-Noise-and-Distortion Ratio (SINAD, or S/N + D)
- Effective Number of Bits (ENOB)
- Signal-to-Noise Ratio (SNR)
- Analog Bandwidth (Full-Power, Small-Signal)
- Spurious Free Dynamic Range (SFDR)
- Two-Tone Intermodulation Distortion
- Multi-tone Intermodulation Distortion
- Noise Power Ratio (NPR)
- Adjacent Channel Leakage Ratio (ACLR)
- Noise Figure
- Settling Time, Overvoltage Recovery Time

**Figure 6-127: Quantifying data converter dynamic performance**

## *Integral and DNL Distortion Effects*

One of the first things to realize when examining the nonlinearities of data converters is that the transfer function of a data converter has artifacts which do not occur in conventional linear devices such as op amps or gain blocks. The overall integral nonlinearity of an ADC is due to the integral nonlinearity of the front end and SHA as well as the overall integral nonlinearity in the ADC transfer function. However, *DNL is due exclusively to the encoding process* and may vary considerably dependent on the ADC encoding architecture. Overall integral nonlinearity produces distortion products whose amplitude varies as a function of the input signal amplitude. For instance, second-order intermodulation products increase 2 dB for every 1 dB increase in signal level, and third-order products increase 3 dB for every 1 dB increase in signal level.

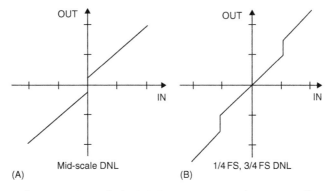

**Figure 6-128: Typical ADC/DAC DNL errors (exaggerated)**

The DNL in the ADC transfer function produces distortion products which not only depend on the amplitude of the signal but the positioning of the DNL errors along the ADC transfer function. Figure 6-128 shows two ADC transfer functions having DNL. Figure 6-128(A) shows an error which occurs at mid-scale. Therefore, for both large and small signals, the signal crosses through this point producing a distortion product which is relatively independent of the signal amplitude. Figure 6-128(B) shows another ADC transfer function which has DNL errors at 1/4 and 3/4 full-scale. Signals which are above ½ scale peak-to-peak will exercise these codes and produce distortion, while those less than ½ scale peak-to-peak will not.

Most high speed ADCs are designed so that DNL is spread across the entire ADC range. Therefore, for signals which are within a few dB of full-scale, the overall integral nonlinearity of the transfer function determines the distortion products. For lower level signals, however, the harmonic content becomes dominated by the DNLs and does not generally decrease proportionally with decreases in signal amplitude.

## Harmonic Distortion, Worst Harmonic, Total Harmonic Distortion, Total Harmonic Distortion Plus Noise

There are a number of ways to quantify the distortion of an ADC. An FFT analysis can be used to measure the amplitude of the various harmonics of a signal. The harmonics of the input signal can be distinguished from other distortion products by their location in the frequency spectrum. Figure 6-129 shows a 7 MHz input signal sampled at 20 MSPS and the location of the first 9 harmonics. Aliased harmonics of $f_a$ fall at frequencies equal to $|\pm K f_s \pm n f_a|$, where n is the order of the harmonic, and $K = 0, 1, 2, 3, \ldots$. The second and third harmonics are generally the only ones specified on a data sheet because they tend to be the largest, although some data sheets may specify the value of the *worst* harmonic.

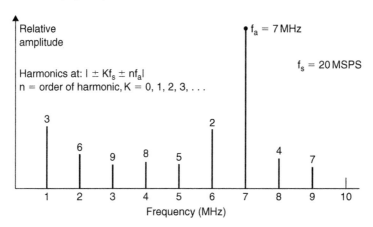

**Figure 6-129: Location of distortion products: Input signal = 7 MHz, sampling rate = 20 MSPS**

*Harmonic distortion* is normally specified in dBc (decibels below *carrier*), although at audio frequencies it may be specified as a percentage. Harmonic distortion is generally specified with an input signal near full-scale (generally 0.5–1 dB below full-scale to prevent clipping), but it can be specified at any level. For signals much lower than full-scale, other distortion products due to the DNL of the converter (not direct harmonics) may limit performance.

THD is the ratio of the RMS value of the fundamental signal to the mean value of the root-sum-square (RSS) of its harmonics (generally, only the first five are significant). THD of an ADC is also generally specified with the input signal close to full-scale, although it can be specified at any level.

THD + N is the ratio of the RMS value of the fundamental signal to the mean value of the RSS of its harmonics plus all noise components (excluding DC). The bandwidth over which the noise is measured must be specified. In the case of an FFT, the bandwidth is DC to $f_s/2$. (If the bandwidth of the measurement is DC to $f_s/2$, THD + N is equal to SINAD—see below.)

## Signal-to-Noise and Distortion Ratio, Signal-to-Noise Ratio and Effective Number of Bits

Signal-to-noise-and-distortion ratio (SINAD) and SNR deserve careful attention, because there is still some variation between ADC manufacturers as to their precise meaning. Signal-to-noise-and-distortion (SINAD, or S/(N + D) is the ratio of the RMS signal amplitude to the mean value of the RSS of all other spectral components, *including harmonics*, but excluding DC (see Figure 6-130). SINAD is a good indication of the overall dynamic performance of an ADC as a function of input frequency because it includes all components which make up noise (including thermal noise) and distortion. It is often plotted for various input amplitudes. SINAD is equal to THD + N if the bandwidth for the noise measurement is the same. A typical plot for the AD9226 12-bit, 65 MSPS ADC is shown in Figure 6-131.

- SINAD (Signal-to-Noise-and-Distortion Ratio):
  - The ratio of the RMS signal amplitude to the mean value of the root-sum-squares (RSS) of all other spectral components, including harmonics, but excluding DC.
- ENOB (Effective Number of Bits):

$$ENOB = \frac{SINAD - 1.76\,dB}{6.02}$$

- SNR (Signal-to-Noise Ratio, or Signal-to-Noise Ratio Without Harmonics):
  - The ratio of the RMS signal amplitude to the mean value of the root-sum-squares (RSS) of all other spectral components, excluding the first 5 harmonics and DC.

**Figure 6-130: SINAD, ENOB, and SNR**

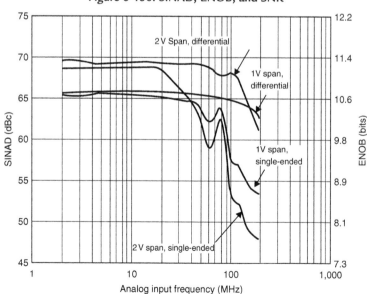

**Figure 6-131: AD9226 12-bit, 65 MSPS ADC SINAD and ENOB for various input full-scale spans (range)**

The SINAD plot shows where the AC performance of the ADC degrades due to high frequency distortion and is usually plotted for frequencies well above the Nyquist frequency so that performance in undersampling applications can be evaluated. SINAD is often converted to ENOB using the relationship for the theoretical SNR of an ideal N-bit ADC: SNR = 6.02N + 1.76 dB. The equation is solved for N, and the value of SINAD is substituted for SNR:

$$\text{ENOB} = \frac{\text{SINAD} - 1.76\,\text{dB}}{6.02} \qquad (6\text{-}21)$$

Signal-to-noise ratio (SNR, or *SNR-without-harmonics*) is calculated the same as SINAD except that the signal harmonics are excluded from the calculation, leaving only the noise terms. In practice, it is only necessary to exclude the first five harmonics since they dominate. The SNR plot will degrade at high frequencies, but not as rapidly as SINAD because of the exclusion of the harmonic terms.

Many current ADC data sheets somewhat loosely refer to SINAD as SNR, so the engineer must be careful when interpreting these specifications.

## Analog Bandwidth

The analog bandwidth of an ADC is that frequency at which the spectral output of the *fundamental* swept frequency (as determined by the FFT analysis) is reduced by 3 dB. It may be specified for either a small signal (SSBW—*small signal bandwidth*), or a full-scale signal (FPBW—*full power bandwidth*), so there can be a wide variation in specifications between manufacturers.

The small signal bandwidth will be larger than the FPBW. This issue is one of slew rate for the analog portion of the converter. This is similar to the bandwidth specifications of an op amp.

Like an amplifier, the analog bandwidth specification of a converter does not imply that the ADC maintains good distortion performance up to its bandwidth frequency. In fact, the SINAD (or ENOB) of most ADCs will begin to degrade considerably before the input frequency approaches the actual 3 dB bandwidth frequency. Figure 6-132 shows ENOB and full-scale frequency response of an ADC with an FPBW of 1 MHz, however, the ENOB begins to drop rapidly above 100 kHz.

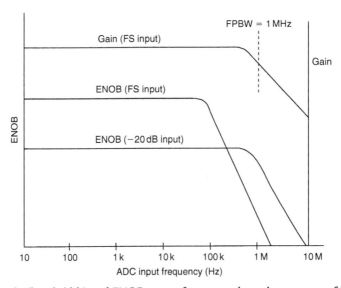

Figure 6-132: ADC gain (bandwidth) and ENOB versus frequency shows importance of ENOB specification

In some systems, notably video applications the bandwidth is specified to the level at which the level is reduced by 0.1 dB.

## *Spurious Free Dynamic Range*

Probably the most significant specification for an ADC used in a communications application is its SFDR. The SFDR specification is to ADCs what the third-order intercept specification is to mixers and low noise amplifiers (LNAs). SFDR of an ADC is defined as the ratio of the RMS signal amplitude to the RMS value of the *peak spurious spectral content* measured over the bandwidth of interest. Unless otherwise stated, the bandwidth is assumed to be the Nyquist bandwidth DC to $f_s/2$.

Occasionally the frequency spectrum is divided into an *in-band* region (containing the signals of interest) and an *out-of-band* region (signals here are filtered out digitally). In this case there may be an *in-band SFDR* specification and an *out-of-band SFDR* specification, respectively.

SFDR is generally plotted as a function of signal amplitude and may be expressed relative to the signal amplitude (dBc) or the ADC full-scale (dBFS) as shown in Figure 6-133.

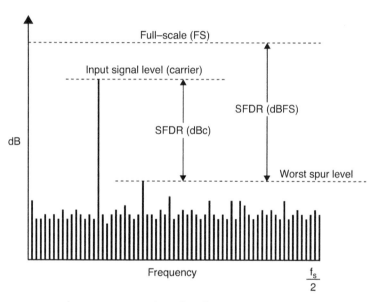

Figure 6-133: Spurious free dynamic range (SFDR)

For a signal near full-scale, the peak spectral spur is generally determined by one of the first few harmonics of the fundamental. However, as the signal falls several dB below full-scale, other spurs generally occur which are not direct harmonics of the input signal. This is because of the DNL of the ADC transfer function as discussed earlier. Therefore, SFDR considers *all* sources of distortion, regardless of their origin.

The AD6645 is a 14-bit, 80 MSPS wideband ADC designed for communications applications where high SFDR is important. The single-tone SFDR for a 69.1 MHz input and a sampling frequency of 80 MSPS is shown in Figure 6-134. Note that a minimum of 89-dBc SFDR is obtained over the entire first Nyquist zone (DC to 40 MHz).

SFDR as a function of signal amplitude is shown in Figure 6-135 for the AD6645. Notice that over the entire range of signal amplitudes, the SFDR is greater than 90 dBFS. The abrupt changes in the SFDR plot

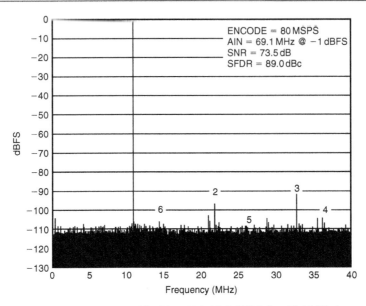

Figure 6-134: AD6645 14-bit, 80 MSPS ADC SFDR for 69.1 MHz input

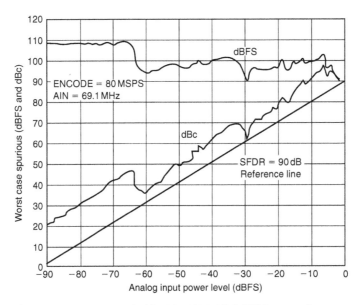

Figure 6-135: AD6645 14-bit, 80 MSPS ADC SFDR versus input
power level for 69.1 MHz input

are due to the DNLs in the ADC transfer function. The nonlinearities correspond to those shown in Figure 6-128(B), and are offset from mid-scale such that input signals less than about 65 dBFS do not exercise any of the points of increased DNL. It should be noted that the SFDR can be improved by injecting a small out-of-band dither signal—at the expense of a slight degradation in SNR.

SFDR is generally much greater than the ADCs theoretical N-bit SNR (6.02N + 1.76 dB). For example, the AD6645 is a 14-bit ADC with an SFDR of 90 dBc and a typical SNR of 73.5 dB (the theoretical SNR for 14-bits is 86 dB). This is because there is a fundamental distinction between noise and distortion measurements. The process gain of the FFT (33 dB for a 4096-point FFT) allows frequency spurs well below the noise floor to be observed. Adding extra resolution to an ADC may serve to increase its SNR but may or may not increase its SFDR.

## Two-Tone Intermodulation Distortion

Two-tone intermodulation distortion (IMD) is measured by applying two spectrally pure sinewaves to the ADC at frequencies $f_1$ and $f_2$, usually relatively close together. The amplitude of each tone is set slightly more than 6 dB below full-scale so that the ADC does not clip when the two tones add in-phase. The location of the second- and third-order products are shown in Figure 6-136. Notice that the second-order products fall at frequencies which can be removed by digital filters. However, the third-order products $2f_2 - f_1$ and $2f_1 - f_2$ are close to the original signals and are more difficult to filter. Unless otherwise specified, two-tone IMD refers to these third-order products. The value of the IMD product is expressed in dBc relative to the value of *either* of the two original tones, and not to their sum.

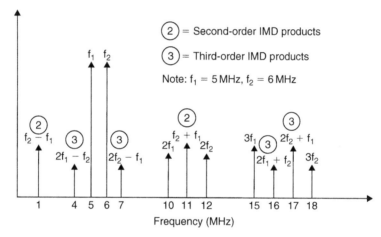

Figure 6-136: Second- and third-order intermodulation products for f1 = 5 MHz, f2 = 6 MHz

Note, however, that if the two tones are close to $f_s/4$, then the aliased third harmonics of the fundamentals can make the identification of the actual $2f_2 - f_1$ and $2f_1 - f_2$ products difficult. This is because the third harmonic of $f_s/4$ is $3f_s/4$, and the alias occurs at $f_s - 3f_s/4 = f_s/4$. Similarly, if the two tones are close to $f_s/3$, the aliased second harmonics may interfere with the measurement. The same reasoning applies here; the second harmonic of $f_s/3$ is $2f_s/3$, and its alias occurs at $f_s - 2f_s/3 = f_s/3$.

## Multi-Tone SFDR

Two-tone and multi-tone SFDR are often measured in communications applications. The larger number of tones more closely simulates the wideband frequency spectrum of cellular telephone systems such as AMPS or GSM. Figure 6-137 shows the two-tone intermodulation performance of the AD6645 14-bit, 80 MSPS ADC. The input tones are at 55.25 MHz and 56.25 MHz and are located in the second Nyquist zone.

The aliased tones therefore occur at 23.75 MHz and 24.75 MHz in the first Nyquist zone. High SFDR increases the receiver's ability to capture small signals in the presence of larger ones, and prevents the small signals from being masked by the intermodulation products of the larger ones. Figure 6-138 shows the AD6645 two-tone SFDR as a function of input signal amplitude for the same input frequencies.

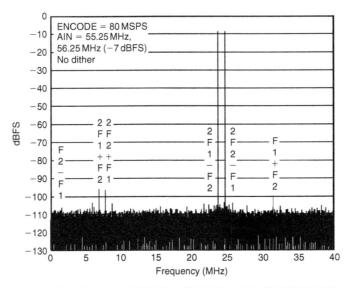

Figure 6-137: Two-tone SFDR for AD6645 14-bit, 80 MSPS ADC, input tones: 55.25 MHz and 56.25 MHz

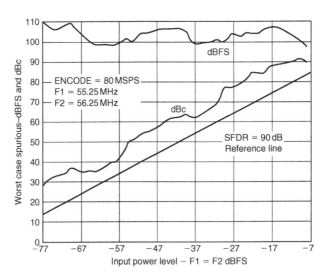

Figure 6-138: Two-tone SFDR versus input amplitude for AD6645 14-bit, 80 MSPS ADC

## Second- and Third-Order Intercept Points, 1 dB Compression Point

Third-order IMD products are especially troublesome in multi-channel communications systems where the channel separation is constant across the frequency band. Third-order IMD products can mask out small signals in the presence of larger ones.

In amplifiers, it is a common practice to specify the third-order IMD products in terms of the *third-order intercept* point, as shown in Figure 6-139. Two spectrally pure tones are applied to the system. The output signal power in a single tone (in dBm) as well as the relative amplitude of the third-order products (referenced to a single tone) is plotted as a function of input signal power. The fundamental is shown by the *slope = 1* curve in the diagram. If the system nonlinearity is approximated by a power series expansion, it can be shown that second-order IMD amplitudes increase 2 dB for every 1 dB of signal increase, as represented by the *slope = 2* curve in the diagram.

Similarly, the third-order IMD amplitudes increase 3 dB for every 1 dB of signal increase, as indicated by the *slope = 3* plotted line. With a low level two-tone input signal, and two data points, one can draw the second- and third-order IMD lines as they are shown in Figure 6-139 (using the principle that a point and a slope define a straight line).

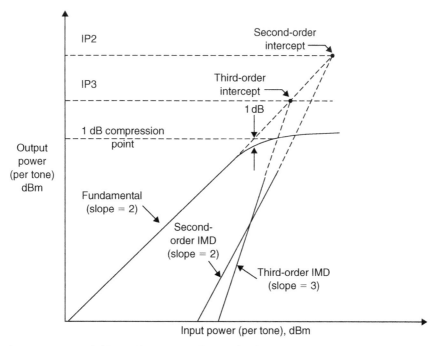

**Figure 6-139: Definition of intercept points and 1 dB compression points for amplifiers**

Once the input reaches a certain level, however, the output signal begins to soft-limit, or compress. A parameter of interest here is the *1 dB compression point*. This is the point where the output signal is compressed 1 dB from an ideal input/output transfer function. This is shown in Figure 6-139 within the region where the ideal slope = 1 line becomes dotted, and the actual response exhibits compression (solid).

Nevertheless, both the second- and third-order intercept lines may be extended, to intersect the (dotted) extension of the ideal output signal line. These intersections are called the *second-* and *third-order intercept points*, respectively, or IP2 and IP3. These power level values are usually referenced to the output power of the device delivered to a matched load (usually, but not necessarily 50 Ω) expressed in dBm.

It should be noted that IP2, IP3, and the 1 dB compression point are all a function of frequency, and as one would expect, the distortion is worse at higher frequencies.

For a given frequency, knowing the third-order intercept point allows calculation of the approximate level of the third-order IMD products as a function of output signal level.

The concept of *second- and third-order intercept points* is not valid for an ADC, because the distortion products do not vary in a predictable manner (as a function of signal amplitude). The ADC does not gradually begin to compress signals approaching full-scale (there is no 1 dB compression point); it acts as a *hard limiter* as soon as the signal exceeds the ADC input range, thereby suddenly producing extreme amounts of distortion because of clipping. On the other hand, for signals much below full-scale, the distortion floor remains relatively constant and is independent of signal level. This is shown graphically in Figure 6-140.

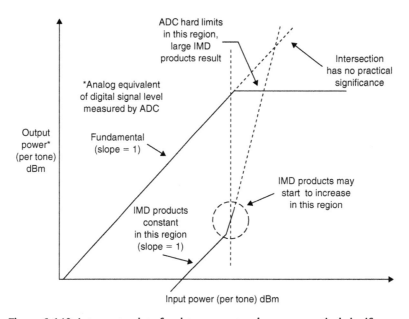

**Figure 6-140: Intercept points for data converters have no practical significance**

The IMD curve in Figure 6-140 is divided into three regions. For low level input signals, the IMD products remain relatively constant regardless of signal level. This implies that as the input signal increases 1 dB, the ratio of the signal to the IMD level will increase 1 dB also. When the input signal is within a few dB of the ADC FSR, the IMD may start to increase (but it might not in a very well-designed ADC). The exact level at which this occurs is dependent on the particular ADC under consideration—some ADCs may not exhibit significant increases in the IMD products over their full input range, however most will. As the input signal continues to increase beyond full-scale, the ADC should as an ideal limiter, and the IMD products become very large.

For these reasons, the second- and third-order IMD intercept points are not specified for ADCs. It should be noted that essentially the same arguments apply to DACs. In either case, the single- or multi-tone SFDR specification is the most accepted way to measure data converter distortion.

## Wideband CDMA Adjacent Channel Power Ratio and Adjacent Channel Leakage Ratio

A wideband code division multiple access (WCDMA) channel has a bandwidth of approximately 3.84 MHz, and channel spacing is 5 MHz. The ratio in dBc between the measured power within a channel relative to its adjacent channel is defined as the *adjacent channel power ratio* (ACPR).

The ratio in dBc between the measured power within the channel bandwidth relative to the noise level in an adjacent empty carrier channel is defined as *adjacent channel leakage ratio* (ACLR).

Figure 6-141 shows a single WCDMA channel centered at 140 MHz sampled at a frequency of 76.8 MSPS using the AD6645. This is a good example of undersampling (direct IF-to-digital conversion). The signal lies within the third Nyquist zone: $3f_s/2 - 2f_s$ (115.2–153.6 MHz). The aliased signal within the first Nyquist zone is therefore centered at $2f_s - f_a = 153.6 - 140 = 13.6$ MHz. The diagram also shows the location of the aliased harmonics. For example, the second harmonic of the input signal occurs at $2 \times 140 = 280$ MHz, and the aliased component occurs at $4f_s - 2f_a = 4 \times 76.8 - 280 = 307.2 - 280 = 27.2$ MHz.

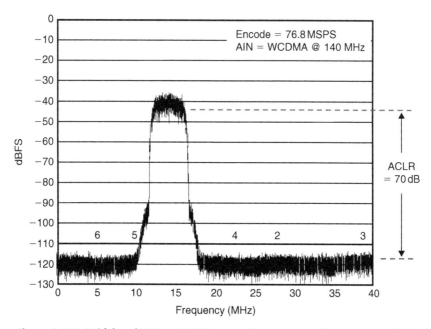

**Figure 6-141: Wideband CDMA (WCDMA), adjacent channel leakage ratio (ACLR)**

## Noise Power Ratio

Noise power ratio (NPR) has been used extensively to measure the transmission characteristics of frequency division multiple access (FDMA) communications links (see Reference 7). In a typical FDMA system, 4 kHz wide voice channels are "stacked" in frequency bins for transmission over coaxial, microwave, or satellite equipment. At the receiving end, the FDMA data is demultiplexed and returned to 4 kHz individual baseband channels. In an FDMA system having more than approximately 100 channels, the FDMA signal can be approximated by Gaussian noise with the appropriate bandwidth. An individual 4 kHz channel can be measured for "quietness" using a narrowband notch (bandstop) filter and a specially tuned receiver which measures the noise power inside the 4 kHz notch (see Figure 6-142).

NPR measurements are straightforward. With the notch filter out, the RMS noise power of the signal inside the notch is measured by the narrowband receiver. The notch filter is then switched in, and the residual noise inside the slot is measured. The ratio of these two readings expressed in dB is the NPR. Several slot frequencies across the noise bandwidth (low, midband, and high) are tested to characterize the system adequately. NPR measurements on ADCs are made in a similar manner except the analog receiver is replaced by a buffer memory and an FFT processor.

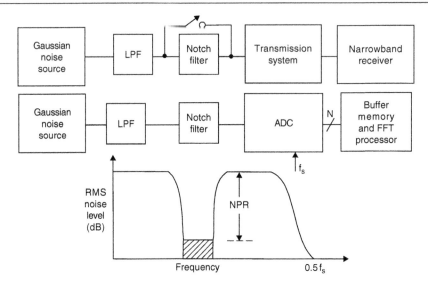

**Figure 6-142: NPR measurements**

The NPR is plotted as a function of RMS noise level referred to the peak range of the system. For very low noise loading level, the undesired noise (in non-digital systems) is primarily thermal noise and is independent of the input noise level. Over this region of the curve, a 1 dB increase in noise loading level causes a 1 dB increase in NPR. As the noise loading level is increased, the amplifiers in the system begin to overload, creating intermodulation products which cause the noise floor of the system to increase. As the input noise increases further, the effects of "overload" noise predominate, and the NPR is reduced dramatically. FDMA systems are usually operated at a noise loading level a few dB below the point of maximum NPR.

In a digital system containing an ADC, the noise within the slot is primarily quantization noise when low levels of noise input are applied. The NPR curve is linear in this region. As the noise level increases, there is a one-for-one correspondence between the noise level and the NPR. At some level, however, "clipping"

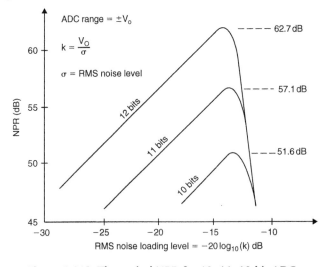

**Figure 6-143: Theoretical NPR for 10, 11, 12-bit ADCs**

*458*

noise caused by the hard-limiting action of the ADC begins to dominate. A theoretical curve for 10-, 11-, and 12-bit ADCs is shown in Figure 6-143 (see Reference 8).

Figure 6-144 shows the maximum theoretical NPR and the noise loading level at which the maximum value occurs for 8–16-bit ADCs. The ADC input range is $2V_O$ peak-to-peak. The RMS noise level is $\sigma$, and the noise loading factor k (crest factor) is defined as $V_O/\sigma$, the peak-to-RMS ratio (k is expressed either as numerical ratio or in dB).

| Bits | k Optimum | k(dB) | Max NPR (dB) |
|------|-----------|-------|--------------|
| 8 | 3.92 | 11.87 | 40.60 |
| 9 | 4.22 | 12.50 | 46.05 |
| 10 | 4.50 | 13.06 | 51.56 |
| 11 | 4.76 | 13.55 | 57.12 |
| 12 | 5.01 | 14.00 | 62.71 |
| 13 | 5.26 | 14.41 | 68.35 |
| 14 | 5.49 | 14.79 | 74.01 |
| 15 | 5.72 | 15.15 | 79.70 |
| 16 | 5.94 | 15.47 | 85.40 |

ADC Range = $\pm V_O$
k = $V_O/\sigma$
$\sigma$ = RMS Noise Level

**Figure 6-144: Theoretical maximum NPR for 8 to 16-bit ADCs**

In multi-channel high frequency communication systems where there is little or no phase correlation between channels, NPR can also be used to simulate the distortion caused by a large number of individual channels, similar to an FDMA system. A notch filter is placed between the noise source and the ADC, and an FFT output is used in place of the analog receiver. The width of the notch filter is set for several MHz as shown in Figure 6-145 for the AD9430 12-bit 170 MSPS ADC. The notch is centered at 19 MHz, and

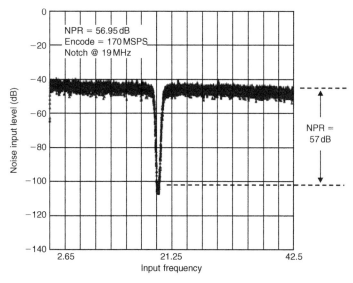

**Figure 6-145: AD9430 12-bit, 170 MSPS ADC NPR measures 57 dB (62.7 dB theoretical)**

the NPR is the "depth" of the notch. An ideal ADC will only generate quantization noise inside the notch, however a practical one has additional noise components due to additional noise and IMD caused by ADC imperfections. Notice that the NPR is about 57 dB compared to 62.7 dB theoretical.

## Noise Factor (F) and Noise Figure (NF)

Noise figure (NF) is a popular specification among RF system designers. It is used to characterize RF amplifiers, mixers, etc., and widely used as a tool in radio receiver design. Many excellent textbooks on communications and receiver design treat NF extensively (e.g., see Reference 9)—it is not the purpose of this discussion to discuss the topic in much detail, but only how it applies to data converters.

Since many wideband operational amplifiers and ADCs are now being used in RF applications, the inevitable day has come where the NF of these devices becomes important. As discussed in Reference 10, in order to determine the NF of an op amp correctly, one must not only know op amp voltage and current noise, but the exact circuit conditions—closed-loop gain, gain-setting resistor values, source resistance, bandwidth, etc. Calculating the NF for an ADC is even more of a challenge as will be seen.

Figure 6-146 shows the basic model for defining the NF of an ADC. The *noise factor*, F, is simply defined as the ratio of the total effective input noise power of the ADC to the amount of that noise power caused by the source resistance alone. Because the impedance is matched, the square of the voltage noise can be used instead of noise power. The *noise figure*, NF, is simply the noise factor expressed in dB, NF = $10 \log_{10} F$.

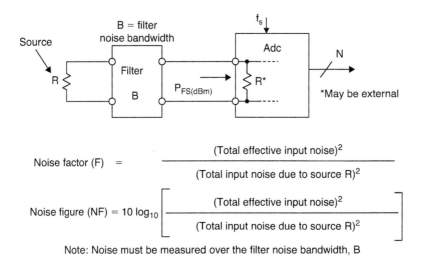

$$\text{Noise factor (F)} = \frac{(\text{Total effective input noise})^2}{(\text{Total input noise due to source R})^2}$$

$$\text{Noise figure (NF)} = 10 \log_{10} \left[ \frac{(\text{Total effective input noise})^2}{(\text{Total input noise due to source R})^2} \right]$$

Note: Noise must be measured over the filter noise bandwidth, B

**Figure 6-146: NF for ADCs: Use with caution!**

This model assumes that the input to the ADC comes from a source having a resistance, R, and that the input is band-limited to $f_s/2$ with a filter having a noise bandwidth equal to $f_s/2$. It is also possible to further band-limit the input signal resulting in oversampling and process gain, and this condition will be discussed shortly.

It is also assumed that the input impedance to the ADC is equal to the source resistance. Many ADCs have a high input impedance, so this termination resistance may be external to the ADC or used in parallel with the internal resistance to produce an equivalent termination resistance equal to R. The full-scale input power is the power of a sinewave whose peak-to-peak amplitude fills the entire ADC input range. The full-scale input sinewave given by the following equation has a peak-to-peak amplitude of $2V_O$ corresponding to the peak-to-peak input range of the ADC:

$$v(t) = V_O \sin 2\pi ft \tag{6-22}$$

The full-scale power in this sinewave is given by:

$$P_{FS} = \frac{\left(V_O / \sqrt{2}\right)^2}{R} = \frac{V_O^2}{2R} \tag{6-23}$$

It is customary to express this power in dBm (referenced to 1 mW) as follows:

$$P_{FS(dBm)} = 10 \log_{10}\left[\frac{P_{FS}}{1\,mW}\right] \tag{6-24}$$

The *noise bandwidth* of a non-ideal brick wall filter is defined as the bandwidth of an ideal brick wall filter which will pass the same noise power as the non-ideal filter. Therefore, the noise bandwidth of a filter is always greater than the 3 dB bandwidth of the filter by a factor which depends on the sharpness of the cutoff region of the filter. Figure 6-147 shows the relationship between the noise bandwidth and the 3 dB bandwidth for Butterworth filters up to five poles. Note that for two poles, the noise bandwidth and 3 dB bandwidth are within 11% of each other, and beyond that the two quantities are essentially equal.

| Number of poles | Noise BW/3 dB BW |
|:---:|:---:|
| 1 | 1.57 |
| 2 | 1.11 |
| 3 | 1.05 |
| 4 | 1.03 |
| 5 | 1.02 |

**Figure 6-147: Relationship between noise bandwidth and 3 dB bandwidth for Butterworth filter**

The first step in the NF calculation is to calculate the effective input noise of the ADC from its SNR. The SNR of the ADC is given for a variety of input frequencies, so be sure and use the value corresponding to the input frequency of interest. Also, make sure that the harmonics are not included in the SNR number—some ADC data sheets may confuse SINAD with SNR. Once the SNR is known, the equivalent input RMS voltage noise can be calculated starting from the equation:

$$SNR = 20 \log_{10}\left[\frac{V_{FS\ RMS}}{V_{NOISE\ RMS}}\right] \tag{6-25}$$

Solving for $V_{NOISE\ RMS}$:

$$V_{NOISE\ RMS} = V_{FS\ RMS} \times 10^{-SNR/20} \tag{6-26}$$

This is the total effective input RMS noise voltage at the carrier frequency measured over the Nyquist bandwidth, DC to $f_s/2$. Note that this noise includes the source resistance noise. These results are summarized in Figure 6-148.

- Start with the SNR of the ADC measured at the carrier frequency (Note: this SNR value does not include the harmonics of the fundamental and is measured over the Nyquist bandwidth, dc to $f_s/2$)

$$SNR = 20\log_{10}\frac{V_{FS\ RMS}}{V_{NOISE\ RMS}}$$

$$V_{NOISE\text{-}RMS} = V_{FS\ RMS}\ 10^{-SNR/20}$$

- This is the total ADC effective input noise at the carrier frequency measured over the Nyquist bandwidth, DC to $f_s/2$

**Figure 6-148: Calculating ADC total effective input noise from SNR**

The next step is to actually calculate the NF. In Figure 6-149 notice that the amount of the input voltage noise due to the source resistance is the voltage noise of the source resistance $(4kTBR)$ divided by two, or $\sqrt{(kTBR)}$ because of the 2:1 attenuator formed by the ADC input termination resistor.

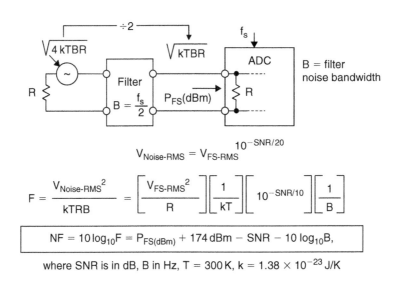

$$V_{Noise\text{-}RMS} = V_{FS\text{-}RMS}\ 10^{-SNR/20}$$

$$F = \frac{V_{Noise\text{-}RMS}^2}{kTRB} = \left[\frac{V_{FS\text{-}RMS}^2}{R}\right]\left[\frac{1}{kT}\right]\left[10^{-SNR/10}\right]\left[\frac{1}{B}\right]$$

$$NF = 10\log_{10}F = P_{FS(dBm)} + 174\,dBm - SNR - 10\log_{10}B,$$

where SNR is in dB, B in Hz, $T = 300\,K$, $k = 1.38 \times 10^{-23}\,J/K$

**Figure 6-149: ADC noise figure in terms of SNR, sampling rate and input power**

The expression for the noise factor F can be written:

$$F = \frac{V_{NOISE\ RMS}^2}{kTRB} = \left[\frac{V_{FS\ RMS}^2}{R}\right]\left[\frac{1}{kT}\right][10^{-SNR/10}]\left[\frac{1}{B}\right] \qquad (6\text{-}27)$$

The NF is obtained by converting F into dB and simplifying:

$$NF = 10_{10}\log F = P_{FS(dBm)} + 174\,dBm - SNR - 10_{10}\log B \qquad (6\text{-}28)$$

where SNR is in dB, B in Hz, $T = 300\,K$, $k = 1.38 \times 10^{-23}\,J/K$.

Oversampling and filtering can be used to decrease the NF as a result of the process gain as has been previously discussed. In this case, the signal bandwidth B is less than $f_s/2$. Figure 6-150 shows the correction factor which results in the following equation:

$$NF = 10_{10} \log F = P_{FS(dBm)} + 174\,dBm - SNR - 10\log_{10}[f_s/2B] - 10\log_{10}B \quad (6\text{-}29)$$

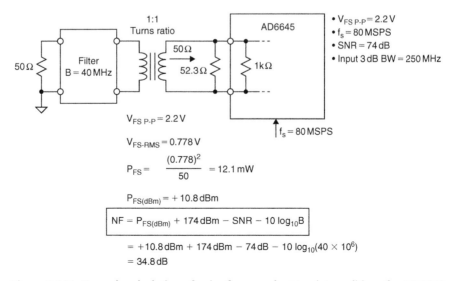

$$NF = P_{FS(dBm)} + 174\,dBm - \underbrace{SNR - 10\log_{10}\left[\frac{f_s/2}{B}\right]}_{\substack{\text{Measured} \\ \text{DC to } f_s/2}} \underbrace{}_{\substack{\text{Process} \\ \text{gain}}} - 10\log_{10}B,$$

where SNR is in dB, B in Hz, T = 300 K, k = $1.38 \times 10^{-23}$ J/K

**Figure 6-150: Effect of oversampling and process gain on ADC noise figure**

Figure 6-151 shows an example NF calculation for the AD6645 14-bit, 80 MSPS ADC. A $52.3\,\Omega$ resistor is added in parallel with the AD6645 input impedance of $1\,k\Omega$ to make the net input impedance $50\,\Omega$. The ADC is operating under Nyquist conditions, and the SNR of 74 dB is the starting point for the calculations using Eq. (6.26). An NF of 34.8 dB is obtained.

$$V_{FS\,P\text{-}P} = 2.2\,V$$

$$V_{FS\text{-}RMS} = 0.778\,V$$

$$P_{FS} = \frac{(0.778)^2}{50} = 12.1\,mW$$

$$P_{FS(dBm)} = +10.8\,dBm$$

$$NF = P_{FS(dBm)} + 174\,dBm - SNR - 10\log_{10}B$$

$$= +10.8\,dBm + 174\,dBm - 74\,dB - 10\log_{10}(40 \times 10^6)$$

$$= 34.8\,dB$$

**Figure 6-151: Example calculation of noise figure under Nyquist conditions for AD6645**

Figure 6-152 shows how using an RF transformer with voltage gain can improve the NF. Figure 6-152(A) shows a 1:1 turns ratio, and the NF (from Figure 6-151) is 34.8. Figure 6-152(B) shows a transformer with a 1:2 turns ratio. The 249 Ω resistor in parallel with the AD6645 internal resistance results in a net input impedance of 200 Ω. The NF is improved by 6 dB because of the "noise-free" voltage gain of the transformer. Figure 6-152(C) shows a transformer with a 1:4 turns ratio. The AD6645 input is paralleled with a 4.02 kΩ resistor to make the net input impedance 800 Ω. The NF is improved by another 6 dB. Transformers with higher turns ratios are not generally practical because of bandwidth and distortion limitations.

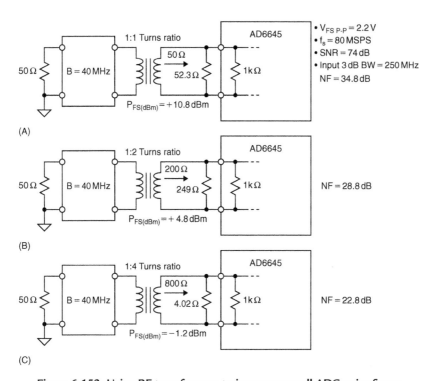

(A)

(B)

(C)

**Figure 6-152: Using RF transformers to improve overall ADC noise figure**

Even with the 1:4 turns ratio transformer, the overall NF for the AD6645 was still 22.8 dB, still relatively high by RF standards. The solution is to provide low noise high gain stages ahead of the ADC. Figure 6-153 shows how the Friis equation is used to calculate the noise factor for cascaded gain stages. Notice that high gain in the first stage reduces the contribution of the noise factor of the second stage—the noise factor of the first stage dominates the overall noise factor.

Figure 6-154 shows the effects of a high gain (25 dB) low noise (NF = 4 dB) stage placed in front of a relatively high NF stage (30 dB)—the NF of the second stage is typical of high performance ADCs. The overall NF is 7.53 dB, only 3.53 dB higher than the first stage NF of 4 dB.

In summary, applying the NF concept to characterize wideband ADCs must be done with extreme caution to prevent misleading results. Simply trying to minimize the NF using the equations can actually increase circuit noise.

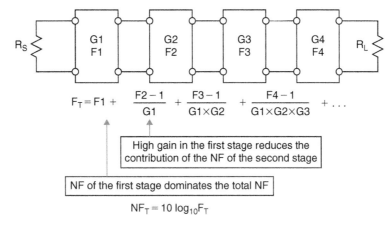

$$F_T = F1 + \frac{F2 - 1}{G1} + \frac{F3 - 1}{G1 \times G2} + \frac{F4 - 1}{G1 \times G2 \times G3} + \ldots$$

High gain in the first stage reduces the contribution of the NF of the second stage

NF of the first stage dominates the total NF

$$NF_T = 10 \log_{10} F_T$$

**Figure 6-153: Cascaded noise figure using the Friis equation**

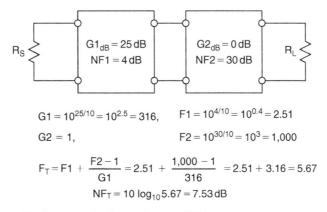

$$G1 = 10^{25/10} = 10^{2.5} = 316, \qquad F1 = 10^{4/10} = 10^{0.4} = 2.51$$

$$G2 = 1, \qquad\qquad\qquad\qquad F2 = 10^{30/10} = 10^3 = 1,000$$

$$F_T = F1 + \frac{F2 - 1}{G1} = 2.51 + \frac{1,000 - 1}{316} = 2.51 + 3.16 = 5.67$$

$$NF_T = 10 \log_{10} 5.67 = 7.53 \, dB$$

- The first stage dominates the overall NF
- It should have the highest gain possible with the lowest NF possible

**Figure 6-154: Example of two-stage cascaded network**

For instance, NF decreases with increasing source resistance according to the calculations, but increased source resistance increases circuit noise. Also, NF decreases with increasing ADC input bandwidth if there is no input filtering. This is also contradictory, because widening the bandwidth increases noise. In both these cases, the circuit noise increases, and the NF decreases. The reason NF decreases is that the source noise makes up a larger component of the total noise (which remains relatively constant because the ADC noise is much greater than the source noise); therefore according to the calculation, NF decreases, but actual circuit noise increases.

It is true that on a stand-alone basis ADCs have relatively high NFs compared to other RF parts such as LNAs or mixers. In the system the ADC should be preceded with low noise gain blocks as shown in the example of Figure 6-154.

## Aperture Time, Aperture Delay Time, and Aperture Jitter

Perhaps the most misunderstood and misused ADC and SHA (or track-and-hold) specifications are those that include the word *aperture*. The most essential dynamic property of an SHA is its ability to disconnect quickly the hold capacitor from the input buffer amplifier as shown in Figure 6-155. The short (but non-zero) interval required for this action is called *aperture time (or sampling aperture)*, $t_a$. The actual value of the voltage that is held at the end of this interval is a function of both the input signal slew rate and the errors introduced by the switching operation itself. Figure 6-155 shows what happens when the hold command is applied with an input signal of two arbitrary slopes labeled as 1 and 2. For clarity, the sample-to-hold pedestal and switching transients are ignored. The value that is finally held is a delayed version of the input signal, averaged over the aperture time of the switch as shown in Figure 6-155. The first-order model assumes that the final value of the voltage on the hold capacitor is approximately equal to the average value of the signal applied to the switch over the interval during which the switch changes from a low to high impedance ($t_a$).

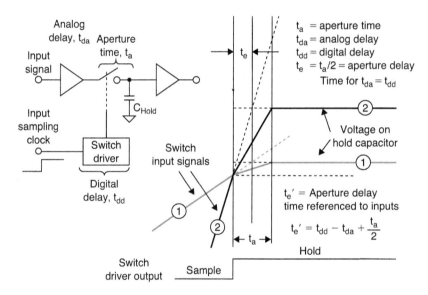

Figure 6-155: Sample-and-hold waveforms and definitions

The model shows that the finite time required for the switch to open ($t_a$) is equivalent to introducing a small delay $t_e$ in the sampling clock driving the SHA. This delay is constant and may either be positive or negative. The diagram shows that the same value of $t_e$ works for the two signals, even though the slopes are different. This delay is called *effective aperture delay time, aperture delay time*, or simply *aperture delay*, $t_e$. In an ADC, the aperture delay time is referenced to the input of the converter, and the effects of the analog propagation delay through the input buffer, $t_{da}$, and the digital delay through the switch driver, $t_{dd}$, must be considered. Referenced to the ADC inputs, aperture time, $t_e$, is defined as the time difference between the analog propagation delay of the front-end buffer, $t_{da}$, and the switch driver digital delay, $t_{dd}$, plus one-half the aperture time, $t_a/2$.

The effective aperture delay time is usually positive, but may be negative if the sum of one-half the aperture time, $t_a/2$, and the switch driver digital delay, $t_{dd}$, is less than the propagation delay through the input buffer, $t_{da}$. The aperture delay specification thus establishes when the input signal is actually sampled with respect to the sampling clock edge.

Aperture delay time can be measured by applying a bipolar sinewave signal to the ADC and adjusting the synchronous sampling clock delay such that the output of the ADC is mid-scale (corresponding to the zero-crossing of the sinewave). The relative delay between the input sampling clock edge and the actual zero-crossing of the input sinewave is the aperture delay time (see Figure 6-156).

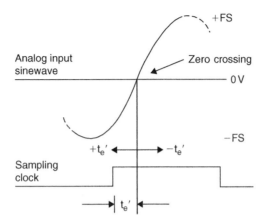

**Figure 6-156: Effective aperture delay time measured with respect to ADC input**

Aperture delay produces no errors (assuming it is relatively short with respect to the hold time), but acts as a fixed delay in either the sampling clock input or the analog input (depending on its sign). However, in simultaneous sampling applications or in direct I/Q demodulation where two or more ADCs must be well matched, variations in the aperture delay between converters can produce errors on fast slewing signals. In these applications, the aperture delay mismatches must be removed by properly adjusting the phases of the individual sampling clocks to the various ADCs.

If, however, there is *sample-to-sample* variation in aperture delay (*aperture jitter*), then a corresponding voltage error is produced as shown in Figure 6-157. This sample-to-sample variation in the instant the switch opens is called *aperture uncertainty*, or *aperture jitter*, and is usually measured in RMS picoseconds. The amplitude of the associated output error is related to the rate of change of the analog input. For any given value of aperture jitter, the aperture jitter error increases as the input dv/dt increases. The effects of phase jitter on the external sampling clock (or the analog input for that matter) produce exactly the same type or error.

The effects of aperture and sampling clock jitter on an ideal ADCs SNR can be predicted by the following simple analysis. Assume an input signal given by:

$$v(t) = V_O \sin 2\pi ft \qquad (6\text{-}30)$$

The rate of change of this signal is given by:

$$dv/dt = 2\pi f V_O \cos 2\pi ft \qquad (6\text{-}31)$$

The RMS value of dv/dt can be obtained by dividing the amplitude, $2\pi f V_O$, by 2:

$$dv/dt|_{RMS} = 2\pi f V_O / \sqrt{2} \qquad (6\text{-}32)$$

Now let $\Delta v_{RMS}$ = the RMS voltage error and $\Delta t$ = the RMS aperture jitter $t_j$, and substitute:

$$\Delta v_{RMS} / t_j = 2\pi f V_O / \sqrt{2} \qquad (6\text{-}33)$$

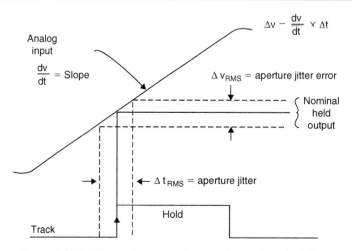

**Figure 6-157: Effects of aperture jitter and sampling clock jitter**

Solving for $\Delta v_{RMS}$:

$$\Delta v_{RMS} = 2\pi f V_0 t_j / \sqrt{2} \qquad (6\text{-}34)$$

The RMS value of the full-scale input sinewave is $V_0/\sqrt{2}$, therefore the RMS signal to RMS noise ratio is given by:

$$SNR = 20\log_{10}\left[\frac{V_0/\sqrt{2}}{\Delta v_{RMS}}\right] = 20\log_{10}\left[\frac{V_0/\sqrt{2}}{2\pi f V_0 t_j/\sqrt{2}}\right] = 20\log_{10}\left[\frac{1}{2\pi f t_j}\right] \qquad (6\text{-}35)$$

This equation assumes an infinite resolution ADC where aperture jitter is the only factor in determining the SNR. This equation is plotted in Figure 6-158 and shows the serious effects of aperture and sampling

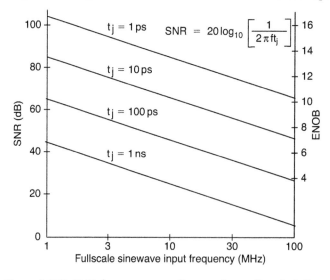

**Figure 6-158: SNR due to aperture jitter and sampling clock jitter**

clock jitter on SNR, especially at higher input/output frequencies. Therefore, extreme care must be taken to minimize phase noise in the sampling/reconstruction clock of any sampled data system.

This care must extend to all aspects of the clock signal: the oscillator itself (e.g., a 555 timer is absolutely inadequate, but even a quartz crystal oscillator can give problems if it uses an active device which shares a chip with noisy logic); the transmission path (these clocks are very vulnerable to interference of all sorts), and phase noise introduced in the ADC or DAC. As discussed, a very common source of phase noise in converter circuitry is aperture jitter in the integral SHA circuitry, however the total RMS jitter will be composed of a number of components—the actual SHA aperture jitter often being the least of them.

## A Simple Equation for the Total SNR of an ADC

A relatively simple equation for the ADC SNR in terms of sampling clock and aperture jitter, DNL, effective input noise, and the number of bits of resolution is shown in Figure 6-159. The equation combines the various error terms on an RSS basis. The average DNL error, $\varepsilon$, is computed from histogram data. This

$$ \text{SNR} = -20\log_{10}\left[ \overbrace{(2\pi \times f_a \times t_{j\,\text{RMS}})^2}^{\substack{\text{Sampling}\\\text{clock jitter}}} + \overbrace{\frac{2}{3}\left[\frac{1+\varepsilon}{2^N}\right]^2}^{\substack{\text{Quantization}\\\text{noise, DNL}}} + \overbrace{\left[\frac{2\times\sqrt{2}\times V_{\text{Noise RMS}}}{2^N}\right]^2}^{\substack{\text{Effective}\\\text{input noise}}} \right]^{\frac{1}{2}} $$

$f_a$ = Analog input frequency of full-scale input sinewave

$t_{j\,\text{RMS}}$ = Combined RMS jitter of internal ADC and external clock

$\varepsilon$ = Average DNL of the ADC (typically 0.41 LSB for AD6645)

$N$ = Number of bits in the ADC

$V_{\text{NOISE RMS}}$ = Effective input noise of ADC (typically 0.9 LSB RMS for AD6645)

If $t_j = 0$, $\varepsilon = 0$, and $V_{\text{NOISE RMS}} = 0$, the above equation reduces to the familiar:

$$ \text{SNR} = 6.02\,N + 1.76\,\text{dB} $$

**Figure 6-159: Relationship between SNR, sampling clock jitter, quantization noise, DNL, and input noise**

equation is used in Figure 6-160 to predict the SNR performance of the AD6645 14-bit, 80 MSPS ADC as a function of sampling clock and aperture jitter.

Two decades or so ago, sampling ADCs were built up from a separate SHA and ADC. Interface design was difficult, and a key parameter was aperture jitter in the SHA. Today, most sampled data systems use *sampling* ADCs which contain an integral SHA. The aperture jitter of the SHA may not be specified as such, but this is not a cause of concern if the SNR or ENOB is clearly specified, since a guarantee of a specific SNR is an implicit guarantee of an adequate aperture jitter specification. However, the use of an additional high performance SHA will sometimes improve the high frequency ENOB of even the best sampling ADC by presenting "DC" to the ADC, and may be more cost-effective than replacing the ADC with a more expensive one.

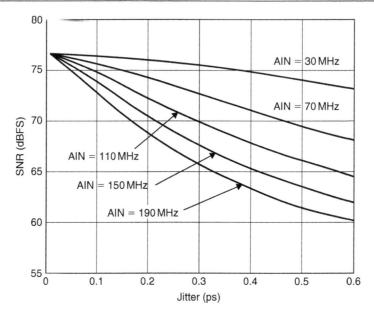

Figure 6-160: AD6645 SNR versus aperture jitter

## ADC Transient Response and Overvoltage Recovery

Most high speed ADCs designed for communications applications are specified primarily in the frequency domain. However, in general purpose data acquisition applications the transient response (or settling time) of the ADC is important. The *transient response* of an ADC is the time required for the ADC to settle to rated accuracy (usually 1 LSB) after the application of a full-scale step input. The typical response of a general purpose 12 bit, 10 MSPS ADC is shown in Figure 6-161, showing a 1 LSB settling time of less than 40 ns. The settling time specification is critical in the typical data acquisition system application where the ADC is being driven by an analog multiplexer as shown in Figure 6-162. The multiplexer output can

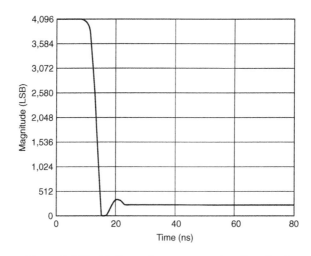

Figure 6-161: ADC transient response (settling time)

*470*

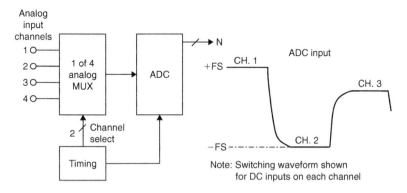

**Figure 6-162: Settling time is critical in multiplexed applications**

deliver a full-scale sample-to-sample change to the ADC input. If both the multiplexer and the ADC have not both settled to the required accuracy, channel-to-channel crosstalk will result, even though only DC or low frequency signals are present on the multiplexer inputs.

Most ADCs have settling times which are less than $1/f_{s\ max}$, even if not specified. However sigma–delta ADCs have a built-in digital filter which can take several output sample intervals to settle. This should be kept in mind when using sigma–delta ADCs in multiplexed applications.

The importance of settling time in multiplexed systems can be seen in Figure 6-163, where the ADC input is modeled as a single-pole filter having a corresponding time constant, $\tau = RC$. The required number of time constants to settle to a given accuracy (1 LSB) is shown. A simple example will illustrate the point.

| Resolution, # of bits | LSB (%FS) | # Of time constants |
|:---:|:---:|:---:|
| 6 | 1.563 | 4.16 |
| 8 | 0.391 | 5.55 |
| 10 | 0.0977 | 6.93 |
| 12 | 0.0244 | 8.32 |
| 14 | 0.0061 | 9.70 |
| 16 | 0.00153 | 11.09 |
| 18 | 0.00038 | 12.48 |
| 20 | 0.000095 | 13.86 |
| 22 | 0.000024 | 15.25 |

**Figure 6-163: Settling time as a function of time constant for various resolutions**

Assume a multiplexed 16-bit data acquisition system uses an ADC with a sampling frequency $f_s = 100$ kSPS. The ADC must settle to 16-bit accuracy for a full-scale step function input in less than $1/f_s = 10\,\mu s$. The chart shows that 11.09 time constants are required to settle to 16-bit accuracy. The input filter time constant must therefore be less than $\tau = 10\,\mu s/11.09 = 900\,ns$. The corresponding risetime $t_r = 2.2\tau = 1.98\,\mu s$. The required ADC full power input bandwidth can now be calculated from $BW = 0.35/t_r = 177\,kHz$. This neglects the settling time of the multiplexer and second-order settling time effects in the ADC.

*Overvoltage recovery time* is defined as that amount of time required for an ADC to achieve a specified accuracy, measured from the time the overvoltage signal re-enters the converter's range, as shown in Figure 6-164. This specification is usually given for a signal which is 50% outside the ADC's input range. Needless to say, the ADC should act as an ideal limiter for out-of-range signals and should produce either the positive full-scale code or the negative full-scale code during the overvoltage condition. Some converters provide over- and under-range flags to allow gain-adjustment circuits to be activated. Care should always be taken to avoid overvoltage signals which will damage an ADC input.

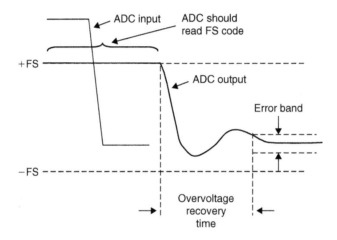

**Figure 6-164: Overvoltage recovery time**

## ADC Sparkle Codes, Metastable States, and Bit Error Rate

A primary concern in the design of many digital communications systems using ADCs is the bit error rate (BER). Unfortunately, ADCs contribute to the BER in ways that are not predictable by simple analysis. This section describes the mechanisms within the ADCs that can contribute to the error rate, ways to minimize the problem, and methods for measuring the BER.

Random noise, regardless of the source, creates a finite probability of errors (deviations from the expected output). Before describing the error code sources, however, it is important to define what constitutes an ADC error code. Noise generated prior to, or inside, the ADC can be analyzed in the traditional manner. Therefore, an ADC error code is any deviation from the expected output that is not attributable to the equivalent input noise of the ADC. Figure 6-165 illustrates an exaggerated output of a low amplitude sinewave applied to an ADC that has error codes. Note that the noise of the ADC creates some uncertainty in the output. These anomalies are not considered error codes, but are simply the result of ordinary noise and quantization. The large errors are more significant and are not expected. These errors are random and so infrequent that an SNR test of the ADC will rarely detect them. These types of errors plagued a few of the early ADCs for video applications, and were given the name *sparkle codes* because of their appearance on a TV screen as small white dots or "sparkles" under certain test conditions. These errors have also been

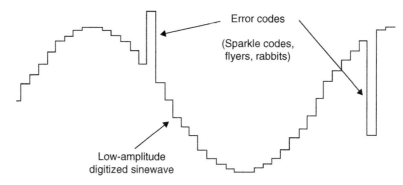

**Figure 6-165: Exaggerated output of ADC showing error codes**

called *rabbits* or *flyers*. In digital communications applications, this type of error increases the overall system BER.

In order to understand the causes of the error codes, we will first consider the case of a simple flash converter. The comparators in a flash converter are latched comparators usually arranged in a master-slave configuration. If the input signal is in the center of the threshold of a particular comparator, that comparator will balance, and its output will take a longer period of time to reach a valid logic level after the application of the latch strobe than the outputs of its neighboring comparators which are being overdriven. This phenomenon is known as *metastability* and occurs when a balanced comparator cannot reach a valid logic level in the time allowed for decoding. If simple binary decoding logic is used to decode the thermometer code, a metastable comparator output may result in a large output code error. Consider the case of a simple 3-bit flash converter shown in Figure 6-166. Assume that the input signal is exactly at the threshold of Comparator 4 and random noise is causing the comparator to toggle between a "1" and a "0" output each time a latch strobe is applied. The corresponding binary output should be interpreted as either 011 or 100. If, however, the comparator output is in a metastable state, the simple binary decoding logic shown may produce binary codes 000, 011, 100, or 111. The codes 000 and 111 represent a one-half scale departure from the expected codes.

The probability of errors due to metastability increases as the sampling rate increases because less time is available for a metastable comparator to settle.

Various measures have been taken in flash converter designs to minimize the metastable state problem. Decoding schemes described in References 12–15 minimize the magnitude of these errors. Optimizing comparator designs for regenerative gain and small time constants is another way to reduce these problems.

Metastable state errors may also appear in successive approximation and subranging ADCs which make use of comparators as building blocks. The same concepts apply, although the magnitudes and locations of the errors may be different.

Establishing the BER of a well-behaved ADC is a difficult, time-consuming task; a single unit can sometimes be tested for days without an error. For example, tests on a typical 8-bit flash converter operating at a sampling rate of 75 MSPS yield a BER of approximately $3.7 \times 10^{-12}$ (1 error per hour) with an error limit of 4 LSBs. Meaningful tests for longer periods of time require special attention to EMI/RFI effects (possibly requiring a shielded screen room), isolated power supplies, etc. Figure 6-167 shows the average time between errors as a function of BER for a sampling frequency of 75 MSPS. This illustrates the difficulty in measuring low BER because the long measurement times increase the probability of power supply transients, noise, etc. causing an error.

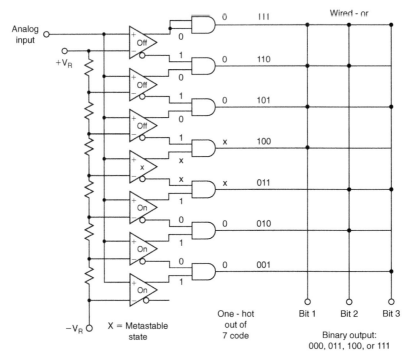

**Figure 6-166: Metastable comparator output states may cause error codes in data converters**

| Bit error rate (BER) | Average time between errors |
|---|---|
| $1 \times 10^{-8}$ | 1.3 seconds |
| $1 \times 10^{-9}$ | 13.3 seconds |
| $1 \times 10^{-10}$ | 2.2 minutes |
| $1 \times 10^{-11}$ | 22 minutes |
| $1 \times 10^{-12}$ | 3.7 hours |
| $1 \times 10^{-13}$ | 1.5 days |
| $1 \times 10^{-14}$ | 15 days |

**Figure 6-167: Average time between errors versus BER when sampling at 75 MSPS**

## DAC Dynamic Performance

The AC specifications which are most likely to be important with DACs are *settling time, glitch impulse area, distortion,* and *SFDR.*

## DAC Settling Time

The settling time of a DAC is the time from a change of digital code to when the output comes within *and remains within* some error band as shown in Figure 6-168. With amplifiers, it is hard to make comparisons of settling time, since their error bands may differ from amplifier to amplifier, but with DACs the error band will almost invariably be $\pm 1$ or $\pm\frac{1}{2}$ LSB.

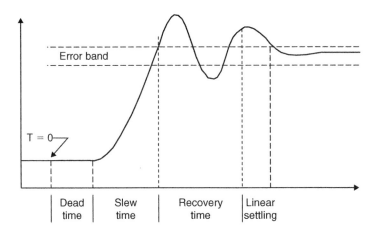

**Figure 6-168: DAC settling time**

The settling time of a DAC is made up of four different periods: the *switching time* or *dead time* (during which the digital switching, but not the output, is changing), the *slewing time* (during which the rate of change of output is limited by the slew rate of the DAC output), the *recovery time* (when the DAC is recovering from its fast slew and may overshoot), and the *linear settling time* (when the DAC output approaches its final value in an exponential or near-exponential manner). If the slew time is short compared to the other three (as is usually the case with current-output DACs), then the settling time will be largely independent of the output step size. On the other hand, if the slew time is a significant part of the total, then the larger the step, the longer the settling time.

Settling time is especially important in video display applications. For example a standard $1024 \times 768$ display updated at a 60 Hz refresh rate must have a pixel rate of $1024 \times 768 \times 60\,\mathrm{Hz} = 47.2\,\mathrm{MHz}$ with no overhead. Allowing 35% overhead time increases the pixel frequency to 64 MHz corresponding to a pixel duration of $1/(64 \times 10^6) = 15.6\,\mathrm{ns}$. In order to accurately reproduce a single fully white pixel located between two black pixels, the DAC settling time should be less than the pixel duration time of 15.6 ns.

Higher resolution displays require even faster pixel rates. For example, a $2048 \times 2048$ display requires a pixel rate of approximately 330 MHz at a 60 Hz refresh rate.

## Glitch Impulse Area

Ideally, when a DAC output changes it should move from one value to its new one monotonically. In practice, the output is likely to overshoot, undershoot, or both (see Figure 6-169). This uncontrolled movement of the DAC output during a transition is known as a *glitch*. It can arise from two mechanisms: capacitive coupling of

475

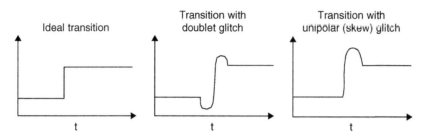

**Figure 6-169: DAC transitions (showing glitch)**

digital transitions to the analog output, and the effects of some switches in the DAC operating more quickly than others and producing temporary spurious outputs.

Capacitive coupling frequently produces roughly equal positive and negative spikes (sometimes called a *doublet* glitch) which more or less cancel in the longer term. The glitch produced by switch timing differences is generally unipolar, much larger, and of greater concern.

Glitches can be characterized by measuring the *glitch impulse area*, sometimes inaccurately called glitch energy. The term *glitch energy* is a misnomer, since the unit for glitch impulse area is volt-seconds (or more probably $\mu$V-seconds or pV-seconds). The *peak glitch area* is the area of the largest of the positive or negative glitch areas. The glitch impulse area is the net area under the voltage-versus-time curve and can be estimated by approximating the waveforms by triangles, computing the areas, and subtracting the negative area from the positive area as shown in Figure 6-170.

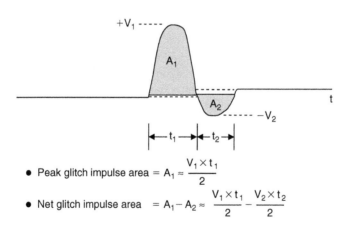

- Peak glitch impulse area $= A_1 \approx \dfrac{V_1 \times t_1}{2}$

- Net glitch impulse area $= A_1 - A_2 \approx \dfrac{V_1 \times t_1}{2} - \dfrac{V_2 \times t_2}{2}$

**Figure 6-170: Calculating net glitch impulse area**

The mid-scale glitch produced by the transition between the codes 0111...111 and 1000...000 is usually the worst glitch. Glitches at other code transition points (such as 1/4 and 3/4 full-scale) are generally less. Figure 6-171 shows the mid-scale glitch for a fast low glitch DAC. The peak and net glitch areas are estimated using triangles as described above. Settling time is measured from the time the waveform leaves the initial 1 LSB error band until it enters and remains within the final 1LSB error band. The step size between the transition regions is also 1 LSB.

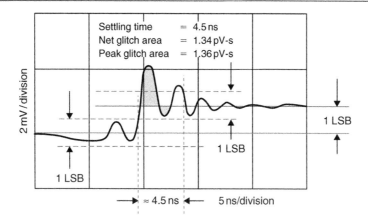

**Figure 6-171: DAC mid-scale glitch shows 1.34 pV-seconds net impulse area and settling time of 4.5 ns**

## DAC SFDR and SNR

DAC settling time is important in applications such as RGB raster scan video display drivers, but frequency-domain specifications such as SFDR are generally more important in communications.

If we consider the spectrum of a waveform reconstructed by a DAC from digital data, we find that in addition to the expected spectrum (which will contain one or more frequencies, depending on the nature of the reconstructed waveform), there will also be noise and distortion products. Distortion may be specified in terms of harmonic distortion, SFDR, IMD, or all of the above. Harmonic distortion is defined as the ratio of harmonics to fundamental when a (theoretically) pure sinewave is reconstructed, and is the most common specification. SFDR is the ratio of the worst spur (usually, but not necessarily, always a harmonic of the fundamental) to the fundamental.

Code-dependent glitches will produce both out-of-band and in-band harmonics when the DAC is reconstructing a digitally generated sinewave as in a direct digital synthesis (DDS) system. The mid-scale glitch occurs twice during a single cycle of a reconstructed sinewave (at each mid-scale crossing), and will therefore produce a second harmonic of the sinewave, as shown in Figure 6-172. Note that the higher order harmonics of the sinewave, which alias back into the Nyquist bandwidth (DC to $f_s/2$), cannot be filtered.

It is difficult to predict the harmonic distortion or SFDR from the glitch area specification alone. Other factors, such as the overall linearity of the DAC, also contribute to distortion. In addition, certain ratios between the DAC output frequency and the sampling clock cause the quantization noise to concentrate at harmonics of the fundamental thereby increasing the distortion at these points.

It is therefore customary to test reconstruction DACs in the frequency domain (using a spectrum analyzer) at various clock rates and output frequencies as shown in Figure 6-173. Typical SFDR for the 16-bit AD9777 Transmit TxDAC™ is shown in Figure 6-174. The clock rate is 160 MSPS, and the output frequency is swept to 50 MHz. As in the case of ADCs, quantization noise will appear as increased harmonic distortion if the ratio between the clock frequency and the DAC output frequency is an integer number. These ratios should be avoided when making the SFDR measurements.

There are nearly infinite combinations of possible clock and output frequencies for a low distortion DAC, and SFDR is generally specified for a limited number of selected combinations. For this reason, Analog

477

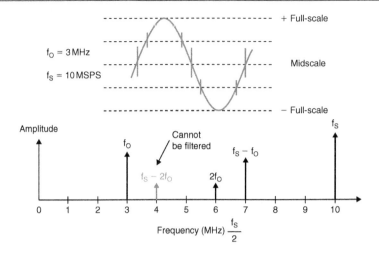

Figure 6-172: Effect of code-dependent glitches on spectral output

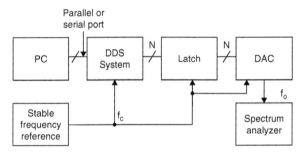

Figure 6-173: Test setup for measuring DAC SFDR

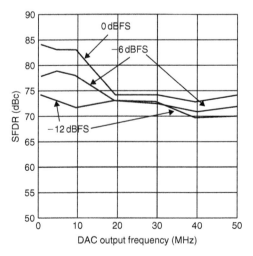

Figure 6-174: AD9777 16-bit TxDAC™ SFDR,
data update rate = 160 MSPS

Devices offers fast turnaround on customer-specified test vectors for the Transmit TxDAC™ family. A test vector is a combination of amplitudes, output frequencies, and update rates specified directly by the customer for SFDR data on a particular DAC.

## *Measuring DAC SNR with an Analog Spectrum Analyzer*

Analog spectrum analyzers are used to measure the distortion and SFDR of high performance DACs. Care must be taken such that the front end of the analyzer is not overdriven by the fundamental signal. If overdrive is a problem, a bandstop filter can be used to filter out the fundamental signal such that the spurious components can be observed.

Spectrum analyzers can also be used to measure the SNR of a DAC provided attention is given to bandwidth considerations. SNR of an ADC is normally defined as the SNR measured over the Nyquist bandwidth DC to $f_s/2$. However, spectrum analyzers have a resolution bandwidth which is less than $f_s/2$—this therefore lowers the analyzer noise floor by the process gain equal to $10 \log_{10}(f_s/2 \text{ BW})$, where BW is the resolution noise bandwidth of the analyzer (see Figure 6-175).

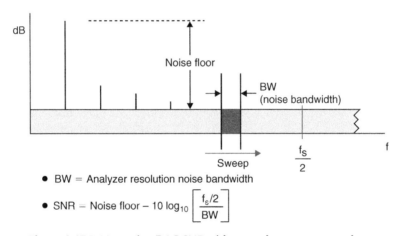

- BW = Analyzer resolution noise bandwidth

- $\text{SNR} = \text{Noise floor} - 10 \log_{10}\left[\dfrac{f_s/2}{\text{BW}}\right]$

**Figure 6-175: Measuring DAC SNR with an analog spectrum analyzer**

It is important that the noise bandwidth (not the 3 dB bandwidth) be used in the calculation, however from Figure 6-147 the error is small assuming that the analyzer narrowband filter is at least two poles. The ratio of the noise bandwidth to the 3 dB bandwidth of a one-pole Butterworth filter is 1.57 (causing an error of 1.96 dB in the process gain calculation). For a two-pole Butterworth filter, the ratio is 1.11 (causing an error of 0.45 dB in the process gain calculation).

## *Other AC Specifications*

Again, there are some specifications that may be encountered that are less common and are listed below.

*Acquisition Time*: The acquisition time of an SHA circuit for a step change is the time required by the output to reach its final value, within a specified error band, after the track command has been given. Included are switch delay time, the slewing interval, and settling time for a specified output voltage change. This spec is less common now that the SHA function has largely been integrated into the ADC.

*Automatic Zero*: To achieve zero stability in many integrating-type converters, a time interval is provided during each conversion cycle to allow the circuitry to compensate for drift errors. The drift error in such converters is substantially zero.

*Channel-to-Channel Isolation*: In multiple DACs, the proportion of analog input signal from one DAC's reference input that appears at the output of the other DAC, expressed logarithmically in dB. See also *Crosstalk.*

*Charge Transfer (or Offset Step)*: The principal component of *sample-to-hold offset* (or *pedestal)*, is the small charge transferred to the storage capacitor via interelectrode capacitance of the switch and stray capacitance when switching to the *hold* mode. The offset step is directly proportional to this charge, viz.,

$$\text{Offset error} = \text{Incremental Charge/Capacitance} = \Delta Q/C$$

It can be reduced somewhat by lightly coupling an appropriate polarity version of the *hold* signal to the capacitor for first-order cancelation. The error can also be reduced by increasing the capacitance, but this increases *acquisition time.* Again a spec that has become less common since the SHA function has largely been integrated into the ADC.

*Crosstalk*: Leakage of signals, usually via capacitance between circuits or channels of a multi-channel system or device, such as a multiplexer, multiple op amp, or multiple DAC. Crosstalk is usually determined by the impedance parameters of the physical circuit, and actual values are frequency-dependent. See also *Channel-to-Channel Isolation.*

Multiple DACs have a *digital crosstalk* specification: the spike (sometimes called a glitch) impulse appearing at the output of one converter due to a change in the digital input code of another of the converters. It is specified in nanovolt- or picovolt-seconds and measured at $V_{REF} = 0\,V$.

*Differential Gain ($\Delta G$)*: A video specification which measures the variation in the amplitude (in percent) of a small amplitude color subcarrier signal as it is swept across the video range from black to white.

*Differential Phase ($\Delta \varphi$)*: A video specification which measures the phase variation (in degrees) of a small amplitude color subcarrier signal as it is swept across the video range from black to white.

*Feedthrough:* Undesirable signal-coupling around switches or other devices that are supposed to be turned off or provide isolation, e.g., *feedthrough error* in an SHA, multiplexer, or MDAC. Feedthrough is variously specified in percent, dB, parts per million, fractions of 1 LSB, or fractions of 1 V, with a given set of inputs, at a specified frequency.

In an MDAC, *feedthrough* error is caused by capacitive coupling from an AC $V_{REF}$ to the output, with all switches off. In an *SHA, feedthrough* is the fraction of the input signal variation or AC input waveform that appears at the output in *hold.* It is caused by stray capacitive coupling from the input to the storage capacitor, principally across the open switch.

*Settling Time—ADC*: The time required, following an analog input step change (usually full-scale), for the digital output of the ADC to reach and remain within a given fraction (usually $\pm\frac{1}{2}$LSB).

## Reference: Data Converter AC Errors

1. W.R. Bennett, "Spectra of Quantized Signals," **Bell System Technical Journal**, Vol. 27, July, 1948, pp. 446–471.

2. B.M. Oliver, J.R. Pierce, and C.E. Shannon, "The Philosophy of PCM," **Proceedings of the IRE**, Vol. 36, November, 1948, pp. 1324–1331.

3. W.R. Bennett, "Noise in PCM Systems," **Bell Labs Record**, Vol. 26, December, 1948, pp. 495–499.

4. H.S. Black and J.O. Edson, "Pulse Code Modulation," **AIEE Transactions**, Vol. 66, 1947, pp. 895–899.

5.   H.S. Black, "Pulse Code Modulation," **Bell Labs Record**, Vol. 25, July, 1947, pp. 265–269.

6.   S. Ruscak and L. Singer, "Using Histogram Techniques to Measure A/D Converter Noise," **Analog Dialogue**, Vol. 29, No. 2, 1995.

7.   M.J. Tant, "The White Noise Book," Marconi Instruments, St. Albans/Hertfordshire, England, July, 1974.

8.   G.A. Gray and G.W. Zeoli, "Quantization and Saturation Noise Due to A/D Conversion," **IEEE Transactions on Aerospace and Electronic Systems**, Vol. 23, January, 1971, pp. 222–223.

9.   K. McClaning and T. Vito, **Radio Receiver Design**, Noble Publishing, 2000, ISBN 1-88-4932-07-X.

10.  W.G. Jung, Editor, **Op Amp Applications**, Analog Devices, Inc., Norwood, Ma, 2002, ISBN 0-916550-26-5, pp. 6.144–6.152.

11.  B. Brannon, **Aperture Uncertainty and ADC System Performance, Application Note AN-501**, Analog Devices, Inc., January 1998. (Available for download at http://www.analog.com.)

12.  C.W. Mangelsdorf, "A 400-MHz Input Flash Converter with Error Correction," **IEEE Journal of Solid-State Circuits**, Vol. 25, No. 1, 1990, pp. 184–191.

13.  C.E. Woodward, "A Monolithic Voltage-Comparator Array for A/D Converters," **IEEE Journal of Solid State Circuits**, Vol. SC-10, No. 6, December, 1975, pp. 392–399.

14.  Y. Akazawa et al., A 400MSPS 8 Bit Flash A/D Converter, **1987 ISSCC Digest of Technical Papers**, pp. 98–99.

15.  Matsuzawa et al., An 8b 600 MHz Flash A/D Converter with Multi-stage Duplex-gray Coding, Symposium VLSI Circuits, **Digest of Technical Papers**, May, 1991, pp. 113–114.

16.  R. Waltman and D. Duff, "Reducing Error Rates in Systems Using ADCs," **Electronics Engineer**, April, 1993, pp. 98–104.

17.  K.W. Cattermole, **"Principles of Pulse Code Modulation,"** American Elsevier Publishing Company, Inc., New York, 1969, ISBN 444-19747-8. (*An excellent tutorial and historical discussion of data conversion theory and practice, oriented towards PCM, but covers practically all aspects. This one is a must for anyone serious about data conversion! Try internet secondhand bookshops such as http://www.abebooks.com for starters.*)

18.  R.A. Witte, "Distortion Measurements Using a Spectrum Analyzer," **RF Design**, September, 1992, pp. 75–84.

19.  W. Kester, "Confused About Amplifier Distortion Specs?," **Analog Dialogue**, Vol. 27, No. 1, 1993, pp. 27–29.

20.  D. Sheingold, Editor, **Analog-to-Digital Conversion Handbook**, 3rd Edition, Prentice-Hall, Norwood, Ma., 1986.

# Timing Specifications

In most cases, the digital signals of a converter are designed to operate at standard levels with standard interface specifications. Common interfaces are parallel or serial (most commonly SPI® or I²C® compatible, but, at the high speed, LVDS is making inroads). The fact that we are compatible with defined standards means that the timing and voltage levels defined in these specifications are met.

The digital signals of a converter are generally divided into one of three groups: address, data, and control.

Common timing specifications include:

- *Logic Low Level*: The voltage level at which the signal is guaranteed to be seen as a logic 0. This level is generally specified at whatever power supply the converter is guaranteed to run at. This means that you will generally see different specification tables for 3 V and 5 V supplies.

- *Logic High Level*: The voltage level at which the signal is guaranteed to be seen as a logic 1. Again, this will be given at the various power supply voltages.

- *Rise Time*: For a step function, the time required for a signal to change from a specified low value to a specified high value. Typically, these values are 10% and 90% of the step height (see Figure 6-176).

- *Fall Time*: For a step function, the time required for a signal to change from a specified high value to a specified low value. Typically, these values are 10% and 90% of the step height (see Figure 6-176).

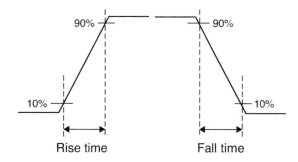

**Figure 6-176: Rise time and fall time**

- *Setup Time*: The time that the data input must be valid before the output latch samples.
- *Hold Time*: The time that data input must be maintained valid after the output samples.
- *Propagation Delay*: The time that takes to the sampled data input to propagate to the output.
- *Pulsewidth High*: The minimum time a pulse must be at the logic high level.
- *Pulsewidth Low*: The minimum time a pulse must be at the logic low level.

Other timing specifications are typically from one signal transition to another. These signals will be defined in the specifications. As an example see Figure 6-177.

**AD6645**

# Switching specifications

(AV$_{CC}$ = 5 V, DV$_{CC}$ = 3.3 V; ENCODE, $\overline{\text{ENCODE}}$, T$_{MIN}$ and T$_{MAX}$ at rated speed grade, (continued) C$_{LOAD}$ = 10 pF, unless otherwise noted.)

| Parameter (Conditions) | Name | Temp | Test Level | AD6645ASQ-80 Min | AD6645ASQ-80 Typ | AD6645ASQ-80 Max | AD6645ASQ-105 Min | AD6645ASQ-105 Typ | AD6645ASQ-105 Max | Unit |
|---|---|---|---|---|---|---|---|---|---|---|
| ENCODE input parameters[1] | | | | | | | | | | |
| Encode period[1] | t$_{ENC}$ | Full | V | | 12.5 | | | 9.5 | | ns |
| Encode pulsewidth high[2] | t$_{ENCH}$ | Full | V | | 6.25 | | | 4.75 | | ns |
| Encode pulsewidth low | t$_{ENCL}$ | Full | V | | 6.25 | | | 4.75 | | ns |
| ENCODE/DataReady | | | | | | | | | | |
| Encode rising to dataready falling | t$_{DR}$ | Full | V | 1.0 | 2.0 | 3.1 | 1.0 | 2.0 | 3.1 | ns |
| Encode rising to dataready rising | t$_{E\_DR}$ | Full | V | | t$_{ENCH}$ + t$_{DR}$ | | | t$_{ENCH}$ + t$_{DR}$ | | ns |
| (50% duty cycle) | | Full | V | 7.3 | 8.3 | 9.4 | 5.7 | 6.75 | 7.9 | ns |
| ENCODE/DATA (D13:0), OVR | | | | | | | | | | |
| ENC to DATA falling low | t$_{E\_FL}$ | Full | V | 2.4 | 4.7 | 7.0 | 2.4 | 4.7 | 7.0 | ns |
| ENC to DATA rising low | t$_{E\_RL}$ | Full | V | 1.4 | 3.0 | 4.7 | 1.4 | 3.0 | 4.7 | ns |
| ENCODE to DATA Delay (Hold time) | t$_{H\_E}$ | Full | V | 1.4 | 3.0 | 4.7 | 1.4 | 3.0 | 4.7 | ns |
| ENCODE to DATA Delay (Setup time) | t$_{S\_E}$ | Full | V | t$_{ENC}$−t$_{E\_FL(max)}$ | | | t$_{ENC}$−t$_{E\_FL(max)}$ | | | ns |
| | | | | | t$_{ENC}$−t$_{E\_FL(typ)}$ | | | t$_{ENC}$−t$_{E\_FL(typ)}$ | | ns |
| | | | | | | t$_{ENC}$−t$_{E\_FL(min)}$ | | | t$_{ENC}$−t$_{E\_FL(min)}$ | ns |
| (50% Duty Cycle) | | Full | V | 5.3 | 7.6 | 10.0 | 2.3 | 4.8 | 7.0 | ns |
| DataReady (DRY[3])/DATA, OVR | | | | | | | | | | |
| DataReady to DATA Delay (Hold time) | t$_{H\_DR}$ | Full | V | | Note 4 | | | Note 4 | | |
| (50% duty cycle) | | | | 6.6 | 7.2 | 7.9 | 5.1 | 5.7 | 6.4 | ns |
| DataReady to DATA Delay (Setup time) | t$_{S\_DR}$ | Full | V | | Note 4 | | | Note 4 | | |
| (50% duty cycle) | | | | 2.1 | 3.6 | 5.1 | 0.6 | 2.1 | 3.5 | ns |
| APERTURE DELAY | t$_A$ | 25°C | V | | −500 | | | −500 | | ps |
| APERTURE UNCERTAINTY (Jitter) | t$_J$ | 25°C | V | | 0.1 | | 0.1 | | | ps rms |

NOTES

[1] Several timing parameters are a function of t$_{ENC}$ AND t$_{ENCH}$.
[2] ENCODE TO DATA Delay (Hold time) is the absolute minimum propagation delay through the analog-to-digital converter, t$_{E\_RL}$ = t$_{H\_E}$.
[3] DRY is an inverted and delayed version of the encode clock. Any change in the duty cycle of the clock will correspondingly change the duty cycle of DRY.
[4] DataReady to DATA Delay (t$_{H\_DR}$ and t$_{S\_DR}$) is calculated relative to rated speed grade and is dependent on t$_{ENC}$ and duty cycle.

Specifications subject to change without notice.

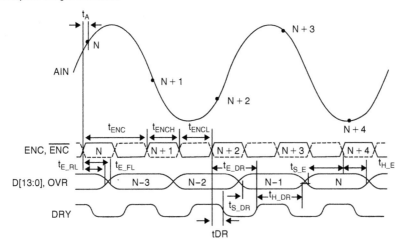

Figure 6-177: Sample timing specifications for the AD6645

Often the output current available is also specified. This will help determine fan out, the number of standard loads that the output can drive. But we will discuss in the circuit board considerations section why it is probably not wise to have the converter drive significant current.

Occasionally a high speed converter has a clock input that is not a standard logic signal level. This is often to make the signal easier to generate and propagate on the PCB. For example, for optimum performance, the AD6645 must be clocked differentially. The encode signal is usually AC coupled into the ENC and *ENC* pins via a transformer or capacitors. These pins are biased internally and require no additional bias. The input impedance of the input and the signal level will be specified (see Figure 6-178).

| Parameter (Conditions) | Temp | Test Level | AD6645ASQ-80 | | | AD6645ASQ-105 | | | Unit |
|---|---|---|---|---|---|---|---|---|---|
| | | | Min | Typ | Max | Min | Typ | Max | |
| ENCODE INPUTS (ENC, ENC) | | | | | | | | | |
| Differential Input Voltage[1] | Full | IV | 0.4 | | | 0.4 | | | $V_{p-p}$ |
| Differential Input Resistance | 25°C | V | | 10 | | | 1  0 | | kΩ |
| Differential Input Capacitance | 25°C | V | | 2.5 | | | | 2.5 | pF |

**Figure 6-178: Encode command specifications for the AD6645**

485

# How to Read a Data Sheet

While there is not an industry standard concerning data sheets, what they cover and what information is included and where that information is located, for the most part, data sheets from various manufacturers generally are similar in construction. In this section we will take a look at several data sheets and try to give a feel for where to find certain information and how to interpret what is found.

As a demonstration we will look at five data sheets:

| | |
|---|---|
| AD6645 | High speed ADC |
| AD9777 | High speed DAC (TxDAC, Interpolating DAC) |
| AD7678 | General Purpose ADC |
| AD5570 | General Purpose DAC |
| AD7730 | $\Sigma\Delta$ ADC |

The part numbers chosen are arbitrary, they were chosen only to give a range of parts.

## The Front Page

This page is designed to give you the basic information you might need to choose the part (referring to Figure 6-179). We can break this up into three sections.

Section 1 is the features. These bullet points are what are considered by the manufacturer to be the more important parameters of the product for its intended application. The targeted applications are typically listed as well.

Section 2 is the product description. This typically covers some of what the manufacturer considers to be the salient features of the op amp.

Section 3 is the functional block diagram. Many times you can get information from the block diagram. In this instance we can see that the ADC has a pipeline architecture, in this case three stage.

## The Specification Tables

There are an unlimited number of conditions possible to measure any given specification. Obviously it is not possible to test all possible conditions. So a representative set of conditions are chosen. The test conditions are specified (Box 1 in Figure 6-179). Occasionally if further clarification or modification of the conditions is required, it is handled as footnotes (Box 2 in Figure 6-179).

On many converters some individual specifications may have multiple entries. This is for different performance levels. It can also be for different temperature ranges (usually commercial, industrial, or military). In this case it is for different speed grades. This can be seen in Figure 6-180.

Note that there are typically three possibilities for the specifications, min, typ, and max. See Figure 6-180 (3). At Analog Devices any specification in the min (minimum) and max (maximum) columns will be guaranteed by test. This can be a direct test, or, in some instances, testing one parameter will guarantee another. A typ (typical) specification is just that, typical. Depending on the particular specification,

487

## Features
SNR = 75 dB, $f_{IN}$ 15 MHz up to 105 MSPS
SNR = 72 dB, $f_{IN}$ 200 MHz up to 105 MSPS
SFDR = 89 dBc, $f_{IN}$ 70 MHz up to 105 MSPS
100 dB Multitone SFDR
IF Sampling to 200 MHz
Sampling jitter 0.1 ps
1.5 W power dissipation
Differential analog inputs
Pin compatible to AD6644
Twos complement digital output format
3.3 V CMOS compatible
DataReady for output latching

## Applications
Multichannel, Multimode Receivers
Base station infrastructure
AMPS, IS-136, CDMA, GSM, WCDMA
Single channel digital receivers
Antenna array processing
Communications instrumentation
Radar, infrared imaging
Instrumentation

**1**

## Product Description
The AD6645 is a high speed, high performance, monolithic 14-bit analog-to-digital converter. All necessary functions, including track-and-hold (T/H) and reference, are included on the chip to provide a complete conversion solution. The AD6645 provides CMOS compatible digital outputs. It is the fourth generation in a wideband ADC family, preceded by the AD9042 (12-bit, 41 MSPS), the AD6640 (12-bit,

65 MSPS, IF sampling), and the AD6644 (14-bit, 40 MSPS/65 MSPS).

Designed for multichannel, multimode receivers, the AD6645 is part of Analog Devices' SoftCell® transceiver chipset. The AD6645 maintains 100 dB multitone, spurious-free dynamic range (SFDR) through the second Nyquist band. This break-through performance eases the burden placed on multimode digital receivers (software radios) that are typically limited by the ADC. Noise performance is exceptional; typical signal-to-noise ratio is 74.5 dB through the first Nyquist band.

The AD6645 is built on Analog Devices' high speed complementary bipolar process (XFCB) and uses an innovative, multipass circuit architecture. Units are available in a thermally enhanced 52-lead PowerQuad 4® (LQFP_PQ4) specified from −40°C to +85°C at 80 MSPS and −10°C to +85°C at 105 MSPS.

**2**

## Product highlights
1. IF sampling
   The AD6645 maintains outstanding AC performance up to input frequencies of 200 MHz, suitable for multicarrier 3G wideband cellular IF sampling receivers.
2. Pin compatibility
   The ADC has the same footprint and pin layout as the AD6644, 14-Bit 40 MSPS/65 MSPS ADC.
3. SFDR Performance and Oversampling
   Multitone SFDR performance of −100 dBc can reduce the requirements of high end RF components and allows the use of receive signal processors such as the AD6620 or AD6624/AD6624A.

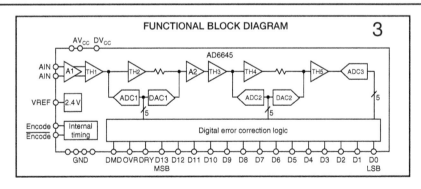

**3**

FUNCTIONAL BLOCK DIAGRAM

One Technology Way, P.O. Box 9106, Norwood, MA 02062-9106, U.S.A.
Tel: 781/329-4700                    www.analog.com
Fax:781/326-8703      © 2003 Analog Devices, Inc. All rights reserved.

Figure 6-179: Example data sheet front page

## AD6645–Specifications

**DC Specifications** (AV$_{CC}$ = 5 V. DV$_{CC}$ = 3.3 V; TMIN and TMAX at rated speed grade, unless otherwise noted.) **1** **3**

| Parameter | Temp | Test Level | AD6645ASQ-80 | | | AD6645ASQ-105 | | | Unit |
|---|---|---|---|---|---|---|---|---|---|
| | | | Min | Typ | Max | Min | Typ | Max | |
| RESOLUTION | | | | 14 | | | | | Bits |
| ACCURACY | | | | | | | | | |
| No missing codes | Full | II | | Guaranteed | | | Guaranteed | | |
| Offset errror | Full | II | −10 | +1.2 | +10 | −10 | +1.2 | +10 | mV |
| Gain error | Full | II | −10 | 0 | +10 | −10 | 0 | +10 | % FS |
| Differential nonlinearity (DNL) | Full | II | −1.0 | ±0.25 | +1.5 | −1.0 | ±0.5 | +1.5 | LSB |
| Integral nonlinearity (INL) | Full | V | | ±0.5 | | | ±1.5 | | LSB |
| TEMPERATURE DRIFT | | | | | | | | | |
| Other error | Full | V | | 1.5 | | | 1.5 | | ppm/°C |
| Gain error | Full | V | | 48 | | | 48 | | ppm/°C |
| POWER SUPPLY REJECTION (PSRR) | 25°C | V | | ±1.0 | | | ±1.0 | | mV/V |
| REFERENCE OUT (VREF)[1] | Full | V | | 2.4 | | | 2.4 | | V |
| ANALOG INPUTS (AIN, $\overline{AIN}$) | | | | | | | | | |
| Differential input voltage range | Full | V | | 2.2 | | | 2.2 | | V p-p |
| Differential input resistance | Full | V | | 1 | | | 1 | | kΩ |
| Differential input capacitance | 25°C | V | | 1.5 | | | 1.5 | | pF |
| POWER SUPPLY | | | | | | | | | |
| Supply voltages | | | | | | | | | |
| AVcc | Full | II | 4.75 | 5.0 | 5.25 | 4.75 | 5.0 | 5.25 | V |
| DVcc | Full | II | 3.0 | 3.3 | 3.6 | 3.0 | 3.3 | 3.6 | V |
| Supply Current | | | | | | | | | |
| I AVcc (AVcc = 5.0 V) | Full | II | | 275 | 320 | | 275 | 320 | mA |
| I DVcc (DVcc = 3.3 V) | Full | II | | 32 | 45 | | 32 | 45 | mA |
| Rise Time[2] | | | | | | | | | |
| AVcc | Full | IV | | | 250 | | | 250 | ms |
| POWER CONSUMPTION | Full | II | | 1.5 | 1.75 | | 1.5 | 1.75 | W |

NOTES
[1]VREF is provided for setting the common-mode offset of a differential amplifier such as the AD8138 when a DC-coupled analog input is required. VREF should be buffered if used to drive additional circuit function. **2**
[2]Specified for DC supplies with linear rise time characteristics.

Specifications subject to change without notice.

**Digital specifications** (AVcc = 5 V, DVcc = 3.3 V; T$_{MIN}$ and T$_{MAX}$ at rated speed grade, unless otherwise noted.) **1**

| Parameter (conditions) | Temp | Test Level | AD6645ASQ-80 | | | AD6645ASQ-105 | | | Unit |
|---|---|---|---|---|---|---|---|---|---|
| | | | Min | Typ | Max | Min | Typ | Max | |
| ENCODE INPUTS (ENC, ENC) | | | | | | | | | |
| Differential input voltage[1] | Full | IV | 0.4 | | | 0.4 | | | V p-p |
| Differential input resistance | 25°C | V | | 10 | | | 10 | | kΩ |
| Differential input capacitance | 25°C | V | | 2.5 | | | 2.5 | | pF |
| LOGIC OUTPUTS (D13-D0, DRY, OVR[2]) | | | | | | | | | |
| Logic compatibility | | | | CMOS | | | CMOS | | |
| Logic 1 voltage (DVcc = 3.3 V)[3] | Full | II | 2.85 | DVcc$^{-2}$ | | 2.85 | DVcc$^{-2}$ | | V |
| Logic 0 voltage (DVcc = 3.3 V)[3] | Full | II | | 0.2 | 0.5 | | 0.2 | 0.5 | V |
| Output coding | | | | Twos Complement | | | Twos Complement | | |
| DMID | Full | V | | DVcc/2 | | | DVcc/2 | | V |

NOTES
[1]All AC specifications tested by driving ENCODE and $\overline{ENCODE}$ differentially. **2**
[2]The functionality of the Overrange bit is specified for a temperature range of 25°C to 85°C only.
[3]Digital output logic levels DVcc = 3.3 V, C$_{LOAD}$ = 10 pF. Capacitive loads >10 pF will degrade performance.

Specifications subject to change without notice.

**Figure 6-180: Example data sheet specification page**

# AC Specifications[1] (AV$_{CC}$ = 5 V, DV$_{CC}$ = 3.3 V; ENCODE, $\overline{ENCODE}$, T$_{MIN}$ and T$_{MAX}$ at rated speed grade, unless otherwise noted.)

| Parameter (Conditions) | | Temp | Test Level | AD6645ASQ-80 | | | AD6645ASQ-105 | | | Unit |
|---|---|---|---|---|---|---|---|---|---|---|
| | | | | Min | Typ | Max | Min | Typ | Max | |
| SNR | | | | | | | | | | |
| Analog input | 15.5 MHz | 25°C | V | | 75.0 | | | 75.0 | | dB |
| @ −1 dBFS | 30.5 MHz | Full | II | 72.5 | 74.5 | | | | | dB |
| | 37.7 MHz | 25°C | I | | | | 72.5 | 74.5 | | dB |
| | 70.0 MHz | Full | II | 72.0 | 73.5 | | 72.0 | 73.5 | | dB |
| | 150.0 MHz | 25°C | V | | 73.0 | | | 73.0 | | dB |
| | 200.0 MHz | 25°C | V | | 72.0 | | | 72.0 | | dB |
| SINAD | | | | | | | | | | |
| Analog input | 15.5 MHz | 25°C | V | | 75.0 | | | 75.0 | | dB |
| @ −1 dBFS | 30.5 MHz | Full | II | 72.5 | 74.5 | | | | | dB |
| | 37.7 MHz | 25°C | I | | | | 72.5 | 74.5 | | dB |
| | 70.0 MHz | Full | V | | 73.0 | | | 73.0 | | dB |
| | 150.0 MHz | 25°C | V | | 68.5 | | | 67.5 | | dB |
| | 200.0 MHz | 25°C | V | | 62.5 | | | 62.5 | | dB |
| Worst Harmonic (Second or third) | | | | | | | | | | |
| Analog input | 15.5 MHz | 25°C | V | | 93.0 | | | 93.1 | | dBc |
| @ −1 dBFS | 30.5 MHz | Full | II | 85.0 | 93.0 | | | | | dBc |
| | 37.7 MHz | 25°C | I | | | | 85.0 | 93.0 | | dBc |
| | 70.0 MHz | Full | V | | 89.0 | | | 87.0 | | dBc |
| | 150.0 MHz | 25°C | V | | 70.0 | | | 70.0 | | dBc |
| | 200.0 MHz | 25°C | V | | 63.5 | | | 63.5 | | dBc |
| Worst Harmonic (Fourth or higher) | | | | | | | | | | |
| Analog input | 15.5 MHz | 25°C | V | | 96.0 | | | 96.0 | | dBc |
| @ −1 dBFS | 30.5 MHz | Full | II | 85.0 | 95.0 | | | | | dBc |
| | 37.7 MHz | 25°C | I | | | | 86.0 | 95.0 | | dBc |
| | 70.0 MHz | Full | V | | 90.0 | | | 90.0 | | dBc |
| | 150.0 MHz | 25°C | V | | 90.0 | | | 90.0 | | dBc |
| | 200.0 MHz | 25°C | V | | 88.0 | | | 88.0 | | dBc |
| Two Tone SFDR | @30.5 MHz[2,3] | 25°C | V | | 100 | | | 98.0 | | dBFS |
| | 55.0 MHz[2,4] | 25°C | V | | 100 | | | 98.0 | | dBFS |
| | 70.0 MHz[2,5] | 25°C | V | | | | | 98.0 | | dBFS |
| Two Tone IMD Rejection[3,4] | | | | | | | | | | |
| F1, F2 @ −7 dBFS | | 25°C | V | | 90 | | | 90 | | dBc |
| Analog input bandwidth | | 25°C | V | | 270 | | | 270 | | MHz |

NOTES

[1] All AC specifications tested by driving ENCODE and $\overline{ENCODE}$ differentially.

[2] Analog input signal power swept from −10 dBFS to −100 dBFS.

[3] F1 = 30.5 MHz, F2 = 31.5 MHz.

[4] F1 = 55.25 MHz, F2 = 56.25 MHz.

[5] F1 = 69.1 MHz, F2 = 71.1 MHz.

Specifications subject to change without notice.

Figure 6-181: Typical AC specifications

**AD6645**

# Switching Specifications (continued)

(AV$_{CC}$ = 5 V, DV$_{CC}$ = 3.3 V; ENCODE, $\overline{ENCODE}$, T$_{MIN}$ and T$_{MAX}$ at rated speed grade, C$_{LOAD}$ = 10 pF, unless otherwise noted.)

| Parameter (Conditions) | Name | Temp | Test Level | AD6645ASQ-80 | | | AD6645ASQ-105 | | | Unit |
|---|---|---|---|---|---|---|---|---|---|---|
| | | | | Min | Typ | Max | Min | Typ | Max | |
| ENCODE input parameters[1] | | | | | | | | | | |
| Encode period[1] | t$_{ENC}$ | Full | V | | 12.5 | | | 9.5 | ns | |
| Encode pulsewidth high[2] | t$_{ENCH}$ | Full | V | | 6.25 | | | 4.75 | | ns |
| Encode pulsewidth low | t$_{ENCL}$ | Full | V | | 6.25 | | | 4.75 | | ns |
| ENCODE/dataready | | | | | | | | | | |
| Encode rising to DataReady falling | t$_{DR}$ | Full | V | 1.0 | 2.0 | 3.1 | 1.0 | 2.0 | 3.1 | ns |
| Encode rising to DataReady rising | t$_{E\_DR}$ | Full | V | | t$_{ENCH}$ + t$_{DR}$ | | | t$_{ENCH}$+t$_{DR}$ | | ns |
| (50% duty cycle) | | Full | V | 7.3 | 8.3 | 9.4 | 5.7 | 6.75 | 7.9 | ns |
| ENCODE/DATA (D13:0), OVR | | | | | | | | | | |
| ENC to DATA falling Low | t$_{E\_FL}$ | Full | V | 2.4 | 4.7 | 7.0 | 2.4 | 4.7 | 7.0 | ns |
| ENC to DATA rising Low | t$_{E\_RL}$ | Full | V | 1.4 | 3.0 | 4.7 | 1.4 | 3.0 | 4.7 | ns |
| ENCODE to DATA Delay (Hold time) | t$_{H\_E}$ | Full | V | 1.4 | 3.0 | 4.7 | 1.4 | 3.0 | 4.7 | ns |
| ENCODE to DATA Delay (Setup time) | t$_{S\_E}$ | Full | V | t$_{ENC}$−t$_{E\_FL(max)}$ | | | t$_{ENC}$−t$_{E\_FL(max)}$ | | | ns |
| | | | | | t$_{ENC}$−t$_{H\_FL(typ)}$ | | | t$_{ENC}$−t$_{E\_FL(typ)}$ | | ns |
| | | | | | | t$_{ENC}$−t$_{H\_FL(min)}$ | | | t$_{ENC}$−t$_{E\_FL(min)}$ | ns |
| (50% duty cycle) | | Full | V | 5.3 | 7.6 | 10.0 | 2.3 | 4.8 | 7.0 | ns |
| DataReady (DRY[3])/DATA, OVR | | | | | | | | | | |
| DataReady to DATA Delay (Hold time) | t$_{H\_DR}$ | Full | V | | Note 4 | | | Note 4 | | |
| (50% duty cycle) | | | | 6.6 | 7.2 | 7.9 | 5.1 | 5.7 | 6.4 | ns |
| DataReady to DATA Delay (Setup time) | t$_{S\_DR}$ | Full | V | | Note 4 | | | Note 4 | | |
| (50% duty cycle) | | | | 2.1 | 3.6 | 5.1 | 0.6 | 2.1 | 3.5 | ns |
| Aperture delay | t$_A$ | 25°C | V | | −500 | | | −500 | | ps |
| Aperture uncertainty (Jitter) | t$_J$ | 25°C | V | | 0.1 | | 0.1 | | | ps rms |

NOTES

[1] Several timing parameters are a function of t$_{ENC}$ and t$_{ENCH}$.
[2] ENCODE TO DATA Delay (Hold Time) is the absolute minimum propagation delay through the analog-to-digital converter, t$_{E\_RL}$ = t$_{H\_E}$.
[3] DRY is an inverted and delayed version of the encode clock. Any change in the duty cycle of the clock will correspondingly change the duty cycle of DRY.
[4] DataReady to DATA Delay (t$_{H\_DR}$ and t$_{S\_DR}$) is calculated relative to rated speed grade and is dependent on t$_{ENC}$ and duty cycle.

Specifications subject to change without notice.

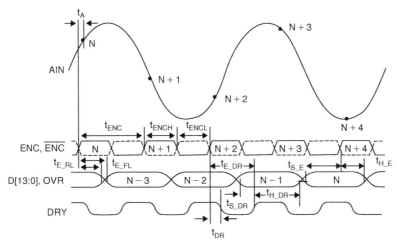

**Figure 6-182: Typical timing specification page**

# Digital Specifications

$(T_{MIN}$ to $T_{MAX}$, AVDD = 3.3 V, CLKVDD = 3.3 V, PLLVDD = 0 V, DVDD = 3.3 V, $I_{OUTFS}$ = 20 mA, unless otherwise noted.)

| Parameter | Min | Typ | Max | Unit |
|---|---|---|---|---|
| Digital inputs | | | | |
| Logic "1" voltage | 2.1 | 3 | | V |
| Logic "0" voltage | | 0 | 0.9 | V |
| Logic "1" current | −10 | | +10 | μA |
| Logic "0" current | −10 | | +10 | μA |
| Input capacitance | | 5 | | pF |
| Clock Inputs | | | | |
| Input voltage range | 0 | | 3 | V |
| Common-mode voltage | 0.75 | 1.5 | 2.25 | V |
| Differential voltage | 0.5 | 1.5 | | V |

Specifications subject to change without notice.

| Parameter | Min | Typ | Max | Unit |
|---|---|---|---|---|
| Serial control bus | | | | |
| Maximum SCLK frequency ($f_{SLCK}$) | 15 | | | MHz |
| Minimum clock pulsewidth high ($t_{PWH}$) | 30 | | | ns |
| Minimum clock pulsewidth low ($t_{PWL}$) | 30 | | | ns |
| Maximum clock rise/fall time | | | 1 | ms |
| Minimum data/chip select setup time ($t_{DS}$) | 25 | | | ns |
| Minimum data hold time ($t_{DH}$) | 0 | | | ns |
| Maximum data valid time ($t_{DV}$) | | | 30 | ns |
| RESET pulsewidth | 1.5 | | | ns |
| Inputs (SDI, SDIO, SCLK, CSB) | | | | |
| Logic "1" voltage | 2.1 | 3 | | V |
| Logic "0" voltage | | 0 | 0.9 | V |
| Logic "1" current | −10 | | +10 | μA |
| Logic "0" current | −10 | | +10 | μA |
| Input capacitance | | 5 | | pF |
| SDIO output | | | | |
| Logic "1" voltage | DRVDD−0.6 | | | V |
| Logic "0" voltage | | | 0.4 | V |
| Logic "1" current | 30 | 50 | | mA |
| Logic "0" current | 30 | 50 | | mA |

**Figure 6-183: Typical timing specification page 2**

AD 9777

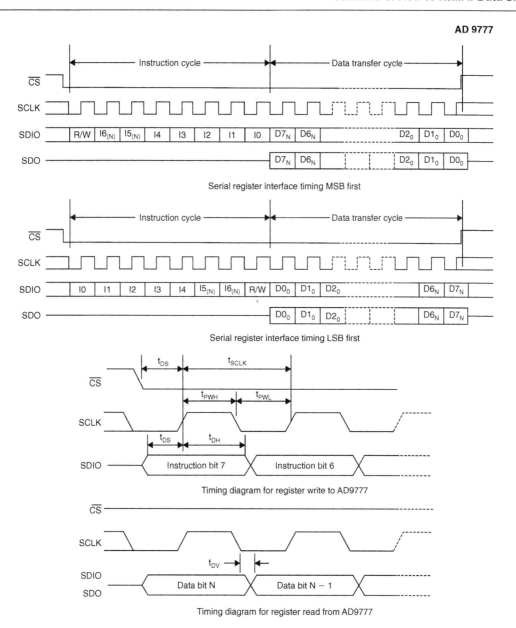

Serial register interface timing MSB first

Serial register interface timing LSB first

Timing diagram for register write to AD9777

Timing diagram for register read from AD9777

**Figure 6-184: Typical timing specifications page 3**

the deviation from the typical can be substantial. And you have no way of knowing what the range of variation on the typ specification is. Sometimes you will find a typ and a min (or max) for the same specification. This tells you that although the test limits are at a particular level (min or max) the typicals tend to run much better than the test limits. For example, in the data sheet example in Figure 6-180, the gain error is guaranteed to be ±10% (full-scale), but the error is typically 0% (FS). When designing, using typs is risky. You are much better off using mins or maxes for error budget analysis.

Testing is one of the most expensive steps in the manufacturing of ICs. Therefore a more highly specified part will typically cost more than a less completely specified part. But, in your system, the higher specified part may be required to guarantee the circuit performance.

As can be seen from the examples in Figures 6-179 and 6-180 and Figures 6-181 to 6-184 the example used (in this case the AD6645) is specified for both DC and AC, as explained in the earlier section on converter specifications. Note that the DC specifications are in terms of absolute levels (volts, amps, etc.) while the AC specifications tend to be in terms of dB.

Also note in Figure 6-181 that the digital signal levels are also specified in terms of voltage levels. They are also specified in terms of time. These are the "switching specifications" of Figure 6-182. These specifications are for individual signals (rise and fall times, pulsewidth high, etc.) as well as between signals (setup and hold times, etc.).

## The Absolute Maximums

There is always a section (usually just after the specification tables) with the absolute maximum ratings (Figure 6-185). These are typically voltage and temperature related.

The process used to fabricate the op amp will typically determine the maximum supply voltage. Maximum input voltages typically are limited to the supply voltages. It should be pointed out that the supply voltage is the instantaneous value, not the average, or ultimate value. So if a converter has voltages on its input but the supply voltage is not present, which could occur during power-up when one section of the system

### Absolute maximum ratings

$T_A = 25°C$, unless otherwise noted.

**Table 4.**

| Parameter | Rating |
|---|---|
| $V_{DD}$ to AGND, AGNDS, DGND | $-0.3\,V, +17\,V$ |
| $V_{SS}$ to AGND, AGNDS, DGND | $+0.3\,V, -17\,V$ |
| AGND, AGNDS to DGND | $-0.3\,V$ to $+0.3\,V$ |
| REFGND to AGND, ADNDS | $V_{SS} - 0.3\,V$ to $V_{DD} + 0.3\,V$ |
| REFIN to AGND, AGNDS | $V_{SS} - 0.3\,V$ to $V_{DD} + 0.3\,V$ |
| REFIN to REFGND | $-0.3\,V$ to $+17\,V$ |
| Digital inputs to DGND | $-0.3\,V$ to $V_{DD} +0.3\,V$ |
| $V_{OUT}$ to AGND, AGNDS | $-0.3\,V$ to $V_{DD} +0.3\,V$ |
| SDO to DGND | $-0.3\,V$ to $+6.5\,V$ |
| Operating temperature range: | $-40°C$ to $+125°C$ |
|   W, Y grades | $-40°C$ to $+125°C$ |
|   A, B grades | $-40°C$ to $+85°C$ |
| Storage temperature range | $-65°C$ to $+50°C$ |
| Maximum junction temperature ($T_J$ Max) | $150°C$ |
| 16-Lead SSOP package | |
|   Power dissipation | $(T_J\,max - T_A)/\theta_{JA}$ |
|   $\theta_{JA}$ thermal impedance | $139°C/W$ |
|   Lead temperature (Soldering 10 s) | $300°C$ |
|   IR reflow, peak temperature | $230°C$ |

Stresses above those listed under Absolute Maximum Ratings may cause permanent damage to the device. This is a stress rating only and functional operation of the device at these or any other conditions above those listed in the operational sections of this specification is not implied. Exposure to absolute maximum rating conditions for extended periods may affect device reliability.

**Figure 6-185: Typical absolute maximum ratings**

is powered but others are not, the converter is overvoltaged, even if when the converter power is applied, everything is within operational limits.

The overriding concern for semiconductor reliability is to keep the junction temperature below 150°C. There will be a $\theta_{ja}$ given for the various package options. This is the thermal resistance from the junction to free air. The units are °C/Watt. To use this information simply determine the dissipation of the package. This would be the quiescent current times the supply voltage. Then take the maximum dissipation generated by the output stages (output current times the difference between the output voltage and the supply voltage). Add these together and you will have the total package dissipation, in watts. Multiply the thermal resistance by the dissipation and you have the temperature rise. You start with the ambient temperature (in °C, typically taken to be 25°C), take the rise calculated above and that will give you the junction temperature. Remember that the ambient temperature should be in operation. Circuits packaged in an enclosure, which is in turn placed in a rack with other equipment, will have an internal ambient temperature that could be significantly above the air temperature in the room that it is located in. This must be considered.

The thermal resistance has two components $\theta_{jc}$ and $\theta_{ca}$. $\theta_{jc}$ is the thermal resistance from the junction to the case. There is not much you can do about that. $\theta_{ca}$ is the thermal resistance from the case to the air. This we can effect relatively easily by adding a heat sink. The thermal resistance adds linearly. Also note that these values are in free air. Moving air allows more cooling, especially with a heatsink.

## The Ordering Guide

Many converters are available in multiple packages and/or multiple temperature ranges. Each of the various combinations of package and temperature range requires a unique part number. This is spelled out in the ordering guide (see Figures 6-186 and 6-187).

Ordering guide

| Model | Maximum INL | No missing code | Temperature range | Package description | Package option | Brand |
|---|---|---|---|---|---|---|
| AD7684ARM | ±6 LSB | 15 bits | −40°C to +85°C | μSOIC-8 | RM-8 | C1M |
| AD7684ARMRL7 | ±6 LSB | 15 bits | −40°C to +85°C | μSOIC-8 | RM-8 (reel) | C1M |
| AD7684BRM | ±3 LSB | 16 bits | −40°C to +85°C | μSOIC-8 | RM-8 | C1D |
| AD7684BRMRL7 | ±3 LSB | 16 bits | −40°C to +85°C | μSOIC-8 | RM-8 (reel) | C1D |
| | | | | | | |
| EVAL-AD7684CB[1] | | | | Evaluation Board | | |
| EVAL-CONTROL BRD2[2] | | | | Controller Board | | |
| EVAL-CONTROL BRD3[2] | | | | Controller Board | | |

NOTES
[1] This board can be used as a standalone evaluation board or in conjunction with the EVAL-CONTROL BRDx for evaluation/demonstration purposes.
[2] These boards allow a PC to control and communicate with all Analog Devices evaluation boards ending in the CB designators.

**Figure 6-186: Ordering guide example 1 (for the AD7684)**

Just as a note, in the case of general purpose ADCs and DACs, the commercial (0–70°C) temperature range has mainly become obsolete. The reason for this is that most circuits yield to the industrial temperature range. It is less expensive to support less part types. Each discrete part number requires a separate test program, separate inventorying, and other supporting documentation. The exception to this rule is for parts

Ordering guide

| Model | Temperature range | Package description | Package option |
|---|---|---|---|
| AD5570ARS | −40°C to +85°C | 16-Lead SSOP | RS-16 |
| AD5570ARS-REEL | −40°C to +85°C | 16-Lead SSOP | RS-16 |
| AD5570ARS-REEL7 | −40°C to +85°C | 16-Lead SSOP | RS-16 |
| AD5570BRS | −40°C to +85°C | 16-Lead SSOP | RS-16 |
| AD5570BRS-REEL | −40°C to +85°C | 16-Lead SSOP | RS-16 |
| AD5570BRS-REEL7 | −40°C to +85°C | 16-Lead SSOP | RS-16 |
| AD5570WRS | −40°C to +125°C | 16-Lead SSOP | RS-16 |
| AD5570WRS-REEL | −40°C to +125°C | 16-Lead SSOP | RS-16 |
| AD5570WRS-REEL7 | −40°C to +125°C | 16-Lead SSOP | RS-16 |
| AD5570YRS | −40°C to +125°C | 16-Lead SSOP | RS-16 |
| AD5570YRS-REEL | −40°C to +125°C | 16-Lead SSOP | RS-16 |
| AD5570YRS-REEL7 | −40°C to +125°C | 16-Lead SSOP | RS-16 |
| Eval-AD5570EB | | Evaluation Board | |

**Figure 6-187: Ordering guides example 2 (for the AD5570)**

designed for a specific application which is, by definition, commercial. An example of this is consumer applications, such as audio. Wider temperature range for these parts offers no advantage.

The industrial temperature range can also mean different things. The standard industrial temperature range is −40°C to 85°C. A common variant on this is what is commonly called the automotive temperature range −55°C to 85°C. 0–100°C is also common.

The military temperature range is −55°C to 125°C.

The "brand" column of the ordering guide is the package marking for small packages. The markings that are customary on the DIP package will not physically fit on the much smaller surface mount packages. For instance, DIP packages would typically include the part number, date code (when the IC was "made," typically when it passes final test), and occasionally some other information. Obviously, the space available for marking on surface mount packages is very much limited. So the three character code is used instead.

## Pin Description

In the pin description, information on the pin function, including optional functionality for multi-purpose pins, is described. Often the descriptions are expanded upon in the main body of the data sheet (see Figures 6-180 through 6-182 and 6-188).

## Defining the Specifications

This section contains a brief description of the specifications. It is, in effect, a subset of the earlier section where we defined the converter specifications. The definitions will be more compact and only those specifications that apply to the particular converter will be defined.

Also in this area specialized specifications will be defined. An example of this might be differential gain and differential phase, which are definitions specific to the video industry.

## Equivalent Circuits

Driving the inputs of a converter, especially at high frequencies, is not a trivial task. Loading of an output pin can also be equally as challenging. Knowing the architecture of circuit connected to that pin may be of assistance in understanding how to interface that pin (see Figure 6-189).

**PIN CONFIGURATION**

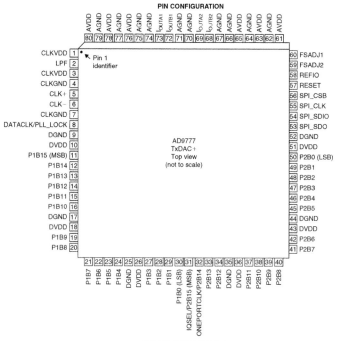

**PIN FUNCTION DESCRIPTIONS**

| Pin Number | Mnemonic | Description |
|---|---|---|
| 1, 3 | CLKVDD | Clock Supply Voltage |
| 2 | LPF | PLL Loop Filter |
| 4,7 | CLKGND | Clock Supply Common |
| 5 | CLK+ | Differential Clock Input |
| 6 | CLK− | Differential Clock Input |
| 8 | DATACLK/PLL_LOCK | With the PLL enabled, this pin indicates the state of the PLL. A read of a Logic "1" indicates the PLL is in the Locked state. Logic "0" indicates the PLL has not achieved lock. This pin may also be programmed to act as either an input or output (Address 02h, Bit 3) DATACLK signal running at the input data rate. |
| 9, 17, 25, 35, 44, 52 | DGND | Digital Common |
| 10, 18, 26, 36, 43, 51 | DVDD | Digital Supply Voltage |
| 11–16, 19–24, 27–30 | P1B15 (MSB) to P1B0 (LSB) | Port 1 Data Inputs |
| 31 | IQSEL/P2B15 (MSB) | In one port mode, IQSEL = 1 followed by a rising edge of the differential input clock will latch the data into the I channel input register. IQSEL = 0 will latch the data into the Q channel input register. In two port mode, this pin becomes the Port 2 MSB. |
| 32 | ONEPORTCLK/P2B14 | With the PLL disabled and the AD9777 in one port mode, this pin becomes a clock output that runs at twice the input data rate of the I and Q channels. This allows the AD9777 to accept and demux interleaved I and Q data to the I and Q input registers. |
| 33, 34, 37–42, 45–50 | P2B13 to P2B0 (LSB) | Port 2 Data Inputs. |
| 53 | SPI_SDO | In the case where SDIO is an input, SDO acts as an output. When SDIO becomes an output, SDO enters a High-Z state. This pin can also be used as an output for the data rate clock. For more information, see the Two Port Data Input Mode section. |
| 54 | SPI_SDIO | Bidirectional Data Pin. Data direction is controlled by Bit 7 of register Address 00h. The default setting for this bit is "0", which sets SDIO as an input. |
| 55 | SPI_CLK | Data input to the SPI port is registered on the rising edge of SPI_CLK. Data output on the SPI port is registered on the falling edge, |
| 56 | SPI_CSB | Chip select/SPI Data Synchronization. On momentary logic high, resets SPI port logic and initializes instruction cycle. |
| 57 | RESET | Logic "1" resets all of the SPI port regiisters, including Address 00h, to their default values. A software reset can also be done by writing a Logic "1" to SPI Register 00h, Bit 5. However, the software reset has no effect on the bits in Address 00h. |
| 58 | REFIO | Reference Output, 1.2 V Nominal |
| 59 | FSADJ2 | Full-Scale Current Adjust, Q Channel |
| 60 | FSADJ1 | Full-Scale Current Adjust, I Channel |
| 61, 63, 65, 76, 78, 80 | AVDD | Analog Supply Voltage |
| 62, 64, 66, 67, 70, 71, 74, 75, 77, 79 | AGND | Analog Common |
| 69, 68 | $I_{OUTA2}$, $I_{OUTB2}$ | Differential DAC Current Outputs, Q Channel |
| 73, 72 | $I_{OUTA1}$, $I_{OUTB1}$ | Differential DAC Current Outputs, I Channel |

Figure 6-188: Typical pin description

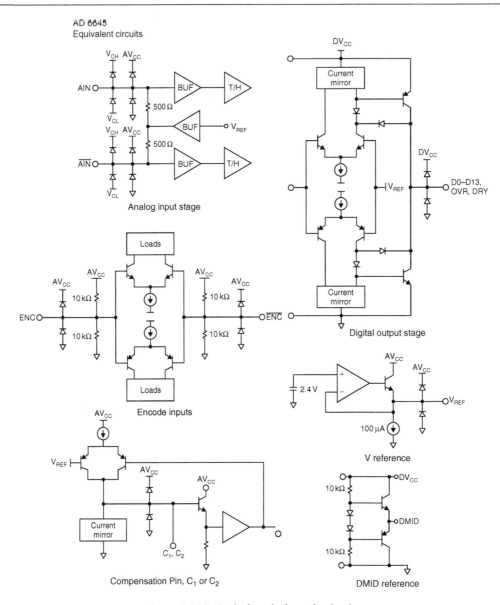

**Figure 6-189: Typical equivalent pin circuits**

Things that might be of interest for an input pin is the input impedance, so that the source impedance can be matched, for instance, and the DC level biasing the pin. Normally this bias is at half the supply voltage (assuming single-supply operation), but this is not always the case.

## The Graphs

Many parameters vary over the operational range of the converter. An example is the variation of SFDR with frequency (see Figures 6-190 to 6-193). So to completely specify the SFDR of a part there would be a specification at particular input frequency, which typically would appear in the specification table, and graphs showing variation with input frequency, sampling rate, and level. The information presented in the graphs is not uniform from vendor to vendor or even from part to part from the same manufacturer. Higher performance parts tend to be more completely specified. For the most part the graphs should be considered typical values.

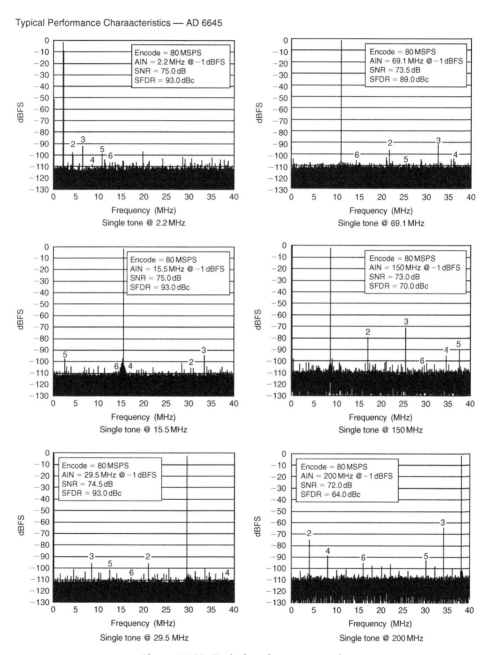

**Figure 6-190: Typical performance graphs**

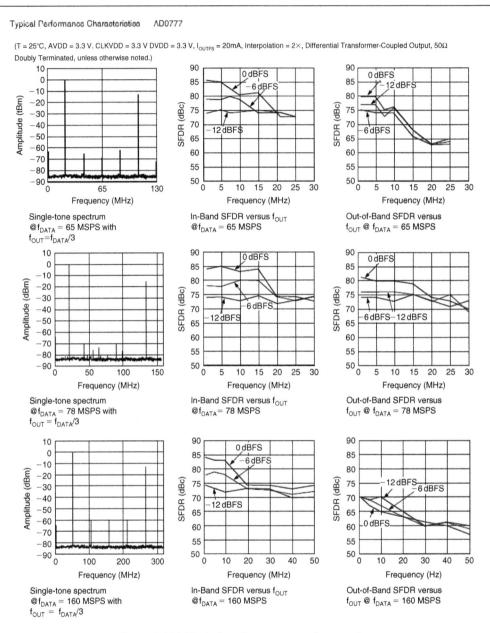

Figure 6-191: Typical performance graphs page 2

## The Main Body

The main body of the data sheet contains detailed information on operations and applications of the converter. Early on at Analog Devices, it was determined that just giving someone an amplifier and letting them go off on their own to try to build whatever it is that they want to build was not the best approach. Therefore, Analog Devices includes application information with the data sheet.

Typical Performance Characteristics AD7678

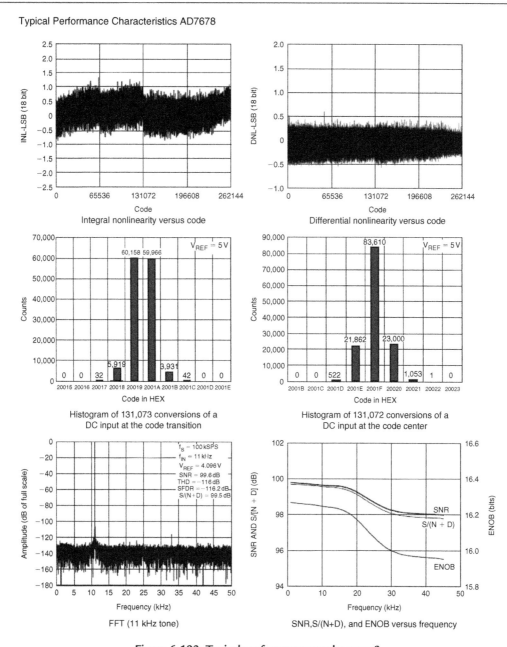

**Figure 6-192: Typical performance graphs page 3**

## Circuit Description

Typically the first part of the main body of the data sheet is the circuit description. Since the topology of the converter can determine the applicability of a particular converter in a particular design, understanding the internal operation of the converter can be very helpful. This is especially true when an understanding of the input structure of a converter may be helpful in designing the driver circuit (Figure 6-194).

Typical Performance Characteristics AD5570

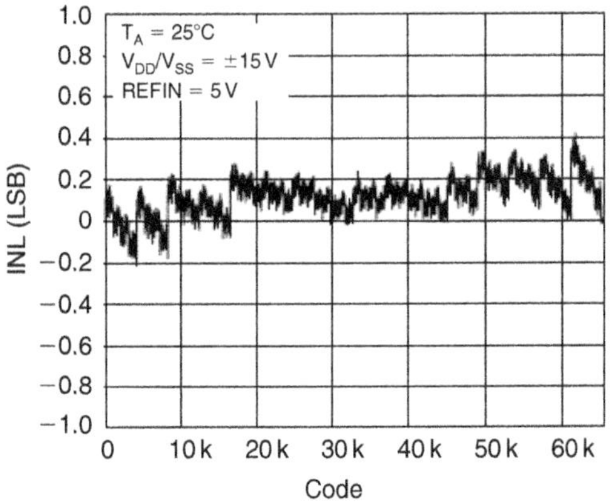

Integral nonlinearity versus code, $V_{DD}/V_{ss} = \pm 15\,V$

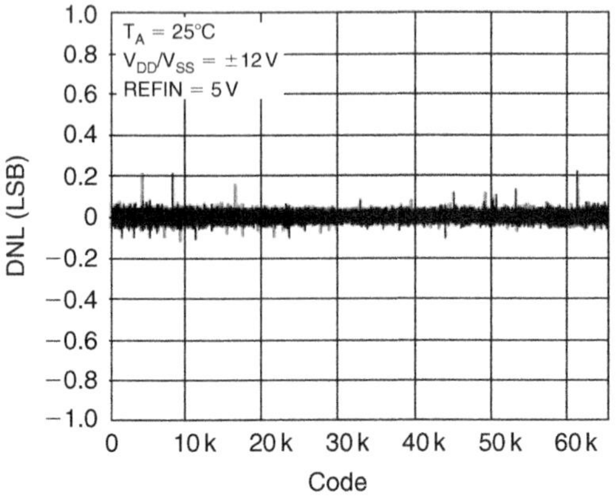

Differential nonlinearity versus code, $V_{DD}/V_{ss} = \pm 12\,V$

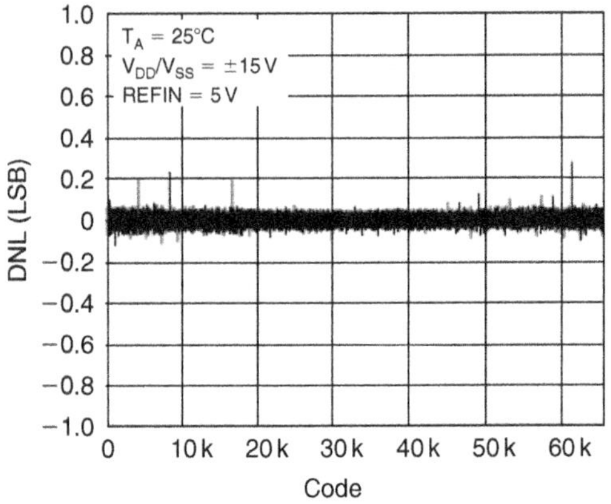

Differential nonlinearity versus code, $V_{DD}/V_{ss} = \pm 15\,V$

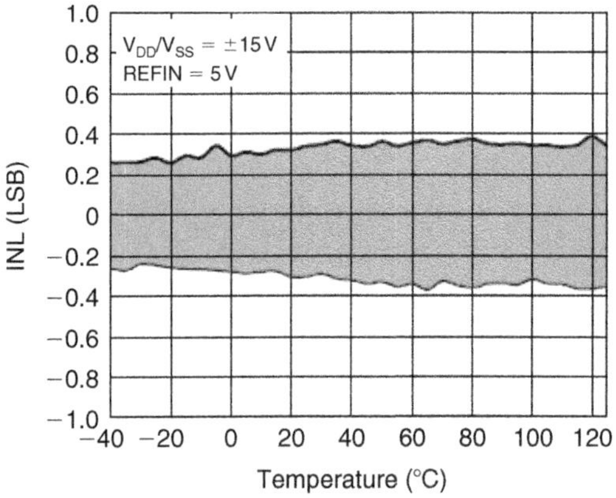

Integral nonlinearity versus Temperature, $\pm 15\,V$ supplies

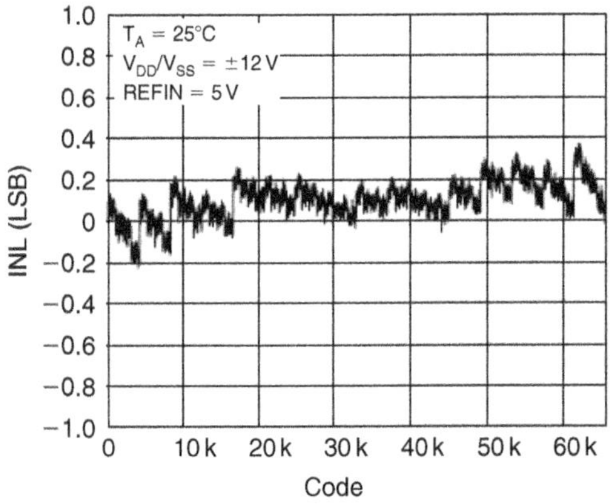

Integral nonlinearity versus code, $V_{DD}/V_{ss} = \pm 12\,V$

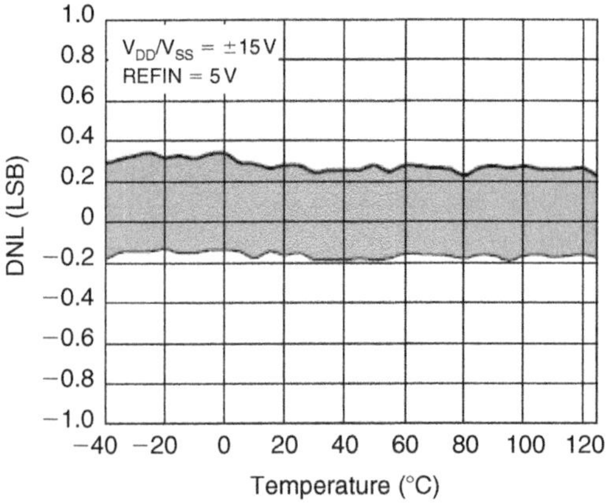

Differential nonlinearity versus temperature, $\pm 15\,V$ supplies

**Figure 6-193: Typical performance graphs page 4**

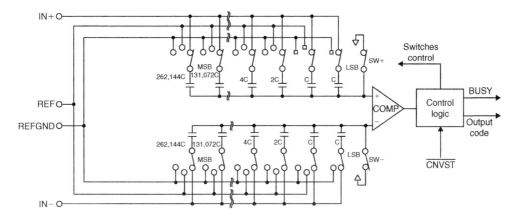

Figure 6-194: Typical circuit diagram

Many converters, such as the AD9777 and the AD7730 that we are using as examples in this section, are really more than just converters. They are more exactly subsystems, containing both converters and support circuitry. The operation of all of the subsections of these circuits is described.

As we said previously, driving an ADC input is not trivial at high speeds. Understanding the input configuration is essential. The same is true for the data outputs. On the DAC side the interface issues are reversed (data in/signal out), but just as important.

## Interface

To use a converter we have to get the data into or out of it. There are basically two different ways to accomplish this, parallel and serial.

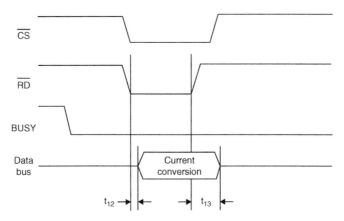

Figure 6-195: Typical parallel interface timing diagram

The parallel interface is relatively straightforward (see Figure 6-195). The only timing issues that may have to be considered are setup and hold times. Obviously, with the advent of low supply voltages, signal levels of the digital interface need to be observed.

In the case of the AD7678, which is an 18-bit converter, there may be an issue interfacing with a 16-bit (or 8 bit) microprocessor bus. The output register logic is flexible enough to allow the 18-bit word to interface to these narrower data busses (see Figure 6-196).

503

| Mode | Mode1 | Mode0 | D0/OB/$\overline{2C}$ | D1/A0 | D2/A1 | D[3] | D[4:9] | D[10:11] | D[12:15] | D[16:17] | Description |
|---|---|---|---|---|---|---|---|---|---|---|---|
| 0 | 0 | 0 | R[0] | R[1] | R[2] | R[3] | R[4:9] | R[10:11] | R[12:15] | R[16:17] | 18-Bit Parallel |
| 1 | 0 | 1 | OB/$\overline{2C}$ | A0:0 | R[2] | R[3] | R[4:9] | R[10:11] | R[12:15] | R[16:17] | 16-Bit High Word |
| 1 | 0 | 1 | OB/$\overline{2C}$ | A0:1 | R[0] | R[1] | All Zeros | | | | 16-Bit Low Word |
| 2 | 1 | 0 | OB/$\overline{2C}$ | A0:0 | A1:0 | All Hi-Z | | R[10:11] | R[12:15] | R[16:17] | 8-Bit HIGH Byte |
| 2 | 1 | 0 | OB/$\overline{2C}$ | A0:0 | A1:1 | All Hi-Z | | R[2:3] | R[4:7] | R[8:9] | 8-Bit MID Byte |
| 2 | 1 | 0 | OB/$\overline{2C}$ | A0:1 | A1:0 | All Hi-Z | | R[0:1] | All Zeros | | 8-Bit LOW Byte |
| 2 | 1 | 0 | OB/$\overline{2C}$ | A0:1 | A1:1 | All Hi-Z | | All Zeros | | R[0:1] | 8-Bit LOW Byte |
| 3 | 1 | 1 | OB/$\overline{2C}$ | All Hi-Z | | | Serial Interface | | | | Serial Interface |

**Table 7. Data bus interface definitions**

R[0:17] is the 18-bit ADC value stored in its output register.

**Figure 6-196: Data bus interface example**

The case of serial interface is typically a bit more complicated (see Figure 6-197). Many times the serial interface conforms to a certain interface standard. You will see that many of the serial interface converters conform to the SPI®, QSPI®, Microwire, or I²C® standards.

For the serial interface the converter could act as a master or a slave. The differentiation is determined by who generates the timing clock. The master typically generates the clock.

The width of the serial clock is variable. The data can be MSB first or LSB first. The time slots that line up with the data bits must be defined. Since there are multiple possibilities, each must be defined.

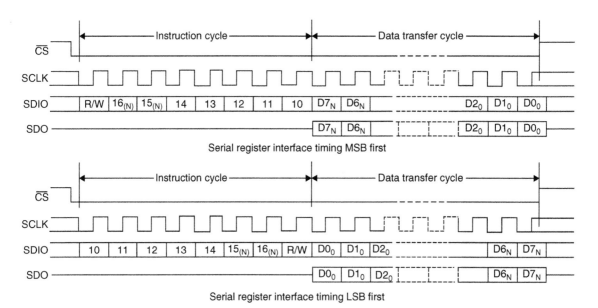

**Figure 6-197: Typical DAC serial timing diagram**

## Communications Register (RS2-RS0 = 0, 0, 0)

The Communications Register is an 8-bit write-only register. All communications to the part must start with a write operation to the Communications Register. The data written to the Communications Register determines whether the next operation is a read or write operation, the type of read operation, and to which register this operation takes place. For single-shot read or write operations, once the subsequent read or write operation to the selected register is complete, the interface returns to where it expects a write operation to the Communications Register. This is the default state of the interface, and on power-up or after a RESET, the AD7730 is in this default state waiting for a write operation to the Communications Register. In situations where the interface sequence is lost, a write operation of at least 32 serial clock cycles with DIN high, returns the AD7730 to this default state by resetting the part. Table VI outlines the bit designations for the Communications Register. CR0 through CR7 indicate the bit location, CR denoting the bits are in the Communications Register. CR7 denotes the first bit of the data stream.

### Table VI. Communications Register

| CR7 | CR6 | CR5 | CR4 | CR3 | CR2 | CR1 | CR0 |
|-----|------|------|------|------|------|------|------|
| $\overline{WEN}$ | ZERO | RW1 | RW0 | ZERO | RS2 | RS1 | RS0 |

| Bit location | Bit mnemonic | Description |
|--------------|--------------|-------------|
| CR7 | $\overline{WEN}$ | Write Enable Bit. A 0 must be written to this bit so the write operation to the Communications Register actually takes place. If a 1 is written to this bit, the part will not clock on to subsequent bits in the register. It will stay at this bit location until a 0 is written to this bit. Once a 0 is written to the WEN bit, the next seven bits will be loaded to the Communications Register. |
| CR6 | ZERO | A zero **must** be written to this bit to ensure correct operation of the AD7730. |
| CR5, CR4 | RW1, RW0 | Read/Write Mode Bits. These two bits determine the nature of the subsequent read/write operation. Table VII outlines the four options. |

### Table VII. Read/write mode

| RWI | RW0 | Read/write mode |
|-----|-----|-----------------|
| 0 | 0 | Single Write to Specified Register |
| 0 | 1 | Single Read of Specified Register |
| 1 | 0 | Start Continuous Read of Specified Register |
| 1 | 1 | Stop Continuous Read Mode |

With 0, 0 written to these two bits, the next operation is a write operation to the register specified by bits RS2, RS1, RS0. Once the subsequent write operation to the specified register has been completed, the part returns to where it is expecting a write operation to the Communications Register. With 0, 1 written to these two bits, the next operation is a read operation of the register specified by bits RS2, RS1, RS0. Once the subsequent read operation to the specified register has been completed, the part returns to where it is expecting a write operation to the Communications Register.

Writing 1,0 to these bits, sets the part into a mode of continuous reads from the register specified by bits RS2, RS1, RS0. The most likely registers with which the user will want to use this function are the Data Register and the Status Register. Subsequent operations to the part will consist of read operations to the specified register without any intermediate writes to the Communications Register. This means that once the next read operation to the specified register has taken place, the part will be in a mode where it is expecting another read from that specified register. The part will remain in this continuous read mode until 30 Hex has been written to the Communications Register.

When 1.1 is written to these bits (and 0 written to bits CR3 through CR0), the continuous read mode is stopped and the part returns to where it is expecting a write operation to the Communications Register. Note, the part continues to look at the DIN line on each SCLK edge during continuous read mode to determine when to stop the continuous read mode. Therefore, the user must be careful not to inadvertently exit the continuous read mode or reset the AD7730 by writing a series of 1s to the part. The easiest way to avoid this is to place a logic 0 on the DIN line while the part is in continuous read mode. Once the part is in continuous read mode, the user should ensure that an integer multiple of 8 serial clocks should have taken place before attempting to take the part out of continuous read mode.

**Figure 6-198: Typical register description (partial)**

## Register Description

Many converters have multiple operational modes. Some have on board circuitry, such as a multiplexer or PGA that must be configured. This requires writing to control registers (see Figure 6-198). Each bit in each word must be defined.

## Application Circuits

Often some typical application circuits (see Figures 6-199 and 6-200) are provided to assist in applying the converters.

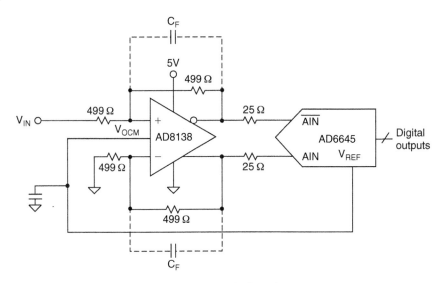

**Figure 6-199: AD6645 typical application circuit**

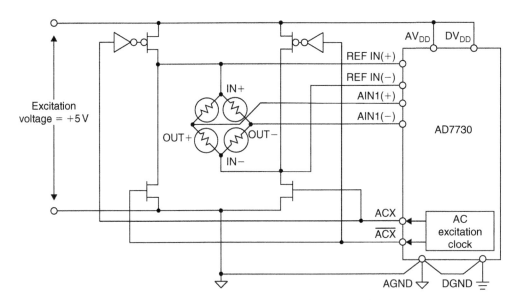

**Figure 6-200: AD7730 typical application circuit**

When looking at the applications circuits, note that the recommended support part numbers, while still valid, may no longer be the best choices. This is because newer parts may have been released since the data sheet was written. Always look to see if a newer part may be better.

## Evaluation Boards

The only way to be sure that your design works is to actually build it. But as has been mentioned several times, the layout of a PCB is as critical as any other part of the design. To that end, manufacturers often make evaluation boards available. This is an advantage for the design engineer, since it relieves him of the responsibility of developing and manufacturing an evaluation board. It also allows him to test a portion of the design before committing to the prototype run.

But the evaluation board serves the manufacturer as well. Since the manufacturer controls the design of the board, he can insure that the evaluation system shows off the part in it best light. The manufacturer will ensure that the performance of the board will not limit the performance of the part. This means a more fair evaluation, since many variables are removed.

The schematic and board layouts are typically presented in the data sheet. Often the Gerber files for the evaluation board are available as well from the manufacturer. A word of warning is in order though. Just cutting and pasting the Gerber files into your design is not enough to ensure optimum performance. Integrating the evaluation board section into the rest of the system is important as well. For instance, what if there is more than one converter in the system? The grounding scheme of the evaluation board, which worked in the instance of the one converter evaluation system, may be inadequate for larger systems.

Evaluation boards are typically part of larger systems for evaluation of a converter. Typically software is included to interface to the part. This software typically runs on a PC and includes a human interface. Evaluation systems are covered in more detail in the chapter on design aids.

## Summary

Not all data sheets for converters (or any other classification of part, for that matter) are the same, not from different manufacturers or even from the same manufacturer. But there are some features which are more or less standard. Knowing what to look for and where to look for it can make the daunting task of part selection a bit easier and possibly more exact.

# Choosing a Data Converter

Often the choice of the data converter is the cornerstone of the entire design. As we have seen in the previous sections, a converter can have many specifications. Now that we have gone over what those specifications mean and how to read a data sheet we are ready to proceed to the next step. How, then, do you determine which converter best suits your needs?

## Determine the Parameters

The most obvious parameters that we may need to specify the converter are the resolution and sample rate. Remember that the resolution of converter and the accuracy may not be the same. Quite often it is really the accuracy that is required.

For ADCs when we think of sample rate, we generally mean the maximum frequency. However, when the sample rate is reduced, the hold time requirements of the hold capacitors in the SHA section of the ADC increase proportionally. This can lead to errors if the sample rate is slow enough for the droop rate of the SHA to allow the sampled voltage to decay till it is out of the error band before the next sample period. While the droop rate of ADCs with internal SHAs is typically not specified, a minimum sample rate will be. This effect is dependent on the architecture of the ADC. Successive approximation ADCs rarely have this problem but pipelined architectures often do.

How the ADC is to be used may also effect the part selection. Pipelined ADCs and ΣΔ converters typically do not have a control signal for starting the conversion. They are designed to convert continuously. This makes them a bit more difficult to use in applications in which the sample must be synchronous. This would include multiplexed applications and those where the sample is to be triggered by an external stimulus. Flash or successive approximation type converters are probably a better choice in these types of applications.

We stated in the specifications section that there are two ways to specify the converter, AC specifications and DC specifications. In general AC specifications tend to be important with continuous sampling, higher speed. DC specifications tend to be more important with single conversion or multiplexed applications, which tend to be lower speed.

What is the frequency range of the input signal? For high frequency applications, is the input frequency band in the first Nyquist zone, or is undersampling to be employed?

*Another point*: Nyquist says that the input frequency can be up to half of the sample rate (for baseband sampling), but the antialiasing filter complexity increases sharply as the upper end of the input frequency band approaches the Nyquist frequency (Fs/2). By using oversampling, moving the sample rate out so that the input frequency band is proportionally smaller, system cost and complexity can be reduced.

The analog considerations for interfacing DACs are typically much less involved. In general, the decisions are whether the DAC should be current out or voltage out. If current out, the DAC will typically dictate the use of a current-to-voltage (I/V) converter. One possible exception is whether the DAC is multiplying, in which case you need to specify the input signal

On the digital side, the primary consideration is whether the data bus is parallel or serial. With the proliferation of low voltage circuits, the voltage level of the interface also needs to be defined. In many cases the data

# High Speed Convertors

April 2004                    THE ANALOG DEVICES SOLUTIONS BULLETIN

## IN THIS ISSUE

### Next-Generation, Dual, High Speed ADCs

Whether you're designing a next-generation wireless communications receiver or a low power data acquisition subsystem, your choice of an A/D convertor solution can be a key element in meeting end system requirements for performance, power, size, and cost.

Analog Devices has developed the next-generation family of dual, high speed ADCs, meeting the most stringent design requirements Ranging from 10 bits to 14 bits, and from 20 MSPS to 65 MSPS (up ro 120 MSPS for 10 bits), this pin compatible family allows for flexibility in design depending on the ADC signal chain requirements, while assuring that performance and power have been optimized.

This dual family builds off the feature-rich AD9238, 12-bit, 20, 40, and 5 MSPS ADCs that includes optimized power consumption, IF sampling capability, and flexible output interface configurations—all in a very space-efficient $9 \times 9$ LFCSP. The AD9216 is the 10-bit companion device that supports speeds from 65 MSPS to 120 MSPS. It is suitable for direct conversion applications, such as in broadband wireless and satellite communications. Extending the family to 14 bits is the AD9248, offered in three speed grades of 20, 40, and 65 MSPS, respectively. The AD9248 gives system designers a low cost convertor alternative to today's wide-ranging choice of receivers.

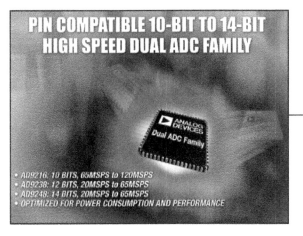

| Part number | Resolution (Bits) | Sample rate (MSPS) | SNR (dB @ 39 MHz) | SFDR (dBc @ 39 MHz) | Power per channel (mW) | Price per channel ($U.S.) |
|---|---|---|---|---|---|---|
| AD9216 | $10 \times 2$ | 65/80/105/120 | 58.0 | 75.0 | 90 | 5.49 |
| AD9238[1] | $12 \times 2$ | 20/40/65 | 70.0 | 85.0 | 90 | 6.57 |
| AD9248[2] | $14 \times 2$ | 20/40/65 | 73.5 | 85.0 | 90 | 14.69 |

[1]Low speed grade.
[2]Also available in LQFP-64.

Visit our website for samples, data sheets, and additional product information.

www.analog.com/bulletins/convertor

**Figure 6-201: Typical solutions bulletin front page**

output level is the same as the power supply, but some converters have a separate power pin which sets the voltage level of the digital interface. While the parallel interface is fairly simple, there are some added questions concerning the serial interface. Does it need to support a standard such as SPI®, I²C®, or LVDS?

If the resolution of the converter is not the same as the data bus width (interfacing a 12-bit converter to an 8 bit parallel bus for instance) the converter will require multiple read/write cycles. Similarly, in a serial interface you may have to specify right or left justified data.

In some converters control words must be written to the converter as well.

As always, what physical environment the converter must operate in is a concern. What is the temperature that the system must operate in? Is there a size limitation? What power supplies are available? Also, high speed converters tend to be relatively high power dissipation devices. Thermal considerations must also be considered.

Part of this process is determining the values for the various parameters. In doing this you should determine an optimum value and an acceptable range. For example, you may have a target value of full 16 bits for the accuracy, but you may be able to live with 2 LSBs of DNL and by loosening this spec, a better overall fit could be made. The temperature range over which the circuit will be required to operate will affect this as well. The physical size of the package and the cost, as always, should be considered. It is good practice to allow a little margin on the specs, if possible, so that aging effects, etc., do not cause the circuit to go out of spec.

## Prioritizing the Parameters

As can be seen from the discussion above, there can be a number of considerations involved in selecting a part. Typically, however, there are one or two that are more important than the rest. It is always a good idea not to overspecify a part. The more specifications that have to be met, the harder it will be to meet all of them.

## Selecting the Part

The last step is to finally select the part. The "brute force" method would be to gather data sheets and randomly start looking at the specs for each of the parts individually. This can quickly get out of hand. There are several tools that make the job much easier.

One such tool is a selection guide. These appear frequently in magazine ads and promotional mailers. The main difficulty with using these guides is that, in many instances, the lists are not all inclusive, but instead are usually focused on specific sub-groups such as new products, single supply, low power, etc. The narrow focus may cause you to miss some otherwise acceptable options. An example of a selection guide, in this case called a solutions bulletin, is given in Figure 6-201.

An alternative is the parametric search engine. Here you enter the relevant parameters for your design. The search will then search the database of parts and it will come up with acceptable alternatives.

# CHAPTER 7
# *Data Converter Support Circuits*

# Voltage References

Reference circuits and linear regulators actually have much in common. In fact, the latter could be functionally described as a reference circuit, but with greater current (or power) output. Accordingly, almost all of the specifications of the two circuit types have great commonality (even though the performance of references is usually tighter with regard to drift, accuracy, etc.). In many cases today the support circuitry is included in the converter package. This is advantageous to the designer since it simplifies the design process and guarantees performance of the system.

## Precision Voltage References

Voltage references have a major impact on the performance and accuracy of analog systems. A $\pm 5\,mV$ tolerance on a 5 V reference corresponds to $\pm 0.1\%$ absolute accuracy which is only 10-bit accuracy. For a 12-bit system, choosing a reference that has a $\pm 1\,mV$ tolerance may be far more cost effective than performing manual calibration, while both high initial accuracy and calibration will be necessary in a system making absolute 16-bit measurements. Note that many systems make *relative* measurements rather than absolute ones, and in such cases the absolute accuracy of the reference is not as important, although noise and short-term stability may be.

Temperature drift or drift due to aging may be an even greater problem than absolute accuracy. The initial error can always be trimmed, but compensating for drift is difficult. Where possible, references should be chosen for temperature coefficient and aging characteristics which preserve adequate accuracy over the operating temperature range and expected lifetime of the system.

Noise in voltage references is often overlooked, but it can be very important in system design. Noise is an instantaneous change in the reference voltage. It is generally specified on data sheets, but system designers frequently ignore the specification and assume that voltage references do not contribute to system noise.

There are two dynamic issues that must be considered with voltage references: their behavior at start-up, and their behavior with transient loads. With regard to the first, always bear in mind that voltage references *do not power-up instantly* (this is true for references inside ADCs and DACs as well as discrete designs). Thus it is rarely possible to turn on an analog-to-digital converters (ADC) and reference, whether internal or external, make a reading, and turn off again within a few microseconds, however attractive such a procedure might be in terms of energy saving.

Regarding the second point, a given reference IC may or may not be well suited for pulse-loading conditions, dependent on the specific architecture. Many references use low power, and therefore low bandwidth, output buffer amplifiers. This makes for poor behavior under fast transient loads, which may degrade the performance of fast ADCs (especially successive approximation and flash ADCs). Suitable decoupling can ease the problem (but some references oscillate with capacitive loads), or an additional external broadband buffer amplifier may be used to drive the node where the transients occur.

## Types of Voltage References

In terms of the functionality of their circuit connection, standard reference ICs are often only available in *series* or *three-terminal* form ($V_{IN}$, Common, $V_{OUT}$), and also in positive polarity only. The series types have the potential advantages of lower and more stable quiescent current, standard pre-trimmed output voltages, and relatively high output current without accuracy loss. *Shunt* or *two-terminal* (i.e., diode-like) references are more flexible regarding operating polarity, but they are also more restrictive as to loading. They can in fact eat up excessive power with widely varying resistor-fed voltage inputs. Also, they sometimes come in non-standard voltages. All of these various factors tend to govern when one functional type is preferred over the other.

Some simple diode-based references are shown in Figure 7-1. In the first of these, a current driven forward-biased diode (or diode-connected transistor) produces a voltage, $V_f = V_{REF}$. While the junction drop is somewhat decoupled from the raw supply, it has numerous deficiencies as a reference. Among them are a strong TC of about $-0.3\%/°C$, some sensitivity to loading, and a rather inflexible output voltage; it is only available in 600 mV jumps.

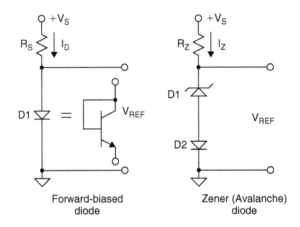

**Figure 7-1: Simple diode reference circuits**

By contrast, these most simple references (as well as all other shunt-type regulators) have a basic advantage, which is the fact that the polarity is readily reversible by flipping connections and reversing the drive current. However, a basic limitation of all shunt regulators is that load current must always be less (usually appreciably less) than the driving current, $I_D$.

In the second circuit of Figure 7-1, a zener or avalanche diode is used, and an appreciably higher output voltage realized. While true *zener* breakdown occurs below 5 V, *avalanche* breakdown occurs at higher voltages and has a positive temperature coefficient. Note that diode reverse breakdown is referred to almost universally today as *zener*, even though it is usually avalanche breakdown. With a D1 breakdown voltage in the 5–8 V range, the net positive TC is such that it equals the negative TC of forward-biased diode D2, yielding a net TC of 100 ppm/°C or less with proper bias current. Combinations of such carefully chosen diodes formed the basis of the early single package "temperature-compensated zener" references, such as the 1N821–1N829 series.

The temperature-compensated zener reference is limited in terms of initial accuracy, since the best TC combinations fall at odd voltages, such as the 1N829's 6.2 V. And, the scheme is also limited for loading, since for best TC the diode current must be carefully controlled. Unlike a fundamentally lower voltage ($<2V$) reference, zener diode-based references must of necessity be driven from voltage sources appreciably higher than 6V levels, so this precludes operation of zener references from 5V system supplies. References based on low TC zener (avalanche) diodes also tend to be noisy, due to the basic noise of the breakdown mechanism. This has been improved greatly with *monolithic* zener types, as is described further below.

## Bandgap References

The development of low voltage (<5V) references based on the bandgap voltage of silicon led to the introduction of various ICs which could be operated on low voltage supplies with good TC performance. The first of these was the LM109 (Reference 1), and a basic bandgap reference cell is shown in Figure 7-2.

This circuit is also called a "$\Delta V_{BE}$" reference because the differing current densities between matched transistors Q1–Q2 produce a $\Delta V_{BE}$ across $R_3$. It works by summing the $V_{BE}$ of Q3 with the amplified $\Delta V_{BE}$ of Q1–Q2, developed across $R_2$. The $\Delta V_{BE}$ and $V_{BE}$ components have opposite polarity TCs; $\Delta V_{BE}$ is proportional to absolute temperature (PTAT), while $V_{BE}$ is complementary to absolute temperature (CTAT). The summed output is $V_R$, and when it is equal to 1.205V (silicon bandgap voltage), the TC is a minimum.

The bandgap reference technique is attractive in IC designs because of several reasons; among these are the relative simplicity, and the avoidance of zeners and their noise. However, very important in these days of ever decreasing system supplies is the fundamental fact that bandgap devices operate at low voltages, i.e., <5V. Not only are they used for standalone IC references, but they are also used within the designs of many other linear ICs such as analog-to-digital converter (ADCs) and digital-to-analog converter (DACs).

However, the basic designs of Figure 7-2 suffer from load and current drive sensitivity, plus the fact that the output needs accurate scaling to more useful levels, i.e., 2.5V, 5V, etc. The load drive issue is best addressed with the use of a buffer amplifier, which also provides convenient voltage scaling to standard levels.

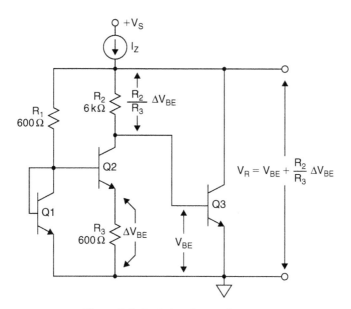

**Figure 7-2: Basic bandgap reference**

An improved three-terminal bandgap reference, the AD580 (introduced in 1974) is shown in Figure 7-3. Popularly called the "Brokaw Cell" (see References 2 and 3), this circuit provides on-chip output buffering, which allows good drive capability and standard output voltage scaling. The AD580 was the first precision bandgap-based IC reference, and variants of the topology have influenced further generations of both

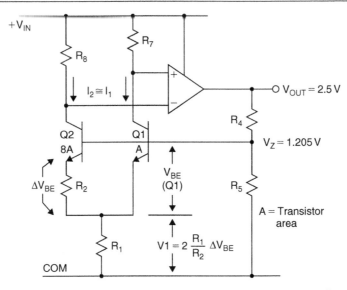

**Figure 7-3: AD580 precision bandgap reference uses Brokaw Cell (1974)**

industry standard references such as the REF01, REF02, and REF03 series, and more recent ADI bandgap parts such as the REF19x series, the AD680, AD780, the AD1582–AD1585 series, the ADR38x series, the ADR39x series, and recent SC-70 and SOT-23 offerings of improved versions of the REF01, REF02, and REF03 (designated ADR01, ADR02, and ADR03).

The AD580 has two 8:1 emitter-scaled transistors Q1–Q2 operating at identical collector currents (and thus 1/8 current densities), by virtue of equal load resistors and a closed loop around the buffer op amp. Due to the resultant smaller $V_{BE}$ of the 8 × area Q2, $R_2$ in series with Q2 drops the $\Delta V_{BE}$ voltage, while $R_1$ (due to the current relationships) drops a PTAT voltage V1:

$$V1 = 2 \times \frac{R_1}{R_2} \times \Delta V_{BE} \qquad (7\text{-}1)$$

The bandgap cell reference voltage $V_Z$ appears at the base of Q1, and is the sum of $V_{BE}$ (Q1) and V1, or 1.205V, the bandgap voltage:

$$V_Z = V_{BE(Q1)} + V1 \qquad (7\text{-}2)$$

$$= V_{BE(Q1)} + 2 \times \frac{R_1}{R_2} \times \Delta V_{BE} \qquad (7\text{-}3)$$

$$= V_{BE(Q1)} + 2 \times \frac{R_1}{R_2} \times \frac{kT}{q} \times \ln \frac{J1}{J2} \qquad (7\text{-}4)$$

$$= V_{BE(Q1)} + 2 \times \frac{R_1}{R_2} \times \frac{kT}{q} \times \ln 8 \tag{7-5}$$

$$= 1.205\,V$$

Note that J1 = current density in Q1, J2 = current density in Q2, and J1/J2 = 8.

However, because of the presence of the $R_4/R_5$ (laser-trimmed) thin-film divider and the op amp, the actual voltage appearing at $V_{OUT}$ can be scaled higher, in the AD580 case 2.5 V. Following this general principle, $V_{OUT}$ can be raised to other practical levels, such as e.g., in the AD584, with taps for precise 2.5, 5, 7.5, and 10 V operations. The AD580 provides up to 10 mA output current while operating from supplies between 4.5 and 30 V. It is available in tolerances as low as 0.4%, with TCs as low as 10 ppm/°C.

Many of the recent developments in bandgap references have focused on smaller package size and cost reduction, to address system needs for smaller, more power efficient and less costly reference ICs. Among these are several recent bandgap-based IC references.

The AD1580 (introduced in 1996) is a shunt mode IC reference which is functionally quite similar to the classic shunt IC reference, the AD589 (introduced in 1980) mentioned above. A key difference is the fact that the AD1580 uses a newer, small geometry process, enabling its availability within the tiny SOT-23 package. The very small size of this package allows use in a wide variety of space limited applications, and the low operating current lends itself to portable battery powered uses. The AD1580 circuit is shown in simplified form in Figure 7-4.

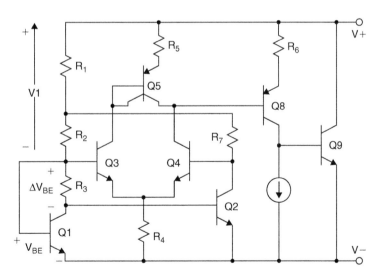

**Figure 7-4: AD1580 1.2 V shunt-type bandgap reference has tiny size in SOT-23 footprint**

In this circuit, like transistors Q1 and Q2 form the bandgap core, and are operated at a current ratio of 5 times, determined by the ratio of $R_7$ to $R_2$. An op amp is formed by the differential pair Q3–Q4, current mirror Q5, and driver/output stage Q8–Q9. In closed-loop equilibrium, this amplifier maintains the bottom ends of $R_2$–$R_7$ at the same potential.

As a result of the closed-loop control described, a basic $\Delta V_{BE}$ voltage is dropped across $R_3$, and a scaled PTAT voltage also appears as V1, which is effectively in series with $V_{BE}$. The nominal bandgap reference voltage of 1.225 V is then the sum of Q1's $V_{BE}$ and V1. The AD1580 is designed to operate at currents as low as 50 μA, also handling maximum currents as high as 10 mA. It is available in grades with voltage tolerances of $\pm1$ or $\pm10$ mV, and with corresponding TCs of 50 or 100 ppm/°C.

The circuit diagram for the series, shown in Figure 7-5, may be recognized as a variant of the basic Brokaw bandgap cell, as described under Figure 7-3. In this case Q1–Q2 form the core, and the overall loop operates to produce the stable reference voltage $V_{BG}$ at the base of Q1. A notable difference here is that the op amp's output stage is designed with push–pull common-emitter stages. This has the effect of requiring an output capacitor for stability, but it also provides the IC with relatively low dropout operation.

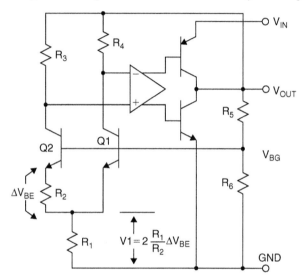

**Figure 7-5: AD1582–AD1585 2.5–5 V series type bandgap references**

The low dropout feature means essentially that $V_{IN}$ can be lowered to as close as several hundred millivolts above the $V_{OUT}$ level without disturbing operation. The push–pull operation also means that this device series can actually both sink and source currents at the output, as opposed to the classic reference operation of sourcing current (only). For the various output voltage ratings, the divider $R_5$–$R_6$ is adjusted for the respective levels.

The AD1582 series is designed to operate with quiescent currents of only 65 μA (maximum), which allows good power efficiency when used in low power systems with varying voltage inputs. The rated output current for the series is 5 mA, and they are available in grades with voltage tolerances of $\pm0.1\%$ or $\pm1\%$ of $V_{OUT}$, with corresponding TCs of 50 or 100 ppm/°C.

Because of stability requirements, devices of the AD1582 series must be used with both an output and input bypass capacitor. Recommended worst-case values for these are shown in the hookup diagram of Figure 7-6. For the electrical values noted, it is likely that tantalum chip capacitors will be the smallest in size.

## Buried Zener References

In terms of the design approaches used within the reference core, the two most popular basic types of IC references consist of the bandgap and buried zener units. Bandgaps have been discussed, but zener-based references warrant some further discussion.

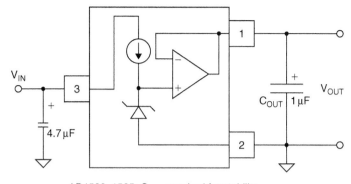

AD1582–1585: C$_{OUT}$ required for stability

ADR380, ADR381: C$_{OUT}$ recommended to absorb transients

**Figure 7-6: AD1582–AD1585 series connection diagram**

In an IC chip, surface operated diode junction breakdown is prone to crystal imperfections and other contamination, thus zener diodes formed at the surface are more noisy and less stable than are *buried* (or subsurface) ones (Figure 7-7). ADI zener-based IC references employ the much preferred buried zener. This improves substantially upon the noise and drift of surface-mode operated zeners (see Reference 4). Buried zener references offer very low temperature drift, down to the 1–2 ppm/°C (AD588 and AD586) (Figure 7-8), and the lowest noise as a percent of full-scale, i.e., $100 \, nV/\sqrt{Hz}$ or less. On the downside, the operating current of zener type references is usually relatively high, typically on the order of several milliamperes. The zener voltage is also relatively high, typically on the order of 5 V. This limits its application in low voltage circuits.

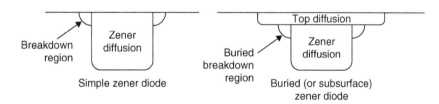

**Figure 7-7: Simple surface zener versus a buried zener**

An important general point arises when comparing noise performance of different references. The best way to do this is to compare the ratio of the noise (within a given bandwidth) to the DC output voltage. For example, a 10 V reference with a $100 \, nV/\sqrt{Hz}$ noise density is 6 dB more quiet in relative terms than is a 5 V reference with the same noise level.

## XFET® References

A third and relatively new category of IC reference core design is based on the properties of junction field effect (JFET) transistors. Somewhat analogous to the bandgap reference for bipolar transistors, the JFET-based reference operates a pair of JFET transistors with different pinchoff voltages, and amplifies the differential output to produce a stable reference voltage. One of the two JFETs uses an extra ion implantation, giving rise to the name XFET® (eXtra implantation junction Field Effect Transistor) for the reference core design.

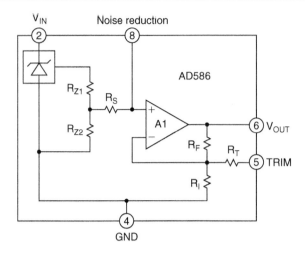

**Figure 7-8: Typical buried zener reference (AD586)**

The basic topology for the XFET reference circuit is shown in Figure 7-9. J1 and J2 are the two JFET transistors, which form the core of the reference. J1 and J2 are driven at the same current level from matched current sources, I1 and I2. To the right, J1 is the JFET with the extra implantation, which causes the difference in the J1–J2 pinchoff voltages to differ by 500 mV. With the pinchoff voltage of two such FETs purposely skewed, a differential voltage will appear between the gates for identical current drive conditions and equal source voltages. This voltage, $\Delta V_P$, is:

$$\Delta V_P = V_{P1} - V_{P2} \tag{7-6}$$

where $V_{P1}$ and $V_{P2}$ are the pinchoff voltages of FETs J1 and J2, respectively.

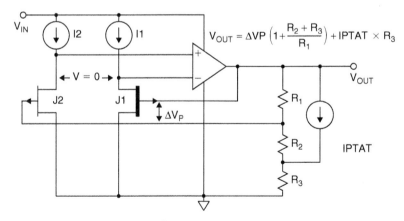

**Figure 7-9: XFET® reference simplified schematic**

Note that, within this circuit, the voltage $\Delta V_P$ exists between the *gates* of the two FETs. We also know that, with the overall feedback loop closed, the op amp axiom of zero input differential voltage will hold the sources of the two JFETs at same potential. These source voltages are applied as inputs to the op amp, the output of which drives feedback divider $R_1$–$R_3$. As this loop is configured, it stabilizes at an output voltage from the $R_1$–$R_2$ tap which does in fact produce the required $\Delta V_P$ between the J1–J2 gates. In essence, the op amp amplifies $\Delta V_P$ to produce $V_{OUT}$, where

$$V_{OUT} = \Delta V_P \left( 1 + \frac{R_2 + R_3}{R_1} \right) + (I_{PTAT})(R_3) \tag{7-7}$$

As can be noted, this expression includes the basic output scaling (leftmost portion of the right terms), plus a rightmost temperature dependent term including $I_{PTAT}$. The $I_{PTAT}$ portion of the expression compensates for a basic negative temperature coefficient of the XFET core, such that the overall net temperature drift of the reference is typically in a range of 3–8 ppm/°C.

The XFET architecture offers performance improvements over bandgap and buried zener references, particularly for systems where operating current is critical, yet drift and noise performance must still be excellent. XFET noise levels are lower than bandgap-based bipolar references operating at an equivalent current, the temperature drift is low and linear at 3–8 ppm/°C (allowing easier compensation when required), and the series has lower hysteresis than bandgaps. Thermal hysteresis is a low 50 ppm over a $-40$ to $+125$°C range, less that half that of a typical bandgap device. Finally, the long-term stability is excellent, typically only 50 ppm/1,000 hours.

Figure 7-10 summarizes the pro and con characteristics of the three reference architectures: bandgap, buried zener, and XFET.

| Bandgap | Buried zener | XFET® |
|---------|--------------|-------|
| <5 V supplies | >5 V supplies | <5 V supplies |
| High noise @ High power | Low noise @ High power | Low noise @ Low power |
| Fair drift and long term stability | Good drift and long term stability | Excellent drift and long term stability |
| Fair hysteresis | Fair hysteresis | Low hysteresis |

**Figure 7-10: Characteristics of reference architectures**

Though modern IC references come in a variety of styles, series operating, fixed output positive types do tend to dominate. They may or may not be low power, low noise, and/or low dropout, or available within a certain package. Of course, in a given application, any one of these differentiating factors can drive a choice, thus it behooves the designer to be aware of all the different devices available.

Figure 7-11 shows the typical schematic for a series type IC positive reference in an 8-pin package. (Note that "(x)" numbers refer to the standard pin for that function). There are several details which are important. Many references allow optional trimming by connecting an external trim circuit to drive the references' *trim* input pin (5). Some bandgap references also have a high impedance PTAT output ($V_{TEMP}$) for temperature sensing (pin 3). The intent here is that no appreciable current be drawn from this pin, but it can be useful for non-loading types of connections such as comparator inputs, to sense temperature thresholds, etc.

Some references have a pin labeled "noise reduction." This may cause some confusion. A capacitor connected to this pin will reduce the noise of the reference cell itself; this cell is typically followed by an internal buffer. The noise of this buffer will not be affected.

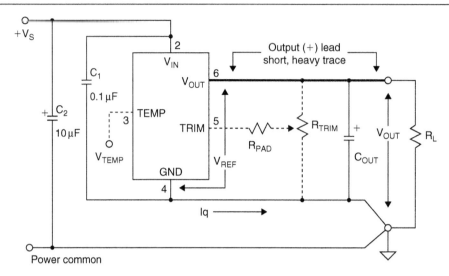

**Figure 7-11: Standard positive output three-terminal reference hookup (8-pin DIP pinout)**

All references should use decoupling capacitors on the input pin (2), but the amount of decoupling (if any) placed on the output (pin 6) depends on the stability of the reference's output op amp with capacitive load. Simply put, there is no hard-and-fast rule for capacitive loads here. For example, some three-terminal types *require* the output capacitor for stability (i.e., REF19x and AD1582–AD1585 series), while with others it is optional for performance improvement (AD780, REF43, ADR29x, ADR43x, AD38x, AD39x, ADR01, ADR02, ADR03). Even if the output capacitor is optional, it may still be required to supply the energy for transient load currents, as presented by some ADC reference input circuits. The safest rule then is that you should use the data sheet to verify what are the specific capacitive loading ground rules for the reference you intend to use, for the load conditions your circuit presents.

## Voltage Reference Specifications

### Tolerance

It is usually better to select a reference with the required value and accuracy and to avoid external trimming and scaling if possible. This allows the best TCs to be realized, as tight tolerances and low TCs usually go hand-in-hand. Tolerances as low as approximately 0.04% can be achieved with the AD586, AD780, REF195, and ADR43x series, while the AD588 is 0.01%. If and when trimming must be used, be sure to use the recommended trim network with no more range than is absolutely necessary. When/if additional external scaling is required, a precision op amp should be used, along with ratio-accurate, low TC tracking thin-film resistors.

### Drift

The XFET and buried zener reference families have the best long-term drift and TC performance. The XFET ADR43x series have TCs as low as 3 ppm/°C. TCs as low as 1–2 ppm/°C are available with the AD586 and AD588 buried zener references, and the AD780 bandgap reference is almost as good at 3 ppm/°C.

The XFET series achieves long-term drifts of 50 ppm/1,000 hours, while the buried zener types come in at 25 ppm/1,000 hours. Note that where a Figure is given for long-term drift, it is usually drift expressed in ppm/1,000 hours. There are 8,766 hours in a year, and many engineers multiply the 1,000 hour Figure by 8.77 to find the annual drift—this is not correct, and can in fact be quite pessimistic. Long-term drift in precision analog circuits is a "random walk" phenomenon and increases with the *square root* of the elapsed time (this supposes that drift is due to random micro-effects in the chip and not some overriding cause such as contamination). The 1 year Figure will therefore be about $\sqrt{8.766} \approx 3$ times the 1,000 hour Figure, and the 10 year value will be roughly 9 times the 1,000 hour value. In practice, things are a little better even than this, as devices tend to stabilize with age.

The accuracy of an ADC or DAC can be no better than that of its reference. Reference temperature drift affects full-scale accuracy as shown in Figure 7-12. This table shows system resolution and the TC required to maintain ½ LSB error over an operating temperature range of 100°C. For example, a TC of about 1 ppm/°C is required to maintain ½ LSB error at 12 bits. For smaller operating temperature ranges, the drift requirement will be less. The last three columns of the table show the voltage value of ½ LSB for popular full-scale ranges.

| Bits | Required drift (ppm/°C) | ½ LSB Weight (mV) 10, 5 and 2.5 V Full-scale ranges | | |
|------|------|------|------|------|
| | | 10 V | 5 V | 2.5 V |
| 8 | 19.53 | 19.53 | 9.77 | 4.88 |
| 9 | 9.77 | 9.77 | 4.88 | 2.44 |
| 10 | 4.88 | 4.88 | 2.44 | 1.22 |
| 11 | 2.44 | 2.44 | 1.22 | 0.61 |
| 12 | 1.22 | 1.22 | 0.61 | 0.31 |
| 13 | 0.61 | 0.61 | 0.31 | 0.15 |
| 14 | 0.31 | 0.31 | 0.15 | 0.08 |
| 15 | 0.15 | 0.15 | 0.08 | 0.04 |
| 16 | 0.08 | 0.08 | 0.04 | 0.02 |

**Figure 7-12: Reference temperature drift requirements for various system accuracies (½ LSB criteria, 100°C span)**

## Supply Range

IC reference supply voltages range from about 3V (or less) above rated output, to as high as 30V (or more) above rated output. Exceptions are devices designed for low dropout, such as the REF19X, AD1582–AD1585, ADR38X, ADR39X series. At low currents, the REF195 can deliver 5V with an input as low as 5.1V (100 mV dropout). Note that due to process limits, some references may have more restrictive maximum voltage input ranges, such as the AD1582–AD1585 series (12V), the ADR29x series (15V), and the ADR43x series (18V).

## Load Sensitivity

Load sensitivity (or output impedance) is usually specified in μV/mA of load current, or mΩ, or ppm/mA. While Figures of 70 ppm/mA or less are quite good (AD780, REF43, REF195, ADR29X, ADR43X), it should be noted that external wiring drops can produce comparable errors at high currents, without care in layout. Load current dependent errors are minimized with short, heavy conductors on the (+) output and on the ground return. For the highest precision, buffer amplifiers and Kelvin sensing circuits (AD588, AD688, ADR39x) are used to ensure accurate voltages at the load.

The output of a buffered reference is the output of an op amp, and therefore the source impedance is a function of frequency. Typical reference output impedance rises at 6 dB/octave from the DC value, and is nominally about $10\,\Omega$ at a few hundred kilohertz. This impedance can be lowered with an external capacitor, provided the op amp within the reference remains stable for such loading.

## Line Sensitivity

Line sensitivity (or regulation) is usually specified in $\mu$V/V, (or ppm/V) of input change, and is typically 25 ppm/V ($-92$ dB) in the REF43, REF195, AD680, AD780, ADR29X, ADR39X, and ADR43X. For DC and very low frequencies, such errors are easily masked by noise.

As with op amps, the line sensitivity (or power supply rejection) of references degrades with increasing frequency, typically 30–50 dB at a few hundred kilohertz. For this reason, the reference input should be highly decoupled (LF and HF). Line rejection can also be increased with a low dropout pre-regulator, such as one of the ADP3300 series parts.

## Noise

Reference noise is not always specified, and when it is, there is not total uniformity on how. For example, some devices are characterized for peak-to-peak noise in a 0.1–10 Hz bandwidth, while others are specified in terms of wideband RMS or peak-to-peak noise over a specified bandwidth. The most useful way to specify noise (as with op amps) is a plot of noise voltage spectral density ($nV/\sqrt{Hz}$) versus frequency.

Low noise references are important in high resolution systems to prevent loss of accuracy. Since white noise is statistical, a given noise density must be related to an equivalent peak-to-peak noise in the relevant bandwidth. Strictly speaking, the peak-to-peak noise in a Gaussian system is infinite (but its probability is infinitesimal). Conventionally, the figure of $6.6 \times$ RMS is used to define a practical peak value— statistically, this occurs less than 0.1% of the time. This peak-to-peak value should be less than $\frac{1}{2}$ LSB in order to maintain required accuracy. If peak-to-peak noise is assumed to be 6 times the RMS value, then for an N-bit system, reference voltage full-scale $V_{REF}$, reference noise bandwidth (BW), the required noise voltage spectral density $E_n$ ($V/\sqrt{Hz}$) is given by:

$$E_n \leq \frac{V_{REF}}{12 \times 2^N \times \sqrt{BW}} \tag{7-8}$$

For a 10 V, 12-bit, 100 kHz system, the noise requirement is a modest 643 $nV/\sqrt{Hz}$. Figure 7-13 shows that increasing resolution and/or lower full-scale references make noise requirements more stringent. The 100 kHz bandwidth assumption is somewhat arbitrary, but the user may reduce it with external filtering, thereby reducing the noise. Most good IC references have noise spectral densities around 100 $nV/\sqrt{Hz}$, so additional filtering is obviously required in most high resolution systems, especially those with low values of $V_{REF}$.

Some references, e.g., the AD587 buried zener type, have a pin designated as the *noise reduction pin* (see data sheet). This pin is connected to a high impedance node preceding the on-chip buffer amplifier. Thus an externally connected capacitor $C_N$ will form a lowpass filter with an internal resistor, to limit the effective noise bandwidth seen at the output. A 1 $\mu$F capacitor gives a 3 dB bandwidth of 40 Hz. Note that

| Bits | Noise density (nv/√Hz) for 10, 5 and 2.5 V Fulls-cale ranges | | |
|---|---|---|---|
| | 10 V | 5 V | 2.5 V |
| 12 | 643 | 322 | 161 |
| 13 | 322 | 161 | 80 |
| 14 | 161 | 80 | 40 |
| 15 | 80 | 40 | 20 |
| 16 | 40 | 20 | 10 |

**Figure 7-13: Reference noise requirements for various system accuracies (½ LSB/100 kHz criteria) scaled references**

this method of noise reduction is by no means universal, and other devices may implement noise reduction differently, if at all. Also note that it does not affect the noise of the buffer amplifier.

There are also general-purpose methods of noise reduction, which can be used to reduce the noise of any reference IC, at any standard voltage level. Note that the DC characteristics of the reference filter will affect the accuracy of the reference.

A useful approach when a non-standard reference voltage is required is to simply buffer and scale a basic low voltage reference diode. With this approach, a potential difficulty is getting an amplifier to work well at such low voltages as 3 V. A workhorse solution is the low power reference and scaling buffer shown in Figure 7-14. Here a low current 1.2 V two-terminal reference diode is used for D1, which can be either a 1.200 V ADR512, 1.235 V AD589, or the 1.225 V AD1580. Resistor $R_1$ sets the diode current in either case,

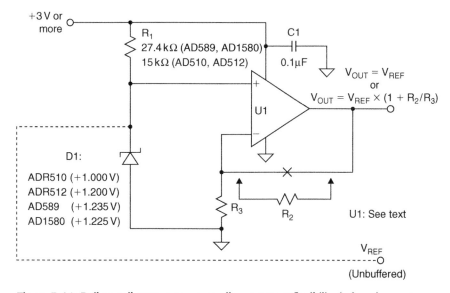

**Figure 7-14: Rail-to-rail output op amps allow greatest flexibility in low dropout references**

527

and is chosen for the diode minimum current requirement at a minimum supply of 2.7V. Obviously, loading on the unbuffered diode must be minimized at the $V_{REF}$ node.

The amplifier U1 both buffers and optionally scales up the nominal 1.0 or 1.2V reference, allowing much higher source/sink output currents. Of course, a higher op amp quiescent current is expended in doing this, but this is a basic tradeoff of the approach.

In Figure 7-14, without gain scaling resistors $R_2$–$R_3$, $V_{OUT}$ is simply equal to $V_{REF}$. With the use of the scaling resistors, $V_{OUT}$ can be set anywhere between a lower limit of $V_{REF}$, and an upper limit of the positive rail, due to the op amp's rail–rail output swing. Also, note that this buffered reference is inherently low dropout, allowing a +4.5V (or more) reference output on a +5V supply, for example. The general expression for $V_{OUT}$ is shown in the Figure, where $V_{REF}$ is the reference voltage.

## Voltage Reference Pulse Current Response

The response of references to dynamic loads is often a concern, especially in applications such as driving ADCs and DACs. Fast changes in load current invariably perturb the output, often outside the rated error band. For example, the reference input to a sigma–delta ADC may be the switched capacitor circuit shown in Figure 7-15. The dynamic load causes current spikes in the reference as the capacitor $C_{IN}$ is charged and discharged. As a result, noise may be induced on the ADC reference circuitry.

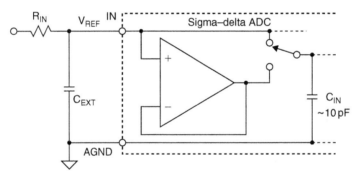

**Figure 7-15: Switched capacitor input of sigma–delta ADC presents a dynamic load to the voltage reference**

Although sigma–delta ADCs have an internal digital filter, transients on the reference input can still cause appreciable conversion errors. Thus it is important to maintain a low noise, transient free potential at the ADC's reference input. Be aware that if the reference source impedance is too high, dynamic loading can cause the reference input to shift by more than 5mV.

A bypass capacitor on the output of a reference may help it to cope with load transients, but many references are unstable with large capacitive loads. Therefore it is quite important to verify that the device chosen will satisfactorily drive the output capacitance required. In any case, the converter reference inputs should always be decoupled—with at least 0.1 µF, and with an additional 5–50 µF if there is any low frequency ripple on its supply.

Since some references misbehave with transient loads, either by oscillating or by losing accuracy for comparatively long periods, it is advisable to test the pulse response of voltage references which may encounter transient loads. A suitable circuit is shown in Figure 7-16. In a typical voltage reference, a step change of 1 mA produces the transients shown. Both the duration of the transient and the amplitude of the ringing *increase* when a 0.01 µF capacitor is connected to the reference output.

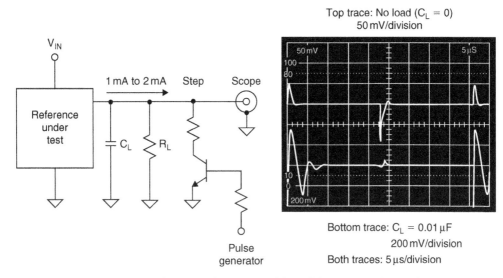

Figure 7-16: Make sure reference is stable with large capacitive loads

As noted above, reference bypass capacitors are useful when driving the reference inputs of successive approximation ADCs. Figure 7-17 illustrates reference voltage settling behavior immediately following the "Start Convert" command. A small capacitor (0.01 µF) does not provide sufficient charge storage to keep the reference voltage stable during conversion, and errors may result. As shown by the bottom trace, decoupling with a ≥1 µF capacitor maintains the reference stability during conversion.

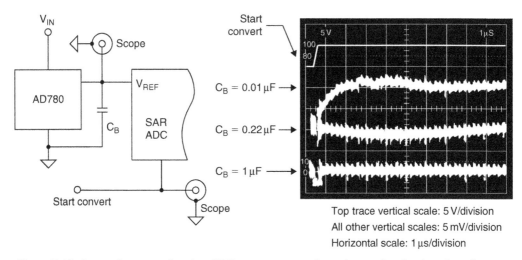

Figure 7-17: Successive approximation ADCs can present a dynamic transient load to the reference

Where voltage references are required to drive large capacitances, it is also critically important to realize that their turn-on time will be prolonged. Experiment may be needed to determine the delay before the reference output reaches full accuracy, but it will certainly be much longer than the time specified on the data sheet for the same reference in a low capacitance loaded state.

## Low Noise References for High Resolution Converters

High resolution converters (both sigma–delta and high speed ones) can benefit from recent improvements in IC references, such as lower noise and the ability to drive capacitive loads. Even though many data converters have internal references, the performance of these references is often compromised because of the limitations of the converter process. In such cases, using an external reference rather than the internal one often yields better overall performance. For example, the AD7710 series of 22-bit ADCs has a 2.5 V internal reference with a 0.1–10 Hz noise of 8.3 μV RMS (2,600 nV/$\sqrt{\text{Hz}}$ ), while the AD780 reference noise is only 0.67 μV RMS (200 nV/$\sqrt{\text{Hz}}$ ). The internal noise of the AD7710 series in this bandwidth is about 1.7 μV RMS. The use of the AD780 increases the effective resolution of the AD7710 from about 20.5 bits to 21.5 bits.

There is one possible but yet quite real problem when replacing the internal reference of a converter with a higher precision external one. The converter in question may have been trimmed during manufacture to deliver its specified performance with a relatively inaccurate internal reference. In such a case, using a more accurate external reference with the converter may actually introduce additional gain error! For example, the early AD574 had a guaranteed uncalibrated gain accuracy of 0.125% when using an internal 10 V reference (which itself had a specified accuracy of only ±1%). It is obvious that if such a device, having an internal reference which is at one end of the specified range, is used with an external reference of exactly 10 V, then its gain will be about 1% in error.

### References: Voltage References

1.  B. Widlar, "New Developments in IC Voltage Regulators," **IEEE Journal of Solid State Circuits**, Vol. SC-6, February, 1971.

2.  P. Brokaw, "A Simple Three-Terminal IC Bandgap Voltage Reference," **IEEE Journal of Solid State Circuits**, Vol. SC-9, December, 1974.

3.  P. Brokaw, "More About the AD580 Monolithic IC Voltage Regulator," **Analog Dialogue**, Vol. 9, No. 1, 1975.

4.  D. Sheingold, **Analog–Digital Conversion Handbook (Section 20.2)**, 3rd Edition, Prentice-Hall, Norwood, MA, 1986.

5.  W. Jung, "Build an Ultra-Low-Noise Voltage Reference," **Electronic Design Analog Applications Issue**, Vol. XX, 1993.

6.  W. Jung, "Getting the Most from IC Voltage References," **Analog Dialogue**, Vol. 28-1, 1994, pp. 13–21.

# Analog Switches and Multiplexers

## Introduction

Solid-state analog switches and multiplexers have become an essential component in the design of electronic systems which require the ability to control and select a specified transmission path for an analog signal. These devices are used in a wide variety of applications including multi-channel data acquisition systems, process control, instrumentation, video systems, etc.

Early complementary-MOS (CMOS) switches and multiplexers were typically designed to handle signal levels up to $\pm 10\,V$ while operating on $\pm 15\,V$ supplies. In 1979, Analog Devices introduced the popular ADG200 series of switches and multiplexers, and in 1988 the ADG201 series was introduced which were fabricated on a proprietary linear-compatible CMOS process ($L^2CMOS$). These devices allowed input signals to $\pm 15\,V$ when operating on $\pm 15\,V$ supplies.

A large number of switches and multiplexers were introduced in the 1980s and 1990s, with the trend toward lower on-resistance, faster switching, lower supply voltages, lower cost, lower power, and smaller surface-mount packages.

Today, analog switches and multiplexers are available in a wide variety of configurations, options, etc., to suit nearly all applications. On-resistances less than $0.5\,\Omega$, picoampere leakage currents, signal bandwidths greater than $1\,GHz$, and single $1.8\,V$ supply operation are now possible with modern CMOS technology.

Although CMOS is by far the most popular IC process today for switches and multiplexers, bipolar processes (with JFETs) and complementary bipolar processes (also with JFET capability) are often used for special applications such as video switching and multiplexing where the high performance characteristics required are not attainable with CMOS. Traditional CMOS switches and multiplexers suffer from several disadvantages at video frequencies. Their switching time is generally not fast enough, and they require external buffering in order to drive typical video loads. In addition, the small variation of the CMOS switch on-resistance with signal level ($R_{ON}$ modulation) can introduce unwanted distortion in differential gain and phase. Multiplexers based on complementary bipolar technology offer better solutions at video frequencies—with obvious power and cost increases above CMOS devices.

## CMOS Switch Basics

The ideal analog switch has no on-resistance, infinite off-impedance, and zero time delay, and can handle large signal and common-mode voltages. Real CMOS analog switches meet none of these criteria. It can be more correctly thought of as a variable resistor which changes from very high to very low resistance. But if we understand the limitations of analog switches, most of these limitations can be overcome.

CMOS switches have an excellent combination of attributes. In its most basic form, the MOSFET transistor is a voltage-controlled resistor. In the "on" state, its resistance can be less than $1\,\Omega$, while in the "off" state, the resistance increases to several hundreds of megaohms, with picoampere leakage currents. CMOS technology is compatible with logic circuitry and can be densely packed in an IC. Its fast switching characteristics are well controlled with minimum circuit parasitics.

MOSFET transistors are bilateral. That is, they can switch positive and negative voltages and conduct positive and negative currents with equal ease. A MOSFET transistor has a voltage-controlled resistance which varies nonlinearly with signal voltage as shown in Figure 7-18.

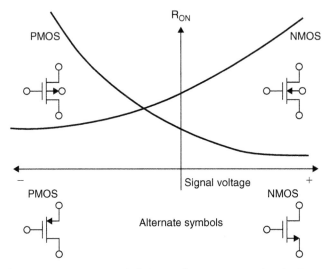

**Figure 7-18: MOSFET switch ON-resistance versus signal voltage**

CMOS yields good P-channel and N-channel MOSFETs. Connecting the PMOS and NMOS devices in parallel forms the basic bilateral CMOS switch of Figure 7-19. This combination reduces the on-resistance, and also produces a resistance which varies much less with signal voltage.

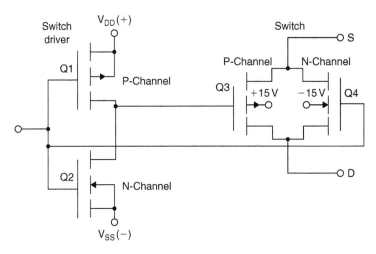

**Figure 7-19: Basic CMOS switch uses complementary pair to minimize $R_{ON}$ variation due to signal swings**

Figure 7-20 shows the on-resistance changing with channel voltage for both N-type and P-type devices. This nonlinear resistance can causes errors in DC accuracy as well as AC distortion. The bilateral CMOS switch solves this problem. On-resistance is minimized, and linearity is also improved. The bottom curve of Figure 7-20 shows the improved flatness of the on-resistance characteristic of the switch.

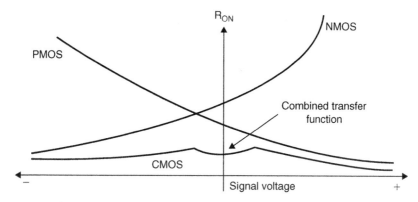

**Figure 7-20: CMOS switch ON-resistance versus signal voltage**

The ADG8xx series of CMOS switches are specifically designed for less than $0.5\,\Omega$ on-resistance and are fabricated on a sub-micron process. These devices can carry currents up to 400 mA, operate on a single 1.8–5.5 V supply, and are rated over an extended temperature range of $-40°C$ to $+125°C$. On-resistance over temperature and input signal level is shown in Figure 7-21.

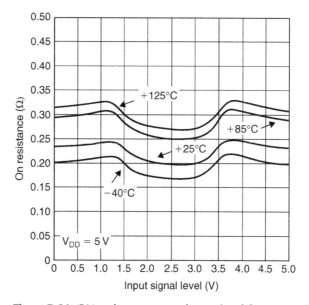

**Figure 7-21: ON-resistance versus input signal for ADG801/ADG802 CMOS switch, $V_{DD}$ = +5 V**

## Error Sources in the CMOS Switch

It is important to understand the error sources in an analog switch. Many affect AC and DC performance, while others only affect AC. Figure 7-22 shows the equivalent circuit of two adjacent CMOS switches. The model includes leakage currents and junction capacitances.

DC errors associated with a single CMOS switch in the on state are shown in Figure 7-23. When the switch is ON, DC performance is affected mainly by the switch on-resistance ($R_{ON}$) and leakage current ($I_{LKG}$). A

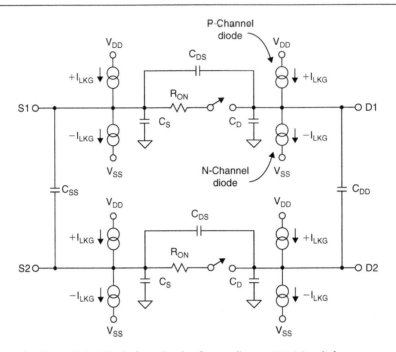

**Figure 7-22: Equivalent circuit of two adjacent CMOS switches**

resistive attenuator is created by the $R_G$–$R_{ON}$–$R_{LOAD}$ combination which produces a gain error. The leakage current, $I_{LKG}$, flows through the equivalent resistance of $R_{LOAD}$ in parallel with the sum of $R_G$ and $R_{ON}$. Not only can $R_{ON}$ cause gain errors—which can be calibrated using a system gain trim—but its variation with applied signal voltage ($R_{ON}$ modulation) can introduce distortion—for which there is no calibration. Low resistance circuits are more subject to errors due to $R_{ON}$, while high resistance circuits are affected by leakage currents. Figure 7-23 also gives equations that show how these parameters affect DC performance.

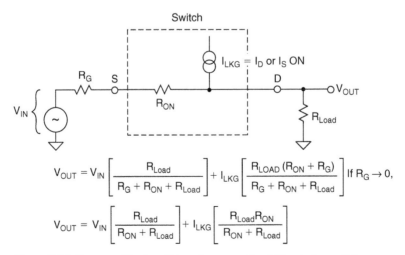

**Figure 7-23: Factors affecting DC performance for ON switch condition:**
$R_{ON}$, $R_{LOAD}$, and $I_{LKG}$

When the switch is OFF, leakage current can introduce errors as shown in Figure 7-24. The leakage current flowing through the load resistance develops a corresponding voltage error at the output.

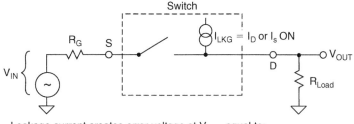

Leakage current creates error voltage at $V_{OUT}$ equal to:
$V_{OUT} = I_{LKG} \times R_{Load}$

**Figure 7-24: Factors affecting DC performance for OFF switch condition: $I_{LKG}$ and $R_{LOAD}$**

Figure 7-25 illustrates the parasitic components that affect the AC performance of CMOS switches. Additional external capacitances will further degrade performance. These capacitances affect feedthrough, crosstalk, and system bandwidth. $C_{DS}$ (drain-to-source capacitance), $C_D$ (drain-to-ground capacitance), and $C_{LOAD}$ all work in conjunction with $R_{ON}$ and $R_{LOAD}$ to form the overall transfer function.

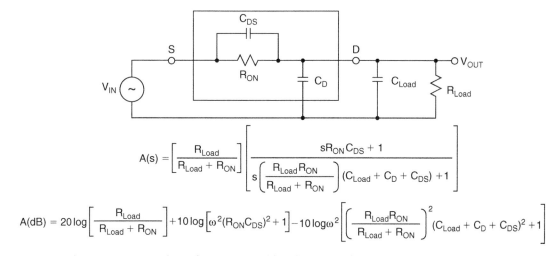

$$A(s) = \left[\frac{R_{Load}}{R_{Load} + R_{ON}}\right] \left[\frac{sR_{ON}C_{DS} + 1}{s\left(\frac{R_{Load}R_{ON}}{R_{Load} + R_{ON}}\right)(C_{Load} + C_D + C_{DS}) + 1}\right]$$

$$A(dB) = 20\log\left[\frac{R_{Load}}{R_{Load} + R_{ON}}\right] + 10\log\left[\omega^2(R_{ON}C_{DS})^2 + 1\right] - 10\log\omega^2\left[\left(\frac{R_{Load}R_{ON}}{R_{Load} + R_{ON}}\right)^2(C_{Load} + C_D + C_{DS})^2 + 1\right]$$

**Figure 7-25: Dynamic performance considerations: transfer accuracy versus frequency**

In the equivalent circuit, $C_{DS}$ creates a frequency zero in the numerator of the transfer function A(s). This zero usually occurs at high frequencies because the switch on-resistance is small. The bandwidth is also a function of the switch output capacitance in combination with $C_{DS}$ and the load capacitance. This frequency pole appears in the denominator of the equation.

The composite frequency domain transfer function may be rewritten as shown in Figure 7-26 which shows the overall Bode plot for the switch in the on state. In most cases, the pole breakpoint frequency occurs first because of the dominant effect of the output capacitance $C_D$. Thus, to maximize bandwidth, a switch should have low input and output capacitance and low on-resistance.

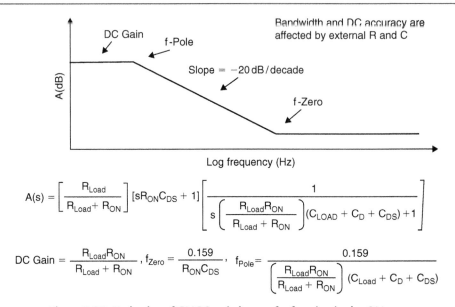

$$A(s) = \left[ \frac{R_{Load}}{R_{Load} + R_{ON}} \right] [sR_{ON}C_{DS} + 1] \left[ \frac{1}{s\left( \frac{R_{Load}R_{ON}}{R_{Load} + R_{ON}} \right)(C_{LOAD} + C_D + C_{DS}) + 1} \right]$$

$$DC\ Gain = \frac{R_{Load}R_{ON}}{R_{Load} + R_{ON}},\ f_{Zero} = \frac{0.159}{R_{ON}C_{DS}},\ f_{Pole} = \frac{0.159}{\left( \frac{R_{Load}R_{ON}}{R_{Load} + R_{ON}} \right)(C_{Load} + C_D + C_{DS})}$$

**Figure 7-26: Bode plot of CMOS switch transfer function in the ON state**

The series-pass capacitance, $C_{DS}$, not only creates a zero in the response in the ON state, it degrades the feedthrough performance of the switch during its OFF state. When the switch is off, $C_{DS}$ couples the input signal to the output load as shown in Figure 7-27.

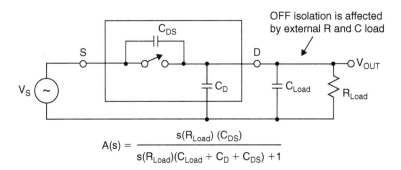

$$A(s) = \frac{s(R_{Load})(C_{DS})}{s(R_{Load})(C_{Load} + C_D + C_{DS}) + 1}$$

**Figure 7-27: Dynamic performance considerations: off-isolation**

Large values of $C_{DS}$ will produce large values of feedthrough, proportional to the input frequency. Figure 7-28 illustrates the drop in OFF-isolation as a function of frequency. The simplest way to maximize the OFF-isolation is to choose a switch that has as small a $C_{DS}$ as possible.

Figure 7-29 shows typical CMOS analog switch OFF-isolation as a function of frequency for the ADG708 eight-channel multiplexer. From DC to several kilohertz, the multiplexer has nearly 90-dB isolation. As the frequency increases, an increasing amount of signal reaches the output. However, even at 10 MHz, the switch shown still has nearly 60 dB of isolation.

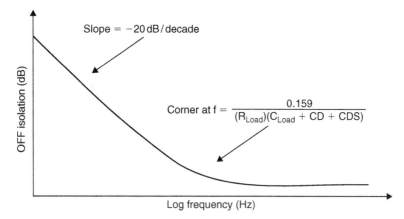

**Figure 7-28: Off-isolation versus frequency**

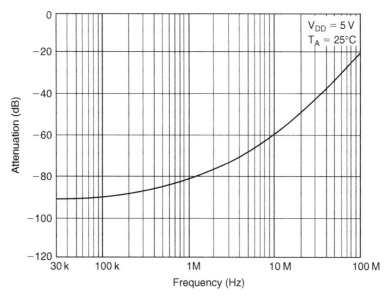

**Figure 7-29: Off-isolation versus frequency for ADG708 eight-channel multiplexer**

Another AC parameter that affects system performance is the charge injection that takes place during switching. Figure 7-30 shows the equivalent circuit of the charge injection mechanism.

When the switch control input is asserted, it causes the control circuit to apply a large voltage change (from $V_{DD}$ to $V_{SS}$, or vice versa) at the gate of the CMOS switch. This fast change in voltage injects a charge into the switch output through the gate-drain capacitance $C_Q$. The amount of charge coupled depends on the magnitude of the gate-drain capacitance.

The charge injection introduces a step change in output voltage when switching as shown in Figure 7-31. The change in output voltage, $\Delta V_{OUT}$, is a function of the amount of charge injected, $Q_{INJ}$ (which is in turn a function of the gate-drain capacitance, $C_Q$), and the load capacitance, $C_L$.

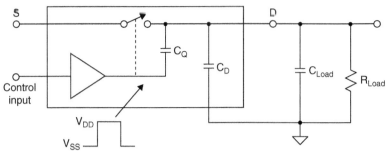

Step waveforms of $\pm(V_{DD} - V_{SS})$ are applied to $C_Q$, the gate capacitance of the output switches

**Figure 7-30: Dynamic performance considerations: charge injection model**

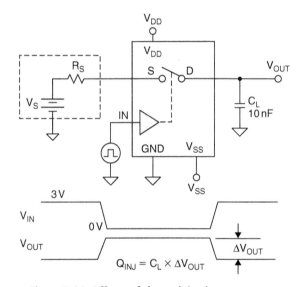

**Figure 7-31: Effects of charge injection on output**

Another problem caused by switch capacitance is the retained charge when switching channels. This charge can cause transients in the switch output, and Figure 7-32 illustrates the phenomenon.

Assume that initially S2 is closed and S1 open. $C_{S1}$ and $C_{S2}$ are charged to $-5\,V$. As S2 opens, the $-5\,V$ remains on $C_{S1}$ and $C_{S2}$, as S1 closes. Thus, the output of amplifier A sees a $-5\,V$ transient. The output will not stabilize until amplifier A's output fully discharges $C_{S1}$ and $C_{S2}$ and settles to $0\,V$. The scope photo in Figure 7-33 depicts this transient. The amplifier's transient load settling characteristics will therefore be an important consideration when choosing the right input buffer.

Crosstalk is related to the capacitances between two switches. This is modeled as the $C_{SS}$ capacitance shown in Figure 7-34.

Figure 7-35 shows typical crosstalk performance of the ADG708 eight-channel CMOS multiplexer.

Finally, the switch itself has a settling time that must be considered. Figure 7-36 shows the dynamic transfer function. The settling time can be calculated, because the response is a function of the switch and circuit

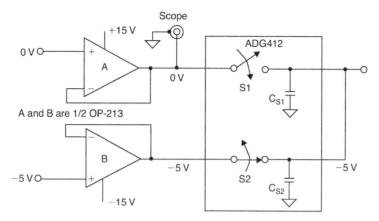

Figure 7-32: Charge coupling causes dynamic settling time transient when multiplexing signals

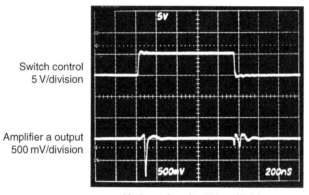

Horizontal scale: 200 ns /division

Figure 7-33: Output of amplifier shows dynamic settling time transient due to charge coupling

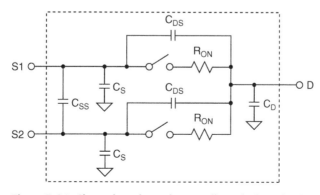

Figure 7-34: Channel-to-channel crosstalk equivalent circuit for adjacent switches

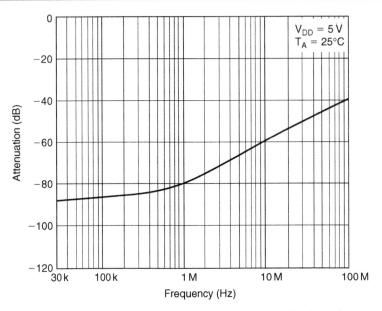

Figure 7-35: Crosstalk versus frequency for ADG708 eight-channel multiplexer

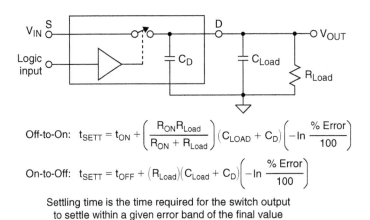

Off-to-On: $t_{SETT} = t_{ON} + \left( \dfrac{R_{ON}R_{Load}}{R_{ON} + R_{Load}} \right)\left( C_{LOAD} + C_D \right)\left( -\ln \dfrac{\% \text{ Error}}{100} \right)$

On-to-Off: $t_{SETT} = t_{OFF} + \left( R_{Load} \right)\left( C_{Load} + C_D \right)\left( -\ln \dfrac{\% \text{ Error}}{100} \right)$

Settling time is the time required for the switch output
to settle within a given error band of the final value

Figure 7-36: Multiplexer settling time

resistances and capacitances. One can assume that this is a single-pole system and calculate the number of time constants required to settle to the desired system accuracy as shown in Figure 7-37.

## Applying the Analog Switch

Switching time is an important consideration in applying analog switches, but switching time should not be confused with settling time. ON and OFF times are simply a measure of the propagation delay from the control input to the toggling of the switch, and are largely caused by time delays in the drive and level-shift circuits (see Figure 7-38). The $t_{ON}$ and $t_{OFF}$ values are generally measured from the 50% point of the control input leading edge to the 90% point of the output signal level.

| Resolution, # of bits | LSB (%Fs) | # of time constants |
|:---:|:---:|:---:|
| 6 | 1.563 | 4.16 |
| 8 | 0.391 | 5.55 |
| 10 | 0.0977 | 6.93 |
| 12 | 0.0244 | 8.32 |
| 14 | 0.0061 | 9.70 |
| 16 | 0.00153 | 11.09 |
| 18 | 0.00038 | 12.48 |
| 20 | 0.000095 | 13.86 |
| 22 | 0.000024 | 15.25 |

**Figure 7-37: Number of time constants required to settle to 1 LSB accuracy for a single-pole system**

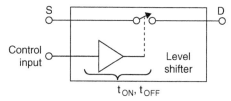

- $t_{ON}$ and $t_{OFF}$ should not be confused with settling time

- $t_{ON}$ and $t_{OFF}$ are simply a measure of the propagation delay from control input to operation of the analog switch. It is caused by time delays in the drive/level-shifter logic circuitry

- $t_{ON}$ and $t_{OFF}$ are measured from the 50% point of the control input to the 90% point of the output signal level

**Figure 7-38: Applying the analog switch: dynamic performance considerations**

We will next consider the issues involved in buffering a CMOS switch or multiplexer output using an op amp. When a CMOS multiplexer switches inputs to an inverting summing amplifier, it should be noted that the on-resistance, and its nonlinear change as a function of input voltage, will cause gain and distortion errors as shown in Figure 7-39. If the resistors are large, the switch leakage current may introduce error. Small resistors minimize leakage current error but increase the error due to the finite value of $R_{ON}$.

To minimize the effect of $R_{ON}$ change due to the change in input voltage, it is advisable to put the multiplexing switches at the op amp summing junction as shown in Figure 7-40. This ensures the switches are only modulated with about $\pm 100\,\text{mV}$ rather than the full $\pm 10\text{V}$—but a separate resistor is required for each input leg.

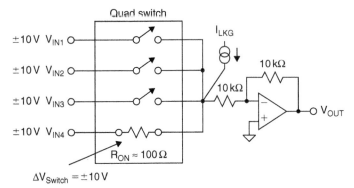

$\Delta V_{Switch} = \pm 10\,V$

- $\Delta R_{ON}$ caused by $\Delta V_{IN}$, degrades linearity of $V_{OUT}$ relative to $V_{IN}$
- $\Delta R_{ON}$ causes overall gain error in $V_{OUT}$ relative to $V_{IN}$

Figure 7-39: Applying the analog switch: unity-gain inverter with switched input

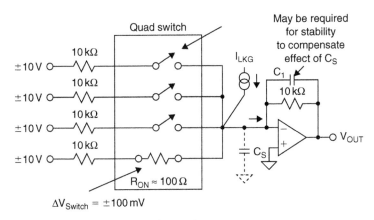

$\Delta V_{Switch} = \pm 100\,mV$

- Switch drives a virtual ground
- Switch sees only $\pm 100\,mV$, not $\pm 10\,V$, minimizes $\Delta R_{ON}$

Figure 7-40: Applying the analog switch: minimizing the influence of $\Delta R_{ON}$

It is important to know how much parasitic capacitance has been added to the summing junction as a result of adding a multiplexer, because any capacitance added to that node introduces phase shift to the amplifier closed-loop response. If the capacitance is too large, the amplifier may become unstable and oscillate. A small capacitance, $C_1$, across the feedback resistor may be required to stabilize the circuit.

The finite value of $R_{ON}$ can be a significant error source in the circuit shown in Figure 7-41. The gain-setting resistors should be at least 1,000 times larger than the switch on-resistance to guarantee 0.1% gain accuracy. Higher values yield greater accuracy but lower bandwidth and greater sensitivity to leakage and bias current.

A better method of compensating for $R_{ON}$ is to place one of the switches in series with the feedback resistor of the inverting amplifier as shown in Figure 7-42. It is a safe assumption that the multiple switches,

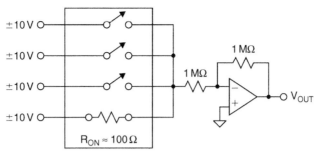

- $\Delta R_{ON}$ is small compared to 1 MΩ switch load
- Effect on transfer accuracy is minimized
- Bias current and leakage current effects are now very important
- Circuit bandwidth degrades

**Figure 7-41: Applying the analog switch: minimizing effects of $\Delta R_{ON}$ using large resistor values**

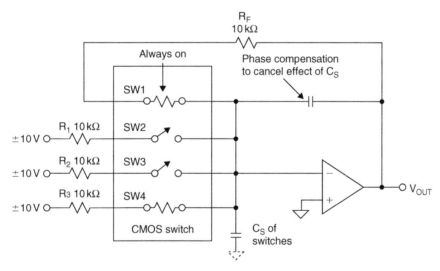

**Figure 7-42: Applying the analog switch: using "dummy" switch in feedback to minimize gain error due to $\Delta R_{ON}$**

fabricated on a single chip, are well matched in absolute characteristics and tracking over temperature. Therefore, the amplifier is closed-loop gain stable at unity gain, since the total feedforward and feedback resistors are matched.

The best multiplexer design drives the non-inverting input of the amplifier as shown in Figure 7-43. The high input impedance of the non-inverting input eliminates the errors due to $R_{ON}$.

CMOS switches and multiplexers are often used with op amps to make programmable gain amplifiers (PGAs). To understand $R_{ON}$'s effect on their performance, consider Figure 7-44, a poor PGA design. A non-inverting op amp has four different gain-set resistors, each grounded by a switch, with an $R_{ON}$ of 100–500 Ω. Even with $R_{ON}$ as low as 25 Ω, the gain of 16 error would be 2.4%, worse than 8-bit accuracy! $R_{ON}$ also changes over temperature, and switch–switch.

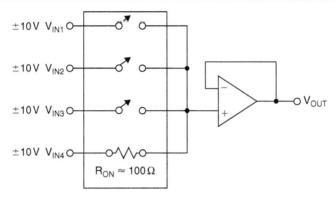

**Figure 7-43: Applying the analog switch: minimizing the influence of $\Delta R_{ON}$ using non-inverting configuration**

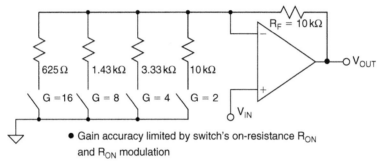

- Gain accuracy limited by switch's on-resistance $R_{ON}$ and $R_{ON}$ modulation
- $R_{ON}$ typically 1–500 $\Omega$ for CMOS or JFET switch
- For $R_{ON}$ = 25 $\Omega$, there is a 2.4% gain error for G = 16
- $R_{ON}$ drift over temperature limits accuracy
- Must use very low $R_{ON}$ switches

**Figure 7-44: A poorly designed PGA using CMOS switches**

To attempt "fixing" this design, the resistors might be increased, but noise and offset could then be a problem. The only way to ensure accuracy with this circuit is to use relays, with virtually no $R_{ON}$. Only then will the few milliohms of relay $R_{ON}$ be a small error vis-à-vis 625 $\Omega$.

It is much better to use a circuit insensitive to $R_{ON}$! In Figure 7-45, the switch is placed in series with the inverting input of an op amp. Since the op amp input impedance is very large, the switch $R_{ON}$ is now irrelevant, and gain is now determined solely by the external resistors. Note that $R_{ON}$ may add a small offset error if op amp bias current is high. If this is the case, it can readily be compensated with an equivalent resistance at $V_{IN}$.

## 1 GHZ CMOS switches

The ADG918/ADG919 are the first switches using a CMOS process to provide high isolation and low insertion loss up to and exceeding 1 GHz. The switches exhibit low insertion loss (0.8 dB) and high off-isolation (37 dB) when transmitting a 1 GHz signal. In high frequency applications with throughput power of +18 dBm or less at 25°C, they are a cost-effective alternative to gallium arsenide (GaAs) switches.

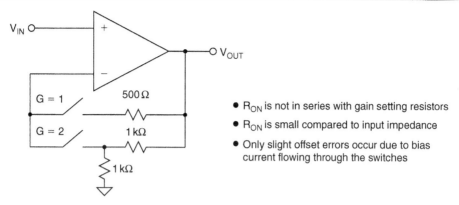

- $R_{ON}$ is not in series with gain setting resistors
- $R_{ON}$ is small compared to input impedance
- Only slight offset errors occur due to bias current flowing through the switches

**Figure 7-45: Alternate PGA configuration minimizes the effects of $R_{ON}$**

A block diagram of the devices is shown in Figure 7-46 along with isolation and loss versus frequency plots given in Figure 7-47.

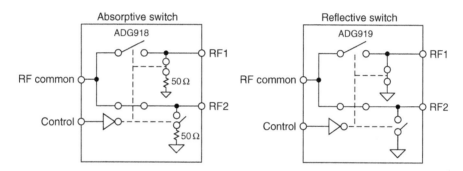

**Figure 7-46: 1-GHz CMOS 1.65–2.75 V 2:1 Mux/SPDT switches**

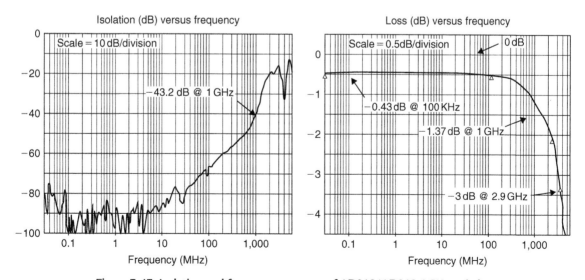

**Figure 7-47: Isolation and frequency response of AD918/AD919 1 GHz switch**

The ADG918 is an absorptive switch with 50 Ω terminated shunt legs that allow impedance matching with the application circuit, while the ADG919 is a reflective switch designed for use where the terminations are external to the chip. Both offer low power consumption (<1 μA), tiny packages (8-lead MSOP and 3 mm × 3 mm leadframe chip scale package), single-pin control voltage levels that are CMOS/LVTTL compatible, making the switches ideal for wireless applications and general-purpose RF switching.

## Video Switches and Multiplexers

In order to meet stringent specifications of bandwidth flatness, differential gain and phase, and 75 Ω drive capability, high speed complementary bipolar processes are more suitable than CMOS processes for video switches and multiplexers. Traditional CMOS switches and multiplexers suffer from several disadvantages at video frequencies. Their switching time (typically 50 ns or so) is not fast enough for today's applications, and they require external buffering in order to drive typical video loads. In addition, the small variation of the CMOS switch on-resistance with signal level ($R_{ON}$ *modulation*) introduces unwanted distortion in differential gain and phase. Multiplexers based on complementary bipolar technology offer a better solution at video frequencies. The tradeoffs, of course, are higher power and cost.

Functional block diagrams of the AD8170/8174/8180/8182 bipolar video multiplexers are shown in Figure 7-48. The AD8183/AD8185 video multiplexer is shown in Figure 7-49. These devices offer a high degree of flexibility and are ideally suited to video applications, with excellent differential gain and phase specifications. Switching time for all devices in the family is 10 ns to 0.1%. The AD8186/8187 are single-supply versions of the AD8183/8185.

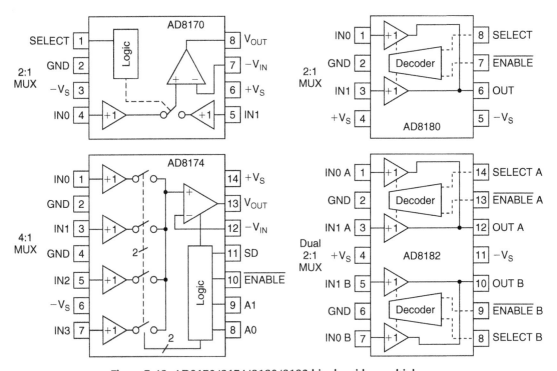

**Figure 7-48: AD8170/8174/8180/8182 bipolar video multiplexers**

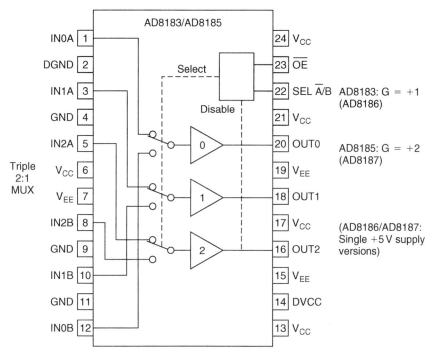

**Figure 7-49: AD8183/AD8185 video multiplexers**

The AD8170/8174 series of muxes include an on-chip current feedback op amp output buffer whose gain can be set externally. Off channel isolation and crosstalk are typically greater than 80 dB for the entire family.

Figure 7-50 shows an application circuit for three AD8170 2:1 muxes, where a single RGB monitor is switched between two RGB computer video sources.

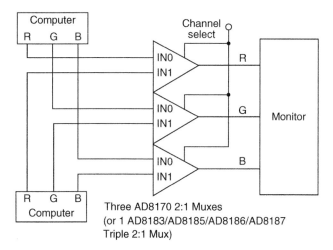

**Figure 7-50: Dual source RGB multiplexer using three 2:1 muxes**

In this setup, the overall effect is that of a three-pole, double-throw switch. The three video sources constitute the three poles, and either the upper or lower of the video sources constitute the two switch states. Note that the circuit can be simplified by using a single AD8183, AD8185, AD8186, or AD8187 triple dual input multiplexer.

The AD8174 or AD8184 4:1 mux is used in Figure 7-51, to allow a single high speed ADC to digitize the RGB outputs of a scanner.

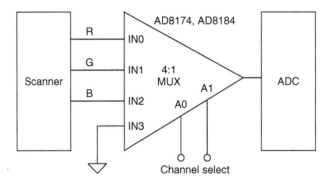

**Figure 7-51: Digitizing RGB signals with one ADC and a 4:1 mux**

The RGB video signals from the scanner are fed in sequence to the ADC, and digitized in sequence, making efficient use of the scanner data with one ADC.

## Video Crosspoint Switches

The AD8116 extends the multiplexer concepts to a fully integrated, 16 × 16 buffered video crosspoint switch matrix (Figure 7-52). A crosspoint switch allows any input to be connected to any output, or combination of outputs. The only limitation is that any output can have no more than one input connected to it.

The 3 dB bandwidth of the AD8116 is greater than 200 MHz, and the 0.1-dB gain flatness extends to 60 MHz. Channel switching time is less than 30 ns to 0.1%. Channel-to-channel crosstalk is −70 dB measured at 5 MHz. Differential gain and phase is 0.01% and 0.01° for a 150 Ω load. Total power dissipation is 900 mW on ±5 V.

The AD8116 includes output buffers that can be put into a high impedance state for paralleling crosspoint stages so that the off channels do not load the output bus. The channel switching is performed via a serial digital control that can accommodate "daisy chaining" of several devices. The AD8116 package is a 128-pin 14 mm × 14 mm LQFP.

Other members of the crosspoint switch family include the AD8108/AD9109 8 × 8 crosspoint switch, the AD8110/AD8111 260 MHz 16 × 8 buffered crosspoint switch, the AD8113 audio/video 60 MHz 16 × 16 crosspoint switch, and the AD8114/AD8115 low cost 225 MHz 16 × 16 crosspoint switch.

## Digital Crosspoint Switches

The AD8152 is a 3.2 Gbps 34 × 34 asynchronous digital crosspoint switch designed for high speed networking (see Figure 7-53). The device operates at data rates up to 3.2 Gbps per port, making it suitable for Sonet/SDH OC-48 with forward error correction (FEC). The AD8152 has digitally programmable current mode outputs that can drive a variety of termination schemes and impedances while maintaining the

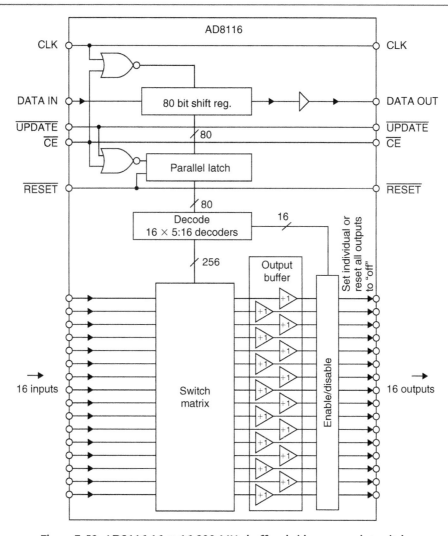

**Figure 7-52: AD8116 16 × 16 200-MHz buffered video crosspoint switch**

correct voltage level and minimizing power consumption. The part operates with a supply voltage as low as +2.5 V, with excellent input sensitivity. The control interface is compatible with LVTTL or CMOS/TTL.

As the lowest power solution of any comparable crosspoint switch, the AD8152 dissipates less than 2 W at 2.5 V supply with all I/Os active and does not require external heat sinks. The low jitter specification of less than 45 ps makes the AD8152 ideal for high speed networking systems. The AD8152's fully differential signal path reduces jitter and crosstalk while allowing the use of smaller single-ended voltage swings. It is offered in a 256-ball SBGA package that operates over the industrial temperature range of 0°C to +85°C.

## Parasitic Latch-up in CMOS Switches and Muxes

Because multiplexers are often at the front end of a data acquisition system, their inputs generally come from remote locations—hence, they are often subjected to overvoltage conditions. Although this topic is treated in more detail in Chapter 11 an understanding of the problem as it relates to CMOS devices is

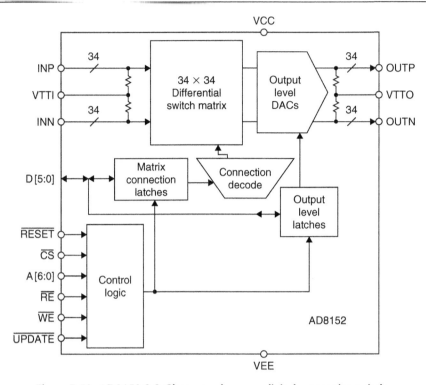

**Figure 7-53: AD8152 3.2-Gbps asynchronous digital crosspoint switch**

particularly important. Although this discussion centers on multiplexers, it is germane to nearly all types of CMOS parts.

Most CMOS analog switches are built using junction-isolated CMOS processes. A cross-sectional view of a single switch cell is shown in Figure 7-54. Parasitic SCR (silicon-controlled rectifier) latch-up can occur if the analog switch terminal has voltages more positive than $V_{DD}$ or more negative than $V_{SS}$. Even a transient situation, such as power-on with an input voltage present, can trigger a parasitic latch-up. If the conduction current is too great (several hundred milliamperes or more), it can damage the switch.

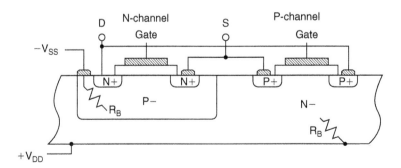

**Figure 7-54: Cross section of a junction-isolation CMOS switch**

The parasitic SCR mechanism is shown in Figure 7-55. SCR action takes place when either terminal of the switch (source or the drain) is either one diode drop more positive than $V_{DD}$ or one diode drop more

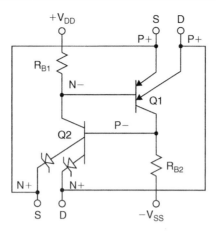

**Figure 7-55: Bipolar transistor equivalent circuit for CMOS switch shows parasitic SCR latch**

negative than $V_{SS}$. In the former case, the $V_{DD}$ terminal becomes the SCR gate input and provides the current to trigger SCR action. In the case where the voltage is more negative than $V_{SS}$, the $V_{SS}$ terminal becomes the SCR gate input and provides the gate current. In either case, high current will flow between the supplies. The amount of current depends on the collector resistances of the two transistors, which can be fairly small.

In general, to prevent the latch-up condition, the inputs to CMOS devices should never be allowed to be more than 0.3 V above the positive supply or 0.3 V below the negative supply. Note that this restriction also applies when the power supplies are off ($V_{DD} = V_{SS} = 0V$), and therefore devices can latch-up if power is applied to a part when signals are present on the inputs. Manufacturers of CMOS devices invariably place this restriction in the data sheet table of absolute maximum ratings. In addition, the input current under overvoltage conditions should be restricted to 5–30 mA, depending on the particular device.

In order to prevent this type of SCR latch-up, a series diode can be inserted into the $V_{DD}$ and $V_{SS}$ terminals as shown in Figure 7-56. The diodes block the SCR gate current. Normally the parasitic transistors Q1 and Q2 have low beta (usually less than 10) and require a comparatively large gate current to fire the SCR. The diodes limit the reverse gate current so that the SCR is not triggered.

If diode protection is used, the analog voltage range of the switch will be reduced by one $V_{BE}$ drop at each rail, and this can be inconvenient when using low supply voltages.

As noted, CMOS switches and multiplexers must also be protected from possible over-current by inserting a series resistor to limit the current to a safe level as shown in Figure 7-57, generally less than 5–30 mA. Because of the resistive attenuator formed by $R_{LOAD}$ and $R_{LIMIT}$, this method works only if the switch drives a relatively high impedance load.

A common method for input protection is shown in Figure 7-58 where Schottky diodes are connected from the input terminal to each supply voltage as shown. The diodes effectively prevent the inputs from exceeding the supply voltage by more than 0.3–0.4 V, thereby preventing latch-up conditions. In addition, if the input voltage exceeds the supply voltage, the input current flows through the external diodes to the supplies, not the device. Schottky diodes can easily handle 50–100 mA of transient current, therefore

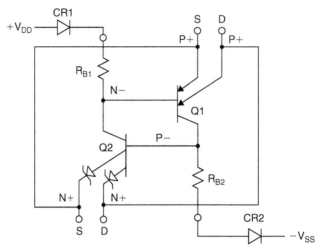

Diodes CR1 and CR2 block base current drive to Q1 and Q2
in the event of overvoltage at S or D.

**Figure 7-56: Diode protection scheme for CMOS switch**

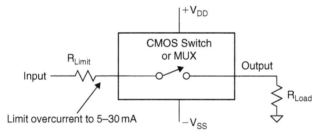

**Figure 7-57: Over-current protection using external resistor**

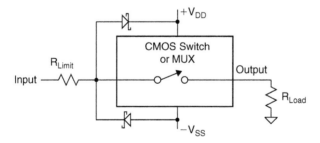

**Figure 7-58: Input protection using external Schottky
diodes**

the $R_{LIMIT}$ resistor can be quite low. It must be remembered that the Schottky diodes will have some
capacitance and leakage current.

Most CMOS devices have internal ESD-protection diodes connected from the inputs to the supply rails,
making the devices less susceptible to latch-up. However, the internal diodes begin conduction at 0.6V, and

have limited current-handling capability, thus adding the external Schottky diodes offers an added degree of protection.

Note that latch-up protection does not provide over-current protection, and vice versa. If both fault conditions can exist in a system, then both protective diodes and resistors should be used.

Analog Devices uses trench-isolation technology to produce its $L^2CMOS$ analog switches. The process reduces the latch-up susceptibility of the device, the junction capacitances, increases switching time and leakage current, and extends the analog voltage range to the supply rails.

Figure 7-59 shows the cross-sectional view of the trench-isolated CMOS structure. The buried oxide layer and the side walls completely isolate the substrate from each transistor junction. Therefore, no reverse-biased PN junction is formed. Consequently the bandwidth-reducing capacitances and the possibility of SCR latch-up are greatly reduced.

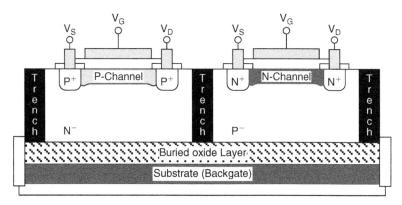

**Figure 7-59: Trench-isolation $L^2$CMOS structure**

The ADG508F, ADG509F, ADG528F, ADG438F, and ADG439F are $\pm15$ V trench-isolated $L^2$CMOS multiplexers which offer "fault protection" for input and output overvoltages between $-40$ and $+55$ V. These devices use a series structure of three MOSFETS in the signal path: an N-channel, followed by a P-channel, followed by an N-channel. In addition, the signal patch becomes a high impedance when the power supplies are turned off. This structure offers a high degree of latch-up and overvoltage protection—at the expense of higher $R_{ON}$ ($\sim300\,\Omega$), and more $R_{ON}$ variation with signal level. For more details of this protection method, refer to the individual product data sheets.

# Sample-and-Hold Circuits

## Introduction and Historical Perspective

The *sample-and-hold amplifier* (SHA) is a critical part of most data acquisition systems. It captures an analog signal and holds it during some operation (most commonly analog–digital conversion). The circuitry involved is demanding, and unexpected properties of commonplace components such as capacitors and printed circuit boards may degrade SHA performance.

Although today the SHA function has become an integral part of the *sampling* ADC, which is the vast majority of ADCs made today, understanding the fundamental concepts governing its operation is essential to understanding ADC dynamic performance.

When the sample-and-hold is in the sample (or track) mode, the output follows the input with only a small voltage offset. There do exist SHAs where the output during the *sample* mode does not follow the input accurately, and the output is only accurate during the *hold* period (such as the AD684, AD781, and AD783). These will not be considered here. Strictly speaking, a sample-and-hold with good tracking performance should be referred to as a *track-and-hold* circuit, but in practice the terms are used interchangeably.

The most common application of an SHA is to maintain the input to an ADC at a constant value during conversion. With many, but not all, types of ADC the input may not change by more than 1 LSB during conversion lest the process be corrupted—this either sets very low input frequency limits on such ADCs, or requires that they be used with an SHA to hold the input during each conversion. See the section in Chapter 6 on successive approximation ADCs.

Integration of the SHA function was made possible by new process developments including high speed complementary bipolar processes and advanced CMOS processes. In fact, the proliferation and popularity of sampling ADCs has been so great that today (2003), one rarely has the need for a separate SHA.

The advantage of a sampling ADC, apart from the obvious ones of smaller size, lower cost, and fewer external components, is that the overall DC and AC performance is fully specified, and the designer need not spend time ensuring that there are no specification, interface, or timing issues involved in combining a discrete ADC and a discrete SHA. This is especially important when one considers dynamic specifications such as SFDR and SNR.

Although the largest applications of SHAs are with ADCs, they are also occasionally used in DAC deglitchers, peak detectors, analog delay circuits, simultaneous sampling systems, and data distribution systems.

## Basic SHA Operation

Regardless of the circuit details or type of SHA in question, all such devices have four major components. The input amplifier, energy-storage device (capacitor), output buffer, and switching circuits are common to all SHAs as shown in the typical configuration of Figure 7-60.

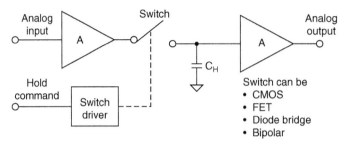

**Figure 7-60: Basic sample-and-hold circuit**

The energy-storage device, the heart of the SHA, is almost always a capacitor. The input amplifier buffers the input by presenting a high impedance to the signal source and providing current gain to charge the hold capacitor. In the *track* mode, the voltage on the hold capacitor follows (or tracks) the input signal (with some delay and bandwidth limiting). In the *hold* mode, the switch is opened, and the capacitor retains the voltage present before it was disconnected from the input buffer. The output buffer offers a high impedance to the hold capacitor to keep the held voltage from discharging prematurely. The switching circuit and its driver form the mechanism by which the SHA is alternately switched between track and hold.

There are four groups of specifications that describe basic SHA operation: track mode, track-to-hold transition, hold mode, hold-to-track transition. These specifications are summarized in Figure 7-61, and some of the SHA error sources are shown graphically in Figure 7-62. Because there are both DC and AC performance implications for each of the four modes, properly specifying an SHA and understanding its operation in a system are a complex matter.

| Sample mode | Sample-to-hold transition | Hold mode | Hold-to-sample transition |
|---|---|---|---|
| Static:<br>• Offset<br>• Gain error<br>• Nonlinearity | Static:<br>• Pedestal<br>• Pedestal nonlinearity | Static:<br>• Droop<br>• Dielectric<br>• Absorption | |
| Dynamic:<br>• Settling time<br>• Bandwidth<br>• Slew rate<br>• Distortion<br>• Noise | Dynamic:<br>• Aperture delay time<br>• Aperture jitter<br>• Switching transient<br>• Settling time | Dynamic:<br>• Feedthrough<br>• Distortion<br>• Noise | Dynamic:<br>• Acquisition time<br>• Switching transient |

**Figure 7-61: Sample-and-hold specifications**

## Specifications

While the specifications of the sample-and-hold part of a sampling converter are not broken out separately, their effect is included in the overall specifications of the converter.

### Track Mode Specifications

Since an SHA in the sample (or track) mode is simply an amplifier, both the static and dynamic specifications in this mode are similar to those of any amplifier. (SHAs which have degraded performance in the track

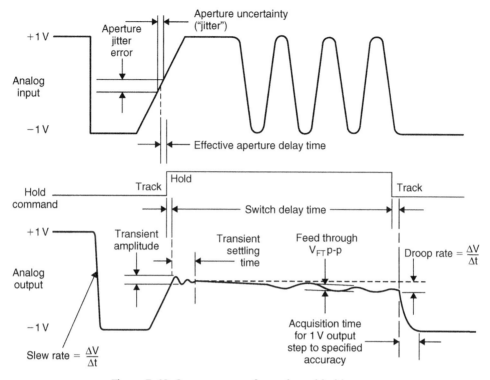

**Figure 7-62: Some sources of sample-and-hold errors**

mode are generally only specified in the hold mode.) The principal track mode specifications are *offset*, *gain*, *nonlinearity*, *bandwidth*, *slew rate*, *settling time*, *distortion*, and *noise*. However, distortion and noise in the track mode are often of less interest than in the hold mode.

## Track-to-Hold Mode Specifications

When the SHA switches from track to hold, there is generally a small amount of charge dumped on the hold capacitor because of non-ideal switches. This results in a hold mode DC offset voltage which is called *pedestal* error as shown in Figure 7-63. If the SHA is driving an ADC, the pedestal error appears as a DC offset voltage which may be removed by performing a system calibration. If the pedestal error is a function of input signal level, the resulting nonlinearity contributes to hold mode distortion.

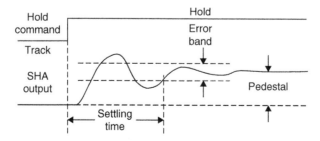

**Figure 7-63: Track-to-hold mode pedestal, transient, and settling time errors**

557

Pedestal errors may be reduced by increasing the value of the hold capacitor with a corresponding increase in acquisition time and a reduction in bandwidth and slew rate.

Switching from track to hold produces a transient, and the time required for the SHA output to settle to within a specified error band is called *hold mode settling time*. Occasionally, the peak amplitude of the switching transient is also specified.

Perhaps the most misunderstood and misused SHA specifications are those that include the word *aperture*. The most essential dynamic property of an SHA is its ability to disconnect quickly the hold capacitor from the input buffer amplifier. The short (but non-zero) interval required for this action is called *aperture time*. The various quantities associated with the internal SHA timing are shown in Figure 7-64.

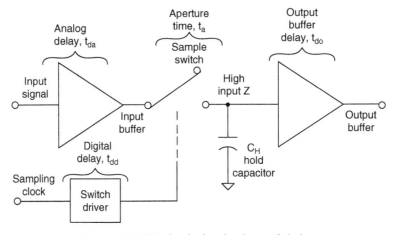

**Figure 7-64: SHA circuit showing internal timing**

The actual value of the voltage that is held at the end of this interval is a function of both the input signal and the errors introduced by the switching operation itself. Figure 7-65 shows what happens when the hold

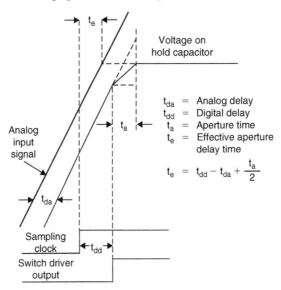

$t_{da}$ = Analog delay
$t_{dd}$ = Digital delay
$t_a$ = Aperture time
$t_e$ = Effective aperture delay time

$$t_e = t_{dd} - t_{da} + \frac{t_a}{2}$$

**Figure 7-65: SHA Waveforms**

command is applied with an input signal of arbitrary slope (for clarity, the sample to hold pedestal and switching transients are ignored). The value that finally gets held is a delayed version of the input signal, averaged over the aperture time of the switch as shown in Figure 7-65. The first-order model assumes that the final value of the voltage on the hold capacitor is approximately equal to the average value of the signal applied to the switch over the interval during which the switch changes from a low to high impedance ($t_a$).

The model shows that the finite time required for the switch to open ($t_a$) is equivalent to introducing a small delay in the sampling clock driving the SHA. This delay is constant and may either be positive or negative. It is called *effective aperture delay time*, *aperture delay time*, or simply *aperture delay* ($t_e$), and is defined as the time difference between the analog propagation delay of the front-end buffer ($t_{da}$) and the switch digital delay ($t_{dd}$) plus one-half the aperture time ($t_a/2$). The effective aperture delay time is usually positive, but may be negative if the sum of one-half the aperture time ($t_a/2$) and the switch digital delay ($t_{dd}$) is less than the propagation delay through the input buffer ($t_{da}$). The aperture delay specification thus establishes when the input signal is actually sampled with respect to the sampling clock edge.

Aperture delay time can be measured by applying a bipolar sinewave signal to the SHA and adjusting the synchronous sampling clock delay such that the output of the SHA is zero during the hold time. The relative delay between the input sampling clock edge and the actual zero-crossing of the input sinewave is the aperture delay time as shown in Figure 7-66.

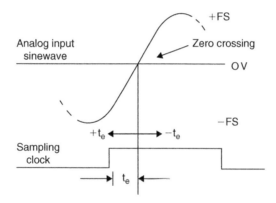

**Figure 7-66: Effective aperture delay time**

Aperture delay produces no errors, but acts as a fixed delay in either the sampling clock input or the analog input (depending on its sign). If there is sample-to-sample variation in aperture delay (*aperture jitter*), then a corresponding voltage error is produced as shown in Figure 7-67. This sample-to-sample variation in the instant the switch opens is called *aperture uncertainty*, or *aperture jitter*, and is usually measured in picoseconds RMS. The amplitude of the associated output error is related to the rate of change of the analog input. For any given value of aperture jitter, the aperture jitter error increases as the input dv/dt increases.

Figure 7-68 shows the effects of total sampling clock jitter on the signal-to-noise ratio (SNR) of a sampled data system. The total RMS jitter will be composed of a number of components, the actual SHA aperture jitter often being the least of them.

## Hold Mode Specifications

During the hold mode there are errors due to imperfections in the hold capacitor, switch, and output amplifier. If a leakage current flows in or out of the hold capacitor, it will slowly charge or discharge, and

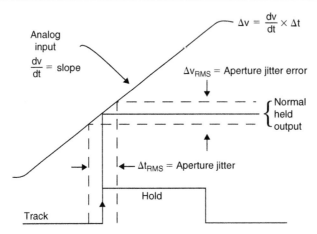

Figure 7-67: Effects of aperture or sampling clock jitter on SHA output

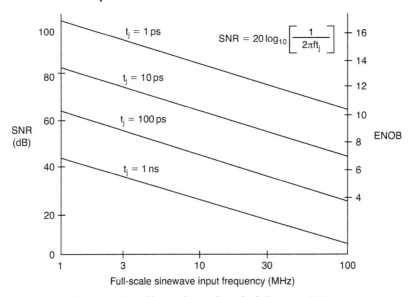

Figure 7-68: Effects of sampling clock jitter on SNR

its voltage will change. This effect is known as *droop* in the SHA output and is expressed in V/μs (Figure 7-69). Droop can be caused by leakage across a dirty PC board if an external capacitor is used, or by a leaky capacitor, but is most usually due to leakage current in semiconductor switches and the bias current of the output buffer amplifier. An acceptable value of droop is where the output of an SHA does not change by more than ½ LSB during the conversion time of the ADC it is driving, although this value is highly dependent on the ADC architecture. Where droop is due to leakage current in reversed-biased junctions (CMOS switches or FET amplifier gates), it will double for every 10°C increase in chip temperature—which means that it will increase a thousand-fold between +25°C and +125°C. Droop can be reduced by increasing the value of the hold capacitor, but this will also increase acquisition time and reduce bandwidth in the track mode. Differential techniques are often used to reduce the effects of droop in modern IC sample-and-hold circuits that are part of the ADC.

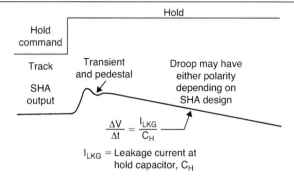

**Figure 7-69: Hold mode droop**

Hold capacitors for SHAs must have low leakage, but there is another characteristic which is equally important: low *dielectric absorption*. If a capacitor is charged, then discharged, and then left open circuit, it will recover some of its charge as shown in Figure 7-70. The phenomenon is known as *dielectric absorption*, and it can seriously degrade the performance of an SHA, since it causes the remains of a previous sample to contaminate a new one, and may introduce random errors of tens or even hundreds of millivolts.

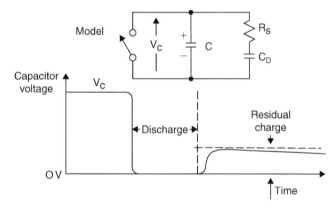

**Figure 7-70: Dielectric absorption**

## Hold-to-Track Transition Specifications

When the SHA switches from hold to track, it must reacquire the input signal (which may have made a full-scale transition during the hold mode). *Acquisition time* is the interval of time required for the SHA to reacquire the signal to the desired accuracy when switching from hold to track. The interval starts at the 50% point of the sampling clock edge, and ends when the SHA output voltage falls within the specified error band (usually 0.1% and 0.01% times are given). Some SHAs also specify acquisition time with respect to the voltage on the hold capacitor, neglecting the delay and settling time of the output buffer. The hold capacitor acquisition time specification is applicable in high speed applications, where the maximum possible time must be allocated for the hold mode. The output buffer settling time must of course be significantly smaller than the hold time.

Acquisition time can be measured directly using modern digital sampling scopes (DSOs) or digital phosphor scopes (DPOs) which are insensitive to large overdrives.

## Internal SHA Circuits for IC ADCs

CMOS ADCs are quite popular because of their low power and low cost. The equivalent input circuit of a typical CMOS ADC using a differential sample-and-hold is shown in Figure 7-71. While the switches

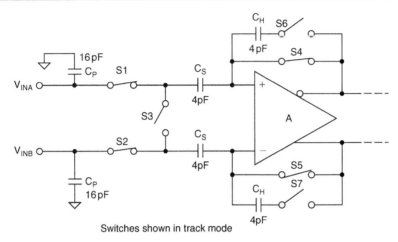

Switches shown in track mode

**Figure 7-71: Simplified input circuit for a typical switched capacitor CMOS sample-and-hold**

are shown in the *track* mode, note that they open/close at the sampling frequency. The 16 pF capacitors represent the effective capacitance of switches S1 and S2, plus the stray input capacitance. The $C_S$ capacitors (4 pF) are the sampling capacitors, and the $C_H$ capacitors are the hold capacitors. Although the input circuit is completely differential, this ADC structure can be driven either single ended or differentially. Optimum performance, however, is generally obtained using a differential transformer or differential op amp drive.

In the *track* mode, the differential input voltage is applied to the $C_S$ capacitors. When the circuit enters the *hold* mode, the voltage across the sampling capacitors is transferred to the $C_H$ hold capacitors and buffered by the amplifier A (the switches are controlled by the appropriate sampling clock phases). When the SHA returns to the *track* mode, the input source must charge or discharge the voltage stored on $C_S$ to a new input voltage. This action of charging and discharging $C_S$, averaged over a period of time and for a given sampling frequency $f_S$, makes the input impedance appear to have a benign resistive component. However, if this action is analyzed within a sampling period ($1/f_S$), the input impedance is dynamic, and certain input drive source precautions should be observed.

The resistive component to the input impedance can be computed by calculating the average charge that is drawn by $C_H$ from the input drive source. It can be shown that if $C_S$ is allowed to fully charge to the input voltage before switches S1 and S2 are opened then the average current into the input is the same as if there were a resistor equal to $1/(C_S f_S)$ connected between the inputs. Since $C_S$ is only a few picofarads, this resistive component is typically greater than several kiloohms for an $f_S = 10$ MSPS.

Figure 7-72 shows a simplified circuit of the input SHA used in the AD9042 12-bit, 41-MSPS ADC introduced in 1995 (Reference 7). The AD9042 is fabricated on a high speed complementary bipolar process, XFCB. The circuit comprises two independent SHAs in parallel for fully differential operation—only one-half the circuit is shown in the Figure. Fully differential operation reduces the error due to droop rate and also reduces second-order distortion. In the track mode, transistors Q1 and Q2 provide unity-gain buffering. When the circuit is placed in the hold mode, the base voltage of Q2 is pulled negative until it is clamped by the diode, D1. The on-chip hold capacitor, $C_H$, is nominally 6 pF. Q3 along with $C_F$ provides output current bootstrapping and reduces the $V_{BE}$ variations of Q2. This reduces third-order signal distortion. Track mode THD is typically $-93$ dB at 20 MHz. In the time domain, full-scale acquisition time to 12-bit accuracy is 8 ns. In the hold mode, signal-dependent pedestal variations are minimized by

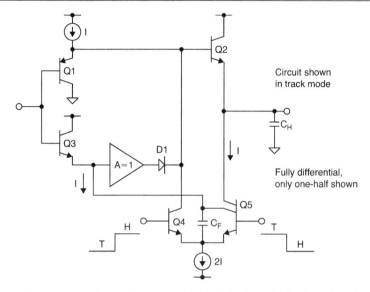

**Figure 7-72: SHA used in AD9042 12-Bit, 41-MSPS ADC introduced in 1995**

the voltage bootstrapping action of Q3 and the A = 1 buffer along with the low feedthrough parasitics of Q2. Hold mode settling time is 5 ns to 12-bit accuracy. Hold mode THD at a clock rate of 50 MSPS and a 20 MHz input signal is −90 dB.

Figure 7-73 shows a simplified schematic of one-half of the differential SHA used in the AD6645 14-bit, 105-MSPS ADC recently introduced (Reference 9 gives a complete description of the ADC including the

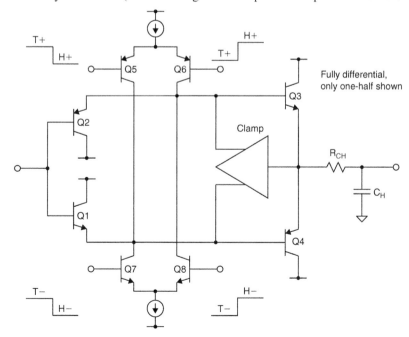

**Figure 7-73: SHA used in AD6645 14-Bit, 105-MSPS ADC**

563

SHA). In the track mode, Q1, Q2, Q3, and Q4 form a complementary emitter follower buffer which drives the hold capacitor, $C_H$. In the hold mode, the polarity of the bases of Q3 and Q4 is reversed and clamped to a low impedance. This turns off Q1, Q2, Q3, and Q4, and results in double isolation between the signal at the input and the hold capacitor. As previously discussed, the clamping voltages are bootstrapped by the held output voltage, thereby minimizing nonlinear effects.

Track mode linearity is largely determined by the $V_{BE}$ modulation of Q3 and Q4 when charging $C_H$. Hold mode linearity depends on track mode linearity plus nonlinear errors in the track-to-hold transitions caused by imbalances in the switching of the base voltages of Q3 and Q4 and the resulting imbalance in charge injection through their base–emitter junctions as they turn off.

## References: Sample-and-Hold Circuits

1. A.H. Reeves, "Electric Signaling System," US Patent 2,272,070, filed, November 22, 1939, issued February 3, 1942. Also French Patent 852,183 issued 1938, and British Patent 538,860 issued 1939 *(the classic patents on PCM including descriptions of a 5-bit, 6-kSPS vacuum tube ADC and DAC).*

2. L.A. Meacham and E. Peterson, "An Experimental Multichannel Pulse Code Modulation System of Toll Quality," **Bell System Technical Journal**, Vol. 27, No. 1, January, 1948, pp. 1–43, *(describes the culmination of much work leading to this 24-channel experimental PCM system. In addition, the article describes a 50-kSPS vacuum tube sample-and-hold based on a pulse transformer driver).*

3. J.R. Gray and S.C. Kitsopoulos, "A Precision Sample-and-Hold Circuit with Subnanosecond Switching," **IEEE Transactions on Circuit Theory**, Vol. CT11, September, 1964, pp. 389–396, *(an excellent description of a solid-state transformer-driven diode bridge SHA, along with a detailed mathematical analysis of the circuit and associated errors).*

4. J.O. Edson and H.H. Henning, "Broadband Codecs for an Experimental 224Mb/s PCM Terminal," **Bell System Technical Journal**, Vol. 44, November, 1965, pp. 1887–1940, *(summarizes experiments on ADCs based on the electron tube coder as well as a bit-per-stage Gray code 9-bit solid state ADC. The electron beam coder was 9-bits at 12MSPS, and represented the fastest of its type).*

5. D.J. Kinniment, D. Aspinall, and D.B.G. Ewards, "High-Speed Analogue–Digital Converter," **IEE Proceedings**, Vol. 113, December 1966, pp. 2061–2069, *(a 7-bit 9-MSPS three-stage pipelined error corrected converter is described based on recirculating through a 3-bit stage three times. Tunnel (Esaki) diodes are used for the individual comparators. The article also shows a proposed faster pipelined 7-bit architecture using three individual 3-bit stages with error correction. The article also describes a fast bootstrapped transformer-driven diode-bridge sample-and-hold circuit).*

6. O.A. Horna, "A 150Mbps A/D and D/A Conversion System," **Comsat Technical Review**, Vol. 2, No. 1, 1972, pp. 39–72, *(a description of a subranging ADC including a detailed analysis of the sample-and-hold circuit).*

7. R. Gosser and F. Murden, "A 12-bit 50MSPS Two-Stage A/D Converter," *1995 ISSCC Digest of Technical Papers*, p. 278 *(a description of the AD9042 error corrected subranging ADC using MagAMP stages for the internal ADCs).*

8. C. Moreland, "An 8-bit 150-MSPS Serial ADC," *1995 ISSCC Digest of Technical Papers*, Vol. 38, p. 272 *(a description of an 8-bit ADC with five folding stages followed by a 3-bit flash converter, including a discussion of the sample-and-hold circuit).*

9. C. Moreland, F. Murden, M. Elliott, J. Young, M. Hensley, and R. Stop, "A 14-bit 100-Msample/s Subranging ADC," **IEEE Journal of Solid State Circuits**, Vol. 35, No. 12, December, 2000, pp. 1791–1798, *(describes the architecture used in the 14-bit, 105MSPS AD6645 ADC and also the sample-and-hold circuit).*

# Clock Generation and Distribution Circuits

Developing a high frequency, high resolution system is a non-trivial task. Any high speed ADC is extremely sensitive to the quality of the sampling clock provided by the user. Since an ADC can be thought of as a sampling mixer, any noise, distortion, or timing jitter on the clock is combined with the desired signal at the ADC output. Clock integrity requirements scale with the analog input frequency and resolution. The higher analog input frequency applications at 14-bit (or higher) resolution are the most stringent. The theoretical SNR of an ADC is limited by the ADC resolution and the jitter on the sampling clock. Considering an ideal ADC of infinite resolution where the step size and quantization error can be ignored, the available SNR can be expressed approximately by:

$$SNR = 20 \times \log_{10}\left(\frac{1}{2\pi f t_j}\right) \tag{7-9}$$

where f is the highest analog frequency being digitized, and $t_j$ is the RMS jitter on the sampling clock. Figure 7-74 shows the required sampling clock jitter as a function of the analog frequency and effective number of bits (ENOB).

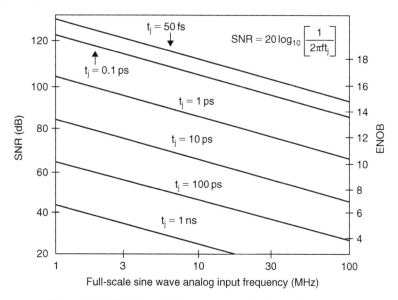

Figure 7-74: ENOB and SNR versus analog input frequency

## Contribution to Overall System Performance

In IF sampling converters, clock purity is of extreme importance. As with the mixing process, the input signal is multiplied by a local oscillator or in this case, a sampling clock. Since multiplication in the time domain is convolution in the frequency domain, the spectrum of the sample clock is convolved with the spectrum of the input signal. Since aperture uncertainty is equivalent to wideband noise on the clock, it shows up as wideband noise in the sampled spectrum as well. And since an ADC is a sampling system, the spectrum is periodic and repeated around the sample rate.

This wideband noise therefore raises the noise floor of the ADC. The theoretical SNR for an ADC, as limited by aperture uncertainty, is determined by the following equation:

$$SNR = -20 \times \log_{10}[(2\pi \, f_{ANALOG} t_{JITTER} RMS)] \tag{7-10}$$

If Eq. (7-10) is evaluated for an analog input of 201 MHz and 0.7 ps RMS "jitter," the theoretical SNR is limited to 61 dB. Therefore, systems that require very high dynamic range and very high analog input frequencies also require a very low jitter encode source. With care phase-locked loops (PLLs) using VCXOs can achieve less than 1 ps RMS jitter, but jitter less than 0.1 ps RMS requires a dedicated low noise crystal oscillator, as discussed in the previous chapter. It should be noted that the jitter of a typical TTL/CMOS gate to about 1–4 ps. Low voltage SiGe reduced swing ECL gate can have about 0.2 ps RMS.

When considering overall system performance, a more generalized equation may be used. This equation builds on the previous equation but includes the effects of thermal noise and differential nonlinearity:

$$SNR = -20 \log_{10} \left[ \underbrace{(2\pi \times f_a \times t_{jRMS})^2}_{\substack{\text{SAMPLING} \\ \text{CLOCKJITTER}}} + \underbrace{\frac{2}{3}\left(\frac{1+\varepsilon}{2^N}\right)^2}_{\substack{\text{QUANTIZATION} \\ \text{NOISE, DNL}}} + \underbrace{\left(\frac{2 \times \sqrt{2} \times V_{NOISE \, RMS}}{2^N}\right)^2}_{\substack{\text{EFFECTIVE} \\ \text{INPUT NOISE}}} \right]^{\frac{1}{2}} \tag{7-11}$$

where $F_a$ is the analog input frequency, $T_{jRMS}$ is the aperture jitter of ADC and external clock, $\varepsilon$ is the average DNL of converter (~0.4 LSB), $V_{NOISERMS}$ is the effective ADC input noise in LSBs, and N is the number of bits.

Although this is a simple equation, it provides much insight into the noise performance that can be expected from a data converter.

## Clock Generation Circuits

Analog Devices has designed dedicated clocking products specifically designed to support the extremely stringent clocking requirements of the highest performance data converters. The first of these is the AD9540. This device features high performance PLL circuitry, including a flexible 200 MHz phase frequency detector (PFD) and a digitally controlled charge pump (CP). The device also provides a low jitter, 655 MHz current mode logic (CML)-mode, positive emitter-coupled, logic (PECL)-compliant output driver with programmable slew rates. External voltage-controlled oscillator (VCO) rates up to 2.7 GHz are supported. Extremely fine tuning resolution (steps less than 2.33 μHz) is another feature supported by this device. Information is loaded into the AD9540 via a serial I/O port that has a device write speed of 25 Mb/s. The AD9540 frequency divider can also be programmed to support a spread spectrum mode of operation.

The block diagram of the AD9540 is shown in Figure 7-75. An overview shows that all the necessary component blocks are present for generating both of the needed clocks. In generating low jitter clocks, it is

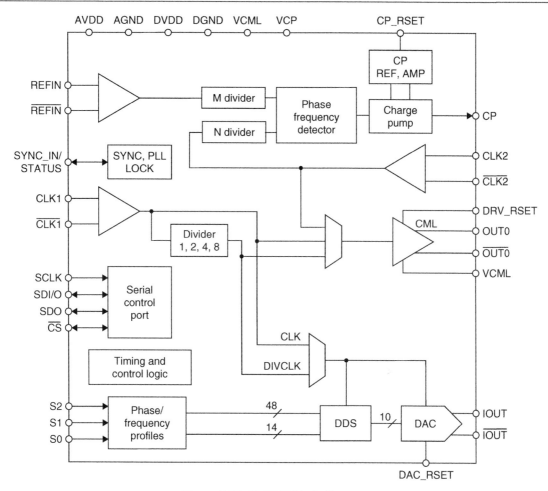

**Figure 7-75: AD9540 block diagram**

almost always preferable to employ a PLL circuit of some sort. In addition to providing frequency stability, PLL circuits offer great noise reduction capability because the loop filter will act as a tracking bandpass filter. Because in most clocking applications a single frequency is required, parameters such as acquisition time and tuning range are not of importance. Therefore performance in these areas can be sacrificed to improve the noise performance of the loop. Specifically, a very narrow range VCO can be selected with a center frequency close to the desired clock rate. As the tuning range is reduced, the gain coefficient for the VCO (Kv) is reduced, and the phase noise of the VCO itself is thereby reduced. Also, the loop filter bandwidth is a concern for designers because there is a tradeoff between loop bandwidth and acquisition time. Generally speaking, the wider the loop bandwidth, the faster the acquisition and lock time of a loop, but more noise from the reference and PFD itself is passed through the loop. In the case of a clocking application, this tradeoff can be used to achieve narrow loop bandwidths, sacrificing settling time in favor of noise suppression through the loop.

The digital clock requires precise frequency and adjustable phase that can be generated from the direct digital synthesizer (DDS) portion of the device. The DDS on the AD9540 offers 48-bit frequency tuning resolution (1.42 Hz, for the maximum clock rate of 400 MHz) and 14-bit phase adjustment (0.022°). The output of a DDS is a reconstructed sinewave, so two additional external circuits are required. First, a bandpass filter at the desired clock rate must filter the reconstructed sinewave. This will remove most sampling artifacts from the output spectrum as well as remove most broadband noise in the DAC output signal. Second, in order to achieve the required slew rates for most clock circuits, an external comparator needs to be inserted into the clock signal path. An excellent choice, used for this example, is the ADCMP563. A simplified block diagram for the resultant circuit is shown in Figure 7-76. Inputs CLK1/CLK1 are shorted to CLK2/CLK2. The device is programmed such that the CML driver gets its input from the undivided input from CLK1, but the DDS is clocked by the divided output (622 MHz divided by 2 = 311 MHz). The drawing shows the crystal oscillator capability of the REF input of the PLL, demonstrating its use with a 38 MHz crystal. The two output clocks are shown at OUT0 (the low jitter 622 MHz clock) and OUT1 (the phase-programmable auxiliary clock). Edge skew (or time delay) in the auxiliary clock is accomplished by programming a phase offset into the DDS, which will change the relative point in time for the complementary input crossing at the comparator.

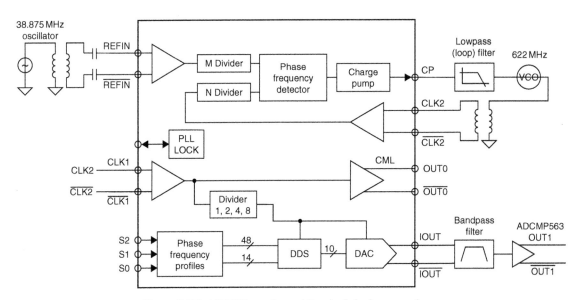

**Figure 7-76: AD9540 configured for dual clock generation**

## PLL Circuitry

The AD9540 includes an RF divider (divide by R), a 48-bit DDS core, a 14-bit programmable delay adjustment, a 10-bit DAC, a PFD, and a programmable current output CP. Incorporating these blocks together, users can generate many useful circuits for clock synthesis.

The RF divider accepts differential or single-ended signals up to 2.7 GHz on the CLK1 input pin. The RF divider also supplies the SYSCLK input to the DDS. Because the DDS operates only up to 400 MSPS, the RF divider must be engaged for any CLK1 signal greater than 400 MHz. The RF divider can be programmed to take values of 1, 2, 4, or 8. The ratio for the divider is programmed in the control register. The output of the divider can be routed to the input of the on-chip CML driver. For lower frequency input signals, it is possible to use the divider to divide the input signal to the CML driver and to use the undivided input of the divider as the SYSCLK input to the DDS, or vice versa. In all cases, the clock to the DDS should not exceed 400 MSPS.

The on-chip PFD has two differential inputs, REFIN (the reference input) and CLK2 (the feedback or oscillator input). These differential inputs can be driven by single-ended signals. When doing so, tie the unused input through a 100 pF capacitor to the analog supply (AVDD). The maximum speed of the PFD inputs is 200 MHz. Each of the inputs has a buffer and a divider (÷M on REFIN and ÷N on CLK2) that operates up to 655 MHz. If the signal exceeds 200 MHz, the divider must be used. The dividers are programmed through the control registers and take any integer value between 1 and 16.

The REFIN input also has the option of engaging an in-line oscillator circuit. Engaging this circuit means that the REFIN input can be driven with a crystal in the frequency range of 20 MHz ≤ REFIN ≤ 30 MHz.

The CP outputs a current in response to an error signal generated in the PFD. The output current is programmed through by placing a resistor (CP_RSET) from the CP_RSET pin to ground.

This sets the CP's reference output current. Also, a programmable scaler multiplies this base value by any integer from 1 to 8, programmable through the CP current scale bits in the Control Function Register 2.

## CML Driver

An on-chip CML driver is also included. This CML driver generates very low jitter clock edges. The outputs of the CML driver are current outputs that drive PECL levels when terminated into a 100 Ω load. The continuous output current of the driver is programmed by attaching a resistor from the DRV_RSET pin to ground (nominally 4.02 kΩ for a continuous current of 7.2 mA). An optional on-chip current programming resistor is enabled by setting a bit in the control register. The rising edge and falling edge slew rates are independently programmable to help control overshoot and ringing by the application of surge current during rising edge and falling edge transitions (see Figure 7-77). There is a default surge

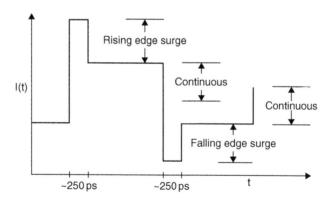

**Figure 7-77: Rising edge and falling edge surge current out of the CML clock driver, as opposed to the steady state continuous current**

current of 7.6 mA on the rising edge and of 4.05 mA on the falling edge. Bits in the control register enable additional rising edge and falling edge surge current, and can disable the default surge current. CML driver can be driven by:

- RF divider input (CLK1 directly to the CML driver)
- RF divider output
- CLK2 input

## DDS and DAC

The precision frequency division within the device is accomplished using DDS technology. The DDS can control the digital phase relationships by clocking a 48-bit accumulator. The incremental value loaded into the accumulator, known as the frequency tuning word, controls the overflow rate of the accumulator. Similar to a sinewave completing a $2\pi$ radian revolution, the overflow of the accumulator is cyclical in nature and generates a fundamental frequency according to:

$$f_0 = \frac{FTW \times (f_S)}{2^{48}} \qquad 0 \le FTW \le 2^{47} \qquad (7\text{-}12)$$

The instantaneous phase of the sinewave is therefore the output of the phase accumulator block. This signal may be phase offset by programming an additive digital phase that is added to each phase sample coming out of the accumulator. These instantaneous phase values are then passed through a phase-to-amplitude conversion (sometimes called an angle-to-amplitude conversion or AAC) block.

This algorithm follows a $\cos(x)$ relationship, where x is the phase coming out of the phase offset block, normalized to $2\pi$. Finally, the amplitude words drive a 10-bit DAC. Because the DAC is a sampled data system, the output is a reconstructed sinewave that needs to be filtered to take high frequency images out of the spectrum. The DAC is a current steering DAC that is AVDD referenced. To get a voltage output, the DAC outputs must be terminated through a load resistor to AVDD, typically $50\,\Omega$. At positive full-scale, $I_{OUT}$ sinks no current and the voltage drop across the load resistor is 0.

However, the $I_{OUT}$ output sinks the DAC's programmed full-scale output current, causing the maximum output voltage drop across the load resistor. At negative full-scale, the situation is reversed, and $I_{OUT}$ sinks the full-scale current (and generates the maximum drop across the load resistor), while $I_{OUT}$ sinks no current (and generates no voltage drop). At mid-scale, the outputs sink equal amounts of current, generating equal voltage drops.

## *Selectable Clock Frequencies and Selectable Edge Delay*

Because the precision driver is implemented using a DDS, it is possible to store multiple clock frequency words to enable externally switchable clock frequencies. The phase accumulator runs at a fixed frequency, according to the active profile clock frequency word. Likewise, any delay applied to the rising and falling edges is a static value that comes from the delay shift word of the active profile. The device has eight different phase/frequency profiles, each with its own 48-bit clock frequency word and 14-bit delay shift word. Profiles are selected by applying their digital value on the clock select (S0, S1, and S2) pins. It is not possible to use the phase offset of one profile and the frequency tuning word of another.

## *Synchronization Modes for Multiple Devices*

In a DDS system, the SYNC_CLK is derived internally from the master system clock, SYSCLK, with a $\div 4$ divider. Because the divider does not power-up to a known state, multiple devices in a system might have staggered clock phase relationships, because each device can potentially generate the SYNC_CLK rising edge from any one of four rising edges of SYSCLK. This ambiguity can be resolved by employing digital synchronization logic to control the phase relationships of the derived clocks among different devices in the system. Note that the synchronization functions included on the AD9540 control only the timing relationships among different digital clocks. They do not compensate for the analog timing delay on the system clock due to mismatched phase relationships on the input clock, CLK1 (see Figure 7-78).

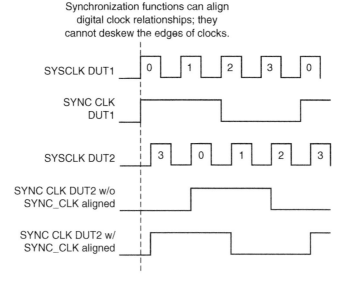

**Figure 7-78: Synchronization functions: capabilities and limitations**

## Automatic Synchronization

In automatic synchronization mode, the device is placed in slave mode and automatically aligns the internal SYNC_CLK to a master SYNC_CLK signal, supplied on the SYNC_IN input. When this bit is enabled, the STATUS is not available as an output; however, an out-of-lock condition can be detected by reading Control Function Register 1 and checking the status of the STATUS_Error bit. The automatic synchronization function is enabled by setting the Control Function Register 1 automatic synchronization bit. To employ this function at higher clock rates (SYNC_CLK > 62.5 MHz, SYSCLK > 250 MHz), the high speed sync enable bit should be set as well.

## Manual Synchronization, Hardware Controlled

In this mode, the user controls the timing relationship of the SYNC_CLK with respect to SYSCLK. When hardware manual synchronization is enabled, the SYNC_IN/STATUS pin becomes a digital input. For each rising edge detected on the SYNC_IN input, the device advances the SYNC_IN rising edge by one SYSCLK period. When this bit is enabled, the STATUS is not available as an output; however, an out-of-lock condition can be detected by reading Control Function Register 1 and checking the status of the STATUS_Error bit. This synchronization function is enabled by setting the hardware manual synchronization enable bit.

## Manual Synchronization, Software Controlled

In this mode, the user controls the timing relationship between SYNC_CLK and SYSCLK through software programming. When the software manual synchronization bit is set high, the SYNC_CLK is advanced by one SYSCLK cycle. Once this operation is complete, the bit is cleared. The user can set this bit repeatedly to advance the SYNC_CLK rising edge multiple times. Because the operation does not use the SYNC_IN/STATUS pin as a SYNC_IN input, the STATUS signal can be monitored on the STATUS pin during this operation.

## Clock Distribution Circuits

In addition to the clock generation circuits Analog Devices make clock distribution circuits such as the A9514 and combination circuits such as the AD9510.

The AD9510 provides a multi-output clock distribution function along with an on-chip PLL core. The design emphasizes low jitter and phase noise to maximize data converter performance. Other applications with demanding phase noise and jitter requirements also benefit from this part.

The PLL section consists of a programmable reference divider (R); a low noise phase frequency detector (PFD); a precision charge pump (CP); and a programmable feedback divider (N).

By connecting an external VCXO or VCO to the CLK2/CLK2B pins, frequencies up to 1.6 GHz may be synchronized to the input reference.

There are eight independent clock outputs. Four outputs are low voltage, positive emitter-coupled, logic (LVPECL) (1.2 GHz maximum), and four are selectable as either low voltage differential (LVDS) (800 MHz maximum) or CMOS (250 MHz maximum) levels.

Each output has a programmable divider that may be bypassed or set to divide by any integer up to 32. The phase of one clock output relative to another clock output may be varied by means of a divider phase select function that serves as a coarse timing adjustment. Two of the LVDS/CMOS outputs feature programmable delay elements with full-scale ranges up to 10 ns of delay. This fine tuning delay block has a 5-bit resolution, giving 32 possible delays from which to choose for each full-scale setting.

## Functional Description

Figure 7-79 shows a block diagram of the AD9510. The chip combines a programmable PLL core with a configurable clock distribution system. A complete PLL requires the addition of a suitable external VCO (or VCXO) and loop filter. This PLL can lock to a reference input signal and produce an output that is related to the input frequency by the ratio defined by the programmable R and N dividers. The PLL reduces the jitter from the external reference signal, depending on the loop bandwidth and the phase noise performance of the VCO (VCXO).

The output from the VCO (VCXO) can be applied to the clock distribution section of the chip, where it can be divided by any integer value from 1 to 32. The duty cycle and relative phase of the outputs can be selected. There are four LVPECL outputs, (OUT0, OUT1, OUT2, and OUT3) and four outputs that can be either LVDS or CMOS level outputs (OUT4, OUT5, OUT6, and OUT7). Two of these outputs (OUT5 and OUT6) can also make use of a variable delay block.

Alternatively, the clock distribution section can be driven directly by an external clock signal, and the PLL can be powered off. Whenever the clock distribution section is used alone, there is no clock clean-up. The jitter of the input clock signal is passed along directly to the distribution section and may dominate at the clock outputs.

### PLL Section

The AD9510 consists of a PLL section and a distribution section. If desired, the PLL section can be used separately from the distribution section.

The AD9510 has a complete PLL core on-chip, requiring only an external loop filter and VCO/VCXO. This PLL is based on the ADF4106, a PLL noted for its superb low phase noise performance. The operation of the AD9510 PLL is nearly identical to that of the ADF4106, offering an advantage to those with experience with the ADF series of PLLs. Differences include the addition of differential inputs at REFIN and CLK2

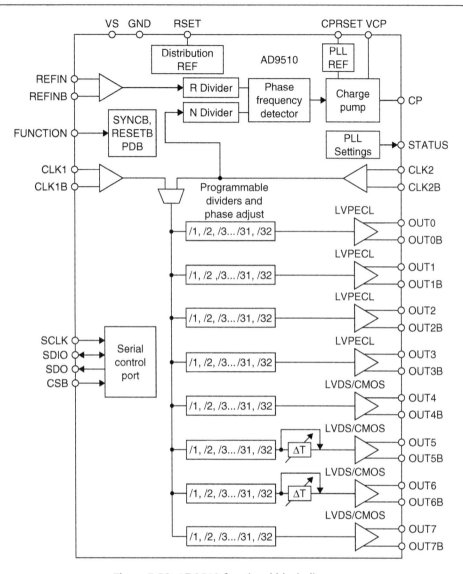

**Figure 7-79: AD9510 functional block diagram**

and a different control register architecture. Also, the prescaler has been changed to allow N as low as 1. The AD9510 PLL implements the digital lock detect feature somewhat differently than the ADF4106 does, offering improved functionality at higher PFD rates.

*PLL Reference Input—REFIN*

The REFIN/REFINB pins can be driven by either a differential or a single-ended signal. These pins are internally self-biased so that they can be AC-coupled via capacitors. It is possible to DC-couple to these inputs. If REFIN is driven single ended, the unused side (REFINB) should be decoupled via a suitable capacitor to a quiet ground. Figure 7-80 shows the equivalent circuit of REFIN.

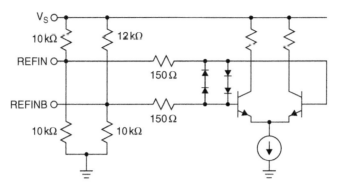

**Figure 7-80: REFIN equivalent circuit**

## VCO/VCXO Clock Input—CLK2

The CLK2 differential input is used to connect an external VCO or VCXO to the PLL. Only the CLK2 input port has a connection to the PLL N divider. This input can receive up to 1.6 GHz. These inputs are internally self-biased and must be AC-coupled via capacitors.

Alternatively, CLK2 may be used as an input to the distribution section.

The default condition is for CLK1 to feed the distribution section (see Figure 7-81).

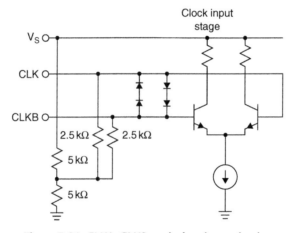

**Figure 7-81: CLK1, CLK2 equivalent input circuit**

## PLL Reference Divider—R

The REFIN/REFINB inputs are routed to reference divider, R, which is a 14-bit counter. R may be programmed to any value from 1 to 16,383 (a value of 0 results in a divide by 1) via its control register. The output of the R divider goes to one of the PFD inputs. The maximum allowable frequency into the PFD must not be exceeded. This means that the REFIN frequency divided by R must be less than the maximum allowable PFD frequency. See Figure 7-80.

## VCO/VCXO Feedback Divider—N (P, A, B)

The N divider is a combination of a prescaler, P (3 bits), and two counters, A (6 bits) and B (13 bits). Although the AD9510's PLL is similar to the ADF4106, the AD9510 has a redesigned prescaler that allows lower values of N. The prescaler has both a dual modulus (DM) and a fixed divide (FD) mode.

When using the prescaler in FD mode, the A counter is not used, and the B counter may need to be bypassed. The DM prescaler modes set some upper limits on the frequency, which can be applied to CLK2.

## A and B Counters

The AD9510 B counter has a bypass mode (B = 1), which is not available on the ADF4106. The B counter bypass mode is valid only when using the prescaler in FD mode. The B counter is bypassed by writing 1 to the B counter bypass bit. The valid range of the B counter is 3 to 8,191. The default after a reset is 0, which is invalid.

Note that the A counter is not used when the prescaler is in FD mode.

Note also that the A/B counters have their own reset bit, which is primarily intended for testing. The A and B counters can also be reset using the R, A, and B counters' shared reset bit.

## Determining Values for P, A, B, and R

When operating the AD9510 in a dual-modulus mode, the input reference frequency, FREF, is related to the VCO output frequency, FVCO:

$$FVCO = \left(\frac{FPRF}{R}\right) \times (PB + A) = FREF \times \frac{N}{R} \qquad (7\text{-}13)$$

When operating the prescaler in FD mode, the A counter is not used and the equation simplifies to:

$$FVCO = \left(\frac{FPRF}{R}\right) \times (PB) = FREF \times \frac{N}{R} \qquad (7\text{-}14)$$

By using combinations of DM and FD modes, the AD9510 can achieve values of N all the way down to N = 1.

## Phase Frequency Detector and Charge Pump

The PFD takes inputs from the R counter and the N counter (N = BP + A) and produces an output proportional to the phase and frequency difference between them. Figure 7-82 shows a simplified

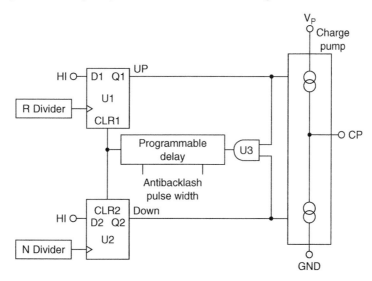

**Figure 7-82: PFD simplified schematic and timing (in lock)**

575

schematic. The PFD includes a programmable delay element that controls the width of the antibacklash pulse. This pulse ensures that there is no dead zone in the PFD transfer function and minimizes phase noise and reference spurs. Two bits in Register 0Dh control the width of the pulse.

### Antibacklash Pulse

The PLL features a programmable antibacklash pulse width. The default antibacklash pulse width is 1.3 ns and normally should not need to be changed. The antibacklash pulse eliminates the dead zone around the phase-locked condition and thereby reduces the potential for certain spurs that could be impressed on the VCO signal.

### STATUS Pin

The output multiplexer on the AD9510 allows access to various signals and internal points on the chip at the STATUS pin. The function of the STATUS pin is controlled by a register.

### PLL Digital Lock Detect

The STATUS pin can display two types of PLL lock detect: digital (DLD) and analog (ALD). Whenever digital lock detect is desired, the STATUS pin provides a CMOS level signal, which can be active high or active low.

The digital lock detect has one of two time windows, as selected by a register. The default requires the signal edges on the inputs to the PFD to be coincident within 9.5 ns to set DLD = true, which then must separate by at least 15 ns to give DLD = false.

The other setting makes these coincidence times 3.5 ns for DLD = true and 7 ns for DLD = false. The DLD may be disabled. If the signal at REFIN goes away while DLD is true, the DLD will not necessarily indicate loss of lock.

## Dividers

Each of the eight clock outputs of the AD9510 has its own divider. The divider can be bypassed to get an output at the same frequency as the input (1×). When a divider is bypassed, it is powered down to save power.

All integer divide ratios from 1 to 32 may be selected. A divide ratio of 1 is selected by bypassing the divider.

Each divider can be configured for divide ratio, phase, and duty cycle. The phase and duty cycle values that can be selected depend on the divide ratio that is chosen.

### Divider Phase Offset

The phase of each output may be selected, depending on the divide ratio chosen. This is selected by writing the appropriate values to the registers which set the phase and start high/low bit for each output. Each divider has a 4-bit phase offset and a start high or low bit.

Following a sync pulse, the phase offset word determines how many fast clock (CLK1 or CLK2) cycles to wait before initiating a clock output edge. The Start H/L bit determines if the divider output starts low or high. By giving each divider a different phase offset, output-to-output delays can be set in increments of the fast clock period, tCLK.

Figure 7-83 shows four dividers, each set for DIV = 4, 50% duty cycle. By incrementing the phase offset from 0 to 3, each output is offset from the initial edge by a multiple of tCLK.

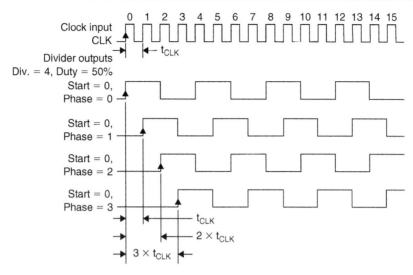

**Figure 7-83: Phase offset—all dividers set for DIV=4, phase set from 0 to 3**

## Delay Block

OUT5 and OUT6 (LVDS/CMOS) include an analog delay element that can be programmed to give variable time delays ($\Delta t$) in the clock signal passing through that output.

The amount of delay that can be used is determined by the frequency of the clock being delayed. The amount of delay can approach one-half cycle of the clock period. For example, for a 10 MHz clock, the delay can extend to the full 10 ns maximum of which the delay element is capable. However, for a 100 MHz clock (with 50% duty cycle), the maximum delay is less than 5 ns (or half of the period).

OUT5 and OUT6 allow a full-scale delay in the range 1–10 ns. The full-scale delay is selected by choosing a combination of ramp current and the number of capacitor. There are 32 fine delay settings for each full-scale (Figure 7-84).

This path adds some jitter greater than that specified for the non-delay outputs. This means that the delay function should be used primarily for clocking digital chips, such as FPGA, ASIC, DUC, and DDC, rather than for data converters. The jitter is higher for long full-scales (~10 ns). This is because the delay

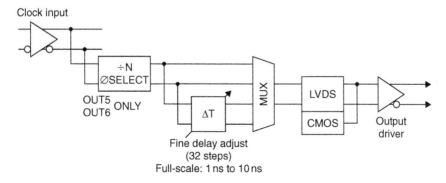

**Figure 7-84: Analog delay (OUT5 and OUT6)**

block uses a ramp and trip points to create the variable delay. A longer ramp means more noise might be introduced.

The clock distribution circuits feature both LVPECL and LVDS outputs that provide differential clock outputs, which enable clock solutions that maximize converter SNR performance. The input requirements of the ADC (differential or single-ended, logic level, termination) should be considered when selecting the best clocking/converter solution.

Whenever single-ended CMOS clocking is used, some of the following general guidelines should be followed. Point-to-point nets should be designed such that a driver has one receiver only on the net, if possible. This allows for simple termination schemes and minimizes ringing due to possible mismatched impedances on the net. Series termination at the source is generally required to provide transmission line matching and/or to reduce current transients at the driver. The value of the resistor is dependent on the board design and timing requirements (typically 10–100 Ω is used). CMOS outputs are limited in terms of the capacitive load or trace length that they can drive. Typically, trace lengths less than 3 inches are recommended to preserve signal rise/fall times and preserve signal integrity (Figure 7-85).

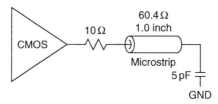

**Figure 7-85: Series termination of CMOS outputs**

Termination at the far end of the PCB trace is a second option. The CMOS outputs typically do not supply enough current to provide a full voltage swing with a low impedance resistive, far-end termination, as shown in Figure 7-86. The far-end termination network should match the PCB trace impedance and provide the desired switching point. The reduced signal swing may still meet receiver input requirements in some applications. This can be useful when driving long trace lengths on less critical nets.

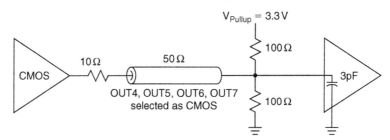

**Figure 7-86: CMOS output with far-end termination**

Because of the limitations of single-ended CMOS clocking, consider using differential outputs when driving high speed signals over long traces. LVPECL and LVDS outputs are better suited for driving long traces where the inherent noise immunity of differential signaling provides superior performance for clocking converters.

## LVPECL Clock Distribution

The LVPECL outputs typically provide the lowest jitter clock signals available from the clock distribution chips. The LVPECL outputs (because they are open emitter) require a DC termination to bias the output transistors. In most applications, a standard LVPECL far-end termination is recommended, as shown in Figure 7-87. The resistor network is designed to match the transmission line impedance ($50\,\Omega$) and the desired switching threshold ($1.3\,V$).

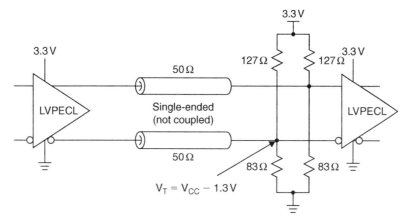

Figure 7-87: LVPECL far-end termination

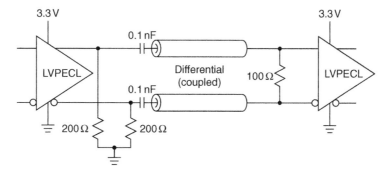

Figure 7-88: LVPECL with parallel transmission line

## LVDS Clock Distribution

LVDS is a second differential output option. LVDS uses a current mode output stage with several user-selectable current levels. The normal value (default) for this current is $3.5\,mA$, which yields $350\,mV$ output swing across a $100\,\Omega$ resistor. The LVDS outputs of the clock chips meet or exceed all ANSI/TIA/EIA-644 specifications. A recommended termination circuit for the LVDS outputs is shown in Figure 7-89.

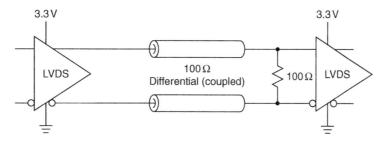

Figure 7-89: LVDS output termination

## Power Management

The power usage of the AD9510 can be managed to use only the power required for the functions that are being used. Unused features and circuitry can be powered down to save power. The following circuit blocks can be powered down, or are powered down when not selected (see the Register Map and Description section):

- The PLL section can be powered down if not needed.
- Any of the dividers are powered down when bypassed—equivalent to divide by one.
- The adjustable delay blocks on OUT5 and OUT6 are powered down when not selected.
- Any output may be powered down. However, LVPECL outputs have both a safe and an off condition. When the LVPECL output is terminated, only the safe shutdown should be used to protect the LVPECL output devices. This still consumes some power.
- The entire distribution section can be powered down when not needed.

Powering down a functional block does not cause the programming information for that block (in the registers) to be lost. This means that blocks can be powered on and off without otherwise having to reprogram the AD9510. However, synchronization is lost. A SYNC must be issued to resynchronize.

## Application to System Debugging

Besides the obvious issues revolving around designing systems to minimize signal degradations, there are several other consequences to these results worth mentioning. These are related to finding the source of mystery spurs and noise.

For instance, if the noise floor rises at the DAC output, it is most likely not caused by clock phase noise. It may be digital coupling into the output circuitry.

If a spur exists in a sampled signal, a good test to see if it comes from the clock is to change the signal amplitude. If it is from the clock it should get proportionally lower.

Analog distortion terms will change at twice (second-order distortion) or three times (third-order distortion) the rate of the signal amplitude change. Spurs due to nonlinearity in the quantizer may not change at all, or if they do change, they will change unpredictably, when the signal amplitude changes. On the other hand, spurs due to the clock will change dB for dB with the signal.

When trying to identify the source of a spur in a sampled data signal, look not only at the explicit spur frequency, which could be caused by a signal directly coupling into the output, but also at the frequency offset from the signal. For example, if a spur is 10 MHz away from the carrier, look to see if there is a 10 MHz oscillator somewhere in the system. If so, this frequency is most likely leaking in through the clock.

### References: Clock Generation and Distribution Circuits

1. B. Brannon, *Sampled Systems and the Effects of Clock Phase Noise and Jitter*, AN-756 Analog Devices, Inc., Norwood, MA.
2. B. Brannon, "Understand the effects of clock jitter and phase noise on sampled systems," **EDN**, December 7, 2004, pp. 87–96.
3. B. Brannon, "Aperture Uncertainty and ADC System Performance," AN-501, Analog Devices Inc.
4. T. Harris, "Generating Multiple Clock Outputs from the AD9540," AN-769, Analog Devices Inc.
5. P. Smith, "Little Known Characteristics of Phase Noise," AN-741, Analog Devices Inc.
6. D. Tuite, "Frequently Asked Questions: CLOCK REQUIREMENTS FOR DATA CONVERTERS," ED Online 9660.
7. J. Keip, "Speedy A/Ds Demand Stable Clocks," EE Times 03/22/2004.

## CHAPTER 8
# Analog Filters

# Introduction

Filters are networks that process signals in a frequency-dependent manner. The basic concept of a filter can be explained by examining the frequency-dependent nature of the impedance of capacitors and inductors. Consider a voltage divider where the shunt leg is a reactive impedance. As the frequency is changed, the value of the reactive impedance changes and the voltage divider ratio changes. This mechanism yields the frequency-dependent change in the input/output transfer function that is defined as the frequency response.

Filters have many practical applications. A simple, single pole, lowpass filter (the integrator) is often used to stabilize amplifiers by rolling off the gain at higher frequencies where excessive phase shift may cause oscillations.

A simple, single pole, highpass filter can be used to block DC offset in high gain amplifiers or single supply circuits. Filters can be used to separate signals, passing those of interest, and attenuating the unwanted frequencies.

An example of this is a radio receiver, where the signal you wish to process is passed through, typically with gain, while attenuating the rest of the signals. In data conversion, filters are also used to eliminate the effects of aliases in A/D systems. They are used in reconstruction of the signal at the output of a D/A as well, eliminating the higher frequency components, such as the sampling frequency and its harmonics, thus smoothing the waveform.

There are a large number of texts dedicated to filter theory. No attempt will be made to go heavily into much of the underlying math—Laplace transforms, complex conjugate poles and the like—although they will be mentioned.

While they are appropriate for describing the effects of filters and examining stability, in most cases examination of the function in the frequency domain is more illuminating.

An ideal filter will have an amplitude response that is unity (or at a fixed gain) for the frequencies of interest (called the *passband*) and zero everywhere else (called the *stopband*). The frequency at which the response changes from passband to stopband is referred to as the *cutoff frequency*.

Figure 8-1(A) shows an idealized lowpass filter. In this filter, the low frequencies are in the passband and the higher frequencies are in the stopband.

The functional complement to the lowpass filter is the highpass filter. Here, the low frequencies are in the stopband, and the high frequencies are in the passband. Figure 8-1(B) shows the idealized highpass filter.

If a highpass filter and a lowpass filter are cascaded, a *bandpass* filter is created. The bandpass filter passes a band of frequencies between a lower cutoff frequency, $f_l$, and an upper cutoff frequency, $f_h$. Frequencies below $f_l$ and above $f_h$ are in the stopband. An idealized bandpass filter is shown in Figure 8-1(C).

A complement to the bandpass filter is the *bandreject*, or *notch* filter. Here, the passbands include frequencies below $f_l$ and above $f_h$. The band from $f_l$ to $f_h$ is in the stopband. Figure 8-1(D) shows a notch response.

The idealized filters defined above, unfortunately, cannot be easily built. The transition from passband to stopband will not be instantaneous, but instead there will be a transition region. Stop band attenuation will not be infinite.

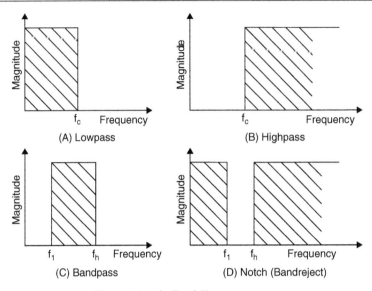

Figure 8-1: Idealized filter responses

The five parameters of a practical filter are defined in Figure 8-2.

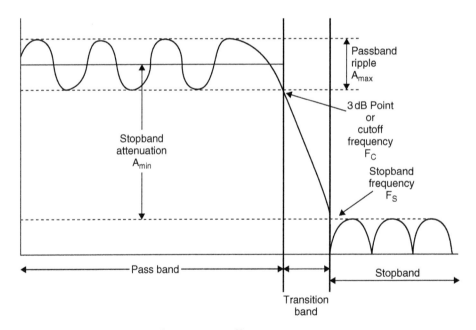

Figure 8-2: Key filter parameters

The *cutoff frequency* ($F_c$) is the frequency at which the filter response leaves the error band (or the $-3\,dB$ point for a Butterworth response filter). The *stopband frequency* ($F_S$) is the frequency at which the minimum attenuation in the stopband is reached. The *passband ripple* ($A_{max}$) is the variation (error band) in the passband response. The *minimum passband attenuation* ($A_{min}$) defines the minimum signal attenuation within the stopband. The steepness of the filter is defined as the *order* (M) of the filter. M is also the number of poles in the transfer function. A pole is a root of the denominator of the transfer function. Conversely, a zero is a root of the numerator of the transfer function. Each pole gives a $-6\,dB$/octave or $-20\,dB$/decade response. Each zero gives a $+6\,dB$/octave or $+20\,dB$/decade response.

Note that not all filters will have all these features. For instance, all-pole configurations (i.e., no zeros in the transfer function) will not have ripple in the stopband. Butterworth and Bessel filters are examples of all-pole filters with no ripple in the passband.

Typically, one or more of the above parameters will be variable. For instance, if you were to design an antialiasing filter for an analog-to-digital convertor (ADC), you will know the cutoff frequency (the maximum frequency that you want to pass), the stopband frequency, (which will generally be the Nyquist frequency ($=1/2$ the sample rate)) and the minimum attenuation required (which will be set by the resolution or dynamic range of the system). You can then go to a chart or computer program to determine the other parameters, such as filter order, $F_0$, and Q, which determine the peaking of the section, for the various sections and/or component values.

It should also be pointed out that the filter will affect the phase of a signal, as well as the amplitude. For example, a single pole section will have a 90° phase shift at the crossover frequency. A pole pair will have a 180° phase shift at the crossover frequency. The Q of the filter will determine the rate of change of the phase. This will be covered more in depth in the next section.

# The Transfer Function

## The s-Plane

Filters have a frequency-dependent response because the impedance of a capacitor or an inductor changes with frequency. Therefore, the complex impedances:

$$Z_L = sL \tag{8-1}$$

and:

$$Z_C = \frac{1}{sC} \tag{8-2}$$

are used to describe the impedance of an inductor and a capacitor, respectively,

$$s = \sigma + j\omega \tag{8-3}$$

where $\sigma$ is the Neper frequency in nepers per second (NP/s) and $\omega$ is the angular frequency in radians per second (rad/s).

By using standard circuit analysis techniques, the transfer equation of the filter can be developed. These techniques include Ohm's law, Kirchoff's voltage and current laws, and superposition, remembering that the impedances are complex. The transfer equation is then:

$$H(s) = \frac{a_m s^m + a_{m-1} s^{m-1} + \cdots + a_1 s + a_0}{b_n s^n + b_{n-1} s^{n-1} + \cdots + b_1 s + b_0} \tag{8-4}$$

Therefore, $H(s)$ is a rational function of s with real coefficients with the degree of m for the numerator and n for the denominator. The degree of the denominator is the order of the filter. Solving for the roots of the equation determines the poles (denominator) and zeros (numerator) of the circuit. Each pole will provide a $-6\,dB$/octave or $-20\,dB$/decade response. Each zero will provide a $+6\,dB$/octave or $+20\,dB$/decade response. These roots can be real or complex. When they are complex, they occur in conjugate pairs. These roots are plotted on the s-plane (complex plane) where the horizontal axis is $\sigma$ (real axis) and the vertical axis is $\omega$ (imaginary axis). How these roots are distributed on the s-plane can tell us many things about the circuit. In order to have stability, all poles must be in the left side of the plane. If we have a zero at the origin, that is a zero in the numerator, the filter will have no response at DC (highpass or bandpass).

Assume an RLC circuit, as in Figure 8-3. Using the voltage divider concept it can be shown that the voltage across the resistor is:

$$H(s) = \frac{V_o}{V_{in}} = \frac{RCs}{LCs^2 + RCs + 1} \tag{8-5}$$

Substituting the component values into the equation yields:

$$H(s) = 10^3 \times \frac{s}{s^2 + 10^3 s + 10^7} \tag{8-6}$$

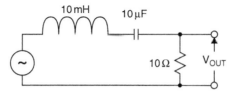

**Figure 8-3: RLC circuit**

Factoring the equation and normalizing gives:

$$H(s) = 10^3 \times \frac{s}{[s - (-0.5 + j\,3.122) \times 10^3] \times [s - (-0.5 - j\,3.122) \times 10^3]} \quad (8\text{-}7)$$

This gives a zero at the origin and a pole pair at:

$$s = (-0.5 \pm j\,3.122) \times 10^3 \quad (8\text{-}8)$$

Next, plot these points on the s-plane as shown in Figure 8-4:

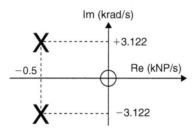

**Figure 8-4: Pole and zero plotted on the s-plane**

The above discussion has a definite mathematical flavor. In most cases, we are more interested in the circuit's performance in real applications. While working in the s-plane is completely valid, I am sure that most of us do not think in terms of Nepers and imaginary frequencies.

## $F_0$ and Q

So if it is not convenient to work in the s-plane, why go through the above discussion? The answer is that the groundwork has been set for two concepts that will be infinitely more useful in practice: $F_0$ and Q.

$F_0$ is the cutoff frequency of the filter. This is defined, in general, as the frequency where the response is down 3 dB from the passband. It can sometimes be defined as the frequency at which it will fall out of the passband. For example, a 0.1 dB Chebyshev filter can have its $F_0$ at the frequency at which the response is down >0.1 dB.

The shape of the attenuation curve (as well as the phase and delay curves, which define the time domain response of the filter) will be the same if the ratio of the actual frequency to the cutoff frequency is examined, rather than just the actual frequency itself. Normalizing the filter to 1 rad/s, a simple system for designing and comparing filters can be developed. The filter is then scaled by the cutoff frequency to determine the component values for the actual filter.

Q is the "quality factor" of the filter. It is also sometimes given as $\alpha$ where:

$$\alpha = \frac{1}{Q} \quad (8\text{-}9)$$

This is commonly known as the *damping ratio.* $\xi$ is sometimes used where:

$$\xi = 2\alpha \qquad\qquad (8\text{-}10)$$

If Q is >0.707, there will be some peaking in the filter response. If the Q is <0.707, rolloff at $F_0$ will be greater; it will have a more gentle slope and will begin sooner. The amount of peaking for a two-pole lowpass filter versus Q is shown in Figure 8-5.

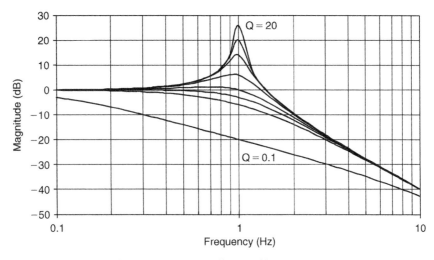

**Figure 8-5: Lowpass filter peaking versus Q**

Rewriting the transfer function H(s) in terms of $\omega_0$ and Q:

$$H(s) = \frac{H_0}{s^2 + \dfrac{\omega_0}{Q}s + \omega_0{}^2} \qquad\qquad (8\text{-}11)$$

where $H_0$ is the passband gain and $\omega_0 = 2\pi F_0$.

This is now the *lowpass prototype* that will be used to design the filters.

## Highpass Filter

Changing the numerator of the transfer equation, H(s), of the lowpass prototype to $H_0 s^2$ transforms the lowpass filter into a highpass filter. The response of the highpass filter is similar in shape to a lowpass, just inverted in frequency.

The transfer function of a highpass filter is then:

$$H(s) = \frac{H_0 s^2}{s^2 + \dfrac{\omega_0}{Q}s + \omega_0{}^2} \qquad\qquad (8\text{-}12)$$

The response of a two-pole highpass filter is illustrated in Figure 8-6.

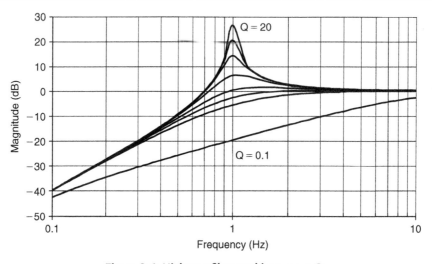

**Figure 8-6: Highpass filter peaking versus Q**

## Bandpass Filter

Changing the numerator of the lowpass prototype to $H_0\omega_0^2$ will convert the filter to a bandpass function. The transfer function of a bandpass filter is then:

$$H(s) = \frac{H_0\omega^2}{s^2 + \dfrac{\omega_0}{Q}s + \omega_0^2} \qquad (8\text{-}13)$$

$\omega_0$ here is the frequency ($F_0 = 2\pi\omega_0$) at which the gain of the filter peaks.

$H_0$ is the circuit gain and is defined:

$$H_0 = H/Q \qquad (8\text{-}14)$$

Q has a particular meaning for the bandpass response. It is the selectivity of the filter. It is defined as:

$$Q = \frac{F_0}{F_H - F_L} \qquad (8\text{-}15)$$

where $F_L$ and $F_H$ are the frequencies where the response is $-3\,dB$ from the maximum.

The bandwidth (BW) of the filter is described as:

$$BW = F_H - F_L \qquad (8\text{-}16)$$

It can be shown that the resonant frequency ($F_0$) is the geometric mean of $F_L$ and $F_H$, which means that $F_0$ will appear half way between $F_L$ and $F_H$ on a logarithmic scale.

$$F_0 = \sqrt{F_H\, F_L} \qquad (8\text{-}17)$$

Also, note that the skirts of the bandpass response will always be symmetrical around $F_0$ on a logarithmic scale.

The response of a bandpass filter to various values of Q are shown in Figure 8-7.

A word of caution is appropriate here. Bandpass filters can be defined in two different ways. The narrowband case is the classic definition that we have shown above.

In some cases, however, if the high and low cutoff frequencies are widely separated, the bandpass filter is constructed out of separate highpass and lowpass sections. Widely separated in this context means separated by at least 2 octaves ($\times 4$ in frequency). This is the wideband case.

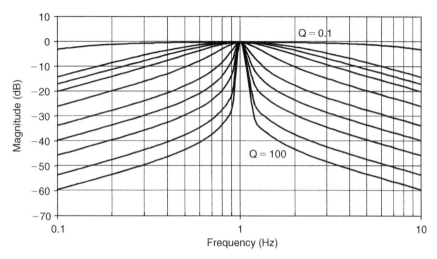

**Figure 8-7: Bandpass filter peaking versus Q**

## Bandreject (Notch) Filter

By changing the numerator to $s^2 + \omega_z^2$, we convert the filter to a bandreject or notch filter. As in the bandpass case, if the corner frequencies of the bandreject filter are separated by more than an octave (the wideband case), it can be built out of separate lowpass and highpass sections. We will adopt the following convention: a narrowband bandreject filter will be referred to as a *notch* filter and the wideband bandreject filter will be referred to as *bandreject* filter.

A notch (or bandreject) transfer function is:

$$H(s) = \frac{H_0(s^2 + \omega_z^2)}{s^2 + \dfrac{\omega_0}{Q}s + \omega_0^2} \tag{8-18}$$

There are three cases of the notch filter characteristics. These are illustrated in Figure 8-8. The relationship of the pole frequency, $\omega_0$, and the zero frequency, $\omega_z$, determines if the filter is a standard notch, a lowpass notch, or a highpass notch.

If the zero frequency is equal to the pole frequency a standard notch exists. In this instance, the zero lies on the $j\omega$-plane where the curve that defines the pole frequency intersects the axis.

A lowpass notch occurs when the zero frequency is greater than the pole frequency. In this case, $\omega_z$ lies outside the curve of the pole frequencies. What this means in a practical sense is that the filter's response below $\omega_z$ will be greater than the response above $\omega_z$. This results in an elliptical lowpass filter.

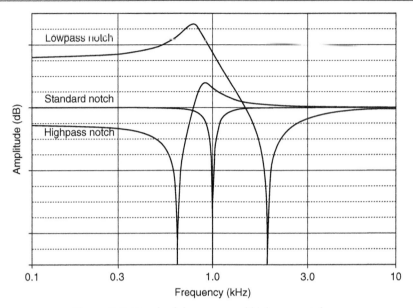

Figure 8-8: Standard, lowpass, and highpass notches

A highpass notch filter occurs when the zero frequency is less than the pole frequency. In this case, $\omega_z$ lies inside the curve of the pole frequencies. What this means in a practical sense is that the filters response below $\omega_z$ will be less than the response above $\omega_z$. This results in an elliptical highpass filter.

The variation of the notch width with Q is shown in Figure 8-9.

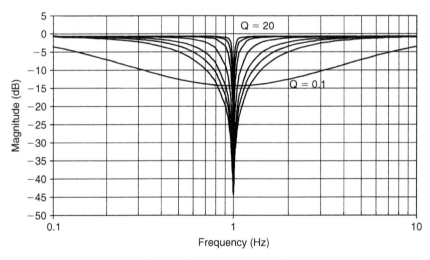

Figure 8-9: Notch filter width versus frequency for various Q values

## Allpass Filter

There is another type of filter that leaves the amplitude of the signal intact but introduces phase shift. This type of filter is called an *allpass*. The purpose of this filter is to add phase shift (delay) to the response of the circuit. The amplitude of an allpass is unity for all frequencies. The phase response, however, changes from 0° to 360° as the frequency is swept from 0 to ∞. The purpose of an allpass filter is to provide phase equalization, typically in pulse circuits. It also has application in single side band, suppressed carrier (SSB–SC) modulation circuits.

The transfer function of an allpass filter is:

$$H(s) = \frac{s^2 - \dfrac{\omega_0}{Q} s + \omega_0^2}{s^2 + \dfrac{\omega_0}{Q} s + \omega_0^2} \tag{8-19}$$

Note that an allpass transfer function can be synthesized as:

$$H_{AP} = H_{LP} - H_{BP} + H_{HP} = 1 - 2H_{BP} \tag{8-20}$$

Figure 8-10 compares the various filter types.

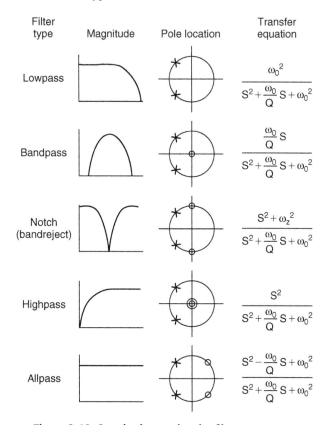

| Filter type | Magnitude | Pole location | Transfer equation |
|---|---|---|---|
| Lowpass | | | $\dfrac{\omega_0^2}{S^2 + \dfrac{\omega_0}{Q} S + \omega_0^2}$ |
| Bandpass | | | $\dfrac{\dfrac{\omega_0}{Q} S}{S^2 + \dfrac{\omega_0}{Q} S + \omega_0^2}$ |
| Notch (bandreject) | | | $\dfrac{S^2 + \omega_z^2}{S^2 + \dfrac{\omega_0}{Q} S + \omega_0^2}$ |
| Highpass | | | $\dfrac{S^2}{S^2 + \dfrac{\omega_0}{Q} S + \omega_0^2}$ |
| Allpass | | | $\dfrac{S^2 - \dfrac{\omega_0}{Q} S + \omega_0^2}{S^2 + \dfrac{\omega_0}{Q} S + \omega_0^2}$ |

Figure 8-10: Standard second-order filter responses

## Phase Response

As mentioned earlier, a filter will change the phase of the signal as well as the amplitude. The question is, does this make a difference? Fourier analysis indicates that a square wave is made up of a fundamental frequency and odd-order harmonics. The magnitude and phase responses, of the various harmonics are precisely defined. If the magnitude or phase relationships are changed, then the summation of the harmonics will not add back together properly to give a square wave. It will instead be distorted, typically showing overshoot and ringing or a slow rise time. This would also hold for any complex waveform.

Each pole of a filter will add 45° of phase shift at the corner frequency. The phase will vary from 0° (well below the corner frequency) to 90° (well beyond the corner frequency). The start of the change can be more than a decade away. In multipole filters, each of the poles will add phase shift, so that the total phase shift will be multiplied by the number of poles (180° total shift for a two-pole system, 270° for a three pole system, etc.).

The phase response of a single pole, lowpass filter is:

$$\phi(\omega) = -\arctan \frac{\omega}{\omega_0} \tag{8-21}$$

The phase response of a lowpass pole pair is:

$$\phi(\omega) = -\arctan \left[\frac{1}{\alpha}\left(2\frac{\omega}{\omega_0} + \sqrt{4-\alpha^2}\right)\right]$$
$$-\arctan \left[\frac{1}{\alpha}\left(2\frac{\omega}{\omega_0} - \sqrt{4-\alpha^2}\right)\right] \tag{8-22}$$

For a single pole highpass filter the phase response is:

$$\phi(\omega) = \frac{\pi}{2} - \arctan \frac{\omega}{\omega_0} \tag{8-23}$$

The phase response of a highpass pole pair is:

$$\phi(\omega) = \pi - \arctan \left[\frac{1}{\alpha}\left(2\frac{\omega}{\omega_0} + \sqrt{4-\alpha^2}\right)\right]$$
$$- \arctan \left[\frac{1}{\alpha}\left(2\frac{\omega}{\omega_0} - \sqrt{4-\alpha^2}\right)\right] \tag{8-24}$$

The phase response of a bandpass filter is:

$$\phi(\omega) = \frac{\pi}{2} - \arctan \left(\frac{2Q\omega}{\omega_0} + \sqrt{4Q^2-1}\right)$$
$$-\arctan \left(\frac{2Q\omega}{\omega_0} - \sqrt{4Q^2-1}\right) \tag{8-25}$$

The variation of the phase shift with frequency due to various values of Q is shown in Figure 8-11 (for lowpass, highpass, bandpass, and allpass) and in Figure 8-12 (for notch).

It is also useful to look at the change of phase with frequency. This is the group delay of the filter. A flat (constant) group delay gives best phase response, but, unfortunately, it also gives the least amplitude discrimination. The group delay of a single lowpass pole is:

$$\tau(\omega) = -\frac{d\phi(\omega)}{d\omega} = \frac{\cos^2\phi}{\omega_0} \tag{8-26}$$

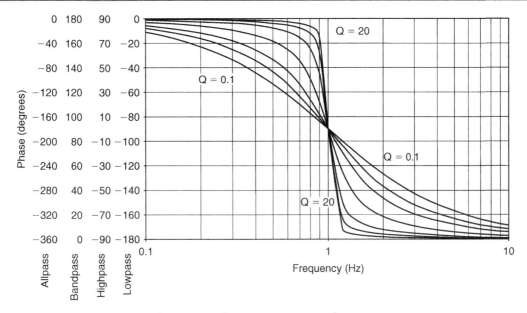

**Figure 8-11: Phase response versus frequency**

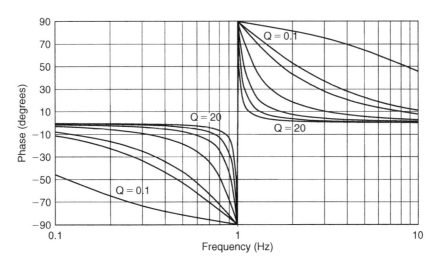

**Figure 8-12: Notch filter phase response**

For the lowpass pole pair it is:

$$\tau(\omega) = \frac{2 \sin^2 \phi}{\alpha \omega_0} - \frac{\sin 2\phi}{2\omega} \tag{8-27}$$

For the single highpass pole it is:

$$\tau(\omega) = -\frac{d\phi(\omega)}{d\omega} = \frac{\sin^2 \phi}{\omega_0} \tag{8-28}$$

**595**

For the highpass pole pair it is:

$$\tau(\omega) = \frac{2\sin^2\phi}{\alpha\omega_0} - \frac{\sin 2\phi}{2\omega} \tag{8-29}$$

And for the bandpass pole pair it is:

$$\tau(\omega) = \frac{2Q\, 2\cos^2\phi}{\alpha\omega_0} + \frac{\sin 2\phi}{2\omega} \tag{8-30}$$

## The Effect of Nonlinear Phase

A waveform can be represented by a series of frequencies of specific amplitude, frequency, and phase relationships. For example, a square wave is:

$$F(t) = A\left(\frac{1}{2} + \frac{2}{\pi}\sin\omega t + \frac{2}{3\pi}\sin 3\omega t + \frac{2}{5\pi}\sin 5\omega t + \frac{2}{7\pi}\sin 7\omega t + \cdots\right) \tag{8-31}$$

If this waveform were passed through a filter, the amplitude and phase response of the filter to the various frequency components of the waveform could be different. If the phase delays were identical, the waveform would pass through the filter undistorted. If, however, the different components of the waveform were changed due to different amplitude and phase response of the filter to those frequencies, they would no longer add up in the same manner. This would change the shape of the waveform. These distortions would manifest themselves in what we typically call overshoot and ringing of the output.

Not all signals will be composed of harmonically related components. An amplitude modulated (AM) signal, for instance, will consist of a carrier and two sidebands at $\pm$ the modulation frequency. If the filter does not have the same delay for the various waveform components, then "envelope delay" will occur and the output wave will be distorted.

Linear phase shift results in constant group delay since the derivative of a linear function is a constant.

# *Time Domain Response*

Up until now the discussion has been primarily focused on the frequency domain response of filters. The time domain response can also be of concern, particularly under transient conditions. Moving between the time domain and the frequency domain is accomplished by the use of the Fourier and Laplace transforms. This yields a method of evaluating performance of the filter to a non-sinusoidal excitation.

The transfer function of a filter is the ratio of the output to input time functions. It can be shown that the impulse response of a filter defines its bandwidth. The time domain response is a practical consideration in many systems, particularly communications, where many modulation schemes use both amplitude and phase information.

## *Impulse Response*

The impulse function is defined as an infinitely high, infinitely narrow pulse, with an area of unity. This is, of course, impossible to realize in a physical sense. If the impulse width is much less than the rise time of the filter, the resulting response of the filter will give a reasonable approximation actual impulse response of the filter response.

The impulse response of a filter, in the time domain, is proportional to the bandwidth of the filter in the frequency domain. The narrower the impulse, the wider the bandwidth of the filter. The pulse amplitude is equal to $\omega_c/\pi$, which is also proportional to the filter bandwidth, the height being taller for wider bandwidths. The pulse width is equal to $2\pi/\omega_c$, which is inversely proportional to bandwidth. It turns out that the product of the amplitude and the bandwidth is a constant.

It would be a non-trivial task to calculate the response of a filter without the use of Laplace and Fourier transforms. The Laplace transform converts multiplication and division to addition and subtraction, respectively. This takes equations, which are typically loaded with integration and/or differentiation, and turns them into simple algebraic equations, which are much easier to deal with. The Fourier transform works in the opposite direction.

The details of these transforms will not be discussed here. However, some general observations on the relationship of the impulse response to the filter characteristics will be made.

It can be shown, as stated, that the impulse response is related to the bandwidth. Therefore, amplitude discrimination (the ability to distinguish between the desired signal from other, out of band signals and noise) and time response are inversely proportional. That is to say that the filters with the best amplitude response are the ones with the worst-time response. For all-pole filters, the Chebyshev filter gives the best amplitude discrimination, followed by the Butterworth and then the Bessel.

If the time domain response were ranked, the Bessel would be best, followed by the Butterworth and then the Chebyshev. Details of the different filter responses will be discussed in the next section.

The impulse response also increases with increasing filter order. Higher filter order implies greater bandlimiting, therefore degraded time response. Each section of a multistage filter will have its own impulse response, and the total impulse response is the accumulation of the individual responses. The degradation in the time response can also be related to the fact that as frequency discrimination is increased, the Q of the individual sections tends to increase. The increase in Q increases the overshoot and ringing of the individual sections, which implies longer time response.

## Step Response

The step response of a filter is the integral of the impulse response. Many of the generalities that apply to the impulse response also apply to the step response. The slope of the rise time of the step response is equal to the peak response of the impulse. The product of the bandwidth of the filter and the rise time is a constant. Just as the impulse has a function equal to unity, the step response has a function equal to 1/s. Both of these expressions can be normalized, since they are dimensionless.

The step response of a filter is useful in determining the envelope distortion of a modulated signal. The two most important parameters of a filter's step response are the overshoot and ringing. Overshoot should be minimal for good pulse response. Ringing should decay as fast as possible, so as not to interfere with subsequent pulses.

Real life signals typically are not made up of impulse pulses or steps, so the transient response curves do not give a completely accurate estimation of the output. They are, however, a convenient figure of merit so that the transient responses of the various filter types can be compared on an equal footing.

Since the calculations of the step and impulse response are mathematically intensive, they are most easily performed by computer. Many CAD (computer aided design) software packages have the ability to calculate these responses. Several of these responses are also collected in the next section.

# Standard Responses

There are many transfer functions that may satisfy the attenuation and/or phase requirements of a particular filter. The one that you choose will depend on the particular system. The importance of the frequency domain response versus the time domain response must be determined. Also, both of these considerations might be traded off against filter complexity, and thereby cost.

## Butterworth

The Butterworth filter is the best compromise between attenuation and phase response. It has no ripple in the passband or the stopband, and because of this it is sometimes called a maximally flat filter. The Butterworth filter achieves its flatness at the expense of a relatively wide transition region from passband to stopband, with average transient characteristics.

The normalized poles of the Butterworth filter fall on the unit circle (in the s-plane). The pole positions are given by:

$$-\sin\frac{(2K-1)\pi}{2n} + j\cos\frac{(2K-1)\pi}{2n} \quad K = 1, 2, ...., n \tag{8-32}$$

where K is the pole pair number, and n is the number of poles.

The poles are spaced equidistant on the unit circle, which means the angles between the poles are equal.

Given the pole locations, $\omega_0$ and $\alpha$ (or Q) can be determined. These values can then be used to determine the component values of the filter. The design tables for passive filters use frequency and impedance normalized filters. They are normalized to a frequency of 1 rad/s and impedance of $1\,\Omega$. These filters can be denormalized to determine actual component values. This allows the comparison of the frequency domain and/or time domain responses of the various filters on equal footing. The Butterworth filter is normalized for a $-3\,dB$ response at $\omega_0 = 1$.

The values of the elements of the Butterworth filter are more practical and less critical than many other filter types. The frequency response, group delay, impulse response, and step response are shown in Figure 8-15. The pole locations and corresponding $\omega_0$ and $\alpha$ terms are tabulated in Figure 8-26.

## Chebyshev

The Chebyshev (or Chevyshev, Tschebychev, Tschebyscheff or Tchevysheff, depending on how you translate from Russian) filter has a smaller transition region than the same-order Butterworth filter, at the expense of ripples in its passband. This filter gets its name because the Chebyshev filter minimizes the height of the maximum ripple, which is the Chebyshev criterion.

Chebyshev filters have 0 dB relative attenuation at DC. Odd-order filters have an attenuation band that extends from 0 dB to the ripple value. Even-order filters have a gain equal to the passband ripple. The number of cycles of ripple in the passband is equal to the order of the filter.

The poles of the Chebyshev filter can be determined by moving the poles of the Butterworth filter to the right, forming an ellipse. This is accomplished by multiplying the real part of the pole by $k_r$ and the imaginary part by $k_I$. The values $k_r$ and $k_I$ can be computed by:

$$K_r = \sinh A \tag{8-33}$$

$$K_I = \cosh A \tag{8-34}$$

where:

$$A = \frac{1}{n} \sinh^{-1} \frac{1}{\varepsilon} \tag{8-35}$$

where n is the filter order and:

$$\varepsilon = \sqrt{10^R - 1} \tag{8-36}$$

where:

$$R = \frac{R_{dB}}{10} \tag{8-37}$$

where:

$$R_{dB} = \text{passband ripple in dB} \tag{8-38}$$

The Chebyshev filters are typically normalized so that the edge of the ripple band is at $\omega_0 = 1$. The 3 dB bandwidth is given by:

$$A_{3\,dB} = \frac{1}{n} \cosh^{-1}\left(\frac{1}{\varepsilon}\right) \tag{8-39}$$

This is tabulated in Table 8-1.

Table 8-1: 3 dB Bandwidth to ripple bandwidth for Chebyshev filters

| Order | 0.01 dB | 0.1 dB | 0.25 dB | 0.5 dB | 1 dB |
|-------|---------|--------|---------|--------|------|
| 2     | 3.30362 | 1.93432 | 1.59814 | 1.38974 | 1.21763 |
| 3     | 1.87718 | 1.38899 | 1.25289 | 1.16749 | 1.09487 |
| 4     | 1.46690 | 1.21310 | 1.13977 | 1.09310 | 1.05300 |
| 5     | 1.29122 | 1.13472 | 1.08872 | 1.05926 | 1.03381 |
| 6     | 1.19941 | 1.09293 | 1.06134 | 1.04103 | 1.02344 |
| 7     | 1.14527 | 1.06800 | 1.04495 | 1.03009 | 1.01721 |
| 8     | 1.11061 | 1.05193 | 1.03435 | 1.02301 | 1.01316 |
| 9     | 1.08706 | 1.04095 | 1.02711 | 1.01817 | 1.01040 |
| 10    | 1.07033 | 1.03313 | 1.02194 | 1.01471 | 1.00842 |

The frequency response, group delay, impulse response and step response are cataloged in Figures 8-16 to 8-20 on following pages, for various values of passband ripple (0.01, 0.1, 0.25, 0.5, and 1 dB). The pole locations and corresponding $\omega_0$ and $\alpha$ terms for these values of ripple are tabulated in Figures 8-27 to 8-31.

## Bessel

Butterworth filters have fairly good amplitude and transient behavior. The Chebyshev filters improve on the amplitude response at the expense of transient behavior. The Bessel filter is optimized to obtain better transient response due to a linear phase (i.e., constant delay) in the passband. This means that there will be relatively poorer frequency response (less amplitude discrimination).

The poles of the Bessel filter can be determined by locating all of the poles on a circle and separating their imaginary parts by:

$$\frac{2}{n} \qquad (8\text{-}40)$$

where n is the number of poles. Note that the top and bottom poles are distanced by where the circle crosses the jω-axis by:

$$\frac{1}{n} \qquad (8\text{-}41)$$

or half the distance between the other poles.

The frequency response, group delay, impulse response, and step response for the Bessel filters are cataloged in Figure 8-21. The pole locations and corresponding $\omega_0$ and $\alpha$ terms for the Bessel filter are tabulated in Figure 8-32.

## Linear Phase with Equiripple Error

The linear phase filter offers linear phase response in the passband, over a wider range than the Bessel, and superior attenuation far from cutoff. This is accomplished by letting the phase response have ripples, similar to the amplitude ripples of the Chebyshev. As the ripple is increased, the region of constant delay extends further into the stopband. This will also cause the group delay to develop ripples, since it is the derivative of the phase response. The step response will show slightly more overshoot than the Bessel and the impulse response will show a bit more ringing.

It is difficult to compute the pole locations of a linear phase filter. Pole locations are taken from the Williams book (see Reference 2), which, in turn, comes from the Zverev book (see Reference 1).

The frequency response, group delay, impulse response, and step response for linear phase filters of 0.05° ripple and 0.5° ripple are given in Figures 8-22 and 8-23. The pole locations and corresponding $\omega_0$ and $\alpha$ terms are tabulated in Figures 8-33 and 8-34.

## Transitional Filters

A transitional filter is a compromise between a Gaussian filter, which is similar to a Bessel, and the Chebyshev. A transitional filter has nearly linear phase shift and smooth, monotonic rolloff in the passband. Above the passband there is a break point beyond which the attenuation increases dramatically compared to the Bessel, and especially at higher values of n.

Two transition filters have been tabulated. These are the Gaussian to 6 dB and Gaussian to 12 dB.

The Gaussian to 6 dB filter has better transient response than the Butterworth in the passband. Beyond the break point, which occurs at $\omega = 1.5$, the rolloff is similar to the Butterworth.

The Gaussian to 12 dB filter's transient response is much better than Butterworth in the passband. Beyond the 12 dB break point, which occurs at $\omega = 2$, the attenuation is less than the Butterworth.

As is the case with the linear phase filters, pole locations for transitional filters do not have a closed form method for computation. Again, pole locations are taken from Williams's book (see Reference 2). These were derived from iterative techniques.

The frequency response, group delay, impulse response, and step response for Gaussian to 12 and 6 dB are shown in Figures 8-24 and 8-25. The pole locations and corresponding $\omega_0$ and $\alpha$ terms are tabulated in Figures 8-35 and 8-36.

## Comparison of All-Pole Responses

The responses of several all-pole filters, namely the Bessel, Butterworth, and Chebyshev (in this case of 0.5 dB ripple) will now be compared. An eight-pole filter is used as the basis for the comparison. The responses have been normalized for a cutoff of 1 Hz. Comparing Figures 8-13 and 8-14, it is easy to see the tradeoffs in the response types. Moving from Bessel through Butterworth to Chebyshev, notice that the amplitude discrimination improves as the transient behavior gets progressively poorer.

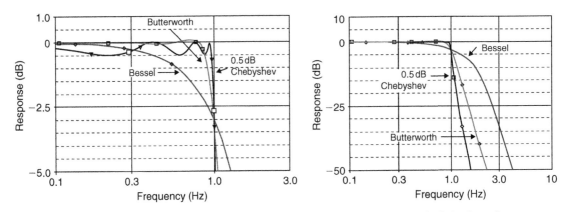

Figure 8-13: Comparison of amplitude response of Bessel, Butterworth, and Chebyshev Filters

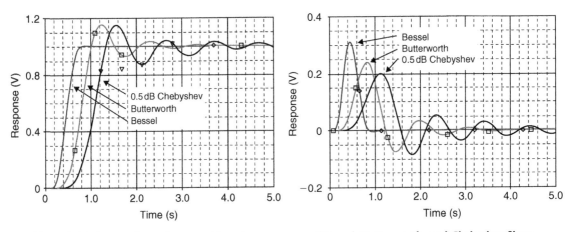

Figure 8-14: Comparison of step and impulse responses of Bessel, Butterworth, and Chebyshev filters

## Elliptical

The previously mentioned filters are all-pole designs, which mean that the zeros of the transfer function (roots of the numerator) are at one of the two extremes of the frequency range (0 or $\infty$). For a lowpass filter, the zeros are at $f = \infty$. If finite frequency transfer function zeros are added to poles an elliptical filter (sometimes referred to as Cauer filters) is created. This filter has a shorter transition region than the Chebyshev filter because it allows ripple in both the stopband and passband. It is the addition of zeros in the stopband that causes ripple in the stopband but gives a much higher rate of attenuation, the most possible for a given number of poles. There will be some "bounceback" of the stopband response between the zeros. This is the stopband ripple. The elliptical filter also has degraded time domain response.

Since the poles of an elliptic filter are on an ellipse, the time response of the filter resembles that of the Chebyshev.

An elliptic filter is defined by the parameters shown in Figure 8-2, those being $A_{max}$, the maximum ripple in the passband; $A_{min}$, the minimum attenuation in the stopband; $F_c$, the cutoff frequency, which is where the frequency response leaves the passband ripple; and $F_S$, the stopband frequency, where the value of $A_{max}$ is reached.

An alternate approach is to define a filter order n, the modulation angle, $\theta$, which defines the rate of attenuation in the transition band, where:

$$\theta = \sin^{-1} \frac{1}{F_s} \tag{8-42}$$

and $\rho$ which determines the passband ripple, where:

$$\rho = \sqrt{\frac{\varepsilon^2}{1 + \varepsilon^2}} \tag{8-43}$$

where $\varepsilon$ is the ripple factor developed for the Chebyshev response, and the passband ripple is:

$$R_{dB} = -10 \log(1 - \rho^2) \tag{8-44}$$

Some general observations can be made. For a given filter order n, and $\theta$, $A_{min}$ increases as the ripple is made larger. Also, as $\theta$ approaches $90°$, $F_S$ approaches $F_C$. This results in extremely short transition region, which means sharp rolloff. This comes at the expense of lower $A_{min}$.

As a side note, $\rho$ determines the input resistance of a passive elliptical filter, which can then be related to the voltage standing wave ratio (VSWR).

Because of the number of variables in the design of an elliptic filter, it is difficult to provide the type of tables provided for the previous filter types. Several CAD packages can provide the design values. Alternatively several sources, such as Williams's (see Reference 2), provide tabulated filter values. These tables classify the filter by

$$C \; n \; \rho \; \theta$$

where the C denotes Cauer. Elliptical filters are sometime referred to as Cauer filters after the network theorist Wilhelm Cauer.

## Maximally Flat Delay With Chebyshev Stopband

Bessel type (Bessel, linear phase with equiripple error and transitional) filters give excellent transient behavior, but less than ideal frequency discrimination. Elliptical filters give better frequency discrimination, but degraded transient response. A maximally flat delay with Chebyshev stopband filter takes a Bessel type function and adds transmission zeros. The constant delay properties of the Bessel type filter in the passband are maintained, and the stopband attenuation is significantly improved. The step response exhibits no overshoot or ringing, and the impulse response is clean, with essentially no oscillatory behavior. Constant group delay properties extend well into the stopband for increasing n.

As with the elliptical filter, numeric evaluation is difficult. Williams's book (see Reference 2) tabulates passive prototypes normalized component values.

## Inverse Chebyshev

The Chebyshev response has ripple in the passband and a monotonic stopband. The inverse Chebyshev response can be defined that has a monotonic passband and ripple in the stopband. The inverse Chebyshev has better passband performance than even the Butterworth. It is also better than the Chebyshev, except very near the cutoff frequency. In the transition band, the inverse Chebyshev has the steepest rolloff. Therefore, the inverse Chebyshev will meet the $A_{min}$ specification at the lowest frequency of the three. In the stopband there will, however, be response lobes which have a magnitude of:

$$\frac{\varepsilon^2}{(1 - \varepsilon)} \tag{8-45}$$

where $\varepsilon$ is the ripple factor defined for the Chebyshev case. This means that deep into the stopband, both the Butterworth and Chebyshev will have better attenuation, since they are monotonic in the stopband. In terms of transient performance, the inverse Chebyshev lies midway between the Butterworth and the Chebyshev.

The inverse Chebyshev response can be generated in three steps. First take a Chebyshev lowpass filter. Then subtract this response from 1. Finally, invert in frequency by replacing $\omega$ with $1/\omega$.

These are by no means all the possible transfer functions, but they do represent the most common.

## Using the Prototype Response Curves

In the following pages, the response curves and the design tables for several of the lowpass prototypes of the all-pole responses will be cataloged. All the curves are normalized to a $-3\,dB$ cutoff frequency of 1 Hz. This allows direct comparison of the various responses. In all cases, the amplitude response for 2- through 10-pole cases for the frequency range of 0.1–10 Hz. will be shown. Then a detail of the amplitude response in the 0.1–2 Hz passband will be shown. The group delay from 0.1 to 10 Hz and the impulse response and step response from 0 to 5 seconds will also be shown.

To use these curves to determine the response of real life filters, they must be denormalized. In the case of the amplitude responses, this is simply accomplished by multiplying the frequency axis by the desired cutoff frequency $F_C$. To denormalize the group delay curves, we divide the delay axis by $2\pi F_C$, and multiply the frequency axis by $F_C$, as before. Denormalize the step response by dividing the time axis by $2\pi F_C$. Denormalize the impulse response by dividing the time axis by $2\pi F_C$ and multiplying the amplitude axis by the same amount.

For a highpass filter, simply invert the frequency axis for the amplitude response. In transforming a lowpass filter into a highpass (or bandreject) filter, the transient behavior is not preserved. Zverev (see Reference 1) provides a computational method for calculating these responses.

In transforming a lowpass into a narrowband bandpass, the 0 Hz axis is moved to the center frequency $F_0$. It stands to reason that the response of the bandpass case around the center frequency would then match the lowpass response around 0 Hz. The frequency response curve of a lowpass filter actually mirrors itself around 0 Hz, although we generally do not concern ourselves with negative frequency.

To denormalize the group delay curve for a bandpass filter, divide the delay axis by $\pi BW$, where BW is the 3 dB bandwidth in hertz. Then multiply the frequency axis by BW/2. In general, the delay of the

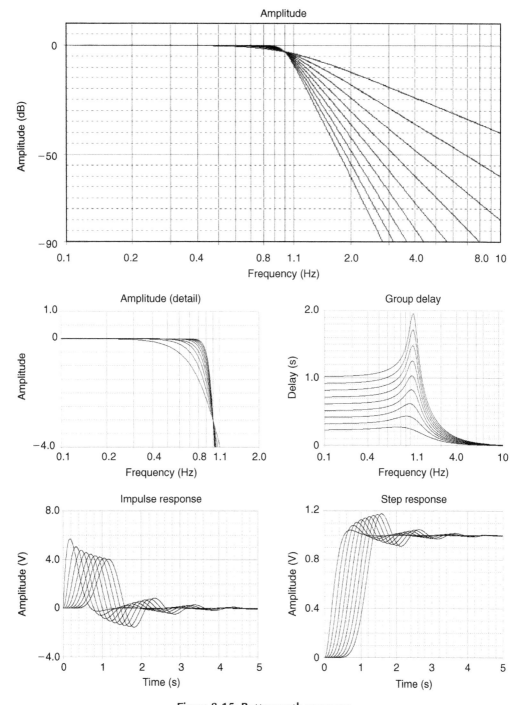

**Figure 8-15: Butterworth response**

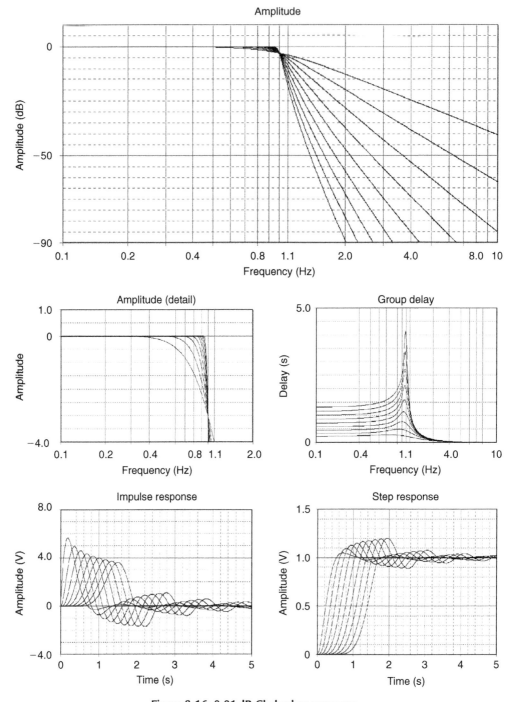

Figure 8-16: 0.01 dB Chebyshev response

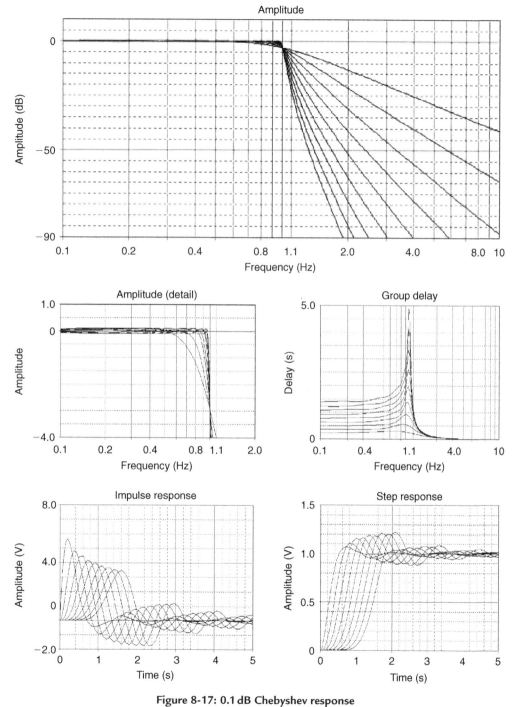

**Figure 8-17: 0.1 dB Chebyshev response**

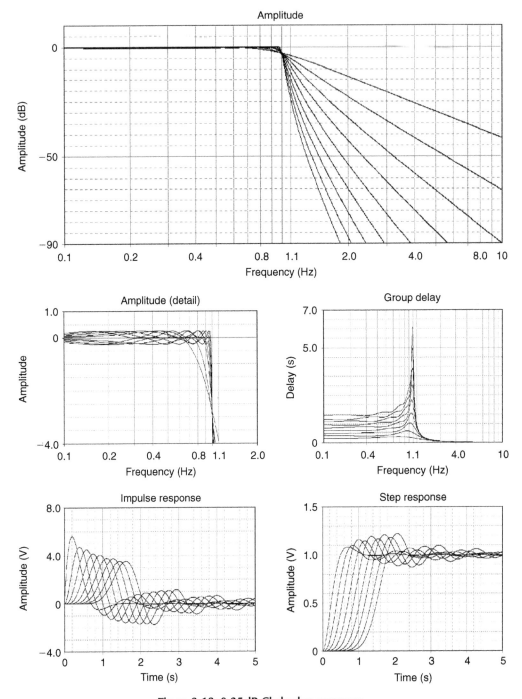

**Figure 8-18: 0.25 dB Chebyshev response**

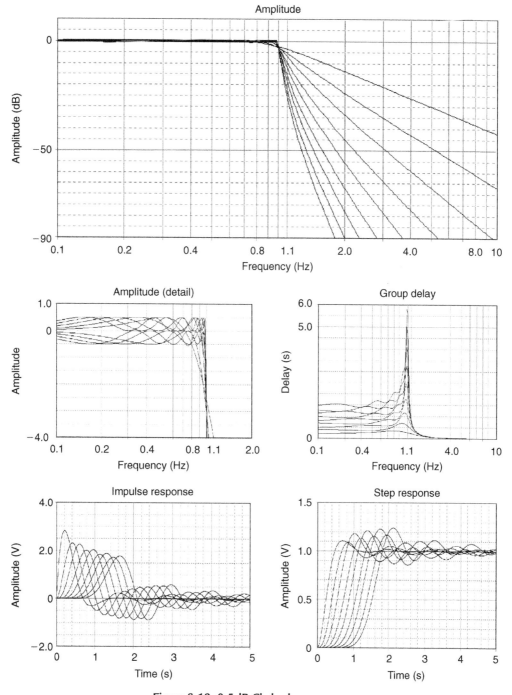

**Figure 8-19: 0.5 dB Chebyshev response**

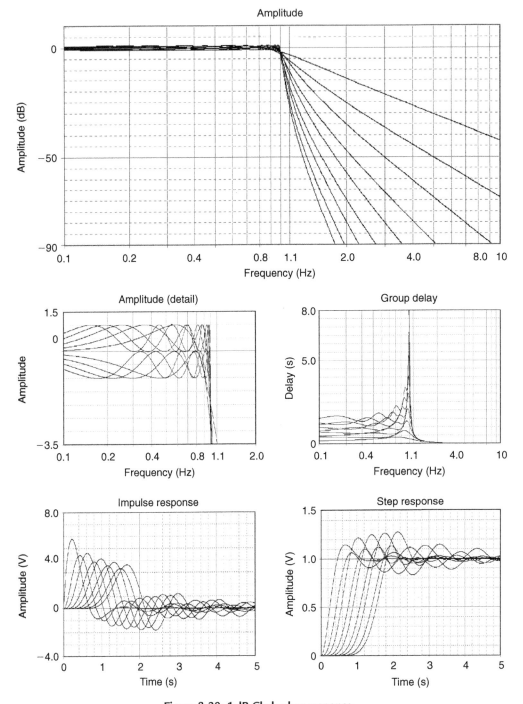

**Figure 8-20: 1 dB Chebyshev response**

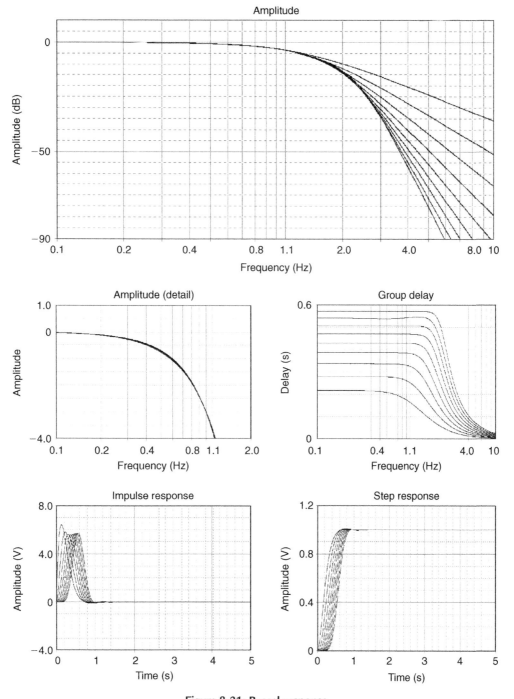

**Figure 8-21: Bessel response**

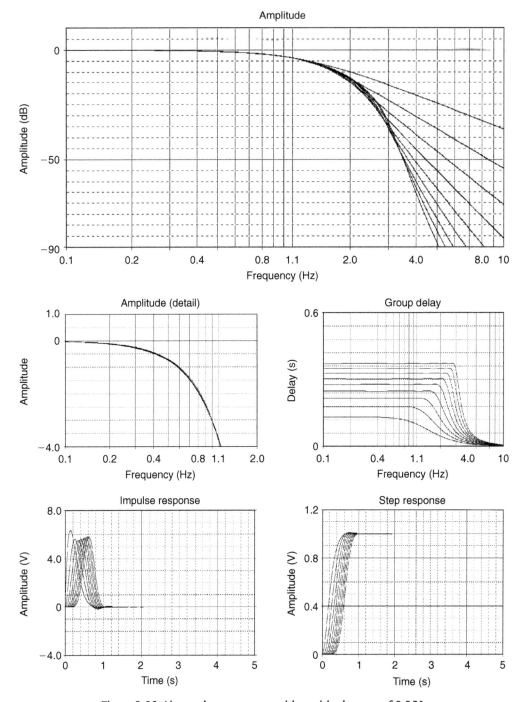

Figure 8-22: Linear phase response with equiripple error of 0.05°

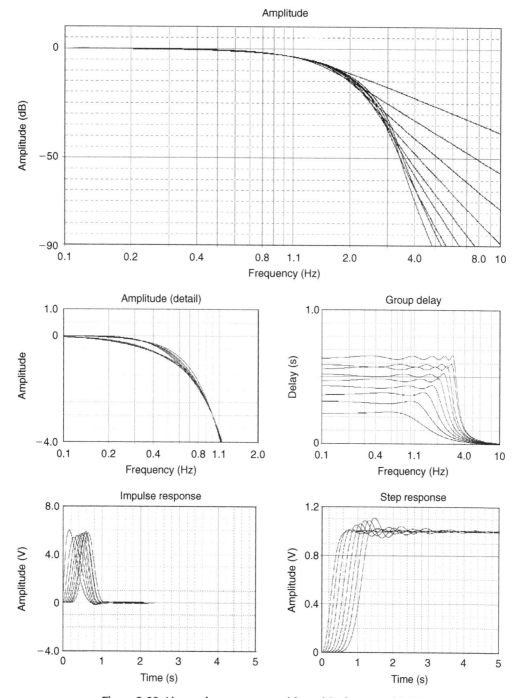

**Figure 8-23: Linear phase response with equiripple error of 0.5°**

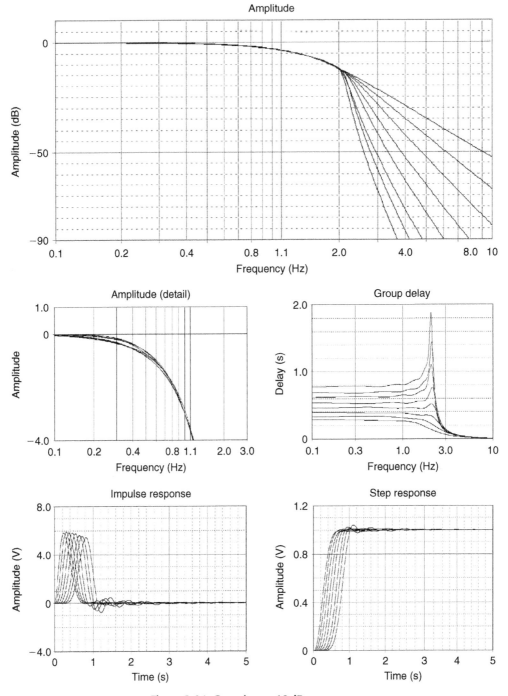

**Figure 8-24: Gaussian to 12 dB response**

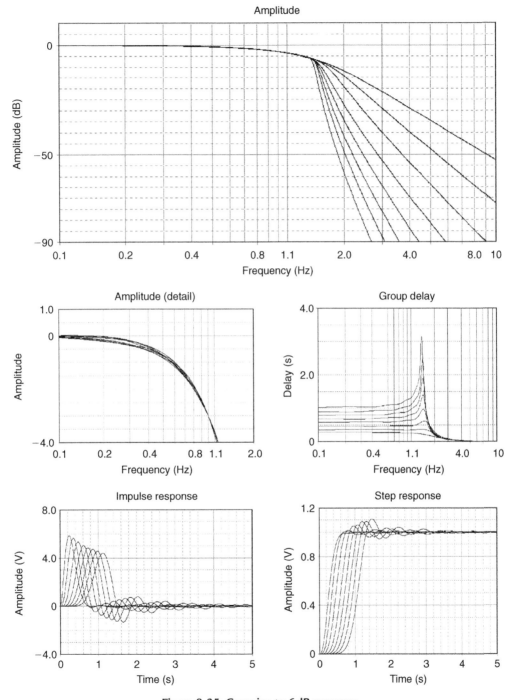

**Figure 8-25: Gaussian to 6 dB response**

| Order | Section | Real part | Imaginary part | $F_0$ | α | Q | −3 dB frequency | Peaking frequency | Peaking level |
|---|---|---|---|---|---|---|---|---|---|
| 2 | 1 | 0.7071 | 0.7071 | 1.0000 | 1.4142 | 0.7071 | 1.0000 | | |
| 3 | 1 | 0.5000 | 0.8660 | 1.0000 | 1.0000 | 1.0000 | | 0.7071 | 1.2493 |
| | 2 | 1.0000 | | 1.0000 | | | 1.0000 | | |
| 4 | 1 | 0.9239 | 0.3827 | 1.0000 | 1.8478 | 0.5412 | 0.7195 | | |
| | 2 | 0.3827 | 0.9239 | 1.0000 | 0.7654 | 1.3065 | | 0.8409 | 3.0102 |
| 5 | 1 | 0.8090 | 0.5878 | 1.0000 | 1.6180 | 0.6180 | 0.8588 | | |
| | 2 | 0.3090 | 0.9511 | 1.0000 | 0.6180 | 1.6182 | | 0.8995 | 4.6163 |
| | 3 | 1.0000 | | 1.0000 | | | 1.0000 | | |
| 6 | 1 | 0.9659 | 0.2588 | 1.0000 | 1.9319 | 0.5176 | 0.6758 | | |
| | 2 | 0.7071 | 0.7071 | 1.0000 | 1.4142 | 0.7071 | 1.0000 | | |
| | 3 | 0.2588 | 0.9659 | 1.0000 | 0.5176 | 1.9319 | | 0.9306 | 6.0210 |
| 7 | 1 | 0.9010 | 0.4339 | 1.0000 | 1.8019 | 0.5550 | 0.7449 | | |
| | 2 | 0.6235 | 0.7818 | 1.0000 | 1.2470 | 0.8019 | | 0.4717 | 0.2204 |
| | 3 | 0.2225 | 0.9749 | 1.0000 | 0.4450 | 2.2471 | | 0.9492 | 7.2530 |
| | 4 | 1.0000 | | 1.0000 | | | 1.0000 | | |
| 8 | 1 | 0.9808 | 0.1951 | 1.0000 | 1.9616 | 0.5098 | 0.6615 | | |
| | 2 | 0.8315 | 0.5556 | 1.0000 | 1.6629 | 0.6013 | 0.8295 | | |
| | 3 | 0.5556 | 0.8315 | 1.0000 | 1.1112 | 0.9000 | | 0.6186 | 0.6876 |
| | 4 | 0.1951 | 0.9808 | 1.0000 | 0.3902 | 2.5628 | | 0.9612 | 8.3429 |
| 9 | 1 | 0.9397 | 0.3420 | 1.0000 | 1.8794 | 0.5321 | 0.7026 | | |
| | 2 | 0.7660 | 0.6428 | 1.0000 | 1.5320 | 0.6527 | 0.9172 | | |
| | 3 | 0.5000 | 0.8660 | 1.0000 | 1.0000 | 1.0000 | | 0.7071 | 1.2493 |
| | 4 | 0.1737 | 0.9848 | 1.0000 | 0.3474 | 2.8785 | | 0.9694 | 9.3165 |
| | 5 | 1.0000 | | 1.0000 | | | 1.0000 | | |
| 10 | 1 | 0.9877 | 0.1564 | 1.0000 | 1.9754 | 0.5062 | 0.6549 | | |
| | 2 | 0.8910 | 0.4540 | 1.0000 | 1.7820 | 0.5612 | 0.7564 | | |
| | 3 | 0.7071 | 0.7071 | 1.0000 | 1.4142 | 0.7071 | 1.0000 | | |
| | 4 | 0.4540 | 0.8910 | 1.0000 | 0.9080 | 1.1013 | | 0.7667 | 1.8407 |
| | 5 | 0.1564 | 0.9877 | 1.0000 | 0.3128 | 3.1970 | | 0.9752 | 10.2023 |

**Figure 8-26: Butterworth design table**

bandpass filter at $F_0$ will be twice the delay of the lowpass prototype with the same bandwidth at 0 Hz. This is due to the fact that the lowpass to bandpass transformation results in a filter with order 2n, even though it is typically referred to as having the same order as the lowpass filter from which it is derived. This approximation holds for narrowband filters. As the bandwidth of the filter is increased, some distortion of the curve occurs. The delay becomes less symmetrical, peaking below $F_0$.

The envelope of the response of a bandpass filter resembles the step response of the lowpass prototype. More exactly, it is almost identical to the step response of a lowpass filter having half the bandwidth. To determine the envelope response of the bandpass filter, divide the time axis of the step response of the lowpass prototype by πBW, where BW is the 3 dB bandwidth. The previous discussions of overshoot, ringing, etc. can now be applied to the carrier envelope.

The envelope of the response of a narrowband bandpass filter to a short burst of carrier (i.e., where the burst width is much less than the rise time of the denormalized step response of the bandpass filter) can be determined by denormalizing the impulse response of the lowpass prototype. To do this, multiply the amplitude axis and divide the time axis by πBW, where BW is the 3 dB bandwidth. It is assumed that the carrier frequency is high enough so that many cycles occur during the burst interval.

While the group delay, step, and impulse curves cannot be used directly to predict the distortion to the waveform caused by the filter, they are useful figures of merit when used to compare filters.

| Order | Section | Real part | Imaginary part | $F_0$ | α | Q | −3 dB frequency | Peaking frequency | Peaking level |
|---|---|---|---|---|---|---|---|---|---|
| 2 | 1 | 0.6743 | 0.7075 | 0.9774 | 1.3798 | 0.7247 | | 0.2142 | 0.0100 |
| 3 | 1 | 0.4233 | 0.8663 | 0.9642 | 0.8780 | 1.1389 | | 0.7558 | 2.0595 |
|   | 2 | 0.8467 | | 0.8467 | | | 0.8467 | | |
| 4 | 1 | 0.6762 | 0.3828 | 0.7770 | 1.7405 | 0.5746 | 0.6069 | | |
|   | 2 | 0.2801 | 0.9241 | 0.9656 | 0.5801 | 1.7237 | | 0.8806 | 5.1110 |
| 5 | 1 | 0.5120 | 0.5879 | 0.7796 | 1.3135 | 0.7613 | | 0.2889 | 0.0827 |
|   | 2 | 0.1956 | 0.9512 | 0.9711 | 0.4028 | 2.4824 | | 0.9309 | 8.0772 |
|   | 3 | 0.6328 | | 0.6328 | | | 0.6328 | | |
| 6 | 1 | 0.5335 | 0.2588 | 0.5930 | 1.7995 | 0.5557 | 0.4425 | | |
|   | 2 | 0.3906 | 0.7072 | 0.8079 | 0.9670 | 1.0342 | | 0.5895 | 1.4482 |
|   | 3 | 0.1430 | 0.9660 | 0.9765 | 0.2929 | 3.4144 | | 0.9554 | 10.7605 |
| 7 | 1 | 0.4393 | 0.4339 | 0.6175 | 1.4229 | 0.7028 | 0.6136 | | |
|   | 2 | 0.3040 | 0.7819 | 0.8389 | 0.7247 | 1.3798 | | 0.7204 | 3.4077 |
|   | 3 | 0.1085 | 0.9750 | 0.9810 | 0.2212 | 4.5208 | | 0.9689 | 13.1578 |
|   | 4 | 0.4876 | | 0.4876 | | | 0.4876 | | |
| 8 | 1 | 0.4268 | 0.1951 | 0.4693 | 1.8190 | 0.5498 | 0.3451 | | |
|   | 2 | 0.3168 | 0.5556 | 0.6396 | 0.9907 | 1.0094 | | 0.4564 | 1.3041 |
|   | 3 | 0.2418 | 0.8315 | 0.8659 | 0.5585 | 1.7906 | | 0.7956 | 5.4126 |
|   | 4 | 0.0849 | 0.9808 | 0.9845 | 0.1725 | 5.7978 | | 0.9771 | 15.2977 |
| 9 | 1 | 0.3686 | 0.3420 | 0.5028 | 1.4661 | 0.6821 | 0.4844 | | |
|   | 2 | 0.3005 | 0.6428 | 0.7096 | 0.8470 | 1.1807 | | 0.5682 | 2.3008 |
|   | 3 | 0.1961 | 0.8661 | 0.8880 | 0.4417 | 2.2642 | | 0.8436 | 7.3155 |
|   | 4 | 0.0681 | 0.9848 | 0.9872 | 0.1380 | 7.2478 | | 0.9824 | 17.2249 |
|   | 5 | 0.3923 | | 0.3923 | | | 0.3923 | | |
| 10 | 1 | 0.3522 | 0.1564 | 0.3854 | 1.8279 | 0.5471 | 0.2814 | | |
|   | 2 | 0.3178 | 0.454 | 0.5542 | 1.1469 | 0.8719 | | 0.3242 | 0.5412 |
|   | 3 | 0.2522 | 0.7071 | 0.7507 | 0.6719 | 1.4884 | | 0.6606 | 3.9742 |
|   | 4 | 0.1619 | 0.891 | 0.9056 | 0.3576 | 2.7968 | | 0.8762 | 9.0742 |
|   | 5 | 0.0558 | 0.9877 | 0.9893 | 0.1128 | 8.8645 | | 0.9861 | 18.9669 |

Figure 8-27: 0.01 dB Chebyshev design table

| Order | Section | Real part | Imaginary part | $F_0$ | α | Q | −3 dB frequency | Peaking frequency | Peaking level |
|---|---|---|---|---|---|---|---|---|---|
| 2 | 1 | 0.6104 | 0.7106 | 0.9368 | 1.3032 | 0.7673 | | 0.3638 | 0.0999 |
| 3 | 1 | 0.3490 | 0.8684 | 0.9359 | 0.7458 | 1.3408 | | 0.7952 | 3.1978 |
|   | 2 | 0.6970 | | 0.6970 | | | 0.6970 | | |
| 4 | 1 | 0.2177 | 0.9254 | 0.9507 | 0.4580 | 2.1834 | | 0.8994 | 7.0167 |
|   | 2 | 0.5257 | 0.3833 | 0.6506 | 1.6160 | 0.6188 | 0.5596 | | |
| 5 | 1 | 0.3842 | 0.5884 | 0.7027 | 1.0935 | 0.9145 | | 0.4457 | 0.7662 |
|   | 2 | 0.1468 | 0.9521 | 0.9634 | 0.3048 | 3.2812 | | 0.9407 | 10.4226 |
|   | 3 | 0.4749 | | 0.4749 | | | 0.4749 | | |
| 6 | 1 | 0.3916 | 0.2590 | 0.4695 | 1.6682 | 0.5995 | 0.3879 | | |
|   | 2 | 0.2867 | 0.7077 | 0.7636 | 0.7509 | 1.3316 | | 0.6470 | 3.1478 |
|   | 3 | 0.1049 | 0.9667 | 0.9724 | 0.2158 | 4.6348 | | 0.9610 | 13.3714 |
| 7 | 1 | 0.3178 | 0.4341 | 0.5380 | 1.1814 | 0.8464 | | 0.2957 | 0.4157 |
|   | 2 | 0.2200 | 0.7823 | 0.8126 | 0.5414 | 1.8469 | | 0.7507 | 5.6595 |
|   | 3 | 0.0785 | 0.9755 | 0.9787 | 0.1604 | 6.2335 | | 0.9723 | 15.9226 |
|   | 4 | 0.3528 | | 0.3528 | | | 0.3528 | | |
| 8 | 1 | 0.3058 | 0.1952 | 0.3628 | 1.6858 | 0.5932 | 0.2956 | | |
|   | 2 | 0.2529 | 0.5558 | 0.6106 | 0.8283 | 1.2073 | | 0.4949 | 2.4532 |
|   | 3 | 0.1732 | 0.8319 | 0.8497 | 0.4077 | 2.4531 | | 0.8137 | 7.9784 |
|   | 4 | 0.0608 | 0.9812 | 0.9831 | 0.1237 | 8.0819 | | 0.9793 | 18.1669 |
| 9 | 1 | 0.2622 | 0.3421 | 0.4310 | 1.2166 | 0.8219 | | 0.2197 | 0.3037 |
|   | 2 | 0.2137 | 0.6430 | 0.6776 | 0.6308 | 1.5854 | | 0.6064 | 4.4576 |
|   | 3 | 0.1395 | 0.8663 | 0.8775 | 0.3180 | 3.1450 | | 0.8550 | 10.0636 |
|   | 4 | 0.0485 | 0.9852 | 0.9864 | 0.0982 | 10.1795 | | 0.9840 | 20.1650 |
|   | 5 | 0.2790 | | 0.2790 | | | 0.2790 | | |
| 10 | 1 | 0.2493 | 0.1564 | 0.2943 | 1.6942 | 0.5902 | 0.2382 | | |
|   | 2 | 0.2249 | 0.4541 | 0.5067 | 0.8876 | 1.1266 | | 0.3945 | 1.9880 |
|   | 3 | 0.1785 | 0.7073 | 0.7295 | 0.4894 | 2.0434 | | 0.6844 | 6.4750 |
|   | 4 | 0.1146 | 0.8913 | 0.8986 | 0.2551 | 3.9208 | | 0.8839 | 11.9386 |
|   | 5 | 0.0395 | 0.9880 | 0.9888 | 0.0799 | 12.5163 | | 0.9872 | 21.9565 |

Figure 8-28: 0.1 dB Chebyshev design table

| Order | Section | Real part | Imaginary part | $F_0$ | $\alpha$ | Q | −3 dB frequency | Peaking frequency | Peaking level |
|---|---|---|---|---|---|---|---|---|---|
| 2 | 1 | 0.5621 | 0.7154 | 0.9098 | 1.2356 | 0.8093 | | 0.4425 | 0.2502 |
| 3 | 1 | 0.3062 | 0.8712 | 0.9234 | 0.6632 | 1.5079 | | 0.8156 | 4.0734 |
| | 2 | 0.6124 | | 0.6124 | | | 0.6124 | | |
| 4 | 1 | 0.4501 | 0.3840 | 0.5916 | 1.5215 | 0.6572 | 0.5470 | | |
| | 2 | 0.1865 | 0.9272 | 0.9458 | 0.3944 | 2.5356 | | 0.9082 | 8.2538 |
| 5 | 1 | 0.3247 | 0.5892 | 0.6727 | 0.9653 | 1.0359 | | 0.4917 | 1.4585 |
| | 2 | 0.1240 | 0.9533 | 0.9613 | 0.2580 | 3.8763 | | 0.9452 | 11.8413 |
| | 3 | 0.4013 | | 0.4013 | | | 0.4013 | | |
| 6 | 1 | 0.3284 | 0.2593 | 0.4184 | 1.5697 | 0.6371 | 0.3730 | | |
| | 2 | 0.2404 | 0.7083 | 0.7480 | 0.6428 | 1.5557 | | 0.6663 | 4.3121 |
| | 3 | 0.0880 | 0.9675 | 0.9715 | 0.1811 | 5.5205 | | 0.9635 | 14.8753 |
| 7 | 1 | 0.2652 | 0.4344 | 0.5090 | 1.0421 | 0.9596 | | 0.3441 | 1.0173 |
| | 2 | 0.1835 | 0.7828 | 0.8040 | 0.4565 | 2.1908 | | 0.7610 | 7.0443 |
| | 3 | 0.0655 | 0.9761 | 0.9783 | 0.1339 | 7.4679 | | 0.9739 | 17.4835 |
| | 4 | 0.2944 | | 0.2944 | | | 0.2944 | | |
| 8 | 1 | 0.2543 | 0.1953 | 0.3206 | 1.5862 | 0.6304 | 0.2822 | | |
| | 2 | 0.2156 | 0.5561 | 0.5964 | 0.7230 | 1.3832 | | 0.5126 | 3.4258 |
| | 3 | 0.1441 | 0.8323 | 0.8447 | 0.3412 | 2.9309 | | 0.8197 | 9.4683 |
| | 4 | 0.0506 | 0.9817 | 0.9830 | 0.1029 | 9.7173 | | 0.9804 | 19.7624 |
| 9 | 1 | 0.2176 | 0.3423 | 0.4056 | 1.0730 | 0.9320 | | 0.2642 | 0.8624 |
| | 2 | 0.1774 | 0.6433 | 0.6673 | 0.5317 | 1.8808 | | 0.6184 | 5.8052 |
| | 3 | 0.1158 | 0.8667 | 0.8744 | 0.2649 | 3.7755 | | 0.8589 | 11.6163 |
| | 4 | 0.0402 | 0.9856 | 0.9864 | 0.0815 | 12.2659 | | 0.9848 | 21.7812 |
| | 5 | 0.2315 | | 0.2315 | | | 0.2315 | | |
| 10 | 1 | 0.2065 | 0.1565 | 0.2591 | 1.5940 | 0.6274 | 0.2267 | | |
| | 2 | 0.1863 | 0.4543 | 0.4910 | 0.7588 | 1.3178 | | 0.4143 | 3.0721 |
| | 3 | 0.1478 | 0.7075 | 0.7228 | 0.4090 | 2.4451 | | 0.6919 | 7.9515 |
| | 4 | 0.0949 | 0.8915 | 0.8965 | 0.2117 | 4.7236 | | 0.8864 | 13.5344 |
| | 5 | 0.0327 | 0.9883 | 0.9888 | 0.0661 | 15.1199 | | 0.9878 | 23.5957 |

Figure 8-29: 0.25 dB Chebyshev design table

| Order | Section | Real part | Imaginary part | $F_0$ | $\alpha$ | Q | −3 dB frequency | Peaking frequency | Peaking level |
|---|---|---|---|---|---|---|---|---|---|
| 2 | 1 | 0.5129 | 0.7225 | 1.2314 | 1.1577 | 0.8638 | | 0.7072 | 0.5002 |
| 3 | 1 | 0.2683 | 0.8753 | 1.0688 | 0.5861 | 1.7061 | | 0.9727 | 5.0301 |
| | 2 | 0.5366 | | 0.6265 | | | 0.6265 | | |
| 4 | 1 | 0.3872 | 0.3850 | 0.5969 | 1.4182 | 0.7051 | 0.5951 | | |
| | 2 | 0.1605 | 0.9297 | 1.0313 | 0.3402 | 2.9391 | | 1.0010 | 9.4918 |
| 5 | 1 | 0.2767 | 0.5902 | 0.6905 | 0.8490 | 1.1779 | | 0.5522 | 2.2849 |
| | 2 | 0.1057 | 0.9550 | 1.0178 | 0.2200 | 4.5451 | | 1.0054 | 13.2037 |
| | 3 | 0.3420 | | 0.3623 | | | 0.3623 | | |
| 6 | 1 | 0.2784 | 0.2596 | 0.3963 | 1.4627 | 0.6836 | 0.3827 | | |
| | 2 | 0.2037 | 0.7091 | 0.7680 | 0.5522 | 1.8109 | | 0.7071 | 5.5025 |
| | 3 | 0.0746 | 0.9687 | 1.0114 | 0.1536 | 6.5119 | | 1.0055 | 16.2998 |
| 7 | 1 | 0.2241 | 0.4349 | 0.5040 | 0.9161 | 1.0916 | | 0.3839 | 1.7838 |
| | 2 | 0.1550 | 0.7836 | 0.8228 | 0.3881 | 2.5767 | | 0.7912 | 8.3880 |
| | 3 | 0.0553 | 0.9771 | 1.0081 | 0.1130 | 8.8487 | | 1.0049 | 18.9515 |
| | 4 | 0.2487 | | 0.2562 | | | 0.2562 | | |
| 8 | 1 | 0.2144 | 0.1955 | 0.2968 | 1.4779 | 0.6767 | 0.2835 | | |
| | 2 | 0.1817 | 0.5565 | 0.5989 | 0.6208 | 1.6109 | | 0.5361 | 4.5815 |
| | 3 | 0.1214 | 0.8328 | 0.8610 | 0.2885 | 3.4662 | | 0.8429 | 10.8885 |
| | 4 | 0.0426 | 0.9824 | 1.0060 | 0.0867 | 11.5305 | | 1.0041 | 21.2452 |
| 9 | 1 | 0.1831 | 0.3425 | 0.3954 | 0.9429 | 1.0605 | | 0.2947 | 1.6023 |
| | 2 | 0.1493 | 0.6436 | 0.6727 | 0.4520 | 2.2126 | | 0.6374 | 7.1258 |
| | 3 | 0.0974 | 0.8671 | 0.8884 | 0.2233 | 4.4779 | | 0.8773 | 13.0759 |
| | 4 | 0.0338 | 0.9861 | 1.0046 | 0.0686 | 14.5829 | | 1.0034 | 23.2820 |
| | 5 | 0.1949 | | 0.1984 | | | 0.1984 | | |
| 10 | 1 | 0.1736 | 0.1566 | 0.2338 | 1.4851 | 0.6734 | 0.2221 | | |
| | 2 | 0.1566 | 0.4545 | 0.4807 | 0.6515 | 1.5349 | | 0.4257 | 4.2087 |
| | 3 | 0.1243 | 0.7078 | 0.7186 | 0.3459 | 2.8907 | | 0.6968 | 9.3520 |
| | 4 | 0.0798 | 0.8919 | 0.8955 | 0.1782 | 5.6107 | | 0.8883 | 15.0149 |
| | 5 | 0.0275 | 0.9887 | 0.9891 | 0.0556 | 17.9833 | | 0.9883 | 25.1008 |

Figure 8-30: 0.5 dB Chebyshev design table

| Order | Section | Real part | Imaginary part | $F_0$ | $\alpha$ | Q | −3 dB frequency | Peaking frequency | Peaking level |
|---|---|---|---|---|---|---|---|---|---|
| 2 | 1 | 0.4508 | 0.7351 | 0.8623 | 1.0456 | 0.9564 | | 0.5806 | 0.9995 |
| 3 | 1 | 0.2257 | 0.8822 | 0.9106 | 0.4957 | 2.0173 | | 0.8528 | 6.3708 |
|   | 2 | 0.4513 | | 0.4513 | | | 0.4513 | | |
| 4 | 1 | 0.3199 | 0.3868 | 0.5019 | 1.2746 | 0.7845 | | 0.2174 | 0.1557 |
|   | 2 | 0.1325 | 0.9339 | 0.9433 | 0.2809 | 3.5594 | | 0.9245 | 11.1142 |
| 5 | 1 | 0.2265 | 0.5918 | 0.6337 | 0.7149 | 1.3988 | | 0.5467 | 3.5089 |
|   | 2 | 0.0865 | 0.9575 | 0.9614 | 0.1800 | 5.5559 | | 0.9536 | 14.9305 |
|   | 3 | 0.2800 | | 0.2800 | | | 0.2800 | | |
| 6 | 1 | 0.2268 | 0.2601 | 0.3461 | 1.3144 | 0.7608 | | 0.1273 | 0.0813 |
|   | 2 | 0.1550 | 0.7106 | 0.7273 | 0.4262 | 2.3462 | | 0.6935 | 7.6090 |
|   | 3 | 0.0608 | 0.9707 | 0.9726 | 0.1249 | 8.0036 | | 0.9688 | 18.0827 |
| 7 | 1 | 0.1819 | 0.4354 | 0.4719 | 0.7710 | 1.2971 | | 0.3956 | 2.9579 |
|   | 2 | 0.1259 | 0.7846 | 0.7946 | 0.3169 | 3.1558 | | 0.7744 | 10.0927 |
|   | 3 | 0.0449 | 0.9785 | 0.9795 | 0.0918 | 10.8982 | | 0.9775 | 20.7563 |
|   | 4 | 0.2019 | | 0.2019 | | | 0.2019 | | |
| 8 | 1 | 0.1737 | 0.1956 | 0.2616 | 1.3280 | 0.7530 | | 0.0899 | 0.0611 |
|   | 2 | 0.1473 | 0.5571 | 0.5762 | 0.5112 | 1.9560 | | 0.5373 | 6.1210 |
|   | 3 | 0.0984 | 0.8337 | 0.8395 | 0.2344 | 4.2657 | | 0.8279 | 12.6599 |
|   | 4 | 0.0346 | 0.9836 | 0.9842 | 0.0702 | 14.2391 | | 0.9830 | 23.0750 |
| 9 | 1 | 0.1482 | 0.3427 | 0.3734 | 0.7938 | 1.2597 | | 0.3090 | 2.7498 |
|   | 2 | 0.1208 | 0.6442 | 0.6554 | 0.3686 | 2.7129 | | 0.6328 | 8.8187 |
|   | 3 | 0.0788 | 0.8679 | 0.8715 | 0.1809 | 5.5268 | | 0.8643 | 14.8852 |
|   | 4 | 0.0274 | 0.9869 | 0.9873 | 0.0555 | 18.0226 | | 0.9865 | 25.1197 |
|   | 5 | 0.1577 | | 0.1577 | | | 0.1577 | | |
| 10 | 1 | 0.1403 | 0.1567 | 0.2103 | 1.3341 | 0.7496 | | 0.0698 | 0.0530 |
|   | 2 | 0.1266 | 0.4546 | 0.4721 | 0.5363 | 1.8645 | | 0.4366 | 5.7354 |
|   | 3 | 0.1005 | 0.7084 | 0.7155 | 0.2809 | 3.5597 | | 0.7012 | 11.1147 |
|   | 4 | 0.0645 | 0.8926 | 0.8949 | 0.1441 | 6.9374 | | 0.8903 | 16.8466 |
|   | 5 | 0.0222 | 0.9895 | 0.9897 | 0.0449 | 22.2916 | | 0.9893 | 26.9650 |

Figure 8-31: 1 dB Chebyshev design table

| Order | Section | Real part | Imaginary part | $F_0$ | $\alpha$ | Q | −3 dB frequency | Peaking frequency | Peaking level |
|---|---|---|---|---|---|---|---|---|---|
| 2 | 1 | 1.1050 | 0.6368 | 1.2754 | 1.7328 | 0.5771 | 1.0020 | | |
| 3 | 1 | 1.0509 | 1.0025 | 1.4524 | 1.4471 | 0.6910 | 1.4185 | | |
|   | 2 | 1.3270 | | 1.3270 | | | 1.3270 | | |
| 4 | 1 | 1.3596 | 0.4071 | 1.4192 | 1.9160 | 0.5219 | 0.9705 | | |
|   | 2 | 0.9877 | 1.2476 | 1.5912 | 1.2414 | 0.8055 | | 0.7622 | 0.2349 |
| 5 | 1 | 1.3851 | 0.7201 | 1.5611 | 1.7745 | 0.5635 | 1.1876 | | |
|   | 2 | 0.9606 | 1.4756 | 1.7607 | 1.0911 | 0.9165 | | 1.1201 | 0.7768 |
|   | 3 | 1.5069 | | 1.5069 | | | 1.5069 | | |
| 6 | 1 | 1.5735 | 0.3213 | 1.6060 | 1.9596 | 0.5103 | 1.0638 | | |
|   | 2 | 1.3836 | 0.9727 | 1.6913 | 1.6361 | 0.6112 | 1.4323 | | |
|   | 3 | 0.9318 | 1.6640 | 1.9071 | 0.9772 | 1.0234 | | 1.3786 | 1.3851 |
| 7 | 1 | 1.6130 | 0.5896 | 1.7174 | 1.8784 | 0.5324 | 1.2074 | | |
|   | 2 | 1.3797 | 1.1923 | 1.8235 | 1.5132 | 0.6608 | 1.6964 | | |
|   | 3 | 0.9104 | 1.8375 | 2.0507 | 0.8879 | 1.1262 | | 1.5961 | 1.9850 |
|   | 4 | 1.6853 | | 1.6853 | | | 1.6853 | | |
| 8 | 1 | 1.7627 | 0.2737 | 1.7838 | 1.9763 | 0.5060 | 1.1675 | | |
|   | 2 | 0.8955 | 2.0044 | 2.1953 | 0.8158 | 1.2258 | | 1.7932 | 2.5585 |
|   | 3 | 1.3780 | 1.3926 | 1.9591 | 1.4067 | 0.7109 | | 0.2011 | 0.0005 |
|   | 4 | 1.6419 | 0.8256 | 1.8378 | 1.7868 | 0.5597 | 1.3849 | | |
| 9 | 1 | 1.8081 | 0.5126 | 1.8794 | 1.9242 | 0.5197 | 1.2774 | | |
|   | 2 | 1.6532 | 1.0319 | 1.9488 | 1.6966 | 0.5894 | 1.5747 | | |
|   | 3 | 1.3683 | 1.5685 | 2.0815 | 1.3148 | 0.7606 | | 0.7668 | 0.0807 |
|   | 4 | 0.8788 | 2.1509 | 2.3235 | 0.7564 | 1.3220 | | 1.9632 | 3.0949 |
|   | 5 | 1.8575 | | 1.8575 | | | 1.8575 | | |
| 10 | 1 | 1.9335 | 0.2451 | 1.9490 | 1.9841 | 0.5040 | 1.2685 | | |
|   | 2 | 1.8467 | 0.7335 | 1.9870 | 1.8587 | 0.5380 | 1.4177 | | |
|   | 3 | 1.6661 | 1.2246 | 2.0678 | 1.6115 | 0.6205 | 1.7848 | | |
|   | 4 | 1.3648 | 1.7395 | 2.2110 | 1.2346 | 0.8100 | | 1.0785 | 0.2531 |
|   | 5 | 0.8686 | 2.2991 | 2.4580 | 0.7067 | 1.4150 | | 2.1291 | 3.5944 |

Figure 8-32: Bessel design table

| Order | Section | Real part | Imaginary part | $F_0$ | $\alpha$ | Q | −3 dB frequency | Peaking frequency | Peaking level |
|---|---|---|---|---|---|---|---|---|---|
| 2 | 1 | 1.0087 | 0.6680 | 1.2098 | 1.6675 | 0.5997 | 0.9999 | | |
| 3 | 1 | 0.8541 | 1.0725 | 1.3710 | 1.2459 | 0.8026 | | 0.6487 | 0.2232 |
| | 2 | 1.0459 | | 1.0459 | | | 1.0459 | | |
| 4 | 1 | 0.9648 | 0.4748 | 1.0753 | 1.7945 | 0.5573 | 0.8056 | | |
| | 2 | 0.7448 | 1.4008 | 1.5865 | 0.9389 | 1.0650 | | 1.1864 | 1.6286 |
| 5 | 1 | 0.8915 | 0.8733 | 1.2480 | 1.4287 | 0.6999 | 1.2351 | | |
| | 2 | 0.6731 | 1.7085 | 1.8363 | 0.7331 | 1.3641 | | 1.5703 | 3.3234 |
| | 3 | 0.9430 | | 0.9430 | | | 0.9430 | | |
| 6 | 1 | 0.8904 | 0.4111 | 0.9807 | 1.8158 | 0.5507 | 0.7229 | | |
| | 2 | 0.8233 | 1.2179 | 1.4701 | 1.1201 | 0.8928 | | 0.8975 | 0.6495 |
| | 3 | 0.6152 | 1.9810 | 2.0743 | 0.5932 | 1.6859 | | 1.8831 | 4.9365 |
| 7 | 1 | 0.8425 | 0.7791 | 1.1475 | 1.4684 | 0.6810 | 1.1036 | | |
| | 2 | 0.7708 | 1.5351 | 1.7177 | 0.8975 | 1.1143 | | 1.3276 | 1.9162 |
| | 3 | 0.5727 | 2.2456 | 2.3175 | 0.4942 | 2.0233 | | 2.1713 | 6.3948 |
| | 4 | 0.8615 | | 0.8615 | | | 0.8615 | | |
| 8 | 1 | 0.8195 | 0.3711 | 0.8996 | 1.8219 | 0.5489 | 0.6600 | | |
| | 2 | 0.7930 | 1.1054 | 1.3604 | 1.1658 | 0.8578 | | 0.7701 | 0.4705 |
| | 3 | 0.7213 | 1.8134 | 1.9516 | 0.7392 | 1.3528 | | 1.6638 | 3.2627 |
| | 4 | 0.5341 | 2.4761 | 2.5330 | 0.4217 | 2.3713 | | 2.4178 | 7.6973 |
| 9 | 1 | 0.7853 | 0.7125 | 1.0604 | 1.4812 | 0.6751 | 1.0102 | | |
| | 2 | 0.7555 | 1.4127 | 1.6020 | 0.9432 | 1.0602 | | 1.1937 | 1.6005 |
| | 3 | 0.6849 | 2.0854 | 2.1950 | 0.6241 | 1.6024 | | 1.9697 | 4.5404 |
| | 4 | 0.5060 | 2.7133 | 2.7601 | 0.3667 | 2.7274 | | 2.6657 | 8.8633 |
| | 5 | 0.7983 | | 0.7983 | | | 0.7983 | | |
| 10 | 1 | 0.7592 | 0.3413 | 0.8324 | 1.8241 | 0.5482 | 0.6096 | | |
| | 2 | 0.7467 | 1.0195 | 1.2637 | 1.1818 | 0.8462 | | 0.6941 | 0.4145 |
| | 3 | 0.7159 | 1.6836 | 1.8295 | 0.7826 | 1.2778 | | 1.5238 | 2.8507 |
| | 4 | 0.6475 | 2.3198 | 2.4085 | 0.5377 | 1.8598 | | 2.2276 | 5.7152 |
| | 5 | 0.4777 | 2.9128 | 2.9517 | 0.3237 | 3.0895 | | 2.8734 | 9.9130 |

Figure 8-33: Linear phase with equiripple error of 0.05° design table

| Order | Section | Real part | Imaginary part | $F_0$ | $\alpha$ | Q | −3 dB frequency | Peaking frequency | Peaking level |
|---|---|---|---|---|---|---|---|---|---|
| 2 | 1 | 0.8590 | 0.6981 | 1.1069 | 1.5521 | 0.6443 | 1.0000 | | |
| 3 | 1 | 06969 | 1.1318 | 1.3292 | 1.0486 | 0.9536 | | 0.8918 | 0.9836 |
| | 2 | 0.8257 | | 0.8257 | | | 0.8257 | | |
| 4 | 1 | 0.7448 | 0.5133 | 0.9045 | 1.6468 | 0.6072 | 0.7597 | | |
| | 2 | 0.6037 | 1.4983 | 1.6154 | 0.7475 | 1.3379 | | 1.3713 | 3.1817 |
| 5 | 1 | 0.6775 | 0.9401 | 1.1588 | 1.1693 | 0.8552 | | 0.6518 | 0.4579 |
| | 2 | 0.5412 | 1.8256 | 1.9041 | 0.5684 | 1.7592 | | 1.7435 | 5.2720 |
| | 3 | 0.7056 | | 0.7056 | | | 0.7056 | | |
| 6 | 1 | 0.6519 | 0.4374 | 0.7850 | 1.6608 | 0.6021 | 0.6522 | | |
| | 2 | 0.6167 | 1.2963 | 1.4355 | 0.8592 | 1.1639 | | 1.1402 | 2.2042 |
| | 3 | 0.4893 | 2.0982 | 2.1545 | 0.4542 | 2.2016 | | 2.0404 | 7.0848 |
| 7 | 1 | 0.6190 | 0.8338 | 1.0385 | 1.1922 | 0.8388 | | 0.5586 | 0.3798 |
| | 2 | 0.5816 | 1.6455 | 1.7453 | 0.6665 | 1.5004 | | 1.5393 | 4.0353 |
| | 3 | 0.4598 | 2.3994 | 2.4431 | 0.3764 | 2.6567 | | 2.3549 | 8.6433 |
| | 4 | 0.6283 | | 0.6283 | | | 0.6283 | | |
| 8 | 1 | 0.5791 | 0.3857 | 0.6958 | 1.6646 | 0.6007 | 0.5764 | | |
| | 2 | 0.5665 | 1.1505 | 1.2824 | 0.8835 | 1.1319 | | 1.0014 | 2.0187 |
| | 3 | 0.5303 | 1.8914 | 1.9643 | 0.5399 | 1.8521 | | 1.8155 | 5.6819 |
| | 4 | 0.4148 | 2.5780 | 2.6112 | 0.3177 | 3.1475 | | 2.5444 | 10.0703 |
| 9 | 1 | 0.5688 | 0.7595 | 0.9489 | 1.1989 | 0.8341 | | 0.5033 | 0.3581 |
| | 2 | 0.5545 | 1.5089 | 1.6076 | 0.6899 | 1.4496 | | 1.4033 | 3.7748 |
| | 3 | 0.5179 | 2.2329 | 2.2922 | 0.4519 | 2.2130 | | 2.1720 | 7.1270 |
| | 4 | 0.4080 | 2.9028 | 2.9313 | 0.2784 | 3.5923 | | 2.8740 | 11.1925 |
| | 5 | 0.5728 | | 0.5728 | | | 0.5728 | | |
| 10 | 1 | 0.5249 | 0.3487 | 0.6302 | 1.6659 | 0.6003 | 0.5215 | | |
| | 2 | 0.5193 | 1.0429 | 1.1650 | 0.8915 | 1.1217 | | 0.9044 | 1.9598 |
| | 3 | 0.5051 | 1.7264 | 1.7988 | 0.5616 | 1.7806 | | 1.6509 | 5.3681 |
| | 4 | 0.4711 | 2.3850 | 2.4311 | 0.3876 | 2.5802 | | 2.3380 | 8.3994 |
| | 5 | 0.3708 | 2.9940 | 3.0169 | 0.2458 | 4.0681 | | 2.9709 | 12.2539 |

Figure 8-34: Linear phase with equiripple error of 0.5° design table

| Order | Section | Real part | Imaginary part | $F_0$ | $\alpha$ | Q | −3 dB frequency | Peaking frequency | Peaking level |
|---|---|---|---|---|---|---|---|---|---|
| 3 | 1 | 0.9360 | 1.2168 | 1.5352 | 1.2194 | 0.8201 | | 0.7775 | 0.2956 |
| | 2 | 0.9360 | | 0.9360 | | | 0.9360 | | |
| 4 | 1 | 0.9278 | 1.6995 | 1.9363 | 0.9583 | 1.0435 | | 1.4239 | 1.5025 |
| | 2 | 0.9192 | 0.5560 | 1.0743 | 1.7113 | 0.5844 | 0.8582 | | |
| 5 | 1 | 0.8075 | 0.9973 | 1.2832 | 1.2585 | 0.7946 | | 0.5853 | 0.1921 |
| | 2 | 0.7153 | 0.2053 | 0.7442 | 1.9224 | 0.5202 | 0.5065 | | |
| | 3 | 0.8131 | | 0.8131 | | | 0.8131 | | |
| 6 | 1 | 0.7019 | 0.4322 | 0.8243 | 1.7030 | 0.5872 | 0.6627 | | |
| | 2 | 0.6667 | 1.2931 | 1.4549 | 0.9165 | 1.0911 | | 1.1080 | 1.7809 |
| | 3 | 0.4479 | 2.1363 | 2.1827 | 0.4104 | 2.4366 | | 2.0888 | 7.9227 |
| 7 | 1 | 0.6155 | 0.7703 | 0.9860 | 1.2485 | 0.8010 | | 0.4632 | 0.2168 |
| | 2 | 0.5486 | 1.5154 | 1.6116 | 0.6808 | 1.4689 | | 1.4126 | 3.8745 |
| | 3 | 0.2905 | 2.1486 | 2.1681 | 0.2680 | 3.7318 | | 2.1289 | 11.5169 |
| | 4 | 0.6291 | | 0.6291 | | | 0.6291 | | |
| 8 | 1 | 0.5441 | 0.3358 | 0.6394 | 1.7020 | 0.5876 | 0.5145 | | |
| | 2 | 0.5175 | 0.9962 | 1.1226 | 0.9220 | 1.0846 | | 0.8512 | 1.7432 |
| | 3 | 0.4328 | 1.6100 | 1.6672 | 0.5192 | 1.9260 | | 1.5507 | 5.9962 |
| | 4 | 0.1978 | 2.0703 | 2.0797 | 0.1902 | 5.2571 | | 2.0608 | 14.4545 |
| 9 | 1 | 0.4961 | 0.6192 | 0.7934 | 1.2505 | 0.7997 | | 0.3705 | 0.2116 |
| | 2 | 0.4568 | 1.2145 | 1.2976 | 0.7041 | 1.4203 | | 1.1253 | 3.6221 |
| | 3 | 0.3592 | 1.7429 | 1.7795 | 0.4037 | 2.4771 | | 1.7055 | 8.0594 |
| | 4 | 0.1489 | 2.1003 | 2.1056 | 0.1414 | 7.0704 | | 2.0950 | 17.0107 |
| | 5 | 0.5065 | | 0.5065 | | | 0.5065 | | |
| 10 | 1 | 0.4535 | 0.2794 | 0.5327 | 1.7028 | 0.5873 | 0.4283 | | |
| | 2 | 0.4352 | 0.8289 | 0.9362 | 0.9297 | 1.0756 | | 0.7055 | 1.6904 |
| | 3 | 0.3886 | 1.3448 | 1.3998 | 0.5552 | 1.8011 | | 1.2874 | 5.4591 |
| | 4 | 0.2908 | 1.7837 | 1.8072 | 0.3218 | 3.1074 | | 1.7598 | 9.9618 |
| | 5 | 0.1136 | 2.0599 | 2.0630 | 0.1101 | 9.0802 | | 2.0568 | 19.1751 |

Figure 8-35: Gaussian to 12 dB design table

| Order | Section | Real part | Imaginary part | $F_0$ | $\alpha$ | Q | −3 dB frequency | Peaking frequency | Peaking level |
|---|---|---|---|---|---|---|---|---|---|
| 3 | 1 | 0.9622 | 1.2214 | 1.5549 | 1.2377 | 0.8080 | | 0.7523 | 0.2448 |
| | 2 | 0.9776 | 0.5029 | 1.0994 | 1.7785 | 0.5623 | 0.8338 | | |
| 4 | 1 | 0.7940 | 0.5029 | 0.9399 | 1.6896 | 05919 | 0.7636 | | |
| | 2 | 0.6304 | 1.5407 | 1.6647 | 0.7574 | 1.3203 | | 1.4058 | 3.0859 |
| 5 | 1 | 0.6190 | 0.8254 | 1.0317 | 1.1999 | 0.8334 | | 0.5460 | 0.3548 |
| | 2 | 0.3559 | 1.5688 | 1.6087 | 0.4425 | 2.2600 | | 1.5279 | 7.3001 |
| | 3 | 0.6650 | | 0.6650 | | | 0.6650 | | |
| 6 | 1 | 0.5433 | 0.3431 | 0.6426 | 1.6910 | 0.5914 | 0.5215 | | |
| | 2 | 0.4672 | 0.9991 | 1.1029 | 0.8472 | 1.1804 | | 0.8831 | 2.2992 |
| | 3 | 0.2204 | 1.5067 | 1.5227 | 0.2895 | 3.4545 | | 1.4905 | 10.8596 |
| 7 | 1 | 0.4580 | 0.5932 | 0.7494 | 1.2223 | 0.8182 | | 0.3770 | 0.2874 |
| | 2 | 0.3649 | 1.1286 | 1.1861 | 0.6153 | 1.6253 | | 1.0680 | 4.6503 |
| | 3 | 0.1522 | 1.4938 | 1.5015 | 0.2027 | 4.9328 | | 1.4860 | 13.9067 |
| | 4 | 0.4828 | | 0.4828 | | | 0.4828 | | |
| 8 | 1 | 0.4222 | 0.2640 | 0.4979 | 1.6958 | 0.5897 | 0.4026 | | |
| | 2 | 0.3833 | 0.7716 | 0.8616 | 0.8898 | 1.1239 | | 0.6697 | 1.9722 |
| | 3 | 0.2678 | 1.2066 | 1.2360 | 0.4333 | 2.3076 | | 1.1765 | 7.4721 |
| | 4 | 0.1122 | 1.4798 | 1.4840 | 0.1512 | 6.6134 | | 1.4755 | 16.4334 |
| 9 | 1 | 0.3700 | 0.4704 | 0.5985 | 1.2365 | 0.8088 | | 0.2905 | 0.2480 |
| | 2 | 0.3230 | 0.9068 | 0.9626 | 0.6711 | 1.4901 | | 0.8473 | 3.9831 |
| | 3 | 0.2309 | 1.2634 | 1.2843 | 0.3596 | 2.7811 | | 1.2421 | 9.0271 |
| | 4 | 0.0860 | 1.4740 | 1.4765 | 0.1165 | 8.5804 | | 1.4715 | 18.6849 |
| | 5 | 0.3842 | | 0.3812 | | | 0.3812 | | |
| 10 | 1 | 0.3384 | 0.2101 | 0.3983 | 1.6991 | 0.5885 | 0.3212 | | |
| | 2 | 0.3164 | 0.6180 | 0.6943 | 0.9114 | 1.0972 | | 0.5309 | 1.8164 |
| | 3 | 0.2677 | 0.9852 | 1.0209 | 0.5244 | 1.9068 | | 0.9481 | 5.9157 |
| | 4 | 0.1849 | 1.2745 | 1.2878 | 0.2871 | 3.4825 | | 1.2610 | 10.9284 |
| | 5 | 0.0671 | 1.4389 | 1.4405 | 0.0931 | 10.7401 | | 1.4373 | 20.6296 |

Figure 8-36: Gaussian to 6 dB design table

# Frequency Transformations

Until now, only filters using the lowpass configuration have been examined. In this section, transforming the lowpass prototype into the other configurations—highpass, bandpass, bandreject (notch), and allpass—will be discussed.

## Lowpass to Highpass

The lowpass prototype is converted to highpass filter by scaling by $1/s$ in the transfer function. In practice, this amounts to capacitors becoming inductors with a value $1/C$, and inductors becoming capacitors with a value of $1/L$ for passive designs. For active designs, resistors become capacitors with a value of $1/R$, and capacitors become resistors with a value of $1/C$. This applies only to frequency setting resistor, not those only used to set gain.

Another way to look at the transformation is to investigate the transformation in the s-plane. The complex pole pairs of the lowpass prototype are made up of a real part, $\alpha$, and an imaginary part, $\beta$. The normalized highpass poles are the given by:

$$\alpha_{HP} = \frac{\alpha}{\alpha^2 + \beta^2} \tag{8-46}$$

and:

$$\beta_{HP} = \frac{\beta}{\alpha^2 + \beta^2} \tag{8-47}$$

A simple pole, $\alpha_0$, is transformed to:

$$\alpha_{\omega,HP} = \frac{1}{\alpha_0} \tag{8-48}$$

Lowpass zeros, $\omega_{Z,LP}$, are transformed by:

$$\omega_{Z,HP} = \frac{1}{\omega_{Z,LP}} \tag{8-49}$$

In addition, a number of zeros equal to the number of poles are added at the origin.

After the normalized lowpass prototype poles and zeros are converted to highpass, they are then denormalized in the same way as the lowpass, i.e., by frequency and impedance.

As an example a three-pole 1 dB Chebyshev lowpass filter will be converted to a highpass filter.

From the design tables of the last section:

$$\alpha_{LP1} = 0.2257$$
$$\beta_{LP1} = 0.8822$$
$$\alpha_{LP2} = 0.4513$$

This will transform to:

$$\alpha_{HP1} = 0.2722$$
$$\beta_{HP1} = 1.0639$$
$$\alpha_{HP2} = 2.2158$$

Which then becomes:

$$F_{01} = 1.0982$$
$$\alpha = 0.4958$$
$$Q = 2.0173$$
$$F_{02} = 2.2158$$

A worked out example of this transformation will appear in a latter section.

A highpass filter can be considered to be a lowpass filter turned on its side. Instead of a flat response at DC, there is a rising response of n $\times$ (20 dB/decade), due to the zeros at the origin, where n is the number of poles. At the corner frequency a response of n $\times$ ($-20$ dB/decade) due to the poles is added to the above rising response. This results in a flat response beyond the corner frequency.

## *Lowpass to Bandpass*

Transformation to the bandpass response is a little more complicated. Bandpass filters can be classified as either wideband or narrowband, depending on the separation of the poles. If the corner frequencies of the bandpass are widely separated (by more than 2 octaves), the filter is wideband and is made up of separate lowpass and highpass sections, which will be cascaded. The assumption made is that with the widely separated poles, interaction between them is minimal. This condition does not hold in the case of a narrowband bandpass filter, where the separation is less than 2 octaves. We will be covering the narrowband case in this discussion.

As in the highpass transformation, start with the complex pole pairs of the lowpass prototype, $\alpha$ and $\beta$. The pole pairs are known to be complex conjugates. This implies symmetry around DC (0 Hz). The process of transformation to the bandpass case is one of mirroring the response around DC of the lowpass prototype to the same response around the new center frequency $F_0$.

This clearly implies that the number of poles and zeros is doubled when the bandpass transformation is done. As in the lowpass case, the poles and zeros below the real axis are ignored. So an nth-order lowpass prototype transforms into an nth-order bandpass, even though the filter order will be 2n. An nth-order bandpass filter will consist of n sections, versus n/2 sections for the lowpass prototype. It may be convenient to think of the response as n poles up and n poles down.

The value of $Q_{BP}$ is determined by:

$$Q_{BP} = \frac{F_0}{BW} \tag{8-50}$$

where BW is the bandwidth at some level, typically $-3$ dB.

A transformation algorithm was defined by Geffe ( Reference 16) for converting lowpass poles into equivalent bandpass poles.

Given the pole locations of the lowpass prototype:

$$-\alpha \pm j\beta \tag{8-51}$$

and the values of $F_0$ and $Q_{BP}$, the following calculations will result in two sets of values for Q and frequencies, $F_H$ and $F_L$, which define a pair of bandpass filter sections.

$$C = \alpha^2 + \beta^2 \tag{8-52}$$

$$D = \frac{2\alpha}{Q_{BP}} \tag{8-53}$$

$$E = \frac{C}{Q_{BP}^2} + 4 \tag{8-54}$$

$$G = \sqrt{E^2 - 4D^2} \tag{8-55}$$

$$Q = \sqrt{\frac{E + G}{2D^2}} \tag{8-56}$$

Observe that the Q of each section will be the same.

The pole frequencies are determined by:

$$M = \frac{\alpha Q}{Q_{BP}} \tag{8-57}$$

$$W = M + \sqrt{M^2 - 1} \tag{8-58}$$

$$F_{BP1} = \frac{F_0}{W} \tag{8-59}$$

$$F_{BP2} = WF_0 \tag{8-60}$$

Each pole pair transformation will also result in 2 zeros that will be located at the origin.

A normalized lowpass real pole with a magnitude of $\alpha_0$ is transformed into a bandpass section where:

$$Q = \frac{Q_{BP}}{\alpha_0} \tag{8-61}$$

and the frequency is $F_0$.

Each single pole transformation will also result in a zero at the origin.

Elliptical function lowpass prototypes contain zeros as well as poles. In transforming the filter, the zeros must be transformed as well. Given the lowpass zeros at $\pm j\omega_Z$, the bandpass zeros are obtained as follows:

$$M = \frac{\alpha Q}{Q_{BP}} \tag{8-62}$$

$$W = M + \sqrt{M^2 - 1} \tag{8-63}$$

$$F_{BP1} = \frac{F_0}{W} \tag{8-64}$$

$$F_{BP2} = WF_0 \tag{8-65}$$

Since the gain of a bandpass filter peaks at $F_{BP}$ instead of $F_0$, an adjustment in the amplitude function is required to normalize the response of the aggregate filter. The gain of the individual filter section is given by:

$$A_R = A_0 \sqrt{1 + Q^2 \left( \frac{F_0}{F_{BP}} - \frac{F_{BP}}{F_0} \right)^2}$$  (8-66)

where:

$A_0$ = gain at filter center frequency,

$A_R$ = filter section gain at resonance,

$F_0$ = filter center frequency, and

$F_{BP}$ = filter section resonant frequency.

Again using a three-pole 1 dB Chebyshev as an example:

$$\alpha_{LP1} = 0.2257$$
$$\beta_{LP1} = 0.8822$$
$$\alpha_{LP2} = 0.4513$$

A 3 dB bandwidth of 0.5 Hz with a center frequency of 1 Hz is arbitrarily assigned. Then:

$$Q_{BP} = 2$$

Going through the calculations for the pole pair, the intermediate results are:

$$C = 0.829217 \quad D = 0.2257$$
$$E = 4.2073 \quad G = 4.18302$$
$$M = 1.0247 \quad W = 1.245$$

and:

$$F_{BP1} = 0.80322 \quad F_{BP2} = 1.24499$$
$$Q_{BP1} = Q_{BP2} = 9.0749$$
$$\text{Gain} = 4.1318$$

And for the single pole:

$$F_{BP3} = 1 \quad Q_{BP3} = 4.431642$$
$$\text{Gain} = 1$$

Again a full example will be worked out in a latter section.

## Lowpass to Bandreject (Notch)

As in the bandpass case, a bandreject filter can be either wideband or narrowband, depending on whether or not the poles are separated by 2 octaves or more. To avoid confusion, the following convention will be adopted. If the filter is wideband, it will be referred to as a bandreject filter. A narrowband filter will be referred to as a notch filter.

One way to build a notch filter is to construct it as a bandpass filter whose output is subtracted from the input $(1 - BP)$. Another way is with cascaded lowpass and highpass sections, especially for the bandreject (wideband) case. In this case, the sections are in parallel, and the output is the difference.

Just as the bandpass case is a direct transformation of the lowpass prototype, where DC is transformed to $F_0$, the notch filter can be first transformed to the highpass case, and then DC, which is now a zero, is transformed to $F_0$.

A more general approach would be to convert the poles directly. A notch transformation results in two pairs of complex poles and a pair of second-order imaginary zeros from each lowpass pole pair.

First, the value of $Q_{BR}$ is determined by:

$$Q_{BR} = \frac{F_0}{BW} \tag{8-67}$$

where BW is the bandwidth at $-3\,dB$.

Given the pole locations of the lowpass prototype:

$$-\alpha \pm j\beta \tag{8-68}$$

and the values of $F_0$ and $Q_{BR}$, the following calculations will result in two sets of values for Q and frequencies, $F_H$ and $F_L$, which define a pair of notch filter sections.

$$C = \alpha^2 + \beta^2 \tag{8-69}$$

$$D = \frac{\alpha}{Q_{BR}C} \tag{8-70}$$

$$E = \frac{\beta}{Q_{BR}C} \tag{8-71}$$

$$F = E^2 - D^2 + 4 \tag{8-72}$$

$$G = \sqrt{\frac{F}{2} + \sqrt{\frac{F2}{4} + D^2E^2}} \tag{8-73}$$

$$H = \frac{DE}{G} \tag{8-74}$$

$$K = \frac{1}{2}\sqrt{(D+H)^2 + (E+G)^2} \tag{8-75}$$

$$Q = \frac{K}{D+H} \tag{8-76}$$

The pole frequencies are given by:

$$F_{BR1} = \frac{F_0}{K} \tag{8-77}$$

$$F_{BR2} = KF_0 \tag{8-78}$$

$$F_Z = F_0 \tag{8-79}$$

$$F_0 = \sqrt{F_{BR1} \times F_{BR2}} \tag{8-80}$$

where $F_0$ is the notch frequency and the geometric mean of $F_{BR1}$ and $F_{BR2}$.

A simple real pole, $\alpha_0$, transforms to a single section having a Q given by:

$$Q = Q_{BR}\alpha_0 \tag{8-81}$$

with a frequency $F_{BR} = F_0$. There will also be transmission zero at $F_0$.

In some instances, such as the elimination of the power line frequency (hum) from low level sensor measurements, a notch filter for a specific frequency may be designed.

Assuming that an attenuation of A dB is required over a bandwidth of B, then the required Q for a single frequency notch is determined by:

$$Q = \frac{\omega_0}{B\sqrt{10^{0.1A} - 1}} \qquad (8\text{-}82)$$

For transforming a lowpass prototype, a three-pole 1 dB Chebyshev is again used as an example:

$$\alpha_{LP1} = 0.2257$$
$$\beta_{LP1} = 0.8822$$
$$\alpha_{LP2} = 0.4513$$

A 3 dB bandwidth of 0.1 Hz. with a center frequency of 1 Hz is arbitrarily assigned. Then:

$$Q_{BR} = 10$$

Going through the calculations for the pole pair yields the intermediate results:

$$
\begin{array}{ll}
C = 0.829217 & D = 0.027218 \\
E = 0.106389 & F = 4.01058 \\
G = 2.002643 & H = 0.001446 \\
\multicolumn{2}{c}{K = 1.054614}
\end{array}
$$

and:

$$F_{BR1} = 0.94821 \qquad F_{BR2} = 1.0546$$
$$Q_{BR1} = Q_{BR2} = 36.7918$$

and for the single pole:

$$F_{BP3} = 1 \qquad Q_{BP3} = 4.4513$$

Once again a full example will be worked out in a latter section.

## Lowpass to Allpass

The transformation from lowpass to allpass involves adding a zero in the right-hand side of the s-plane corresponding to each pole in the left-hand side.

In general, however, the allpass filter is usually not designed in this manner. The main purpose of the allpass filter is to equalize the delay of another filter. Many modulation schemes in communications use some form or another of quadrature modulation, which processes both the amplitude and phase of the signal.

Allpass filters add delay to flatten the delay curve without changing the amplitude. In most cases, a closed form of the equalizer is not available. Instead the amplitude filter is designed and the delay calculated or measured. Then graphical means or computer programs are used to figure out the required sections of equalization.

Each section of the equalizer gives twice the delay of the lowpass prototype due to the interaction of the zeros. A rough estimate of the required number of sections is given by:

$$n = 2\Delta_{BW}\Delta_T + 1 \qquad (8\text{-}83)$$

where $\Delta_{BW}$ is the bandwidth of interest in hertz and $\Delta_T$ is the delay distortion over $\Delta_{BW}$ in seconds.

# *Filter Realizations*

Now that it has been decided what to build, it must be decided how to build it. That means that it is necessary to decide which of the filter topologies to use. Filter design is a two step process where it is determined what is to be built (the filter transfer function) and then how to build it (the topology used for the circuit).

In general, filters are built out of one-pole sections for real poles, and two-pole sections for pole pairs. While you can build a filter out of three-pole, or higher order sections, the interaction between the sections increases, and therefore, component sensitivities go up.

It is better to use buffers to isolate the various sections. In addition, it is assumed that all filter sections are driven from a low impedance source. Any source impedance can be modeled as being in series with the filter input.

In all of the design equation figures, the following convention will be used:

$H$ = circuit gain in the passband or at resonance,

$F_0$ = cutoff or resonant frequency in hertz,

$\omega_0$ = cutoff or resonant frequency in rad/s,

$Q$ = circuit "quality factor." Indicates circuit peaking, and

$\alpha = 1/Q$ = damping ratio.

It is unfortunate that the symbol $\alpha$ is used for damping ratio. It is not the same as the $\alpha$ that is used to denote pole locations ($\alpha \pm j\beta$). The same issue occurs for Q. It is used for the *circuit* quality factor and also the *component* quality factor, which are not the same thing.

The circuit Q is the amount of peaking in the circuit. This is a function of the angle of the pole to the origin in the s-plane. The component Q is the amount of losses in what should be lossless reactances. These losses are the parasitics of the components; dissipation factor, leakage resistance, equivalent series resistance (ESR), etc. in capacitors, and series resistance and parasitic capacitances in inductors.

## *Single Pole RC*

The simplest filter building block is the passive RC section. The single pole can be either lowpass or highpass. Odd-order filters will have a single pole section.

The basic form of the lowpass RC section is shown in Figure 8-37(A). It is assumed that the load impedance is high ($>\times 10$), so that there is no loading of the circuit. The load will be in parallel with the shunt arm of the filter. If this is not the case, the section will have to be buffered with an op amp. A lowpass filter can be transformed to a highpass filter by exchanging the resistor and the capacitor. The basic form of the highpass filter is shown in Figure 8-37(B). Again it is assumed that load impedance is high.

The pole can also be incorporated into an amplifier circuit. Figure 8-38(A) shows an amplifier circuit with a capacitor in the feedback loop. This forms a lowpass filter since as frequency is increased, the effective feedback impedance decreases, which causes the gain to decrease.

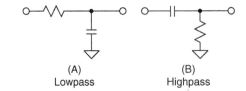

(A)
Lowpass

(B)
Highpass

**Figure 8-37: Single pole sections**

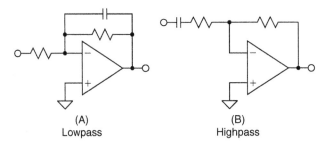

(A)
Lowpass

(B)
Highpass

**Figure 8-38: Single pole active filter blocks**

Figure 8-38(B) shows a capacitor in series with the input resistor. This causes the signal to be blocked at DC. As the frequency is increased from DC, the impedance of the capacitor decreases and the gain of the circuit increases. This is a highpass filter.

The design equations for single pole filters appear in Figure 8-66.

## Passive LC Section

While not strictly a function that uses op amps, passive filters form the basis of several active filters topologies and are included here for completeness.

As in active filters, passive filters are built up of individual subsections. Figure 8-39 shows lowpass filter sections. The full section is the basic two-pole section. Odd-order filters use one-half section which is a single pole section. The m-derived sections, shown in Figure 8-40, are used in designs requiring transmission zeros as well as poles.

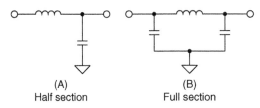

(A)
Half section

(B)
Full section

**Figure 8-39: Passive filter blocks (lowpass)**

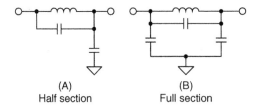

(A)
Half section

(B)
Full section

**Figure 8-40: Passive filter blocks (lowpass m-derived)**

630

A lowpass filter can be transformed into a highpass (see Figures 8-41 and 8-42) by simply replacing capacitors with inductors with reciprocal values and vice versa so:

$$L_{HP} = \frac{1}{C_{LP}} \tag{8-84}$$

and:

$$C_{HP} = \frac{1}{L_{LP}} \tag{8-85}$$

Transmission zeros are also reciprocated in the transformation so:

$$\omega_{Z,HP} = \frac{1}{\omega_{Z,LP}} \tag{8-86}$$

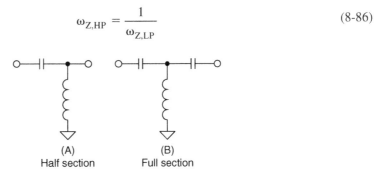

(A)
Half section

(B)
Full section

**Figure 8-41: Passive filter blocks (highpass)**

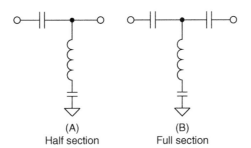

(A)
Half section

(B)
Full section

**Figure 8-42: Passive filter blocks (highpass m-derived)**

The lowpass prototype is transformed to bandpass and bandreject filters as well by using the table in Figure 8-43.

For a passive filter to operate, the source and load impedances must be specified. One issue with designing passive filters is that in multipole filters each section is the load for the preceding sections and also the source impedance for subsequent sections, so component interaction is a major concern. Because of this, designers typically make use of tables, such as in Williams's book (Reference 2).

## Integrator

Any time that you put a frequency-dependent impedance in a feedback network the inverse frequency response is obtained. For example, if a capacitor, which has a frequency-dependent impedance that decreases with increasing frequency, is put in the feedback network of an op amp, an integrator is formed, as in Figure 8-44.

| Lowpass branch | Bandpass configuration | Circuit values |
|---|---|---|
| C | L / C | $C = \dfrac{1}{\omega_0^2 L}$ |
| L | L C | $L = \dfrac{1}{\omega_0^2 C}$ |
| La / Cb | La Ca / Lb / Cb | $Ca = \dfrac{1}{\omega_0^2 La}$ $Lb = \dfrac{1}{\omega_0^2 Cb}$ |
| L1 C2 | L1 C1 C2 / L2 | $C1 = \dfrac{1}{\omega_0^2 L1}$ $L2 = \dfrac{1}{\omega_0^2 C2}$ |
| Highpass branch | Bandreject configuration | Circuit values |

**Figure 8-43: Lowpass → bandpass and highpass → bandreject transformation**

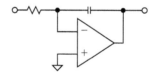

**Figure 8-44: Integrator**

The integrator has high gain (i.e., the open-loop gain of the op amp) at DC. An integrator can also be thought of as a lowpass filter with a cutoff frequency of 0 Hz.

## General Impedance Converter (GIC)

Figure 8-45 is the block diagram of a GIC. The impedance of this circuit is:

$$Z = \frac{Z1Z3Z5}{Z2Z4} \tag{8-87}$$

By substituting one or two capacitors into appropriate locations (the other locations being resistors), several impedances can be synthesized (see Reference 25).

One limitation of this configuration is that the lower end of the structure must be grounded.

## Active Inductor

Substituting a capacitor for Z4 and resistors for Z1, Z2, Z3, and Z5 in the GIC results in an impedance given by:

$$Z_{11} = \frac{sCR_1R_3R_5}{R_2} \tag{8-88}$$

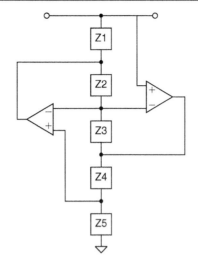

Figure 8-45: General impedance converter

By inspection it can be shown that this is an inductor with a value of:

$$L = \frac{CR_1R_3R_5}{R_2}$$

(8-89)

This is just one way to simulate an inductor as shown in Figure 8-46.

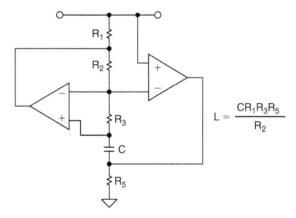

$$L = \frac{CR_1R_3R_5}{R_2}$$

Figure 8-46: Active inductor

## Frequency-Dependent Negative Resistor (FDNR)

By substituting capacitors for two of the Z1, Z3, or Z5 elements, a structure known as a FDNR is generated. The impedance of this structure is:

$$Z_{11} = \frac{sC^2R_2R_4}{R_5}$$

(8-90)

This impedance, which is called a D element, has the value:

$$D = C^2 R_4 \tag{8-91}$$

Assuming:

$$C_1 = C_2 \quad \text{and} \quad R_2 = R_5 \tag{8-92}$$

The three possible versions of the FDNR are shown in Figure 8-47.

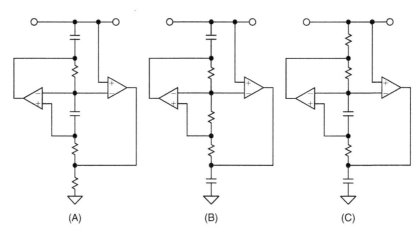

**Figure 8-47: FDNR blocks**

There is theoretically no difference in these three blocks, and so they should be interchangeable. In practice though there may be some differences; circuit (a) is sometimes preferred because it is the only block to provide a return path for the amplifier bias currents.

For the FDNR filter (see Reference 24), the passive realization of the filter is used as the basis of the design. As in the passive filter, the FDNR filter must then be denormalized for frequency and impedance. This is typically done before the conversion by 1/s. First take the denormalized passive prototype filter and transform the elements by 1/s. This means that inductors, whose impedance is equal to sL, transform into a resistor with an impedance of L. A resistor of value R becomes a capacitor with an impedance of R/s; and a capacitor of impedance 1/sC transforms into a frequency-dependent resistor, D, with an impedance of $1/s^2 C$. The transformations involved with the FDNR configuration and the GIC implementation of the D element are shown in Figure 8-48. We can apply this transformation to lowpass, highpass, bandpass, or notch filters, remembering that the FDNR block must be restricted to shunt arms.

A worked out example of the FDNR filter is included in the next section.

A perceived advantage of the FDNR filter in some circles is that there are no op amps in the direct signal path, which can add noise and/or distortion, however small, to the signal. It is also relatively insensitive to component variation. These advantages of the FDNR come at the expense of an increase in the number of components required.

## Sallen–Key

The Sallen–Key configuration, also known as a voltage control voltage source (VCVS), was first introduced in 1955 by R.P. Sallen and E.L. Key of MIT's Lincoln Labs (see Reference 14). It is one of the most widely

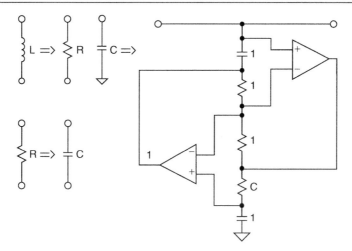

**Figure 8-48: 1/s transformation**

used filter topologies and is shown in Figure 8-49. One reason for this popularity is that this configuration shows the least dependence of filter performance on the performance of the op amp. This is due to the fact that the op amp is configured as an amplifier, as opposed to an integrator, which minimizes the gain-bandwidth requirements of the op amp. This infers that for a given op amp, you will be able to design a higher frequency filter than with other topologies since the op amp gain-bandwidth product will not limit the performance of the filter as it would if it were configured as an integrator. The signal phase through the filter is maintained (non-inverting configuration).

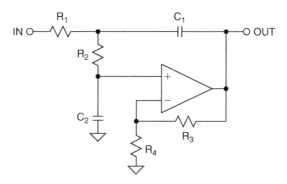

**Figure 8-49: Sallen–Key lowpass filter**

Another advantage of this configuration is that the ratio of the largest resistor value to the smallest resistor value and the ratio of the largest capacitor value to the smallest capacitor value (component spread) are low, which is good for manufacturability. The frequency and Q terms are somewhat independent, but they are very sensitive to the gain parameter. The Sallen–Key is very Q-sensitive to element values, especially for high Q sections. The design equations for the Sallen–Key lowpass are shown in Figure 8-67.

There is a special case of the Sallen–Key lowpass filter. If the gain is set to 2, the capacitor values, as well as the resistor values, will be the same.

While the Sallen–Key filter is widely used, a serious drawback is that the filter is not easily tuned, due to interaction of the component values on $F_0$ and Q.

To transform the lowpass into the highpass we simply exchange the capacitors and the resistors in the frequency determining network (i.e., not the amp gain resistors). This is shown in Figure 8-50. The comments regarding sensitivity of the filter given above for the lowpass case apply to the highpass case as well. The design equations for the Sallen–Key highpass are shown in Figure 8-68.

The bandpass case of the Sallen–Key filter has a limitation (see Figure 8-51). The value of Q will determine the gain of the filter, i.e., it cannot be set independent, as in the lowpass or highpass cases. The design equations for the Sallen–Key bandpass are shown in Figure 8-69.

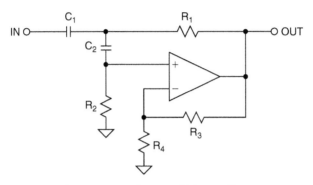

Figure 8-50: Sallen–Key highpass filter

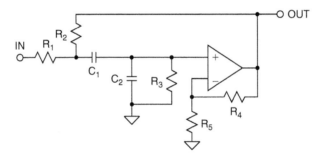

Figure 8-51: Sallen–Key bandpass filter

A Sallen–Key notch filter may also be constructed, but it has a large number of undesirable characteristics. The resonant frequency, or the notch frequency, cannot be adjusted easily due to component interaction. As in the bandpass case, the section gain is fixed by the other design parameters, and there is a wide spread in component values, especially capacitors. Because of this and the availability of easier to use circuits, it is not covered here.

## Multiple Feedback

The multiple feedback filter uses an op amp as an integrator as shown in Figure 8-52. Therefore, the dependence of the transfer function on the op amp parameters is greater than in the Sallen–Key realization. It is hard to generate high Q, high frequency sections due to the limitations of the open-loop gain of the op amp. A rule-of-thumb is that the open-loop gain of the op amp should be at least 20 dB ($\times$10) above the amplitude response at the resonant (or cutoff) frequency, including the peaking caused by the Q of the filter. The peaking due to Q will cause an amplitude, $A_0$:

$$A_0 = HQ \tag{8-93}$$

where H is the gain of the circuit. The multiple feedback filter will invert the phase of the signal. This is equivalent to adding the resulting 180° phase shift to the phase shift of the filter itself.

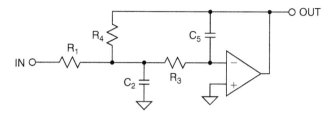

**Figure 8-52: Multiple feedback lowpass**

The maximum to minimum component value ratios are higher in the multiple feedback case than in the Sallen–Key realization. The design equations for the multiple feedback lowpass are given in Figure 8-70.

Comments made about the multiple feedback lowpass case apply to the highpass case as well (see Figure 8-53). Note that we again swap resistors and capacitors to convert the lowpass case to the highpass case. The design equations for the multiple feedback highpass are given in Figure 8-71.

The design equations for the multiple feedback bandpass case (see Figure 8-54 ) are given in Figure 8-72.

This circuit is widely used in low Q (<20) applications. It allows some tuning of the resonant frequency, $F_0$, by making $R_2$ variable. Q can be adjusted (with $R_5$) as well, but this will also change $F_0$.

Tuning of $F_0$ can be accomplished by monitoring the output of the filter with the horizontal channel of an oscilloscope, with the input to the filter connected to the vertical channel. The display will be a Lissajous pattern. This pattern will be an ellipse that will collapse to a straight line at resonance, since the phase shift will be 180°. You could also adjust the output for maximum output, which will also occur at resonance, but this is usually not as precise, especially at lower values of Q where there is a less pronounced peak.

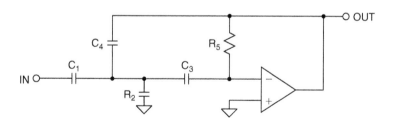

**Figure 8-53: Multiple feedback highpass**

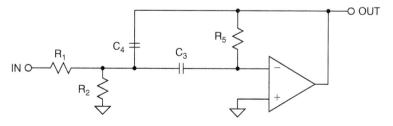

**Figure 8-54: Multiple feedback bandpass**

## State Variable

The state variable realization (see Reference 11) is shown in Figure 8-55, along with the design equations in Figure 8-73. This configuration offers the most precise implementation, at the expense of many more circuit elements. All three major parameters (gain, Q, and $\omega_0$) can be adjusted independently, and lowpass, highpass, and bandpass outputs are available simultaneously. Note that the lowpass and highpass output are inverted in phase while the bandpass output maintains the phase. The gain of each of the outputs of the filter is also independently variable. With an added amplifier section summing the lowpass and highpass sections the notch function can also be synthesized. By changing the ratio of the summed sections, lowpass notch, standard notch, and highpass notch functions can be realized. A standard notch may also be realized by subtracting the bandpass output from the input with the added op amp section. An allpass filter may also be built with the four amplifier configuration by subtracting the bandpass output from the input. In this instance, the bandpass gain must equal 2.

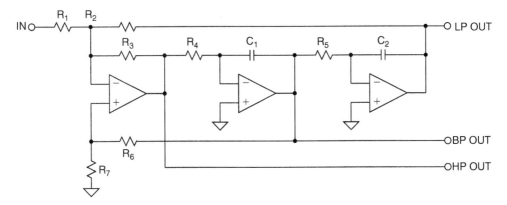

**Figure 8-55: State variable filter**

Since all parameters of the state variable filter can be adjusted independently, component spread can be minimized. Also, variations due to temperature and component tolerances are minimized. The op amps used in the integrator sections will have the same limitations on op amp gain bandwidth as described in the multiple feedback section.

Tuning the resonant frequency of a state variable filter is accomplished by varying $R_4$ and $R_5$. While you do not have to tune both, if you are varying over a wide range it is generally preferable. Holding $R_1$ constant, tuning $R_2$ sets the lowpass gain and tuning $R_3$ sets the highpass gain. Bandpass gain and Q are set by the ratio of $R_6$ and $R_7$.

Since the parameters of a state variable filter are independent and tunable, it is easy to add electronic control of frequency, Q and $\omega_0$. This adjustment is accomplished by using an analog multiplier, multiplying digital-to-analog converters (MDACs) or digital pots, as shown in one of the examples in a later section. For the integrator sections adding the analog multiplier or MDAC effectively increases the time constant by dividing the voltage driving the resistor, which, in turn, provides the charging current for the integrator capacitor. This in effect raises the resistance and, in turn, the time constant. The Q and gain can be varied by changing the ratio of the various feedback paths. A digital pot will accomplish the same feat in a more direct manner, by directly changing the resistance value. The resultant tunable filter offers a great deal of utility in measurement and control circuitry. A worked out example is given in Section 8-8 of this chapter.

## Biquadratic (Biquad)

A close cousin of the state variable filter is the biquad as shown in Figure 8-56. The name of this circuit was first used by J. Tow in 1968 (Reference 11) and later by L.C. Thomas in 1971 (see Reference 12). The name derives from the fact that the transfer function is a quadratic function in both the numerator and the denominator. Hence, the transfer function is a biquadratic function. This circuit is a slight rearrangement of the state variable circuit. One significant difference is that there is not a separate highpass output. The bandpass output inverts the phase. There are two lowpass outputs, one in phase and one out of phase. With the addition of a fourth amplifier section, highpass, notch (lowpass, standard, and highpass), and allpass filters can be realized. The design equations for the biquad are given in Figure 8-74.

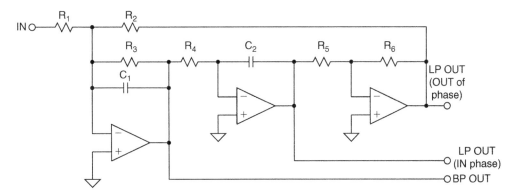

**Figure 8-56: Biquad filter**

Referring to Figure 8-74 the allpass case of the biquad, $R_8 = R_9/2$ and $R_7 = R_9$. This is required to make the terms in the transfer function line up correctly. For the highpass output, the input, bandpass, and second lowpass outputs are summed. In this case, the constraints are that $R_1 = R_2 = R_3$ and $R_7 = R_8 = R_9$.

Like the state variable, the biquad filter is tunable. Adjusting $R_3$ will adjust the Q. Adjusting $R_4$ will set the resonant frequency. Adjusting $R_1$ will set the gain. Frequency would generally be adjusted first followed by Q and then gain. Setting the parameters in this manner minimizes the effects of component value interaction.

## Dual Amplifier Bandpass (DABP)

The DABP filter structure is useful in designs requiring high Qs and high frequencies. Its component sensitivity is small, and the element spread is low. A useful feature of this circuit is that the Q and resonant frequency can be adjusted more or less independently.

Referring to Figure 8-57, the resonant frequency can be adjusted by $R_2$. $R_1$ can then be adjusted for Q. In this topology, it is useful to use dual op amps. The match of the two op amps will lower the sensitivity of Q to the amplifier parameters.

It should be noted that the DABP has a gain of 2 at resonance. If lower gain is required, resistor $R_1$ may be split to form a voltage divider. This is reflected in the addendum to the design equations of the DABP, Figure 8-75.

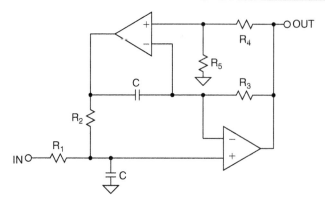

**Figure 8-57: DABP filter**

## Twin-T Notch

The twin T is widely used as a general purpose notch circuit as shown in Figure 8-58. The passive implementation of the twin T (i.e., with no feedback) has a major shortcoming of having a Q that is fixed at 0.25. This issue can be rectified with the application of positive feedback to the reference node. The amount of the signal feedback, set by the $R_4/R_5$ ratio, will determine the value of Q of the circuit, which, in turn, determines the notch depth. For maximum notch depth, the resistors $R_4$ and $R_5$ and the associated op amp can be eliminated. In this case, the junction of $C_3$ and $R_3$ will be directly connected to the output.

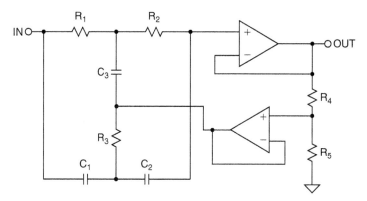

**Figure 8-58: Twin-T notch filter**

Tuning is not easily accomplished. Using standard 1% components a 60 dB notch is as good as can be expected, with 40–50 dB being more typical.

The design equations for the twin T are given in Figure 8-76.

## Bainter Notch

A simple notch filter is the Bainter circuit (see Reference 21). It is composed of simple circuit blocks with two feedback loops as shown in Figure 8-59. Also, the component sensitivity is very low.

This circuit has several interesting properties. The Q of the notch is not based on component matching as it is in every other implementation, but is instead only dependent on the gain of the amplifiers. Therefore, the notch depth will not drift with temperature, aging, and other environmental factors. The notch frequency may shift, but not the depth.

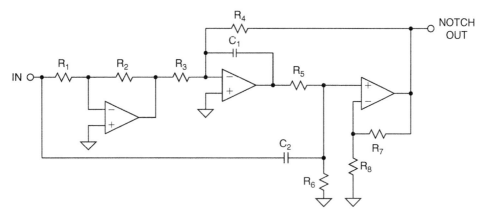

Figure 8-59: Bainter notch filter

Amplifier open-loop gain of $10^4$ will yield a $Q_z$ of >200. It is capable of orthogonal tuning with minimal interaction. $R_6$ tunes Q and $R_1$ tunes $\omega_z$. Varying $R_3$ sets the ratio of $\omega_0/\omega_z$, produces lowpass notch ($R_4 > R_3$), notch ($R_4 = R_3$), or highpass notch ($R_4 < R_3$).

The design equations of the Bainter circuit are given in Figure 8-77.

## Boctor Notch

The Boctor circuits (see References 22, 23), while moderately complicated, use only one op amp. Due to the number of components, there is a great deal of latitude in component selection. These circuits also offer low sensitivity and the ability to tune the various parameters more or less independently.

There are two forms, a lowpass notch (Figure 8-60) and a highpass notch (Figure 8-61). For the lowpass case, the preferred order of adjustment is to tune $\omega_0$ with $R_4$, then $Q_0$ with $R_2$, next $Q_z$ with $R_3$ and finally $\omega_z$ with $R_1$.

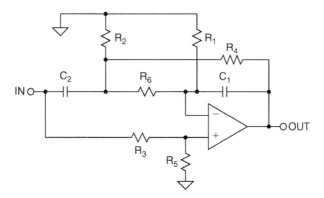

Figure 8-60: Boctor lowpass notch filter

In order for the components to be realizable we must define a variable, k1, such that:

$$\frac{\omega_0^2}{\omega_z^2} < k1 < 1 \qquad (8\text{-}94)$$

The design equations are given in Figure 8-78 for the lowpass case and in Figure 8-79 for the highpass case.

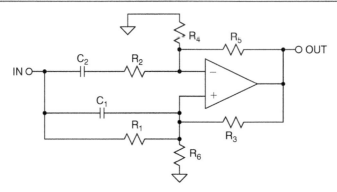

**Figure 8-61: Boctor highpass filter**

In the highpass case, circuit gain is required and it applies only when:

$$Q < \frac{1}{1 - \frac{\omega_z^2}{\omega_0^2}} \tag{8-95}$$

But a highpass notch can be realized with one amplifier and only 2 capacitors, which can be of the same value. The pole and zero frequencies are completely independent of the amplifier gain. The resistors can be trimmed so that even 5% capacitors can be used.

## "1 – Bandpass" Notch

As mentioned in the state variable and biquad sections, a notch filter can be built as $1 - \text{BP}$. The bandpass section can be any of the all-pole bandpass realizations discussed above, or any others. Keep in mind whether the bandpass section is inverting as shown in Figure 8-62 (such as the multiple feedback circuit) or non-inverting as shown in Figure 8-63 (such as the Sallen–Key), since we want to subtract, not add, the bandpass output from the input.

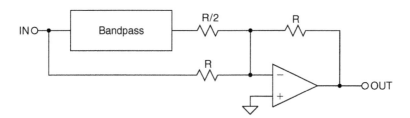

**Figure 8-62: 1 – BP filter for inverting bandpass configurations**

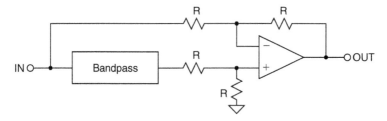

**Figure 8-63: 1 – BP filter for non-inverting bandpass configurations**

It should be noted that the gain of the bandpass amplifier must be taken into account in determining the resistor values. Unity gain bandpass would yield equal values.

## First-Order Allpass

The general form of a first-order allpass filter is shown in Figure 8-64. If the function is a simple RC highpass (Figure 8-64(A)), the circuit will have a phase shift that goes from $-180°$ at 0 Hz and $0°$ at high frequency. It will be $-90°$ at $\omega = 1/RC$. The resistor may be made variable to allow adjustment of the delay at a particular frequency.

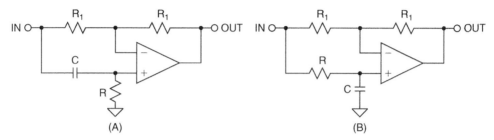

**Figure 8-64: First-order allpass filters**

If the function is changed to a lowpass function (Figure 8-64(B)), the filter is still a first-order allpass and the delay equations still hold, but the signal is inverted, changing from $0°$ at DC to $-180°$ at high frequency.

## Second-Order Allpass

A second-order allpass circuit shown in Figure 8-65 was first described by Delyiannis (see Reference 17). The main attraction of this circuit is that it only requires one op amp.

Remember also that an allpass filter can also be realized as $1-2$ BP.

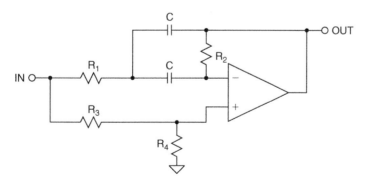

**Figure 8-65: Second-order allpass filter**

We may use any of the all-pole realizations discussed above to build the filter, but you need to be aware of whether the BP inverts the phase or not. We must also be aware that the gain of the BP section must be 2. To this end, the DABP structure is particularly useful, since its gain is fixed at 2.

Figures 8-66 to 8-81 summarize design equations for various active filter realizations. In all cases, H, $\omega_0$, Q, and $\alpha$ are given, being taken from the design tables.

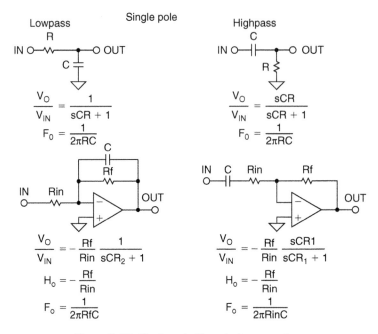

**Figure 8-66: Single pole filter design equations**

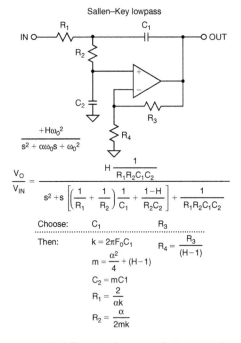

**Figure 8-67: Sallen–Key lowpass design equations**

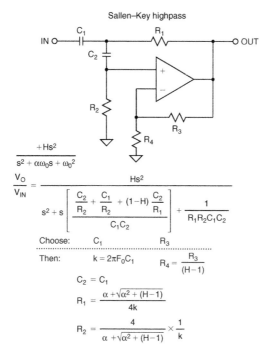

Sallen–Key highpass

$$\dfrac{+Hs^2}{s^2 + \alpha\omega_0 s + \omega_0^2}$$

$$\dfrac{V_O}{V_{IN}} = \dfrac{Hs^2}{s^2 + s\left[\dfrac{\dfrac{C_2}{R_2} + \dfrac{C_1}{R_2} + (1-H)\dfrac{C_2}{R_1}}{C_1 C_2}\right] + \dfrac{1}{R_1 R_2 C_1 C_2}}$$

Choose:  $C_1$  $R_3$

Then:  $k = 2\pi F_0 C_1$  $R_4 = \dfrac{R_3}{(H-1)}$

$C_2 = C_1$

$R_1 = \dfrac{\alpha + \sqrt{\alpha^2 + (H-1)}}{4k}$

$R_2 = \dfrac{4}{\alpha + \sqrt{\alpha^2 + (H-1)}} \times \dfrac{1}{k}$

**Figure 8-68: Sallen–Key highpass design equations**

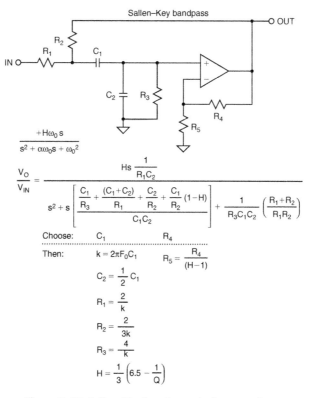

Sallen–Key bandpass

$$\dfrac{+H\omega_0 s}{s^2 + \alpha\omega_0 s + \omega_0^2}$$

$$\dfrac{V_O}{V_{IN}} = \dfrac{Hs \dfrac{1}{R_1 C_2}}{s^2 + s\left[\dfrac{\dfrac{C_1}{R_3} + \dfrac{(C_1 + C_2)}{R_1} + \dfrac{C_2}{R_2} + \dfrac{C_1}{R_2}(1-H)}{C_1 C_2}\right] + \dfrac{1}{R_3 C_1 C_2}\left(\dfrac{R_1 + R_2}{R_1 R_2}\right)}$$

Choose:  $C_1$  $R_4$

Then:  $k = 2\pi F_0 C_1$  $R_5 = \dfrac{R_4}{(H-1)}$

$C_2 = \dfrac{1}{2} C_1$

$R_1 = \dfrac{2}{k}$

$R_2 = \dfrac{2}{3k}$

$R_3 = \dfrac{4}{k}$

$H = \dfrac{1}{3}\left(6.5 - \dfrac{1}{Q}\right)$

**Figure 8-69: Sallen–Key bandpass design equations**

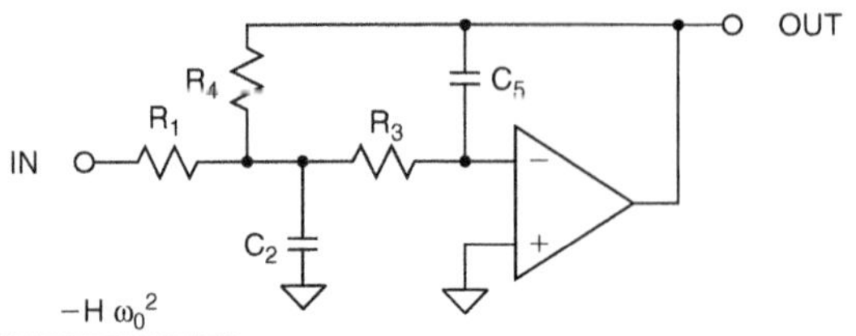

$$\frac{-H\,\omega_0{}^2}{s^2 + \alpha\,\omega_0 s + \omega_0{}^2}$$

$$\frac{V_O}{V_{IN}} = \frac{-H\,\dfrac{1}{R_1 R_3 C_2 C_5}}{s^2 + s\,\dfrac{1}{C_2}\left(\dfrac{1}{R_1} + \dfrac{1}{R_3} + \dfrac{1}{R_4}\right) + \dfrac{1}{R_3 R_4 C_2 C_5}}$$

Choose:    $C_5$
  Then:    $k = 2\pi F_0 C_5$

$$C_2 = \frac{4}{\alpha^2}\,(H + 1)\,C_5$$

$$R_1 = \frac{\alpha}{2H\,k}$$

$$R_3 = \frac{\alpha}{2(H + 1)k}$$

$$R_4 = \frac{\alpha}{2k}$$

**Figure 8-70: Multiple feedback lowpass design equations**

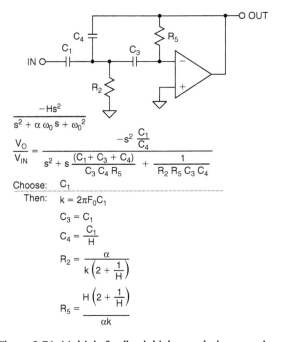

$$\frac{-Hs^2}{s^2 + \alpha\,\omega_0\,s + \omega_0{}^2}$$

$$\frac{V_O}{V_{IN}} = \frac{-s^2\,\dfrac{C_1}{C_4}}{s^2 + s\,\dfrac{(C_1 + C_3 + C_4)}{C_3 C_4 R_5} + \dfrac{1}{R_2 R_5 C_3 C_4}}$$

Choose:    $C_1$
  Then:    $k = 2\pi F_0 C_1$

$$C_3 = C_1$$

$$C_4 = \frac{C_1}{H}$$

$$R_2 = \frac{\alpha}{k\left(2 + \dfrac{1}{H}\right)}$$

$$R_5 = \frac{H\left(2 + \dfrac{1}{H}\right)}{\alpha k}$$

**Figure 8-71: Multiple feedback highpass design equations**

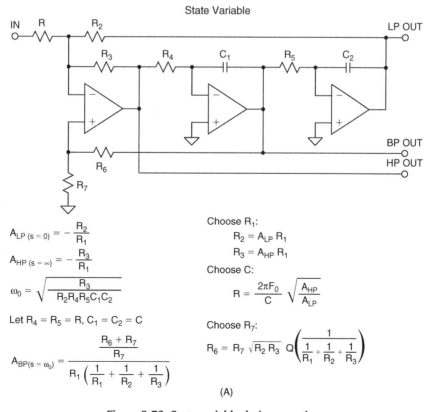

Multiple feedback bandpass

$$\frac{-H\,\omega_0 s}{s^2 + \alpha\,\omega_0 s + \omega_0{}^2}$$

$$\frac{V_O}{V_{IN}} = \frac{-s\,\dfrac{1}{R_1 C_4}}{s^2 + s\,\dfrac{(C_3 + C_4)}{C_3 C_4 R_5} + \dfrac{1}{R_5 C_3 C_4}\left(\dfrac{1}{R_1} + \dfrac{1}{R_2}\right)}$$

Choose:   $C_3$

Then:   $k = 2\pi F_0 C_3$

$C_4 = C_3$

$R_1 = \dfrac{1}{Hk}$

$R_2 = \dfrac{1}{(2Q - H)\,k}$

$R_5 = \dfrac{2Q}{k}$

**Figure 8-72: Multiple feedback bandpass design equations**

State Variable

$$A_{LP\,(s\,=\,0)} = -\frac{R_2}{R_1}$$

$$A_{HP\,(s\,=\,\infty)} = -\frac{R_3}{R_1}$$

$$\omega_0 = \sqrt{\frac{R_3}{R_2 R_4 R_5 C_1 C_2}}$$

Let $R_4 = R_5 = R$, $C_1 = C_2 = C$

$$A_{BP(s\,=\,\omega_0)} = \frac{\dfrac{R_6 + R_7}{R_7}}{R_1\left(\dfrac{1}{R_1} + \dfrac{1}{R_2} + \dfrac{1}{R_3}\right)}$$

Choose $R_1$:

$R_2 = A_{LP}\,R_1$

$R_3 = A_{HP}\,R_1$

Choose C:

$$R = \frac{2\pi F_0}{C}\sqrt{\frac{A_{HP}}{A_{LP}}}$$

Choose $R_7$:

$$R_6 = R_7\,\sqrt{R_2 R_3}\;Q\left(\frac{1}{\dfrac{1}{R_1} + \dfrac{1}{R_2} + \dfrac{1}{R_3}}\right)$$

(A)

**Figure 8-73: State variable design equations**

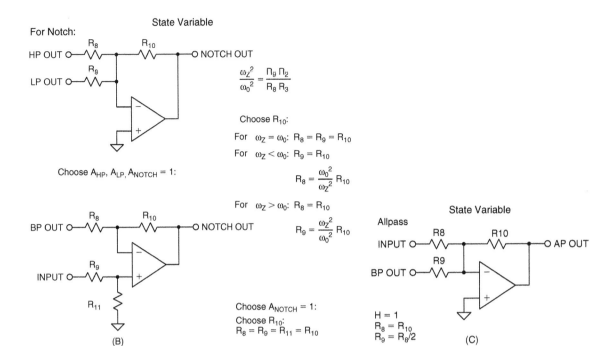

For Notch:

State Variable

$$\frac{\omega_Z{}^2}{\omega_0{}^2} = \frac{R_9\, R_2}{R_8\, R_3}$$

Choose $A_{HP}$, $A_{LP}$, $A_{NOTCH} = 1$:

Choose $R_{10}$:

For $\omega_Z = \omega_0$: $R_8 = R_9 = R_{10}$

For $\omega_Z < \omega_0$: $R_9 = R_{10}$

$$R_8 = \frac{\omega_0{}^2}{\omega_Z{}^2} R_{10}$$

For $\omega_Z > \omega_0$: $R_8 = R_{10}$

$$R_9 = \frac{\omega_Z{}^2}{\omega_0{}^2} R_{10}$$

Choose $A_{NOTCH} = 1$:
Choose $R_{10}$:
$R_8 = R_9 = R_{11} = R_{10}$

(B)

State Variable

Allpass

$H = 1$
$R_8 = R_{10}$
$R_9 = R_8/2$

(C)

Figure 8-73: Continued

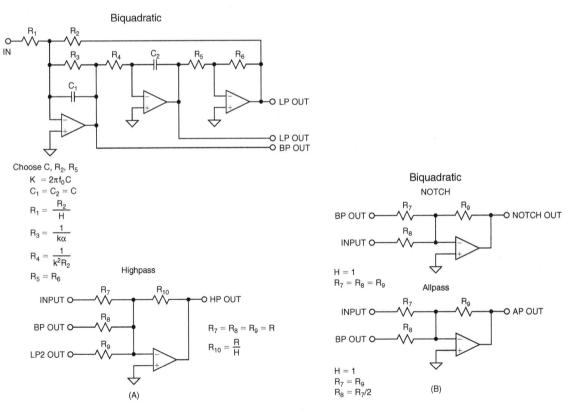

Biquadratic

Choose C, $R_2$, $R_5$

$K = 2\pi f_0 C$

$C_1 = C_2 = C$

$$R_1 = \frac{R_2}{H}$$

$$R_3 = \frac{1}{k\alpha}$$

$$R_4 = \frac{1}{k^2 R_2}$$

$R_5 = R_6$

Highpass

$R_7 = R_8 = R_9 = R$

$$R_{10} = \frac{R}{H}$$

(A)

Biquadratic

NOTCH

$H = 1$
$R_7 = R_8 = R_9$

Allpass

$H = 1$
$R_7 = R_9$
$R_8 = R_7/2$

(B)

Figure 8-74: Biquad design equations

*648*

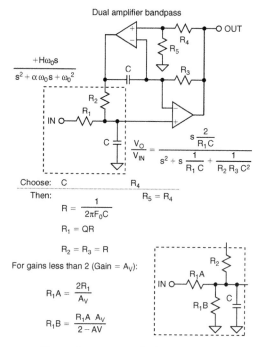

Dual amplifier bandpass

$$\dfrac{+H\omega_0 s}{s^2 + \alpha\,\omega_0 s + \omega_0{}^2}$$

$$\dfrac{V_O}{V_{IN}} = \dfrac{s\dfrac{2}{R_1 C}}{s^2 + s\dfrac{1}{R_1 C} + \dfrac{1}{R_2 R_3 C^2}}$$

Choose:    C             $R_4$

Then:                $R_5 = R_4$

$$R = \dfrac{1}{2\pi F_0 C}$$

$$R_1 = QR$$

$$R_2 = R_3 = R$$

For gains less than 2 (Gain = $A_V$):

$$R_1 A = \dfrac{2R_1}{A_V}$$

$$R_1 B = \dfrac{R_1 A\ A_V}{2 - AV}$$

Figure 8-75: DABP design equations

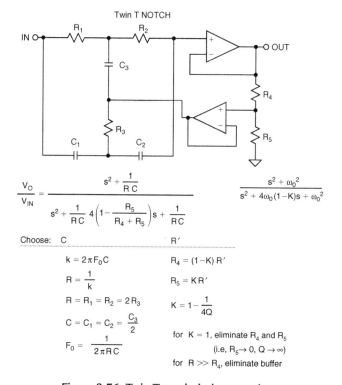

Twin T NOTCH

$$\dfrac{V_O}{V_{IN}} = \dfrac{s^2 + \dfrac{1}{R\,C}}{s^2 + \dfrac{1}{R\,C}\ 4\left(1 - \dfrac{R_5}{R_4 + R_5}\right)s + \dfrac{1}{R\,C}}$$

$$\dfrac{s^2 + \omega_0{}^2}{s^2 + 4\omega_0(1 - K)s + \omega_0{}^2}$$

Choose:    C                  R′

$$k = 2\pi F_0 C \qquad\qquad R_4 = (1 - K)\,R'$$

$$R = \dfrac{1}{k} \qquad\qquad\qquad R_5 = K\,R'$$

$$R = R_1 = R_2 = 2R_3 \qquad K = 1 - \dfrac{1}{4Q}$$

$$C = C_1 = C_2 = \dfrac{C_3}{2}$$

                       for K = 1, eliminate $R_4$ and $R_5$

$$F_0 = \dfrac{1}{2\pi R C}$$

                           (i.e, $R_5 \rightarrow 0$, $Q \rightarrow \infty$)

                       for R ≫ $R_4$, eliminate buffer

Figure 8-76: Twin-T notch design equations

649

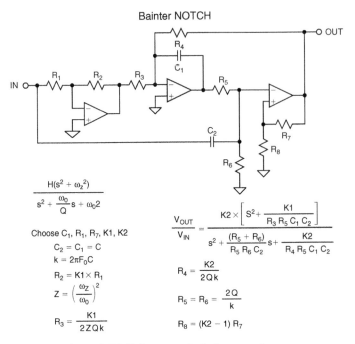

## Bainter NOTCH

$$\frac{H(s^2 + \omega_z{}^2)}{s^2 + \dfrac{\omega_0}{Q}s + \omega_0{}2}$$

Choose $C_1$, $R_1$, $R_7$, K1, K2

$$C_2 = C_1 = C$$
$$k = 2\pi F_0 C$$
$$R_2 = K1 \times R_1$$
$$Z = \left(\frac{\omega_z}{\omega_0}\right)^2$$
$$R_3 = \frac{K1}{2ZQk}$$

$$\frac{V_{OUT}}{V_{IN}} = \frac{K2 \times \left[S^2 + \dfrac{K1}{R_3 R_5 C_1 C_2}\right]}{s^2 + \dfrac{(R_5 + R_6)}{R_5 R_6 C_2}s + \dfrac{K2}{R_4 R_5 C_1 C_2}}$$

$$R_4 = \frac{K2}{2Qk}$$
$$R_5 = R_6 = \frac{2Q}{k}$$
$$R_8 = (K2 - 1)\,R_7$$

**Figure 8-77: Bainter notch design equations**

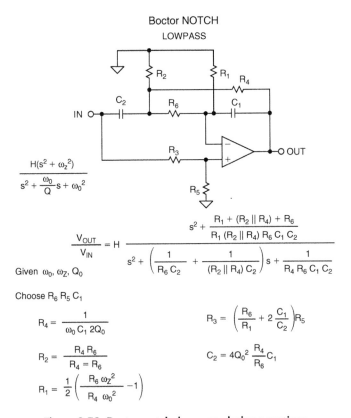

## Boctor NOTCH
### LOWPASS

$$\frac{H(s^2 + \omega_z{}^2)}{s^2 + \dfrac{\omega_0}{Q}s + \omega_0{}^2}$$

$$\frac{V_{OUT}}{V_{IN}} = H\,\frac{s^2 + \dfrac{R_1 + (R_2 \| R_4) + R_6}{R_1 (R_2 \| R_4) R_6 C_1 C_2}}{s^2 + \left(\dfrac{1}{R_6 C_2} + \dfrac{1}{(R_2 \| R_4) C_2}\right)s + \dfrac{1}{R_4 R_6 C_1 C_2}}$$

Given $\omega_0$, $\omega_z$, $Q_0$

Choose $R_6$ $R_5$ $C_1$

$$R_4 = \frac{1}{\omega_0 C_1 \, 2Q_0}$$

$$R_2 = \frac{R_4 R_6}{R_4 = R_6}$$

$$R_1 = \frac{1}{2}\left(\frac{R_6 \omega_z{}^2}{R_4 \; \omega_0{}^2} - 1\right)$$

$$R_3 = \left(\frac{R_6}{R_1} + 2\frac{C_1}{C_2}\right)R_5$$

$$C_2 = 4Q_0{}^2\,\frac{R_4}{R_6}C_1$$

**Figure 8-78: Boctor notch, lowpass, design equations**

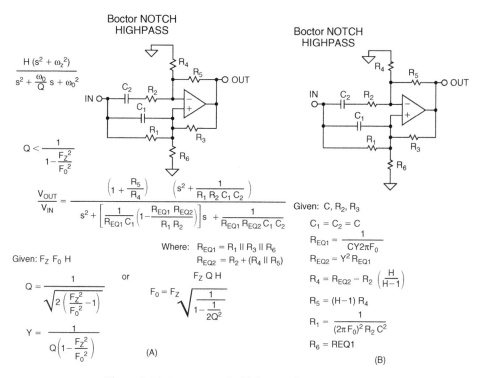

Boctor NOTCH
HIGHPASS

$$\frac{H(s^2 + \omega_z{}^2)}{s^2 + \dfrac{\omega_0}{Q} s + \omega_0{}^2}$$

$$Q < \frac{1}{1 - \dfrac{F_z{}^2}{F_0{}^2}}$$

$$\frac{V_{OUT}}{V_{IN}} = \frac{\left(1 + \dfrac{R_5}{R_4}\right)\left(s^2 + \dfrac{1}{R_1 R_2 C_1 C_2}\right)}{s^2 + \left[\dfrac{1}{R_{EQ1} C_1}\left(1 - \dfrac{R_{EQ1} R_{EQ2}}{R_1 R_2}\right)\right] s + \dfrac{1}{R_{EQ1} R_{EQ2} C_1 C_2}}$$

Where: $R_{EQ1} = R_1 \parallel R_3 \parallel R_6$
$R_{EQ2} = R_2 + (R_4 \parallel R_5)$

Given: $F_z$ $F_0$ $H$

$$Q = \frac{1}{\sqrt{2\left(\dfrac{F_z{}^2}{F_0{}^2} - 1\right)}}$$

or

$$F_0 = F_z \sqrt{\frac{1}{1 - \dfrac{1}{2Q^2}}}$$

$F_z Q H$

$$Y = \frac{1}{Q\left(1 - \dfrac{F_z{}^2}{F_0{}^2}\right)}$$

(A)

Boctor NOTCH
HIGHPASS

Given: $C, R_2, R_3$

$C_1 = C_2 = C$

$$R_{EQ1} = \frac{1}{CY2\pi F_0}$$

$R_{EQ2} = Y^2 R_{EQ1}$

$$R_4 = R_{EQ2} - R_2\left(\frac{H}{H-1}\right)$$

$R_5 = (H-1) R_4$

$$R_1 = \frac{1}{(2\pi F_0)^2 R_2 C^2}$$

$R_6 = R_{EQ1}$

(B)

**Figure 8-79: Boctor notch, highpass, design equations**

First order ALLPASS

$$\frac{V_O}{V_{IN}} = \frac{s - \dfrac{1}{RC}}{s + \dfrac{1}{RC}}$$

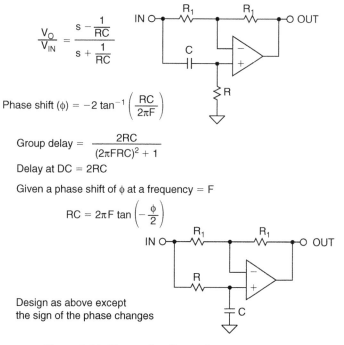

Phase shift $(\phi) = -2 \tan^{-1}\left(\dfrac{RC}{2\pi F}\right)$

Group delay $= \dfrac{2RC}{(2\pi FRC)^2 + 1}$

Delay at DC $= 2RC$

Given a phase shift of $\phi$ at a frequency $= F$

$$RC = 2\pi F \tan\left(-\frac{\phi}{2}\right)$$

Design as above except
the sign of the phase changes

**Figure 8-80: First-order allpass design equations**

Second-order ALLPASS

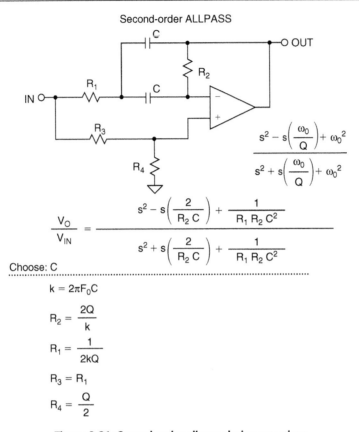

$$\frac{s^2 - s\left(\dfrac{\omega_0}{Q}\right) + \omega_0{}^2}{s^2 + s\left(\dfrac{\omega_0}{Q}\right) + \omega_0{}^2}$$

$$\frac{V_O}{V_{IN}} = \frac{s^2 - s\left(\dfrac{2}{R_2\,C}\right) + \dfrac{1}{R_1\,R_2\,C^2}}{s^2 + s\left(\dfrac{2}{R_2\,C}\right) + \dfrac{1}{R_1\,R_2\,C^2}}$$

Choose: C

$$k = 2\pi F_0 C$$

$$R_2 = \frac{2Q}{k}$$

$$R_1 = \frac{1}{2kQ}$$

$$R_3 = R_1$$

$$R_4 = \frac{Q}{2}$$

**Figure 8-81: Second-order allpass design equations**

# Practical Problems in Filter Implementation

In the previous sections, filters were dealt with as mathematical functions. The filter designs were assumed to have been implemented with "perfect" components. When the filter is built with real-world components design tradeoffs must typically be made.

In building a filter with an order greater than two, multiple second and/or first-order sections are used. The frequencies and Qs of these sections must align precisely or the overall response of the filter will be affected. For example, the antialiasing filter design example in the next section is a fifth-order Butterworth filter, made up of a second-order section with a frequency ($F_0$) = 1 and a Q = 1.618, a second-order section with a frequency ($F_0$) = 1 and a Q = 0.618, and a first-order section with a frequency ($F_0$) = 1 (for a filter normalized to 1 rad/s). If the Q or frequency response of any of the sections is off slightly, the overall response will deviate from the desired response. It may be close, but it will not be exact. As is typically the case with engineering, a decision must be made as to what tradeoffs should be made. For instance, do we really need a particular response exactly? Is there a problem if there is a little more ripple in the passband? Or if the cutoff frequency is at a slightly different frequency? These are the types of questions that face a designer, and will vary from design to design.

## Passive Components (Resistors, Capacitors, and Inductors)

Passive components are the first problem. When designing filters, the calculated values of components will most likely not be available commercially. Resistors, capacitors, and inductors come in standard values. While custom values can be ordered, the practical tolerance will probably still be ±1% at best. An alternative is to build the required value out of a series and/or parallel combination of standard values. This increases the cost and size of the filter. Not only is the cost of components increased, so are the manufacturing costs, both for loading and tuning the filter. Furthermore, success will be still limited by the number of parts that are used, their tolerance, and their tracking, both over temperature and time.

A more practical way is to use a circuit analysis program to determine the response using standard values. The program can also evaluate the effects of component drift over temperature. The values of the sensitive components are adjusted using parallel combinations where needed, until the response is within the desired limits. Many of the higher end filter CAD programs include this feature.

The resonant frequency and Q of a filter are typically determined by the component values. Obviously, if the component value is drifting, the frequency and the Q of the filter will drift which, in turn, will cause the frequency response to vary. This is especially true in higher order filters.

Higher order implies higher Q sections. Higher Q sections mean that component values are more critical, since the Q is typically set by the ratio of two or more components, typically capacitors.

In addition to the initial tolerance of the components, you must also evaluate effects of temperature/time drift. The temperature coefficients (TCs) of the various components may be different in both magnitude and sign. Capacitors, especially, are difficult in that not only do they drift, but the TC is also a function of temperature, as shown in Figure 8-82. This represents the TC of a (relatively) poor film capacitor, which might be typical for a polyester or polycarbonate type. *Linear TC* in film capacitors can be found in the

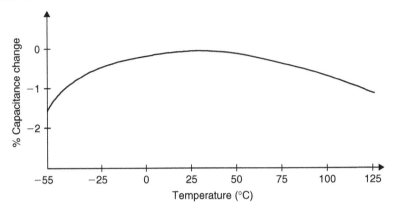

**Figure 8-82: A poor film capacitor temperature coefficient**

polystyrene, polypropylene, and Teflon dielectrics. In these types, TC is on the order of 100–200 ppm/°C, and if necessary, this can be compensated with a complementary TC elsewhere in the circuit.

The lowest TC dielectrics are NPO (or COG) ceramic (±30 ppm/°C), and polystyrene (−120 ppm/°C). Some capacitors, mainly the plastic film types, such as polystyrene and polypropylene, also have a limited temperature range.

While there is infinite choice of the values of the passive components for building filters, in practice there are physical limits. Capacitor values below 10 pF and above 10 μF are not practical. Electrolytic capacitors should be avoided. Electrolytic capacitors are typically very leaky. A further potential problem is if they are operated without a polarizing voltage, they become nonlinear when the AC voltage reverse biases them. Even with a DC polarizing voltage, the AC signal can reduce the instantaneous voltage to 0V or below. Large values of film capacitors are physically very large.

Resistor values of less than 100 Ω should be avoided, as should values over 1 MΩ. Very low resistance values (under 100 Ω) can require a great deal of drive current and dissipate a great deal of power. Both of these should be avoided. And low values and very large values of resistors may not be as readily available. Very large values tend to be more prone to parasitics since smaller capacitances will couple more easily into larger impedance levels. Noise also increases with the square root of the resistor value. Larger value resistors also will cause larger offsets due to the effects of the amplifier bias currents.

Parasitic capacitances due to circuit layout and other sources affect the performance of the circuit. They can form between two traces on a PC board (on the same side or opposite side of the board), between leads of adjacent components, and just about everywhere else you can (and in most cases cannot) think of. These capacitances are usually small, so their effect is greater at high impedance nodes. Thus, they can be controlled most of the time by keeping the impedance of the circuits down. Remember that the effects of stray capacitance are frequency-dependent, being worse at high frequencies because the impedance drops with increasing frequency.

Parasitics are not just associated with outside sources. They are also present in the components themselves.

A capacitor is more than just a capacitor in most instances. A real capacitor has inductance (from the leads and other sources) and resistance as shown in Figure 8-83. This resistance shows up in the specifications as leakage and poor power factor. Obviously, we would like capacitors with very low leakage and good power factor (see Figure 8-84).

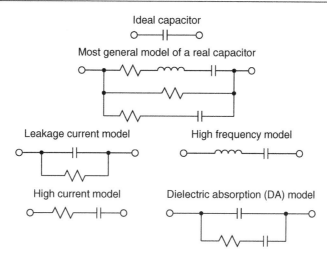

Figure 8-83: Capacitor equivalent circuit

|  | Type | Advantages | Disadvantages |
|---|---|---|---|
| Discrete | Carbon Composition | Lowest cost<br>High power/Small case size<br>Wide range of values | Poor tolerance (5%)<br>Poor temperature coefficient<br>(1500 ppm/°C) |
|  | Wirewound | Excellent tolerance (0.01%)<br>Excellent TC (1ppm/°C)<br>High power | Reactance is a problem<br>Large case size<br>Most expensive |
|  | Metal film | Good tolerance (0.1%)<br>Good TC (<1 to 100ppm/°C)<br>Moderate cost<br>Wide range of values<br>Low voltage coefficient | Must be stabilized with burn-in<br>Low Power |
|  | Bulk metal or metal foil | Excellent tolerance (to 0.005%)<br>Excellent TC (to <1ppm/°C)<br>Low reactance<br>Low voltage coefficient | Low power<br>Very expensive |
|  | High megohm | Very high values ($10^8$ to $10^{14}\Omega$)<br>Only choice for some circuits | High voltage coefficient<br>(200ppm/V)<br>Fragile glass case (Needs special handling)<br>Expensive |
| Networks | Thick film | Low cost<br>High power<br>Laser-trimmable<br>Readily available | Fair matching (0.1%)<br>Poor TC (>100ppm/°C)<br>Poor tracking TC (10ppm/°C) |
|  | Thin film | Good matching (<0.01%)<br>Good TC (<100ppm/°C)<br>Good tracking TC (2ppm/°C)<br>Moderate cost<br>Laser-trimmable<br>Low capacitance<br>Suitable for hybrid IC substrate | Often large geometry<br>Limited values and configurations |

Figure 8-84: Resistor comparison chart

**655**

| Type | Typical Da | Advantages | Disadvantages |
|---|---|---|---|
| Polyatyrene | 0.001% to 0.02% | Inexpensive<br><br>Low DA<br>Good stability<br>($-120$ppm/°C) | Damaged by temperature<br>$>-35$°C<br>Large<br>High inductance<br>Vendors limited |
| Polypropylene | 0.001% to 0.02% | Inexpensive<br><br>Low DA<br>Stable ($-200$ppm/°C)<br>Wide range of values | Damaged by temperature<br>$>-105$°C<br>Large<br>High inductance |
| Teflon | 0.009% to 0.02% | Low DA available<br>Good stability<br>Operational above $-125$°C<br>Wide range of values | Expensive<br>Large<br>High inductance |
| Polycarbonate | 0.1% | Good stability<br>Low cost<br>Wide temperature range<br>Wide range of values | Large<br>DA limits to 8-bit applications<br>High inductance |
| Polyenter | 0.3% to 0.5% | Moderate stability<br>Low cost<br>Wide temperature range<br>Low inductance (standard film) | Large<br>DA limits to 8-bit applications<br>High inductance (conventional) |
| NPO Ceramic | <0.1% | Small case size<br><br>Inexpensive, many vendors<br>Good stability (30 ppm/°C)<br>1% values available<br>Low inductance (chip) | DA generally low (may not be specified)<br><br>Low maximum values (10 µF) |
| Monolithic Ceramic (High K) | >0.2% | Low inductance (chip)<br>Wide range of values | Poor stability<br>Poor Da<br>High voltage coefficient |
| Mica | >0.009% | Low less at HF<br>Low inductance<br>Good stability<br>1% values available | Quite large<br>Low maximum values (10µF)<br>Expensive |
| Aluminum Electrolytic | Very high | Large values<br>High currents<br>High voltages<br>Small size | High leakage<br>Usually polarized<br>Poor stability, accuracy<br>Inductive |
| Tantalum Electrolytic | Very high | Small size<br>Large values<br>Medium inductance | High leakage<br>Usually polarized<br>Expensive<br>Poor stability, accuracy |

**Figure 8-85: Capacitor comparison chart**

In general, it is best to use plastic film (preferably Teflon or polystyrene) or mica capacitors and metal film resistors, both of moderate to low values in our filters.

One way to reduce component parasitics is to use surface mounted devices. Not having leads means that the lead inductance is reduced. Also, being physically smaller allows more optimal placement. A disadvantage is that not all types of capacitors are available in surface mount. Ceramic capacitors are popular surface mount types, and of these, the NPO family has the best characteristics for filtering. Ceramic capacitors may also be prone to microphonics. Microphonics occurs when the capacitor turns into a motion sensor, similar to a strain gauge, and turns vibration into an electrical signal, which is a form of noise.

Resistors also have parasitic inductances due to leads and parasitic capacitance. The various qualities of resistors are compared in Figure 8-85.

## Limitations of Active Elements (Op Amps) in Filters

The active element of the filter will also have a pronounced effect on the response.

In developing the various topologies (multiple feedback, Sallen–Key, state variable, etc.), the active element was always modeled as a "perfect" op amp. That is to say it has:

- infinite gain,

- infinite input impedance,

- zero output impedance.

None of these varies with frequency. While amplifiers have improved a great deal over the years, this model has not yet been realized.

The most important limitation of the amplifier has to do with its gain variation with frequency. All amplifiers are band-limited. This is due mainly to the physical limitations of the devices with which the amplifier is constructed. Negative feedback theory tells us that the response of an amplifier must be first order ($-6\,$dB per octave) when the gain falls to unity in order to be stable. To accomplish this, a real pole is usually introduced in the amplifier so the gain rolls off to $<1$ by the time the phase shift reaches $180°$ (plus some phase margin, hopefully). This rolloff is equivalent to that of a single pole filter. So in simplistic terms, the transfer function of the amplifier is added to the transfer function of the filter to give a composite function. How much the frequency-dependent nature of the op amp affects the filter is dependent on which topology is used as well as the ratio of the filter frequency to the amplifier bandwidth.

The Sallen–Key configuration, for instance, is the least dependent on the frequency response of the amplifier. All that is required is for the amplifier response to be flat to just past the frequency where the attenuation of the filter is below the minimum attenuation required. This is because the amplifier is used as a gain block. Beyond cutoff, the attenuation of the filter is reduced by the rolloff of the gain of the op amp. This is because the output of the amplifier is phase shifted, which results in incomplete nulling when fed back to the input. There is also an issue with the output impedance of the amplifier rising with frequency as the open-loop gain rolls off. This causes the filter to lose attenuation.

The state variable configuration uses the op amps in two modes, as amplifiers and as integrators. As amplifiers, the constraint on frequency response is basically the same as for the Sallen–Key, which is flat out to the minimum attenuation frequency. As an integrator, however, more is required. A good rule-of-thumb is that the open-loop gain of the amplifier must be greater than 10 times the closed-loop gain (including peaking from the Q of the circuit). This should be taken as the absolute minimum requirement What this means is that there must be 20 dB loop gain, minimum. Therefore, an op amp with 10 MHz unity gain bandwidth is the minimum required to make a 1 MHz integrator. What happens is that the effective Q of the circuit increases as loop gain decreases. This phenomenon is called Q enhancement. The mechanism for Q enhancement is similar to that of slew rate limitation. Without sufficient loop gain, the op amp virtual ground is no longer at ground. In other words, the op amp is no longer behaving as an op amp. Because of this, the integrator no longer behaves like an integrator.

The multiple feedback configuration also places heavy constraints on the active element. Q enhancement is a problem in this topology as well. As the loop gain falls, the Q of the circuit increases, and the parameters of the filter change. The same rule-of-thumb as used for the integrator also applies to the multiple feedback topology (loop gain should be at least 20 dB). The filter gain must also be factored into this equation.

In the FDNR realization, the requirements for the op amps are not as clear. To make the circuit work, we assume that the op amps will be able to force the input terminals to be the same voltage. This implies that the loop gain be a minimum of 20 dB at the resonant frequency.

Also it is generally considered to be advantageous to have the two op amps in each leg matched. This is easily accomplished using dual op amps. It is also a good idea to have low bias current devices for the op amps, so FET input op amps should be used, all other things being equal.

In addition to the frequency-dependent limitations of the op amp, several of its other parameters may be important to the filter designer.

One is input impedance. We assume in the "perfect" model that the input impedance is infinite. This is required so that the input of the op amp does not load the network around it. This means that we probably want to use FET amplifiers with high impedance circuits.

There is also a small frequency-dependent term to the input impedance, since the effective impedance is the real input impedance multiplied by the loop gain. This usually is not a major source of error, since the network impedance of a high frequency filter should be low.

## Distortion Resulting from Input Capacitance Modulation

Another subtle effect can be noticed with FET input amps. The input capacitance of a FET changes with the applied voltage. When the amplifier is used in the inverting configuration, such as with the multiple feedback configuration, the applied voltage is held to 0 V. Therefore, there is no capacitance modulation. However, when the amplifier is used in the non-inverting configuration, such as in the Sallen–Key circuit, this form of distortion can exist.

There are two ways to address this issue. The first is to keep the equivalent impedance low. The second is to balance the impedance seen by the inputs. This is accomplished by adding a network into the feedback leg of the amplifier which is equal to the equivalent input impedance. Note that this will only work for a unity gain application.

As an example, which is taken from the OP176 data sheet, a 1 kHz highpass Sallen–Key filter is shown (Figure 8-86). Figure 8-87 shows the distortion for the uncompensated version (curve A1) as well as with the compensation (curve A2). Also shown is the same circuit with the impedances scaled up by a factor of 10 (B1 uncompensated, B2 compensated). Note that the compensation improves the distortion, but not as much as having low impedance to start with.

Similarly, the op amp output impedance affects the response of the filter. The output impedance of the amplifier is divided by the loop gain; therefore, the output impedance will rise with increasing frequency.

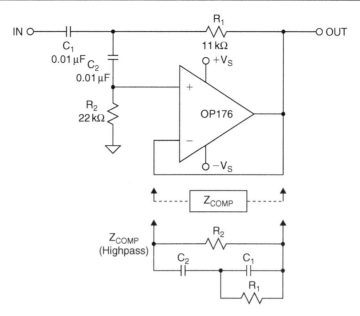

**Figure 8-86: Compensation for input capacitance voltage modulation**

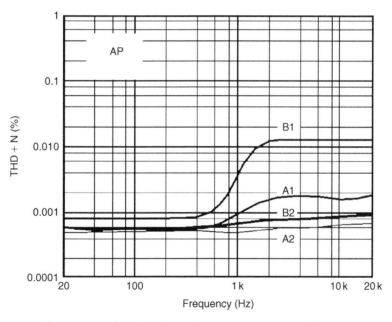

**Figure 8-87: Distortion due to input capacitance modulation**

This may have an effect with high frequency filters if the output impedance of the stage driving the filter becomes a significant portion of the network impedance.

The fall of loop gain with frequency can also affect the distortion of the op amp, since there is less loop gain available for correction. In the multiple feedback configuration the feedback loop is also frequency dependent, which may further reduce the feedback correction, resulting in increased distortion. This effect

*659*

is counteracted somewhat by the reduction of distortion components in the filter network (assuming a lowpass or bandpass filter).

All of the discussion so far is based on using classical voltage feedback op amps. Current feedback, or transimpedance, op amps offer improved high frequency response, but are unusable in any topologies discussed except the Sallen–Key. The problem is that capacitance in the feedback loop of a current feedback amplifier usually causes it to become unstable. Also, most current feedback amplifiers will only drive a small capacitive load. Therefore, it is difficult to build classical integrators using current feedback amplifiers. Some current feedback op amps have an external pin that may be used to configure them as a very good integrator, but this configuration does not lend itself to classical active filter designs.

Current feedback integrators tend to be non-inverting, which is not acceptable in the state variable configuration. Also, the bandwidth of a current feedback amplifier is set by its feedback resistor, which would make the multiple feedback topology difficult to implement. Another limitation of the current feedback amplifier in the multiple feedback configuration is the low input impedance of the inverting terminal. This would result in loading of the filter network. Sallen–Key filters are possible with current feedback amplifiers, since the amplifier is used as a non-inverting gain block. New topologies that capitalize on the current feedback amplifier's superior high frequency performance and compensate for its limitations will have to be developed.

## Q Peaking and Q Enhancement

The last thing that you need to be aware of is exceeding the dynamic range of the amplifier. Qs over 0.707 will cause peaking in the response of the filter (see Figures 8-5 to 8-7). For high Q's, this could cause overload of the input or output stages of the amplifier with a large input. Note that relatively small values of Q can cause significant peaking. The Q times the gain of the circuit must stay under the loop gain (plus some margin, again, 20 dB is a good starting point). This holds for multiple amplifier topologies as well. Be aware of internal node levels, as well as input and output levels. As an amplifier overloads, its effective Q decreases, so the transfer function will appear to change even if the output appears undistorted. This shows up as the transfer function changing with increasing input level.

We have been dealing mostly with lowpass filters in our discussions, but the same principles are valid for highpass, bandpass, and bandreject as well. In general, things like Q enhancement and limited gain/bandwidth will not affect highpass filters, since the resonant frequency will hopefully be low in relation to the cutoff frequency of the op amp. Remember, though, that the highpass filter will have a lowpass section, by default, at the cutoff frequency of the amplifier. Bandpass and bandreject (notch) filters will be affected, especially since both tend to have high values of Q.

The general effect of the op amp's frequency response on the filter Q is shown in Figure 8-88.

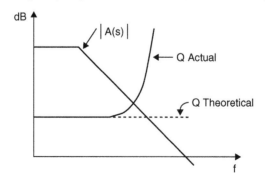

**Figure 8-88: Q enhancement**

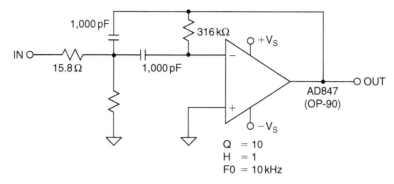

**Figure 8-89: 1 kHz multiple feedback bandpass filter**

As an example of the Q enhancement phenomenon, consider the Spice simulation of a 10 kHz bandpass multiple feedback filter with Q = 10 and gain = 1, using a good high frequency amplifier (the AD847) as the active device. The circuit diagram is shown in Figure 8-89. The open-loop gain of the AD847 is greater than 70 dB at 10 kHz as shown in Figure 8-91(A). This is well over the 20 dB minimum, so the filter works as designed as shown in Figure 8-90.

We now replace the AD847 with an OP-90. The OP-90 is a DC precision amplifier and so has a limited bandwidth. In fact, its open-loop gain is less than 10 dB at 10 kHz (see Figure 8-91(B)). This is not to imply that the AD847 is in all cases better than the OP-90. It is a case of misapplying the OP-90.

From the output for the OP-90, also shown in Figure 8-90, we see that the magnitude of the output has been reduced, and the center frequency has shifted downward.

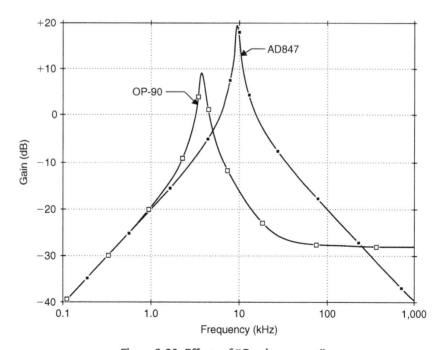

**Figure 8-90: Effects of "Q enhancement"**

**661**

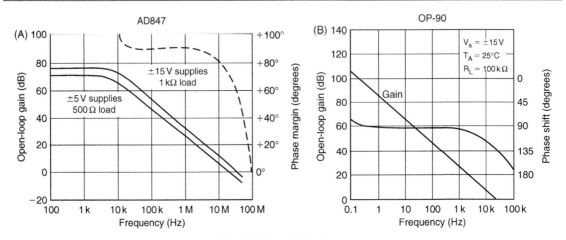

Figure 8-91: AD847 and OP-90 Bode plots

# Design Examples

Several examples will now be worked out to demonstrate the concepts previously discussed.

## Antialias Filter

As an example, passive and active antialias filters will now be designed based on a common set of specifications. The active filter will be designed in four ways: Sallen–Key, multiple feedback, state variable, and FDNR.

The specifications for the filter are given as follows:

*   The cutoff frequency will be 8 kHz.

*   The stopband attenuation will be 72 dB. This corresponds to a 12-bit system.

*   Nyquist frequency of 50 kSPS.

*   The Butterworth filter response is chosen in order to give the best compromise between attenuation and phase response.

Consulting the Butterworth response curves (Figure 8-14, reproduced below in Figure 8-92), we see that for a frequency ratio of 6.25 (50 kSPS/8 kSPS), a filter order of 5 is required.

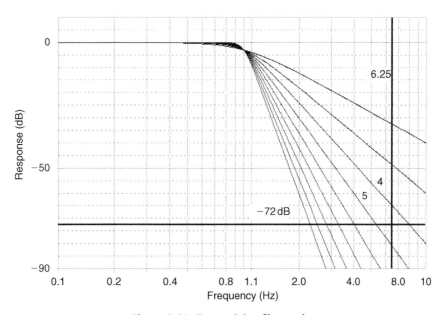

**Figure 8-92: Determining filter order**

663

Now consulting the Butterworth design table (Figure 8-25), the normalized poles of a fifth-order Butterworth filter are:

| Stage | $F_0$ | $\alpha$ |
|-------|-------|----------|
| 1 | 1.000 | 1.618 |
| 2 | 1.000 | 0.618 |
| 3 | 1.000 | – |

The last stage is a real (single) pole, thus the lack of an $\alpha$ value. It should be noted that this is not necessarily the order of implementation in hardware. In general, you would typically put the real pole last and put the second-order sections in order of decreasing $\alpha$ (increasing Q) as we have done here. This will avoid peaking due to high Q sections possibly overloading internal nodes. Another feature of putting the single pole at the end is to bandlimit the noise of the op amps. This is especially true if the single pole is implemented as a passive filter.

For the passive design, we will choose the zero input impedance configuration. While "classic" passive filters are typically double terminated, i.e., with termination on both source and load ends, we are concerned with voltage transfer not power transfer so the source termination will not be used. From the design table (see Reference 2, p. 313), we find the normalized values for the filter (see Figure 8-93).

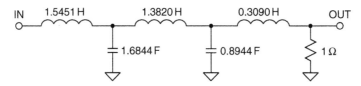

**Figure 8-93: Normalized passive filter implementation**

These values are normalized for a 1 rad/s filter with a 1 Ω termination. To scale the filter we divide all reactive elements by the desired cutoff frequency, 8 kHz ($=50,265$ rad/s, $=2\pi 8 \times 10^3$). This is commonly referred to as the frequency scale factor (FSF). We also need to scale the impedance.

For this example, an arbitrary value of 1,000 Ω is chosen. To scale the impedance, we multiply all resistor and inductor values and divide all capacitor values by this magnitude, which is commonly referred to as the impedance-scaling factor (Z).

After scaling, the circuit looks like Figure 8-94.

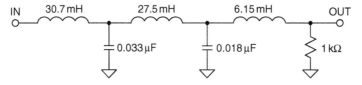

**Figure 8-94: Passive filter implementation**

For the Sallen–Key active filter, we use the design equations shown in Figure 8-49. The values for $C_1$ in each section are arbitrarily chosen to give reasonable resistor values. The implementation is shown in Figure 8-95.

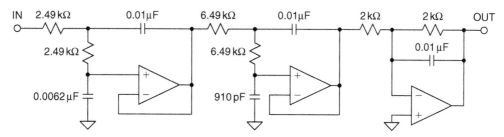

**Figure 8-95: Sallen–Key implementation**

The exact values have been rounded to the nearest standard value. For most active realization to work correctly, it is required to have a zero-impedance driver, and a return path for DC due to the bias current of the op amp. Both of these criteria are approximately met when you use an op amp to drive the filter.

In the above example, the single pole has been built as an active circuit. It would have been just as correct to configure it as a passive RC filter. The advantage to the active section is lower output impedance, which may be an advantage in some applications, notably driving an ADC input that uses a switched capacitor structure.

This type of input is common on sigma delta ADCs as well as many other complementary-MOS process (CMOS) type of converters. It also eliminates the loading effects of the input impedance of the following stage on the passive section.

Figure 8-96 shows a multiple feedback realization of our filter. It was designed using the equations in Figure 8-52. In this case, the last section is a passive RC circuit.

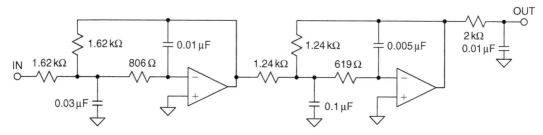

**Figure 8-96: Multiple feedback implementation**

An optional buffer could be added after the passive section, if desired. This would give many of the advantages outlined above, except for bandlimiting the noise of the output amp. By using one of the above two filter realizations, we have both an inverting and a non-inverting design.

The state variable filter, shown in Figure 8-97, was designed with the equations in Figure 8-55. Again, we have rounded the resistor values to the nearest standard 1% value.

Obviously, this filter implementation has many more parts than either the Sallen–Key or the multiple feedback. The rational for using this circuit is that stability is improved and the individual parameters are independently adjustable.

The FDNR realization of this filter is shown in Figure 8-98.

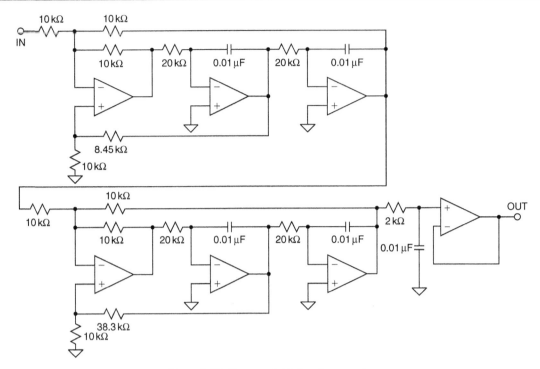

Figure 8-97: State variable implementation

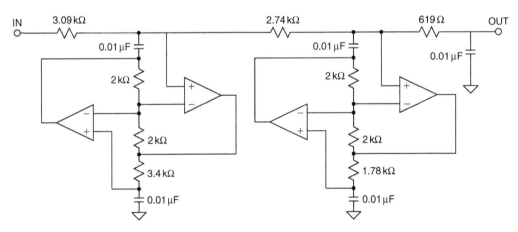

Figure 8-98: FDNR implementation

In the conversion process from passive to FDNR, the D element is normalized for a capacitance of 1 F. We then scale the filter to a more reasonable value (0.01 μF in this case).

In all of the above implementations, standard values were used instead of the calculated values. Any variation from the ideal values will cause a shift in the filter response characteristic, but often the effects are minimal. The computer can be used to evaluate these variations on the overall performance and determine if they are acceptable.

To examine the effect of using standard values, take the Sallen–Key implementation. Figure 8-99 shows the response of each of the three sections of the filter. While the Sallen–Key was the filter used, the results from any of the other implementations will give similar results.

Figure 8-100 then shows the effect of using standard values instead of calculated values. Notice that the general shape of the filter remains the same, just slightly shifted in frequency. This investigation was done only for the standard value of the resistors. To understand the total effect of component tolerance the same type of calculations would have to be done for the tolerance of all the components and also for their temperature and aging effects.

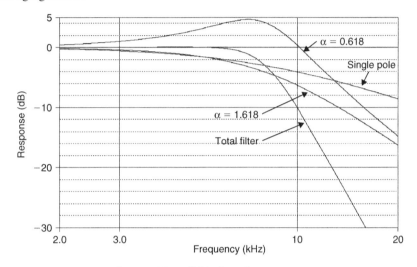

Figure 8-99: Individual section response

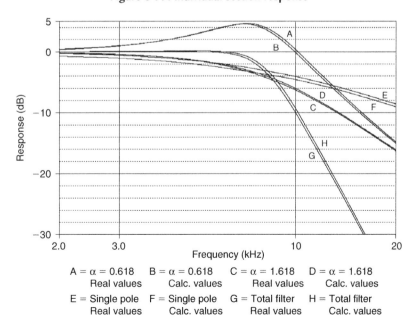

A = α = 0.618    B = α = 0.618    C = α = 1.618    D = α = 1.618
    Real values       Calc. values       Real values       Calc. values

E = Single pole   F = Single pole   G = Total filter   H = Total filter
    Real values       Calc. values       Real values       Calc. values

Figure 8-100: Effect of using standard value resistors

In active filter applications using op amps, the DC accuracy of the amplifier is often critical to optimal filter performance. The amplifier's offset voltage will be passed by the lowpass filter and may be amplified to produce excessive output offset. For low frequency applications requiring large value resistors, bias currents flowing through these resistors will also generate an output offset voltage.

In addition, at higher frequencies, an op amp's dynamics must be carefully considered. Here, slew rate, bandwidth, and open-loop gain play a major role in op amp selection. The slew rate must be fast as well as symmetrical to minimize distortion.

## Transformations

In the next example, the transformation process will be investigated.

As mentioned earlier, filter theory is based on a lowpass prototype, which is then manipulated into the other forms. In these examples, the prototype that will be used is a 1 kHz, three-pole, 0.5 dB Chebyshev filter. A Chebyshev was chosen because it would show more clearly if the responses were not correct; a Butterworth would probably be too forgiving in this instance. A three-pole filter was chosen so that a pole pair and a single pole would be transformed.

The pole locations for the LP prototype were taken from Figure 8-30. They are:

| Stage | $\alpha$ | $\beta$ | $F_0$ | $\alpha$ |
|-------|----------|---------|-------|----------|
| 1 | 0.2683 | 0.8753 | 1.0688 | 0.5861 |
| 2 | 0.5366 | – | 0.6265 | – |

The first stage is the pole pair and the second stage is the single pole. Note the unfortunate convention of using $\alpha$ for 2 entirely separate parameters. The $\alpha$ and $\beta$ on the left are the pole locations in the s-plane. These are the values that are used in the transformation algorithms. The $\alpha$ on the right is 1/Q, which is what the design equations for the physical filters want to see.

The Sallen–Key topology will be used to build the filter. The design equations in Figure 8-67 (pole pair) and Figure 8-66 (single pole) were then used to design the filter. The schematic is shown in Figure 8-101.

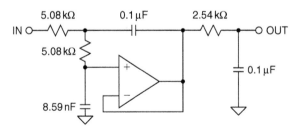

**Figure 8-101: Lowpass prototype**

Using the equation string described in Section 8-8, the filter is now transformed into a highpass filter. The results of the transformation are:

| Stage | $\alpha$ | $\beta$ | $F_0$ | $\alpha$ |
|-------|----------|---------|-------|----------|
| 1 | 0.3201 | 1.0443 | 0.9356 | 0.5861 |
| 2 | 1.8636 | – | 1.596 | – |

A word of caution is warranted here. Since the convention of describing a Chebyshev filter is to quote the end of the error band instead of the 3 dB frequency, the $F_0$ must be divided (for highpass) by the ratio of ripple band to 3 dB bandwidth (Table 8-1, Section 8-4).

The Sallen–Key topology will again be used to build the filter. The design equations in Figure 8-68 (pole pair) and Figure 8-66 (single pole) were then used to design the filter. The schematic is shown in Figure 8-102.

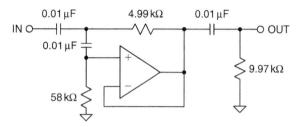

**Figure 8-102: Highpass transformation**

Figure 8-103 shows the response of the lowpass prototype and the highpass transformation. Note that they are symmetric around the cutoff frequency of 1 kHz. Also note that the error band is at 1 kHz, not the $-3$ dB point, which is characteristic of Chebyshev filters.

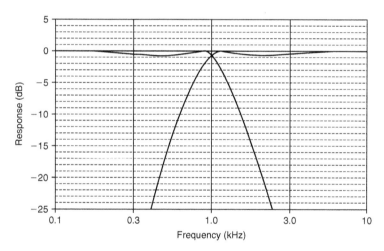

**Figure 8-103: Lowpass and highpass response**

The lowpass prototype is now converted to a bandpass filter. The equation string outlined in Section 8-5 is used for the transformation. Each pole of the prototype filter will transform into a pole pair. Therefore, the three-pole prototype, when transformed, will have six poles (three-pole pairs). In addition, there will be six zeros at the origin.

Part of the transformation process is to specify the 3 dB bandwidth of the resultant filter. In this case this bandwidth will be set to 500 Hz. The results of the transformation yield:

| Stage | $F_0$ | Q | $A_0$ |
|---|---|---|---|
| 1 | 804.5 | 7.63 | 3.49 |
| 2 | 1,243 | 7.63 | 3.49 |
| 3 | 1,000 | 3.73 | 1 |

The reason for the gain requirement for the first two stages is that their center frequencies will be attenuated relative to the center frequency of the total filter. Since the resultant Q's are moderate (less than 20) the multiple feedback topology will be chosen. Figure 8-72 was then used to design the filter sections.

Figure 8-104 is the schematic of the filter and Figure 8-105 shows the filter response.

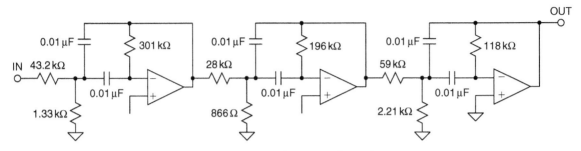

**Figure 8-104: Bandpass transformation**

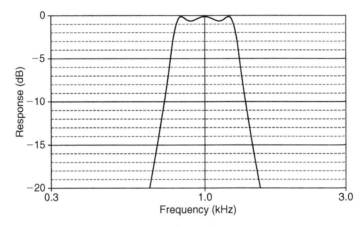

**Figure 8-105: Bandpass filter response**

Note that again there is symmetry around the center frequency. Also the 800 Hz bandwidth is not 250 Hz either side of the center frequency (arithmetic symmetry). Instead the symmetry is geometric, which means that for any two frequencies ($F_1$ and $F_2$) of equal amplitude are related by:

$$F_0 = \sqrt{F_1 \times F_2} \qquad (8\text{-}96)$$

Lastly, the prototype will be transformed into a bandreject filter. For this the equation string in Section 8-5 is used. Again, each pole of the prototype filter will transform into a pole pair. Therefore, the three-pole prototype, when transformed, will have six poles (three-pole pairs).

As in the bandpass case, part of the transformation process is to specify the 3 dB bandwidth of the resultant filter. Again in this case, this bandwidth will be set to 500 Hz. The results of the transformation yield:

| Stage | $F_0$ | Q | $F_{oz}$ |
|-------|-------|------|-------|
| 1 | 763.7 | 6.54 | 1,000 |
| 2 | 1,309 | 6.54 | 1,000 |
| 3 | 1,000 | 1.07 | 1,000 |

Note that there are three cases of notch filters required. There is a standard notch ($F_0 = F_Z$, Section 8-3), a lowpass notch ($F_0 < F_Z$, Section 8-1) and a highpass notch ($F_0 > F_Z$, Section 8-2). Since there is a requirement for all three types of notches, the Bainter notch is used to build the filter. The filter is designed using Figure 8-77. The gain factors K1 and K2 are arbitrarily set to 1. Figure 8-106 is the schematic of the filter.

The response of the filter is shown in Figure 8-107 and in detail in Figure 8-108. Again, note the symmetry around the center frequency. Again the frequencies have geometric symmetry.

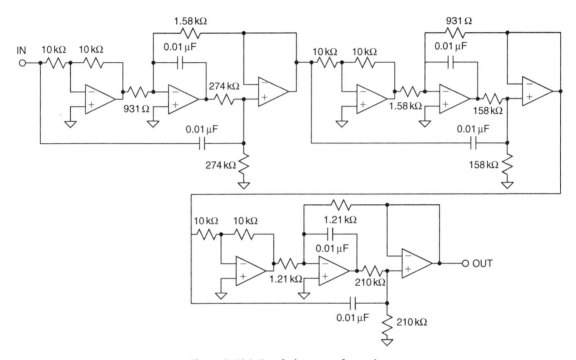

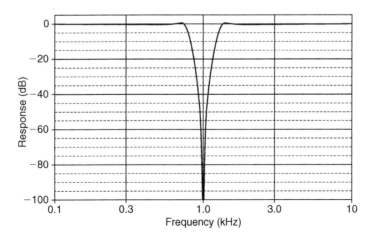

**Figure 8-106: Bandreject transformation**

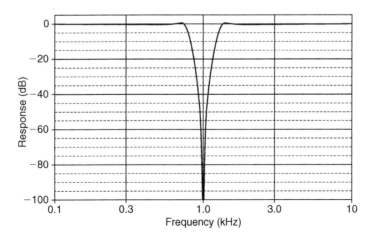

**Figure 8-107: Bandreject response**

*671*

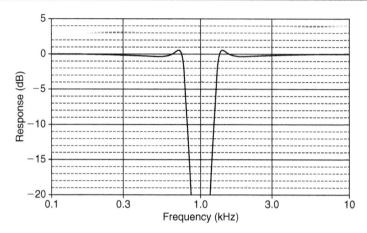

**Figure 8-108: Bandreject response (detail)**

## CD Reconstruction Filter

This design was done for a magazine article describing a high quality outboard D/A converter for use with digital audio sources (see Reference 26).

A reconstruction filter is required on the output of a D/A converter because, despite the name, the output of a D/A converter is not really an analog voltage but instead a series of steps. The converter will put out a discrete voltage, which it will then hold until the next sample is asserted. The filter's job is to remove the high frequency components, smoothing out the waveform. This is why the filter is sometimes referred to as a smoothing filter. This also serves to eliminate the aliases of the conversion process. The "standard" in the audio industry is to use a third-order Bessel function as the reconstruction filter. The reason to use a Bessel filter is that it has the best phase response. This helps to preserve the phase relationship of the individual tones in the music. The price for this phase "goodness" is that the amplitude discrimination is not as good as some other filter types. If we assume that we are using $8 \times$ oversampling of the 48 kSPS data stream in the D/A converter then the aliases will appear at 364 kHz ($8 \times 48$k $-20$k). The digital filter that is used in the interpolation process will eliminate the frequencies between 20 and 364 kHz. If we assume that the band edge is 30 kHz, then we have a frequency ratio of approximately 12 ($364 \div 30$). We use 30 kHz as the band edge, rather than 20 kHz to minimize the rolloff due to the filter in the passband. In fact, the complete design for this filter includes a shelving filter to compensate for the passband rolloff. Extrapolating from Figure 8-20, a third-order Bessel will only provide on the order of 55 dB attenuation at $12 \times F_0$. This is only about 9-bit accuracy.

By designing the filter as seventh order, and by designing it as a linear phase with equiripple error of $0.05°$ we can increase the stopband attenuation to about 120 dB at $12 \times F_0$. This is close to the 20-bit system that we are hoping for.

The filter will be designed as a FDNR type. This is an arbitrary decision. Reasons to choose this topology are its low sensitivities to component tolerances and the fact that the op amps are in the shunt arms rather than in the direct signal path.

The first step is to find the passive prototype. To do this, use the charts in the Williams book. We then get the circuit shown in Figure 8-109(A). Next, perform a translation in the s-plane. This gives the circuit shown in Figure 8-109(B). This filter is scaled for a frequency of 1 Hz and an impedance level of 1 Ω. The D structure of the converted filter is replaced by a GIC structure that can be physically realized. The filter is then denormalized by frequency (30 kHz) and impedance (arbitrarily chosen to be 1 kΩ). This gives a FSF of $1.884 \times 10^5$ ($=2\pi(3 \times 10^4)$). Next, arbitrarily choose a value of 1 nF for the capacitor. This gives an impedance-scaling factor (Z) of 5305 ($=(C_{OLD}/C_{NEW})/FSF$).

Then multiply the resistor values by Z. This results in the resistors that had the normalized value of 1 Ω will now have a value of 5.305 kΩ. For the sake of simplicity adopt the standard value of 5.36 kΩ. Working backwards, this will cause the cutoff frequency to change to 29.693 kHz. This slight shift of the cutoff frequency will be acceptable.

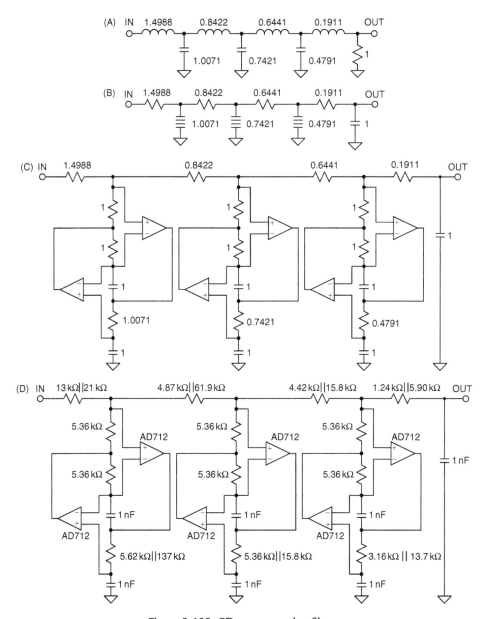

**Figure 8-109: CD reconstruction filter**

The FSF is then recalculated with the new center frequency and this value is used to denormalize the rest of the resistors. The D structure of the converted filter is replaced by a GIC structure that can be physically realized (Figure 8-109(C)). The final schematic is shown in Figure 8-109(D).

The performance of the filter is shown in Figure 8-110(A–D).

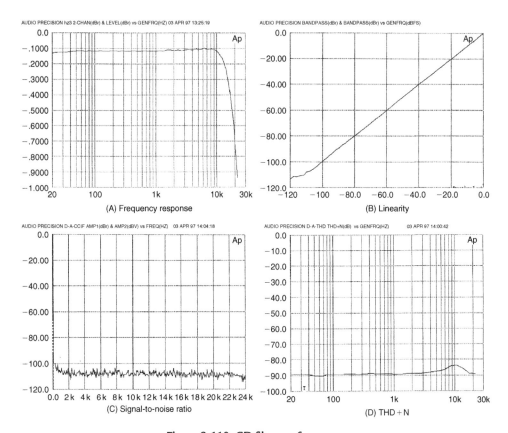

Figure 8-110: CD filter performance

## Digitally Programmable State Variable Filter

One of the attractive features of the state variable filter is that the parameters (gain, cutoff frequency, and "Q") can be individually adjusted. This attribute can be exploited to allow control of these parameters.

To start, the filter is reconfigured slightly. The resistor divider that determines Q ($R_6$ and $R_7$ of Figure 8-84) is changed to an inverting configuration. The new filter schematic is shown in Figure 8-111. Then the resistors $R_1$–$R_4$ (of Figure 8-111) are replaced by CMOS multiplying DACs. Note that $R_5$ is implemented as the feedback resistor implemented in the DAC. The schematic of this circuit is shown in Figure 8-112.

The AD7528 is an 8-bit dual MDAC. The AD825 is a high speed FET input op amp. Using these components the frequency range can be varied from around 550 Hz to around 150 kHz (Figure 8-113). The Q can be varied from approximately 0.5 to over 12.5 (Figure 8-114). The gain of circuit can be varied from 0 to −48 dB (Figure 8-115).

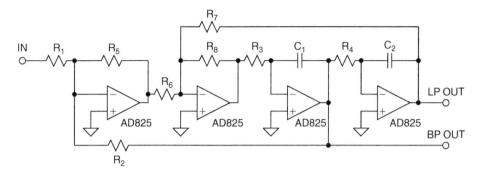

**Figure 8-111: Redrawn state variable filter**

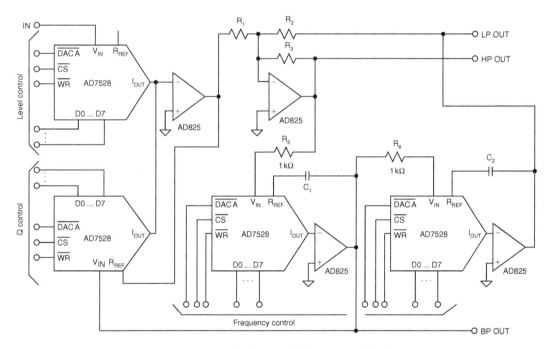

**Figure 8-112: Digitally controlled state variable filter**

The operation of the DACs in controlling the parameters can be best thought of as the DACs changing the effective resistance of the resistors. This relationship is:

$$\text{DAC equivalent resistance} = \frac{256 \times \text{DAC resistance}}{\text{DAC code (decimal)}} \qquad (8\text{-}97)$$

This, in effect, varies the resistance from $11\,\text{k}\Omega$ to $2.8\,\text{M}\Omega$ for the AD7528.

One limitation of this design is that the frequency is dependent on the ladder resistance of the DAC. This particular parameter is not controlled. DACs are trimmed so that the ratios of the resistors, not their absolute values, are controlled. In the case of the AD7528, the typical value is $11\,\text{k}\Omega$. It is specified

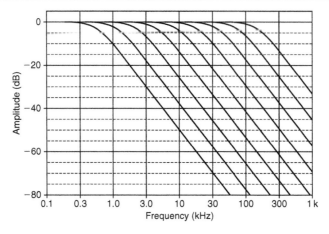

Figure 8-113: Frequency response versus DAC control word

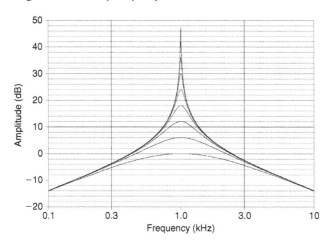

Figure 8-114: Q variation versus DAC control word

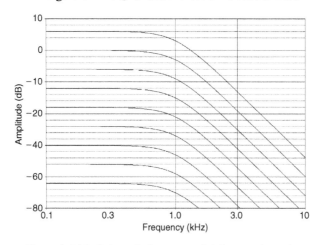

Figure 8-115: Gain variation versus DAC control word

as 8 kΩ min and 15 kΩ max. A simple modification of the circuit can eliminate this issue. The cost is two more op amps (Figure 8-116). In this case, the effective resistor value is set by the fixed resistors rather than the DAC's resistance. Since there are two integrators the extra inversions caused by the added op amps cancel.

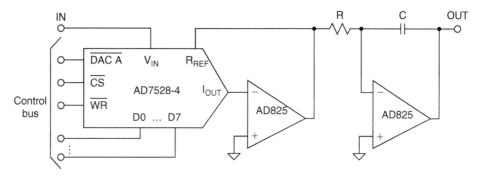

Figure 8-116: Improved digitally variable integrator

As a side note, the multiplying DACs could be replaced by analog multipliers. In this case, the control would obviously be an analog rather than a digital signal. We also could just as easily have used a digital pot in place of the MDACs. The difference is that instead of increasing the effective resistance, the value of the pot would be the maximum.

## 60 Hz Notch Filter

A very common problem in instrumentation is that of interference of the telemetry that is to be measured. One of the primary sources of this interference is the power line. This is particularly true of high impedance circuits. Another path for this noise is ground loops. One possible solution is to use a notch filter to remove the 60 Hz component. Since this is a single frequency interference, the twin-T circuit will be used.

Since the maximum attenuation is desired and the minimum notch width is desired, the maximum Q of the circuit is desired. This means the maximum amount of positive feedback is used ($R_5$ open and $R_4$ shorted). Due to the high impedance of the network, a FET input op amp is used.

The filter is designed using Figure 8-78. The schematic is shown in Figure 8-117 and the response in Figure 8-118.

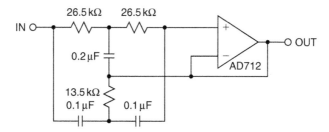

Figure 8-117: 60 Hz twin-T notch filter

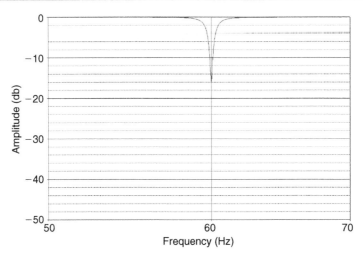

**Figure 8-118: 60 Hz notch filter response**

## References: Design Examples

1. A.I. Zverev, **Handbook of Filter Synthesis**, John Wiley & Sons, New York, 1967.

2. A.B. Williams, **Electronic Filter Design Handbook**, McGraw-Hill, New York, 1981, ISBN: 0-07-070430-9.

3. M.E. Van Valkenburg, **Analog Filter Design**, Rinehart & Winston, Holt, 1982.

4. M.E. Van Valkenburg, **Introduction to Modern Network Synthesis**, John Wiley & Sons, New York, 1960.

5. A.I. Zverev and H.J. Blinchikoff, **Filtering in the Time and Frequency Domain**, John Wiley & Sons, New York, 1976.

6. S. Franco, **Design with Operational Amplifiers and Analog Integrated Circuits**, McGraw-Hill, New York, 1988, ISBN: 0-07-021799-8.

7. W. Cauer, **Synthesis of Linear Communications Networks**, McGraw-Hill, New York, 1958.

8. A. Budak, **Passive and Active Network Analysis and Synthesis**, Houghton Mifflin Company, Boston, MA, 1974.

9. L.P. Huelsman and P.E. Allen, **Introduction to the Theory and Design of Active Filters**, McGraw-Hill, New York, 1980, ISBN: 0-07-030854-3.

10. R.W. Daniels, **Approximation Methods for Electronic Filter Design**, McGraw-Hill, New York, 1974.

11. J. Tow, "Active RC Filters—A State-Space Realization," **Proceedings of the IEEE**, Vol. 56, 1968, pp. 1137–1139.

12. L.C. Thomas, "The Biquad: Part I—Some Practical Design Considerations," **IEEE Transactions on Circuits and Systems**, Vol. CAS-18, 1971, pp. 350–357.

13. L.C. Thomas, "The Biquad: Part I—A Multipurpose Active Filtering System," **IEEE Transactions on Circuits and Systems**, Vol. CAS-18, 1971, pp. 358–361.

14. R.P. Sallen and E.L. Key, "A Practical Method of Designing RC Active Filters," **IRE Transactions on Circuit Theory**, Vol. CT-2, 1999, pp. 74–85.

15. P.R. Geffe, "How to Build High-Quality Filters out of Low-Quality Parts," **Electronics**, November 1976, pp. 111–113.

16. P.R. Geffe, "Designers Guide to Active Bandpass Filters," **EDN**, April 5, 1974, pp. 46–52.

17. T. Delyiannis, "High-Q Factor Circuit with Reduced Sensitivity," **Electronics Letters**, December 4, 1968, p. 577.

18. J.J. Friend, "A Single Operational-Amplifier Biquadratic Filter Section," **IEEE ISCT Digest Technical Papers**, 1970, p. 189.

19. L. Storch, "Synthesis of Constant-Time-Delay Ladder Networks Using Bessel Polynomials," **Proceedings of the IRE**, Vol. 42, 1954, pp. 1666–1675.

20. K.W. Henderson and W.H. Kautz, "Transient Response of Conventional Filters," **IRE Transactions on Circuit Theory**, Vol. CT-5, 1958, pp. 333–347.

21. J.R. Bainter, "Active Filter Has Stable Notch and Response Can be Regulated," **Electronics**, October 2, 1975, pp. 115–117.

22. S.A. Boctor, "Single Amplifier Functionally Tunable Low-Pass Notch Filter," **IEEE Transactions on Circuits and Systems**, Vol. CAS-22, 1975, pp. 875–881.

23. S.A. Boctor, "A Novel Second-Order Canonical RC-Active Realization of High-Pass-Notch Filter," **Proceedings of the 1974 IEEE International Symposium Circuits and Systems**, pp. 640–644.

24. L.T. Burton, "Network Transfer Function Using the Concept of Frequency Dependant Negative Resistance," **IEEE Transactions on Circuit Theory**, Vol. CT-16, 1969, pp. 406–408.

25. L.T. Burton and D. Trefleaven, "Active Filter Design Using General Impedance Converters," **EDN**, 1973, pp. 68–75.

26. H. Zumbahlen, "A New Outboard DAC, Part 2," **Audio Electronics**, Vol. 1, January 1997, pp. 26–32, 42.

27. M. Williamsen, "Notch-Filter Design," **Audio Electronics**, Vol. 1, January 2000, pp. 10–17.

28. W. Jung, *Bootstrapped IC Substrate Lowers Distortion in JFET Op Amps*, Analog Devices AN232.

29. H. Zumbahlen, *Passive and Active Filtering*, Analog Devices AN281.

30. P. Toomey and W. Hunt, *AD7528 Dual 8-Bit CMOS DAC*, Analog Devices AN318.

31. W. Slattery, *8th Order Programmable Lowpass Analog Filter Using 12-Bit DACs*, Analog Devices AN209.

32. *CMOS DAC Application Guide*, Analog Devices. ADI, Norwood, MA.

# CHAPTER 9

# *Power Management*

## Chapter Introduction

All electronic systems require power supplies to operate. This part of the system design is often overlooked. But, as discussed in the amplifier section on PSRR (power supply rejection ratio), the AC component of the power supply (noise) may add to the output of the amplifier. And the PSRR has a strong frequency dependence, falling as frequency increases. The supply voltage(s) must be kept as quiet as possible for optimum performance of high performance circuits. Local decoupling helps—but it is not sufficient.

*Power management* broadly refers to the generation and control of regulated voltages required to operate an electronic system. It encompasses much more than just power supply design. Today's systems require power supply design be integrated with the system design in order to maintain high efficiency. In addition, distributed power supply systems require localized regulators at the PC board level, thereby requiring the design engineer to master at least the basics of both switching and linear regulators.

Integrated circuit (IC) components such as switching regulators, linear regulators, switched capacitor voltage converters, and voltage references are typical elements of power management.

Historically, the standard for supply voltages was $\pm 15\,V$. In recent years the trend is toward lower supply voltages. This is partially due to the processes used to manufacture integrated circuits. Circuit speeds have increased. One of the enabling technologies of this increase is the reduction of size of the transistors used in the process. These smaller feature sizes imply lower breakdown voltages, which, in turn, indicate lower supply voltages.

Another trend in power management is the often misguided attempt to operate the analog and digital circuitry on the same supply, eliminating one of the supplies. While the supply voltages may be the same, the low noise requirement is at odds with the often very noisy digital supply.

While reducing the supply voltage has the very desirable effect of reducing the power dissipation of digital circuits, lowering the supply of linear circuits limits the dynamic range of the signal. And, unfortunately, lowering the signal swing (the upper end of the dynamic range) by lowering the supply voltage does not imply that the noise level (the lower end of the signal dynamic range) will be reduced by a like amount.

Another trend in power management is the adoption of unipolar (single) supplies. This is often done in the attempt to eliminate the negative supply. Often, a negative supply is cheaper and provides better performance. This is due to the extra circuitry required for level shifting and AC coupling required by a single supply.

# Linear Voltage Regulators

## Linear Regulator Basics

Linear IC voltage regulators have long been standard power system building blocks. After an initial introduction in 5 V logic voltage regulator form, they have since expanded into other standard voltage levels spanning from less than 1 to 24 V, handling output currents from as low as 25 mA (or less) to as high as 5 A (or more). For several good reasons, linear style IC voltage regulators have been valuable system components since the early days. One reason is the relatively low noise characteristic vis-à-vis the switching type of regulator. Others are a low parts count and overall simplicity compared to discrete solutions. Because of their power losses, these linear regulators have also been known for being relatively inefficient. Early generation devices (of which many are still available) required an unregulated input 2 V or more above the regulated output voltage, making them lossy in power terms. Typical maximum efficiency for a 5 V supply is 71%, meaning 29% of the power is dissipated in the regulator.

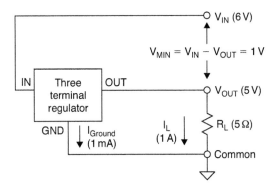

**Figure 9-1: A basic three-terminal voltage regulator**

Figure 9-1 also allows a more detailed analysis of power losses in the regulator. There are two components to power which are dissipated in the regulator, one a function of $V_{IN} - V_{OUT}$ and $I_L$, plus a second which is a function of $V_{IN}$ and $I_{Ground}$. If we call the total power $P_D$, this then becomes:

$$P_D = (V_{IN} - V_{OUT})(I_L) + (V_{IN})(I_{Ground}) \qquad (9\text{-}1)$$

A more detailed look within a typical regulator block diagram reveals a variety of elements, as is shown in Figure 9-2.

In this diagram virtually all of the elements shown are fundamentally necessary, the exceptions being the shutdown control and saturation sensor functions (shown dotted). While these are present on many current regulators, the shutdown feature is relatively new as a standard function, and certainly is not part of standard three-terminal regulators. When present, shutdown control is a logic level controllable input.

The optional error output, $\overline{\text{ERR}}$, is useful within a system to detect regulator overload, such as saturation of the pass device, thermal overload, etc. The remaining functions shown are always part of an IC power regulator. While this diagram shows the blocks in a conceptual way, actual implementation may vary.

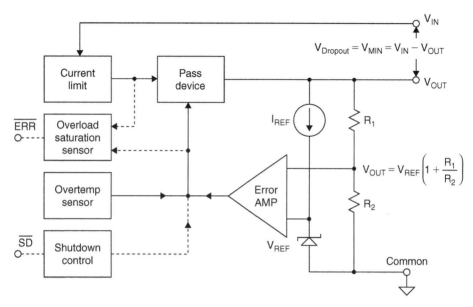

**Figure 9-2: Block diagram of a voltage regulator**

In operation, a voltage reference block produces a stable voltage $V_{REF}$, which is almost always a bandgap-based voltage, typically $\sim 1.2\,$V. This voltage is presented to one input of an error amplifier, with the other input connected to the $V_{OUT}$ sensing divider, $R_1$ and $R_2$. The error amplifier drives the pass device, which in turn controls the output. The resulting regulated voltage is then simply:

$$V_{OUT} = V_{REF}\left(1 + \frac{R_1}{R_2}\right) \tag{9-2}$$

When standby power is critical, several design steps will be taken. The resistor values of the divider will be high, the error amplifier and pass device driver will be low power, and the reference current $I_{REF}$ will also be low. By these means the regulator's unloaded standby current can be reduced to a mA or less using bipolar technology, and to only a few microamperes in complementary-MOS (CMOS) parts. In regulators which offer a shutdown mode, the shutdown state standby current may be reduced to $1\,\mu$A or less.

Nearly all regulators will have some means of current limiting and over temperature sensing, to protect the pass device against failure. Current limiting is usually implemented by a series sensing resistor for high current parts, or alternately by a more simple drive current limit to a controlled $\beta$ pass device (which achieves the same end). For higher voltage circuits, this current limiting may also be combined with voltage limiting, to provide complete load line control for the pass device. All power regulator devices will also have some means of sensing excessive temperature, usually by means of a fixed reference voltage and a $V_{BE}$-based sensor monitoring chip temperature. When the die temperature exceeds a dangerous level (above $\sim 150°$C), this can be used to shutdown the chip, by removing the drive to the pass device. In some cases an error flag output may be provided to warn of this shutdown (and also loss of regulation from some other means).

## Pass Devices and Their Associated Tradeoffs

The discussion thus far has not treated the pass device in any detail. In practice, this major part of the regulator can actually take on quite a number of alternate forms. Precisely which type of pass device is chosen has a major influence on almost all major regulator performance issues. Most notable among these is the dropout voltage, $V_{MIN}$.

It is difficult to fully compare all of the devices from their schematic representations, since they differ in so many ways beyond their applicable dropout voltages. For this reason, the chart of Figure 9-4 is useful.

This chart compares the various pass elements in greater detail, allowing easy comparison between the device types, dependent on which criteria is most important. Note that columns A–E correspond to the schematics of Figure 9-3. Note also that the pro/con comparison items are in *relative* terms, as opposed to a hard specification limit for any particular pass device type.

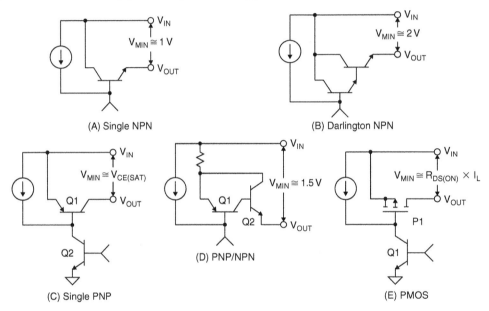

Figure 9-3: Pass devices useful in voltage regulators

| A<br>Single<br>NPN | B<br>Darlington<br>NPN | C<br>Single<br>PNP | D<br>PNP/NPN | E<br>PMOS |
|---|---|---|---|---|
| $V_{MIN} \sim 1\,V$ | $V_{MIN} \sim 2\,V$ | $V_{MIN} \sim 0.1\,V$ | $V_{MIN} \sim 1.5\,V$ | $V_{MIN} \sim R_{DS(ON)} \times I_L$ |
| $I_L < 1\,A$ | $I_L > 1\,A$ | $I_L < 1\,A$ | $I_L > 1\,A$ | $I_L > 1\,A$ |
| Follower | Follower | Inverter | Inverter | Inverter |
| Low $Z_{OUT}$ | Low $Z_{OUT}$ | High $Z_{OUT}$ | High $Z_{OUT}$ | High $Z_{OUT}$ |
| Wide BW | Wide BW | Narrow BW | Narrow BW | Narrow BW |
| $C_L$ Immune | $C_L$ Immune | $C_L$ Sensitive | $C_L$ Sensitive | $C_L$ Sensitive |

Figure 9-4: Pros and cons of voltage regulator pass devices

For example, it can be seen that the all NPN pass devices of columns A and B have the attributes of a follower circuit, which allows high bandwidth and provides relative immunity to cap loading because of the characteristic low $Z_{OUT}$.

All of the three connections C/D/E have the characteristic of high output impedance, and require an output capacitor for stability. The fact that the output cap is part of the regulator frequency compensation is a most basic application point, and one which needs to be clearly understood by the regulator user. This factor, denoted by "$C_L$ sensitive," makes regulators using them generally critical as to the exact $C_L$ value, as well as its ESR (equivalent series resistance). Typically this type of regulator must be used only with a specific size as well as type of output capacitor, where the ESR is stable and predictable with respect to both time and temperature to fully guarantee regulator stability. Some recent Analog Devices LDO IC circuit developments have eased this burden on the part of the regulator user a great deal and will be discussed below in further detail.

The classic LM309 5 V/1 A three-terminal regulator (see Reference 1) was the originator in a long procession of linear regulators. This circuit is shown in much simplified form in Figure 9-5, with current limiting and over temperature details omitted. This IC type is still in standard production today, not just in original form, but in family derivatives such as the 7805, 7815, etc., and their various low and medium current alternates. Using a Darlington pass connection for Q18–Q19, the design does not have low dropout (LDO) characteristics ($\sim$1.5 V typical minimum), or for low quiescent current ($\sim$5 mA). It is however relatively immune to instability issues, due to the internal compensation of $C_1$, and the buffering of the emitter follower output. This helps make it easy to apply.

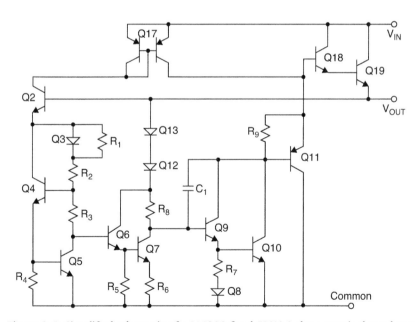

**Figure 9-5: Simplified schematic of a LM309 fixed 5 V/1 A three-terminal regulator**

Later developments in references and three-terminal regulation techniques led to the development of the voltage adjustable regulator. The original IC to employ this concept was the LM317 (see Reference 2),

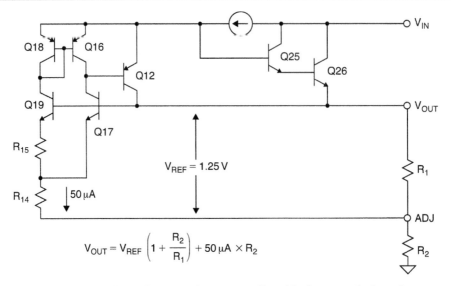

**Figure 9-6: Simplified schematic of a LM317 adjustable three-terminal regulator**

which is shown in simplified schematic form in Figure 9-6. Note that this design does not use the same $\Delta V_{BE}$ form of reference as in the LM309. Instead, Q17–Q19, etc. are employed as a form of a Brokaw bandgap reference cell (Reference 3).

This adjustable regulator bootstraps the reference cell transistors Q17–Q19 and the error amplifier transistors Q16–Q18. The output of the error amplifier drives Darlington pass transistors Q25–Q26 through buffer Q12. The basic reference cell produces a fixed voltage of 1.25 V, which appears between the $V_{OUT}$ and ADJ pins of the IC as shown. External scaling resistors $R_1$ and $R_2$ set up the desired output voltage, which is:

$$V_{OUT} = V_{REF}\left(1 + \frac{R_2}{R_1}\right) + 50\,\mu A \times R_2 \tag{9-3}$$

As can be noted, the voltage output is a scaling of $V_{REF}$ with $R_2$ and $R_1$, plus a small voltage component which is a function of the $50\,\mu A$ reference cell current. Typically, the $R_1$ and $R_2$ values are chosen to draw $>5\,mA$, making the offset current term relatively small by comparison. The design is internally compensated, and in many applications will not necessarily need an output bypass capacitor to ensure stability. But in most cases you will still want to use one—see the section on decoupling.

Like the LM309 fixed voltage regulator, the LM317 series has relatively high dropout voltage, due to the use of Darlington pass transistors. It is also not a low power IC (quiescent current typically 3.5 mA). The strength of this regulator lies in the wide range of user voltage adaptability it allows.

## Low Dropout Regulator Architectures

In many systems it is desirable to have a linear regulator with low input–output differential. This allows for reduced power dissipation. It also allows for declining input voltage (such as a discharging battery). This is known as a low dropout regulator (LDO). As has been shown thus far, all LDO pass devices have the fundamental characteristics of operating in an inverting mode. This allows the regulator circuit to regulate down to the pass device saturation (but if saturation is reached, the circuit is no longer a regulator), and thus LDO. A by-product of this mode of operation is that this type of topology will necessarily be more susceptible to stability issues. This is due to the higher output impedance of the inverting pass devices, relative to the follower configurations. This higher impedance, combined with the impedance of the output capacitor, can move the second pole of the system too far in, causing instability and possible oscillations. These basic points give rise to some of the more difficult issues with regard to LDO performance. In fact, these points influence both the design and the application of LDOs to a very large degree, and in the end, determine how they are differentiated in the performance arena.

A traditional LDO architecture is shown in Figure 9-7, and is generally representative of actual parts employing either a PNP pass device as shown, or alternately, a PMOS device. There are both DC and AC design and application issues to be resolved with this architecture, which are now discussed.

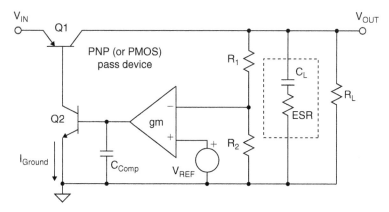

**Figure 9-7: Traditional LDO architecture**

In DC terms, perhaps the major issue is the type of pass device used, which influences dropout voltage and ground current. If a lateral PNP device is used for Q1, the $\beta$ will be low, sometimes only on the order of 10 or so. Since Q1 is driven from the collector of Q2, the relatively high base current demanded by a lateral PNP results in relatively high emitter current in Q2, or a high $I_{Ground}$. For a typical lateral PNP-based regulator operating with a 5 V/150 mA output, $I_{Ground}$ will be typically $\sim$18 mA, and can be as high as 40 mA. To compound the problem of high $I_{Ground}$ in PNP LDOs, there is also the "spike" in $I_{Ground}$, as the regulator tries to regulate down to the dropout region. Under such conditions, the output voltage is out of tolerance, and the regulation loop requires higher drive to the pass device, in an unsuccessful attempt to maintain loop regulation. This results in a substantial spike upward in $I_{Ground}$, which is typically internally limited by the regulator's saturation control circuits.

PMOS pass devices do not demonstrate a similar current spike in $I_{Ground}$, since they are voltage controlled. But, while devoid of the $I_{Ground}$ spike, PMOS pass devices do have some problems of their own. Problem number one is that high quality, low $R_{ON}$, low threshold PMOS devices generally are not compatible with many IC processes. This makes the best technical choice for a PMOS pass device an external part, driven from the collector of Q2 in the figure. This introduces the term "LDO controller," where the LDO architecture is completed by an external pass device. While in theory NMOS pass devices would offer lower $R_{ON}$ choice options, they also demand a boosted voltage supply to turn on, making them impractical for a simple LDO. PMOS pass devices are widely available in both low $R_{ON}$ and low threshold forms, with current levels up to several amperes. They offer the potential of the lowest dropout of any device, since dropout can always be lowered by picking a lower $R_{ON}$ part.

The dropout voltage of lateral PNP pass devices is typically around 300 mV at 150 mA, with a maximum of 600 mV. These performance levels are considerably better in regulators using vertical PNPs, which have a typical $\beta$ of ~150 at currents of 200 mA. This corresponds to an $I_{Ground}$ of 1.3 mA at the 200 mA output current. The dropout voltage of vertical PNPs is also an improvement vis-à-vis that of the lateral PNP regulator, and is typically 180 mV at 200 mA, with a maximum of 400 mV.

There are also major AC performance issues to be dealt with in the LDO architecture of Figure 9-7. This topology has an inherently high output impedance, due to the operation of the PNP pass device in a common-emitter (or common source with a PMOS device) mode. In either case, this factor causes the regulator to appear as a high source impedance to the load.

The internal compensation capacitor of the regulator, $C_{COMP}$, forms a fixed frequency pole, in conjunction with the $g_m$ of the error amplifier. In addition, load capacitance $C_L$ forms an output pole, in conjunction with $R_L$. This particular pole, because it is a second (and sometimes variable) pole of a two-pole system, is the source of a major LDO application problem. The $C_L$ pole can strongly influence the overall frequency response of the regulator, in ways that are both useful as well as detrimental. Depending on the relative positioning of the two poles in the frequency domain, along with the relative value of the ESR of capacitor $C_L$, it is quite possible that the stability of the system can be compromised for certain combinations of $C_L$ and ESR. Note that $C_L$ is shown here as a real capacitor, which is actually composed of a pure capacitance plus the series parasitic resistance ESR.

If the two poles of such a system are widely separated in terms of frequency, stability may not be a serious problem. The emitter follower output of a classic regulator like the LM309 is an example with widely separated pole frequencies, as the very low $Z_{OUT}$ of the NPN follower pushes the output pole due to load capacitance far out in frequency, where it does little harm. The internal compensation capacitance ($C_1$ of Figure 9-5) then forms part of a *dominant pole*, which reduces loop gain to below unity at the much higher frequencies where the second (output) pole does occur. Thus stability is not necessarily compromised by load capacitance in this type of regulator.

By their nature, however, LDOs simply cannot afford the luxury of emitter follower outputs; they must instead operate with pass devices capable of saturation. Thus, given the existence of two or more poles (one or more internal and another formed by external capacitance) there is the potential for the cumulative phase shift to exceed 180° before the gain drops to unity, thus causing oscillations. The potential for instability under certain output loading conditions is, for better or worse, a fact of life for most LDO topologies.

However, the very output capacitor which gives rise to the instability can, in certain circumstances, also be the solution to the same instability. This seemingly paradoxical situation can be appreciated by realizing that all practical capacitors are actually a series combination of the capacitance $C_L$ and the parasitic resistance, ESR. While load resistance $R_L$ and $C_L$ do form a pole, $C_L$ and its ESR also form a zero. The effect of the zero is to mitigate the de-stabilizing effect of $C_L$ for certain conditions. For example, if the pole and zero in question are appropriately placed in frequency relative to the internal regulator poles, some of the deleterious effects can be made to essentially cancel. The basic challenge with this approach to stability is simply that the capacitor's ESR, being a parasitic term, is not well controlled. As a result, LDOs which depend on output pole-zero compensation schemes must very carefully limit the capacitor ESR to certain *zones*, such as shown by Figure 9-8.

A zoned ESR chart such as this is meant to guide the user of an LDO in picking an output capacitor which confines ESR to the stable region, i.e., the central zone, for all operating conditions. Note that this generic chart is not intended to portray any specific device, just the general pattern, but finding a capacitor that guarantees minimum and maximum ESR (especially over temperature). This effectively means that general purpose aluminum electrolytics are prohibited from use, since they deteriorate (increase) in terms of ESR

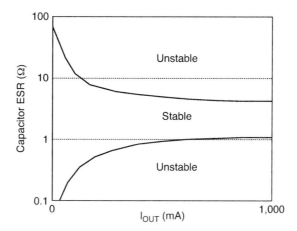

**Figure 9-8: Zoned load capacitor ESR can make LDO applications a nightmare**

at cold temperatures. Very low ESR types such as OS-CON or multi-layer ceramic units have ESRs which are too low for use. While they could in theory be padded up into the stable zone with external resistance, this would hardly be a practical solution. This leaves tantalum types as the best all around choice for LDO output use. Finally, since a large capacitor value is likely to be used to maximize stability, this effectively means that the solution for an LDO such as Figure 9-8 must use a more expensive and physically large tantalum capacitor. This is not desirable if small size is a major design criterion.

## The anyCAP® LDO Regulator Family

Some novel modifications to the basic LDO architecture of Figure 9-7 allow major improvements in terms of both DC and AC performance. These developments are shown schematically in Figure 9-9, which is a simplified diagram of the Analog Devices ADP330X series LDO regulator family. These regulators are also known as the anyCAP® family, so named for their relative insensitivity to the output capacitor in terms of both size and ESR. They are available in power efficient packages such as the Thermal Coastline (discussed below), in both standalone LDO and LDO controller forms, and also in a wide span of output voltage options.

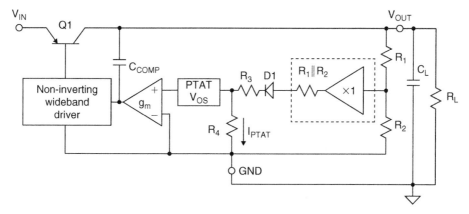

**Figure 9-9: ADP330X anyCAP® topology features improved DC and AC performance over traditional LDOs**

## Design Features Related to DC Performance

One of the key differences in the ADP330X series is the use of a high gain vertical PNP pass device, with all of the advantages described above with Figure 9-9 (Reference 6). This allows the typical dropout voltages for the series to be on the order of 1 mV/mA for currents of 200 mA or less.

It is important to note that the topology of this LDO is distinctly different from that of the generic form in Figure 9-7, as there is no obvious $V_{REF}$ block. The reason for this is the fact that the ADP330X series uses what is termed a "merged" amplifier-reference design. The operation of the integral amplifier and reference scheme illustrated in Figure 9-9 can be described as follows.

In this circuit, $V_{REF}$ is defined as a reference voltage existing at the output of a zero impedance divider of ratio $R_1$ and $R_2$. In the figure, this is depicted symbolically by the (dotted) unity gain buffer amplifier fed by $R_1$ and $R_2$, which has an output of $V_{REF}$. This reference voltage feeds into a series connection of (dotted) $R_1$ and $R_2$, then actual components D1, $R_3$, $R_4$, etc.

The error amplifier, shown here as a $g_m$ stage, is actually a PNP input differential stage with the two transistors of the pair operated at different current densities, so as to produce a predictable proportional to absolute temperature (PTAT) offset voltage. Although shown here as a separate block $V_{OS}$, this offset voltage is inherent to a bipolar pair for such operating conditions. The PTAT $V_{OS}$ causes a current $I_{PTAT}$ to flow in $R_4$, which is simply:

$$I_{PTAT} = \frac{V_{OS}}{R_4} \qquad (9\text{-}4)$$

Note that this current also flows in series connected $R_4$, $R_3$, and the Thevenin resistance of the divider, $R_1 \parallel R_2$, so:

$$V_{PTAT} = I_{PTAT}\,(R_3 + R_4 + R_1 \parallel R_2) \qquad (9\text{-}5)$$

The *total* voltage defined as $V_{REF}$ is the sum of two component voltages:

$$V_{REF} = V_{PTAT} + V_{D_1} \qquad (9\text{-}6)$$

where the $I_{PTAT}$ scaled voltages across $R_3$, $R_4$, and $R_1 \parallel R_2$ produce a net PTAT voltage $V_{PTAT}$, and the diode voltage $V_{D1}$ is a complementary to absolute temperature (CTAT) voltage. As in a standard bandgap reference, the PTAT and CTAT components add up to a temperature stable reference voltage of 1.25V. In this case however, the reference voltage is not directly accessible, but instead it exists in the virtual form described above. It acts as it would be seen at the output of a zero impedance divider of a numeric ratio of $R_1/R_2$, which is then fed into the $R_3$–D1 series string through a Thevenin resistance of $R_1 \parallel R_2$ in series with D1.

With the closed loop regulator at equilibrium, the voltage at the virtual reference node will be:

$$V_{REF} = V_{OUT} \left( \frac{R_2}{R_1 + R_2} \right) \qquad (9\text{-}7)$$

With minor re-arrangement, this can be put into the standard form to describe the regulator output voltage as:

$$V_{OUT} = V_{REF} \left( 1 + \frac{R_1}{R_2} \right) \qquad (9\text{-}8)$$

In the various devices of the ADP330X series, the $R_1$–$R_2$ divider is adjusted to produce standard output voltages of 2.7, 3.0, 3.2, 3.3, and 5.0V.

As can be noted from this discussion, unlike a conventional reference setup, there is no power wasting reference current such as used in a conventional regulator topology ($I_{REF}$ of Figure 9-2). In fact, Figure 9-9 regulator behaves as if the entire error amplifier has simply an offset voltage of $V_{REF}$ volts, as seen at the output of a conventional $R_1$–$R_2$ divider.

## Design Features Related to AC Performance

While the above described DC performance enhancements of the ADP330X series are worthwhile, the most dramatic improvements come in areas of AC related performance. These improvements are in fact the genesis of the anyCAP® series name.

Capacitive loading and the potential instability it brings are a major deterrent to easily applying LDOs. While LDO goals prevent the use of emitter follower type outputs, and so preclude their desirable buffering effect against cap loading, there is an alternative technique of providing load immunity. One method of reducing susceptibility against variation in a particular amplifier response pole is called *pole splitting* (see Reference 8). It refers to an amplifier compensation method whereby two response poles are shifted in such a way so as to make one a dominant, lower frequency pole. In this manner the secondary pole (which in this case is the $C_L$ related output pole) becomes much less of a major contributor to the net AC response. This has the desirable effect of greatly de-sensitizing the amplifier to variations in the output pole.

## The anyCAP® Pole-Splitting Topology

Returning to the anyCAP® series topology, (Figure 9-9) it can be noted that in this case $C_{COMP}$ is isolated from the pass device's base (and thus input ripple variations), by the wideband non-inverting driver. But insofar as frequency compensation is concerned, because of this buffer's isolation, $C_{COMP}$ still functions as a modified pole-splitting capacitor (see Reference 9), and it does provide the benefits of a buffered, $C_L$ independent single-pole response. The regulator's frequency response is dominated by the internal compensation, and becomes relatively immune to the value and ESR of load capacitor $C_L$. Thus the name anyCAP® for the series is apt, as the design is tolerant of virtually any output capacitor type. $C_L$ can be as low as $0.47\,\mu F$, and it can also be a multi-layer ceramic capacitor (MLCC) type. This allows a very small physical size for the entire regulation function, such as when a SOT-23 packaged anyCAP® LDO is used, e.g., the ADP3300 device.

The ADP3300 and other anyCAP® series devices maintain regulation over a wide range of load, input voltage, and temperature conditions. However, when the regulator is overloaded or entering the dropout

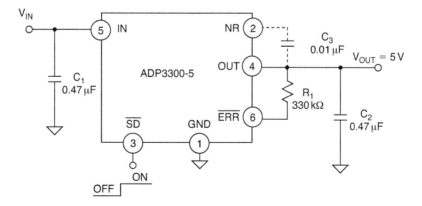

**Figure 9-10: A basic ADP3000 50 mA LDO regulator**

region (e.g., by a reduction in the input voltage) the open collector $\overline{\text{ERR}}$ pin becomes active, by going to a *low* or conducting state. Once set, the $\overline{\text{ERR}}$ pin's internal hysteresis keeps the output low, until some margin of operating range is restored. In the circuit of Figure 9-10, $R_1$ is a pull-up resistor for the $\overline{\text{ERR}}$ output, $E_{OUT}$. This resistor can be eliminated if the load being driven provides a pull-up current.

The $\overline{\text{ERR}}$ function can also be activated by the regulator's over temperature protection circuit, which trips at 165°C. These internal current and thermal limits are intended to protect the device against accidental overload conditions. For normal operation, device power dissipation should be externally limited by means of heat sinking, air flow, etc. so that junction temperatures will not exceed 125°C.

A capacitor, $C_3$, connected between pins 2 and 4, can be used for an optional noise reduction (NR) feature. This is accomplished by AC bypassing a portion of the regulator's internal scaling divider, which has the effect of reducing the output noise ~10 dB. When this option is exercised, only low leakage 10–100 nF capacitors should be used. Also, input and output capacitors should be changed to 1 and 4.7 µF values respectively, for lowest noise and the best overall performance. Note that the NR pin is internally connected to a high impedance node, so connections to it should be carefully done to avoid noise. PC traces and pads connected to this pin should be as short and small as possible.

## LDO Regulator Thermal Considerations

To determine a regulator's power dissipation, calculate it as follows:

$$P_D = (V_{IN} - V_{OUT})(I_L) + (V_{IN})(I_{Ground}) \qquad (9\text{-}9)$$

where $I_L$ and $I_{Ground}$ are load and ground current, and $V_{IN}$ and $V_{OUT}$ are the input and output voltages, respectively. Assuming $I_L = 50$ mA, $I_{Ground} = 0.5$ mA, $V_{IN} = 8$ V, and $V_{OUT} = 5$ V, the device power dissipation is:

$$P_D = (8 - 5)(0.05) + (8)(0.0005) = 0.150 + 0.004 = 0.154 \text{ W} \qquad (9\text{-}10)$$

To determine the regulator's temperature rise, $\Delta T$, calculate it as follows:

$$\Delta T = T_J - T_A = P_D \times \theta_{JA} = 0.154 \text{ W} \times 165°C/W = 25.4°C \qquad (9\text{-}11)$$

With a maximum junction temperature of 125°C, this yields a calculated maximum safe ambient operating temperature of 125–25.4°C, or just under 100°C. Since this temperature is in excess of the device's rated temperature range of 85°C, the device will then be operated conservatively at an 85°C (or less) maximum ambient temperature.

These general procedures can be used for other devices in the series, substituting the appropriate $\theta_{JA}$ for the applicable package, and applying the remaining operating conditions.

In addition, layout and printed circuit boards (PCB) design can have a significant influence on the power dissipation capabilities of power management ICs. This is due to the fact that the surface mount packages used with these devices rely heavily on thermally conductive traces or pads, to transfer heat away from the package. Appropriate PC layout techniques should then be used to remove the heat due to device power dissipation. The following general guidelines will be helpful in designing a board layout for lowest thermal resistance in SOT-23 and SO-8 packages:

1.  *PC board traces with large cross sectional areas remove more heat. For optimum results, use large area PCB patterns with wide and heavy (2 ounces) copper traces, placed on the uppermost side of the PCB.*

2.  *Electrically connect dual $V_{IN}$ and $V_{OUT}$ pins in parallel, as well as to the corresponding $V_{IN}$ and $V_{OUT}$ large area PCB lands.*

3. *In cases where maximum heat dissipation is required, use double-sided copper planes connected with multiple vias.*

4. *Where possible, increase the thermally conducting surface area(s) openly exposed to moving air, so that heat can be removed by convection (or forced air flow, if available).*

5. *Do not use solder mask or silkscreen on the heat dissipating traces, as they increase the net thermal resistance of the mounted IC package.*

A real life example visually illustrates a number of the above points far better than words can. It is shown in Figure 9-11, a photo of the ADP3300 1.5 inch square evaluation PCB. The boxed area on the board represents the actual active circuit area.

**Figure 9-11: Size does make a difference: ADP3300 evaluation board**

Recent developments in packaging have led to much improved thermal performance for power management ICs. The anyCAP® LDO regulator family capitalizes on this most effectively, using a thermally improved leadframe as the basis for all 8 pin devices. This package is called a "Thermal Coastline" design, and is shown in Figure 9-12. The foundation of the improvement in heat transfer is related to two key parameters of the leadframe design, distance, and width. The payoff comes in the reduced thermal resistance of the leadframe based on the Thermal Coastline, only 90°C/W versus 160°C/W for a standard SO-8 package. The increased dissipation of the Thermal Coastline allows the anyCAP® series of SO-8 regulators to support more than 1 W of dissipation at 25°C.

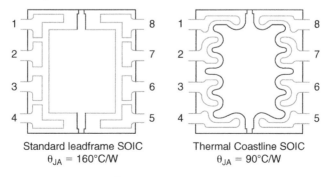

Standard leadframe SOIC $\theta_{JA} = 160°C/W$

Thermal Coastline SOIC $\theta_{JA} = 90°C/W$

**Figure 9-12: anyCAP® series regulators in SO-8 use thermal coastline packages**

Additional insight into how the new leadframe increases heat transfer can be appreciated by Figure 9-13. In this figure, it can be noted how the spacing of the Thermal Coastline paddle and leads shown on the right is reduced, while the width of the lead ends is increased, versus the standard leadframe, on the left.

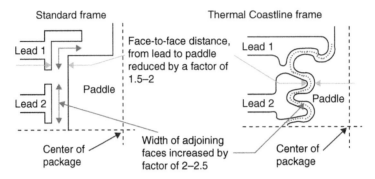

Figure 9-13: Details of the Thermal Coastline package

## Regulator Controller Differences

A basic difference between the regulator controller and a standalone regulator is the removal of the pass device from the regulator chip. This design step has both advantages and disadvantages. A positive is that the external PMOS pass device can be chosen for the exact size, package, current rating, and power handling which are most useful to the application. This approach allows the same basic controller IC to be useful for currents of several hundred milliamperes to more than 10A, simply by choice of the FET. Also, since the regulator controller IC's $I_{Ground}$ of 800 µA results in very little power dissipation, its thermal drift will be enhanced. On the downside, there are two packages now used to make up the regulator function. And, current limiting (which can be made completely integral to a standalone IC LDO regulator) is now a function which must be split between the regulator controller IC and an external sense resistor. This step also increases the dropout voltage of the LDO regulator controller somewhat, by about 50 mV.

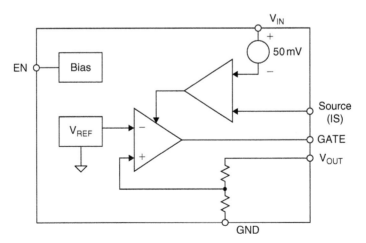

Figure 9-14: Functional block diagram of anyCAP series LDO regulator controller

A functional diagram of the ADP3310 regulator controller is shown in Figure 9-14. The basic error amplifier, reference, and scaling divider of this circuit are similar to the standalone anyCAP® regulator, and will not be described in detail. The regulator controller version does share the same cap load immunity of the standalone versions, and also has a shutdown function, similarly controlled by the EN (enable) pin.

The main differences in the regulator controller IC architecture is the buffered output of the amplifier, which is brought out on the GATE pin, to drive the external PMOS FET. In addition, the current limit sense amplifier has a built-in 50 mV threshold voltage, and is designed to compare the voltage between the $V_{IN}$ and IS pins. When this voltage exceeds 50 mV, the current limit sense amplifier takes over control of the loop, by shutting down the error amplifier and limiting output current to the preset level.

## A Basic 5 V/1 A LDO Regulator Controller

An LDO regulator controller is easy to use, since a PMOS FET, a resistor, and two relatively small capacitors (one at the input, one at the output) are all that are needed to form an LDO regulator. The general configuration is shown by Figure 9-15, LDO suitable as a 5 V/1 A regulator operating from a $V_{IN}$ of 6 V, using the ADP3310-5 controller IC.

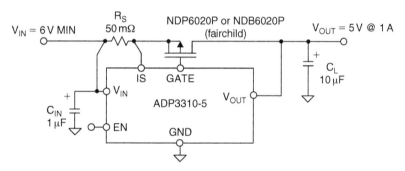

**Figure 9-15: A basic ADP3310 PMOS FET 1 A LDO regulator controller**

This regulator is stable with virtually any good quality output capacitor used for $C_L$ (as is true with the other anyCAP® devices). The actual $C_L$ value required and its associated ESR depend on the $g_m$ and capacitance of the external PMOS device. In general, a 10 µF capacitor at the output is sufficient to ensure stability for load currents up to 10 A. Larger capacitors can also be used, if high output surge currents are present. In such cases, low ESR capacitors such as OS-CON electrolytics are preferred, because they offer lowest ripple on the output. For less demanding requirements, a standard tantalum or aluminum electrolytic can be adequate. When an aluminum electrolytic is used, it should be qualified for adequate performance over temperature. The input capacitor, $C_{IN}$, is only necessary when the regulator is several inches or more distant from the raw DC filter capacitor. However, since it is small physically, it is usually prudent to use it in most instances. It should be located close to the $V_{IN}$ pin of the regulator. Note also the current-sensing resistor, $R_S$. This will be discussed in a following section.

## Selecting the Pass Device

The type and size of the pass transistor are determined by a set of requirements for threshold voltage, input–output voltage differential, load current, power dissipation, and thermal resistance. An actual PMOS pass device selected must satisfy all of these electrical requirements, plus physical and thermal parameters. There are a number of manufacturers offering suitable devices in packages ranging from SO-8 up through TO-220 in size.

To ensure that the maximum available drive from the controller will adequately drive the FET under worst case conditions of temperature range and manufacturing tolerances, the maximum drive from the controller ($V_{GS(DRIVE)}$) to the pass device must be determined. This voltage is calculated as follows:

$$V_{GS(DRIVE)} = V_{IN} - V_{BE} - (I_{L(MAX)})(R_S) \qquad (9\text{-}12)$$

where $V_{IN}$ is the minimum input voltage, $I_{L(MAX)}$ is the maximum load current, $R_S$ is the sense resistor, and $V_{BE}$ is a voltage internal to the ADP3310 ($\sim$0.5 at high temperature, 0.9 cold, and 0.7 V at room temperature). Note that since $I_{L(MAX)} \times R_S$ will be no more than 75 mV, and $V_{BE}$ at cold temperature $\cong$0.9 V, this equation can be further simplified to:

$$V_{GS(DRIVE)} \cong V_{IN} - 1\,V \qquad (9\text{-}13)$$

In Figure 9-15 example, $V_{IN} = 6\,V$ and $V_{OUT} = 5\,V$, so $V_{GS(DRIVE)}$ is $6 - 1 = 5\,V$.

It should be noted that the above two equations apply to FET drive voltages which are *less* than the typical gate-to-source clamp voltage of 8 V (built into the ADP3310, for the purposes of FET protection).

An overall goal of the design is to then select an FET which will have an $R_{DS(ON)}$ sufficiently low so that the resulting dropout voltage will be less than $V_{IN} - V_{OUT}$, which in this case is 1 V. For the NDP6020P used in Figure 2-43 (see Reference 10), this device achieves an $R_{DS(ON)}$ of 70 m$\Omega$ (maximum) with a $V_{GS}$ of 2.7 V, a voltage drive appreciably less than the ADP3310's $V_{GS(DRIVE)}$ of 5 V. The dropout voltage $V_{MIN}$ of this regulator configuration is the sum of two series voltage drops, the FET's drop plus the drop across $R_S$, or

$$V_{MIN} = I_{L(MAX)} (R_{DS(ON)} + R_S) \qquad (9\text{-}14)$$

In the design here, the two resistances are roughly comparable to one another, so the net $V_{MIN}$ will be $1\,A \times (50 + 70\,m\Omega) = 120\,mV$.

## Thermal Design

The maximum allowable thermal resistance between the FET junction and the highest expected ambient temperature must be taken into account, to determine the type of FET package and heat sink used (if any).

Using 2 ounces copper PCB material and one square inch of copper PCB land area as a heat sink, it is possible to achieve a net thermal resistance, $\theta_{JA}$, for mounted SO-8 devices on the order of 60°C/W or less. Such data are available for SO-8 power FETs (see Reference 11). There are also a variety of larger packages with lower thermal resistance than the SO-8, but still useful with surface mount techniques. Examples are the DPAK and D²PAK, etc.

For higher power dissipation applications, corresponding to thermal resistance of 50°C/W or less, a bolt-on external heat sink is required to satisfy the $\theta_{JA}$ requirement. Compatible package examples would be the TO-220 family, which is used with the NDP6020P example of Figure 2-43.

## Sensing Resistors for LDO Controllers

Current limiting in the ADP3310 controller is achieved by choosing an appropriate external current sense resistor, $R_S$, which is connected between the controller's $V_{IN}$ and IS (source) pins. An internally derived 50 mV current limit threshold voltage appears between these pins to establish a comparison threshold for current limiting. This 50 mV determines the threshold where current limiting begins. For a continuous current limiting, a foldback mode is established, with dissipation controlled by reducing the gate drive. The net effect is that the ultimate current limit level is a factor of 2/3 of maximum. The foldback limiting reduces the power dissipated in the pass transistor substantially.

To choose a sense resistor for a maximum output current $I_L$, $R_S$ is calculated as follows:

$$R_S = \frac{0.05}{K_F \, I_L} \tag{9-15}$$

In this expression, the nominal 50 mV current limit threshold voltage appears in the numerator. In the denominator appears a scaling factor $K_F$, which can be either 1.0 or 1.5, plus the maximum load current, $I_L$. For example, if a scaling factor of 1.0 is to be used for a 1 A $I_L$, the $R_S$ calculation is straightforward, and 50 m$\Omega$ is the correct $R_S$ value.

However, to account for uncertainties in the threshold voltage and to provide a more conservative output current margin, a scaling factor of $K_F = 1.5$ can alternately be used. When this approach is used, the same 1 A $I_L$ load conditions will result in a 33 m$\Omega$ $R_S$ value. In essence, the use of the 1.5 scaling factor takes into account the foldback scheme's reduction in output current, allowing higher current in the limit mode.

The simplest and least expensive sense resistor for high current applications such as Figure 9-15 is a copper PCB trace controlled in both thickness and width. Both the temperature dependence of copper and the relative size of the trace must be taken into account in the resistor design. The temperature coefficient (TC) of resistivity for copper has a positive TC of +0.39%/°C. This natural copper TC, in conjunction with the controller's PTAT-based current limit threshold voltage, can provide for a current limit characteristic which is simple and effective over temperature.

The table of Figure 9-16 provides resistance data for designing PCB copper traces with various PCB copper thickness (or weight), in ounces of copper per square foot area. To use this information, note that the center column contains a resistance coefficient, which is the conductor resistance in milliohms/inch, divided by the trace width, W. For example, the first entry, for 1/2 ounce copper is 0.983 m$\Omega$/inch/W. So, for a reference trace width of 0.1 inch, the resistance would be 9.83 m$\Omega$/inch. Since these are all linear relationships, everything scales for wider/skinnier traces, or for differing copper weights. As an example, to design a 50 m$\Omega$ $R_S$ for the circuit of Figure 9-15 using 1/2 ounce copper, a 2.54 inches length of a 0.05 inch wide PCB trace could be used.

| Copper thickness | Resistance coefficient, milliohms/inch/ W (trace width W in inches) | Reference 0.1 inch wide trace, milliohms/inch |
|---|---|---|
| 1/2 ounces/feet$^2$ | 0.983/W | 9.83 |
| 1 ounces/feet$^2$ | 0.491/W | 4.91 |
| 2 ounces/feet$^2$ | 0.246/W | 2.46 |
| 3 ounces/feet$^2$ | 0.163/W | 1.63 |

**Figure 9-16: Printed circuit copper resistance**

To minimize current limit sense voltage errors, the two connections to $R_S$ should be made four-terminal style, as is noted in Figure 9-15. It is not absolutely necessary to actually use four-terminal style resistors, except for the highest current levels. However, as a minimum, the heavy currents flowing in the source circuit of the pass device should not be allowed to flow in the ADP3310 sense pin traces. To minimize such errors, the $V_{IN}$ connection trace to the ADP3310 should connect close to the body of $R_S$ (or the resistor's input sense terminal), and the IS connection trace should also connect close to the resistor body (or the resistor's output sense terminal). Four-terminal wiring is increasingly important for output currents of 1A or more.

Alternately, an appropriate selected sense resistor such as surface mount sense devices available from resistor vendors can be used (see Reference 13). Sense resistor $R_S$ may not be needed in all applications, if a current limiting function is provided by the circuit feeding the regulator. For circuits that do not require current limiting, the IS and $V_{IN}$ pins of the ADP3310 must be tied together.

## PCB Layout Issues

For best voltage regulation, place the load as close as possible to the controller device's $V_{OUT}$ and GND pins. Where the best regulation is required, the $V_{OUT}$ trace from the ADP3310 and the pass device's drain connection should connect to the positive load terminal via separate traces. This step (Kelvin sensing) will keep the heavy load currents in the pass device's drain out of the feedback-sensing path, and thus maximize output accuracy. Similarly, the unregulated input common should connect to the common side of the load via a separate trace from the ADP3310 GND pin.

### References: Linear Voltage Regulators

1.  B. Widlar, "New Developments in IC Voltage Regulators," **IEEE Journal of Solid State Circuits**, Vol. SC-6, February, 1971.

2.  R.C. Dobkin, "3-Terminal Regulator is Adjustable," National Semiconductor AN-181, March, 1977.

3.  P. Brokaw, "A Simple Three-Terminal IC Bandgap Voltage Reference," **IEEE Journal of Solid State Circuits**, Vol. SC-9, December, 1974.

4.  F. Goodenough, "Linear Regulator Cuts Dropout Voltage," **Electronic Design**, April 16, 1987.

5.  C. Simpson, "LDO Regulators Require Proper Compensation," **Electronic Design**, November 4, 1996.

6.  F. Goodenough, "Vertical-PNP-Based Monolithic LDO Regulator Sports Advanced Features," **Electronic Design**, May 13, 1996.

7.  F. Goodenough, "Low Dropout Regulators Get Application Specific," **Electronic Design**, May 13, 1996.

8.  J. Solomon, "The Monolithic Op Amp: A Tutorial Study," **IEEE Journal of Solid State Circuits**, Vol. SC-9, No. 6, December 1974.

9.  R.J. Reay and G.T.A. Kovacs, "An Unconditionally Stable Two-Stage CMOS Amplifier," **IEEE Journal of Solid State Circuits**, Vol. SC-30, No. 5, May 1995.

10. "NDP6020P/NDB6020P P-Channel Logic Level Enhancement Mode Field Effect Transistor," Fairchild Semiconductor data sheet, September 1997, http://www.fairchildsemi.com.

11. A. Li, et al., "Maximum Power Enhancement Techniques for SO-8 Power MOSFETs," Fairchild Semiconductor application note AN1029, April 1996, http://www.fairchildsemi.com.

12. R. Blattner, Wharton McDaniel, "Thermal Management in On-Board DC-to-DC Power Conversion," Temic application note, http://www.temic.com.

13. "S" series surface mount current sensing resistors, KRL/Bantry Components, 160 Bouchard Street, Manchester, NH, 03103-3399, (603) 668-3210.

# Switch Mode Regulators

## Introduction

The trend toward lower power, lower weight, portable equipment has driven the technology and the requirement for converting power efficiently. Switch mode power converters, often referred to simply as "switchers," offer a versatile way of achieving this goal. Modern IC switching regulators are small, flexible, and allow either step-up (boost) or step-down (buck) operation. Some topologies operate in both modes.

The most basic switcher topologies require only one transistor which is essentially used as a switch, one diode, one inductor, a capacitor across the output, and for practical but not fundamental reasons, another one across the input. A practical converter, however, requires a control section comprised several additional elements, such as a voltage reference, error amplifier, comparator, oscillator, and switch driver, and may also include optional features like current limiting and shutdown capability.

Depending on the power level, modern IC switching regulators may integrate the entire converter except for the main magnetic element(s) (usually a single inductor) and the input/output capacitors. Often, a diode, the one which is an essential element of basic switcher topologies, cannot be integrated either. In any case, the complete power conversion for a switcher cannot be as integrated as is a linear regulator. The requirement of a magnetic element means that system designers should not be inclined to think of switching regulators as simply "drop in" solutions. This presents a challenge to switching regulator manufacturers to provide careful design guidelines, commonly used application circuits (using off-the-shelf components where possible), and plenty of design assistance. As the power levels increase, ICs tend to grow in complexity because it becomes more critical to optimize the control flexibility and precision. Also, since the switches begin to dominate the size of the die, it becomes more cost effective to remove them and integrate only the controller.

The primary limitations of switching regulators, as compared to linear regulators, are the generation of input and output voltage noise, EMI/RFI emissions, and the comparatively stringent requirements of the external support components. Although switching regulators do not necessarily require transformers, they do use inductors, and magnetic theory is not generally well understood by design engineers. However, manufacturers of switching regulators generally offer applications support in this area by offering design software and complete data sheets with recommended parts lists for the external inductor as well as capacitors and switching elements.

One unique advantage of switching regulators lies in their ability to convert a given supply voltage with a known voltage range to virtually any desired output voltage, with no "first-order" limitations on efficiency. This is true regardless of whether the output voltage is higher or lower than the input voltage—the same or the opposite polarity. Consider the basic components of a switcher, as stated above. The inductor and capacitor are, ideally, reactive elements which dissipate no power. The transistor is effectively, ideally, a switch in that it is either "on," thus having no voltage dropped across it while current flows through it, or "off," thus having no current flowing through it while there is voltage across it. Since either voltage or current is always zero, the power dissipation is zero, thus, ideally, the switch dissipates no power. Finally, there is the diode, which has a finite voltage drop while current flows through it, and thus dissipates *some* power. But even that can be substituted for by a synchronized switch, called a "synchronous rectifier," so that it ideally dissipates no power. Practical efficiencies can exceed 90%.

Switchers also offer the advantage that, since they inherently require a magnetic element, it is often a simple matter to "tap" an extra winding onto that element and, often with just a diode and capacitor, generate a reasonably well regulated additional output. If more outputs are needed, more such taps can be used. Since the tap winding requires no electrical connection, it can be isolated from other circuitry, or made to "float" atop other voltages. Note that only one of the outputs would be "regulated." The others track according to the ratio of the taps.

Of course, real components create inefficiencies. Inductors have resistance, and their magnetic cores are not ideal either, so they dissipate power. Capacitors have resistance, and as current flows in and out of them, they dissipate power as well. Transistors, bipolar or field effect, are not ideal switches, and have a voltage drop when they are turned on, plus they cannot be switched instantly, and thus dissipate power while they are turning on or off.

As we shall soon see, switchers create ripple currents in their input and output capacitors. Those ripple currents create voltage ripple and noise on the converter's input and output due to the resistance, inductance, and finite capacitance of the capacitors used. That is the *conducted* part of the noise. Then there are often ringing voltages in the converter, parasitic inductances in components and PCB traces, and an inductor which creates a magnetic field which it cannot perfectly contain within its core—all contributors to *radiated* noise. Noise is an inherent by-product of a switcher and must be controlled by proper component selection, PCB layout, and, if that is not sufficient, additional input or output filtering or shielding.

## Inductor and Capacitor Fundamentals

In order to understand switching regulators, the fundamental energy storage capabilities of inductors and capacitors must be fully understood. When a voltage is applied to an ideal inductor (see Figure 9-17), the current builds up linearly over time at a rate equal to V/L, where V is the applied voltage, and L is the value of the inductance. This energy is stored in the inductor's magnetic field, and if the switch is opened quickly, the magnetic field collapses, and the inductor voltage goes to a large instantaneous value until the field has fully collapsed.

When a current is applied to an ideal capacitor, the capacitor is gradually charged, and the voltage builds up linearly over time at a rate equal to I/C, where I is the applied current, and C is the value of the capacitance. Note that the voltage across an ideal capacitor cannot change instantaneously (Figure 9-17).

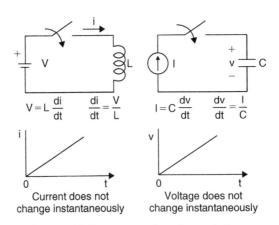

$$V = L\frac{di}{dt} \qquad \frac{di}{dt} = \frac{V}{L} \qquad I = C\frac{dv}{dt} \qquad \frac{dv}{dt} = \frac{I}{C}$$

Current does not change instantaneously

Voltage does not change instantaneously

**Figure 9-17: Energy transfer using an inductor**

Of course, there is no such thing as an ideal inductor or capacitor. Real inductors have stray winding capacitance, series resistance, and can saturate for large currents. Real capacitors have series resistance and inductance and may break down under large voltages. Nevertheless, the fundamentals of the ideal inductor and capacitor are fundamental in understanding the operation of switching regulators.

An inductor can be used to transfer energy between two voltage sources as shown in Figure 9-18. While energy transfer could occur between two voltage sources with a resistor connected between them, the energy transfer would be inefficient due to the power loss in the resistor, and the energy could only be transferred from the higher to the lower value source. In contrast, an inductor ideally returns all the energy that is stored in it, and with the use of properly configured switches, the energy can flow from any one source to another, regardless of their respective values and polarities.

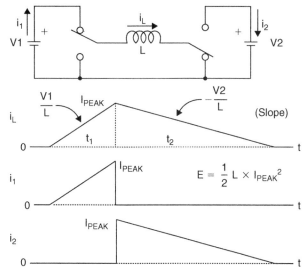

**Figure 9-18: Inductor and capacitor fundamentals**

When the switches are initially placed in the position shown, the voltage V1 is applied to the inductor, and the inductor current builds up at a rate equal to V1/L. The peak value of the inductor current at the end of the interval $t_1$ is:

$$I_{PEAK} = \frac{V1}{L} t_1 \tag{9-16}$$

The average power transferred to the inductor during the interval $t_1$ is:

$$P_{AVG} = \frac{1}{2} I_{PEAK} V1 \tag{9-17}$$

The energy transferred during the interval $t_1$ is:

$$E = P_{AVG} t_1 = \frac{1}{2} I_{PEAK} V1 t_1 \tag{9-18}$$

Solving Eq. (9-16) for $t_1$ and substituting into Eq. (9-18) yields:

$$E = \frac{1}{2} L \times I_{PEAK}^2 \tag{9-19}$$

When the switch positions are reversed, the inductor current continues to flow into the load voltage V2, and the inductor current decreases at a rate $-V_{OUT} + V_F/L$. At the end of the interval $t_2$, when the inductor current has decreased to zero, all of the energy previously stored in the inductor has been transferred into the load. The figure shows the current waveforms for the inductor, the input current $I_1$, and the output current $I_2$. The ideal inductor dissipates no power, so there is no power loss in this transfer, assuming ideal circuit elements. This fundamental method of energy transfer forms the basis for all switching regulators.

## Ideal Step-Down (Buck) Converter

The basic topology of an ideal step-down (buck) converter is shown in Figure 9-19. The actual integrated circuit switching regulator contains the switch control circuit and may or may not include the switch (depending on the output current requirement). The inductor, diode, and load bypass capacitor are external.

The output voltage is sensed and then regulated by the switch control circuit. There are several methods for controlling the switch, but for now assume that the switch is controlled by a pulse width modulator (PWM) operating at a fixed frequency, f.

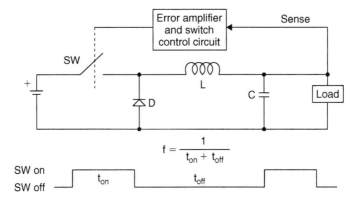

**Figure 9-19: Basic step-down (buck) converter**

The actual waveforms associated with the buck converter are shown in Figure 9-20. When the switch is on, the voltage $V_{IN} - V_{OUT}$ appears across the inductor (neglecting the voltage drop across the inductor), and the inductor current increases with a slope of $(V_{IN} - V_{OUT})/L$ (see Figure 9-20(B)). When the switch turns off, current continues to flow through the inductor in the same direction and into the load (remember that the current cannot change instantaneously in an inductor). The diode providing the return current path, called a "freewheeling" diode in this application, completes the current path broken by opening the switch. It also clamps the $V_D$ as the inductor tries to pull current out of the node. The voltage across the inductor is now $V_{OUT} + V_F$, but the polarity has reversed. Therefore, the inductor current decreases with a slope equal to $-V_{OUT}/L$. Note that the inductor current is equal to the output current in a buck converter.

The diode and switch currents are shown in Figure 9-20(C) and (D), respectively, and the inductor current is the sum of these waveforms. Also note by inspection that the instantaneous input current equals the switch current. Note, however, that the average input current is less than the average output current. In a practical regulator, both the switch and the diode have voltage drops across them during their conduction which creates internal power dissipation and a loss of efficiency, but these voltages will be neglected for now. It is also assumed that the output capacitor, C, is large enough so that the output voltage does not change significantly during the switch on- or off-times.

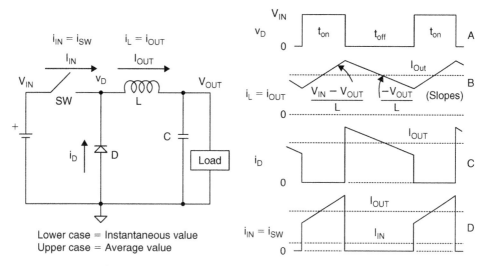

**Figure 9-20: Basic step-down (buck) converter waveforms**

There are several important things to note about these waveforms. First is that ideal components have been assumed, i.e., the input voltage source has zero impedance, the switch has zero on-resistance and zero turn-on and turn-off times. It is also assumed that the inductor does not saturate and that the diode is ideal with no forward drop.

Also note that the output current is continuous, while the input current is pulsating. Obviously, this has implications regarding input and output filtering. If one is concerned about the voltage ripple created on the power source which supplies a buck converter, the input filter capacitor (not shown) is generally more critical that the output capacitor with respect to equivalent series resistance/inductance (ESR/ESL).

If a steady-state condition exists (see Figure 9-21), the basic relationship between the input and output voltage may be derived by inspecting the inductor current waveform and writing:

$$\frac{V_{IN} - V_{OUT}}{L} t_{on} = \frac{V_{OUT}}{L} t_{off} \tag{9-20}$$

Solving for $V_{OUT}$:

$$V_{OUT} = V_{IN} \frac{t_{on}}{t_{on} + t_{off}} = V_{IN}D \tag{9-21}$$

where D is the switch *duty ratio* (more commonly called *duty cycle*), defined as the ratio of the switch on-time ($t_{on}$) to the total switch cycle time ($t_{on} + t_{off}$).

This is the classic equation relating input and output voltage in a buck converter which is operating with *continuous* inductor current, defined by the fact that the inductor current never goes to zero.

Notice that this relationship is independent of the inductor value L as well as the switching frequency $1/(t_{on} + t_{off})$ and the load current. Decreasing the inductor value, however, will result in a larger peak-to-peak output ripple current, while increasing the value results in smaller ripple. There are many other tradeoffs involved in selecting the inductor, and these will be discussed in a later section.

In this simple model, line and load regulation (of the output voltage) is achieved by varying the duty cycle using a PWM operating at a fixed frequency, f. The PWM is in turn controlled by an error amplifier—an

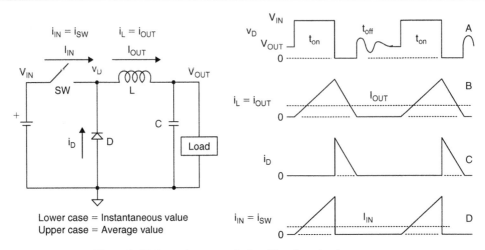

**Figure 9-21: Input/output relationships for a buck converter**

amplifier which amplifies the "error" between the measured output voltage and a reference voltage. As the input voltage increases, the duty cycle decreases; and as the input voltage decreases, the duty cycle increases. Note that while the average inductor current changes proportionally to the output current, the duty cycle does not change. Only dynamic changes in the duty cycle are required to modulate the inductor current to the desired level; then the duty cycle returns to its steady-state value. In a practical converter, the duty cycle might increase slightly with load current to counter the increase in voltage drops in the circuit, but would otherwise follow the ideal model.

This discussion so far has assumed the regulator is in the *continuous mode* of operation, defined by the fact that the inductor current never goes to zero. If, however, the output load current is decreased, there comes a point where the inductor current will go to zero between cycles, and the inductor current is said to be *discontinuous*. It is necessary to understand the implications of this operating mode as well, since many switchers must supply a wide dynamic range of output current, where this phenomenon is unavoidable. Waveforms for discontinuous operation are shown in Figure 9-22.

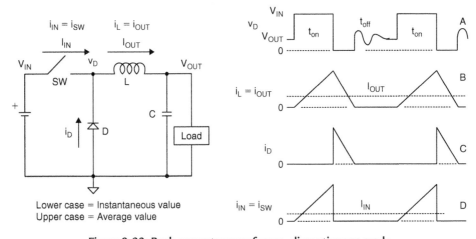

**Figure 9-22: Buck converter waveforms—discontinuous mode**

Behavior during the switch on-time is identical to that of the continuous mode of operation. However, during the switch off-time, there are two regions of unique behavior. First, the inductor current ramps down at the same rate as it does during continuous mode, but then the inductor current goes to zero. When it reaches zero, the current tries to reverse but cannot find a path through the diode any longer. So the voltage on the input side of the inductor (same as the diode and switch junction) jumps up to $V_{OUT}$ such that the inductor has no voltage across it, and the current can remain at zero.

Because the impedance at diode node ($V_D$) is high, ringing occurs due to the inductor, L, resonating with the stray capacitance which is the sum of the diode capacitance, $C_D$, and the switch capacitance, $C_{SW}$. The oscillation is damped by stray resistances in the circuit, and occurs at a frequency given by:

$$f_o = \frac{1}{2\pi\sqrt{L(C_D + C_{SW})}} \tag{9-22}$$

A circuit devoted simply to dampening resonances via power dissipation is called a *snubber*. If the ringing generates EMI/RFI problems, it may be damped with a suitable RC snubber. However, this will cause additional power dissipation and reduced efficiency.

If the load current of a standard buck converter is low enough, the inductor current becomes discontinuous. The current at which this occurs can be calculated by inspection of the waveform shown in Figure 9-23. This waveform is drawn showing the inductor current going to exactly zero at the end of the switch off-time. Under these conditions, the average output current is:

$$I_{OUT} = \frac{I_{PEAK}}{2} \tag{9-23}$$

We have already shown that the peak inductor current is:

$$I_{PEAK} = \frac{V_{IN} - V_{OUT}}{L}\, t_{on} \tag{9-24}$$

Thus, discontinuous operation will occur if:

$$I_{OUT} < \frac{V_{IN} - V_{OUT}}{2L}\, t_{on} \tag{9-25}$$

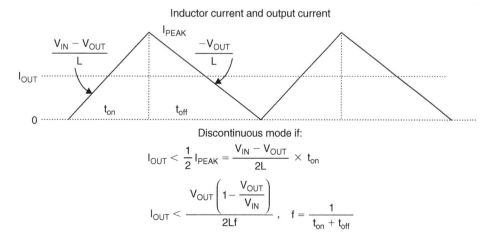

Figure 9-23: Buck converter point of discontinuous operation

However, $V_{OUT}$ and $V_{IN}$ are related by:

$$V_{OUT} = V_{IN}\,D = V_{IN}\,\frac{t_{on}}{t_{on} + t_{off}} \tag{9-26}$$

Solving for $t_{on}$:

$$t_{on} = \frac{V_{OUT}}{V_{IN}}(t_{on} + t_{off}) = \frac{V_{OUT}}{V_{IN}}\,\frac{1}{f} \tag{9-27}$$

Substituting this value for $t_{on}$ into the previous equation for $I_{OUT}$:

$$I_{OUT} < \frac{V_{OUT}\left(1 - \dfrac{V_{OUT}}{V_{IN}}\right)}{2Lf} \tag{9-28}$$

(Criteria for discontinuous operation—buck converter.)

## Ideal Step-Up (Boost) Converter

The basic step-up (boost) converter circuit is shown in Figure 9-24. During the switch on-time, the current builds up in the inductor. When the switch is opened, the energy stored in the inductor is transferred to the load through the diode.

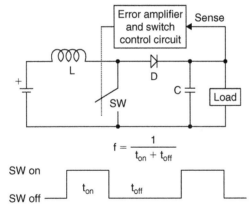

Figure 9-24: Basic step-up (boost) converter

The actual waveforms associated with the boost converter are shown in Figure 9-25. When the switch is on, the voltage $V_{IN}$ appears across the inductor, and the inductor current increases at a rate equal to $V_{IN}/L$. When the switch is opened, a voltage equal to $V_{OUT} - V_{IN}$ appears across the inductor, current is supplied to the load, and the current decays at a rate equal to $(V_{OUT} - V_{IN})/L$. The inductor current waveform is shown in Figure 9-25(B).

Note that in the boost converter, the input current is continuous, while the output current (Figure 9-25(D)) is pulsating. This implies that filtering the output of a boost converter is more difficult than that of a buck converter. (Refer back to the previous discussion of buck converters.) Also note that the input current is the sum of the switch and diode current.

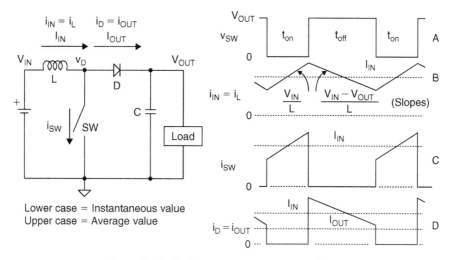

Figure 9-25: Basic step-up converter waveforms

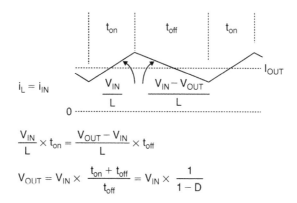

$$\frac{V_{IN}}{L} \times t_{on} = \frac{V_{OUT} - V_{IN}}{L} \times t_{off}$$

$$V_{OUT} = V_{IN} \times \frac{t_{on} + t_{off}}{t_{off}} = V_{IN} \times \frac{1}{1 - D}$$

Figure 9-26: Input/output relationship for a boost converter

If a steady-state condition exists (see Figure 9-26), the basic relationship between the input and output voltage may be derived by inspecting the inductor current waveform and writing:

$$\frac{V_{IN}}{L} t_{on} = \frac{V_{OUT} - V_{IN}}{L} t_{off} \tag{9-29}$$

Solving for $V_{OUT}$:

$$V_{OUT} = V_{IN} \frac{t_{on} + t_{off}}{t_{off}} = V_{IN} \frac{1}{1 - D} \tag{9-30}$$

This discussion so far has assumed the boost converter is in the *continuous mode* of operation, defined by the condition that the inductor current never goes to zero. If, however, the output load current is decreased, there comes a point where the inductor current will go to zero between cycles, and the inductor current is

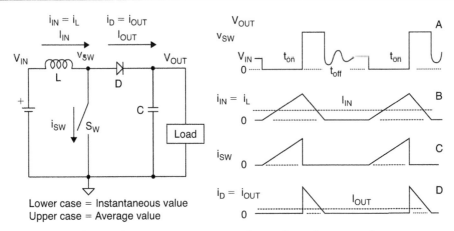

**Figure 9-27: Boost converter waveform—discontinuous mode**

said to be *discontinuous*. It is necessary to understand this operating mode as well, since many switchers must supply a wide dynamic range of output current, where this phenomenon is unavoidable.

Discontinuous operation for the boost converter is similar to that of the buck converter. Figure 9-27 shows the waveforms. Note that when the inductor current goes to zero, ringing occurs at the switch node at a frequency $f_o$ given by:

$$f_o = \frac{1}{2\pi\sqrt{L(C_D + C_{SW})}} \tag{9-31}$$

The inductor, L, resonates with the stray switch capacitance and diode capacitance, $C_{SW} + C_D$ as in the case of the buck converter. The ringing is dampened by circuit resistances, and, if needed, a snubber.

The current at which a boost converter becomes discontinuous can be derived by inspecting the inductor current (same as input current) waveform of Figure 9-28.

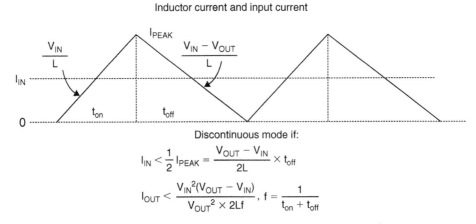

**Figure 9-28: Boost converter point of discontinuous operation**

The average input current at the point of discontinuous operation is:

$$I_{IN} = \frac{I_{PEAK}}{2} \tag{9-32}$$

Discontinuous operation will occur if:

$$I_{IN} < \frac{I_{PEAK}}{2} \tag{9-33}$$

However,

$$I_{IN} = \frac{I_{PEAK}}{2} = \frac{V_{OUT} - V_{IN}}{2L} t_{off} \tag{9-34}$$

Also,

$$V_{IN} I_{IN} = V_{OUT} I_{OUT} \tag{9-35}$$

and therefore:

$$I_{OUT} = \frac{V_{IN}}{V_{OUT}} I_{IN} = \frac{V_{IN}}{V_{OUT}} \frac{(V_{OUT} - V_{IN})}{2L} t_{off} \tag{9-36}$$

However,

$$\frac{V_{OUT}}{V_{IN}} = \frac{1}{1 - D} = \frac{1}{1 - \dfrac{t_{on}}{t_{on} + t_{off}}} = \frac{t_{on} + t_{off}}{t_{off}} \tag{9-37}$$

Solving for $t_{off}$:

$$t_{off} = \frac{V_{IN}}{V_{OUT}} (t_{on} + t_{off}) = \frac{V_{IN}}{f \ V_{OUT}} \tag{9-38}$$

Substituting this value for $t_{off}$ into the previous expression for $I_{OUT}$, the criteria for discontinuous operation of a boost converter is established:

$$I_{OUT} < \frac{V_{IN}{}^2 (V_{OUT} - V_{IN})}{V_{OUT}{}^2 \times 2Lf} \tag{9-39}$$

(Criteria for discontinuous operation—boost converter.)

The basic buck and boost converter circuits can work equally well for negative inputs and outputs as shown in Figure 9-29. Note that the only difference is that the polarities of the input voltage and the diode have been reversed. In practice, however, not many IC buck and boost regulators or controllers will work with negative inputs. In some cases, external circuitry can be added in order to handle negative inputs and outputs. Rarely are regulators or controllers designed specifically for negative inputs or outputs. In any case, data sheets for the specific ICs will indicate the degree of flexibility allowed.

## Buck–Boost Topologies

The simple buck converter can only produce an output voltage which is less than the input voltage, while the simple boost converter can only produce an output voltage greater than the input voltage. There are many applications where more flexibility is required. This is especially true in battery-powered

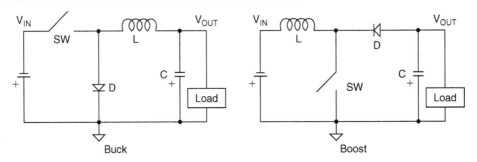

Figure 9-29: Negative in, negative out buck and boost converter

applications, where the fully charged battery voltage starts out greater than the desired output (the converter must operate in the buck mode), but as the battery discharges, its voltage becomes less than the desired output (the converter must then operate in the boost mode).

A *buck–boost* converter is capable of producing an output voltage which is either greater than or less than the absolute value of the input voltage. A simple buck–boost converter topology is shown in Figure 9-30. The input voltage is positive, and the output voltage is negative. When the switch is on, the inductor current builds up. When the switch is opened, the inductor supplies current to the load through the diode. Obviously, this circuit can be modified for a negative input and a positive output by reversing the polarity of the diode.

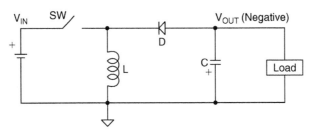

The absolute value of the output can be less than or greater than the absolute value of the input.

Figure 9-30: Buck–boost converter #1 $+V_{IN}$, $-V_{OUT}$

A second buck–boost converter topology is shown in Figure 9-31. This circuit allows both the input and output voltage to be positive. When the switches are closed, the inductor current builds up. When the switches open, the inductor current is supplied to the load through the current path provided by D1 and D2. A fundamental disadvantage to this circuit is that it requires two switches and two diodes. As in the previous circuits, the polarities of the diodes may be reversed to handle negative input and output voltages.

Another way to accomplish the buck–boost function is to cascade two switching regulators, a boost regulator followed by a buck regulator as shown in Figure 9-32. The example shows some practical voltages in a battery-operated system. The input from the four AA cells can range from 6 V (charged) to about 3.5 V (discharged). The intermediate voltage output of the boost converter is 8 V, which is always greater than

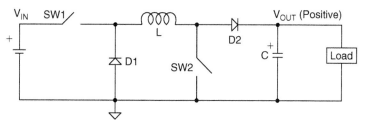

The absolute value of the output can be less than
or greater than the absolute value of the input.

**Figure 9-31: Buck–boost converter #2 +V$_{IN}$, −V$_{OUT}$**

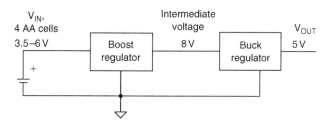

**Figure 9-32: Cascaded buck–boost regulators (example voltages)**

the input voltage. The buck regulator generates the desired 5 V from the 8 V intermediate voltage. The total efficiency of the combination is the product of the individual efficiencies of each regulator and can be greater than 85% with careful design.

An alternate topology is to use a buck regulator followed by a boost regulator. This approach, however, has the disadvantage of pulsating currents on both the input and output and a higher current at the intermediate voltage output.

## Other Non-Isolated Switcher Topologies

The coupled-inductor single-ended primary inductance converter (SEPIC) topology is shown in Figure 9-33. This converter uses a transformer with the addition of capacitor C$_C$ which couples additional energy to the load. If the turns ratio (N = the ratio of the number of primary turns to the number of secondary turns) of the transformer in the SEPIC converter is 1:1, the capacitor serves only to recover the energy in the leakage inductance (i.e., that energy which is not perfectly coupled between the windings) and deliver it to the load. In that case, the relationship between input and output voltage is given by:

$$V_{OUT} = V_{IN} \frac{D}{1 - D} \tag{9-40}$$

For non-unity turns ratios the input/output relationship is highly nonlinear due to transfer of energy occurring via both the coupling between the windings and the capacitor C$_C$. For that reason, it is not analyzed here.

This converter topology often makes an excellent choice in non-isolated battery-powered systems for providing both the ability to step up or down the voltage, and, unlike the boost converter, the ability to have zero voltage at the output when desired.

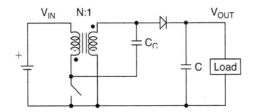

**Figure 9-33: Single-ended primary inductance converter (SEPIC)**

The Zeta and Cük converters, not shown, are two examples of non-isolated converters which require capacitors to deliver energy from input to output, i.e., rather than just to store energy or deliver only recovered leakage energy, as the SEPIC can be configured via a 1:1 turns ratio. Because capacitors capable of delivering energy efficiently in such converters tend to be bulky and expensive, these converters are not frequently used.

## Isolated Switching Regulator Topologies

The switching regulators discussed so far have direct galvanic connections between the input and output. Transformers can be used to supply galvanic isolation as well as allowing the buck–boost function to be easily performed. However, adding a transformer to the circuit creates a more complicated and expensive design as well as increasing the physical size.

The basic *flyback* buck–boost converter circuit is shown in Figure 9-34. It is derived from the buck–boost converter topology. When the switch is on, the current builds up in the primary of the transformer and energy is stored in the magnetic core. When the switch is opened, the current reverts to the secondary winding and flows through the diode delivering the stored energy into the load. The relationship between the input and output voltage is determined by the turns ratio, N, and the duty cycle, D, per the following equation:

$$V_{OUT} = \frac{V_{IN}}{N} \frac{D}{1-D} \tag{9-41}$$

One advantage of the flyback topology is that the transformer provides galvanic isolation as well as acting like an inductor (the transformer is more appropriately referred to as a coupled inductor in this application). A disadvantage of the flyback converter is the high energy which must be stored in the transformer in the form of DC current in the windings. This requires larger cores than would be necessary if the transformer just passed energy (instead of also acting as an inductor).

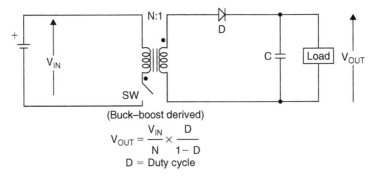

(Buck–boost derived)

$$V_{OUT} = \frac{V_{IN}}{N} \times \frac{D}{1-D}$$

D = Duty cycle

**Figure 9-34: Isolated topology: flyback converter**

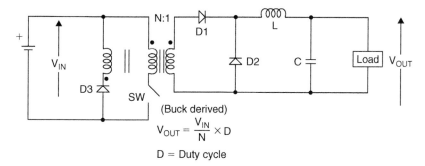

$$V_{OUT} = \frac{V_{IN}}{N} \times D$$

D = Duty cycle

**Figure 9-35: Isolated topology: forward converter**

The basic *forward* converter topology is shown in Figure 9-35. It is derived from the buck converter. This topology avoids the problem of having to store large amounts of energy in the transformer core. However, the circuit is more complex and requires an additional magnetic element (a transformer), an inductor, an additional transformer winding, plus three diodes. When the switch is on, current builds up in the primary winding and also in the secondary winding, where it is transferred to the load through diode D1. When the switch is on, the current in the inductor flows out of D1 from the transformer and is reflected back to the primary winding according to the turns ratio. Additionally, a current (called a *magnetization current*) builds up in the primary due to the input voltage applied across the primary inductance, called the *magnetizing inductance*, and flows in the primary winding. When the switch is opened, the current in the inductor continues to flow through the load via the return path provided by diode D2. The load current is no longer reflected into the transformer, but the magnetizing current induced in the primary still requires a return path so that the transformer can be *reset*. Hence the extra *reset* winding and diode are needed.

The relationship between the input and output voltage is given by:

$$V_{OUT} = \frac{V_{IN}}{N}\, D \tag{9-42}$$

There are many other possible isolated switching regulator topologies which use transformers; however, the balance of this section will focus on non-isolated topologies because of their wider application in portable and distributed power systems.

## Switch Modulation Techniques

Important keys to understanding switching regulators are the various methods used to control the switch. For simplicity of analysis, the examples previously discussed used a simple fixed frequency PWM technique. There can be two other standard variations of the PWM technique: variable frequency constant on-time and variable frequency constant off-time.

In the case of a buck converter, using a fixed off-time ensures that the peak-to-peak output ripple current in the inductor current remains constant as the input voltage varies. This is illustrated in Figure 9-36, where the output current is shown for two conditions of input voltage. Note that as the input voltage increases, the slope during the on-time increases, but the on-time decreases, thereby causing the frequency to increase. Fixed off-time control techniques are popular for buck converters where a wide input voltage range must be accommodated. The ADP1147 family implements this switch modulation technique.

In the case of a boost converter, however, neither input ramp slopes nor output ramp slopes are solely a function of the output voltage (see Figure 9-35), so there is no inherent advantage in the variable frequency

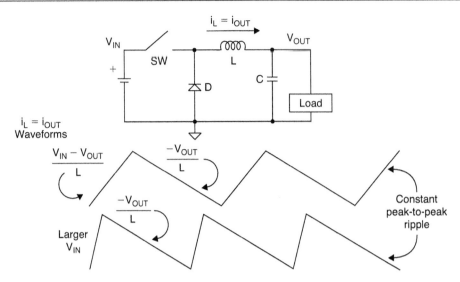

**Figure 9-36: Control of a buck converter using fixed off-time, variable frequency PWM**

constant off-time modulation method with respect to maintaining constant output ripple current. Still, that modulation method tends to allow for less ripple current variation than does fixed frequency, so it is sometimes used.

In the case where very low duty cycles are needed, e.g., under short circuit conditions, sometimes the limitation of a minimum achievable duty cycle is encountered. In such cases, in order to maintain a steady-state condition and prevent runaway of the switch current, a pulse skipping function must be implemented to reduce effective duty cycle. This might take the form of a current monitoring circuit which detects that the switch current is excessive. So either a fixed frequency cycle is skipped without turning on the switch, or the off-time is extended in some way to delay the turn-on.

The pulse skipping technique for a fixed frequency controller can be applied even to operation at *normal* duty cycles. Such a switch modulation technique is then referred to as *pulse burst modulation* (PBM). At its simplest, this technique simply gates a fixed frequency, fixed duty cycle oscillator to be applied to the switch or not. The duty cycle of the oscillator sets the maximum achievable duty cycle for the converter, and smaller duty cycles are achieved over an average of a multiplicity of pulses by skipping oscillator cycles. This switch modulation method accompanies a simple control method of using a hysteretic comparator to monitor the output voltage versus a reference and decides whether to use the oscillator to turn on the switch for that cycle or not. The hysteresis of the comparator tends to give rise to several cycles of switching followed by several cycles of not switching. Hence, the resulting switching signal is characterized by pulses which tend to come in bursts—hence the name for the modulation technique.

There are at least two inherent fundamental drawbacks of the PBM switch modulation technique. First, the constant variation of the duty cycle between zero and maximum produces high ripple currents and accompanying losses. Second, there is an inherent generation of subharmonic frequencies with respect to the oscillator frequency. This means that the noise spectrum is not well controlled, and often audible frequencies can be produced. This is often apparent in higher power converters which use pulse skipping to maintain short circuit current control. An audible noise can often be heard under such a condition, due to the large magnetic elements acting like speaker voice coils. For these reasons, PBM is seldom used at power levels above $\sim 10\,\text{W}$, but for its simplicity, it is often preferred below that power level, but above a power level or with a power conversion requirement where charge pumps are not well suited.

## Control Techniques

Though often confused with or used in conjunction with discussing the switch modulation technique, the control technique refers to what parameters of operation are monitored and how they are processed to control the modulation of the switch. The specific way in which the switch is modulated can be thought of separately, and was just presented in the previous section.

In circuits using PBM for switch modulation, the control technique typically used is a voltage-mode (VM) hysteretic control. In this implementation the switch is controlled by monitoring the output voltage and modulating the switch such that the output voltage oscillates between two hysteretic limits. The ADP3000 switching regulator is an example of a regulator which combines these modulation and control techniques.

Figure 9-37 shows the most basic control technique for use with PWM is *VM control*. Here, the output voltage is the only parameter used to determine how the switch will be modulated. An error amplifier (first mentioned in the Buck Converter section) monitors the output voltage; its error is amplified with the required frequency compensation for maintaining stability of the control loop; and the switch is modulated directly in proportion to that amplifier output.

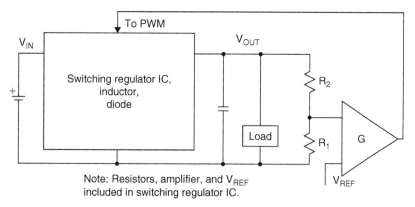

**Figure 9-37: Voltage feedback for PWM control**

The output voltage is divided down by a ratio-matched resistor divider and drives one input of an amplifier, G. A precision reference voltage ($V_{REF}$) is applied to the other input of the amplifier. The output of the amplifier in turn controls the duty cycle of the PWM. It is important to note that the resistor divider, amplifier, and reference are actually part of the switching regulator IC, but are shown externally in the diagram for clarity. The output voltage is set by the resistor divider ratio and the reference voltage:

$$V_{OUT} = V_{REF} + \left(1 + \frac{R_2}{R_1}\right) \tag{9-43}$$

The internal resistor ratios and the reference voltage are set to produce standard output voltage options such as 12, 5, 3.3, or 3 V. In some regulators, the resistor divider can be external, allowing the output voltage to be adjusted.

A simple modification of VM control is voltage *feedforward*. This technique adjusts the duty cycle automatically as the input voltage changes so that the feedback loop does not have to make an adjustment (or as much of an adjustment). Voltage feedforward can even be used in the simple PBM regulators. Feedforward is especially useful in applications where the input voltage can change suddenly or, perhaps due to current limit protection limitations, it is desirable to limit the maximum duty cycle to lower levels when the input voltage is higher.

In switchers, the VM control loop needs to be compensated to provide stability, considering that the voltage being controlled by the modulator is the average voltage produced at the switched node, whereas the actual output voltage is filtered through the switcher's LC filter. The phase shift produced by the filter can make it difficult to produce a control loop with a fast response time.

A popular way to circumvent the problem produced by the LC filter phase shift is to use current-mode (CM) control as shown in Figure 9-38. In CM control, it is still desirable, of course, to regulate the output voltage. Thus, an error amplifier (G1) is still required. However, the switch modulation is no longer controlled directly by the error amplifier. Instead, the inductor current is sensed, amplified by G2, and used to modulate the switch in accordance with the command signal from the (output voltage) error amplifier. It should be noted that the divider network, $V_{REF}$, G1, and G2 are usually part of the IC switching regulator itself, rather than external as shown in the simplified diagram.

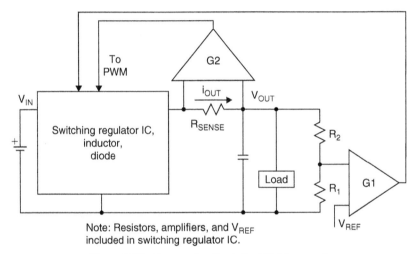

Note: Resistors, amplifiers, and $V_{REF}$ included in switching regulator IC.

**Figure 9-38: Current feedback for PWM control**

The CM control system uses feedback from both the output voltage and output current. Recall that at the beginning of each PWM cycle, the switch turns on, and the inductor current begins to rise. The inductor current develops a voltage across the small sense resistor, $R_{SENSE}$, which is amplified by G2 and fed back to the PWM controller to turn off the switch. The output voltage, sensed by amplifier G1 and also fed back to the PWM controller, sets the level at which the peak inductor current will terminate the switch on-time. Since it is inductor current that turns off the switch (and thereby sets the duty cycle) this method is commonly referred to as *CM control*, even though there are actually two feedback control loops: the fast responding current loop, and the slower responding output voltage loop. Note that inductor current is being controlled on a pulse-by-pulse basis, which simplifies protection against switch over-current and inductor saturation conditions.

In essence, then, in CM control, rather than controlling the average voltage which is applied to the LC filter as in VM control, the inductor current is controlled directly on a cycle-by-cycle basis. The only phase shift remaining between the inductor current and the output voltage is that produced by the impedance of the output capacitor(s). The correspondingly lower phase shift in the output filter allows the loop response to be faster while still remaining stable. Also, instantaneous changes in input voltage are immediately reflected in the inductor current, which provides excellent line transient response. The obvious disadvantage of CM control is the requirement of sensing current and, if needed, an additional amplifier. With increasingly higher performance requirements in modern electronic equipment, the performance advantage of CM control typically outweighs the cost of implementation. Also, some sort of current limit protection is often required, whatever the control technique. Thus it tends to be necessary to implement some sort of current sensing even in VM-controlled systems.

Now even though we speak of a CM controller as essentially controlling the inductor current, more often than not the switch current is controlled instead, since it is more easily sensed (especially in a switching regulator) and it is a representation of the inductor current for at least the on-time portion of the switching cycle. Rather than actually controlling the average switch current, which is not the same as the average inductor current anyway, it is often simpler to control the peak current—which is the same for both the switch and the inductor in all the basic topologies. The error between the average inductor current and the peak inductor current produces a nonlinearity within the control loop. In most systems, that is not a problem. In other systems, a more precise current control is needed, and in such a case, the inductor current is sensed directly and amplified and frequency compensated for the best response.

Other control variations are possible, including *valley* rather than peak control, *hysteretic current* control, and even *charge* control—a technique whereby the integral of the inductor current (i.e., charge) is controlled. That eliminates even the phase shift of the output capacitance from the loop, but presents the problem that instantaneous current is not controlled, and therefore short circuit protection is not inherent in the system. All techniques offer various advantages and disadvantages. Usually the best tradeoff between performance and cost/simplicity is peak current control—as used by the ADP1147 family. This family also uses the current sense output to control a *sleep,* or *power saving mode* of operation, to maintain high efficiency for low output currents.

## Gated Oscillator (PBM) Control Example

All of the PWM techniques discussed thus far require some degree of feedback loop compensation. This can be especially tricky for boost converters, where there is more phase shift between the switch and the output voltage.

As previously mentioned, a technique which requires no feedback compensation uses a fixed frequency gated oscillator as the switch control (see Figure 9-39). This method is often (incorrectly) referred to as the pulse frequency modulation (PFM) mode, but is more correctly called *pulse burst modulation (PBM)* or *gated-oscillator* control.

The output voltage ($V_{OUT}$) is divided by the resistive divider ($R_1$ and $R_2$) and compared against a reference voltage, $V_{REF}$. The comparator hysteresis is required for stability and also affects the output voltage ripple. When the resistor divider output voltage drops below the comparator threshold ($V_{REF}$ minus the hysteresis voltage), the comparator starts the gated oscillator. The switcher begins switching again which then causes the output voltage to increase until the comparator threshold is reached ($V_{REF}$ plus the hysteresis voltage), at which time the oscillator is turned off. When the oscillator is off, quiescent current drops to a very low value (e.g., 95 µA in the ADP1073) making PBM controllers very suitable for battery-powered applications.

A simplified output voltage waveform is shown in Figure 9-40 for a PBM buck converter. Note that the comparator hysteresis voltage multiplied by the reciprocal of the attenuation factor primarily determines the peak-to-peak output voltage ripple (typically between 50 and 100 mV). It should be noted that the actual

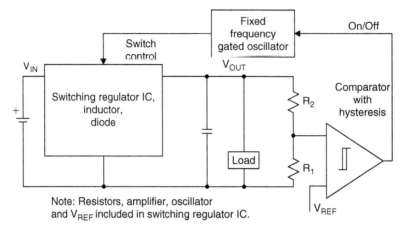

**Figure 9-39: Switch control using gated oscillator (pulse burst modulation, PBM)**

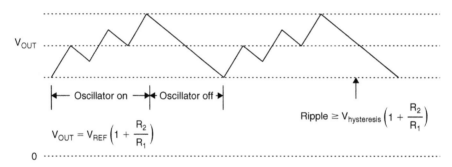

**Figure 9-40: Representative output voltage waveform for gated oscillator-controlled (PBM) buck regulators**

output voltage ripple waveform can look quite different from that shown in Figure 3-30 depending on the design and whether the converter is a buck or boost.

A practical switching regulator IC using the PBM approach is the ADP3000, which has a fixed switching frequency of 400 kHz and a fixed duty cycle of 80%. This device is a versatile step-up/step-down converter. It can deliver an output current of 100 mA in a 5 to 3 V step-down configuration and 180 mA in a 2 to 3.3 V step-up configuration. Input supply voltage can range between 2 and 12 V in the boost mode, and up to 30 V in the buck mode. It should be noted that when the oscillator is turned off, the internal switch is opened so that the inductor current does not continue to increase.

In the gated-oscillator method, the comparator hysteresis serves to stabilize the feedback loop making the designs relatively simple. The disadvantage, of course, is that the peak-to-peak output voltage ripple can never be less than the comparator hysteresis multiplied by the reciprocal of the attenuation factor:

$$\text{Output Ripple} \geq V_{\text{hysteresis}} \left( \frac{R_2}{R_1} \right) \qquad (9\text{-}44)$$

Because the gated-oscillator (PBM)-controlled switching regulator operates with a fixed duty cycle, output regulation is achieved by changing the number of "skipped pulses" as a function of load current and voltage. From this perspective, PBM-controlled switchers tend to operate in the "discontinuous" mode under light load conditions. Also, the maximum average duty cycle is limited by the built-in duty cycle of the oscillator. Once the required duty cycle exceeds that limit, no pulse skipping occurs, and the device will lose regulation.

One disadvantage of the PBM switching regulator is that the frequency spectrum of the output ripple is "fuzzy" because of the burst-mode of operation. Frequency components may fall into the audio band, so proper filtering of the output of such a regulator is mandatory.

Selection of the inductor value is also more critical in PBM regulators. Because the regulation is accomplished with a burst of fixed duty cycle pulses (i.e., higher than needed on average) followed by an extended off time, the energy stored in the inductor during the burst of pulses must be sufficient to supply the required energy to the load. If the inductor value is too large, the regulator may never start up, or may have poor transient response and inadequate line and load regulation. On the other hand, if the inductor value is too small, the inductor may saturate during the charging time, or the peak inductor current may exceed the maximum rated switch current. However, devices such as the ADP3000 incorporate on-chip overcurrent protection for the switch. An additional feature allows the maximum peak switch current to be set with an external resistor, thereby preventing inductor saturation. Techniques for selecting the proper inductor value will be discussed in a following section.

## Diode and Switch Considerations

So far, we have based our discussions around an ideal lossless switching regulator having ideal circuit elements. In practice, the diode, switch, and inductor all dissipate power which leads to less than 100% efficiency.

Figure 9-41 shows typical buck and boost converters, where the switch is part of the IC. The process is bipolar, and this type of transistor is used as the switching element. The ADP3000 and its relatives (ADP1108, ADP1109, ADP1110, ADP1111, ADP1073, ADP1173) use this type of internal switch.

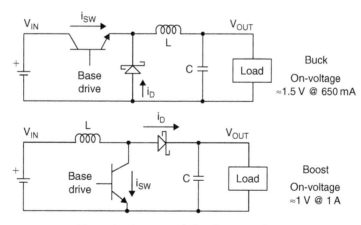

**Figure 9-41: NPN switches in IC regulators**

The diode is external to the IC and must be chosen carefully. Current flows through the diode during the off-time of the switching cycle. This translates into an average current which causes power dissipation because of the diode forward voltage drop. The power dissipation can be minimized by selecting a Schottky diode with a low forward drop (0.5 V), such as the 1N5818 type. It is also important that the diode capacitance and recovery time be low to prevent additional power loss due to charging current, and this is also afforded by the Schottky diode. Power dissipation can be approximated by multiplying the average diode current by the forward voltage drop.

The drop across the NPN switch also contributes to internal power dissipation. The power (neglecting switching losses) is equal to the average switch current multiplied by the collector–emitter on-state voltage. In the case of the ADP3000 series, it is 1.5 V at the maximum rated switch current of 650 mA (when operating in the buck mode).

In the boost mode, the NPN switch can be driven into saturation, so the on-state voltage is reduced, and thus, so is the power dissipation. Note that in the case of the ADP3000, the saturation voltage is about 1 V at the maximum rated switch current of 1 A.

In examining the two configurations, it would be logical to use a PNP switching transistor in the buck converter and an NPN transistor in the boost converter in order to minimize switch voltage drop. However, the PNP transistors available on processes which are suitable for IC switching regulators generally have poor performance, so the NPN transistor must be used for both topologies.

In addition to lowering efficiency by their power dissipation, the switching transistors and the diode also affect the relationship between the input and output voltage. The equations previously developed assumed zero switch and diode voltage drops. Rather than re-deriving all the equations to account for these drops, we will examine their effects on the inductor current for a simple buck and boost converter operating in the continuous mode as shown in Figure 9-42.

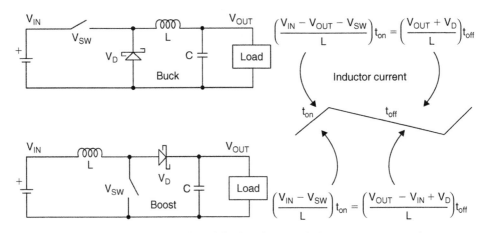

**Figure 9-42: Effects of switch and diode voltage on inductor current equations**

In the buck converter, the voltage applied to the inductor when the switch is on is equal to $V_{IN} - V_{OUT} - V_{SW}$, where $V_{SW}$ is the approximate average voltage drop across the switch. When the switch is off, the inductor current is discharged into a voltage equal to $V_{OUT} + V_D$, where $V_D$ is the approximate average forward drop across the diode. The basic inductor equation used to derive the relationship between the input and output voltage becomes:

$$\left(\frac{V_{IN} - V_{OUT} - V_{SW}}{L}\right) t_{on} = \left(\frac{V_{OUT} + V_D}{L}\right) t_{off} \tag{9-45}$$

In the actual regulator circuit, negative feedback will force the duty cycle to maintain the correct output voltage, but the duty cycle will also be affected by the switch and the diode drops to a lesser degree.

When the switch is on in a boost converter, the voltage applied to the inductor is equal to $V_{IN} - V_{SW}$. When the switch is off, the inductor current discharges into a voltage equal to $V_{OUT} - V_{IN} + V_D$. The basic inductor current equation becomes:

$$\left( \frac{V_{IN} - V_{SW}}{L} \right) t_{on} = \left( \frac{V_{OUT} - V_{IN} + V_D}{L} \right) t_{off} \qquad (9\text{-}46)$$

From the above equations, the basic relationships between input voltage, output voltage, duty cycle, switch, and diode drops can be derived for the buck and boost converters (Figures 9-43 and 9-44).

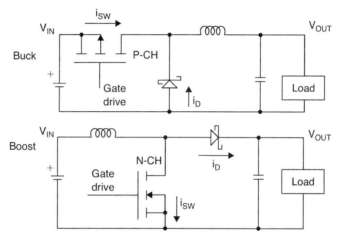

**Figure 9-43: Power MOSFET switches**

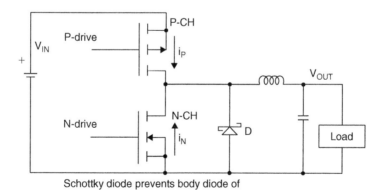

Schottky diode prevents body diode of
N-channel MOSFET from conducting during "deadtime."

**Figure 9-44: Buck converter with synchronous switch using P- and
N-channel MOSFETS**

## *Inductor Considerations*

The selection of the inductor used in a switching regulator is probably the most difficult part of the design. Fortunately, manufacturers of switching regulators supply a wealth of applications information, and standard off-the-shelf inductors from well-known and reliable manufacturers are quite often recommended on the switching regulator data sheet. However, it is important for the design engineer to understand at least some of the fundamental issues relating to inductors. This discussion, while by no means complete, will give some insight into the relevant magnetics issues.

Selecting the actual value for the inductor in a switching regulator is a function of many parameters. Fortunately, in a given application the exact value is generally not all that critical, and equations supplied on the data sheets allow the designer to calculate a minimum and maximum acceptable value. That's the easy part.

Unfortunately, there is more to a simple inductor than its inductance! Figure 9-45 shows an equivalent circuit of a real inductor and also some of the many considerations that go into the selection process. To further complicate the issue, most of these parameters interact, thereby making the design of an inductor truly more of an art than a science.

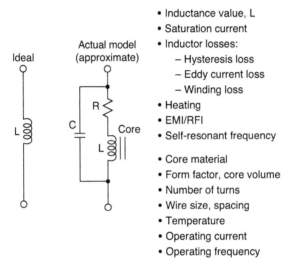

- Inductance value, L
- Saturation current
- Inductor losses:
    - Hysteresis loss
    - Eddy current loss
    - Winding loss
- Heating
- EMI/RFI
- Self-resonant frequency

- Core material
- Form factor, core volume
- Number of turns
- Wire size, spacing
- Temperature
- Operating current
- Operating frequency

**Figure 9-45: Inductor considerations**

Probably the easiest inductor problem to solve is selecting the proper value. In most switching regulator applications, the exact value is not very critical, so approximations can be used with a high degree of confidence.

The heart of a switching regulator analysis involves a thorough understanding of the inductor current waveform. Figure 9-46 shows an assumed inductor current waveform (which is also the output current) for a buck converter, such as the ADP3000, which uses the gated-oscillator PBM switch modulation technique. Note that this waveform represents a worst case condition from the standpoint of storing energy in the inductor, where the inductor current starts from zero on each cycle. In high output current applications, the inductor current does not return to zero, but ramps up until the output voltage comparator senses that the oscillator should be turned off, at which time the current ramps down until the comparator turns the oscillator on again. This assumption about the worst case waveform is necessary because in a simple PBM regulator, the oscillator duty cycle remains constant regardless of input voltage or output load current. Selecting the inductor value using this assumption will always ensure that there is enough energy stored in the inductor to maintain regulation:

*It should be emphasized that the following inductance calculations for the PBM buck and boost regulators should be used only as a starting point, and larger or smaller values may actually be required depending on the specific regulator and the input/output conditions.*

Output and inductor current

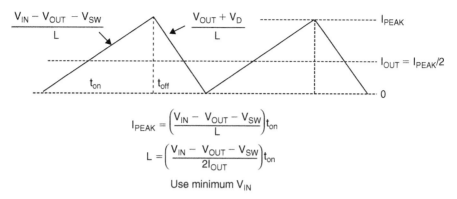

Figure 9-46: Calculating L for a buck converter: gated oscillator (PBM) type

The peak current is easily calculated from the slope of the positive-going portion of the ramp:

$$I_{PEAK} = \left( \frac{V_{IN} - V_{OUT} - V_{SW}}{L} \right) t_{on} \tag{9-47}$$

This equation can then be solved for L:

$$L = \left( \frac{V_{IN} - V_{OUT} - V_{SW}}{I_{PEAK}} \right) t_{on} \tag{9-48}$$

However, the average output current, $I_{OUT}$ is equal to $I_{PEAK}/2$, and therefore $I_{PEAK} = 2I_{OUT}$. Substituting this value for $I_{PEAK}$ into the previous equation yields:

$$L = \left( \frac{V_{IN} - V_{OUT} - V_{SW}}{2I_{OUT}} \right) t_{on} \tag{9-49}$$

(L for buck PBM converter.)

The minimum expected value of $V_{IN}$ should be used in order to minimize the inductor value and maximize its stored energy. If $V_{IN}$ is expected to vary widely, an external resistor can be added to the ADP3000 to limit peak current and prevent inductor saturation at maximum $V_{IN}$.

A similar analysis can be carried out for a boost PBM regulator as shown in Figure 9-47.

We make the same assumptions about the inductor current, but note that the output current shown on the diagram is pulsating and not continuous. The output current, $I_{OUT}$, can be expressed in terms of the peak current, $I_{PEAK}$, and the duty cycle, D, as:

$$I_{OUT} = \frac{I_{PEAK}}{2} (1 - D) \tag{9-50}$$

Solving for $I_{PEAK}$ yields:

$$I_{PEAK} = \frac{2I_{OUT}}{1 - D} \tag{9-51}$$

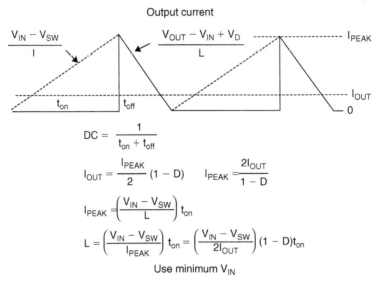

Figure 9-47: Calculating L for a boost converter: gated oscillator (PBM) type

However, $I_{PEAK}$ can also be expressed in terms of $V_{IN}$, $V_{SW}$, L, and $t_{on}$:

$$I_{PEAK} = \left(\frac{V_{IN} - V_{SW}}{L}\right) t_{on} \qquad (9\text{-}52)$$

which can be solved for L:

$$L = \left(\frac{V_{IN} - V_{SW}}{I_{PEAK}}\right) t_{on} \qquad (9\text{-}53)$$

Substituting the previous expression for $I_{PEAK}$ yields:

$$L = \left(\frac{V_{IN} - V_{SW}}{2I_{OUT}}\right)(1 - D)t_{on} \qquad (9\text{-}54)$$

(L for boost PBM converter.)

The minimum expected value of $V_{IN}$ should be used in order to ensure sufficient inductor energy storage under all conditions. If $V_{IN}$ is expected to vary widely, an external resistor can be added to the ADP3000 to limit peak current and prevent inductor saturation at maximum $V_{IN}$:

> *The above equations will only yield approximations to the proper inductor value for the PBM-type regulators and should be used only as a starting point. An exact analysis is difficult and highly dependent on the regulator and input/output conditions. However, there is considerable latitude with this type of regulator, and other analyses may yield different results but still fall within the allowable range for proper regulator operation.*

Calculating the proper inductor value for PWM regulators is more straightforward. Figure 9-48 shows the output and inductor current waveform for a buck PWM regulator operating in the continuous mode.

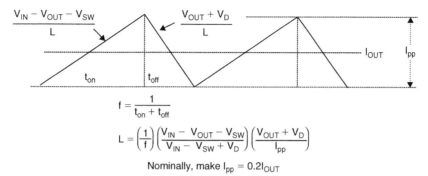

Output and inductor current, continuous mode

$$f = \frac{1}{t_{on} + t_{off}}$$

$$L = \left(\frac{1}{f}\right)\left(\frac{V_{IN} - V_{OUT} - V_{SW}}{V_{IN} - V_{SW} + V_D}\right)\left(\frac{V_{OUT} + V_D}{I_{pp}}\right)$$

Nominally, make $I_{pp} = 0.2 I_{OUT}$

**Figure 9-48: Calculating L for a buck converter: constant frequency PWM type**

It is accepted design practice to design for a peak-to-peak ripple current, $I_{pp}$, which is between 10% and 30% of the output current, $I_{OUT}$. We will assume that $I_{pp} = 0.2 \times I_{OUT}$.

By inspection, we can write:

$$\left(\frac{V_{IN} - V_{OUT} - V_{SW}}{L}\right)t_{on} = \left(\frac{V_{OUT} + V_D}{L}\right)t_{off} \tag{9-55}$$

or

$$t_{off} = \left(\frac{V_{IN} - V_{OUT} - V_{SW}}{V_{OUT} + V_D}\right)t_{on} \tag{9-56}$$

However, the switching frequency, f, is given by:

$$f = \frac{1}{t_{on} + t_{off}} \tag{9-57}$$

or

$$t_{off} = \frac{1}{f} - t_{on} \tag{9-58}$$

Substituting this expression for $t_{off}$ in the previous equation for $t_{off}$ and solving for $t_{on}$ yields:

$$t_{on} = \frac{1}{f}\left(\frac{V_{OUT} + V_D}{V_{IN} - V_{SW} + V_D}\right) \tag{9-59}$$

However,

$$I_{pp} = \left(\frac{V_{IN} - V_{OUT} - V_{SW}}{L}\right)t_{on} \tag{9-60}$$

Combining the last two equations and solving for L yields:

$$L = \left(\frac{1}{f}\right)\left(\frac{V_{IN} - V_{OUT} - V_{SW}}{V_{IN} - V_{SW} + V_D}\right)\left(\frac{V_{OUT} + V_D}{I_{pp}}\right) \tag{9-61}$$

(L for buck PWM converter, constant frequency.)

As indicated earlier, choose $I_{pp}$ to be nominally $0.2 \times I_{OUT}$ and solve the equation for L. Calculate L for the minimum and maximum expected value of $V_{IN}$ and choose a value halfway between. System requirements may dictate a larger or smaller value of $I_{pp}$, which will inversely affect the inductor value.

A variation of the buck PWM constant frequency regulator is the buck PWM regulator with variable frequency and constant off-time (e.g., ADP1148).

A diagram of the output and inductor current waveform is shown in Figure 9-49 for the continuous mode.

Output and inductor current, continuous mode

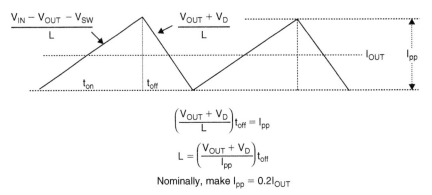

$$\left(\frac{V_{OUT} + V_D}{L}\right) t_{off} = I_{pp}$$

$$L = \left(\frac{V_{OUT} + V_D}{I_{pp}}\right) t_{off}$$

Nominally, make $I_{pp} = 0.2 I_{OUT}$

**Figure 9-49: Calculating L for a buck converter: constant off-time, variable frequency PWM type**

The calculations are very straightforward, since the peak-to-peak amplitude of the ripple current is constant:

$$I_{pp} = \left(\frac{V_{OUT} + V_D}{L}\right) t_{off} \tag{9-62}$$

Solving for L:

$$L = \left(\frac{V_{OUT} + V_D}{I_{pp}}\right) t_{off} \tag{9-63}$$

(L for buck PWM constant off-time, variable frequency converter.)

Again, choose $I_{pp} = 0.2 \times I_{OUT}$, or whatever the system requires.

The final example showing the inductance calculation is for the boost PWM constant frequency regulator. The inductor (and input) current waveform is shown in Figure 9-50.

The analysis is similar to that of the constant frequency buck PWM regulator.

By inspection of the inductor current, we can write:

$$\left(\frac{V_{IN} - V_{SW}}{L}\right) t_{on} = \left(\frac{V_{OUT} - V_{IN} + V_D}{L}\right) t_{off} \tag{9-64}$$

or

$$t_{off} = \left(\frac{V_{IN} - V_{SW}}{V_{OUT} - V_{IN} + V_D}\right) t_{on} \tag{9-65}$$

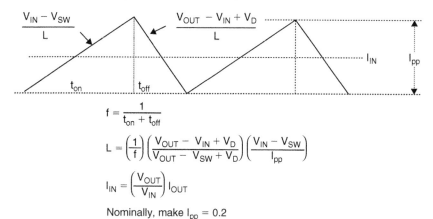

Figure 9-50: Calculating L for a boost converter: constant frequency PWM type

However, the switching frequency, f, is given by:

$$f = \frac{1}{t_{on} + t_{off}}$$ (9-66)

or

$$t_{off} = \frac{1}{f} - t_{on}$$ (9-67)

Substituting this expression for $t_{off}$ in the previous equation for $t_{off}$ and solving for $t_{on}$ yields:

$$t_{on} = \frac{1}{f}\left(\frac{V_{OUT} - V_{IN} + V_D}{V_{OUT} - V_{SW} + V_D}\right)$$ (9-68)

However,

$$I_{pp} = \left(\frac{V_{IN} - V_{SW}}{L}\right)t_{on}$$ (9-69)

Combining the last two equations and solving for L yields:

$$L = \left(\frac{1}{f}\right)\left(\frac{V_{OUT} - V_{IN} + V_D}{V_{OUT} - V_{SW} + V_D}\right)\left(\frac{V_{IN} - V_{SW}}{I_{pp}}\right)$$ (9-70)

(L for boost PWM, constant frequency converter.)

For the boost converter, the inductor (input) current, $I_{IN}$, can be related to the output current, $I_{OUT}$, by:

$$I_{IN} = \left(\frac{V_{OUT}}{V_{IN}}\right)I_{OUT}$$ (9-71)

Nominally, make $I_{pp} = 0.2I_{IN}$.

Note that for the boost PWM, even though the input current is continuous, while the output current pulsates, we still base the inductance calculation on the peak-to-peak inductor ripple current.

As was previously suggested, the actual selection of the inductor value in a switching regulator is probably the easiest part of the design process. Choosing the proper type of inductor is much more complicated as the following discussions will indicate.

Fundamental magnetic theory says that if a current passes through a wire, a magnetic field will be generated around the wire (right-hand rule). The strength of this field is measured in ampere-turns per meter, or *oersteds* and is proportional to the current flowing in the wire. The magnetic field strength produces a *magnetic flux density* (B, measured in webers per square meter, or *gauss*).

Using a number of turns of wire to form a coil increases the magnetic flux density for a given current. The effective inductance of the coil is proportional to the ratio of the magnetic flux density to the field strength.

This simple air core inductor is not very practical for the values of inductance required in switching regulators because of wiring resistance, interwinding capacitance, sheer physical size, and other factors. Therefore, in order to make a reasonable inductor, the wire is wound around some type of ferromagnetic core having a high *permeability*. Core permeability is often specified as a relative permeability which is basically the increase in inductance which is obtained when the inductor is wound on a core instead of just air. A relative permeability of 1,000, for instance, will increase inductance by 1,000:1 above that of an equivalent air core.

Figure 9-51 shows magnetic flux density, B, versus inductor current for the air core and also ferromagnetic cores. Note that B is linear with respect to H for the air core inductor, i.e., the inductance remains constant regardless of current.

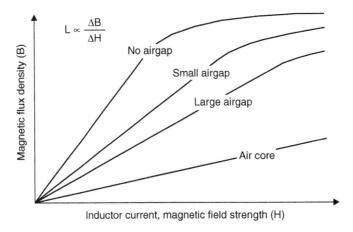

**Figure 9-51: Magnetic flux density versus inductor current**

The addition of a ferromagnetic core increases the slope of the curve and increases the effective inductance, but at some current level, the inductor core will saturate (i.e., the inductance is drastically reduced). It is obvious that inductor saturation can wreak havoc in a switching regulator, and can even burn out the switch if it is not current limited.

This effect can be reduced somewhat while still maintaining higher inductance than an air core by the addition of an air gap in the ferromagnetic core. The air gap reduces the slope of the curve, but provides a wider linear operating range of inductor current. Air gaps do have their problems, however, and one of them is the tendency of the air-gapped inductor to radiate high frequency energy more than a non-gapped inductor. Proper design and manufacturing techniques, however, can be used to minimize this EMI problem, so air-gapped cores are popular in many applications.

The effects of inductor core saturation in a switcher can be disastrous to the switching elements as well as lowering efficiency and increasing noise. Figure 9-52 shows a normal inductor current waveform in a switching regulator as well as a superimposed waveform showing the effects of core saturation. Under normal conditions the slope is linear for both the charge and discharge cycle. If saturation occurs, however, the inductor current increases exponentially, corresponding to the drop in effective inductance. It is therefore important in all switching regulator designs to determine the peak inductor current expected under the worst case conditions of input voltage, load current, duty cycle, etc. This worst case peak current must be less than the peak current rating of the inductor. Notice that when inductor literature does not have a "DC current" rating, or shows only an "AC amps" rating, such inductors are often prone to saturation.

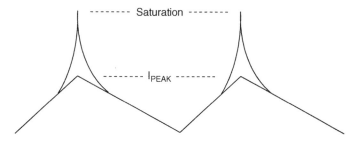

**Figure 9-52: Effects of saturation on inductor current**

From a simplified design standpoint, the effects or presence of inductor saturation can best be observed with a scope and a current probe. If a current probe is not available, a less direct but still effective method is to measure the voltage across a small sense resistor in series with the inductor. The resistor value should be $1\,\Omega$ or less (depending on the inductor current), and the resistor must be sized to dissipate the power. In most cases, a $1\,\Omega$, $1\,W$ resistor will work for currents up to a few hundred milliamperes, and a $0.1\,\Omega$, $10\,W$ resistor is good for currents up to $10\,A$.

Another inductor consideration is its loss. Ideally, an inductor should dissipate no power. However, in a practical inductor, power is dissipated in the form of hysteresis loss, eddy current loss, and winding loss. Figure 9-53 shows a typical B/H curve for an inductor. The enclosed area swept out by the B/H curve during one complete operating cycle is the hysteresis loss exhibited by the core during that cycle. Hysteresis loss is a function of core material, core volume, operating frequency, and the maximum flux density during each cycle. The second major loss within the core is eddy current loss. This loss is caused by the flow of circulating magnetic currents within the core material caused by rapid transitions in the magnetic flux density. It is also dependent on the core material, core volume, operating frequency, and flux density.

In addition to core loss, there is winding loss, the power dissipated in the DC resistance of the winding. This loss is a function of the wire size, core volume, and the number of turns.

In a switching regulator application, excessive loss will result in a loss of efficiency and high inductor operating temperatures.

Fortunately, inductor manufacturers have simplified the design process by specifying maximum peak current, maximum continuous current, and operating frequency range and temperature for their inductors. If the designer derates the maximum peak and continuous current levels by a factor of 20% or so, the inductor

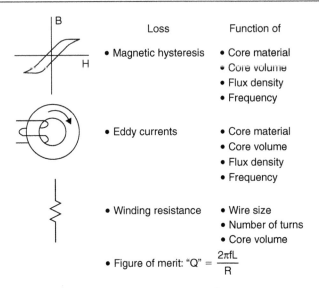

Figure 9-53: Inductor power losses

should be satisfactory for the application. If these simple guidelines are observed, then the designer can be reasonable confident that the major sources of efficiency losses will be due to other parts of the regulator, i.e., the switch ($I^2R$, gate charge, on-voltage), the diode (on-voltage), and the quiescent power dissipation of the regulator itself.

One method to ensure that the inductor losses do not significantly degrade the regulator performance is to measure the Q of the inductor at the switching frequency. If the Q is greater than about 25, then the losses should be insignificant.

There are many possible choices in inductor core materials: ferrite, molypermalloy (MPP) ferrite, powdered iron, etc. High efficiency converters generally cannot accommodate the core loss found in the low cost powdered iron cores, forcing the use of more expensive ferrite, molypermalloy (MPP), or "Kool Mμ"® cores.

Ferrite core material saturates "hard," which causes the inductance to collapse abruptly when the peak current is exceeded. This results in a sharp increase in inductor ripple current.

MPP from Magnetics, Inc., is a very good, low loss core material for toroids, but is more expensive than ferrite. A reasonable compromise from the same manufacturer is "Kool Mμ."

The final consideration is the inductor self-resonant frequency. A practical example would be an inductor of $10\,\mu H$ which has an equivalent distributed capacitance of $5\,pF$. The self-resonant frequency can be calculated as follows:

$$f_{resonance} = \frac{1}{2\pi}\sqrt{\frac{1}{LC}} = 22\,\text{MHz} \tag{9-72}$$

The switching frequency of the regulator should be at least 10 times less than the resonant frequency. In most practical designs with switching frequencies less than $1\,\text{MHz}$ this will always be the case, but a quick calculation is a good idea.

## Capacitor Considerations

Capacitors play a critical role in switching regulators by acting as storage elements for the pulsating currents produced by the switching action. Although not shown on the diagrams previously, all switching regulators need capacitors on their inputs as well as their outputs for proper operation. The capacitors must have very low impedance at the switching frequency as well as the high frequencies produced by the pulsating current waveforms.

Recall the input and output current waveforms for the simple buck converter shown in Figure 9-54. Note that the input current to the buck converter is pulsating, while the output is continuous. Obviously, the input capacitor $C_{IN}$ is critical for proper operation of the regulator. It must maintain the input at a constant voltage during the switching spikes. This says that the impedance of the capacitor must be very low at high frequencies, much above the regulator switching frequency. The load capacitor is also critical in that its impedance will determine the peak-to-peak output voltage ripple, but its impedance at high frequencies is not as critical due to the continuous nature of the output current waveform.

The situation is reversed in the case of the boost converter shown in Figure 9-55. Here the input waveform is continuous, while the output waveform is pulsating. The output capacitor must have good low and high frequency characteristics in order to minimize the output voltage ripple. Boost converters are often followed by a post filter to remove the high frequency switching noise.

Switching regulator capacitors are generally of the electrolytic type because of the relatively large values required. An equivalent circuit for an electrolytic capacitor is shown in Figure 9-56. In addition to the capacitance value itself, the capacitor has some ESR and ESL. It is useful to make a few assumptions and examine the approximate response of the capacitor to a fast current step input. For the sake of the discussion, assume the input current switches from 0 to 1 A in 100 ns. Also, assume that the ESR is 0.2 $\Omega$

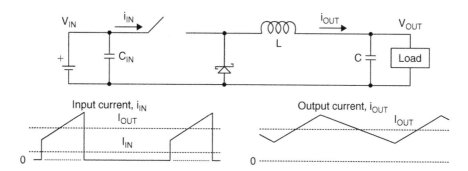

Figure 9-54: Buck converter input and output current waveforms

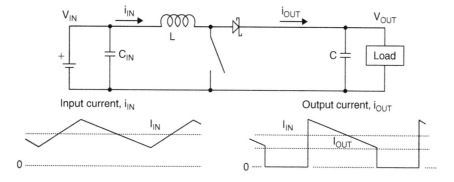

Figure 9-55: Boost converter input and output current waveforms

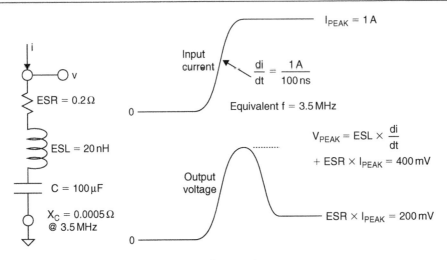

$I_{PEAK} = 1 A$

Input current

$$\frac{di}{dt} = \frac{1 A}{100 ns}$$

Equivalent $f = 3.5 MHz$

$ESR = 0.2 \Omega$

$ESL = 20 nH$

$$V_{PEAK} = ESL \times \frac{di}{dt}$$
$$+ ESR \times I_{PEAK} = 400 mV$$

$C = 100 \mu F$

Output voltage

$ESR \times I_{PEAK} = 200 mV$

$X_C = 0.0005 \Omega$
@ 3.5 MHz

**Figure 9-56: Response of a capacitor to a current step**

and that the ESL is 20 nH. ESR and ESL vary widely between manufacturers and are also dependent on body style (through hole versus surface mount), but these values will serve to illustrate the point.

Assume that the actual value of the capacitor is large enough so that its reactance is essentially a short circuit with respect to the step function input. For example, 100 μF at 3.5 MHz (the equivalent frequency of a 100 ns risetime pulse) has a reactance of $1/2\pi fC = 0.0005 \Omega$. In this case, the output voltage ripple is determined exclusively by the ESR and ESL of the capacitor, not the actual capacitor value itself.

These waveforms show the inherent limitations of electrolytic capacitors used to absorb high frequency switching pulses. In a practical system, the high frequency components must be attenuated by low inductance ceramic capacitors with low ESL or by the addition of an LC filter.

Figure 9-57 shows the impedance versus frequency for a typical 100 μF electrolytic capacitor having an ESR of 0.2 Ω and an ESL of 20 nH. At frequencies below about 10 kHz, the capacitor is nearly ideal. Between 10 kHz and 1 MHz (the range of switching frequencies for most IC switching regulators!) the impedance is limited by the ESR to 0.2 Ω. Above about 1 MHz the capacitor behaves like an inductor due to the ESL of 20 nH. These values, although they may vary somewhat depending on the actual type

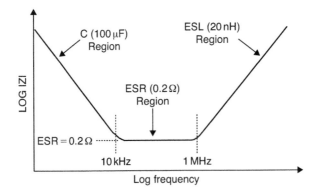

**Figure 9-57: Typical electrolytic capacitor impedance versus frequency**

of electrolytic capacitor (aluminum general purpose, aluminum switching type, tantalum, or organic semiconductor), are representative and illustrate the importance of understanding the limitations of capacitors in switching regulators.

From the electrolytic capacitor impedance characteristic, it is clear that the ESR and ESL of the output capacitor will determine the peak-to-peak output voltage ripple caused by the switching regulator output ripple current.

In most electrolytic capacitors, ESR degrades noticeably at low temperature, by as much as a factor of 4–6 times at $-55°C$ versus the room temperature value. For circuits where ESR is critical to performance, this can lead to problems. Some specific electrolytic types do address this problem; for example, within the HFQ switching types, the $-10°C$ ESR at 100 kHz is no more than 2 times that at room temperature. The OS-CON electrolytics have an ESR versus temperature characteristic which is relatively flat.

There are generally three classes of capacitors useful in 10 kHz–100 MHz frequency range, broadly distinguished as the generic dielectric types; *electrolytic*, *film*, and *ceramic*. These can in turn can be further sub-divided. A thumbnail sketch of capacitor characteristics is shown in the chart of Figure 9-58.

| | Aluminum electrolytic (general purpose) | Aluminum electrolytic (switching type) | Tantalum electrolytic | OS-CON electrolytic | Polyester (stacked film) | Ceramic (multilayer) |
|---|---|---|---|---|---|---|
| Size | 100 µF | 120 µF | 120 µF | 100 µF | 1 µF | 0.1 µF |
| Rated voltage | 25 V | 25 V | 20 V | 20 V | 400 V | 50 V |
| ESR | 0.6 Ω @ 100 kHz | 0.18 Ω @ 100 kHz | 0.12 Ω @ 100 kHz | 0.02 Ω @ 100 kHz | 0.11 Ω @ 1 MHz | 0.12 Ω @ 1 MHz |
| Operating frequency (*) | ≅100 kHz | ≅500 kHz | ≅1 MHz | ≅1 MHz | ≅10 MHz | ≅1 GHz |

(*) Upper frequency strongly size and package dependent

**Figure 9-58: Capacitor selection guide**

The *electrolytic* family provides an excellent, cost-effective low frequency component, because of the wide range of values, a high capacitance-to-volume ratio, and a broad range of working voltages. It includes *general purpose aluminum electrolytic* types, available in working voltages from below 10 V up to about 500 V, and in size from one to several thousand µF (with proportional case sizes). All electrolytic capacitors are polarized, and thus cannot withstand more than a volt or so of reverse bias without damage. They also have relatively high leakage currents (up to tens of microamperes, and strongly dependent on design specifics).

A subset of the general electrolytic family includes *tantalum* types, generally limited to voltages of 100 V or less, with capacitance of 500 µF or less (Reference 21). In a given size, tantalums exhibit a higher capacitance-to-volume ratio than do general purpose electrolytics, and have both a higher frequency range and lower ESR. They are generally more expensive than standard electrolytics, and must be carefully applied with respect to surge and ripple currents.

A subset of aluminum electrolytic capacitors is the *switching* type, designed for handling high pulse currents at frequencies up to several hundred kHz with low losses (Reference 22). This capacitor type competes directly with tantalums in high frequency filtering applications, with the advantage of a broader range of values.

A more specialized high performance aluminum electrolytic capacitor type uses an organic semiconductor electrolyte (Reference 23). The *OS-CON* capacitors feature appreciably lower ESR and higher frequency range than do other electrolytic types, with an additional feature of low temperature ESR degradation.

*Film* capacitors are available in very broad value ranges and different dielectrics, including polyester, polycarbonate, polypropylene, and polystyrene. Because of the low dielectric constant of these films, their volumetric efficiency is quite low, and a 10 μF/50 V polyester capacitor (for example) is actually a handful. Metalized (as opposed to foil) electrodes do help to reduce size, but even the highest dielectric constant units among film types (polyester, polycarbonate) are still larger than any electrolytic, even using the thinnest films with the lowest voltage ratings (50 V). Where film types excel is in their low dielectric losses, a factor which may not necessarily be a practical advantage for filtering switchers. For example, ESR in film capacitors can be as low as 10 mΩ or less, and the behavior of films generally is very high in terms of Q. In fact, this can cause problems of spurious resonance in filters, requiring damping components.

Typically using a wound layer-type construction, film capacitors can be inductive, which can limit their effectiveness for high frequency filtering. Obviously, only non-inductively made film caps are useful for switching regulator filters. One specific style which is non-inductive is the *stacked film* type, where the capacitor plates are cut as small overlapping linear sheet sections from a much larger wound drum of dielectric/plate material. This technique offers the low inductance attractiveness of a plate sheet style capacitor with conventional leads (see References 22–24). Obviously, minimal lead length should be used for best high frequency effectiveness. Very high current polycarbonate film types are also available, specifically designed for switching power supplies, with a variety of low inductance terminations to minimize ESL (Reference 25).

Dependent on their electrical and physical size, film capacitors can be useful at frequencies to well above 10 MHz. At the highest frequencies, only stacked film types should be considered. Some manufacturers are now supplying film types in leadless surface mount packages, which eliminates the lead length inductance.

*Ceramic* is often the capacitor material of choice above a few MHz, due to its compact size, low loss, and availability up to several μF in the high-K dielectric formulations (X7R and Z5U), at voltage ratings up to 200 V (see ceramic families of Reference 21). NP0 (also called COG) types use a lower dielectric constant formulation, and have nominally zero TC, plus a low voltage coefficient (unlike the less stable high-K types). NP0 types are limited to values of 0.1 μF or less, with 0.01 μF representing a more practical upper limit.

Multi-layer ceramic "chip caps" are very popular for bypassing/filtering at 10 MHz or more, simply because their very low inductance design allows near optimum radio frequency (RF) bypassing. For smaller values, ceramic chip caps have an operating frequency range to 1 GHz. For high frequency applications, a useful selection can be ensured by selecting a value which has a self-resonant frequency *above* the highest frequency of interest.

The *ripple current* rating of electrolytic capacitors must not be ignored in switching regulator applications because, unlike linear regulators, switching regulators subject capacitors to large AC currents. AC currents can cause heating in the dielectric material and change the temperature-dependent characteristics of the capacitor. Also, the capacitor is more likely to fail at the higher temperatures produced by the ripple current. Fortunately, most manufacturers provide ripple current ratings, and this problem can be averted if understood.

Calculating the exact ripple current can be tedious, especially with complex switching regulator waveforms. Simple approximations can be made, however, which are sufficiently accurate. Consider first the buck converter input and output currents (refer to Figure 9-59). The RMS input capacitor ripple current can be approximated by a square wave having a peak-to-peak amplitude equal to $I_{OUT}$. The RMS value of this

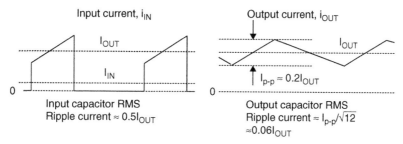

**Figure 9-59: Buck converter input and output capacitors RMS ripple current approximations**

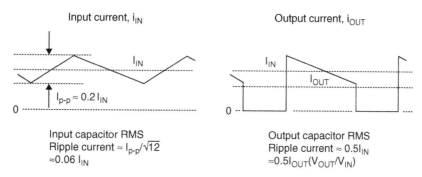

**Figure 9-60: Boost converter input and output capacitors RMS ripple current approximations**

square wave is therefore $I_{OUT}/2$. The output capacitor current waveform can be approximated by a sawtooth waveform having a peak-to-peak amplitude of $0.2 \times I_{OUT}$. The RMS value of this sawtooth is therefore approximately $0.2 \times I_{OUT}/12$, or $0.06 \times I_{OUT}$.

Similarly for a boost converter (see waveforms shown in Figure 9-60), the input capacitor RMS ripple current is $0.06 \times I_{IN}$, and the RMS output current ripple is $0.5 \times I_{IN}$. These boost converter expressions can also be expressed in terms of the output current, $I_{OUT}$, using the relationship, $I_{IN} = I_{OUT}(V_{OUT}/V_{IN})$. In any case, the minimum expected value of input voltage should be used which will result in the largest value of input current.

In practice, a safety factor of 25% should be added to the above approximations for further derating. In practical applications, especially those using surface mount components, it may be impossible to meet the capacitance value, ESR, and ripple current requirement using a single capacitor. Paralleling a number of equal value capacitors is a viable option which will increase the effective capacitance and reduce ESR, ESL. In addition, the ripple current is divided between the individual capacitors.

Several electrolytic capacitor manufacturers offer low ESR surface mount devices including the AVX TPS series (Reference 27), and the Sprague 595D series (Reference 28). Low ESR through-hole electrolytic capacitors are the HFQ series from Panasonic (Reference 29) and the OS-CON series from Sanyo (Reference 30).

## Switching Regulator Output Filtering

In order to minimize switching regulator output voltage ripple it is often necessary to add additional filtering. In many cases, this is more efficient than simply adding parallel capacitors to the main output capacitor to reduce ESR.

Output ripple current in a boost converter is pulsating, while that of a buck converter is a sawtooth. In any event, the high frequency components in the output ripple current can be removed with a small inductor (2–10 µH or so followed by a low ESR capacitor). Figure 9-61 shows a simple LC filter on the output of a switching regulator whose switching frequency is f. Generally the actual value of the filter capacitor is not as important as its ESR when filtering the switching frequency ripple. For instance, the reactance of a 100 µF capacitor at 100 kHz is approximately 0.016 Ω, which is much less than available ESRs.

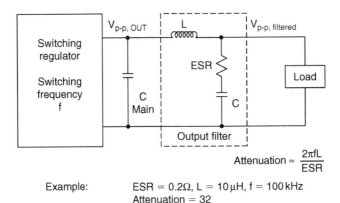

$$\text{Attenuation} \approx \frac{2\pi f L}{ESR}$$

Example:    ESR = 0.2 Ω, L = 10 µH, f = 100 kHz
            Attenuation = 32

**Figure 9-61: Switching regulator output filtering**

The capacitor ESR and the inductor reactance attenuate the ripple voltage by a factor of approximately $2\pi fL/ESR$. The example shown in Figure 9-61 uses a 10 µH inductor and a capacitor with an ESR of 0.2 Ω. This combination attenuates the output ripple by a factor of about 32.

The inductor core material is not critical, but it should be rated to handle the load current. Also, its DC resistance should be low enough so that the load current does not cause a significant voltage drop across it.

## Switching Regulator Input Filtering

The input ripple current in a buck converter is pulsating, while that of a boost converter is a sawtooth. Additional filtering may be required to prevent the switching frequency and the other higher frequency components from affecting the main supply ripple current.

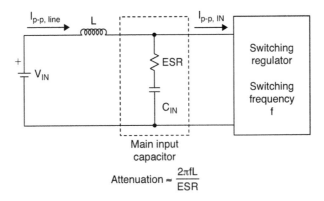

$$\text{Attenuation} \approx \frac{2\pi f L}{ESR}$$

**Figure 9-62: Switching regulator input filtering**

738

This is easily accomplished by the addition of a small inductor in series with the main input capacitor of the regulator as shown in Figure 9-62. The reactance of the inductor at the switching frequency forms a divider with the ESR of the input capacitor. The inductor will block both low and high frequency components from the main input voltage source. The attenuation of the ripple current at the switching frequency, f, is approximately $2\pi fL/ESR$.

### References: Switch Mode Regulators

1.  I.M. Gottlieb, **Power Supplies, Switching Regulators, Inverters, and Converters**, 2nd Edition, McGraw-Hill (TAB Books), New York, 1994.

2.  M. Brown, **Practical Switching Power Supply Design**, Academic Press, Burlington, MA, 1990.

3.  M. Brown, **Power Supply Cookbook**, Butterworth-Heinemann, Oxford, 1994.

4.  J.D. Lenk, **Simplified Design of Switching Power Supplies**, Butterworth-Heinemann, Oxford, 1995.

5.  K. Billings, **Switchmode Power Supply Handbook**, McGraw-Hill, New York, 1989.

6.  G. Chryssis, **High-Frequency Switching Power Supplies: Theory and Design**, 2nd Edition, McGraw-Hill, New York, 1989.

7.  A.I. Pressman, **Switching Power Supply Design**, McGraw-Hill, New York, 1991.

8.  "Tantalum Electrolytic and Ceramic Families," Kemet Electronics, Box 5828, Greenville, SC, 29606, 803-963-6300.

9.  Type HFQ Aluminum Electrolytic Capacitor and Type V Stacked Polyester Film Capacitor, Panasonic, 2 Panasonic Way, Secaucus, NJ, 07094, 201-348-7000.

10. OS-CON Aluminum Electrolytic Capacitor 93/94 Technical Book, Sanyo, 3333 Sanyo Road, Forest City, AK, 72335, 501-633-6634.

11. I. Clelland, "Metalized Polyester Film Capacitor Fills High Frequency Switcher Needs," **PCIM**, June, 1992.

12. Type 5MC Metallized Polycarbonate Capacitor, Electronic Concepts, Inc., Box 1278, Eatontown, NJ, 07724, 908-542-7880.

13. W. Jung and D. Marsh, "Picking Capacitors, Parts 1 and 2," Audio, February, March, 1980.

### Capacitor Manufacturers

14. AVX Corporation, 801 17th Avenue S., Myrtle Beach, SC 29577, 803-448-9411.

15. Sprague, 70 Pembroke Road, Concord, NH 03301, 603-224-1961.

16. Panasonic, 2 Panasonic Way, Secaucus, NJ 07094, 201-392-7000.

17. Sanyo Corporation, 2001 Sanyo Avenue, San Diego, CA 92173, 619-661-6835.

18. Kemet Electronics, Box 5828, Greenville, SC 29606, 803-963-6300.

## Inductor Manufacturers

19.   Coiltronics, 6000 Park of Commerce Blvd., Boca Raton, FL 33487, 407-241-7876.

20.   Sumida, 5999 New Wilke Road. Suite 110, Rolling Meadow, IL 60008, 847-956-0666.

21.   Pulse Engineering, 12220 World Trade Drive, San Diego, CA 92128, 619-674-8100.

22.   Gowanda Electronics, 1 Industrial Place, Gowanda, NY 14070, 716-532-2234.

23.   Coilcraft, 1102 Silver Lake Road, Cary, IL 60013, 847-639-2361.

24.   Dale Electronics, Inc., E. Highway 50, P.O. Box 180, Yankton, SD 57078, 605-665-9301.

25.   Hurricane Electronics Lab, 331 N. 2260 West, P.O. Box 1280, Hurricane, UT 84737, 801-635-2003.

## Core Manufacturers

26.   Magnetics, P.O. Box 391, Butler, PA 16003, 412-282-8282.

## MOSFET Manufacturers

27.   International Rectifier, 233 Kansas Street, El Segundo, CA 90245, 310-322-3331.

28.   Motorola Semiconductor, 3102 North 56th Street, MS56-126, Phoenix, AZ 85018, 800-521-6274.

29.   Siliconix Inc., 2201 Laurelwood Road, P.O. Box 54951, Santa Clara, CA 95056, 408-988-8000.

## Schottky Diode Manufacturers

30.   General Instrument, Power Semiconductor Division, 10 Melville Park Road, Melville, NY 11747, 516-847-3000.

31.   International Rectifier, 233 Kansas Street, El Segundo, CA 90245, 310-322-3331.

32.   Motorola Semiconductor, 3102 North 56th Street, MS56-126, Phoenix, AZ 85018, 800-521-6274.

# Switched Capacitor Voltage Converters

## Introduction

In the previous section, we saw how inductors can be used to transfer energy and perform voltage conversions. This section examines switched capacitor voltage converters which accomplish energy transfer and voltage conversion using capacitors.

The two most common switched capacitor voltage converters are the *voltage inverter* and the *voltage doubler* circuit shown in Figure 9-63. In the voltage inverter, the charge pump capacitor, $C_1$, is charged to the input voltage during the first half of the switching cycle. During the second half of the switching cycle, its voltage is inverted and applied to capacitor $C_2$ and the load. The output voltage is the negative of the input voltage, and the average input current is approximately equal to the output current. The switching frequency impacts the size of the external capacitors required, and higher switching frequencies allow the use of smaller capacitors. The duty cycle—defined as the ratio of charging time for $C_1$ to the entire switching cycle time—is usually 50%, because that generally yields the optimal charge transfer efficiency.

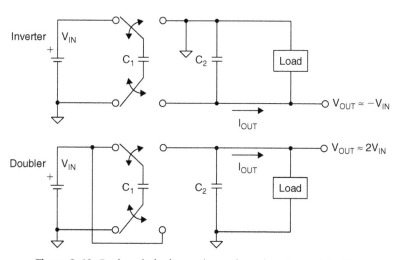

Figure 9-63: Basic switched capacitor voltage inverter and doubler

After initial start-up transient conditions and when a steady-state condition is reached, the charge pump capacitor has to supply only a small amount of charge to the output capacitor on each switching cycle. The amount of charge transferred depends on the load current and the switching frequency. During the time the pump capacitor is charged by the input voltage, the output capacitor $C_2$ must supply the load current. The load current flowing out of $C_2$ causes a droop in the output voltage which corresponds to a component of output voltage ripple. Higher switching frequencies allow smaller capacitors for the same amount of droop. There are, however, practical limitations on the switching speeds and switching losses, and switching frequencies are generally limited to a few hundred kHz.

The voltage doubler works similarly to the inverter; however, the pump capacitor is placed in series with the input voltage during its discharge cycle, thereby accomplishing the voltage doubling function. In the voltage doubler, the average input current is approximately twice the average output current.

The basic inverter and doubler circuits provide no output voltage regulation; however, techniques exist to add regulated capability and have been implemented in the ADP3603/ADP3604/ADP3605/ADP3607.

There are certain advantages and disadvantages of using switched capacitor techniques rather than inductor-based switching regulators (Figure 9-64). An obvious key advantage is the elimination of the inductor and the related magnetic design issues. In addition, these converters typically have relatively low noise and minimal radiated EMI. Application circuits are simple, and usually only two or three external capacitors are required. Because there is no need for an inductor, the final PCB component height can generally be made smaller than a comparable switching regulator. This is important in many applications such as display panels.

Switched capacitor inverters are low cost and compact and are capable of achieving efficiencies greater than 90%. Obviously, the current output is limited by the size of the capacitors and the current carrying capacity of the switches. Typical IC switched capacitor inverters have maximum output currents of about 150 mA maximum.

- No inductors!
- Minimal radiated EMI
- Simple implementation: Only 2 external capacitors (plus an input capacitor if required)
- Efficiency > 90% achievable
- Optimized for doubling or inverting supply voltage—efficiency degrades for other output voltages
- Low cost, compact, low profile (height)
- Parts with voltage regulation are available: ADP3603/ADP3604/ADP3605/ADP3607

**Figure 9-64: Advantages of switched capacitor voltage converters**

Switched capacitor voltage converters do not maintain high efficiency for a wide range of ratios of input to output voltages, unlike their switching regulator counterparts. Because the input to output current ratio is scaled according to the basic voltage conversion (i.e., doubled for a doubler, inverted for an inverter) regardless of whether or not regulation is used to reduce the doubled or inverted voltage, any output voltage magnitude less than $2 \times V_{IN}$ for a doubler or less than $|V_{IN}|$ for an inverter will result in additional power dissipation within the converter, and efficiency will be degraded proportionally.

The voltage inverter is useful where a relatively low current negative voltage is required in addition to the primary positive voltage. This may occur in a single supply system where only a few high performance parts require the negative voltage. Similarly, voltage doublers are useful in low current applications where a voltage greater than the primary supply voltage is required.

## Charge Transfer Using Capacitors

A fundamental understanding of capacitors (theoretical and real) is required in order to master the subtleties of switched capacitor voltage converters. Figure 9-65 shows the theoretical capacitor and its real world counterpart. If the capacitor is charged to a voltage V, then the total charge stored in the capacitor, q, is given by q = CV. Real capacitors have ESR and ESL as shown in the diagram, but these parasitics do not affect the ability of the capacitor to store charge. They can, however, have a large effect on the overall efficiency of the switched capacitor voltage converter.

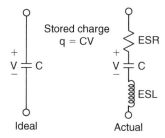

**Figure 9-65: Stored charge in a capacitor**

If an ideal capacitor is charged with an ideal voltage source as shown in Figure 9-66(A), the capacitor charge buildup occurs instantaneously, corresponding to a unit impulse of current. A practical circuit (Figure 9-66(B)) will have resistance in the switch ($R_{SW}$) as well as the ESR of the capacitor. In addition, the capacitor has an ESL. The charging current path also has an effective series inductance which can be minimized with proper component layout techniques. These parasitics serve to limit the peak current, and also increase the charge transfer time as shown in the diagram. Typical switch resistances can range from 1 to 50$\Omega$, and ESRs between 50 and 200 m$\Omega$. Typical capacitor values may range from about 0.1 to 10 µF, and typical ESL values 1 to 5 nH. Although the equivalent RLC circuit of the capacitor can be underdamped or overdamped, the relatively large switch resistance generally makes the final output voltage response overdamped.

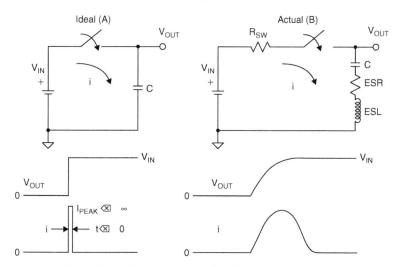

**Figure 9-66: Charging a capacitor from a voltage source**

The law of conservation of charge states that if two capacitors are connected together, the total charge on the parallel combination is equal to the sum of the original charges on the capacitors. Figure 9-67 shows two capacitors, $C_1$ and $C_2$, each charged to voltages V1 and V2, respectively. When the switch is closed, an impulse of current flows, and the charge is redistributed. The total charge on the parallel combination of the two capacitors is $q_T = C_1 \times V1 + C_2 \times V2$. This charge is distributed between the two capacitors, so the new voltage, $V_T$, across the parallel combination is equal to $qT/(C_1 + C_2)$, or

$$V_T = \frac{q_T}{C_1 + C_2} = \frac{C_1\,V1 + C_2\,V2}{C_1 + C_2} = \left(\frac{C_1}{C_1 + C_2}\right)V1 + \left(\frac{C_2}{C_1 + C_2}\right)V2 \qquad (9\text{-}73)$$

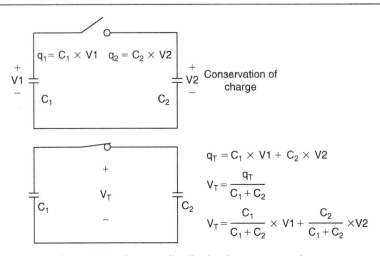

**Figure 9-67: Charge redistribution between capacitors**

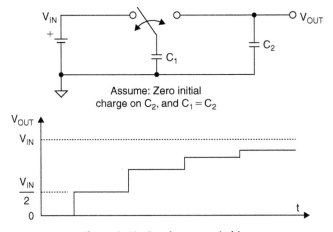

**Figure 9-68: Continuous switching**

This principle may be used in the simple charge pump circuit shown in Figure 9-68. Note that this circuit is neither a doubler nor inverter, but only a voltage replicator. The pump capacitor is $C_1$, and the initial charge on $C_2$ is zero. The pump capacitor is initially charged to $V_{IN}$. When it is connected to $C_2$, the charge is redistributed, and the output voltage is $V_{IN}/2$ (assuming $C_1 = C_2$). On the second transfer cycle, the output voltage is pumped to $V_{IN}/2 + V_{IN}/4$. On the third transfer cycle, the output voltage is pumped to $V_{IN}/2 + V_{IN}/4 + V_{IN}/8$. The waveform shows how the output voltage exponentially approaches $V_{IN}$.

Figure 9-69 shows a pump capacitor, $C_1$, switched continuously between the source, V1, and $C_2$ in parallel with the load. The conditions shown are after a steady-state condition has been reached. The charge transferred each cycle is $\Delta q = C_1(V1 - V2)$. This charge is transferred at the switching frequency, f. This corresponds to an average current (current = charge transferred per unit time) of:

$$I = f \, \Delta q = f \, C_1(V1 - V2) \qquad (9\text{-}74)$$

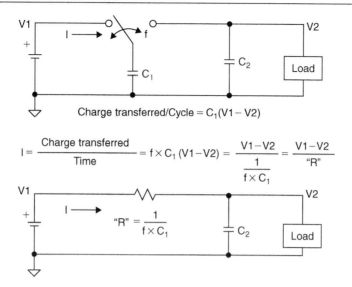

**Figure 9-69: Continuous switching, steady state**

or

$$I = \frac{V1 - V2}{\frac{1}{f}\, C_1} \tag{9-75}$$

Notice that the quantity, $1/f \cdot C_1$, can be considered an equivalent resistance, "R," connected between the source and the load. The power dissipation associated with this virtual resistance, "R," is essentially forced to be dissipated in the switch on-resistance and the capacitor ESR, regardless of how low those values are reduced. (It should be noted that capacitor ESR and the switch on-resistance cause additional power losses as will be discussed shortly.)

In a typical switched capacitor voltage inverter, a capacitance of $10\,\mu F$ switched at $100\,kHz$ corresponds to "R" $= 1\,\Omega$. Obviously, minimizing "R" by increasing the frequency minimizes power loss in the circuit. However, increasing switching frequency tends to increase switching losses. The optimum switched capacitor operating frequency is therefore highly process and device dependent. Therefore, specific recommendations are given in the data sheet for each device.

### Unregulated Switched Capacitor Inverter and Doubler Implementations

An unregulated switched capacitor inverter implementation is shown in Figure 9-70. Notice that the SPDT switches (shown in previous diagrams) actually comprise two SPST switches. The control circuit consists of an oscillator and the switch drive signal generators. Most IC switched capacitor inverters and doublers contain all the control circuits as well as the switches and the oscillator. The pump capacitor, $C_1$, and the load capacitor, $C_2$, are external. Not shown in the diagram is a capacitor on the input which is generally required to ensure low source impedance at the frequencies contained in the switching transients.

The switches used in IC switched capacitor voltage converters may be CMOS or bipolar as shown in Figure 9-71. Standard CMOS processes allow low on-resistance MOSFET switches to be fabricated along with the oscillator and other necessary control circuits. Bipolar processes can also be used, but add cost and increase power dissipation.

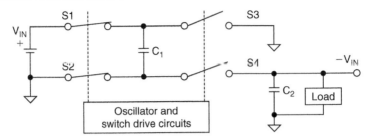

**Figure 9-70: Switched capacitor voltage inverter**

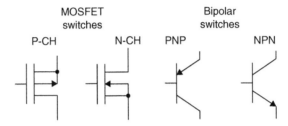

**Figure 9-71: Switches used in voltage converters**

## Voltage Inverter and Doubler Dynamic Operation

The steady-state current and voltage waveforms for a switched capacitor voltage inverter are shown in Figure 9-72. The average value of the input current waveform (A) must be equal to $I_{OUT}$. When the pump capacitor is connected to the input, a charging current flows. The initial value of this charging current depends on the initial voltage across $C_1$, the ESR of $C_1$, and the resistance of the switches. The switching frequency, switch resistance, and the capacitor ESRs generally limit the peak amplitude of the charging current to less than $2.5 \times I_{OUT}$. The charging current then decays exponentially as $C_1$ is charged. The waveforms in Figure 9-72 assume that the time constant due to capacitor $C_1$, the switch resistance, and the ESR of $C_1$ is several times greater than the switching period ($1/f$). Smaller time constants will cause the peak currents to increase as well as increase the slopes of the charge/discharge waveforms. Long time constants cause longer start-up times and require larger and more costly capacitors. For the conditions shown in Figure 9-72(A), the peak value of the input current is only slightly greater than $2 \times I_{OUT}$.

The output current waveform of $C_1$ is shown in Figure 9-72(B). When $C_1$ is connected to the output capacitor, the step change in the output capacitor current is approximately $2 \times I_{OUT}$. This current step therefore creates an output voltage step equal to $2 \times I_{OUT} \times ESR_{C_2}$ as shown in Figure 9-72(C). After the step change, $C_2$ charges linearly by an amount equal to $I_{OUT}/2f \times C_2$. When $C_1$ is connected back to the input, the ripple waveform reverses direction as shown in the diagram. The total peak-to-peak output ripple voltage is therefore:

$$V_{RIPPLE} \approx 2I_{OUT} \, ESR_{C_2} + \frac{I_{OUT}}{2f \, C_2} \qquad (9\text{-}76)$$

The current and voltage waveforms for a simple voltage doubler are shown in Figure 9-73 and are similar to those of the inverter. Typical voltage ripple for practical switched capacitor voltage inverter/doublers range from 25 to 100 mV, but can be reduced by filtering techniques.

Note that the input current waveform has an average value of $2 \times I_{OUT}$ because $V_{IN}$ is connected to $C_1$ during $C_1$'s charge cycle and to the load during $C_1$'s discharge cycle. The expression for the ripple voltage is identical to that of the voltage inverter.

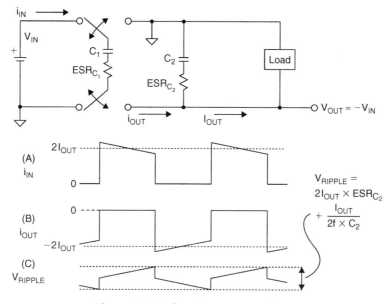

**Figure 9-72: Voltage inverter waveforms**

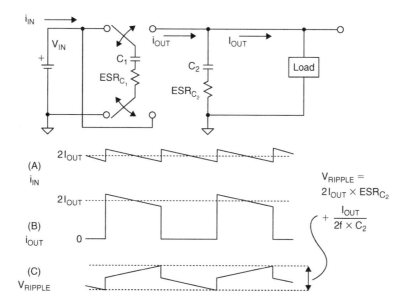

**Figure 9-73: Voltage doubler waveforms**

## Switched Capacitor Voltage Converter Power Losses

The various sources of power loss in a switched capacitor voltage inverter are shown in Figure 9-74. In addition to the inherent switched capacitor resistance, "R" = $1/f \times C_1$, there are resistances associated with each switch, as well as the ESRs of the capacitors. The quiescent power dissipation, $I_q \times V_{IN}$, must also be included, where $I_q$ is the quiescent current drawn by the IC itself.

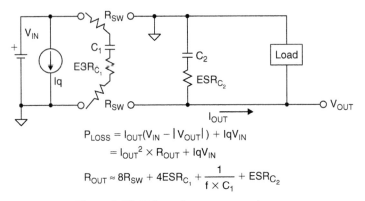

$$P_{LOSS} = I_{OUT}(V_{IN} - |V_{OUT}|) + I_q V_{IN}$$
$$= I_{OUT}{}^2 \times R_{OUT} + I_q V_{IN}$$
$$R_{OUT} \approx 8R_{SW} + 4ESR_{C_1} + \frac{1}{f \times C_1} + ESR_{C_2}$$

**Figure 9-74: Voltage inverter power losses**

The power dissipated in the switching arm is first calculated. When $C_1$ is connected to $V_{IN}$, a current of $2 \times I_{OUT}$ flows through the switch resistances ($2R_{SW}$) and the ESR of $C_1$, $ESR_{C_1}$. When $C_1$ is connected to the output, a current of $2 \times I_{OUT}$ continues to flow through $C_1$, $2R_{SW}$, and $ESR_{C_1}$. Therefore, there is always an RMS current of $2 \times I_{OUT}$ flowing through these resistances, resulting in a power dissipation in the switching arm of:

$$P_{SW} = (2 I_{OUT})^2 + (2 R_{SW} + ESR_{C_1}) = I_{OUT}{}^2 (8R_{SW} + 4ESR_{C_1}) \qquad (9\text{-}77)$$

In addition to these purely resistive losses, an RMS current of $I_{OUT}$ flows through the "resistance" of the switched capacitor, $C_1$, yielding an additional loss of:

$$P_{C_1} = I_{OUT}{}^2 : R_{C_1} = I_{OUT2} \frac{1}{f C_1} \qquad (9\text{-}78)$$

The RMS current flowing through $ESR_{C_2}$ is $I_{OUT}$, yielding a power dissipation of:

$$P_{ESR_{C_2}} = I_{OUT}{}^2 \, ESR_{C_2} \qquad (9\text{-}79)$$

Adding all the resistive power dissipations to the quiescent power dissipation yields:

$$P_{LOSS} = I_{OUT}{}^2 \left( 8R_{SW} + 4ESR_{C_1} + ESR_{C_2} + \frac{1}{f C_1} \right) + I_q V_{IN} \qquad (9\text{-}80)$$

All of the resistive losses can be grouped into an equivalent $R_{OUT}$ as shown in the diagram:

$$R_{OUT} \approx 8R_{SW} + 4ESR_{C_1} + \frac{1}{f} C_1 + ESR_{C_2} \qquad (9\text{-}81)$$

Typical values for switch resistances are between $1 - 20\,\Omega$, and ESRs between 50 and $200\,\text{m}\Omega$. The values of $C_1$ and f are generally chosen such that the term, $1/f \cdot C_1$, is less than $1\,\Omega$. For instance, $10\,\mu\text{F}$ at $100\,\text{kHz}$ yields "R" $= 1\,\Omega$. The dominant sources of power loss in most inverters are therefore the switch resistances and the ESRs of the pump capacitor and output capacitor.

The ADP3603/ADP3604/ADP3605/ADP3607 series regulators have a shutdown control pin which can be asserted when load current is not required. When activated, the shutdown feature reduces quiescent current to a few tens of microamperes.

Power losses in a voltage doubler circuit are shown in Figure 9-75, and the analysis is similar to that of the inverter.

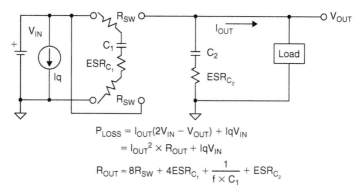

$$P_{LOSS} = I_{OUT}(2V_{IN} - V_{OUT}) + I_qV_{IN}$$
$$= I_{OUT}^2 \times R_{OUT} + I_qV_{IN}$$
$$R_{OUT} \approx 8R_{SW} + 4ESR_{C_1} + \frac{1}{f \times C_1} + ESR_{C_2}$$

**Figure 9-75: Voltage doubler power losses**

## Regulated Output Switched Capacitor Voltage Converters

Adding regulation to the simple switched capacitor voltage converter greatly enhances its usefulness in many applications. There are three general techniques for adding regulation to a switched capacitor converter. The most straightforward is to follow the switched capacitor inverter/doubler with a LDO linear regulator. The LDO provides the regulated output and also reduces the ripple of the switched capacitor converter. This approach, however, adds complexity and reduces the available output voltage by the dropout voltage of the LDO.

Another approach to regulation is to vary the duty cycle of the switch control signal with the output of an error amplifier which compares the output voltage with a reference. This technique is similar to that used in inductor-based switching regulators and requires the addition of a PWM and appropriate control circuitry. However, this approach is highly nonlinear and requires long time constants (i.e., lossy components) in order to maintain good regulation control.

By far the simplest and most effective method for achieving regulation in a switched capacitor voltage converter is to use an error amplifier to control the on-resistance of one of the switches as shown in Figure 9-76, a block diagram of the ADP3603/ADP3604/ADP3605 voltage inverters. These devices offer a regulated $-3$V output for an input voltage of $+4.5$ to $+6$V. The output is sensed and fed back into the device via the $V_{SENSE}$ pin. Output regulation is accomplished by varying the on-resistance of one of the MOSFET switches as shown by control signal labeled "$R_{ON}$ CONTROL" in the diagram. This signal accomplishes the switching of the MOSFET as well as controlling the on-resistance.

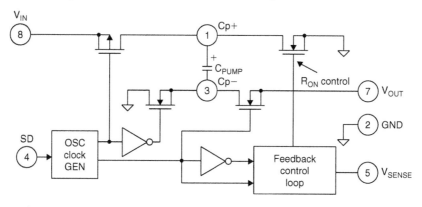

**Figure 9-76: ADP3603/3604/3605 regulated $-3$V output voltage inverters**

A typical application circuit for the ADP3603/ADP3604/ADP3605 series is shown in Figure 9-77. In the normal mode of operation, the SHUTDOWN pin should be connected to ground. The 10 μF capacitors should have ESRs of less than 150 mΩ, and values of 4.7 μF can be used at the expense of slightly higher output ripple voltage. The equations for ripple voltage shown in Figure 9-72 also apply to the ADP3603/ADP3604/ADP3605. Using the values shown, typical ripple voltage ranges from 25 to 60 mV as the output current varies over its allowable range.

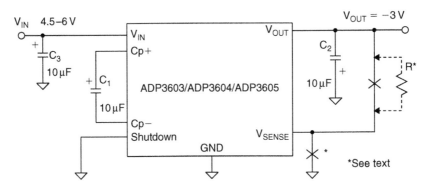

Figure 9-77: ADP3603/3604/3605 application circuit for −3 V operation

The regulated output voltage of the ADP3603/ADP3604/ADP3605 series can varied between −3 V and −$V_{IN}$ by connecting a resistor between the output and the $V_{SENSE}$ pin as shown in the diagram. Regulation will be maintained for output currents up to about 30 mA. The value of the resistor is calculated from the following equation:

$$V_{OUT} = -\left( \frac{R}{5\,k\Omega} + 3\,V \right) \quad (9\text{-}82)$$

The devices can be made to operate as standard inverters providing an unregulated output voltage if the $V_{SENSE}$ pin is simply connected to ground.

A typical application circuit is shown in Figure 9-78. The Schottky diode connecting the input to the output is required for proper operation during start-up and shutdown. If $V_{SENSE}$ is connected to ground, the devices operate as unregulated voltage doublers.

The output voltage of each device can be adjusted with an external resistor. The equation which relates output voltage to the resistor value for the ADP3607 is given by:

$$V_{OUT} = \frac{R}{9.5\,k\Omega} + 1\,V \quad \text{for } V_{OUT} < 2V_{IN} \quad (9\text{-}83)$$

The ADP3607 should be operated with an output voltage of at least 3 V in order to maintain regulation.

Although the ADP3607-5 is optimized for an output voltage of 5 V, its output voltage can be adjusted between 5 V and 2 × $V_{IN}$ with an external resistor using the equation:

$$V_{OUT} = \frac{2R}{9.5\,k\Omega} + 5\,V \quad \text{for } V_{OUT} < 2V_{IN} \quad (9\text{-}84)$$

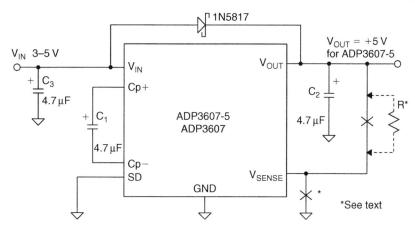

**Figure 9-78: ADP3607 application circuit**

When using either the ADP3607 or the ADP3607-5 in the adjustable mode, the output current should be no greater than 30 mA in order to maintain good regulation.

The circuit shown in Figure 9-79 generates a regulated 12 V output from a 5 V input using the ADP3607-5 in a voltage tripler application. Operation is as follows. First assume that the $V_{SENSE}$ pin of the ADP3607-5 is grounded and that the resistor R is not connected. The output of the ADP3607-5 is an unregulated voltage equal to $2 \times V_{IN}$. The voltage at the Cp+ pin of the ADP3607-5 is a square wave with a minimum value of $V_{IN}$ and a maximum value of $2 \times V_{IN}$. When the voltage at Cp+ is $V_{IN}$, capacitor $C_2$ is charged to $V_{IN}$ (less the D1 diode drop) from $V_{OUT1}$ via diode D1. When the voltage at Cp+ is $2 \times V_{IN}$, the output capacitor $C_4$ is charged to a voltage $3 \times V_{IN}$ (less the diode drops of D1 and D2). The final unregulated output voltage of the circuit, $V_{OUT2}$, is therefore approximately $3 \times V_{IN} - 2 \times V_D$, where $V_D$ is the Schottky diode voltage drop.

The addition of the feedback resistor, R, ensures that the output is regulated for values of $V_{OUT2}$ between $2 \times V_{IN} - 2 \times V_D$ and $3 \times V_{IN} - 2 \times V_D$. Choosing $R = 33.2\,k\Omega$ yields an output voltage $V_{OUT2}$ of $+12\,V$ for a nominal input voltage of $+5\,V$. Regulation is maintained for output currents up to approximately 20 mA.

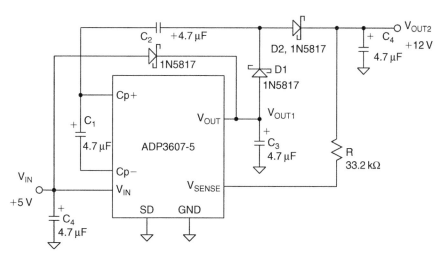

**Figure 9-79: Regulated +12 V from a +5 V input**

*751*

# Passive Components

## Chapter Introduction

When designing precision analog circuits, it is critical that users avoid the pitfall of poor passive component choice. In fact, the wrong passive component can derail even the best op amp or data converter application. This section includes discussion of some basic traps of choosing passive components.

So, you have spent good money for a precision op amp or data converter, only to find that, when plugged into your board, the device does not meet spec. Perhaps the circuit suffers from drift, poor frequency response, and oscillations—or simply does not achieve expected accuracy. Well, before you blame the device, you should closely examine your passive components—including capacitors, resistors, potentiometers, and yes, even the printed circuit boards. In these areas, subtle effects of tolerance, temperature, parasitics, aging, and user assembly procedures can unwittingly sink your circuit. And all too often these effects go unspecified (or underspecified) by passive component manufacturers.

In general, if you use data converters having 12 bits or more of resolution, or high precision op amps, pay very close attention to passive components. Consider the case of a 12-bit DAC, where 1/2 LSB corresponds to 0.012% of full scale, or only 122 ppm. A host of passive component phenomena can accumulate errors far exceeding this! But, buying the most expensive passive components will not necessarily solve your problems either. Often, a *correct* 25-cent capacitor yields a better-performing, more cost-effective design than a premium-grade (expensive) part. With a few basics, understanding and analyzing passive components may prove rewarding, albeit not easy.

## SECTION 10-1

# *Capacitors*

## Basics

A capacitor is a passive electronic component that stores energy in the form of an electrostatic field. In its simplest form, a capacitor consists of two conducting plates separated by an insulating material called the dielectric. The capacitance is directly proportional to the surface areas of the plates, and is inversely proportional to the separation between the plates. Capacitance also depends on the dielectric constant of the substance separating the plates.

Capacitive reactance is defined as:

$$X_c = 1/\omega C = 1/2\,\pi f C \qquad (10\text{-}1)$$

where $X_C$ is the capacitive reactance, $\omega$ is the angular frequency, f is the frequency in Hertz, and C is the capacitance.

Capacitive reactance is the negative imaginary component of impedance.

The complex impedance of an inductor is then given by:

$$Z = 1/j\omega C = 1/j2\,\pi fC \tag{10-2}$$

where j is the imaginary number.

$$j = \sqrt{-1} \tag{10-3}$$

## Dielectric Types

There are many different types of capacitors, and an understanding of their individual characteristics is absolutely mandatory to the design of practical circuits. A thumbnail sketch of capacitor characteristics is shown in the chart of Figure 10-1. Background and tutorial information on capacitors can be found in Reference 2 and many vendor catalogs.

With any dielectric, a major potential filter loss element is ESR (equivalent series resistance), the net parasitic resistance of the capacitor. ESR provides an ultimate limit to filter performance, and requires more than casual consideration, because it can vary with both frequency and temperature in some types. Another capacitor loss element is ESL (equivalent series inductance). ESL determines the frequency where the net impedance of the capacitor switches from a capacitive to inductive characteristic. This varies from as low as 10 kHz in some electrolytics to as high as 100 MHz or more in chip ceramic types. Both ESR and ESL are minimized when a leadless package is used, and all capacitor types discussed here are available in surface mount packages, which are preferable for high speed uses.

The *electrolytic* family provides an excellent, cost-effective low-frequency filter component, because of the wide range of values, a high capacitance-to-volume ratio, and a broad range of working voltages. It includes *general purpose aluminum electrolytic* types, available in working voltages from below 10 V up to about 500 V, and in size from 1 to several thousand μF (with proportional case sizes). All electrolytic capacitors are polarized, and thus cannot withstand more than a volt or so of reverse bias without damage. They have relatively high leakage currents (this can be tens of μA, but is strongly dependent on specific family design, electrical size, and voltage rating versus applied voltage). However, this is not likely to be a major factor for basic filtering applications.

Also included in the electrolytic family are *tantalum* types, which are generally limited to voltages of 100 V or less, with capacitance of 500 μF or less. In a given size, tantalums exhibit a higher capacitance-to-volume ratios than do the general purpose electrolytics, and have both a higher frequency range and lower ESR. They are generally more expensive than standard electrolytics, and must be carefully applied with respect to surge and ripple currents.

A subset of aluminum electrolytic capacitors is the *switching* type, which is designed and specified for handling high pulse currents at frequencies up to several hundred kHz with low losses. This type of capacitor competes directly with the tantalum type in high frequency filtering applications and has the advantage of a much broader range of available values.

More recently, high performance aluminum electrolytic capacitors using an organic semiconductor electrolyte have appeared. These *OS-CON* families of capacitors feature appreciably lower ESR and higher frequency range than do the other electrolytic types, with an additional feature of low-temperature ESR degradation.

| Type | Typical DA | Advantages | Disadvantages |
|------|-----------|------------|---------------|
| Polystyrene | 0.001% – 0.02% | Inexpensive<br>Low DA<br>Good stability (~120 ppm/°C) | Damaged by temperature > +85°C<br>Large<br>High inductance<br>Vendors limited |
| Polypropylene | 0.001% – 0.02% | Inexpensive<br>Low DA<br>Stable (~200 ppm/°C)<br>Wide range of values | Damaged by temperature > +105°C<br>Large<br>High inductance |
| Teflon | 0.003% – 0.02% | Low DA available<br>Good stability<br>Operational above +125°C<br>Wide range of values | Expensive<br>Large<br>High inductance |
| Polycarbonate | 0.1% | Good stability<br>Low cost<br>Wide temperature range<br>Wide range of values | Large<br>DA limits to 8-bit applications<br>High inductance |
| Polyester | 0.3% – 0.5% | Moderate stability<br>Low cost<br>Wide temperature range<br>Low inductance (stacked film) | Large<br>DA limits to 8-bit applications<br>High inductance (conventional) |
| NP0 ceramic | <0.1% | Small case size<br>Inexpensive, many vendors<br>Good stability (30 ppm/°C)<br>1% values available<br>Low inductance (chip) | DA generally low (may not be specified)<br>Low maximum values (≤10 nF) |
| Monolithic ceramic (High K) | >0.2% | Low inductance (chip)<br>Wide range of values | Poor stability<br>Poor DA<br>High voltage coefficient |
| Mica | >0.003% | Low loss at HF<br>Low inductance<br>Good stability<br>1% values available | Quote large<br>Low maximum values (≤10 nF)<br>Expensive |
| Aluminum electrolytic | Very high | Large values<br>High currents<br>High voltages<br>Small size | High leakage<br>Usually polarized<br>Poor stability, accuracy<br>Inductive |
| Tantalum electrolytic | Very high | Small size<br>Large values<br>Medium inductance | High leakage<br>Usually polarized<br>Expensive<br>Poor stability, accuracy |

**Figure 10-1: Capacitor comparison chart**

*Film* capacitors are available in very broad ranges of values and an array of dielectrics, including polyester, polycarbonate, polypropylene, and polystyrene. Because of the low dielectric constant of these films, their volumetric efficiency is quite low, and a 10 μF/50 V polyester capacitor (for example) is actually a handful. Metalized (as opposed to foil) electrodes do help to reduce size, but even the highest dielectric constant units among film types (polyester, polycarbonate) are still larger than any electrolytic, even using the thinnest films with the lowest voltage ratings (50 V). Where film types excel is in their low dielectric losses, a factor which may not necessarily be a practical advantage for filtering switchers. For example, ESR in film capacitors can be as low as $10\,\mathrm{m}\Omega$ or less, and the behavior of films generally is very high in terms of Q. In fact, this can cause problems of spurious resonance in filters, requiring damping components.

Typically using a wound layer-type construction, film capacitors can be inductive, which can limit their effectiveness for high frequency filtering. Obviously, only non-inductively made film caps are useful for switching regulator filters. One specific style which is non-inductive is the *stacked-film* type, where the capacitor plates are cut as small overlapping technique offers the low inductance attractiveness of a plate sheet style capacitor with conventional leads. Obviously, minimal lead length should be used for best high frequency effectiveness. Very high current polycarbonate film types are also available, specifically designed for switching power supplies, with a variety of low inductance terminations to minimize ESL. Linear sheet sections from a much larger wound drum of dielectric/plate material.

Typically using a wound layer-type construction, film capacitors can be inductive, which can limit their effectiveness for high frequency filtering. Obviously, only non-inductively made film caps are useful for switching regulator filters. One specific style which is non-inductive is the *stacked-film* type, where the capacitor plates are cut as small overlapping linear sheet sections from a much larger wound drum of dielectric/plate material. This technique offers the low inductance attractiveness of a plate sheet style capacitor with conventional leads. Obviously, minimal lead length should be used for best high frequency effectiveness. Very high current polycarbonate film types are also available, specifically designed for switching power supplies, with a variety of low inductance terminations to minimize ESL.

Dependent on their electrical and physical size, film capacitors can be useful at frequencies to well above $10\,\mathrm{MHz}$. At very high frequencies, only stacked-film types should be considered. Some manufacturers are also supplying film types in leadless surface mount packages, which eliminate the lead length inductance.

Ceramic is often the capacitor material of choice above a few MHz, due to its compact size and low loss. But the characteristics of ceramic dielectrics vary widely. Some types are better than others for various applications, especially power supply decoupling. Ceramic dielectric capacitors are available in values up to several µF in the high-K dielectric formulations of X7R and Z5U, at voltage ratings up to $200\,\mathrm{V}$. NP0 (also called COG) types use a lower dielectric constant formulation, and have nominally zero TC, plus a low voltage coefficient (unlike the less stable high-K types). The NP0 types are limited in available values to $0.1\,\mu\mathrm{F}$ or less, with $0.01\,\mu\mathrm{F}$ representing a more practical upper limit.

Multilayer ceramic "chip caps" are increasingly popular for bypassing and filtering at $10\,\mathrm{MHz}$ or more, because their very low inductance design allows near optimum RF bypassing. In smaller values, ceramic chip caps have an operating frequency range to $1\,\mathrm{GHz}$. For these and other capacitors for high frequency applications, a useful value can be ensured by selecting a value which has a self-resonant frequency *above* the highest frequency of interest.

All capacitors have some finite ESR. In some cases, the ESR may actually be helpful in reducing resonance peaks in filters, by supplying "free" damping. For example, in general purpose, tantalum, and switching type electrolytics, a broad series resonance region is noted in an impedance versus frequency plot. This occurs where |Z| falls to a minimum level, which is nominally equal to the capacitor's ESR at that frequency. In an example below, this low Q resonance is noted to encompass quite a wide frequency range, several octaves in fact. Contrasted to the very high Q sharp resonances of film and ceramic caps, this low Q behavior can be useful in controlling resonant peaks.

In most electrolytic capacitors, ESR degrades noticeably at low temperature, by as much as a factor of 4–6 times at $-55^{\circ}\mathrm{C}$ versus the room temperature value. For circuits where a high level of ESR is critical, this can lead to problems. Some specific electrolytic types do address this problem, for example within the HFQ switching types, the $-10°C$ ESR at $100\,kHz$ is no more than $2\times$ that at room temperature. The OSCON electrolytics have an ESR versus temperature characteristic which is relatively flat.

Figure 10-2 illustrates the high frequency impedance characteristics of a number of electrolytic capacitor types, using nominal $100\,\mu F/20\,V$ samples. In these plots, the impedance, $|Z|$, versus frequency over the $20\,Hz$–$200\,kHz$ range is displayed using a high resolution 4-terminal setup. Shown in this display are performance samples for a $100\,\mu F/25\,V$ general purpose aluminum unit, a $120\,\mu F/25\,V$ HFQ unit, a $100\,\mu F/20\,V$ tantalum bead type, and a $100\,\mu F/20\,V$ OS-CON unit (lowest curve at right). While the HFQ and tantalum samples are close in $100\,kHz$ impedance, the general purpose unit is about 4 times worse. The OS-CON unit is nearly an order of magnitude lower in $100\,kHz$ impedance than the tantalum and switching electrolytic types.

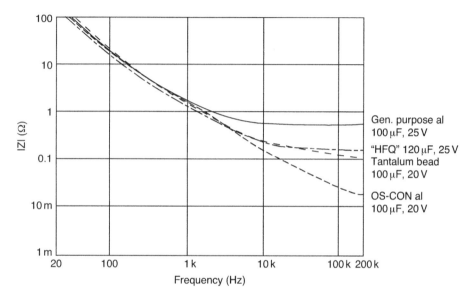

**Figure 10-2: Impedance Z($\Omega$) versus frequency for $100\,\mu F$ electrolytic capacitors (AC current = $50\,mA$ RMS)**

As noted above, all real capacitors have parasitic elements which limit their performance. As an insight into why the impedance curves of Figure 10-2 appear the way they do, a (simplified) model for a $100\,\mu F/20\,V$ tantalum capacitor is shown in Figure 10-3.

The electrical network representing this capacitor is shown, and it models the ESR and ESL components with simple R and L elements, plus a $1\,M\Omega$ shunt resistance. While this simple model ignores temperature and dielectric absorption effects which occur in the real capacitor, it is still sufficient for this discussion.

When driven with a constant level of AC current swept from $10\,Hz$ to $100\,MHz$, the voltage across this capacitor model is proportional to its net impedance, which is shown in Figure 10-4.

At low frequencies the net impedance is almost purely capacitive, as noted by the $100\,Hz$ impedance of $15.9\,\Omega$. At the bottom of this "bathtub" curve, the net impedance is determined by ESR, which is shown to be $0.12\,\Omega$ at $125\,kHz$. Above about $1\,MHz$ this capacitor becomes inductive, and impedance is dominated by the effect of ESL. While this particular combination of capacitor characteristics has been chosen purposely to correspond to the tantalum sample used with Figure 10-4, it is also true that all electrolytics will display impedance curves which are similar in general shape. The minimum impedance will vary with the ESR, and the inductive region will vary with ESL (which in turn is strongly affected by package

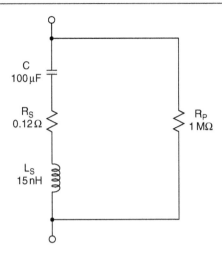

**Figure 10-3: Simplified spice model for a leaded 100 μF/20 V tantalum electrolytic capacitor**

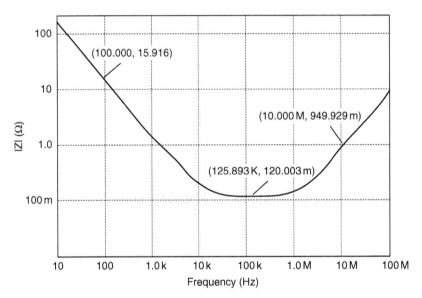

**Figure 10-4: The100 μF/20 V tantalum capacitor simplified model impedance (Ω) versus frequency (Hz)**

style). The simulation curve of Figure 10-4 can be considered as an extension of the 100 μF/20 V tantalum capacitor curve from Figure 10-2.

## Tolerance, Temperature, and Other Effects

In general, precision capacitors are expensive and—even then—not necessarily easy to buy. In fact, choice of capacitance is limited both by the range of available values, and also by tolerances. In terms of size, the better-performing capacitors in the film families tend to be limited in practical terms to 10 μF or less (for dual reasons of size and expense). In terms of low value tolerance, ±1% is possible for NP0 ceramic and some film devices, but with possibly unacceptable delivery times. Many film capacitors can be made available with tolerances of less than $\pm 1\%$, but on a special order basis only.

Most capacitors are sensitive to temperature variations. DF, DA, and capacitance value are all functions of temperature. For some capacitors, these parameters vary approximately linearly with temperature, in others they vary quite nonlinearly. Although it is usually not important for sample-hold (SH) applications, an excessively large *temperature coefficient* (TC, measured in ppm/°C) can prove harmful to the performance of precision integrators, voltage-to-frequency converters, and oscillators. NP0 ceramic capacitors, with TCs as low as 30 ppm/°C, are the best for stability, with polystyrene and polypropylene next best, with TCs in the 100–200 ppm/°C range. On the other hand, when capacitance stability is important, one should stay away from types with TCs of more than a few hundred ppm/°C, or in fact any TC which is nonlinear.

A capacitor's maximum working temperature should also be considered, in light of the expected environment. Polystyrene capacitors, for instance, melt near 85°C, compared to Teflon's ability to survive temperatures up to 200°C.

Sensitivity of capacitance and DA to applied voltage, expressed as *voltage coefficient*, can also hurt capacitor performance within a circuit application. Although capacitor manufacturers do not always clearly specify voltage coefficients, the user should always consider the possible effects of such factors. For instance, when maximum voltages are applied, some high-K ceramic devices can experience a decrease in capacitance of 50% or more. This is an inherent distortion producer, making such types unsuitable for signal path filtering, for example, and better suited for supply bypassing. Interestingly, NP0 ceramics, the stable dielectric subset from the wide range of available ceramics, do offer good performance with respect to voltage coefficient.

Similarly, the capacitance and dissipation factor (DF) of many types vary significantly with frequency, mainly as a result of a variation in dielectric constant. In this regard, the better dielectrics are polystyrene, polypropylene, and Teflon.

## Parasitics

Most designers are generally familiar with the range of capacitors available. But the mechanisms by which both static and dynamic errors can occur in precision circuit designs using capacitors are sometimes easy to forget, because of the tremendous variety of types available. These include dielectrics of glass, aluminum foil, solid tantalum and tantalum foil, silver mica, ceramic, Teflon, and the film capacitors, including polyester, polycarbonate, polystyrene, and polypropylene types. In addition to the traditional leaded packages, many of these are now also offered in surface mount styles.

Figure 10-5 is a workable model of a non-ideal capacitor. The nominal capacitance, C, is shunted by a resistance $R_P$, which represents *insulation resistance* or leakage. A second resistance, $R_S$—*equivalent series resistance*, or ESR—appears in series with the capacitor and represents the resistance of the capacitor leads and plates.

Note that capacitor phenomena are not that easy to separate out. The matching of phenomena and models is for convenience in explanation. Inductance, L—the *equivalent series inductance*, or ESL—models the inductance of the leads and plates. Finally, resistance $R_{DA}$ and capacitance $C_{DA}$ together form a simplified model of a phenomenon known as *dielectric absorption*, or DA. It can ruin fast and slow circuit dynamic performance. In a real capacitor $R_{DA}$ and $C_{DA}$ extend to include multiple parallel sets. These parasitic RC elements can act to degrade timing circuits substantially, and the phenomenon is discussed further below.

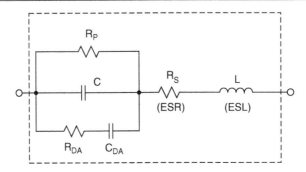

**Figure 10-5: A non-ideal capacitor equivalent circuit includes parasitic elements**

## Dielectric Absorption

Dielectric absorption, which is also known as "soakage" and sometimes as "dielectric hysteresis"—is perhaps the least understood and potentially most damaging of various capacitor parasitic effects. Upon discharge, most capacitors are reluctant to give up all of their former charge, due to this memory consequence.

Figure 10-6 illustrates this effect. On the left of the diagram, after being charged to the source potential of V volts at time $t_0$, the capacitor is shorted by the switch S1 at time $t_1$, discharging it. At time $t_2$, the capacitor is then open-circuited; a residual voltage slowly builds up across its terminals and reaches a nearly constant value. This error voltage is due to DA, and is shown in the right figure, a time/voltage representation of the charge/discharge/recovery sequence. Note that the recovered voltage error is proportional to both the original charging voltage V, as well as the rated DA for the capacitor in use.

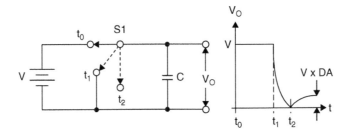

**Figure 10-6: A residual open-circuit voltage after charge/discharge characterizes capacitor dielectric absorption**

Standard techniques for specifying or measuring dielectric absorption are few and far between. Measured results are usually expressed as the percentage of the original charging voltage that reappears across the capacitor. Typically, the capacitor is charged for a long period, then shorted for a shorter established time. The capacitor is then allowed to recover for a specified period, and the residual voltage is then measured (see Reference 8 for details). While this explanation describes the basic phenomenon, it is important to note that real-world capacitors vary quite widely in their susceptibility to this error, with their rated DA ranging from well below to above 1%, the exact number being a function of the dielectric material used.

In practice, DA makes itself known in a variety of ways. Perhaps an integrator refuses to reset to zero, a voltage-to-frequency converter exhibits unexpected nonlinearity, or an SH exhibits varying errors. This last manifestation can be particularly damaging in a data-acquisition system, where adjacent channels may be at voltages which differ by nearly full scale, as shown below.

Figure 10-7 illustrates the case of DA error in a simple SH. On the left, switches S1 and S2 represent an input multiplexer and SH switch, respectively. The multiplexer output voltage is $V_X$, and the sampled voltage held on C is $V_Y$, which is buffered by the op amp for presentation to an ADC. As can be noted by the timing diagram on the right, a DA error voltage, $\epsilon$, appears in the hold mode, when the capacitor is effectively open circuit. This voltage is proportional to the difference of voltages V1 and V2, which, if at opposite extremes of the dynamic range, exacerbates the error. As a practical matter, the best solution for good performance in terms of DA in a SH is to use only the best capacitor.

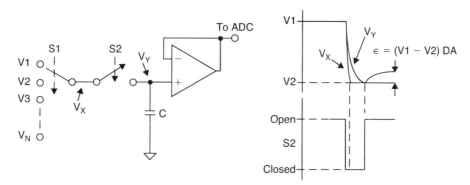

**Figure 10-7: Dielectric absorption induces errors in SH applications**

The DA phenomenon is a characteristic of the dielectric material itself, although inferior manufacturing processes or electrode materials can also affect it. DA is specified as a percentage of the charging voltage. It can range from a low of 0.02% for Teflon, polystyrene, and polypropylene capacitors, up to a high of 10% or more for some electrolytics. For some time frames, the DA of polystyrene can be as low as 0.002%.

Common high-K ceramics and polycarbonate capacitor types display typical DA on the order of 0.2%; it should be noted this corresponds to 1/2 LSB at only 8 bits! Silver mica, glass, and tantalum capacitors typically exhibit even larger DA, ranging from 1.0% to 5.0%, with those of polyester devices failing in the vicinity of 0.5%. As a rule, if the capacitor spec sheet does not specifically discuss DA *within your time frame and voltage range*, exercise caution! Another type with lower *specified* DA is likely a better choice.

DA can produce long tails in the transient response of fast-settling circuits, such as those found in high-pass active filters or AC amplifiers. In some devices used for such applications, Figure 10-5's $R_{DA}$–$C_{DA}$ model of DA can have a time constant of milliseconds. Much longer time constants are also quite usual. In fact, several paralleled $R_{DA}$–$C_{DA}$ circuit sections with a wide range of time constants can model some devices.

In fast-charge, fast-discharge applications, the behavior of the DA mechanism resembles "analog memory"; the capacitor in effect tries to remember its previous voltage.

In some designs, you can compensate for the effects of DA if it is simple and easily characterized, and you are willing to do custom tweaking. In an integrator, for instance, the output signal can be fed back through a suitable compensation network, tailored to cancel the circuit equivalent of the DA by placing a negative impedance effectively in parallel. Such compensation has been shown to improve SH circuit performance by factors of 10 or more (Reference 6).

## Capacitor Parasitics and Dissipation Factor

In Figure 10-5, a capacitor's leakage resistance, $R_P$, the effective series resistance, $R_S$, and effective series inductance, L, act as parasitic elements, which can degrade an external circuit's performance. The effects of these elements are often lumped together and defined as a DF.

A capacitor's leakage is the small current that flows through the dielectric when a voltage is applied. Although modeled as a simple insulation resistance ($R_P$) in parallel with the capacitor, the leakage actually is nonlinear with voltage. Manufacturers often specify leakage as a megohm-microfarad product, which describes the dielectric's self-discharge time constant, in seconds. It ranges from a low of 1 second or less for high-leakage capacitors, such as electrolytic devices, to the 100s of seconds for ceramic capacitors. Glass devices exhibit self-discharge time-constants of 1,000 or more; but the best leakage performance is shown by Teflon and the film devices (polystyrene, polypropylene), with time constants exceeding 1,000,000 megohm-microfarads. For such a device, external leakage paths—created by surface contamination of the device's case or in the associated wiring or physical assembly—can overshadow the internal dielectric-related leakage.

Equivalent series inductance, ESL (Figure 10-5) arises from the inductance of the capacitor leads and plates, which, particularly at the higher frequencies, can turn a capacitor's normally capacitive reactance into an inductive reactance. Its magnitude strongly depends on construction details within the capacitor. Tubular wrapped-foil devices display significantly more lead inductance than molded radial-lead configurations. Multilayer ceramic and film-type devices typically exhibit the lowest series inductance, while ordinary tantalum and aluminum electrolytics typically exhibit the highest. Consequently, standard electrolytic types, if used alone, usually prove insufficient for *high-speed* local bypassing applications. Note however that there also are more specialized aluminum and tantalum electrolytics available, which may be suitable for higher speed uses. These are the types generally designed for use in switch-mode power supplies, which are covered more completely in a following section.

Manufacturers of capacitors often specify effective series impedance by means of impedance versus frequency plots. Not surprisingly, these curves show graphically a predominantly capacitive reactance at low frequencies, with rising impedance at higher frequencies because of the effect of series inductance.

Effective series resistance, ESR (resistor $R_S$ of Figure 10-5), is made up of the resistance of the leads and plates. As noted, many manufacturers lump the effects of ESR, ESL, and leakage into a single parameter called *dissipation factor*, or DF. DF measures the basic inefficiency of the capacitor. Manufacturers define it as the ratio of the energy lost to energy stored per cycle by the capacitor. The ratio of ESR to total capacitive reactance—at a specified frequency—approximates the DF, which turns out to be equivalent to the reciprocal of the figure of merit, Q. Stated as an approximation, $Q \approx 1/DF$ (with DF in numeric terms). For example, a DF of 0.1% is equivalent to a fraction of 0.001; thus, the inverse in terms of Q would be 1000.

DF often varies as a function of both temperature and frequency. Capacitors with mica and glass dielectrics generally have DF values from 0.03% to 1.0%. For ordinary ceramic devices, DF ranges from a low of 0.1% to as high as 2.5% at room temperature. And electrolytics usually exceed even this level. The film capacitors are the best as a group, with DFs of less than 0.1%. Stable-dielectric ceramics, notably the NP0 (also called COG) types, have DF specs comparable to films (more below).

## Assemble Critical Components Last

The designer's worries do not end with the design process. Some common printed circuit assembly techniques can prove ruinous to even the best designs. For instance, some commonly used cleaning solvents can infiltrate certain electrolytic capacitors—those with rubber end caps are particularly susceptible.

Even worse, some of the film capacitors, polystyrene in particular, actually melt when contacted by some solvents. Rough handling of the leads can damage still other capacitors, creating random or even intermittent circuit problems. Etched-foil types are particularly delicate in this regard. To avoid these difficulties it may be advisable to mount especially critical components as the last step in the board assembly process, if possible.

Designers should also consider the natural failure mechanisms of capacitors. Metallized film devices, for instance, often self-heal. They initially fail due to conductive bridges that develop through small perforations in the dielectric film. But the resulting fault currents can generate sufficient heat to destroy the bridge, thus returning the capacitor to normal operation (at a slightly lower capacitance). Of course, applications in high-impedance circuits may not develop sufficient current to clear the bridge, so the designer must be wary here.

Tantalum capacitors also exhibit a degree, of self-healing, but—unlike film capacitors—the phenomenon depends on the temperature at the fault location rising slowly. Therefore, tantalum capacitors self-heal best in high impedance circuits which limit the surge in current through the capacitor's defect. Use caution therefore, when specifying tantalums for high-current applications.

Electrolytic capacitor life often depends on the rate at which capacitor fluids seep through end caps. Epoxy end seals perform better than rubber seals, but an epoxy sealed capacitor can explode under severe reverse-voltage or overvoltage conditions. Finally, *all* polarized capacitors must be protected from exposure to voltages outside their specifications.

# Resistors and Potentiometers

## Basics

Designers have a broad range of resistor technologies to choose from, including carbon composition, carbon film, bulk metal, metal film, and both inductive and non-inductive wirewound types. As perhaps the most basic—and presumably most trouble-free—of components, resistors are often overlooked as error sources in high performance circuits.

Yet, an improperly selected resistor can subvert the accuracy of a 12-bit design by developing errors well in excess of 122 ppm (1/2 LSB). When did you last read a resistor data sheet? You would be surprised what can be learned from an informed review of data.

Consider the simple circuit of Figure 10-8, showing a non-inverting op amp where the 100× gain is set by $R_1$ and $R_2$. The TCs of these two resistors are a somewhat obvious source of error. Assume the op amp gain errors to be negligible, and that the resistors are perfectly matched to a 99/1 ratio at +25°C. If, as noted, the resistor TCs differ by only 25 ppm/°C, the gain of the amplifier changes by 250 ppm for a 10°C temperature change. This is about a 1 LSB error in a 12-bit system, and a major disaster in a 16-bit system.

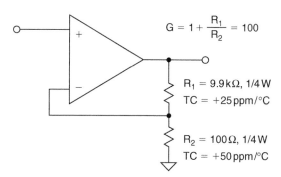

$$G = 1 + \frac{R_1}{R_2} = 100$$

$R_1 = 9.9\,\text{k}\Omega,\ 1/4\,\text{W}$
$TC = +25\,\text{ppm/°C}$

$R_2 = 100\,\Omega,\ 1/4\,\text{W}$
$TC = +50\,\text{ppm/°C}$

- Temperature change of 10°C causes gain change of 250 ppm.
- This is 1LSB in a 12 bit system and a disaster in a 16 bit system.

**Figure 10-8: Mismatched resistor TCs can induce temperature-related gain errors**

Temperature changes, however, can limit the accuracy of the Figure 10-8 amplifier in several ways. In this circuit (as well as many op amp circuits with component-ratio defined gains), the *absolute* TC of the resistors is less important—*as long as they track one another in ratio*. But even so, some resistor types simply are not suitable for precise work. For example, *carbon composition* units—with TCs of approximately 1,500 ppm/°C, will not work. Even if the TCs could be matched to an unlikely 1%, the resulting 15 ppm/°C differential still proves inadequate—an 8°C shift creates a 120 ppm error.

Many manufacturers offer metal film and bulk metal resistors, with absolute TCs ranging between $\pm 1$ and $\pm 100\,\text{ppm/}°\text{C}$. Beware, though; TCs can vary a great deal, particularly among discrete resistors from different batches. To avoid this problem, more expensive matched resistor pairs are offered by some manufacturers, with TCs that track one another to within $2–10\,\text{ppm/}°\text{C}$. Low-priced thin-film networks have good relative performance and are widely used.

Suppose, as shown in Figure 10-9, $R_1$ and $R_2$ are 1/4W resistors with identical $25\,\text{ppm/}°\text{C}$ TCs. Even when the TCs are identical, there can still be significant errors! When the signal input is zero, the resistors dissipate no heat. But, if it is 100mV, there is 9.9V across $R_1$, which then dissipates 9.9mW. It will experience a temperature rise of $1.24°\text{C}$ (due to a $125°\text{C/W}$ 1/4W resistor thermal resistance). This $1.24°\text{C}$ rise causes a resistance change of 31ppm, and thus a corresponding gain change. But $R_2$, with only 100mV

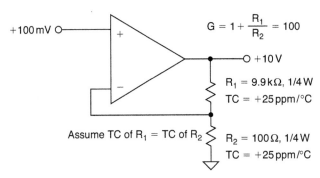

$$G = 1 + \frac{R_1}{R_2} = 100$$

+100mV

+10V

$R_1 = 9.9\,\text{k}\Omega$, 1/4W
TC $= +25\,\text{ppm/}°\text{C}$

Assume TC of $R_1$ = TC of $R_2$

$R_2 = 100\,\Omega$, 1/4W
TC $= +25\,\text{ppm/}°\text{C}$

- $R_1$, $R_2$ Thermal resistance $= 125°\text{C/W}$.
- Temperature of $R_1$ will rise by $1.24°\text{C}$, $P_D = 9.9\,\text{mW}$.
- Temperature rise of $R_2$ is negligible, $P_D = 0.1\,\text{mW}$.
- Gain is altered by 31 ppm, or 1/2 LSB @ 14 bits.

**Figure 10-9: Uneven power dissipation between resistors with identical TCs can also introduce temperature-related gain errors**

across it, is only heated a negligible $0.0125°\text{C}$. The resulting 31ppm net gain error represents a full-scale error of 1/2 LSB at 14 bits, and is a disaster for a 16-bit system.

Even worse, the effects of this resistor self-heating also create easily calculable *nonlinearity errors*. In the Figure 10-9 example, with 1/2 the voltage input, the resulting self-heating error is only 15ppm. In other words, the stage gain is not constant at 1/2 and full scale (nor is it so at other points), as long as uneven temperature shifts exist between the gain-determining resistors. This is by no means a worst-case example; physically smaller resistors would give worse results, due to higher associated thermal resistance.

These, and similar errors, are avoided by selecting critical resistors that are accurately matched for both value and TC, are well derated for power, and have tight thermal coupling between those resistors where matching is important. This is best achieved by using a resistor network on a single substrate—such a network either may be within an IC, or may be a separately packaged thin-film resistor network.

When the circuit resistances are very low ($\leq 10\,\Omega$), *interconnection stability* also becomes important. For example, while often overlooked as an error, the resistance TC of typical copper wire or printed circuit traces can add errors. The TC of copper is typically $\sim 3{,}900\,\text{ppm/}°\text{C}$. Thus a precision $10\,\Omega$, $10\,\text{ppm/}°\text{C}$ wirewound resistor with $0.1\,\Omega$ of copper interconnect effectively becomes a $10.1\,\Omega$ resistor with a TC of nearly $50\,\text{ppm/}°\text{C}$.

One final consideration applies mainly to designs that see widely varying ambient temperatures: a phenomenon known as *temperature retrace* describes the change in resistance which occurs after a specified number of cycles of exposure to low and high ambients with constant internal dissipation. Temperature retrace can exceed 10 ppm/°C, even for some of the better thin-film components.

In summary, to design resistance-based circuits for minimum temperature-related errors, consider the points noted in Figure 10-10 (along with their cost):

- Closely match resistance TCs.
- Use resistors with low absolute TCs.
- Use resistors with low thermal resistance (higher power ratings, larger cases).
- Tightly couple matched resistors thermally (use standard common-substrate networks).
- For large ratios consider using stepped attenuators.

**Figure 10-10: Important points for minimizing temperature-related errors in resistors**

## Resistor Parasitics

Resistors can exhibit significant levels of parasitic inductance or capacitance, especially at high frequencies. Manufacturers often specify these parasitic effects as a reactance error, in % or ppm, based on the ratio of the difference between the impedance magnitude and the DC resistance, to the resistance, at one or more frequencies.

Wirewound resistors are especially susceptible to difficulties. Although resistor manufacturers offer wirewound components in either normal or non-inductively wound form, even non-inductively wound resistors create headaches for designers. These resistors still appear slightly inductive (of the order of 20 μH) for values below 10 kΩ. Above 10 kΩ the same style resistors actually exhibit 5 pF of shunt capacitance.

These parasitic effects can raise havoc in dynamic circuit applications. Of particular concern are applications using wirewound resistors with values both greater than 10 kΩ. Here it is not uncommon to see peaking, or even oscillation. These effects become more evident at low-kHz frequency ranges.

Even in low-frequency circuit applications, parasitic effects in wirewound resistors can create difficulties. Exponential settling to 1 ppm may take 20 time constants or more. The parasitic effects associated with wirewound resistors can significantly increase net circuit settling time to beyond the length of the basic time constants.

Unacceptable amounts of parasitic reactance are often found even in resistors that are not wirewound. For instance, some metal-film types have significant interlead capacitance, which shows up at high frequencies. In contrast, when considering this end–end capacitance, carbon resistors do the best at high frequencies.

## Thermoelectric Effects

Another more subtle problem with resistors is the *thermocouple effect*, also sometimes referred to as *thermal EMF*. Wherever there is a junction between two different metallic conductors, a thermoelectric voltage results.

The thermocouple effect is widely used to measure temperature. However, in any low level precision op amp circuit it is also a potential source of inaccuracy, since wherever two different conductors meet, a thermocouple is formed (whether we like it or not). In fact, in many cases, it can easily produce the dominant error within an otherwise precision circuit design.

Parasitic thermocouples will cause errors when and if the various junctions forming the parasitic thermocouples are at different temperatures. With two junctions present on each side of the signal being processed within a circuit, by definition we have formed at least one thermocouple pair. If the two junctions of this thermocouple pair are at different temperatures, there will be a net temperature dependent error voltage produced. Conversely, if the two junctions of a parasitic thermocouple pair are kept at an identical temperature, then the net error produced will be zero, as the voltages of the two thermocouples effectively will be canceled.

This is a critically important point, since in practice we cannot avoid connecting dissimilar metals together to build an electronic circuit. But, what we can do is carefully control temperature differentials across the circuit, such that the undesired thermocouple errors cancel one another.

The effect of such parasitics is very hard to avoid. To understand this, consider a case of making connections *with copper wire only*. In this case, even a junction formed by different copper wire alloys can have a thermoelectric voltage which is a small fraction of $1 \mu V/°C$! And, taking things a step further, even such apparently benign components as resistors contain parasitic thermocouples, with potentially even stronger effects.

For example, consider the resistor model shown in Figure 10-11. The two connections between the resistor material and the leads form thermocouple junctions, T1 and T2. This thermocouple EMF can be as high as $400 \mu V/°C$ for some carbon composition resistors, and as low as $0.05 \mu V/°C$ for specially constructed resistors. Ordinary metal film resistors (RN-types) are typically about $20 \mu V/°C$.

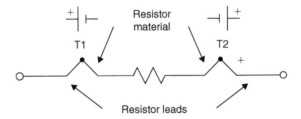

Typical resistor thermocouple EMFs

- Carbon composition      $\approx 400 \mu V/°C$
- Metal film      $\approx 20 \mu V/°C$
- Evenohm or manganin wirewound      $\approx 2 \mu V/°C$
- RCD components HP-series      $\approx 0.05 \mu V/°C$

**Figure 10-11: Every resistor contains two thermocouples, formed between the leads and resistance element**

Note that these thermocouple effects are relatively unimportant for AC signals. Even for DC-only signals, they will nicely cancel one another, if, as noted above, the entire resistor is at a uniform temperature. However, if there is significant power dissipation in a resistor, or if its orientation with respect to a heat source is non-symmetrical, this can cause one of its ends to be warmer than the other, causing a net thermocouple error voltage. Using ordinary metal film resistors, an end-to-end temperature differential of 1°C causes a thermocouple voltage of about $20 \mu V$. This error level is quite significant compared to the offset voltage drift of a precision op amp like the OP177, and extremely significant when compared to chopper-stabilized op amps, with their drifts of $<1 \mu V/°C$.

Figure 10-12 shows how resistor orientation can make a difference in the net thermocouple voltage. In the left diagram, standing the resistor on end in order to conserve board space will invariably cause a temperature gradient across the resistor, especially if it is dissipating any significant power. In contrast, placing the resistor flat on the PC board as shown at the right will generally eliminate the gradient. An exception might occur, if there is end-to-end resistor airflow. For such cases, orienting the resistor axis perpendicular to the airflow will minimize this source of error, since this tends to force the resistor ends to the same temperature.

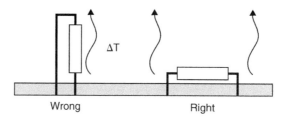

**Figure 10-12: The effects of thermocouple EMFs generated by resistors can be minimized by orientation that normalizes the end temperatures**

Note that this line of thinking should be extended, to include orientation of resistors on a vertically mounted PC board. In such cases, natural convection air currents tend to flow upward across the board. Again, the resistor thermal axis should be perpendicular to convection, to minimize thermocouple effects. With tiny surface mount resistors, the thermocouple effects can be less problematic, due to tighter thermal coupling between the resistor ends.

In general, designers should strive to avoid thermal gradients on or around critical circuit boards. Often this means thermally isolating components that dissipate significant amounts of power. Thermal turbulence created by large temperature gradients can also result in dynamic noise-like low-frequency errors.

## Voltage Sensitivity, Failure Mechanisms, and Aging

Resistors are also plagued by changes in value as a function of applied voltage. The deposited-oxide high-megohm type components are especially sensitive, with voltage coefficients ranging from 1 ppm/V to more than 200 ppm/V. This is another reason to exercise caution in such precision applications as high-voltage dividers.

The normal failure mechanism of a resistor can also create circuit difficulties, if not carefully considered beforehand. For example, carbon composition resistors fail safely, by turning into open circuits. Consequently, in some applications, these components can play a useful secondary role, as a fuse. Replacing such a resistor with a carbon-film type can possibly lead to trouble, since carbon-films can fail as short circuits. (Metal-film components usually fail as open circuits.)

All resistors tend to change slightly in value with age. Manufacturers specify long-term stability in terms of change—ppm/year. Values of 50 or 75 ppm/year are not uncommon among metal film resistors. For critical applications, metal-film devices should be burned-in for at least 1 week at rated power. During burn-in, resistance values can shift by up to 100 or 200 ppm. Metal film resistors may need 4–5,000 operational hours for full stabilization, especially if deprived of a burn-in period (Figure 10-13).

| | Type | Advantages | Disadvantages |
|---|---|---|---|
| **Discrete** | Carbon Composition | Lowest cost<br>High power/small case size<br>Wide range of values | Poor tolerance (5%)<br>Poor temperature coefficient (1,500 ppm/°C) |
| | Wirewound | Excellent tolerance (0.1%)<br>Excellent TC (1 ppm/°C)<br>High power | Reactance is a problem<br>Large case size<br>Most expensive |
| | Metal film | Good tolerance (0.1%)<br>Good TC (<1 to 100 ppm/°C)<br>Moderate cost<br>Wide range of values<br>Low voltage coefficient | Must be stabilized with burn-in<br>Low power |
| | Bulk metal or Metal foil | Excellent tolerance (to 0.005%)<br>Excellent TC (to <1 ppm/°C)<br>Low reactance<br>Low voltage coefficient | Low power<br>Very expensive |
| | High megohm | Very high values ($10^8$ to $10^{14}\,\Omega$)<br>Only choice for some circuits | High voltage coefficient (200 ppm/V)<br>Fragile glass case (needs special handling)<br>Expensive |
| **Networks** | Thick film | Low cost<br>High power<br>Laser-trimmable<br>Readily available | Fair matching (0.1%)<br>Poor TC (>100 ppm/°C)<br>Poor tracking TC (10 ppm/°C) |
| | Thin film | Good matching (<0.01%)<br>Good TC (<100 ppm/°C)<br>Good tracking TC (2 ppm/°C)<br>Moderate cost<br>Laser-trimmable<br>Low capacitance<br>Suitable for hybrid IC substrate | Often large geometry<br>Limited values and Configurations |

**Figure 10-13: Resistor comparison chart**

## Resistor Excess Noise

Most designers have some familiarity with thermal, or Johnson noise, which occurs in resistors. But a less widely recognized secondary noise phenomenon is associated with resistors, and it is called *excess noise*. It can prove particularly troublesome in precision op amp and converter circuits, as it is evident only when current passes through a resistor.

To review briefly, thermal noise results from thermally induced random vibration of charge resistor carriers. Although the average current from the vibrations remains zero, instantaneous charge motions result in an instantaneous voltage across the terminals.

Excess noise, on the other hand, occurs primarily when DC flows in a discontinuous medium—for example the conductive particles of a carbon composition resistor. The current flows unevenly through the compressed carbon granules, creating microscopic particle-to-particle "arcing." This phenomenon gives rise to a 1/f noise-power spectrum, in addition to the thermal noise spectrum. In other words, the excess spot noise voltage increases as the inverse square-root of frequency.

Excess noise often surprises the unwary designer. Resistor thermal noise and op amp input noise set the noise floor in typical op amp circuits. Only when voltages appear across input resistors and causes current to flow does the excess noise become a significant—and often dominant—factor. In general, carbon composition resistors generate the most excess noise. As the conductive medium becomes more uniform, excess noise becomes less significant. Carbon film resistors do better, with metal film, wirewound and bulk-metal-film resistors doing better yet.

Manufacturers specify excess noise in terms of a noise index—the number of microvolts of rms noise in the resistor in each decade of frequency per volt of DC drop across the resistor. The index can rise to 10 dB (3 μV per DC volt per decade of bandwidth) or more. Excess noise is most significant at low frequencies, while above 100 kHz thermal noise predominates.

## *Potentiometers*

Trimming potentiometers (trimpots) can suffer from most of the phenomena that plague fixed resistors. In addition, users must also remain vigilant against some hazards unique to these components.

For instance, many trimpots are not sealed, and can be severely damaged by board washing solvents, and even by excessive humidity. Vibration—or simply extensive use—can damage the resistive element and wiper terminations. Contact noise, TCs, parasitic effects, and limitations on adjustable range can all hamper trimpot circuit operation. Furthermore, the limited resolution of wirewound types and the hidden limits to resolution in cermet and plastic types (hysteresis, incompatible material TCs, slack) make obtaining and maintaining precise circuit settings anything but an "infinite resolution" process. Given this background, two rules are suggested for the potential trimpot user. Rule 1: Use infinite care and infinitesimal adjustment range to avoid infinite frustration when applying manual trimpots. Rule 2: *Consider the elimination of manual trimming potentiometers altogether, if possible!* A number of digitally addressable potentiometers (RDACs) are now available for direct application in similar circuit functions as classic trimpots. There are also many low cost multi-channel voltage output DACs expressly designed for system voltage trimming.

# *Inductors*

## *Basics*

An inductor is a passive electronic component that stores energy in the form of a magnetic field. In its simplest form, an inductor consists of a wire loop or coil. The inductance is directly proportional to the number of turns in the coil. Inductance also depends on the radius of the coil and on the type of material around which the coil is wound.

Inductive reactance is defined as:

$$X_L =_- \omega L = 2\pi f L \qquad (10\text{-}4)$$

where $X_L$ is the inductive reactance, $\omega$ is the angular frequency, f is the frequency in Hertz, and L is the inductance.

Inductive reactance is the positive imaginary component of impedance.

The complex impedance of an inductor is then given by:

$$Z = j\omega L = j2\pi f L \qquad (10\text{-}5)$$

where j is the imaginary number.

$$j = \sqrt{-1} \qquad (10\text{-}6)$$

Inductors had been rare in circuit design, especially at lower frequencies, due of the physical size. Since inductance is the inverse mathematical function of capacitance, inductors are sometimes synthesized by placing a capacitor in the feedback network of an op amp (see Figure 10-14). This technique is obviously limited to frequencies where the open-loop gain of the op amp is sufficient to support this operation. It also may not be practical at high current levels.

Inductors are popularity in RF circuits since the small inductance values make them physically small. Passive LC filters are used at RF frequencies since active filters are not practical.

Switch-mode power supplies are probably the common place to find inductors today. This is covered in depth in the section on switch-mode regulators in the power and ground section.

It is only relatively recently that manufacturers developed the ability to fabricate inductors in monolithic semiconductor processes. This is beyond the scope of this book though.

## *Ferrites*

*Ferrites* (non-conductive ceramics manufactured from the oxides of nickel, zinc, manganese, or other compounds) are extremely useful for decoupling in power supply filters. At low frequencies ($<100\,\text{kHz}$), ferrites are inductive; thus, they are useful in lowpass LC filters. Above 100 kHz, ferrites becomes resistive,

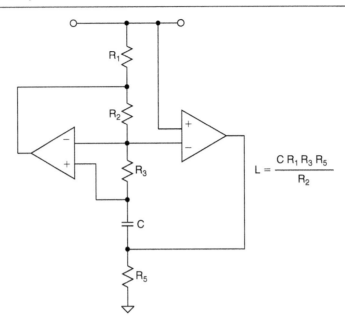

$$L = \frac{C\, R_1\, R_3\, R_5}{R_2}$$

**Figure 10-14: A synthetic inductor**

an important characteristic in high-frequency filter designs. Ferrite impedance is a function of material, operating frequency range, DC bias current, number of turns, size, shape, and temperature.

Several ferrite manufacturers offer a wide selection of ferrite materials from which to choose, as well as a variety of packaging styles for the finished network. The most simple form is the *bead* of ferrite material, a cylinder of the ferrite which is simply slipped over the power supply lead to the stage being decoupled. Alternately, the *leaded ferrite bead* is the same bead, mounted by adhesive on a length of wire, and used simply as a component. More complex beads offer multiple holes through the cylinder for increased decoupling, plus other variations. Surface mount bead styles are also available.

Recently, PSpice ferrite models for Fair-Rite materials have become available that allow ferrite impedance to be estimated (Reference 12). These models have been designed to match measured impedances rather than theoretical impedances.

A ferrite's impedance is dependent on a number of interdependent variables and is difficult to quantify analytically, thus selecting the proper ferrite is not straightforward. However, knowing the following system characteristics will make selection easier. First, determine the frequency range of the noise to be filtered. A spectrum analyzer is useful here. Second, the expected temperature range of the filter should be known, because ferrite impedance varies with temperature. Third, the DC bias current flowing through the ferrite must be known, to ensure that the ferrite does not saturate. Although models and other analytical tools may prove useful, the general guidelines given above, coupled with some experimentation with the actual filter connected to the supply output under system load conditions, should ultimately lead to the selection of the proper ferrite.

Sizing of the ferrite for current is especially important; if the saturation current of the ferrite is reached, it loses it inductive properties. This will obviously limit the ferrite's usefulness in decoupling applications. This saturation current should be sized for the peak current requirements of the circuit (with some added margin, of course).

## References: Inductors

1. W. Jung and D. Marsh, "Picking Capacitors, Parts 1 & 2," *Audio*, February, March, 1980.

2. *Tantalum Electrolytic and Ceramic Capacitor Families*, Kemet Electronics, Box 5928, Greenville, SC, 29606, (803) 963–6300.

3. *Type HFQ Aluminum Electrolytic Capacitor and Type V Stacked Polyester Film Capacitor*, Panasonic, 2 Panasonic Way, Secaucus, NJ, 07094, (201) 348–7000.

4. *OS-CON Aluminum Electrolytic Capacitor 93/94 Technical Book*, Sanyo, 3333 Sanyo Road, Forrest City, AK, 72335, (501) 633–6634.

5. I. Clelland, "Metalized Polyester Film Capacitor Fills High Frequency Switcher Needs," **PCIM**, Vol. June, 1992.

6. *Type 5MC Metallized Polycarbonate Capacitor*, Electronic Concepts, Inc., Box 1278, Eatontown, NJ, 07724, (908) 542–7880.

7. H. Ott, **Noise Reduction Techniques in Electronic Systems**, 2d Edition, Wiley, Hoboken, NJ, 1988.

8. *Fair-Rite Linear Ferrites Catalog, Fair-Rite Products*, Box J, Wallkill, NY, 12886, (914) 895–2055.

9. *Type EXCEL Leaded Ferrite Bead EMI Filter, and Type EXC L Leadless Ferrite Bead*, Panasonic, 2 Panasonic Way, Secaucus, NJ, 07094, (201) 348–7000.

10. S. Hageman, "Use Ferrite Bead Models to Analyze EMI Suppression," The Design Center Source, *MicroSim Newsletter*, January, 1995.

11. J.W. Miller, *Type 5250 and 6000-101 K Chokes*, 306 E. Alondra Blvd., Gardena, CA, 90247, (310) 515–1720.

12. *Tantalum Electrolytic Capacitor SPICE Models*, Kemet Electronics, Box 5928, Greenville, SC, 29606, (803) 963–6300.

13. Eichhoff Electronics, Inc., 205 Hallene Road, Warwick, RI., 02886, (401) 738–1440.

## CHAPTER 11

# Overvoltage Effects on Analog Integrated Circuits

# *Overvoltage Effects*

One of the most commonly asked applications questions is: "What happens if external voltages are applied to an analog integrated circuit with the supplies turned off?" This question describes situations that can take on many different forms: from lightning strikes on cables which propagate very high transient voltages into signal conditioning circuits, to walking across a carpet and then touching a printed circuit board (PCB) full of sensitive precision circuits. Regardless of the situation, the general issue is the effect of overvoltage stress (and, in some cases, abuse) on analog integrated circuits. The discussion which follows will be limited in general to operational amplifiers, because it is these devices that most often interface to the outside world. The principles developed here can and should be applied to all analog integrated circuits which are required to condition or digitize analog waveforms. These devices include (but are not limited to) instrumentation amplifiers, analog comparators, sample-and-hold amplifiers, analog switches and multiplexers, and analog-to-digital converters.

## *Amplifier Input Stage Overvoltage*

In real-world signal conditioning, sensors are often used in hostile environments where faults can and do occur. When these faults take place, signal conditioning circuitry can be exposed to large voltages which exceed the power supplies. The likelihood for damage is quite high, even though the components' power supplies may be turned on. Published specifications for operational amplifier absolute maximum ratings state that applied input signal levels should never exceed the power supplies by more than 0.3 V or, in some devices, 0.7 V. Exceeding these levels exposes amplifier input stages to potentially destructive fault currents which flow through internal metal traces and parasitic P–N junctions to the supplies. Without some type of current limiting, unprotected input differential pairs (BJTs or FETs) can be destroyed in a matter of microseconds. There are, however, some devices with built-in circuitry that can provide protection beyond the supply voltages, but in general, absolute maximum ratings must still be observed.

Although more recent vintage operational amplifiers designed for single-supply or rail-to-rail operation are now including information with regard to input stage overvoltage effects, there are very many amplifiers available today without such information provided by the manufacturer. In those cases, the circuit designer using these components must ascertain the input stage current–voltage characteristic of the device in question before steps can be taken to protect it. All amplifiers will conduct current to the positive/negative supply, provided the applied input voltage exceeds some internal threshold. This threshold is device dependent, and can range from 0.7 to 30 V, depending on the internal construction of the input stage. Regardless of the threshold level, externally generated fault currents should be limited to no more than ±5 mA.

Many factors contribute to the current–voltage characteristic of an amplifier's input stage: internal differential clamping diodes, current-limiting series resistances, substrate potential connections, and differential input stage topologies (BJTs or FETs). Input protection diodes used as differential input clamps are typically constructed from the base–emitter junctions of NPN transistors. These diodes usually form a parasitic P–N junction to the negative supply when the applied input voltage exceeds the negative supply. Current-limiting series resistances used in the input stages of operational amplifiers can be fabricated from three types of material: N- or P-type diffusions, polysilicon, or thin films (e.g., SiCr ). Polysilicon and thin-film resistors are fabricated over thin layers of oxide which provide an insulating barrier to the substrate; as such, they do not exhibit any parasitic P–N junctions to either supply. Diffused resistors, on the other hand, exhibit P–N junctions to the supplies because they are constructed from either P- or N-type diffusion regions. The substrate potential of the amplifier is the most critical component, for it will determine the sensitivity of an amplifier's input current–voltage characteristic to supply voltage (Figure 11-1).

- Input should not exceed absolute maximum ratings (usually specified with respect to supply voltages)
- A common specification requires the input signal $<|V_s| \pm 0.3\,V$
- Input voltage should be held near zero in the absence of supplies
- Input stage conduction current needs to be limited (rule of thumb: $\leq 5\,mA$)
- Avoid reverse bias junction breakdown in input stage base–emitter junctions
- Differential and common–mode ratings may differ
- No two amplifiers are exactly the same
- Some op amps contain input protection (voltage clamps, current limits, or both), but absolute maximum ratings must still be observed

**Figure 11-1: Input stage overvoltage**

The configuration of the amplifier's input stage also plays a large role in the current–voltage characteristic of the amplifier. Input differential pairs of operational amplifiers are constructed from either bipolar transistors (NPN or PNP) or field-effect transistors (junction or MOS, N- or P-channel). While the bipolar input differential pairs do not have any direct path to either supply, FET differential pairs do. For example, an N-channel junction field effect (JFET) forms a parasitic P–N junction between its backgate and the P-substrate that energizes when $V_{IN} + 0.7\,V < V_{NEG}$. As mentioned previously, many manufacturers of analog integrated circuits do not provide any details with regard to the behavior of the device's input structure. Either simplified schematics are not provided or, if they are shown, the behavior of the input stage under an overvoltage condition is omitted. Therefore, other measures must be taken in order to identify the conduction paths.

A standard transistor curve tracer can be configured to determine the current–voltage characteristic of any amplifier regardless of input circuit topology. Both amplifier supply pins are connected to ground, and the collector output drive is connected to one of the amplifier's inputs. The curve tracer applies a DC ramp voltage and measures the current flowing through the input stage. In the event that a transistor curve tracer is not available, a DC voltage source and a multimeter can be substituted for the curve tracer. A $10\,k\Omega$ resistor should be used between the DC voltage source and the amplifier input for additional protection. Ammeter readings from the multimeter at each applied DC voltage will yield the same result as that produced by the curve tracer. Although either input can be tested (both inputs should), it is recommended that the unused input be left open; otherwise, additional junctions could come into play and would complicate matters further. Evaluations of current feedback amplifier input stages are more difficult because of the lack of symmetry between the inputs. As a result, both inputs should be characterized for their individual current–voltage characteristics.

Once the input current–voltage characteristic has been determined for the device in question, the next step is to determine the minimum level of resistance required to limit fault currents to $\pm 5\,mA$. Equation (11-1) illustrates the computation for $R_s$ when the input overvoltage level is known:

$$R_s = \frac{V_{IN(MAX)} - V_{SUPPLY}}{5\,mA} \tag{11-1}$$

The worst-case condition for overvoltage would be when the power supplies are initially turned off or disconnected. In this case, $V_{SUPPLY}$ is equal to zero. For example, if the input overvoltage could reach 100 V under some type of fault condition, then the external resistor should be no smaller than 20 k$\Omega$. Most operational amplifier applications only require protection at one input; however, there are a few configurations (e.g., difference amplifiers ) where both inputs can be subjected to overvoltage and both must be protected. The need for protection on both inputs is much more common with instrumentation amplifiers (Figure 11-2).

- Junctions may be <u>forward biased</u> if the current is limited
- In general a safe current limit is 5 mA
- <u>Reverse bias</u> junction breakdown is damaging regardless of the current level
- When in doubt, protect with external diodes and series resistances
- Curve tracers can be used to check the overvoltage characteristics of a device
- Simplified equivalent circuits in data sheets do not tell the entire story!!!

**Figure 11-2: Overvoltage effects**

## *Amplifier Output Voltage Phase Reversal*

Some operational amplifiers exhibit output voltage phase reversal when one or both of their inputs exceeds their input common-mode voltage range. Phase reversal is usually associated with JFET (N- or P-channel) input amplifiers, but some bipolar devices (especially single-supply amplifiers operating as unity-gain followers) may also be susceptible. In the vast majority of applications, output voltage phase reversal does not harm the amplifier nor the circuit in which the amplifier is used. Although a number of operational amplifiers suffer from phase reversal, it is rarely a problem in system design. However, in servo loop applications, this effect can be quite hazardous. Fortunately, this is only a temporary condition. Once the amplifier's inputs return to within their normal operating common-mode range, output voltage phase reversal ceases. It may still be necessary to consult the amplifier manufacturer, since phase reversal information rarely appears on device data sheets (Figure 11-3).

- Sometimes occurs in FET and bipolar input (especially single-supply) op amps when input exceeds common mode range
- Does not harm amplifier, but may be disastrous in servo systems!
- Not usually specified on data sheet, so amplifier must be checked
- Easily prevented:
    BiFETs:    Add appropriate input series resistance (determined empirically, unless provided in data sheet)
    Bipolars:   Use Schottky diode clamps to the supply rails.

**Figure 11-3: Beware of amplifier output phase reversal**

In bipolar-FET (BiFET) operational amplifiers, phase reversal may be prevented by adding an appropriate resistance in series with the amplifier's input to limit the current. Bipolar input devices can be protected by using a Schottky diode to clamp the input to within a few hundred millivolts of the negative rail. For a complete description of the output voltage phase reversal effect, please consult Reference 1.

Rail-to-rail operational amplifiers present a special class of problems to the integrated circuit designer, because these types of devices should not exhibit any abnormal behavior throughout the entire input common-mode range. In fact, it is desirable that devices used in these applications also not exhibit any abnormal behavior if the applied input voltages exceed the power-supply range. One of the more recent vintage rail-to-rail input/output operational amplifiers, the OPX91 family (the OP191, the OP291, and the OP491), includes additional components that prevent overvoltage and damage to the device. As shown in Figure 11-4, the input stage of the OPX91 devices use six diodes and two resistors to clamp the input terminals to each other and to the supplies. D1 and D2 are base–emitter NPN diodes which are used to protect the bases of Q1–Q2 and Q3–Q4 against avalanche breakdown when the applied differential input voltage to the device exceeds 0.7 V. Diodes D3–D6 are diodes formed from substrate PNP transistors that clamp the applied input voltages on the OPX91 to the supply rails.

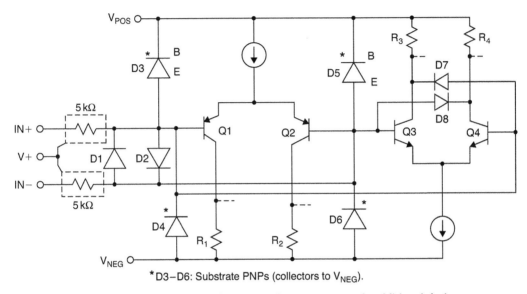

*D3–D6: Substrate PNPs (collectors to $V_{NEG}$).

**Figure 11-4: A closer look at the OP-X91 input stage reveals additional devices**

An interesting benefit from using substrate PNPs as clamp diodes is that their collectors are connected to the negative supply; thus, when the applied input voltage exceeds either supply rail, the diodes energize, and the fault currents are diverted directly to the supply and not through or into the device's input stage. There are also 5 kΩ resistors in series with each of the inputs to the OPX91 to limit the fault current through D1 and D2 when the differential input voltage exceeds 0.7 V (Figure 11-5). Note that these 5 kΩ resistors are P-type diffusions placed inside an n-well, which is then connected to the positive supply. When the applied input voltage exceeds the positive supply, some of the fault current generated is also diverted to $V_{POS}$ and away from the input stage. As a result of these measures, the input overvoltage characteristic of the OPX91 is well behaved as shown in Figure 11-6. Note that the combination of the 5 kΩ resistors and clamp diodes safely limits the input current to less than 2 mA, even when the inputs of the device exceed the supply rails by 10 V.

As an added safety feature, an additional pair of diodes is used in the input stage across Q3 and Q4 to prevent subsequent stages internal to the OPX91 from collapsing (i.e., forced into cutoff). If these stages were forced into cutoff, then the amplifier would undergo output voltage phase reversal when the inputs exceeded the positive input common-mode voltage. An illustration of the diodes' effectiveness is shown in Figure 11-6. Here, the OPX91 family can safely handle a 20 Vp-p input signal on ±5 V supplies without exhibiting any sign of output voltage phase reversal or other anomalous behavior. With these amplifiers, no external clamping diodes are required.

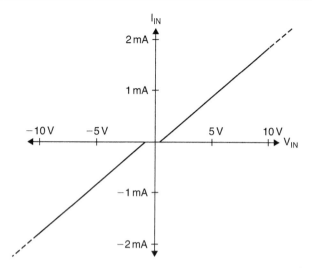

Figure 11-5: Internal 5 kΩ resistors plus input clamp
diodes combine to protect OP-X91 devices against
overvoltage

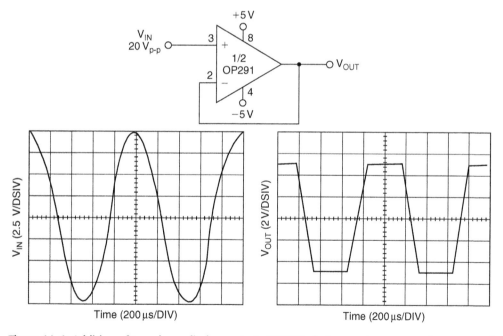

**Figure 11-6: Addition of two clamp diodes protects OP-X91 devices against output phase reversal**

For those amplifiers where external protection is clearly required against both overvoltage abuse and output phase reversal, a common technique is to use a series resistance, $R_s$, to limit fault current, and Schottky diodes to clamp the input signal to the supplies, as shown in Figure 11-7.

The external input series resistance, $R_s$, will be provided by the manufacturer of the amplifier, or determined empirically by the user with the method previously shown in Figure 11-2 and Eq. (11-1).

*785*

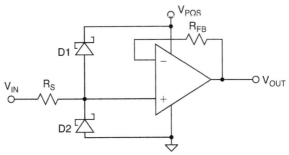

Value for $R_S$ provided by manufacturer or determined empirically.

$R_{FB}$ may be required for high bias current devices.

D1 and D2 can be Schottky diodes (check their capacitance and leakage current first).

**Figure 11-7: Generalized external protection schemes against input overvoltage abuse and output voltage phase reversal in single-supply op amps**

More often than not, the value of this resistor will provide enough protection against output voltage phase reversal, as well as limiting the fault current through the Schottky diodes.

It is evident that whenever resistance is added in series with an amplifier's input, its offset and noise performance will be affected. The effects of this series resistance on circuit noise can be calculated using the following equation:

$$E_{n,total} = \sqrt{(e_{n,op\ amp})^2 + (e_{n,R_s})^2 + (R_s \times i_{n,op\ amp})^2} \qquad (11\text{-}2)$$

The thermal noise of the resistor, the voltage noise due to amplifier noise current flowing through the resistor, and the input noise voltage of the amplifier are added together (in root-sum-square manner, since the noise voltages are uncorrelated) to determine the total input noise and may be compared with the input voltage noise in the absence of the protection resistor.

A protection resistor in series with an amplifier input will also produce a voltage drop due to the amplifier bias current flowing through it. This drop appears as an increase in the circuit offset voltage (and, if the bias current changes with temperature, offset drift). In amplifiers where bias currents are approximately equal, a resistor in series with each input will tend to balance the effect and reduce the error. The effects of this additional series resistance on the circuit's overall offset voltage can be calculated:

$$V_{os(total)} = V_{os} + I_b R_s \qquad (11\text{-}3)$$

For the case where $R_{FB}$ $R_s$ or where the source impedance levels are balanced, then the total circuit offset voltage can be expressed as:

$$V_{os(total)} = V_{os} + I_{os} R_s \qquad (11\text{-}4)$$

To limit the additional noise of $R_{FB}$, it can be shunted with a capacitor.

When using external clamp diodes to protect operational amplifier inputs, the effects of diode junction capacitance and leakage current should be evaluated in the application. Diode junction capacitance and

$R_s$ will add an additional pole in the signal path, and diode leakage currents will double for every 10°C rise in ambient temperature. Therefore, low leakage diodes should be used such that, at the highest ambient temperature for the application, the total diode leakage current is less than one-tenth of the input bias current for the device at that temperature. Another issue with regard to the use of Schottky diodes is the change in their forward voltage drop as a function of temperature. These diodes do not, in fact, limit the signal to ±0.3 V at all ambient temperatures, but if the Schottky diodes are at the same temperature as the op amp, they will limit the voltage to a safe level, even if they do not limit it at all times to within the data sheet rating. This is true if overvoltage is only possible at turn-on, when the diodes and the op amp will always be at the same temperature. If the op amp is warm when it is repowered, however, steps must be taken to ensure that diodes and op amp are at the same temperature.

# Electrostatic Discharge

## Understanding and Protecting Integrated Circuits from Electrostatic Discharge

Integrated circuits can be damaged by the high voltages and high peak currents that can be generated by electrostatic discharge (ESD). Precision analog circuits, which often feature very low bias currents, are more susceptible to damage than common digital circuits, because the traditional input protection structures which protect against ESD damage also increase input leakage.

The keys to eliminating ESD damage are: (1) awareness of the sources of ESD voltages and (2) understanding the simple handling steps that will discharge potential voltages safely.

The basic definitions relating to ESD are given in Figure 11-8. Notice that the *ESD failure threshold* level relates to any of the IC data sheet limits, and not simply to a *catastrophic failure* of the device. Also, the limits apply to each pin of the IC, not just to the input and output pins.

- ESD (electrostatic discharge):
  - A single fast, high current transfer of electrostatic charge.
  - Direct contact between two objects at different potentials.
  - A high electrostatic field between two objects when they are in close proximity.
- ESD failure threshold:
  - The highest voltage level at which all pins on a device can be subjected to ESD zaps without failing any 25°C data sheet limits.

**Figure 11-8: ESD definitions**

The generation of static electricity caused by rubbing two substances together is called the *triboelectric effect*. Static charge can be generated either by dissimilar materials (e.g., rubber-soled shoes moving across a rug) or by separating similar materials (e.g., pulling transparent tape off of a roll).

A wide variety of common human activities can create high electrostatic charge. Some examples are given in Figure 11-9. The values shown will occur with a fairly high relative humidity. Low humidity, such as can occur indoors during cold weather, can generate voltages 10 times (or more) greater than the values shown.

In an effort to standardize the testing and classification of integrated circuits for ESD robustness, ESD models have been developed (Figure 11-10). These models attempt to simulate the source of ESD voltage. The assumptions underlying the three commonly used models are different, so results are not directly comparable.

The most often encountered ESD model is the human body model (HBM). This model simulates the approximate resistance and capacitance of a human body with a simple RC network. The capacitor is charged through a high voltage power supply (HVPS) and then discharged (using a high voltage switch) through a series resistor. The RC values for different individuals will, of course, vary. However, the HBM has been standardized by MIL-STD-883 Method 3015 Electrostatic Discharge Sensitivity Classification, which specifies R–C combinations of 1.5 kΩ and 100 pF. (R, C, and L values for all three ESD models are shown in Figure 11-11.)

- Person walks across a typical carpet.
  - 1,000–1,500 V generated
- Person walks across a typical vinyl floor.
  - 150–250 V generated
- Person handles instructions protected by clear plastic covers.
  - 400–600 V generated
- Person handles polyethylene bags.
  - 1,000–1,200 V generated
- Person pours polyurethane foam into a box.
  - 1,200–1,500 V generated
- An IC slides down a grounded handler chute.
  - 50–500 V generated
- An IC slides down an open conductive shipping tube.
  - 25–250 V generated

*Note*: Above values can occur in a high ($\approx$60%) RH environment. For low RH ($\approx$30%), generated voltages can be >10 times those listed above!

**Figure 11-9: Examples of ESD generation**

- Three Models:
  - Human Body Model (HBM)
  - Machine Model (MM)
  - Charged Device Model (CDM)
- Model Correlation:
  - Low – assumptions are different

**Figure 11-10: Modeling electrostatic potential**

- Human Body Model (HBM)
  Simulates the discharge event that occurs when a person charged to either a positive or negative potential touches an IC at a different potential.
  $$\text{RLC:} \qquad R = 1.5\,\text{k}\Omega, \qquad L \approx 0\,\text{nH}, \qquad C = 100\,\text{pF}$$
- Machine Model (MM)
  Non-real-world Japanese model based on worst-case HBM.
  $$\text{RLC:} \qquad R \approx 0\,\Omega, \qquad L \approx 500\,\text{nH}, \qquad C = 200\,\text{pF}$$
- Charged Device Model (CDM)
  Simulates the discharge that occurs when a pin on an IC, charged to either a positive or negative potential, contacts a conductive surface at a different (usually ground) potential.
  $$\text{RLC:} \qquad R = 1\,\Omega \qquad L \approx 0\,\text{nH}, \qquad C = 1\text{–}20\,\text{pF}$$

**Figure 11-11: ESD models applicable to ICs**

The machine model (MM) is a worst-case HBM. Rather than using an *average* value for resistance and capacitance of the human body, the MM assumes a worst-case value of 200 pF and 0 Ω. The 0 Ω output resistance of the MM is also intended to simulate the discharge from a charged conductive object (e.g., a charged device under test (DUT) socket on an automatic test system) to an IC pin, which is how the MM earned its name. However, the MM does not simulate many known real-world ESD events. Rather, it models the ESD event resulting from an ideal voltage source (in other words, with no resistance in the discharge path). EIAJ Specification ED-4701 Test Method C-111 Condition A and ESD Association Specification S5.2 provide guidelines for MM testing.

The charged device model (CDM) originated at AT&T. This model differs from the HBM and the MM, in that the source of the ESD energy is the IC itself. The CDM assumes that the integrated circuit die, bond wires, and leadframe are charged to some potential (usually positive with respect to ground). One or more of the IC pins then contacts ground, and the stored charge rapidly discharges through the leadframe and bond wires. Typical examples of triboelectric charging followed by a CDM discharge include:

1.  An IC slides down a handler chute and then a corner pin contacts a grounded stop bar.

2.  An IC slides down an open conductive shipping tube and then a corner pin contacts a conductive surface.

The basic concept of the CDM is different from the HBM and MM in two ways. First, the CDM simulates a charged IC discharging to ground, while the HBM and MM both simulate a charged source discharging into the IC. Thus, current flows out of the IC during CDM testing, and into the IC during HBM and MM testing. The second difference is that the capacitor in the CDM is the capacitance of the package, while the HBM and MM use a fixed external capacitor.

Unlike the HBM and MM, CDM ESD thresholds may vary for the same die in different packages. This occurs because the DUT capacitance is a function of the package. For example, the capacitance of an 8-pin package is different from the capacitance of a 14-pin package. CDM capacitance values can vary from about 1 to 20 pF. The device capacitance is discharged through a 1 Ω resistor.

Schematic representations of the three models are shown in Figure 11-12. Notice that C1 in the HBM and MM are external capacitors, while $C_{PKG}$ in the CDM is the internal capacitance of the DUT.

The HBM discharge waveform is a predicable unipolar RC pulse, while the MM discharge shows ringing because of the parasitic inductance in the discharge path (typically 500 nH). Ideally, the CDM waveform is also a single unipolar pulse, but the parasitic inductance in series with the 1 Ω resistor slows the rise time and introduces some ringing.

The significant features of each ESD model are summarized in Figure 11-13. The peak currents shown for each model are based on a test voltage of 400 V. Peak current is lowest for the HBM because of the relatively high discharge resistance. The CDM discharge has low energy because device capacitance is only in the range of 1–20 pF, but peak current is high. The MM has the highest energy discharge, because it has the highest capacitance value (power $= 0.5 CV^2$).

Figure 11-14 compares 400 V discharge waveforms of the CDM, MM, and HBM, with the same current and time scales.

The CDM waveform corresponds to the shortest known real-world ESD event. The waveform has a rise time of <1 ns, with the total duration of the CDM event only about 2 ns. The CDM waveform is essentially unipolar, although some ringing occurs at the end of the pulse that results in small negative-going peaks. The very short duration of the overall CDM event results in an overall discharge of relatively low energy, but peak current is high.

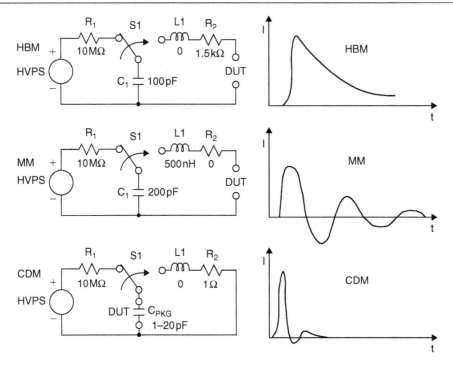

**Figure 11-12: Schematic representations of ESD models and typical discharge waveforms**

| Model | HBM | MM | Socketed CDM |
|---|---|---|---|
| Simulate | Human body | Machine | Charged device |
| Origin | US Military, late 1960s | Japan, 1976 | AT&T, 1974 |
| Real world? | Yes | Generally | Yes |
| RC | 1.5 k$\Omega$, 100 pF | 0 $\Omega$, 200 pF | 1 $\Omega$, 1–20 pF |
| Rise time | <10 ns (6–9 ns typ) | 14 ns* | 400 ps** |
| Ipeak at 400 V | 0.27 A | 5.8A* | 2.1 A** |
| Energy | Moderate | High | Low |
| Package dependent | No | No | Yes |
| Standard | MIL-STD-883 Method 3015 | ESD Association std. S5.2; EIAJ Std. ED-4701, Method C-111 | ESD Association Draft Std. DS5.3 |

\* These values per ESD association std. S5.2. EIAJ std. ED-4701, Method C-111 includes no waveform specifications.
\*\* These values are for the direct charging (socketed) method.

**Figure 11-13: Comparison of HBM, MM, and CDM ESD models**

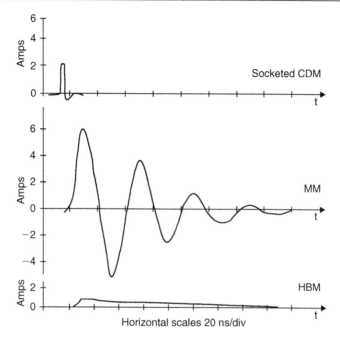

**Figure 11-14: Relative comparison of 400 V HBM, MM, and CDM discharges**

The MM waveform consists of both positive- and negative-going sinusoidal peaks, with a resonance frequency of 10–15 MHz. The initial MM peak has a typical rise time of 14 ns, and the total pulse duration is about 150 ns. The multiple high current, moderate duration peaks of the MM result in an overall discharge energy that is by far the highest of the three models for a given test voltage.

The rise time for the unipolar HBM waveform is typically 6–9 ns, and the waveform decays exponentially toward 0 V with a fall time of approximately 150 ns (Method 3015 requires a rise time of <10 ns and a delay time of 150 ± 20 ns, with decay time defined as the time for the waveform to drop from 100% to 36.8% of peak current). The peak current for the HBM is 400 V/1,500 Ω or 0.267 A, which is much lower than is produced by 400 V CDM and MM events. However, the relatively long duration of the total HBM event still results in an overall discharge of moderately high energy.

As previously noted, the MM waveform is bipolar while HBM and CDM waveforms are primarily unipolar. However, HBM and CDM testing is done with both positive and negative polarity pulses. Thus all three models stress the IC in both directions.

MIL-STD-883 Method 3015 classifies ICs for ESD failure threshold. The classification limits, shown in Figure 11-15, are derived using the HBM shown in Figure 11-13. Method 3015 also mandates a marking method to denote the ESD classification. All military grade Class 1 and 2 devices have their packages marked with one or two "Δ" symbols, respectively, while Class 3 devices (with a failure threshold >4 kV) do not have any ESD marking. Commercial and industrial grade IC packages may not be marked with any ESD classification symbol.

Notice that the Class 1 limit includes all devices which do not pass a 2 kV threshold. However, a Class 1 rating does not imply that all devices within that class will pass 1,999 V. In any event, the emphasis must

| HBM ESD class | Failuro threshold | Marking |
|:---:|:---:|:---:|
| 1 | <2 kV | Δ |
| 2 | 2–<4 kV | Δ Δ |
| 3 | >4 kV | None |

*Note*: Commercial and industrial ICs are not marked for ESD.

**Figure 11-15: Classifying and marking ICs for ESD per MIL-883C, Method 3015**

be placed on eliminating ESD exposure, not on attempting to decide how much ESD exposure is "safe" (Figure 11-16).

- ESD failure mechanisms:
  - Dielectric or junction damage
  - Surface charge accumulation
  - Conductor fusing.
- ESD damage can cause:
  - Increased leakage
  - Reduced performance
  - Functional failures of ICs.
- ESD damage is often cumulative:
  - For example, each ESD "Zap" may increase junction damage until, finally, the device fails.

**Figure 11-16: Understanding ESD damage**

A detailed discussion of IC failure mechanisms is beyond the scope of this seminar, but some typical ESD effects are shown in Figure 11-17.

- ESD damage *cannot* be "cured"!
- Circuits cannot be *tweaked, nulled, adjusted,* etc., to compensate for ESD damage.
- ESD damage must be *prevented*!

**Figure 11-17: The most important thing to remember about ESD damage**

For the design engineer or technician, the most common manifestation of ESD damage is a catastrophic failure of the IC. However, exposure to ESD can also cause increased leakage or degrade other parameters. If a device appears not to meet a data sheet specification during evaluation, the possibility of ESD damage should be considered.

Special care should be taken when breadboarding and evaluating ICs. The effects of ESD damage can be cumulative, so repeated mishandling of a device can eventually cause a failure. Inserting and removing ICs from a test socket, storing devices during evaluation, and adding or removing external components on the breadboard should all be done while observing proper ESD precautions. Again, if a device fails during a prototype system development, repeated ESD stress may be the cause.

The key word to remember with respect to ESD is *prevention*. There is no way to undo ESD damage, or to compensate for its effects.

Since ESD damage cannot be undone, the only cure is prevention. Luckily, prevention is a simple two-step process. The first step is recognizing ESD-sensitive products, and the second step is understanding how to handle these products (Figure 11-18).

Two key elements in protecting circuits from ESD damages are:
- Recognizing ESD-sensitive products
- Always handling ESD-sensitive products at a *grounded* workstation

**Figure 11-18: Preventing ESD damage to ICs**

All static-sensitive devices are shipped in protective packaging. ICs are usually contained in either conductive foam or in antistatic tubes. Either way, the container is then sealed in a static-dissipative plastic bag. The sealed bag is marked with a distinctive sticker, such as is shown in Figure 11-19, which outlines the appropriate handling procedures.

All static sensitive devices are sealed in
protective packaging and marked with
special handling instructions

Caution

Sensitive electronic devices

Do not ship or store near strong
electrostatic, electromagnetic, magnetic,
or radioactive fields

Caution

Sensitive electronic devices

Do not open except at
approved field force
protective work station

**Figure 11-19: Recognizing ESD-sensitive devices**

Once ESD-sensitive devices are identified, protection is easy. Obviously, keeping ICs in their original protective packaging as long as possible is the first step. The second step is to discharge potential ESD sources before damage to the IC can occur. The HBM capacitance is only 100 pF, so discharging a potentially dangerous voltage can be done quickly and safely through a high impedance. Even with a source resistance of 10 MΩ, the 100 pF will be discharged in less than 100 ms.

The key component required for safe ESD handling is a workbench with a static-dissipative surface, as shown in Figure 11-20. This surface is connected to ground through a 1 MΩ resistor, which dissipates static

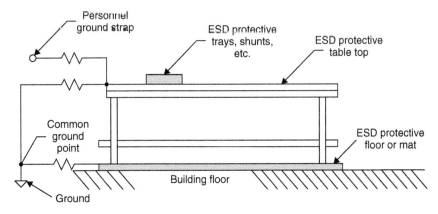

*Note*: Conductive table top sheet resistance » $10^6 \Omega/$

**Figure 11-20: Workstation for handling ESD-sensitive devices**

charge while protecting the user from electrical shock hazards caused by ground faults. If existing bench tops are non-conductive, a static-dissipative mat should be added, along with a discharge resistor.

Notice that the surface of the workbench has a moderately high sheet resistance. It is neither necessary nor desirable to use a low resistance surface (such as a sheet of copper-clad PC board) for the work surface. Remember, the CDM assumes that a high peak current will flow if a charged IC is discharged through a low impedance. This is precisely what happens when a charged IC contacts a grounded copper-clad board. When the same charged IC is placed on the surface shown in Figure 11-20, however, the peak current is not high enough to damage the device.

A conductive wrist strap is also recommended while handling ESD-sensitive devices. The wrist strap ensures that normal tasks, such as peeling tape off of packages, will not cause damage to ICs. Again, a 1 MΩ resistor, from the wrist strap to ground, is required for safety.

When building prototype breadboards or assembling PC boards which contain ESD-sensitive devices, all passive components should be inserted and soldered before the ICs. This procedure minimizes the ESD exposure of the sensitive devices. The soldering iron must, of course, have a grounded tip.

- Analog Devices is committed to helping our customers prevent ESD damage by:
    - Building products with the highest level of ESD protection commensurate with performance requirements
    - Protecting products from ESD during shipment
    - Helping customers to avoid ESD exposure during manufacture

**Figure 11-21: Analog devices commitment**

Protecting ICs from ESD requires the participation of both the IC manufacturer and the customer. IC manufacturers have a vested interest in providing the highest possible level of ESD protection for their products (Figure 11-21). IC circuit designers, process engineers, packaging specialists, and others are constantly looking for new and improved circuit designs, processes, and packaging methods to withstand or shunt ESD energy (Figure 11-22).

Analog devices
- Circuit design and fabrication
    - Design and manufacture products with the highest level of ESD protection
    - Consistent with required analog and digital performance.
- Pack and Ship
    - Pack in static dissipative material. Mark packages with ESD warning.

Customers
- Incoming Inspection
    - Inspect at grounded workstation. Minimize handling.
- Inventory control
    - Store in original ESD-safe packaging. Minimize handling.
- Manufacturing
    - Deliver to work area in original ESD-safe packaging. Open packages only at
    - grounded workstation. Package subassemblies in static dissipative packaging.
- Pack and ship
    - Pack in static dissipative material if required. Replacement or optional boards may require special attention.

**Figure 11-22: ESD protection requires a partnership between the IC supplier and the customer**

A complete ESD protection plan, however, requires more than building-ESD protection into ICs. Users of ICs must also provide their employees with the necessary knowledge of and training in ESD handling procedures.

# EMI/RFI Considerations

Electromagnetic interference (EMI) has become a hot topic in the last few years among circuit designers and systems engineers. Although the subject matter and prior art have been in existence for over the last 50 years or so, the advent of portable and high frequency industrial and consumer electronics has provided a comfortable standard of living for many EMI testing engineers, consultants, and publishers. With the help of *EDN* magazine and Kimmel Gerke Associates, this section will highlight general issues of EMC (electromagnetic compatibility) to familiarize the system/circuit designer with this subject and to illustrate proven techniques for protection against EMI.

## A Primer on EMI Regulations

The intent of this section is to summarize the different types of EMC regulations imposed on equipment manufacturers, both voluntary and mandatory. Published EMC regulations apply at this time only to equipment and systems, and not to components. Thus, EMI *hardened* equipment does not necessarily imply that each of the components used (integrated circuits, especially) in the equipment must also be EMI *hardened*.

### Commercial Equipment

The two driving forces behind commercial EMI regulations are the FCC (Federal Communications Commission) in the US and the VDE (Verband Deutscher Electrotechniker) in Germany. VDE regulations are more restrictive than the FCC's with regard to emissions and radiation, but the European Community will be adding immunity to RF, ESD, and power-line disturbances to the VDE regulations, and will require mandatory compliance in 1996. In Japan, commercial EMC regulations are covered under the VCCI (Voluntary Control Council for Interference) standards and, implied by the name, are much looser than their FCC and VDE counterparts.

All commercial EMI regulations primarily focus on *radiated* emissions, specifically to protect nearby radio and TV receivers, although both FCC and VDE standards are less stringent with respect to *conducted* interference (by a factor of 10 over radiated levels). The FCC Part 15 and VDE 0871 regulations group commercial equipment into two classes: Class A, for all products intended for business environments; and Class B, for all products used in residential applications. For example, Table 11-1 illustrates the

Table 11-1: Radiated emission limits for commercial computer equipment

| Frequency (MHz) | Class A (at 3 m) | Class B (at 3 m) |
|---|---|---|
| 30–88 | 300 µV/m | 100 µV/m |
| 88–216 | 500 µV/m | 150 µV/m |
| 216–1,000 | 700 µV/m | 200 µV/m |

*Source*: Reprinted from *EDN* magazine (January 20, 1994) © CAHNERS PUBLISHING COMPANY 1995, a division of Reed Publishing USA.

electric-field emission limits of commercial computer equipment for both FCC Part 15 and VDE 0871 compliance.

In addition to the already stringent VDE emission limits, the European Community EMC standards (IEC and IEEE) require mandatory compliance to these additional EMI threats: immunity to RF fields, ESD, and power-line disturbances. All equipment/systems marketed in Europe must exhibit an immunity to RF field strengths of 1–10V/m (IEC standard 801-3), ESD (generated by human contact or through material movement) in the range of 10–15kV (IEC standard 801-2), and power-line disturbances of 4kV EFTs (extremely fast transients, IEC standard 801-4) and 6kV lightning surges (IEEE standard C62.41).

## Military Equipment

The defining EMC specification for military equipment is MIL-STD-461 which applies to radiated equipment emissions and equipment susceptibility to interference. Radiated emission limits are very typically 10–100 times more stringent than the levels shown in Table 11-1. Required limits on immunity to RF fields are typically 200 times more stringent (RF field strengths of 5–50mV/m) than the limits for commercial equipment.

## Medical Equipment

Although not yet mandatory, EMC regulations for medical equipment are presently being defined by the FDA (Food and Drug Administration) in the USA and the European Community. The primary focus of these EMC regulations will be on immunity to RF fields, ESD, and power-line disturbances, and may very well be more stringent than the limits spelled out in MIL-STD-461. The primary objective of the medical EMC regulations is to guarantee safety to humans.

## Automotive Equipment

Perhaps the most difficult and hostile environment in which electrical circuits and systems must operate is that found in the automobile. All of the key EMI threats to electrical systems exist here. In addition, operating temperature extremes, moisture, dirt, and toxic chemicals further exacerbate the problem. To complicate matters further, standard techniques (ferrite beads, feed-through capacitors, inductors, resistors, shielded cables, wires, and connectors) used in other systems are not generally used in automotive applications because of the cost of the additional components.

Presently, automotive EMC regulations, defined by the very comprehensive SAE standards J551 and J1113, are not yet mandatory. They are, however, very rigorous. SAE standard J551 applies to vehicle-level EMC specifications, and standard J1113 (functionally similar to MIL-STD-461) applies to all automotive electronic modules. For example, the J1113 specification requires that electronic modules cannot radiate electric fields greater than 300nV/m at a distance of 3m. This is roughly 1,000 times more stringent than the FCC Part 15 Class A specification. In many applications, automotive manufacturers are imposing J1113 RF field immunity limits on *each of the active components* used in these modules. Thus, in the very near future, automotive manufacturers will require that IC products comply with existing EMC standards and regulations.

## EMC Regulations' Impact on Design

In all these applications and many more, complying with mandatory EMC regulations will require careful design of individual circuits, modules, and systems using established techniques for cable shielding, signal and power-line filtering against both small- and large-scale disturbances, and sound multi-layer PCB layouts. The key to success is to incorporate sound EMC principles early in the design phase to avoid time-consuming and expensive redesign efforts.

## Passive Components: Your Arsenal Against EMI

Minimizing the effects of EMI requires that the circuit/system designer be completely aware of the primary arsenal in the battle against interference: *passive components*. To successfully use these components, the designer must understand their non-ideal behavior. For example, Figure 11-23 illustrates the *real* behavior of the passive components used in circuit design. At very high frequencies, wires become transmission lines, capacitors become inductors, inductors become capacitors, and resistors behave as resonant circuits.

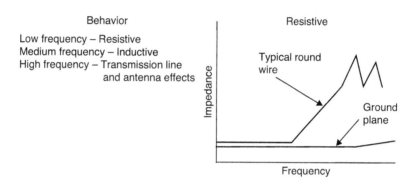

**Figure 11-23: Impedance comparison: wire versus ground plane**

A specific case in point is the frequency response of a simple wire compared to that of a ground plane. In many circuits, wires are used as either power or signal returns, and there is no ground plane. A wire will behave as a very low resistance (<0.02 Ω/feet for 22-gauge wire) at low frequencies, but because of its parasitic inductance of approximately 20 nH/feet, it becomes inductive at frequencies above 160 kHz. Furthermore, depending on size and routing of the wire and the frequencies involved, it ultimately becomes a transmission line with an uncontrolled impedance. From our knowledge of RF, unterminated transmission lines become antennas with gain, as illustrated in Figure 11-26. On the other hand, large area ground planes are much more well behaved, and maintain a low impedance over a wide range of frequencies. With a good understanding of the behavior of *real* components, a strategy can now be developed to find solutions to most EMI problems.

With any problem, a strategy should be developed before any effort is expended trying to solve it. This approach is similar to the scientific method: initial circuit misbehavior is noted, theories are postulated, experiments designed to test the theories are conducted, and results are again noted. This process continues until all theories have been tested and expected results achieved and recorded. With respect to EMI, a problem solving framework has been developed. As shown in Figure 11-24, the model suggested by Kimmel Gerke in Reference 2 illustrates that all three elements (a *source*, a *receptor* or *victim*, and a *path* between the two) must exist in order to be considered an EMI problem. The sources of EMI can take on many forms, and the ever-increasing number of portable instrumentation and personal communications/computation equipment only adds the number of possible sources and receptors.

Interfering signals reach the receptor by *conduction* (the circuit or system interconnections) or *radiation* (parasitic mutual inductance and/or parasitic capacitance). In general, if the frequencies of the interference are less than 30 MHz, the primary means by which interference is coupled is through the *interconnects*.

Any interference problem can be broken down into

- The source of interference
- The receptor of interference
- The path coupling the source to the receptor

| Sources | Paths | Receptors |
|---|---|---|
| Microcontroller<br>• Analog<br>• Digital | Radiated<br>• EM Fields<br>• Crosstalk<br>  Capacitive<br>  Inductive | Microcontroller<br>• Analog<br>• Digital |
| ESD<br>Communications<br>Transmitters<br>Power<br>Disturbances<br>Lightning | Conducted<br>• Signal<br>• Power<br>• Ground | Communications<br>• Receivers<br><br>Other electronic<br>systems |

Reprinted from *EDN* magazine (January 20, 1994) © CAHNERS PUBLISHING COMPANY 1995,
A Division of Reed Publishing USA

**Figure 11-24: A diagnostic framework for EMI**

Between 30 and 300 MHz, the primary coupling mechanism is *cable radiation and connector leakage*. At frequencies greater than 300 MHz, the primary mechanism is *slot and board radiation*. There are many cases where the interference is broadband, and the coupling mechanisms are combinations of the above.

When all three elements exist together, a framework for solving any EMI problem can be drawn from Figure 11-25. There are three types of interference with which the circuit or system designer must contend. The first type of interference is that generated by and emitted from an instrument; this is known as circuit/system *emission* and can be either *conducted* or *radiated*. An example of this would be the personal

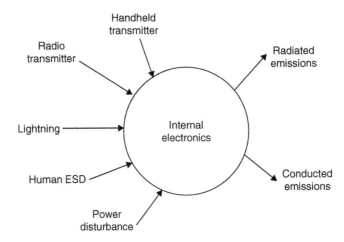

Reprinted from *EDN* magazine (January 20, 1994) © CAHNERS PUBLISHING COMPANY 1995,
A Division of Reed Publishing USA

**Figure 11-25: Three types of interference: emissions, immunity, and internal**

computer. Portable and desktop computers must pass the stringent FCC Part 15 specifications prior to general use.

The second type of interference is circuit or system *immunity*. This describes the behavior of an instrument when it is exposed to large electromagnetic fields, primarily electric fields with an intensity in the range of 1–10V/m at a distance of 3 m. Another term for immunity is *susceptibility*, and it describes circuit/system behavior against radiated or conducted interference.

The third type of interference is *internal*. Although not directly shown on the figure, internal interference can be high speed digital circuitry within the equipment which affects sensitive analog (or other digital circuitry), or noisy power supplies which can contaminate both analog and digital circuits. Internal interference often occurs between digital and analog circuits, or between motors or relays and digital circuits. In mixed-signal environments, the digital portion of the system often interferes with analog circuitry. In some systems, the internal interference reaches such high levels that even very high speed digital circuitry can affect other low speed digital circuitry as well as analog circuits.

In addition to the source-path-receptor model for analyzing EMI-related problems, Kimmel Gerke Associates have also introduced the FAT-ID concept (Reference 2). FAT-ID is an acronym that describes the five key elements inherent in any EMI problem. These five key parameters are: *frequency*, *amplitude*, *time*, *impedance*, and *distance*.

The *frequency* of the offending signal suggests its path. For example, the path of low frequency interference is often the circuit conductors. As the interference frequency increases, it will take the path of least impedance, usually stray capacitance. In this case, the coupling mechanism is radiation.

Time and frequency in EMI problems are interchangeable. In fact, the physics of EMI shows that the time response of signals contains all the necessary information to construct the spectral response of the interference. In digital systems, both the signal rise time and pulse repetition rate produce spectral components according to the following relationship:

$$f_{EMI} = \frac{1}{\pi \, t_{rise}}$$ (11-5)

For example, a pulse having a 1 ns rise time is equivalent to an EMI frequency of over 300MHz. This time–frequency relationship can also be applied to high speed analog circuits, where slew rates in excess of 1,000V/μs and gain-bandwidth products greater than 500MHz are not uncommon.

When this concept is applied to instruments and systems, EMI emissions are again functions of signal rise time and pulse repetition rates. Spectrum analyzers and high speed oscilloscopes used with voltage and current probes are very useful tools in quantifying the effects of EMI on circuits and systems.

Another important parameter in the analysis of EMI problems is the physical dimensions of cables, wires, and enclosures. Cables can behave as either passive antennas (receptors) or very efficient transmitters (sources) of interference. Their physical length and their shield must be carefully examined where EMI is a concern. As previously mentioned, the behavior of simple conductors is a function of length, cross-sectional area, and frequency. Openings in equipment enclosures can behave as slot antennas, thereby allowing EMI energy to affect the internal electronics.

## Radio Frequency Interference

The world is rich in radio transmitters: radio and TV stations, mobile radios, computers, electric motors, garage door openers, electric jackhammers, and countless others. All this electrical activity can affect circuit/system performance and, in extreme cases, may render it inoperable. Regardless of the location and magnitude of the interference, circuits/systems must have a minimum level of immunity to radio frequency interference (RFI). The next section will cover two general means by which RFI can disrupt normal instrument operation: the direct effects of RFI-sensitive analog circuits, and the effects of RFI on shielded cables.

Two terms are typically used in describing the sensitivity of an electronic system to RF fields. In communications, radio engineers define *immunity* to be an instrument's *susceptibility to the applied RFI power density at the unit*. In more general EMI analysis, the *electric-field intensity* is used to describe RFI stimulus. For comparative purposes, Eq. (11-6) can be used to convert electric-field intensity to power density and vice versa:

$$\vec{E}\left(\frac{V}{m}\right) = 61.4 \sqrt{P_T\left(\frac{mW}{cm^2}\right)} \tag{11-6}$$

where E is the electric-field strength, in V/m; and $P_T$ is the transmitted power, in $mW/cm^2$.

From the standpoint of the source-path-receptor model, the *strength of the electric field*, E, surrounding the receptor is a function of *transmitted power*, *antenna gain*, and *distance* from the source of the disturbance. An approximation for the electric-field intensity (for both near- and far-field sources) in these terms is given by Eq. (11-7):

$$\vec{E}\left(\frac{V}{m}\right) = 5.5 \left(\frac{\sqrt{P_T \times G_A}}{d}\right) \tag{11-7}$$

where E is the electric-field intensity, in V/m; $P_T$ is the transmitted power, in $mW/cm^2$; $G_A$ is the antenna gain (numerical); and d is the distance from source, in meters.

For example, a 1 W hand-held radio at a distance of 1 m can generate an electric field of 5.5 V/m, whereas a 10 kW radio transmission station located 1 km away generates a field smaller than 0.6 V/m.

Analog circuits are generally more sensitive to RF fields than digital circuits because analog circuits, operating at high gains, must be able to resolve signals in the microvolt/millivolt region. Digital circuits, on the other hand, are more immune to RF fields because of their larger signal swings and noise margins. As shown in Figure 11-26, RF fields can use inductive and/or capacitive coupling paths to generate noise currents and voltages which are amplified by high impedance analog instrumentation. In many cases, out-of-band noise signals are detected and rectified by these circuits. The result of the RFI rectification is usually unexplained offset voltage shifts in the circuit or in the system.

There are techniques that can be used to protect analog circuits against interference from RF fields (see Figure 11-27). The three general points of RFI coupling are *signal inputs*, *signal outputs*, and *power supplies*. At a minimum, all power-supply pin connections on analog and digital ICs should be decoupled with 0.1 μF ceramic capacitors. As was shown in Reference 3, lowpass filters (LPFs), whose cutoff frequencies are set no higher than 10–100 times the signal bandwidth, can be used at the inputs and the outputs of signal conditioning circuitry to filter noise.

Care must be taken to ensure that the LPFs are effective at the highest RF interference frequency expected. As illustrated in Figure 11-28, real LPFs may exhibit *leakage* at high frequencies. Their inductors can lose their effectiveness due to parasitic capacitance, and capacitors can lose their effectiveness due to parasitic

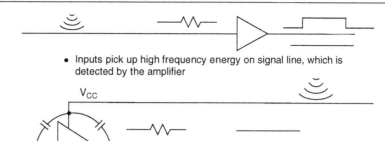

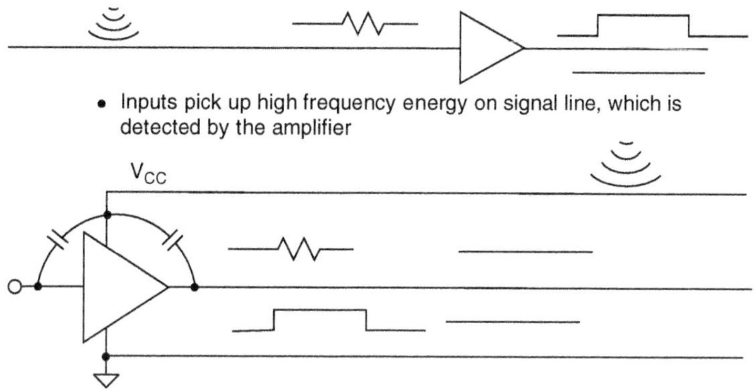

- Inputs pick up high frequency energy on signal line, which is detected by the amplifier
- Output drivers can be jammed, too: energy couples back to input via $V_{CC}$ or signal line and then is detected or amplified

Reprinted from *EDN* magazine (January 20, 1994) © CAHNERS PUBLISHING COMPANY 1995, A Division of Reed Publishing USA

**Figure 11-26: RFI can cause rectification in sensitive analog circuits**

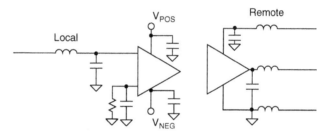

Decouple all voltage supplies to analog chip with high frequency capacitors
Use high frequency filters on all lines that leave the board
Use high frequency filters on the voltage reference if it is not grounded

Reprinted from *EDN* magazine (January 20, 1994) © CAHNERS PUBLISHING COMPANY 1995, A Division of Reed Publishing USA

**Figure 11-27: Keeping RFI away from analog circuits**

Reprinted from *EDN* magazine (January 20, 1994) © CAHNERS PUBLISHING COMPANY 1995, A Division of Reed Publishing USA

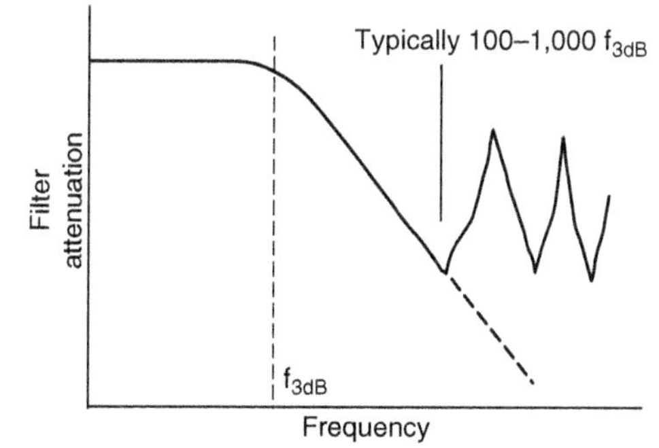

**Figure 11-28: Single low power lowpass filter loses effectiveness at 100–1,000 $f_{3\,dB}$**

inductance. A rule-of-thumb is that a conventional LPF (made up of a single capacitor and inductor) can begin to *leak* when the applied signal frequency is 100–1,000 higher than the filter's cutoff frequency. For example, a 10 kHz LPF would not be considered very efficient at filtering frequencies above 1 MHz.

Rather than using one LPF stage, it is recommended that the interference frequency bands be separated into *low band*, *midband*, and *high band*, and then use individual filters for each band. Kimmel Gerke Associates use the stereo speaker analogy of *woofer-midrange-tweeter* for RFI LPF design illustrated in Figure 11-29. In this approach, low frequencies are grouped from 10 kHz to 1 MHz, midband frequencies are grouped from 1 MHz to 100 MHz, and high frequencies grouped from 100 MHz to 1 GHz. In the case of a shielded cable input/output, the high frequency section should be located close to the shield to prevent high frequency leakage at the shield boundary. This is commonly referred to as *feed-through* protection. For applications where shields are not required at the inputs/outputs, then the preferred method is to locate the high frequency filter section as close to the analog circuit as possible. This is to prevent the possibility of pickup from other parts of the circuit.

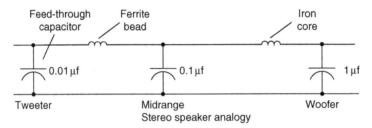

**Figure 11-29: Multistage filters are more effective**

## Ground Reduces Effectiveness

Another cause of filter failure is illustrated in Figure 11-30. If there is any impedance in the ground connection (e.g., a long wire or narrow trace connected to the ground plane), then the high frequency noise uses this impedance path to bypass the filter completely. Filter grounds must be broadband and tied to low impedance points or planes for optimum performance. High frequency capacitor leads should be kept as short as possible, and low inductance surface-mounted ceramic chip capacitors are preferable.

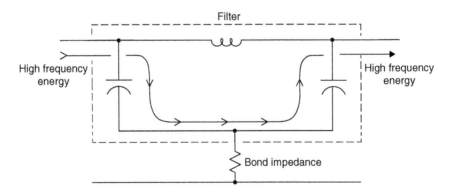

**Figure 11-30: Non-zero (inductive and/or resistive) filter**

In the first part of this discussion on RF immunity, circuit-level techniques were discussed. In this next section, the second strategic concept for RF immunity will be discussed: *all cables behave as antennas*. As shown in Figure 11-31, pigtail terminations on cables very often cause systems to fail radiated emissions tests because high frequency noise has coupled into the cable shield, generally through stray capacitance. If the length of the cable is considered *electrically long* (a concept to be explained later) at the interference frequency, then it can behave as a very efficient quarter-wave antenna. The cable pigtail forms a matching network, as shown in the Figure, to radiate the noise which coupled into the shield. In general, pigtails are only recommended for applications below 10 kHz, such as 50/60 Hz interference protection. For applications where the interference is greater than 10 kHz, shielded connectors, electrically and physically connected to the chassis, should be used. In applications where shielding is not used, filters on input/output signal and power lines work well. Small ferrites and capacitors should be used to filter high frequencies, provided that: (1) the capacitors have short leads and are tied directly to the chassis ground and (2) the filters are physically located close to the connectors to prevent noise pickup.

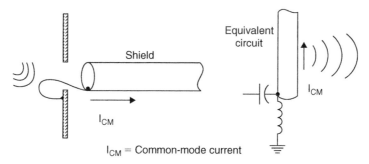

$I_{CM}$ = Common-mode current

**Figure 11-31: "Shielded" cable can carry high frequency current and behaves as an antenna**

The key issues and techniques described in this section on solving RFI-related problems are summarized in Figure 11-32. Some of the issues were not discussed in detail, but are equally important. For a complete treatment of this issue, the interested reader should consult References 1 and 2. The main thrust of this section was to provide the reader with a problem-solving strategy against RFI and to illustrate solutions to commonly encountered RFI problems.

## Solutions for Power-Line Disturbances

The goal of this next section is not to describe in detail all the circuit/system failure mechanisms which can result from power-line disturbances or faults. Nor is it the intent of this section to describe methods by which power-line disturbances can be prevented. Instead, this section will describe techniques that allow circuits and systems to accommodate *transient* power-line disturbances.

Figure 11-33 is an example of a hybrid power transient protection network commonly used in many applications where lightning transients or other power-line disturbances are prevalent. These networks can be designed to provide protection against transients as high as 10 kV and as fast as 10 ns. Gas discharge tubes (crowbars) and large geometry zener diodes (clamps) are used to provide both differential and common-mode protection. Metal-oxide varistors (MOVs) can be substituted for the zener diodes in less critical, or in more compact, designs. Chokes are used to limit the surge current until the gas discharge tubes fire.

- Radio-frequency interference is a serious threat
  - Equipment causes interference to nearby radio and television
  - Equipment upset by nearby transmitters
- RFI-failure modes
  - Digital circuits prime source of emissions
  - Analog circuits more vulnerable to RF than digital circuits
- Two strategic concepts
  - Treat all cables as antennas
  - Determine the most critical circuits
- RFI circuit protection
  - Filters and multilayers boards
  - Multistage filters often needed
- RFI Shielding
  - Slots and seams cause the most problems
- RFI cable protection
  - High-quality shields and connectors needed for RF protection

**Figure 11-32: Summary of RFI problems and protection techniques**

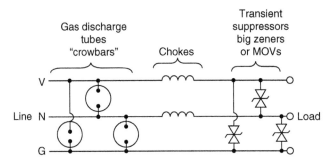

Reprinted from *EDN* magazine (January 20, 1994) © CAHNERS PUBLISHING COMPANY 1995, A Division of Reed Publishing USA

**Figure 11-33: Power-line disturbances can generate EMI**

Commercial EMI filters, as illustrated in Figure 11-34, can be used to filter less catastrophic transients or high frequency interference. These EMI filters provide both common- and differential-mode filtering. An optional choke in the safety ground can provide additional protection against common-mode noise. The value of this choke cannot be too large, however, because its resistance may affect power-line fault clearing.

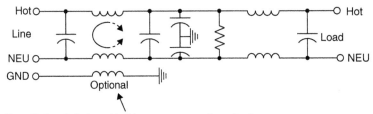

*Note*: Optional choke added for common-mode protection.

Reprinted from *EDN* magazine (January 20, 1994) © CAHNERS PUBLISHING COMPANY 1995, A Division of Reed Publishing USA

**Figure 11-34: Schematic for a commercial power-line filter**

Transformers provide the best common-mode power-line isolation. They provide good protection at low frequencies (<1 MHz), or for transients with rise and fall times greater than 300 ns. Most motor noise and lightning transients are in this range, so isolation transformers work well for these types of disturbances. Although the isolation between input and output is galvanic, isolation transformers do not provide sufficient protection against EFTs (<10 ns) or those caused by high amplitude ESD (1–3 ns). As illustrated in Figure 11-35, isolation transformers can be designed for various levels of differential- or common-mode protection. For differential-mode noise rejection, the Faraday shield is connected to the neutral, and for common-mode noise rejection, the shield is connected to the safety ground.

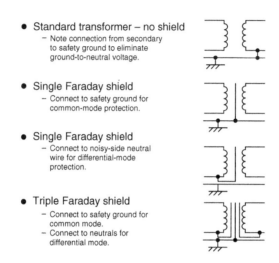

- Standard transformer – no shield
  - Note connection from secondary to safety ground to eliminate ground-to-neutral voltage.

- Single Faraday shield
  - Connect to safety ground for common-mode protection.

- Single Faraday shield
  - Connect to noisy-side neutral wire for differential-mode protection.

- Triple Faraday shield
  - Connect to safety ground for common mode.
  - Connect to neutrals for differential mode.

Reprinted from *EDN* magazine (January 20, 1994) © CAHNERS PUBLISHING COMPANY 1995, A Division of Reed Publishing USA

Figure 11-35: Faraday shields in isolation transformers provide increasing levels of protection

## PCB Design for EMI Protection

This section will summarize general points regarding the most critical portion of the design phase: the PCB layout. It is at this stage where the performance of the system is most often compromised. This is not only true for signal-path performance, but also for the system's susceptibility to EMI and the amount of electromagnetic energy radiated by the system. Failure to implement sound PCB layout techniques will very likely lead to system/instrument EMC failures.

Figure 11-36 is a real-world PCB layout which shows all the paths through which high frequency noise can couple/radiate into/out of the circuit. Although the diagram shows digital circuitry, the same points are applicable to precision analog, high speed analog, or mixed analog/digital circuits. Identifying critical circuits and paths helps in designing the PCB layout for both low emissions and susceptibility to radiated and conducted external and internal noise sources.

A key point in minimizing noise problems in a design is to *choose devices no faster than actually required by the application.* Many designers assume that faster is better: fast logic is better than slow, high bandwidth amplifiers are clearly better than low bandwidth ones, and fast digital-to-analog converters (DACs) and analog-to-digital converters (ADCs) are better, even if the speed is not required by the system. Unfortunately, faster is not better, but worse where EMI is concerned.

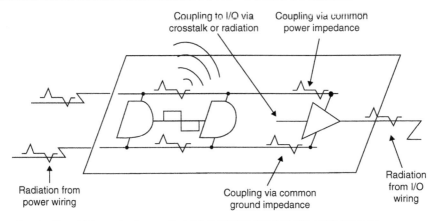

Coupling to I/O via crosstalk or radiation

Coupling via common power impedance

Radiation from power wiring

Coupling via common ground impedance

Radiation from I/O wiring

**Figure 11-36: Methods by which high frequency energy couples and radiates into circuitry via placements**

Many fast DACs and ADCs have digital inputs and outputs with rise and fall times in the nanosecond region. Because of their wide bandwidth, the sampling clock and the digital inputs can respond to any form of high frequency noise, even glitches as narrow as 1–3 ns. These high speed data converters and amplifiers are easy prey for the high frequency noise of microprocessors, digital signal processors, motors, switching regulators, hand-held radios, electric jackhammers, etc. With some of these high speed devices, a small amount of input/output filtering may be required to desensitize the circuit from its EMI/RFI environment. Adding a small ferrite bead just before the decoupling capacitor as shown in Figure 11-37 is very effective in filtering high frequency noise on the supply lines. For those circuits that require bipolar supplies, this technique should be applied to both positive- and negative-supply lines.

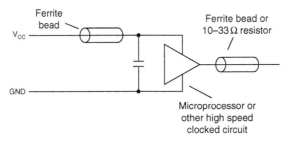

Ferrite bead

$V_{CC}$

GND

Ferrite bead or 10–33 Ω resistor

Microprocessor or other high speed clocked circuit

**Figure 11-37: Power-supply filtering and signal line snubbing greatly reduce EMI emissions**

To help reduce the emissions generated by extremely fast moving digital signals at DAC inputs or ADC outputs, a small resistor or ferrite bead may be required at each digital input/output.

Once the system's critical paths and circuits have been identified, the next step in implementing sound PCB layout is to partition the PCB according to circuit function. This involves the appropriate use of power, ground, and signal planes. Good PCB layouts also isolate critical analog paths from sources of high interference (e.g., I/O lines and connectors). High frequency circuits (analog and digital) should be separated from low frequency ones. Furthermore, automatic signal routing CAD layout software should be used with extreme caution, and critical paths routed by hand.

Properly designed multi-layer PCBs can reduce EMI emissions and increase immunity to RF fields by a factor of 10 or more compared to double-sided boards. A multi-layer board allows a complete layer to be used for the ground plane, whereas the ground plane side of a double-sided board is often disrupted with signal crossovers, etc.

The preferred multi-layer board arrangement is to embed the signal traces between the power and ground planes, as shown in Figure 11-38. These low impedance planes form very high frequency *stripline* transmission lines with the signal traces. The return current path for a high frequency signal on a trace is located directly above and below the trace on the ground/power planes. The high frequency signal is thus contained inside the PCB, thereby minimizing emissions. The embedded signal trace approach has an obvious disadvantage: debugging circuit traces that are hidden from plain view is difficult.

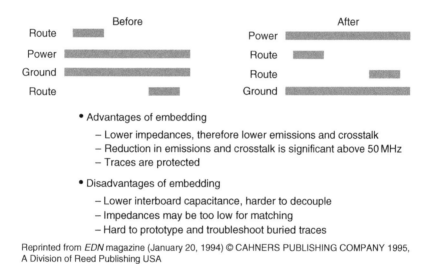

- Advantages of embedding
    - Lower impedances, therefore lower emissions and crosstalk
    - Reduction in emissions and crosstalk is significant above 50 MHz
    - Traces are protected

- Disadvantages of embedding
    - Lower interboard capacitance, harder to decouple
    - Impedances may be too low for matching
    - Hard to prototype and troubleshoot buried traces

Reprinted from *EDN* magazine (January 20, 1994) © CAHNERS PUBLISHING COMPANY 1995, A Division of Reed Publishing USA

**Figure 11-38: "To embed or not to embed" that is the question**

Much has been written about terminating PCB traces in their characteristic impedance to avoid reflections. A good rule-of-thumb to determine when this is necessary is as follows: *Terminate the line in its characteristic impedance when the one-way propagation delay of the PCB track is equal to or greater than one-half the applied signal rise/fall time (whichever edge is faster).* A conservative approach is to use a 2 inch (PCB track length)/nanosecond (rise-, fall-time) criterion. For example, PCB tracks for high speed logic with rise/fall time of 5 ns should be terminated in their characteristic impedance and if the track length is equal to or greater than 10 inches (including any meanders). The 2 inch/ns track length criterion is summarized in Figure 11-39 for a number of logic families.

This same 2 inch/ns rule-of-thumb should be used with analog circuits in determining the need for transmission line techniques. For instance, if an amplifier must output a maximum frequency of $f_{max}$, then the equivalent rise time, $t_r$, can be calculated using the equation $t_r = 0.35/f_{max}$. The maximum PCB track length is then calculated by multiplying the rise time by 2 inch/ns. For example, a maximum output frequency of 100 MHz corresponds to a rise time of 3.5 ns, and a track carrying this signal greater than 7 inches should be treated as a transmission line.

| Digital IC family | $t_r$, $t_f$ (ns) | PCB track length (inches) | PCB track length (cm) |
|---|---|---|---|
| GaAs | 0.1 | 0.2 | 0.5 |
| ECL | 0.75 | 1.5 | 3.8 |
| Schottky | 3 | 6 | 15 |
| Fast | 3 | 6 | 15 |
| AS | 3 | 6 | 15 |
| AC | 4 | 8 | 20 |
| ALS | 6 | 12 | 30 |
| LS | 8 | 16 | 40 |
| TTL | 10 | 20 | 50 |
| HC | 18 | 36 | 90 |

$t_r$ = rise time of signal in ns
$t_f$ = fall time of signal in ns

- For analog signals @ $f_{max}$, calculate $t_r = t_f = 0.35/f_{max}$

Reprinted from *EDN* magazine (January 20, 1994) © CAHNERS PUBLISHING COMPANY 1995, A Division of Reed Publishing USA

**Figure 11-39: Line termination should be used when the length of the PCB trace exceeds 2 inches/ns**

Equation (11-8) can be used to determine the characteristic impedance of a PCB track separated from a power/ground plane by the board's dielectric (microstrip transmission line):

$$Z_0(\Omega) = \frac{87}{\sqrt{\varepsilon_r + 1.41}} \ln\left[\frac{5.98d}{0.89w + t}\right] \tag{11-8}$$

where $\varepsilon_r$ is the dielectric constant of PCB material; d is the thickness of the board between metal layers, in mils; w is the width of metal trace, in mils; and t is the thickness of metal trace, in mils.

The one-way transit time for a single metal trace over a power/ground plane can be determined from Eq. (11-9):

$$t_{pd}(\text{ns/feet}) = 1.017\sqrt{0.475\varepsilon_r + 0.67} \tag{11-9}$$

For example, a standard four-layer PCB board might use 8-mil wide, 1 ounce (1.4 mils) copper traces separated by 0.021 inch FR-4 ($\varepsilon_r = 4.7$) dielectric material. The characteristic impedance and one-way transit time of such a signal trace would be 88 $\Omega$ and 1.7 ns/feet (7 inch/ns), respectively. Transmission lines can be effectively terminated in several ways depending on the application.

Figure 11-40 is a summary of techniques that should be applied to PCB layouts to minimize the effects of EMI, both emissions and immunity.

"All EMI problems begin and end at a circuit"

- Identify critical, sensitive circuits
- Where appropriate, choose ICs no faster than needed
- Consider and implement sound PCB design
- Spend time on the initial layout (by hand, if necessary)
- Power supply decoupling (digital and analog circuits)
- High-speed digital and high-accuracy analog do not mix
- Beware of connectors for input/output circuits
- Text, evaluate, and correct early and often

Reprinted from *EDN* magazine (January 20, 1994) © CAHNERS PUBLISHING COMPANY 1995, A Division of Reed Publishing USA

**Figure 11-40: Circuit board design and EMI**

## *A Review of Shielding Concepts*

The concepts of shielding effectiveness presented next are background material. Interested readers should consult References 1, 3, and 4 cited at the end of the section for more detailed information.

Applying the concepts of shielding requires an understanding of the source of the interference, the environment surrounding the source, and the distance between the source and point of observation (the receptor or victim). If the circuit is operating close to the source (in the near-, or induction-field), then the field characteristics are determined by the source. If the circuit is remotely located (in the far-, or radiation-field), then the field characteristics are determined by the transmission medium.

A circuit operates in a near-field if its distance from the source of the interference is less than the wavelength ($\lambda$) of the interference divided by $2\pi$, or $\pi/2\lambda$. If the distance between the circuit and the source of the interference is larger than this quantity, then the circuit operates in the far-field. For instance, the interference caused by a 1 ns pulse edge has an upper bandwidth of approximately 350 MHz. The wavelength of a 350 MHz signal is approximately 32 inches (the speed of light is approximately 12 inch/ns). Dividing the wavelength by $2\pi$ yields a distance of approximately 5 inches, the boundary between near- and far-field. If a circuit is within 5 inches of a 350 MHz interference source, then the circuit operates in the near-field of the interference. If the distance is greater than 5 inches, the circuit operates in the far-field of the interference.

Regardless of the type of interference, there is a characteristic impedance associated with it. The characteristic, or wave impedance, of a field is determined by the ratio of its electric (or E-) field to its magnetic (or H-) field. In the far-field, the ratio of the electric field to the magnetic field is the characteristic (wave impedance) of free space, given by $Z_0 = 377\,\Omega$. In the near-field, the wave impedance is determined by the nature of the interference and its distance from the source. If the interference source is high current and low voltage (e.g., a loop antenna or a power-line transformer), the field is predominately magnetic and exhibits a wave impedance which is less than $377\,\Omega$. If the source is low current and high voltage (e.g., a rod antenna or a high speed digital switching circuit), then the field is predominately electric and exhibits a wave impedance which is greater than $377\,\Omega$.

Conductive enclosures can be used to shield sensitive circuits from the effects of these external fields. These materials present an impedance mismatch to the incident interference because the impedance of the shield is lower than the wave impedance of the incident field. The effectiveness of the conductive shield depends on two things: first is the loss due to the *reflection* of the incident wave off the shielding material. Second is the loss due to the *absorption* of the transmitted wave *within* the shielding material. Both concepts are illustrated in Figure 11-41. The amount of reflection loss depends on the type of interference and its wave

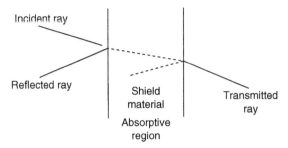

**Figure 11-41: Reflection and absorption are the two principal shielding mechanisms**

impedance. The amount of absorption loss, however, is independent of the type of interference. It is the same for near- and far-field radiation, as well as for electric or magnetic fields.

Reflection loss at the interface between two media depends on the difference in the characteristic impedances of the two media. For electric fields, reflection loss depends on the frequency of the interference and the shielding material. This loss can be expressed in dB, and is given by:

$$R_e(dB) = 322 + 10 \log_{10} \left[ \frac{\sigma_r}{\mu_r f^3 r^2} \right] \tag{11-10}$$

where $\sigma_r$ is the relative conductivity of the shielding material, in Siemens per meter; $\mu_r$ is the relative permeability of the shielding material, in Henries per meter; f is the frequency of the interference; and r is the distance from source of the interference, in meters.

For magnetic fields, the loss depends also on the shielding material and the frequency of the interference. Reflection loss for magnetic fields is given by:

$$R_m(dB) = 14.6 + 10 \log_{10} \left[ \frac{f r^2 \sigma_r}{\mu_r} \right] \tag{11-11}$$

and, for plane waves ( $r > \lambda/2\pi$ ), the reflection loss is given by:

$$R_{pw}(dB) = 168 + 10 \log_{10} \left[ \frac{\sigma_r}{\mu_r f} \right] \tag{11-12}$$

*Absorption* is the second loss mechanism in shielding materials. Wave attenuation due to absorption is given by:

$$A(dB) = 3.34t \sqrt{\sigma_r \mu_r f} \tag{11-13}$$

where t is the thickness of the shield material, in inches. This expression is valid for plane waves and electric and magnetic fields. Since the intensity of a transmitted field decreases exponentially relative to the thickness of the shielding material, the absorption loss in a shield one skin-depth ($\delta$) thick is 9 dB. Since absorption loss is proportional to thickness and inversely proportional to skin depth, increasing the thickness of the shielding material improves shielding effectiveness at high frequencies.

Reflection loss for plane waves in the far-field decreases with increasing frequency because the shield impedance, $Z_s$, increases with frequency. Absorption loss, on the other hand, increases with frequency because skin depth decreases. For electric fields and plane waves, the primary shielding mechanism is reflection loss, and at high frequencies, the mechanism is absorption loss. For these types of interference, high conductivity materials, such as copper or aluminum, provide adequate shielding. At low frequencies, both reflection and absorption loss to magnetic fields is low; thus, it is very difficult to shield circuits from low frequency magnetic fields. In these applications, high permeability materials that exhibit low reluctance provide the best protection. These low reluctance materials provide a magnetic shunt path that diverts the magnetic field away from the protected circuit. Some characteristics of metallic materials commonly used for shielded enclosures are shown in Table 11-2.

**Table 11-2: Impedance and skin depths for various shielding materials**

| Material | Conductivity $\sigma_r$ | Permeability $\mu_r$ | Shield impedance $|Z_s|$ | Skin depth $\delta$ (inch) |
|---|---|---|---|---|
| Cu | 1 | 1 | $3.68E-7 \cdot \sqrt{f}$ | $\dfrac{2.6}{\sqrt{f}}$ |
| Al | 1 | 0.61 | $4.71E-7 \cdot \sqrt{f}$ | $\dfrac{3.3}{\sqrt{f}}$ |
| Steel | 0.1 | 1,000 | $3.68E-5 \cdot \sqrt{f}$ | $\dfrac{0.26}{\sqrt{f}}$ |
| $\mu$ Metal | 0.03 | 20,000 | $3E-4 \cdot \sqrt{f}$ | $\dfrac{0.11}{\sqrt{f}}$ |

$\sigma_o$ is $5.82 \times 10^7$ S/m, $\mu_o$ is $4\pi \times 10^{-7}$ H/m, and $\varepsilon_o$ is $8.85 \times 10^{-12}$ F/m.

*Source:* Reprinted from *EDN* magazine (January 20, 1994) © CAHNERS PUBLISHING COMPANY 1995, a division of Reed Publishing USA.

A properly shielded enclosure is very effective at preventing external interference from disrupting its contents as well as confining any internally generated interference. However, in the real world, openings in the shield are often required to accommodate adjustment knobs, switches, and connectors, or to provide ventilation (see Figure 11-42). Unfortunately, these openings may compromise shielding effectiveness by providing paths for high frequency interference to enter the instrument.

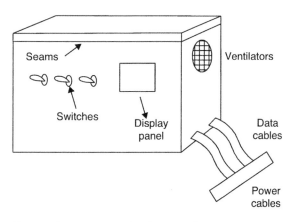

**Figure 11-42: Any opening in an enclosure can act as an EMI waveguide by compromising shielding effectiveness**

*815*

The longest dimension (not the total area) of an opening is used to evaluate the ability of external fields to enter the enclosure, because the openings behave as slot antennas. Equation (11-14) can be used to calculate the shielding effectiveness, or the susceptibility to EMI leakage or penetration, of an opening in an enclosure:

$$\text{Shielding effectiveness (dB)} = 20\log_{10}\left(\frac{\lambda}{2L}\right) \tag{11-14}$$

where $\lambda$ is the wavelength of the interference and L is the maximum dimension of the opening.

Maximum radiation of EMI through an opening occurs when the longest dimension of the opening is equal to one half-wavelength of the interference frequency (0 dB shielding effectiveness). A rule-of-thumb is to keep the longest dimension less than 1/20 wavelength of the interference signal, as this provides 20 dB shielding effectiveness. Furthermore, a few small openings on each side of an enclosure are preferred over many openings on one side. This is because the openings on different sides radiate energy in different directions, and as a result, shielding effectiveness is not compromised. If openings and seams cannot be avoided, then conductive gaskets, screens, and paints alone or in combination should be used judiciously to limit the longest dimension of any opening to less than 1/20 wavelength. Any cables, wires, connectors, indicators, or control shafts penetrating the enclosure should have circumferential metallic shields physically bonded to the enclosure at the point of entry. In those applications where unshielded cables/wires are used, then filters are recommended at the point of shield entry.

## General Points on Cables and Shields

Although covered in more detail later, the improper use of cables and their shields is a significant contributor to both radiated and conducted interference. Rather than developing an entire treatise on these issues, the interested reader should consult References 1, 2, 4, and 5. As illustrated in Figure 11-43 effective cable and enclosure shielding confines sensitive circuitry and signals within the entire shield without compromising shielding effectiveness.

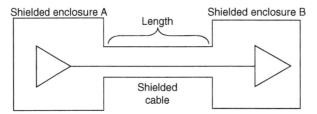

Fully shielded enclosures connected by fully
shielded cable keep all internal circuits and
signal lines inside the shield

• Transition region: 1/20 wavelength

**Figure 11-43: Length of shielded cables determines as "electrically long" or "electronically short" applications**

Depending on the type of interference (pickup/radiated, low/high frequency), proper cable shielding is implemented differently and is very dependent on the length of the cable (Figure 11-44). The first step is to determine whether the length of the cable is *electrically short* or *electrically long* at the frequency of concern. A cable is considered *electrically short* if the length of the cable is less than 1/20 wavelength of

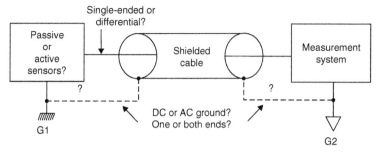

Reprinted from *EDN* magazine (January 20, 1994) © CAHNERS PUBLISHING COMPANY 1995, A Division of Reed Publishing USA

**Figure 11-44: Precision sensors and cable shielding**

the highest frequency of the interference; otherwise it is *electrically long*. For example, at 50/60 Hz, an *electrically short* cable is any cable length less than 150 miles, where the primary coupling mechanism for these low frequency electric fields is capacitive. As such, for any cable length less than 150 miles, the amplitude of the interference will be the same over the entire length of the cable. To protect circuits against low frequency electric-field pickup, only one end of the shield should be returned to a low impedance point. A generalized example of this mechanism is illustrated in Figure 11-45.

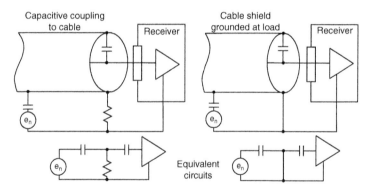

Reprinted from *EDN* magazine (January 20, 1994) © CAHNERS PUBLISHING COMPANY 1995, A Division of Reed Publishing USA

**Figure 11-45: Connect the shield at one point at the load to protect against low frequency (50/60 Hz) threats**

In this example, the shield is grounded at the receiver. An exception to this approach (which will be highlighted again later) is the case where line-level ($>1 V_{RMS}$) audio signals are transmitted over long distances using twisted pair, shielded cables. In these applications, the shield again offers protection against low frequency interference, and an accepted approach is to ground the shield at the driver end (low and high frequency ground) and ground it at the receiver with a capacitor (high frequency ground only).

In those applications where the length of the cable is *electrically long*, or protection against high frequency interference is required, then the preferred method is to connect the cable shield to low impedance points at both ends (direct connection at the driving end, and capacitive connection at the receiver). Otherwise, unterminated transmission lines effects can cause reflections and standing waves along the cable. At frequencies of 10 MHz and above, circumferential (360°) shield bonds and metal connectors are required to maintain low impedance connections to ground.

In summary, for protection against low frequency (<1 MHz), electric-field interference, grounding the shield at one end is acceptable. For high frequency interference (>1 MHz), the preferred method is grounding the shield at both ends, using 360° circumferential bonds between the shield and the connector, and maintaining metal-to-metal continuity between the connectors and the enclosure. Low frequency ground loops can be eliminated by replacing one of the DC shield connections to ground with a low inductance 0.01 µF capacitor. This capacitor prevents low frequency ground loops and shunts high frequency interference to ground.

## EMI Troubleshooting Philosophy

System EMI problems often occur after the equipment has been designed and is operating in the field. More often than not, the original designer of the instrument has retired and is living in Tahiti, so the responsibility of repairing it belongs to someone else who may not be familiar with the product. Figure 11-46 summarizes the EMI problem-solving techniques discussed in this section and should be useful in these situations.

- Diagnose before you fix
- Ask yourself:
  - What are the symptoms?
  - What are the causes?
  - What are the constraints?
  - How will you know you have fixed it?
- Use available models for EMI to identify source – path – victim
- Start at low frequency and work up to high frequency
- EMI doctor's bag of tricks:
  - Aluminum foil
  - Conductive tape
  - Bulk ferrites
  - Power line filters
  - Signal filters
  - Resistors, capacitors, inductors, ferrites
  - Physical separation

Reprinted from *EDN* magazine (January 20, 1994) © CAHNERS PUBLISHING COMPANY 1995, A Division of Reed Publishing USA

**Figure 11-46: EMI troubleshooting philosophy**

### References: EMI/RFI Considerations

1. "EDN's Designer's Guide to Electromagnetic Compatibility," **EDN**, January 20, 1994, material reprinted by permission of Cahners Publishing Company, 1995.

2. **Designing for EMC (Workshop Notes)**, Kimmel Gerke Associates, Ltd., 1994.

3. **Systems Application Guide**, Chapter 1, Analog Devices, Inc., Norwood, MA, 1944, pp. 21–55.

4. H. Ott, **Noise Reduction Techniques in Electronic Systems**, **2nd Edition**, John Wiley & Sons, New York, 1988.

5. R. Morrison, **Grounding and Shielding Techniques in Instrumentation**, **3rd Edition,** John Wiley & Sons, New York, 1986.

6. **Amplifier Applications Guide**, Chapter XI, Analog Devices, Inc., Norwood, MA, 1992, p. 61.

7.  B. Slattery and J. Wynne, "Design and Layout of a Video Graphics System for Reduced EMI," Analog Devices Application Note AN-333.

8.  P. Brokaw, "An IC Amplifier User Guide to Decoupling, Grounding, and Making Things Go Right for a Change," Analog Devices Application Note AN-202.

9.  A. Rich, "Understanding Interference-Type Noise," **Analog Dialogue**, Vol. 16, No. 3, 1982, pp. 16–19.

10. A. Rich, "Shielding and Guarding," **Analog Dialogue**, Vol. 17, No. 1, 1983, pp. 8–13.

11. **EMC Test & Design,** Cardiff Publishing Company, Englewood, CO. An excellent, general purpose trade journal on issues of EMI and EMC.

12. **Amplifier Applications Guide,** Section XI, Analog Devices, Inc., Norwood, MA, 1992, pp. 1–10.

13. **Systems Applications Guide,** Section 1, Analog Devices, Inc., Norwood, MA, 1993, pp. 56–72.

14. **Linear Design Seminar,** Section 1, Analog Devices, Inc., Norwood, MA, 1994, pp. 19–22.

15. **ESD Prevention Manual**, Analog Devices, Inc, ADI, Norwood, MA.

16. **MIL-STD-883 Method 3015, Electrostatic Discharge Sensitivity Classification**. Available from Standardization Document Order Desk, 700 Robbins Ave., Building #4, Section D, Philadelphia, PA 19111–5094.

17. **EIAJ ED-4701 Test Method C-111, Electrostatic Discharges**. Available from the Japan Electronics Bureau, 250 W 34th St., New York, NY 10119, Attn.: Tomoko.

18. **ESD Association Standard S5.2 for Electrostatic Discharge (ESD) Sensitivity Testing: Machine Model (MM) – Component Level**. Available from the ESD Association, Inc., 200 Liberty Plaza, Rome, NY 13440.

19. **ESD Association Draft Standard DS5.3 for Electrostatic Discharge (ESD) Sensitivity Testing – Charged Device Model (CDM) Component Testing**. Available from the ESD Association, Inc., 200 Liberty Plaza, Rome, NY 13440.

20. N. Lyne, "Electrical Overstress Damage to CMOS Converters," **Application Note AN-397**, Analog Devices, 1995.

## CHAPTER 12

# Printed Circuit-Board Design Issues

## Chapter Introduction

Printed circuit boards (PCBs) are by far the most common method of assembling modern electronic circuits. They comprise a sandwich of one or more insulating layers and one or more copper layers which contain the signal traces and the powers and grounds; the design of the layout of PCBs can be as demanding as the design of the electrical circuit.

Most modern systems consist of multilayer boards of anywhere up to eight layers (or sometimes even more). Traditionally, components were mounted on the top layer in holes which extended through all layers. These are referred to as "through-hole" components. More recently, with the near universal adoption of surface mount components, you commonly find components mounted on both the top and the bottom layers.

The design of the PCB can be as important as the circuit design to the overall performance of the final system. We shall discuss in this chapter the partitioning of the circuitry, the problem of interconnecting traces, parasitic components, grounding schemes, and decoupling. All of these are important in the success of a total design.

PCB effects that are harmful to precision circuit performance include leakage resistances, IR voltage drops in trace foils, vias, and ground planes, the influence of stray capacitance, and dielectric absorption (DA). In addition, the tendency of PCBs to absorb atmospheric moisture (*hygroscopicity*) means that changes in humidity often cause the contributions of some parasitic effects to vary from day to day.

In general, PCB effects can be divided into two broad categories—those that most noticeably affect the static or DC operation of the circuit, and those that most noticeably affect dynamic or AC circuit operation, especially at high frequencies.

Another very broad area of PCB design is the topic of grounding. Grounding is a problem area in itself for all analog and mixed-signal designs, and it can be said that simply implementing a PCB-based circuit does not change the fact that proper techniques are required. Fortunately, certain principles of quality grounding, namely the use of ground planes, are intrinsic to the PCB environment. This factor is one of the more significant advantages to PCB-based analog designs, and appreciable discussion in this section is focused on this issue.

Some other aspects of grounding that must be managed include the control of spurious ground and signal return voltages that can degrade performance. These voltages can be due to external signal coupling, common currents, or simply excessive IR drops in ground conductors. Proper conductor routing and sizing, as well as differential signal-handling and ground isolation techniques enable control of such parasitic voltages.

One final area of grounding to be discussed is grounding appropriate for a mixed-signal, analog/digital environment. Indeed, the single issue of quality grounding can influence the entire layout philosophy of a high performance mixed-signal PCB design—as it well should.

# *Partitioning*

Any subsystem or circuit layout operating at high frequency and/or high precision with both analog and digital signals should like to have those signals physically separated as much as possible to prevent crosstalk. This is typically difficult to accomplish in practice.

Crosstalk can be minimized by paying attention to the system layout and preventing different signals from interfering with each other. High level analog signals should be separated from low level analog signals, and both should be kept away from digital signals. TTL and CMOS digital signals have high edge rates, implying frequency components starting with the system clock and going up from there. And most logic families are saturation logic, which has uneven current flow (high transient currents) that can modulate the ground. We have seen elsewhere that in waveform sampling and reconstruction systems the sampling clock (which is a digital signal) is as vulnerable to noise as any analog signal. Noise on the sampling clock manifests itself as phase jitter, which as we have seen in a previous section, translates directly to reduced signal-to-noise ratio (SNR) of the sampled signal. If clock driver packages are used in clock distribution, only one frequency clock should be passed through a single package. Sharing drivers between clocks of different frequencies in the same package will produce excess jitter and crosstalk and degrade performance.

The ground plane can act as a shield where sensitive signals cross. Figure 12-1 shows a good layout for a data acquisition board where all sensitive areas are isolated from each other and signal paths are kept as short as possible. While real life is rarely as simple as this, the principle remains a valid one.

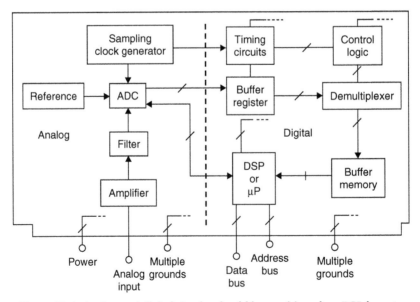

**Figure 12-1: Analog and digital circuits should be partitioned on PCB layout**

There are a number of important points to be considered when making signal and power connections. First of all a connector is one of the few places in the system where all signal conductors must run in parallel—it is therefore imperative to separate them with ground pins (creating a Faraday shield) to reduce coupling between them.

Multiple ground pins are important for another reason: they keep down the ground impedance at the junction between the board and the backplane. The contact resistance of a single pin of a PCB connector is quite low (typically on the order of $10\,m\Omega$) when the board is new—as the board gets older the contact resistance is likely to rise, and the board's performance may be compromised. It is therefore well worthwhile to allocate extra PCB connector pins so that there are many ground connections (perhaps 30–40% of all the pins on the PCB connector should be ground pins). For similar reasons there should be several pins for each power connection.

Manufacturers of high performance mixed-signal ICs, like Analog Devices, often offer evaluation boards to assist customers in their initial evaluations and layout. Analog-to-digital converter (ADC) evaluation boards generally contain an on-board low jitter sampling clock oscillator, output registers, and appropriate power and signal connectors. They may also have additional support circuitry such as the ADC input buffer amplifier and external reference.

The layout of the evaluation board is optimized in terms of grounding, decoupling, and signal routing and can be used as a model when laying out the ADC section of the PC board in a system. The actual evaluation board layout is usually available from the ADC manufacturer in the form of computer CAD files (Gerber files). In many cases, the layout of the various layers appears on the data sheet for the device. It should be pointed out, though, that an evaluation board is an extremely simple system. While some guidelines can be inferred from inspection of the evaluation board layout, the system that you are designing is undoubtedly more complicated. Therefore, direct use of the layout may not be optimum in larger systems.

# *Traces*

## *Resistance of Conductors*

Every engineer is familiar with resistors. But far too few engineers consider that all the wires and PCB traces with which their systems and circuits are assembled are also resistors (as well as inductors, as will be discussed later). In higher precision systems, even these trace resistances and simple wire interconnections can have degrading effects. Copper is *not* a superconductor—and too many engineers appear to think it is!

Figure 12-2 illustrates a method of calculating the sheet resistance R of a copper square, given the length Z, the width X, and the thickness Y.

$$R = \frac{\rho Z}{XY}$$

$\rho$ = Resistivity

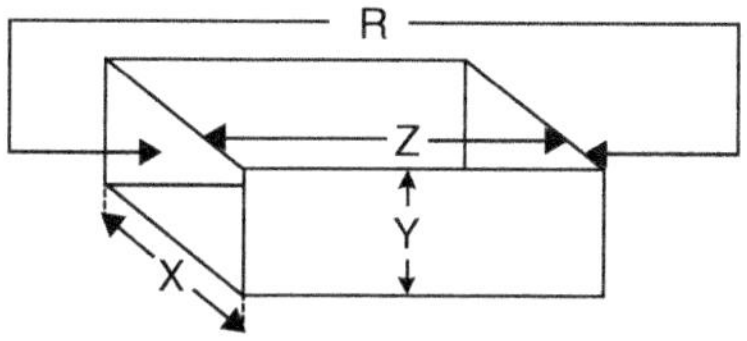

Sheet resistance calculation for
1oz. copper conductor:

$\rho = 1.724 \times 10^{-6} \Omega cm$, Y = 0.0036 cm

$R = 0.48 \dfrac{Z}{X} m\Omega$

$\dfrac{Z}{X}$ = Number of squares

R = Sheet resistance of 1 square (Z=X)
   = 0.48 mΩ/ square

**Figure 12-2: Calculation of sheet resistance and linear resistance for standard copper PCB conductors**

At 25°C the resistivity of pure copper is $1.724 \times 10^{-6} \Omega$/cm. The thickness of standard 1 ounce PCB copper foil is 0.036 mm (0.0014 inches). Using the relations shown, the resistance of such a standard copper element is therefore 0.48 mΩ/square. One can readily calculate the resistance of a linear trace by effectively "stacking" a series of such squares end–end, to make up the line's length. The line length is Z and the width is X, so the line resistance R is simply a product of Z/X and the resistance of a single square, as noted in the figure.

For a given copper weight and trace width, a resistance/length calculation can be made. For example, the 0.25 mm (10 mil) wide traces frequently used in PCB designs equate to a resistance/length of about

19 mΩ/cm (48 mΩ/inch), which is quite large. Moreover, the temperature coefficient of resistance for copper is about 0.4%/°C around room temperature. This is a factor that should not be ignored, in particular within low impedance precision circuits, where the TC can shift the net impedance over temperature.

As shown in Figure 12-3, PCB trace resistance can be a serious error when conditions are not favorable. Consider a 16-bit ADC with a 5 kΩ input resistance, driven through 5 cm of 0.25 mm wide 1 ounce PCB tracks between it and its signal source. The track resistance of nearly 0.1 Ω forms a divider with the 5 kΩ load, creating an error. The resulting voltage drop is a gain error of 0.1/5k (~0.0019%), well over 1 LSB (0.0015% for 16 bits). And this ignores the issue of the return path! It also ignores inductance, which could make the situation worse at high frequencies.

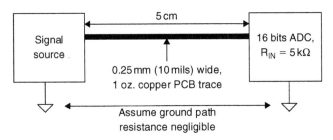

**Figure 12-3: Ohm's law predicts >1 LSB of error due to drop in PCB conductor**

So, when dealing with precision circuits, the point is made that even simple design items such as PCB trace resistance cannot be dealt with casually. There are various solutions that can address this issue, such as wider traces, which may take up excessive space and may not be a viable solution with the smallest packages and with packages with multiple rows of pins, such as a ball grid array (BGA), the use of heavier copper (which may be too expensive), or simply choosing a high input impedance converter. But, the most important thing is to think it all through, avoiding any tendency to overlook items appearing innocuous on the surface.

## Voltage Drop in Signal Leads — "Kelvin" Feedback

The gain error resulting from resistive voltage drop in PCB signal leads is important only with high precision and/or at high resolutions (the Figure 12-3 example), or where large signal currents flow. Where load impedance is constant and resistive, adjusting overall system gain can compensate for the error. In other circumstances, it may often be removed by the use of "Kelvin" or "voltage-sensing" feedback, as shown in Figure 12-4.

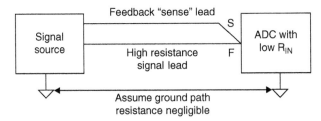

**Figure 12-4: Use of a sense connection moves accuracy to the load point**

In this modification to the case of Figure 12-3 a long resistive PCB trace is still used to drive the input of a high resolution ADC, with low input impedance. In this case, however, the voltage drop in the signal lead does *not* give rise to an error, as feedback is taken directly from the input pin of the ADC, and returned to the driving source. This scheme allows full accuracy to be achieved in the signal presented to the ADC, despite any voltage drop across the signal trace.

The use of separate force (F) and sense (S) connections (often referred to as a "Kelvin connection") at the load removes any errors resulting from voltage drops in the force lead, but, of course, may only be used in systems where there is negative feedback. It is also impossible to use such an arrangement to drive two or more loads with equal accuracy, since feedback may only be taken from one point. Also, in this much-simplified system, errors in the common lead source/load path are ignored, the assumption being that ground path voltages are negligible. In many systems this may not necessarily be the case, and additional steps may be needed, as noted below.

## Signal Return Currents

Kirchoff's law tells us that at any point in a circuit the algebraic sum of the currents is zero. This tells us that all currents flow in circles and, particularly, that the return current must always be considered when analyzing a circuit, as is illustrated in Figure 12-5 (see References 7 and 8).

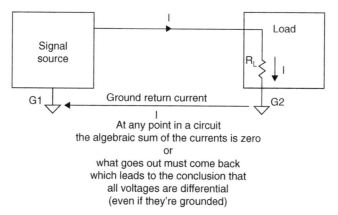

**Figure 12-5: Kirchoff's law helps in analyzing voltage drops around a complete source/load coupled circuit**

In dealing with grounding issues, common human tendencies provide some insight into how the correct thinking about the circuit can be helpful toward analysis. Most engineers readily consider the ground return current "I," only when they are considering a fully differential circuit.

However, when considering the more usual circuit case, where a single-ended signal is referred to "ground," it is common to assume that all the points on the circuit diagram where ground symbols are found are at the same potential. Unfortunately, this happy circumstance just ain't necessarily so!

This overly optimistic approach is illustrated in Figure 12-6 where, if it really should exist, "infinite ground conductivity" would lead to zero ground voltage difference between source ground G1 and load ground G2. Unfortunately this approach is not a wise practice, and when dealing with high precision circuits, it can lead to disasters.

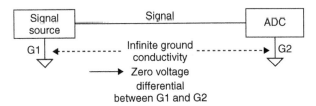

**Figure 12-6: Unlike this optimistic diagram, it is unrealistic to assume infinite conductivity between source/ load grounds in a real-world system**

A more realistic approach to ground conductor integrity includes analysis of the impedance(s) involved and careful attention to minimizing spurious noise voltages.

## Ground Noise and Ground Loops

A more realistic model of a ground system is shown in Figure 12-7. The signal return current flows in the complex impedance existing between ground points G1 and G2 as shown, giving rise to a voltage drop $\Delta V$ in this path. But it is important to note that additional *external* currents, such as $I_{EXT}$, may also flow in this same path. It is critical to understand that such currents may generate uncorrelated noise voltages between G1 and G2 (dependent on the current magnitude and relative ground impedance).

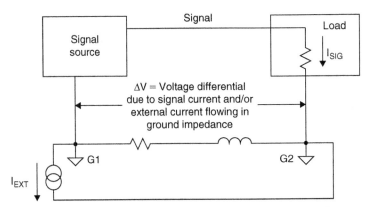

**Figure 12-7: A more realistic source-to-load grounding system view includes consideration of the impedance between G1 and G2, plus the effect of any non-signal-related currents**

Some portion of these undesired voltages may end up being seen at the signal's load end, and they can have the potential to corrupt the signal being transmitted.

It is evident, of course, that other currents can only flow in the ground impedance if there is a current path for them. In this case, severe problems can be caused by a high current circuit sharing an *unlooped* ground return with the signal source.

Figure 12-8 shows just such a common-ground path, shared by the signal source and a high current circuit, which draws a large and varying current from its supply. This current flows in the common-ground return, causing an error voltage $\Delta V$ to develop.

From Figure 12-9, it is also evident that if a ground network contains *loops*, or circular ground conductor patterns (with S1 closed), there is an even greater danger of its being vulnerable to EMFs induced by

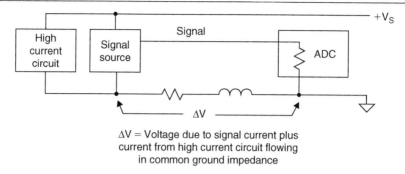

$\Delta V$ = Voltage due to signal current plus current from high current circuit flowing in common ground impedance

**Figure 12-8: Any current flowing through a common-ground impedance can cause errors**

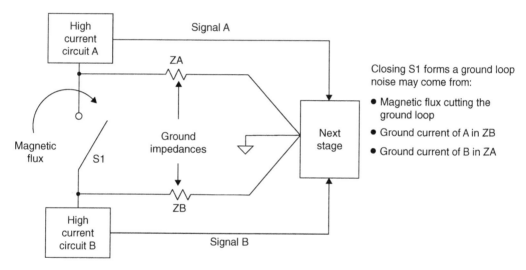

**Figure 12-9: A ground loop**

external magnetic fields. There is also a real danger of ground current related signals "escaping" from the high current areas and causing noise in sensitive circuit regions elsewhere in the system.

For these reasons ground loops are best avoided, by wiring all return paths within the circuit by separate paths back to a common point, i.e., the common ground point toward the mid-right of the diagram. This would be represented by the S1 open condition.

## Ground Isolation Techniques

While the use of ground planes does lower impedance and helps greatly in lowering ground noise, there may still be situations where a prohibitive level of noise exists. In such cases, the use of ground error minimization and isolation techniques can be helpful.

Another illustration of a common-ground impedance coupling problem is shown in Figure 12-10. In this circuit a precision gain of 100 preamp amplifies a low level signal $V_{IN}$, using an AD8551 chopper-stabilized amplifier for best DC accuracy. At the load end, the signal $V_{OUT}$ is measured with respect to G2, the local

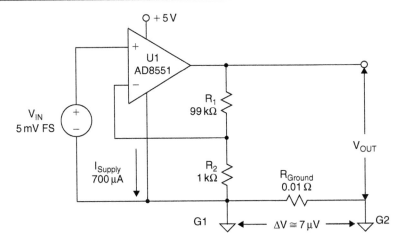

**Figure 12-10: Unless care is taken, even small common-ground currents can degrade precision amplifier accuracy**

ground. Because of the small $700\,\mu A$ $I_{SUPPLY}$ of the AD8551 flowing between G1 and G2, there is a $7\,\mu V$ ground error—about 7 times the typical input offset expected from the op amp!

This error can be avoided simply by routing the negative supply pin current of the op amp back to star ground G2 as opposed to ground G1, by using a separate trace. This step eliminates the G1–G2 path power supply current, and so minimizes the ground leg voltage error. Note that there will be little error developed in the "hot" $V_{OUT}$ lead, so long as the current drain at the load end is small.

In some cases, there may be simply unavoidable ground voltage differences between a source signal and the load point where it is to be measured. Within the context of this "same-board" discussion, this might require rejecting ground error voltages of several tens of millivolts. Or, should the source signal originate from an "off-board" source, then the magnitude of the common-mode voltages to be rejected can easily rise into a several volt range (or even tens of volts).

Fortunately, full signal transmission accuracy can still be accomplished in the face of such high noise voltages, by employing a principle discussed earlier. This is the use of a differential input, *ground isolation* amplifier. The ground isolation amplifier minimizes the effect of ground error voltages between stages by processing the signal in differential fashion, thereby rejecting CM voltages by a substantial margin (typically 60 dB or more).

Two ground isolation amplifier solutions are shown in Figure 12-11. This diagram can alternately employ either the AD629 to handle CM voltages up to $\pm 270\,V$, or the AMP03, which is suitable for CM voltages up to $\pm 20\,V$.

In the circuit, input voltage $V_{IN}$ is referred to G1, but must be measured with respect to G2. With the use of a high CMR unity-gain difference amplifier, the noise voltage $\Delta V$ existing between these two grounds is easily rejected. The AD629 offers a typical CMR of 88 dB, while the AMP03 typically achieves 100 dB. In the AD629, the high CMV rating is done by a combination of high CM attenuation, followed by differential gain, realizing a net differential gain of unity. The AD629 uses the first listed value resistors noted in the figure for $R_1$–$R_5$. The AMP03 operates as a precision four-resistor differential amplifier, using the $25\,k\Omega$ value $R_1$–$R_4$ resistors noted. Both devices are complete, one package solutions to the ground isolation amplifier.

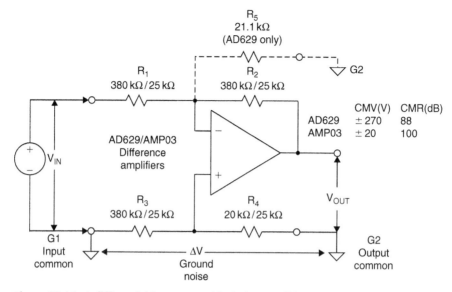

**Figure 12-11: A differential input ground isolating amplifier allows high transmission accuracy by rejecting ground noise voltage between source (G1) and measurement (G2) grounds**

This scheme allows relative freedom from tightly controlling ground drop voltages, or running additional and/or larger PCB traces to minimize such error voltages. Note that it can be implemented either with the fixed gain difference amplifiers shown, or also with a standard in-amp IC, configured for unity gain. The AD623, e.g., also allows single-supply use. In any case, signal polarity is also controllable, by simple reversal of the difference amplifier inputs.

In general terms, transmitting a signal from one point on a PCB to another for measurement or further processing can be optimized by two key interrelated techniques. These are the use of high impedance, differential signal-handling techniques. The high impedance loading of an in-amp minimizes voltage drops, and differential sensing of the remote voltage minimizes sensitivity to ground noise.

When the further signal processing is A/D conversion, these transmission criteria can be implemented *without* adding a differential ground isolation amplifier stage. Simply select an ADC which operates differentially. The high input impedance of the ADC minimizes load sensitivity to the PCB wiring resistance. In addition, the differential input feature allows the output of the source to be sensed directly at the source output terminals (even if single ended). The CMR of the ADC then eliminates sensitivity to noise voltages between the ADC and source grounds.

An illustration of this concept using an ADC with high impedance differential inputs is shown in Figure 12-12. Note that the general concept can be extended to virtually any signal source, driving any load. All loads, even single-ended ones, become differential input by adding an appropriate differential input stage. The differential input can be provided by either a fully developed high Z in-amp, or in many cases a simple subtractor stage op amp, as shown in Figure 12-11.

## Static PCB Effects

Leakage resistance is the dominant static circuit-board effect. Contamination of the PCB surface by flux residues, deposited salts, and other debris can create leakage paths between circuit nodes. Even on

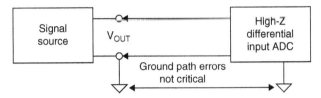

**Figure 12-12: A high impedance differential input ADC also allows high transmission accuracy between source and load**

well-cleaned boards, it is not unusual to find 10 nA or more of leakage to nearby nodes from 15 V supply rails. Nanoamperes of leakage current into the wrong nodes often cause volts of error at a circuit's output; for example, 10 nA into a 10 MΩ resistance causes 0.1 V of error. Unfortunately, the standard op amp pinout places the $-V_S$ supply pin next to the + input, which is often hoped to be at high impedance! To help identify nodes sensitive to the effects of leakage currents ask the simple question: if a spurious current of a few nanoamperes or more were injected into this node, would it matter?

If the circuit is already built, you can localize moisture sensitivity to a suspect node with a classic test. While observing circuit operation, blow on potential trouble spots through a simple soda straw. The straw focuses the breath's moisture, which, with the board's salt content in susceptible portions of the design, disrupts circuit operation upon contact.

There are several means of eliminating simple surface leakage problems. Thorough washing of circuit boards to remove residues helps considerably. A simple procedure includes vigorously brushing the boards with isopropyl alcohol, followed by thorough washing with deionized water and an 85°C bakeout for a few hours. Be careful when selecting board-washing solvents, though. When cleaned with certain solvents, some water-soluble fluxes create salt deposits, exacerbating the leakage problem.

Unfortunately, if a circuit displays sensitivity to leakage, even the most rigorous cleaning can offer only a temporary solution. Problems soon return upon handling, or exposure to foul atmospheres, and high humidity. Some additional means must be sought to stabilize circuit behavior, such as conformal surface coating.

Fortunately, there is an answer to this, namely *guarding*, which offers a fairly reliable and permanent solution to the problem of surface leakage. Well-designed guards can eliminate leakage problems, even for circuits exposed to harsh industrial environments. Two schematics illustrate the basic guarding principle, as applied to typical inverting and non-inverting op amp circuits.

Figure 12-13 illustrates an inverting mode guard application. In this case, the op amp reference input is grounded, so the guard is a grounded ring surrounding all leads to the inverting input, as noted by the dotted line.

Guarding basic principles are simple: *completely* surround sensitive nodes with conductors that can readily sink stray currents, and maintain the guard conductors at the exact potential of the sensitive node (as otherwise the guard will serve as a leakage source rather than a leakage sink). For example, to keep leakage into a node below 1 pA (assuming 1,000 MΩ leakage resistance) the guard and guarded node must be within 1 mV. Generally, the low offset of a modern op amp is sufficient to meet this criterion.

There are important caveats to be noted with implementing a true high quality guard. For traditional through-hole PCB connections, the guard pattern should appear on *both* sides of the circuit board, to be most effective. And, it should also be connected along its length by several vias. Finally, when either justified or required by the system design parameters, do make an effort to include guards in the

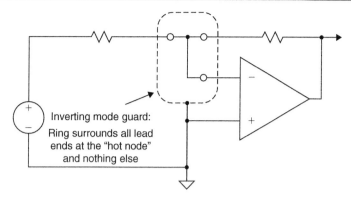

**Figure 12-13: Inverting mode guard encloses all op amp inverting input connections within a grounded guard ring**

PCB design process from the outset—there is little likelihood that a proper guard can be added as an afterthought.

Figure 12-14 illustrates the case for a non-inverting guard. In this instance the op amp reference input is directly driven by the source, which complicates matters considerably. Again, the guard ring completely surrounds all of the input nodal connections. In this instance, however, the guard is driven from the low impedance feedback divider connected to the inverting input.

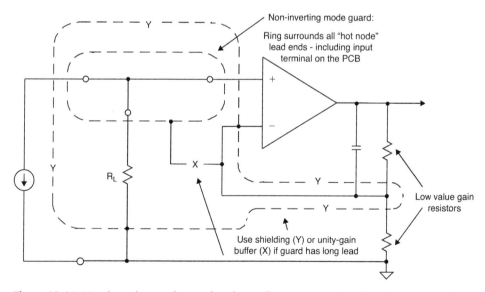

**Figure 12-14: Non-inverting mode guard encloses all op amp non-inverting input connections within a low impedance, driven guard ring**

Usually the guard-to-divider junction will be a direct connection, but in some cases a unity-gain buffer might be used at "X" to drive a cable shield, or also to maintain the lowest possible impedance at the guard ring.

In lieu of the buffer, another useful step is to use an additional, directly grounded screen ring, "Y," which surrounds the inner guard and the feedback nodes as shown. This step costs nothing except some added layout time, and will greatly help buffer leakage effects into the higher impedance inner guard ring.

Of course what has not been addressed to this point is just how the op amp itself gets connected into these guarded islands without compromising performance. The traditional method using a TO-99 metal can package device was to employ double-sided PCB guard rings, with both op amp inputs terminated within the guarded ring.

## Sample MINIDIP and SOIC Op Amp PCB Guard Layouts

Modern assembly practices have favored smaller plastic packages such as 8-pin MINIDIP and SOIC types. Some suggested partial layouts for guard circuits using these packages are shown in the next two figures. While guard traces may also be possible with even more tiny op amp footprints, such as SOT-23 etc., the required trace separations become even more confining, challenging the layout designer as well as the manufacturing processes.

For the ADI "N" style MINIDIP package, Figure 12-15 illustrates how guarding can be accomplished for inverting (A) and non-inverting (B) operating modes. This setup would also be applicable to other op amp devices where relatively high voltages occur at pin 1 or 4. Using a standard 8-pin DIP outline, it can be noted that this package's 0.1 inch pin spacing allows a PC trace (here, the guard trace) to pass between adjacent pins. This is the key to implementing effective DIP package guarding, as it can adequately prevent a leakage path from the $-V_S$ supply at pin 4, or from similar high potentials at pin 1.

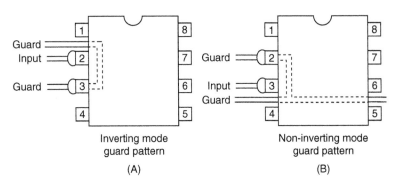

**Figure 12-15: PCB guard patterns for inverting and non-inverting mode op amps using 8 pin MINIDIP (N) package**

For the inverting mode, note that the pin 3 connected and grounded guard traces surround the op amp inverting input (pin 2), and run parallel to the input trace. This guard would be continued out to and around the source and feedback connections of Figure 7-36 (or other similar circuit), including an input pad in the case of a cable. In the non-inverting mode, the guard voltage is the feedback divider voltage to pin 2. This corresponds to the inverting input node of the amplifier, from Figure 12-14.

Note that in both of the cases of Figure 12-15, the guard physical connections shown are only partial—an actual layout would include all sensitive nodes within the circuit. In both the inverting and the non-inverting modes using the MINIDIP or other through-hole style package, the PCB guard traces should be located on both sides of the board, with top and bottom traces connected with several vias.

Things become slightly more complicated when using guarding techniques with the SOIC surface mount ("R") package, as the 0.05 inch pin spacing does not easily allow routing of PCB traces between the pins. But, there is still an effective guarding answer, at least for the inverting case. Figure 12-16 shows guards for the ADI "R" style SOIC package.

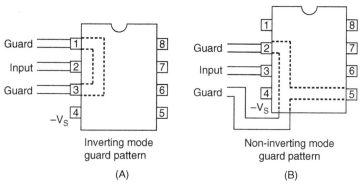

Inverting mode
guard pattern

(A)

Non-inverting mode
guard pattern

(B)

*Note*: Pins 1, 5, and 8 are open on many "R" packaged devices

**Figure 12-16: PCB guard patterns for inverting and non-inverting mode op amps using 8 pin SOIC (R) package**

Note that for many single op amp devices in this SOIC "R" package, pins 1, 5, and 8 are "no connect" pins. Historically these pins were used for offset adjustment and/or frequency compensation. These functions rarely are used in modern op amps. For such instances, this means that these empty locations can be employed in the layout to route guard traces. In the case of the inverting mode (A), the guarding is still completely effective, with the dummy pin 1 and pin 3 serving as the grounded guard trace. This is a fully effective guard without compromise. Also, with SOIC op amps, much of the circuitry around the device will not use through-hole components. So, the guard ring may only be necessary on the op amp PCB side.

In the case of the follower stage (B), the guard trace must be routed around the negative supply at pin 4, and thus pin 4 to pin 3 leakage is not fully guarded. For this reason, a precision high impedance follower stage using an SOIC package op amp is not generally recommended, as guarding is not possible for dual-supply connected devices.

However, an exception to this caveat does apply to the use of a *single-supply* op amp as a non-inverting stage. For example, if the AD8551 is used, pin 4 becomes ground, and some degree of intrinsic guarding is then established by default.

## Dynamic PCB Effects

Although static PCB effects can come and go with changes in humidity or board contamination, problems that most noticeably affect the dynamic performance of a circuit usually remain relatively constant. Short of a new design, washing or any other simple fixes cannot fix them. As such, they can permanently and adversely affect a design's specifications and performance. The problems of stray capacitance, linked to lead and component placement, are reasonably well known to most circuit designers. Since lead placement can be permanently dealt with by correct layout, any remaining difficulty is solved by training assembly personnel to orient components or bend leads optimally.

DA, on the other hand, represents a more troublesome and still poorly understood circuit-board phenomenon. Like DA in discrete capacitors, DA in a PCB can be modeled by a series resistor and capacitor connecting two closely spaced nodes. Its effect is inverse with spacing and linear with length.

As shown in Figure 12-17, the RC model for this effective capacitance ranges from 0.1 to 2.0 pF, with the resistance ranging from 50 to 500 MΩ. Values of 0.5 pF and 100 MΩ are most common. Consequently, circuit-board DA interacts most strongly with high impedance circuits.

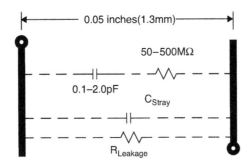

**Figure 12-17: DA plagues dynamic response of PCB-based circuits**

PCB DA most noticeably influences dynamic circuit response, e.g., settling time. Unlike circuit leakage, the effects are not usually linked to humidity or other environmental conditions, but rather, are a function of the board's dielectric properties. The chemistry involved in producing plated-through-holes seems to exacerbate the problem. If your circuits do not meet expected transient response specs, you should consider PCB DA as a possible cause.

Fortunately, there are solutions. As in the case of capacitor DA, external components can be used to compensate for the effect. More importantly, surface guards that totally isolate sensitive nodes from parasitic coupling often eliminate the problem (note that these guards should be duplicated on both sides of the board, in cases of through-hole components). As noted previously, low loss PCB dielectrics are also available.

PCB "hook," similar if not identical to DA, is characterized by variation in effective circuit-board capacitance with frequency (see Reference 1). In general, it affects high impedance circuit transient response where board capacitance is an appreciable portion of the total in the circuit. Circuits operating at frequencies below 10 kHz are the most susceptible. As in circuit-board DA, the board's chemical makeup very much influences its effects.

## Inductance

### Stray Inductance

All conductors are inductive, and at high frequencies, the inductance of even quite short pieces of wire or printed circuit traces may be important. The inductance of a straight wire of length L (mm) and circular cross section with radius R (mm) in free space is given by the first equation shown in Figure 12-18.

The inductance of a strip conductor (an approximation to a PC track) of width W (mm) and thickness H (mm) in free space is given by the second equation in Figure 12-18.

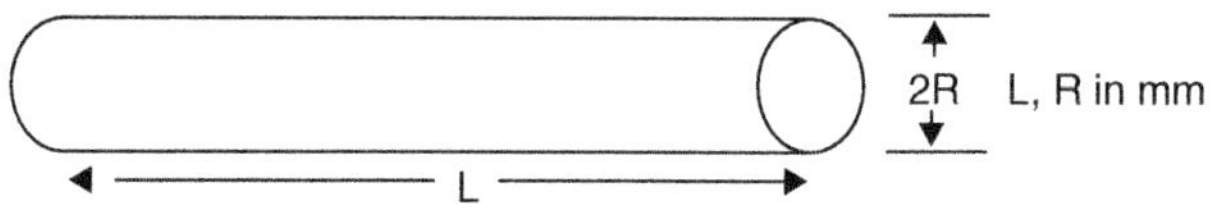

Wire inductance $= 0.0002L \left[ \ln \left( \frac{2L}{R} \right) - 0.75 \right] \mu H$

Example: 1 cm of 0.5 mm o.d. wire has an inductance of 7.26 nH
(2R = 0.5 mm, L = 1 cm)

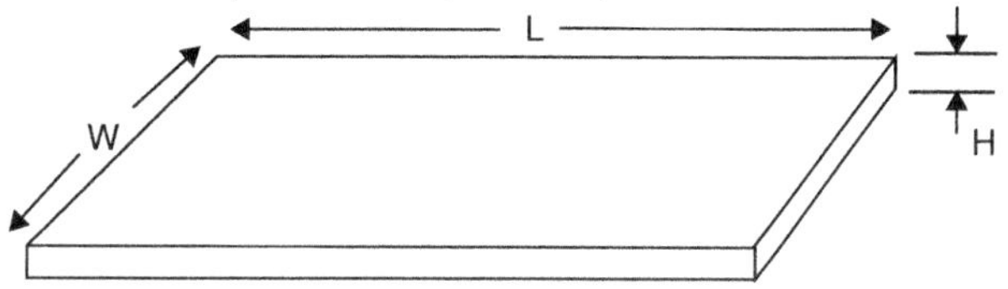

Strip inductance $= 0.0002L \left[ \ln \frac{2L}{(W+H)} + 0.2235 \frac{W+H}{(L)} + 0.5 \right] \mu H$

Example: 1 cm of 0.25 mm PC track has an inductance of 9.59 nH
(H = 0.038 mm, W = 0.25 mm, L = 1 cm)

**Figure 12-18: Wire and strip inductance calculations**

In real systems, both these formulas turn out to be approximate, but they do give some idea of the order of magnitude of inductance involved. They tell us that 1 cm of 0.5 mm o.d. wire has an inductance of 7.26 nH, and 1 cm of 0.25 mm PC track has an inductance of 9.59 nH—these figures are reasonably close to measured results.

At 10 MHz, an inductance of 7.26 nH has an impedance of 0.46 Ω, and so can give rise to 1% error in a 50 Ω system.

### Mutual Inductance

Another consideration regarding inductance is the separation of outward and return currents. As we shall discuss in more detail later, Kirchoff's law tells us that current flows in closed paths—there is always an outward and return path. The whole path forms a single-turn inductor.

This principle is illustrated by the contrasting signal trace routing arrangements of Figure 9-10. If the area enclosed within the turn is relatively large, as in the upper "nonideal" picture, then the inductance (and hence the AC impedance) will also be large. On the other hand, if the outward and return paths are closer together, as in the lower "improved" picture, the inductance will be much smaller.

Note that the nonideal signal routing case of Figure 12-19 has other drawbacks—the large area enclosed within the conductors produces extensive external magnetic fields, which may interact with other circuits, causing unwanted coupling. Similarly, the large area is more vulnerable to interaction with external magnetic fields, which can induce unwanted signals in the loop.

The basic principle is illustrated in Figure 12-20 and is a common mechanism for the transfer of unwanted signals (noise) between two circuits.

As with most other noise sources, as soon as we define the working principle, we can see ways of reducing the effect. In this case, reducing any or all of the terms in the equations in Figure 12-20 reduces the coupling. Reducing the frequency or amplitude of the current causing the interference may be impracticable, but it is frequently possible to reduce the mutual inductance between the interfering and

Nonideal signal trace routing

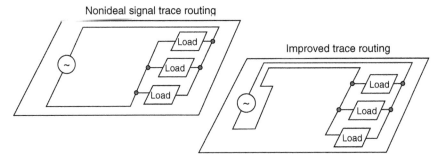

Figure 12-19: Nonideal and improved signal trace routing

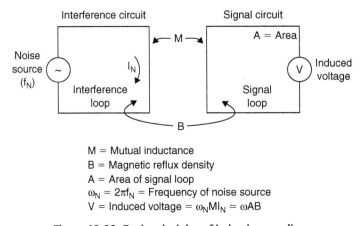

M = Mutual inductance
B = Magnetic reflux density
A = Area of signal loop
$\omega_N = 2\pi f_N$ = Frequency of noise source
V = Induced voltage = $\omega_N M I_N = \omega AB$

Figure 12-20: Basic principles of inductive coupling

interfered with circuits by reducing loop areas on one or both sides and, possibly, increasing the distance between them.

A layout solution is illustrated in Figure 12-21. Here two circuits, shown as Z1 and Z2, are minimized for coupling by keeping each of the loop areas as small as is practical.

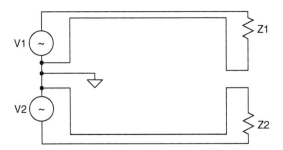

Figure 12-21: Proper signal routing and layout can
reduce inductive coupling

As also illustrated in Figure 12-22, mutual inductance can be a problem in signals transmitted on cables. Mutual inductance is high in ribbon cables, especially when a single return is common to several signal

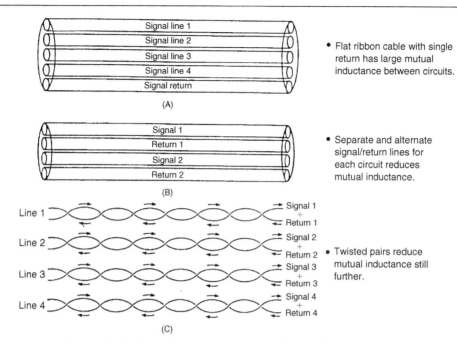

**Figure 12-22: Mutual inductance and coupling within signal cabling**

circuits (A). Separate, dedicated signal and return lines for each signal circuit reduces the problem (B). Using a cable with twisted pairs for each signal circuit as in (C) is even better (but is more expensive and often unnecessary).

Shielding of magnetic fields to reduce mutual inductance is sometimes possible, but is by no means as easy as shielding an electric field with a Faraday shield (following section). HF magnetic fields are blocked by conductive material provided the skin depth in the conductor at the frequency to be screened is much less than the thickness of the conductor, and the screen has no holes (Faraday shields can tolerate small holes, but magnetic screens cannot). LF and DC fields may be screened by a shield made of mu-metal sheet. Mu-metal is an alloy having very high permeability, but it is expensive, its magnetic properties are damaged by mechanical stress, and it will saturate if exposed to too high fields. Its use, therefore, should be avoided where possible.

## Parasitic Effects in Inductors

Although inductance is one of the fundamental properties of an electronic circuit, inductors are far less common as components than are resistors and capacitors. As for precision components, they are even more rare. This is because they are harder to manufacture, less stable, and less physically robust than resistors and capacitors. It is relatively easy to manufacture stable precision inductors with inductances from nH to tens or hundreds of μH, but larger valued devices tend to be less stable, and large.

As we might expect in these circumstances, circuits are designed, where possible, to avoid the use of precision inductors. We find that stable precision inductors are relatively rarely used in precision analog circuitry, except in tuned circuits for high frequency narrowband applications.

Of course, they are widely used in power filters, switching power supplies, and other applications where lack of precision is unimportant (more on this in a following section). The important features of inductors used in such applications are their current carrying and saturation characteristics, and their Q. If an inductor consists of a coil of wire with an air core, its inductance will be essentially unaffected by the current it is carrying. On the other hand, if it is wound on a core of a magnetic material (magnetic alloy or ferrite), its inductance will be nonlinear, since at high currents, the core will start to saturate. The effects of such saturation will reduce the efficiency of the circuitry employing the inductor and are liable to increase noise and harmonic generation.

As mentioned above, inductors and capacitors together form tuned circuits. Since all inductors will also have some stray capacity, all inductors will have a resonant frequency (which will normally be published on their data sheet), and should only be used as precision inductors at frequencies well below this.

## Q or "Quality Factor"

The other characteristic of inductors is their Q (or "quality factor"), which is the ratio of the reactive impedance to the resistance, as indicated in Figure 12-23.

- Q = $2\pi f$ L/R.
- The Q of an inductor or resonant circuit is a measure of the ratio of its reactance to its resistance.
- The resistance is the HF and NOT the DC value.
- The 3 dB bandwidth of a single tuned circuit is Fc/Q where Fc is the center frequency.

**Figure 12-23: Inductor Q or quality factor**

It is rarely possible to calculate the Q of an inductor from its DC resistance, since skin effect (and core losses if the inductor has a magnetic core) ensures that the Q of an inductor at high frequencies is always lower than that predicted from DC values.

Q is also a characteristic of tuned circuits (and of capacitors—but capacitors generally have such high Q values that they may be disregarded, in practice). The Q of a tuned circuit, which is generally very similar to the Q of its inductor (unless it is deliberately lowered by the use of an additional resistor), is a measure of its bandwidth around resonance. LC tuned circuits rarely have Q of much more than 100 (3 dB bandwidth of 1%), but ceramic resonators may have a Q of thousands, and quartz crystals tens of thousands.

## Do Not Overlook Anything

Remember, if your precision op amp or data converter-based design does not meet specification, try not to overlook anything in your efforts to find the error sources. Analyze both active *and* passive components, trying to identify and challenge any assumptions or preconceived notions that may blind you to the facts. Take nothing for granted.

For example, when not tied down to prevent motion, cable conductors, moving within their surrounding dielectrics, can create significant static charge buildups that cause errors, especially when connected to high impedance circuits. Rigid cables, or even costly low noise Teflon-insulated cables, are expensive alternative solutions.

As more and more high precision op amps become available, and system designs call for higher speed and increased accuracy, a thorough understanding of the error sources described in this section (as well in those following) becomes more important.

Some additional discussions of passive components within a succeeding power supply filtering section complement this one. In addition, the very next section on PCB design issues also complements many points within this section. Similar comments apply to the chapter on EMI/RFI.

## Stray Capacitance

When two conductors are not short circuited together, or totally screened from each other by a conducting (Faraday) screen, there is a capacitance between them. So, on any PCB, there will be a large number of capacitors associated with any circuit (which may or may not be considered in models of the circuit). Where high frequency performance matters (and even DC and VLF circuits may use devices with high Ft and therefore be vulnerable to HF instability), it is very important to consider the effects of this stray capacitance.

Any basic textbook will provide formulas for the capacitance of parallel wires and other geometric configurations (see References 9 and 10). The example we need to consider in this discussion is the parallel plate capacitor, often formed by conductors on opposite sides of a PCB. The basic diagram describing this capacitance is shown in Figure 12-24.

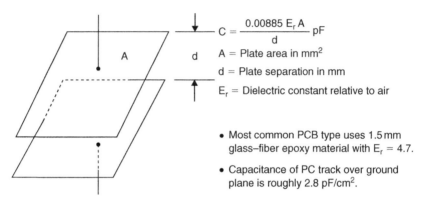

**Figure 12-24: Capacitance of two parallel plates**

Neglecting edge effects, the capacitance of two parallel plates of area A (mm²) and separation d (mm) in a medium of dielectric constant $E_r$ relative to air is:

$$0.00885\,E_r A/d \ pF \quad\quad\quad (12\text{-}1)$$

where $E_r$ is the dielectric constant of the insulator material relative to air, A is the plate area, and D is the distance between the plates.

From this formula, we can calculate that for general-purpose PCB material ($E_r$ = 4.7, d = 1.5 mm), the capacitance between conductors on opposite sides of the board is just under 3 pF/cm². In general, such capacitance will be parasitic, and circuits must be designed so that it does not affect their performance.

While it is possible to use PCB capacitance in place of small discrete capacitors, the dielectric properties of common PCB substrate materials cause such capacitors to behave poorly. They have a rather high temperature coefficient and poor Q at high frequencies, which make them unsuitable for many applications. Boards made with lower loss dielectrics such as Teflon are expensive exceptions to this rule.

## Capacitive Noise and Faraday Shields

There is a capacitance between any two conductors separated by a dielectric (air or vacuum are dielectrics). If there is a change of voltage on one, there will be a movement of charge on the other. A basic model for this is shown in Figure 12-25.

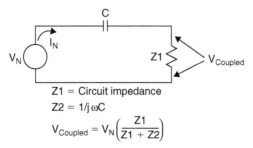

$$Z1 = \text{Circuit impedance}$$
$$Z2 = 1/j\,\omega C$$
$$V_{Coupled} = V_N \left( \frac{Z1}{Z1 + Z2} \right)$$

**Figure 12-25: Capacitive coupling equivalent circuit model**

It is evident that the noise voltage, $V_{COUPLED}$ appearing across Z1, may be reduced by several means, all of which reduce noise current in Z1. They are reduction of the signal voltage $V_N$, reduction of the frequency involved, reduction of the capacitance, or reduction of Z1 itself. Unfortunately however, often none of these circuit parameters can be freely changed, and an alternate method is needed to minimize the interference. The best solution toward reducing the noise coupling effect of C is to insert a grounded conductor, also known as a *Faraday shield*, between the noise source and the affected circuit. This has the desirable effect of reducing Z1 noise current, thus reducing $V_{COUPLED}$.

A Faraday shield model is shown in Figure 12-26. In (A), the function of the shield is noted by how it effectively divides the coupling capacitance, C. In (B) the net effect on the coupled voltage across Z1 is shown. Although the noise current $I_N$ still flows in the shield, most of it is now diverted away from Z1. As a result, the coupled noise voltage $V_{COUPLED}$ across Z1 is reduced.

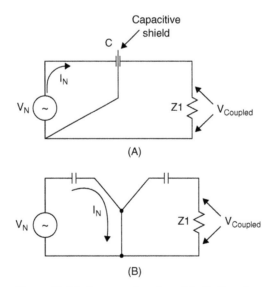

**Figure 12-26: An operational model of a Faraday shield**

A Faraday shield is easily implemented and almost always successful. Thus capacitively coupled noise is rarely an intractable problem. However, to be fully effective, a Faraday shield must completely block the electric field between the noise source and the shielded circuit. It must also be connected so that the displacement current returns to its source, without flowing in any part of the circuit where it can introduce conducted noise.

## Buffering ADCs Against Logic Noise

If we have a high resolution data converter (ADC or DAC, digital-to-analog converters) connected to a high speed data bus which carries logic noise with a 2–5 V/ns edge rate, this noise is easily connected to the converter analog port via stray capacitance across the device. Whenever the data bus is active, intolerable amounts of noise are capacitively coupled into the analog port, thus seriously degrading performance.

This particular effect is illustrated in Figure 12-27, where multiple package capacitors couple noisy edge signals from the data bus into the analog input of an ADC.

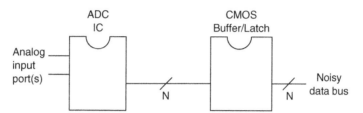

- The output buffer/latch acts as a Faraday shield between "N" lines of a fast, noisy data bus and a high performance ADC

- This measure adds cost, board area, power consumption, reliability reduction, design complexity, and most importantly, improved performance

**Figure 12-27: A high speed ADC IC sitting on a fast data bus couples digital noise into the analog port, thus limiting performance**

Present technology offers no cure for this problem, within the affected IC device itself. The problem also limits performance possible from other broadband monolithic mixed-signal ICs with single-chip analog and digital circuits. Fortunately, this coupled noise problem can be simply avoided, by *not* connecting the data bus directly to the converter.

Instead, *use a CMOS latched buffer as a converter-to-bus interface*, as shown in Figure 12-28. Now the CMOS buffer IC acts as a Faraday shield, and dramatically reduces noise coupling from the digital bus. This solution costs money, occupies board area, reduces reliability (very slightly), consumes power, and complicates the design—but it does improve the SNR of the converter! The designer must decide whether it is worthwhile for individual cases, but in general it is highly recommended.

## High Circuit Impedances Are Susceptible to Noise Pickup

Since low power circuits tend to use high value resistors to conserve power, this tends to make the circuit more susceptible to externally induced radiated noise and conducted noise. Even a small amount of parasitic capacitance can create a significant conduction path for noise to penetrate.

For example, as little as 1 pF of parasitic capacitance allows a 5 V logic transition to cause a large disturbance in a $100 \, \text{k}\Omega$ circuit as illustrated in Figure 12-29.

This serves to illustrate that high impedance circuits are full of potential parasitics which can cause a good paper design to perform poorly when actually implemented. One needs to pay particular attention to the

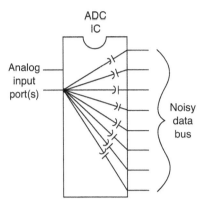

**Figure 12-28: A high speed ADC IC using a CMOS buffer/latch at the output shows enhanced immunity of digital data bus noise**

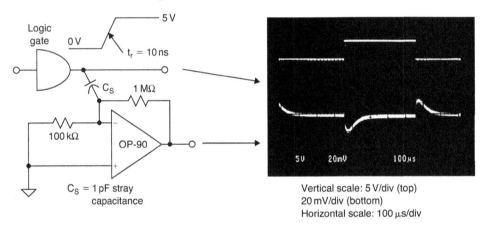

**Figure 12-29: High circuit impedances increase susceptibility to noise pickup**

routing of signals. Interestingly, many high frequency layout techniques for eliminating parasitics can also be applied here for low frequency, low power circuits—for different reasons. While circuit parasitics cause unwanted phase shifts and instabilities in high frequency circuits, the same parasitics pick up unwanted noise in low power precision circuits.

As discussed in the chapter on amplifiers, current feedback amplifiers do not like to have capacitances on their inputs. To that end, ground planes should be cut back from the input pins as shown in Figure 12-30 (A,B), which is an evaluation board for the AD8001 high speed current feedback amplifier. The effect of even small capacitance on the input of a current feedback amplifier is shown in Figure 12-31. Note the ringing on the output.

## Skin Effect

At high frequencies, also consider *skin effect*, where inductive effects cause currents to flow only in the outer surface of conductors. Note that this is in contrast to the earlier discussions of this section on DC resistance of conductors.

*846*

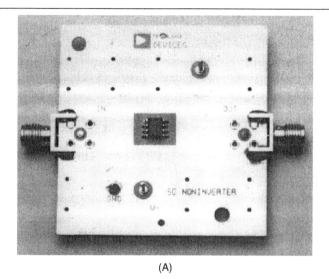

(A)

(B)

**Figure 12-30: AD8001AR (SOIC) evaluation board (A) top
view and (B) bottom view**

The skin effect has the consequence of increasing the resistance of a conductor at high frequencies. Note also that this effect is separate from the increase in impedance due to the effects of the self-inductance of conductors as frequency is increased.

Skin effect is quite a complex phenomenon, and detailed calculations are beyond the scope of this discussion. However, a good approximation for copper is that the skin depth in centimeters is $6.61/\sqrt{f}$ (f in Hz). A summary of the skin effect within a typical PCB conductor foil is shown in Figure 12-32. Note that this copper conductor cross-sectional view assumes looking into the *side* of the conducting trace. Assuming that skin effects become important when the skin depth is less than 50% of the thickness of the conductor, this tells us that for a typical PC foil, we must be concerned about skin effects at frequencies above approximately 12 MHz.

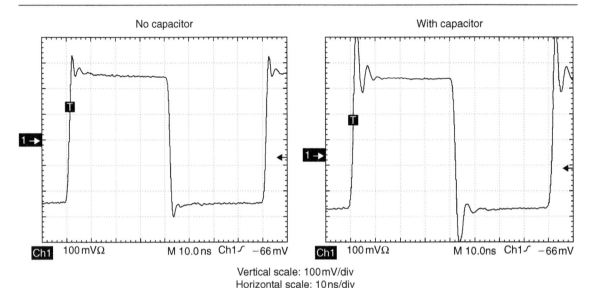

Vertical scale: 100mV/div
Horizontal scale: 10ns/div

**Figure 12-31: Effects of 10 pF stray capacitance on the inverting input on amplifier (AD8001) pulse response**

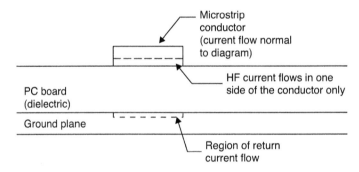

**Figure 12-32: Skin depth in a PCB conductor**

Where skin effect is important, the resistance for copper is $2.6 \times 10^{-7}\sqrt{f}$ $\Omega$/square (f in Hz). This formula is invalid if the skin thickness is greater than the conductor thickness (i.e., at DC or LF).

Figure 12-33 illustrates a case of a PCB conductor with current flow, as separated from the ground plane underneath.

In this diagram, note the (dotted) regions of HF current flow, as reduced by the skin effect. When calculating skin effect in PCBs, it is important to remember that current generally flows in both sides of the PC foil (this is not necessarily the case in microstrip lines, see below), so the resistance per square of PC foil may be half the above value.

## Transmission Lines

We earlier considered the benefits of outward and return signal paths being close together so that inductance is minimized. As shown previously in Figure 7-30, when an HF signal flows in a PC track running over a

• HF current flows only
in thin surface layers

Top

Copper conductor

Bottom

• Skin depth: $6.61/\sqrt{f}$ cm, f in Hz.

• Skin resistance: $2.6 \times 10^{-7}\sqrt{f}$ ohms per square, f in Hz.

• Since skin currents flow on both sides of a PC track, the
value of skin resistance in PCBs must take account of this.

**Figure 12-33: Skin effect with PCB conductor and
ground plane**

ground plane, the arrangement functions as a *microstrip* transmission line, and the majority of the return current flows in the ground plane directly underneath the line.

Figure 12-34 shows the general parameters for a microstrip transmission line, given the conductor width, w, dielectric thickness, h, and the dielectric constant, $E_r$.

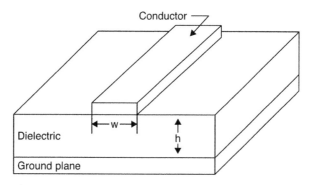

Conductor

w

h

Dielectric

Ground plane

**Figure 12-34: A PCB microstrip transmission line is an
example of a controlled impedance conductor pair**

The characteristic impedance of such a microstrip line will depend on the width of the track and the thickness and dielectric constant of the PCB material. Designs of microstrip lines are covered in more detail later in this chapter.

For most DC and lower frequency applications, the characteristic impedance of PCB traces will be relatively unimportant. Even at frequencies where a track over a ground plane behaves as a transmission line, it is not necessary to worry about its characteristic impedance or proper termination if the free space wavelengths of the frequencies of interest are greater than 10 times the length of the line.

However, at VHF and higher frequencies it is possible to use PCB tracks as microstrip lines within properly terminated transmission systems. Typically the microstrip will be designed to match standard coaxial cable impedances, such as 50, 75, or $100\,\Omega$, simplifying interfacing.

Note that if losses in such systems are to be minimized, the PCB material must be chosen for low high frequency losses. This usually means the use of Teflon or some other comparably low loss PCB material. Often, though, the losses in short lines on cheap glass–fiber board are small enough to be quite acceptable.

## Design PCBs Thoughtfully

Once the system's critical paths and circuits have been identified, the next step in implementing sound PCB layout is to partition the PCB according to circuit function. This involves the appropriate use of power, ground, and signal planes. Good PCB layouts also isolate critical analog paths from sources of high interference (e.g., I/O lines and connectors). High frequency circuits (analog and digital) should be separated from low frequency ones. Furthermore, automatic signal routing CAD layout software should be used with extreme caution. Critical signal paths should be routed by hand, to avoid undesired coupling and/or emissions.

Properly designed multilayer PCBs can reduce EMI emissions and increase immunity to RF fields, by a factor of 10 or more, compared to double-sided boards. A multilayer board allows a complete layer to be used for the ground plane, whereas the ground plane side of a double-sided board is often disrupted with signal crossovers, etc. If the system has separate analog and digital ground and power planes, the analog ground plane should be underneath the analog power plane, and similarly, the digital ground plane should be underneath the digital power plane. There should be no overlap between analog and digital ground planes, nor analog and digital power planes.

## Designing Controlled Impedance Traces on PCBs

A variety of trace geometries are possible with controlled impedance designs, and they may be either integral to or allied to the PCB pattern. In the discussions below, the basic patterns follow those of the IPC, as described in standard 2141 (see Reference 16).

Note that the Figures below use the term "ground plane." It should be understood that this plane is in fact a large area, low impedance *reference* plane. In practice it may actually be either a ground plane or a power plane, both of which are assumed to be at zero AC potential.

The first of these is the simple wire-over-a-plane form of transmission line, also called a *wire microstrip*. A cross-sectional view is shown in Figure 12-35. This type of transmission line might be a signal wire used within a breadboard, for example. It is composed simply of a discrete-insulated wire spaced a fixed distance over a ground plane. The dielectric would be either the insulation wall of the wire or a combination of this insulation and air.

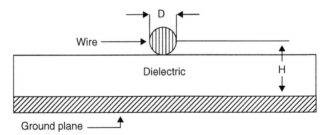

**Figure 12-35: A wire microstrip transmission line with defined impedance is formed by an insulated wire spaced from a ground plane**

The impedance of this line in ohms can be estimated with Eq. (12-2):

$$Z_0(\Omega) = \frac{60}{\sqrt{\varepsilon_r}} \ln\left[\frac{4H}{D}\right] \qquad (12\text{-}2)$$

where D is the conductor diameter, H is the wire spacing above the plane, and $\varepsilon_r$ is the dielectric constant of the material relative to air.

For patterns integral to the PCB, there are a variety of geometric models from which to choose, single ended and differential. These are covered in some detail within IPC standard 2141 (see Reference 16), but information on two popular examples is given here.

Before beginning any PCB-based transmission line design, it should be understood that there are abundant equations, all claiming to cover such designs. In this context, "Which of these are accurate?" is an extremely pertinent question. The unfortunate answer is, *none is perfectly so!* All of the existing equations are approximations, and thus accurate to varying degrees, depending on specifics. The best known and most widely quoted equations are those of Reference 16, but even these come with application caveats.

Reference 17 has evaluated the Reference 16 equations for various geometric patterns against test PCB samples, finding that predicted accuracy varies according to target impedance. Reference 18 also evaluates the Reference 16 equations, offering an alternative and even more complex set (see Reference 19). The equations quoted below are from Reference 16, and are offered here as a starting point for a design, subject to further analysis, testing, and design verification. The bottom line is, study carefully, and take PCB trace impedance equations with a proper dose of salt.

## Microstrip PCB Transmission Lines

For a simple double-sided PCB design where one side is a ground plane, a signal trace on the other side can be designed for controlled impedance. This geometry is known as a *surface microstrip,* or more simply, *microstrip.*

A cross-sectional view of a two-layer PCB illustrates this microstrip geometry as shown in Figure 12-36.

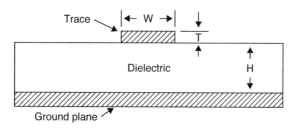

**Figure 12-36: A microstrip transmission line with defined impedance is formed by a PCB trace of appropriate geometry, spaced from a ground plane**

For a given PCB laminate and copper weight, note that all parameters will be predetermined except for W, the width of the signal trace. Equation (12-3) can then be used to design a PCB trace to match the impedance required by the circuit. For the signal trace of width W and thickness T, separated by distance H from a ground (or power) plane by a PCB dielectric with dielectric constant $\varepsilon_r$, the characteristic impedance is:

$$Z_0(\Omega) = \frac{87}{\sqrt{\varepsilon_r + 1.41}} \ln\left[\frac{5.98H}{(0.8W + T)}\right] \qquad (12\text{-}3)$$

Note that in these expressions, measurements are in common dimensions (mils).

These transmission lines will have not only a characteristic impedance, but also capacitance. This can be calculated in terms of pF/inch as shown in Eq. (12-4):

$$C_0(\text{pF/inch}) = \frac{0.67(\varepsilon_r + 1.41)}{\ln[5.98H/(0.8W + T)]} \qquad (12\text{-}4)$$

As an example including these calculations, a two-layer board might use 20-mil wide (W), 1 ounce (T = 1.4) copper traces separated by 10 mil (H) FR-4 ($\varepsilon$ = 4.0) dielectric material. The resulting impedance for this microstrip would be about 50 $\Omega$. For other standard impedances, e.g., the 75 $\Omega$ video standard, adjust "W" to about 8.3 mils.

## Some Microstrip Rules-of-Thumb

This example touches an interesting and quite handy point. Reference 17 discusses a useful rule-of-thumb pertaining to microstrip PCB impedance. For a case of dielectric constant of 4.0 (FR-4), it turns out that when W/H is 2/1, the resulting impedance will be close to 50 $\Omega$ (as in the first example, with W = 20 mils).

Careful readers will note that Eq. (9-21) predicts Z0 to be about 46 $\Omega$, generally consistent with accuracy quoted in Reference 17 (>5%). The IPC microstrip equation is most accurate between 50 and 100 $\Omega$, but is substantially less so for lower (or higher) impedances. Reference 20 gives tabular results of various PCB industry impedance calculator tools.

The propagation delay of the microstrip line can also be calculated, as per Eq. (12-5). This is the one-way transit time for a microstrip signal trace. Interestingly, for a given geometry model, *the delay constant in ns/ feet is a* function *only of the dielectric constant, and not the trace dimensions* (see Reference 21). Note that this is quite a convenient situation. It means that, with a given PCB laminate (and given $\varepsilon_r$), the propagation delay constant is fixed for various impedance lines:

$$t_{pd}(\text{ns/feet}) = 1.017\sqrt{0.475\,\varepsilon_r + 0.67} \qquad (12\text{-}5)$$

This delay constant can also be expressed in terms of ps/in, a form which will be more practical for smaller PCBs. This is:

$$t_{pd}(\text{ps/inch}) = 85\sqrt{0.475\,\varepsilon_r + 0.67} \qquad (12\text{-}6)$$

Thus for an example PCB dielectric constant of 4.0, it can be noted that a microstrip's delay constant is about 1.63 ns/feet, or 136 ps/inch. These two additional rules-of-thumb can be useful in designing the timing of signals across PCB trace runs.

## Symmetric Stripline PCB Transmission Lines

A method of PCB design preferred from many viewpoints is a multi-layer PCB. This arrangement *embeds* the signal trace between a power and a ground plane, as shown in the cross-sectional view of Figure 9-142. The low impedance AC ground planes and the embedded signal trace form a *symmetric stripline* transmission line (Figure 12-37).

As can be noted from the figure, the return current path for a high frequency signal trace is located directly above and below the signal trace on the ground/power planes. The high frequency signal is thus contained

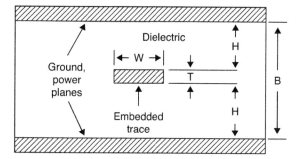

**Figure 12-37: A symmetric stripline transmission line with defined impedance is formed by a PCB trace of appropriate geometry embedded between equally spaced ground and/or power planes**

entirely inside the PCB, minimizing emissions, and providing natural shielding against incoming spurious signals.

The characteristic impedance of this arrangement is again dependent on geometry and the $\varepsilon_r$ of the PCB dielectric. An expression for Z0 of the stripline transmission line is:

$$Z0(\Omega) = \frac{60}{\sqrt{\varepsilon_r}} \ln \left[ \frac{1.9(B)}{(0.8W + T)} \right] \qquad (12\text{-}7)$$

Here, all dimensions are again in mils, and B is the spacing between the two planes. In this symmetric geometry, note that B is also equal to 2H + T. Reference 17 indicates that the accuracy of this Reference 16 equation is typically on the order of 6%.

Another handy rule-of-thumb for the symmetric stripline in an $\varepsilon_r = 4.0$ case is to make B a multiple of W, in the range of 2–2.2. This will result in a stripline impedance of about 50 $\Omega$. Of course this rule is based on a further approximation, by neglecting T. Nevertheless, it is still useful for ballpark estimates.

The symmetric stripline also has a characteristic capacitance, which can be calculated in terms of pF/in:

$$C_0(\text{pF/inch}) = \frac{1.41(\varepsilon_r)}{\ln [3.81H//(0.8W + T)]} \qquad (12\text{-}8)$$

The propagation delay of the symmetric stripline is shown in Eq. (12-9):

$$t_{pd}(\text{ns/feet}) = 1.017 \sqrt{\varepsilon_r} \qquad (12\text{-}9)$$

or, in terms of ps:

$$t_{pd}(\text{ps/inch}) = 85 \sqrt{\varepsilon_r} \qquad (12\text{-}10)$$

For a PCB dielectric constant of 4.0, it can be noted that the symmetric stripline's delay constant is almost exactly 2 ns/feet, or 170 ps/inch.

## Some Pros and Cons of Embedding Traces

The above discussions allow the design of PCB traces of defined impedance, either on a surface layer or embedded between layers. There of course are many other considerations beyond these impedance issues.

Embedded signals do have one major and obvious disadvantage—the debugging of the hidden circuit traces is difficult to impossible. Some of the pros and cons of embedded signal traces are summarized in Figure 12-38.

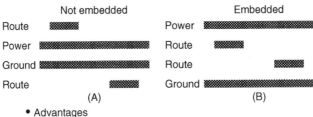

Figure 12-38: The pros and cons of not embedding versus embedding of signal traces in multi-layer PCB designs

Multi-layer PCBs can be designed *without* the use of embedded traces, as is shown in the left-most cross-sectional example (Figure 12-38(A)). This embedded case (Figure 12-38(B)) could be considered as a doubled two-layer PCB design (i.e., four copper layers overall). The routed traces at the top form a microstrip with the power plane, while the traces at the bottom form a microstrip with the ground plane. In this example, the signal traces of both outer layers are readily accessible for measurement and troubleshooting purposes. But, the arrangement does nothing to take advantage of the shielding properties of the planes.

This non-embedded arrangement will have greater emissions and susceptibility to external signals, vis-à-vis the embedded case , which uses the embedding, and does take full advantage of the planes. As in many other engineering efforts, the decision of embedded versus not-embedded for the PCB design becomes a tradeoff, in this case one of reduced emissions versus ease of testing.

## Dealing with High Speed Logic

Much has been written about terminating PCB traces in their characteristic impedance, to avoid signal reflections. A good rule-of-thumb to determine when this is necessary is as follows: *Terminate the transmission line in its characteristic impedance when the one-way propagation delay of the PCB track is equal to or greater than one-half the applied signal rise/fall time (whichever edge is faster)*. For example, a 2 inch microstrip line over an $E_r = 4.0$ dielectric would have a delay of ~270 ps. Using the above rule strictly, termination would be appropriate whenever the signal rise time is <~500 ps. A more conservative rule is to use a 2 inch (PCB track length)/nanosecond (rise/fall time) rule. If the signal trace exceeds this trace length/speed criterion, then termination should be used.

For example, PCB tracks for high speed logic with rise/fall time of 5 ns should be terminated in their characteristic impedance if the track length is equal to or greater than 10 inches (where measured length *includes* meanders).

As an example of what can be expected today in modern systems, Figure 12-39 shows typical rise/fall times for several logic families including the SHARC digital signal processors (DSPs) operating on +3.3 V supplies. As would be expected, the rise/fall times are a function of load capacitance.

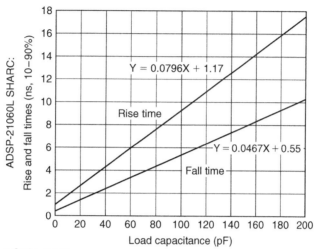

- GaAs: 0.1 ns
- ECL: 0.75 ns
- ADI SHARC DSPs: 0.5 ns to 1 ns (operating on +3.3 V supply)

**Figure 12-39: Typical DSP output rise times and fall times**

In the analog domain, it is important to note that this same 2 inch/ns rule-of-thumb should also be used with op amps and other circuits, to determine the need for transmission line techniques. For instance, if an amplifier must output a maximum frequency of $f_{MAX}$, then the equivalent rise time $t_r$ is related to this $f_{MAX}$. This limiting rise time, $t_r$, can be calculated as:

$$t_r = 0.35/f_{MAX} \qquad (12\text{-}11)$$

The maximum PCB track length is then calculated by multiplying $t_r$ by 2 inch/ns. For example, a maximum frequency of 100 MHz corresponds to a rise time of 3.5 ns, so a 7 inch or more track carrying this signal should be treated as a transmission line.

The best ways to keep sensitive analog circuits from being affected by fast logic are to physically separate the two by the PCB layout, and to use no faster logic family than is dictated by system requirements. In some cases, this may require the use of several logic families in a system. An alternative is to use series resistance or ferrite beads to slow down the logic transitions where highest speed is not required. Figure 12-40 shows two methods.

In the first, the series resistance and the input capacitance of the gate form a lowpass filter. Typical CMOS input capacitance is 5–10 pF. Locate the series resistor close to the driving gate. The resistor minimizes

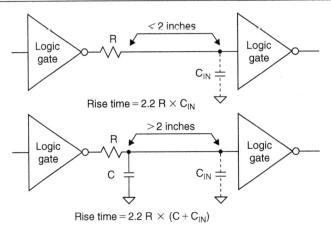

**Figure 12-40: Damping resistors slow down fast logic edges to minimize EMI/RFI problems**

transient currents and may eliminate the necessity of using transmission line techniques. The value of the resistor should be chosen such that the rise and fall times at the receiving gate are fast enough to meet system requirements but no faster. Also, make sure that the resistor is not so large that the logic levels at the receiver are out of specification because of the voltage drop caused by the source and sink currents which flow through the resistor. The second method is suitable for longer distances (>2 inches), where additional capacitance is added to slow down the edge speed. Notice that either one of these techniques increases delay and increases the rise/fall time of the original signal. This must be considered with respect to the overall timing budget, and the additional delay may not be acceptable.

Figure 12-41 shows a situation where several DSPs must connect to a single point, as would be the case when using read or write strobes bidirectionally connected from several DSPs. Small damping resistors shown in Figure 12-41(A) can minimize ringing provided the length of separation is less than about 2 inches.

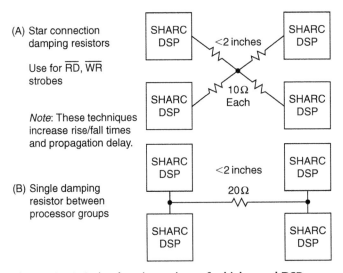

**Figure 12-41: Series damping resistors for high speed DSP interconnections**

This method will also increase rise/fall times and propagation delay. If two groups of processors must be connected, a single resistor between the pairs of processors as shown in Figure 12-41(B) can serve to damp out ringing.

The only way to preserve 1 ns or less rise/fall times over distances greater than about 2 inches without ringing is to use transmission line techniques. Figure 12-42 shows two popular methods of termination: end termination and source termination. The end termination method (Figure 12-42(A)) terminates the cable at its terminating point in the characteristic impedance of the microstrip transmission line. Although higher impedances can be used, $50\,\Omega$ is popular because it minimizes the effects of the termination impedance mismatch due to the input capacitance of the terminating gate (usually 5–10 pF).

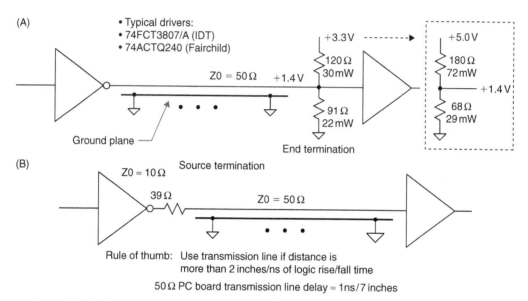

**Figure 12-42: Termination techniques for controlled impedance microstrip transmission lines**

In Figure 12-42(A), the cable is terminated in a Thevenin impedance of $50\,\Omega$ terminated to $+1.4\,V$ (the midpoint of the input logic threshold of 0.8 V and 2.0 V). This requires two resistors (91 and $120\,\Omega$), which add about 50 mW to the total quiescent power dissipation to the circuit. Figure 12-42(A) also shows the resistor values for terminating with a $+5\,V$ supply ($68\,\Omega$ and $180\,\Omega$). Note that 3.3 V logic is much more desirable in line driver applications because of its symmetrical voltage swing, faster speed, and lower power. Drivers are available with less than 0.5 ns time skew, source and sink current capability greater than 25 mA, and rise/fall times of about 1 ns. Switching noise generated by 3.3 V logic is generally less than 5 V logic because of the reduced signal swings and lower transient currents.

The source termination method, shown in Figure 12-42(B), absorbs the reflected waveform with an impedance equal to that of the transmission line. This requires about $39\,\Omega$ in series with the internal output impedance of the driver, which is generally about $10\,\Omega$. This technique requires that the end of the transmission line be terminated in an open circuit; therefore, no additional fanout is allowed. The source termination method adds no additional quiescent power dissipation to the circuit.

Figure 12-43 shows a method for distributing a high speed clock to several devices. The problem with this approach is that there is a small amount of time skew between the clocks because of the propagation delay

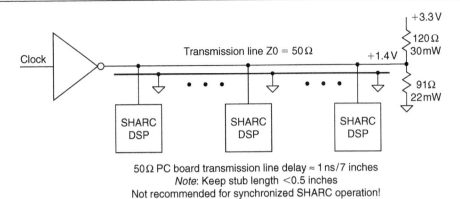

**Figure 12-43: Clock distribution using end-of-line termination**

of the microstrip line (approximately 1 ns/7 inch). This time skew may be critical in some applications. It is important to keep the stub length to each device less than 0.5 inch in order to prevent mismatches along the transmission line.

The clock distribution method shown in Figure 12-44 minimizes the clock skew to the receiving devices by using source terminations and making certain the length of each microstrip line is equal. There is no extra quiescent power dissipation as would be the case using end termination resistors.

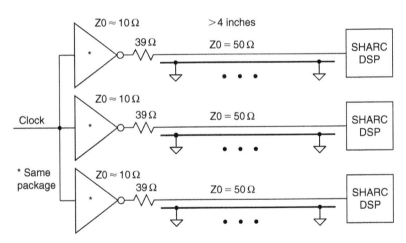

**Figure 12-44: Preferred method of clock distribution using source terminated transmission lines**

Figure 12-45 shows how source terminations can be used in bidirectional link port transmissions between SHARC DSPs. The output impedance of the SHARC driver is approximately 17 Ω, and therefore a 33 Ω series resistor is required on each end of the transmission line for proper source termination.

The method shown in Figure 12-46 can be used for bidirectional transmission of signals from several sources over a relatively long transmission line. In this case, the line is terminated at both ends, resulting in a DC load impedance of 25 Ω. SHARC drivers are capable of driving this load to valid logic levels.

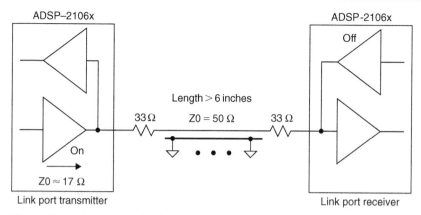

**Figure 12-45: Source termination for bidirectional transmission between SHARC DSPs**

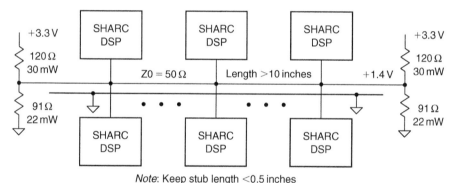

*Note*: Keep stub length <0.5 inches
Not recommended for clocks in synchronized SHARC operation!

**Figure 12-46: Single transmission line terminated at both ends**

Emitter-coupled logic (ECL) has long been known for low noise and its ability to drive terminated transmission lines with rise/fall times less than 2 ns. The family presents a constant load to the power supply, and the low level differential outputs provide a high degree of common-mode rejection. However, ECL dissipates lots of power.

## Low Voltage Differential Signaling

Recently, low-voltage-differential-signaling (LVDS) logic has attained widespread popularity because of similar characteristics, but with lower amplitudes and lower power dissipation than ECL. The defining LVDS specification can be found in the References. The LVDS logic swing is typically 350 mV peak-to-peak centered about a common-mode voltage of +1.2 V. A typical driver and receiver configuration is shown in Figure 12-47. The driver consists of a nominal 3.5 mA current source with polarity switching provided by PMOS and NMOS transistors as in the case of the AD9430 12-bit, 170/210-MSPS ADC. The output voltage of the driver is nominally 350 mV peak-to-peak at each output, and can vary between 247 and 454 mV. The output current can vary between 2.47 and 4.54 mA. The LVDS receiver is terminated

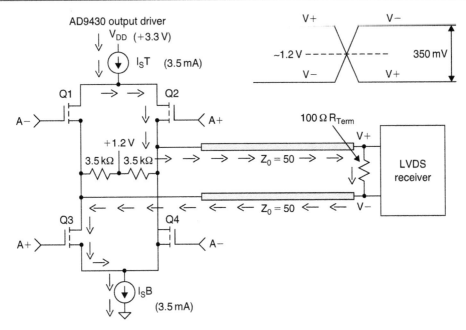

**Figure 12-47: LVDS driver and receiver**

in a $100\,\Omega$ line-to-line. According to the LVDS specification, the receiver must respond to signals as small as $100\,$mV, over a common-mode voltage range of $50\,$mV to $+2.35\,$V. The wide common-mode receiver voltage range is to accommodate ground voltage differences up to $\pm1\,$V between the driver and receiver.

The LVDS edge speed is defined as the 20–90% rise/fall time (as opposed to 10–90% for CMOS logic) and specified to be less than $0.3\,t_{ui}$, where $t_{ui}$ is the inverse of the data signaling rate. For a 210-MSPS sampling rate, $t_{ui} = 4.76\,$ns, and the 20–80% rise/fall time must be less than $0.3 \times 4.76 = 1.43\,$ns. For the AD9430, the rise/fall time is nominally $0.5\,$ns.

LVDS outputs for high performance ADCs should be treated differently than standard LVDS outputs used in digital logic. While standard LVDS can drive 1–10 m in high speed digital applications (dependent on data rate), it is not recommended to let a high performance ADC drive that distance. It is recommended to keep the output trace lengths short (<2 inch), minimizing the opportunity for any noise coupling onto the outputs from the adjacent circuitry, which may get back to the analog inputs. The differential output traces should be routed close together, maximizing common-mode rejection, with the $100\,\Omega$ termination resistor close to the receiver. Users should pay attention to PCB trace lengths to minimize any delay skew. A typical differential microstrip PCB trace cross section is shown in Figure 12-48 along with some recommended layout guidelines.

LVDS also offers some benefits in reduced EMI. The EMI fields generated by the opposing LVDS currents tend to cancel each other (for matched edge rates). In high speed ADCs, LVDS offers simpler timing constraints compared to demultiplexed CMOS outputs at similar data rates. A demultiplexed data bus requires a synchronization signal that is not required in LVDS. In demuxed CMOS buses, a clock equal to one-half the ADC sample rate is needed, adding cost and complexity, which is not required in LVDS.

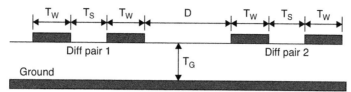

- Keep $T_W$, $T_S$, and D constant over the trace length
- Keep $T_S \sim < 2T_W$
- Avoid use of vias if possible
- Keep $D > 2T_S$
- Avoid 90° bends if possible
- Design $T_W$ and $T_G$ for $\sim 50\ \Omega$

Figure 12-48: Microstrip PCB layout for two pairs of LVDS signals

## References: Traces

1. W. Doeling, W. Mark, T. Tadewald, and P. Reichenbacher, "Getting Rid of Hook: The Hidden PC-Board Capacitance," **Electronics**, October 12, 1978, pp. 111–117.

2. A. Rich, "Shielding and Guarding," **Analog Dialogue**, Vol. 17, No. 1, 1983, p. 8.

3. R. Morrison, **Grounding and Shielding Techniques, 4th Edition**, John Wiley, Inc., New York, 1998, ISBN: 0471245186.

4. H. W. Ott, **Noise Reduction Techniques in Electronic Systems**, 2nd Edition, John Wiley, Inc., New York, 1988, ISBN: 0-471-85068-3.

5. P. Brokaw, "An IC Amplifier User's Guide to Decoupling, Grounding and Making Things Go Right for a Change," **Analog Devices AN202**.

6. P. Brokaw, "Analog Signal-Handling for High Speed and Accuracy," **Analog Devices AN342**.

7. P. Brokaw and J. Barrow, "Grounding for Low- and High-Frequency Circuits," **Analog Devices AN345**.

8. J. Barrow, "Avoiding Ground Problems in High Speed Circuits," **RF Design**, July 1989.

9. B. I. Bleaney and B. Bleaney, **Electricity and Magnetism**, Oxford at the Clarendon Press, 1957, pp 23, 24, and 52.

10. G. W. A. Dummer and H. Nordenberg, **Fixed and Variable Capacitors**, McGraw-Hill, New York, 1960, pp 11–13.

11. W. C. Rempfer, "Get all the Fast ADC Bits You Pay For," **Electronic Design, Special Analog Issue**, June 24, 1996, p. 44.

12. M. Sauerwald, "Keeping Analog Signals Pure in a Hostile Digital World," **Electronic Design, Special Analog Issue**, June 24, 1996, p. 57.

13. J. Grame and B. Baker, "Design Equations Help Optimize Supply Bypassing for Op Amps," **Electronic Design, Special Analog Issue**, June 24, 1996, p. 9.

14. J. Grame and B. Baker, "Fast Op Amps Demand More Than a Single-Capacitor Bypass," **Electronic Design, Special Analog Issue**, November 18, 1996, p. 9.

15. W. Kester and J. Bryant, "Grounding in High Speed Systems," **High Speed Design Techniques**, Analog Devices, Chapter 7, 1996, p. 7–27.

16. J. S. Pattavina, "Bypassing PC Boards: Thumb Your Nose at Rules of Thumb," **EDN**, October 22, 1998, p. 149.

17. H. W. Johnson and M. Graham, **High-Speed Digital Design**, PTR Prentice Hall, 1993, ISBN: 0133957241.

18. W. Kester, "A Grounding Philosophy for Mixed-Signal Systems," **Electronic Design Analog Applications Issue**, June 23, 1997, p. 29.

19. R. Morrison, **Solving Interference Problems in Electronics**, John Wiley, New York, 1995.

20. C. D. Motchenbacher and J. A. Connelly, **Low Noise Electronic System Design**, John Wiley, New York, 1993.

21. Crystal Oscillators: MF Electronics, 10 Commerce Drive, New Rochelle, NY, 10801, 914-576-6570.

22. Crystal Oscillators: Wenzel Associates, Inc., 2215 Kramer Lane, Austin, Texas USA 78758, 512-835-2038, http://www.wenzel.com.

23. M. Montrose, **EMC and the Printed Circuit Board**, IEEE Press, New York, 1999, (IEEE Order Number PC5756).

# *Grounding*

In this section we discuss grounding. This is undoubtedly one of the most difficult subjects in system design. While the basic concepts are relatively simple, implementation is very involved.

For linear systems the ground is the reference against which we base our signal. Unfortunately, it has also become the return path for the power supply current in unipolar supply systems. Improper application of grounding strategies can destroy high accuracy linear system performance.

Grounding is an issue for all analog designs, and it can be said that implementing a PCB-based circuit does not change the fact that proper implementation is essential. Fortunately, certain principles of quality grounding, namely the use of ground planes, are intrinsic to the PCB environment. This factor is one of the more significant advantages to PCB-based analog designs, and appreciable discussion of this section is focused on this issue.

Some other aspects of grounding that must be managed include the control of spurious ground and signal return voltages that can degrade performance. These voltages can be due to external signal coupling, common currents, or simply excessive IR drops in ground conductors. Proper conductor routing and sizing, as well as differential signal-handling and ground isolation techniques enable control of such parasitic voltages.

One final area of grounding to be discussed is grounding appropriate for a mixed-signal, analog/digital environment. Indeed, the single issue of quality grounding can influence the entire layout philosophy of a high performance mixed-signal PCB design—as it well should.

Today's signal processing systems generally require mixed-signal devices such as ADCs and DACs as well as fast DSPs. Requirements for processing analog signals having wide dynamic ranges increase the importance of high performance ADCs and DACs. Maintaining wide dynamic range with low noise in hostile digital environments is dependent on using good high speed circuit design techniques, including proper signal routing, decoupling, and grounding.

In the past, "high precision, low speed" circuits have generally been viewed differently than so-called high speed circuits. With respect to ADCs and DACs, the sampling (or update) frequency has generally been used as the distinguishing speed criterion. However, the following two examples show that in practice, most of today's signal processing ICs are really "high speed," and must therefore be treated as such in order to maintain high performance. This is certainly true of DSPs, and also true of ADCs and DACs.

All sampling ADCs (ADCs with an internal sample-and-hold circuit) suitable for signal processing applications operate with relatively high speed clocks with fast rise and fall times (generally a few nanoseconds) and must be treated as high speed devices, even though throughput rates may appear low. For example, a medium speed 12-bit successive approximation (SAR) ADC may operate on 10 MHz internal clock, while the sampling rate is only 500 kSPS.

Sigma–delta ($\Sigma$–$\Delta$) ADCs also require high speed clocks because of their high oversampling ratios. Even high resolution, so-called low frequency $\Sigma$–$\Delta$ industrial measurement ADCs (having throughputs of 10 Hz to 7.5 kHz) operate on 5 MHz or higher clocks and offer resolution to 24 bits (e.g., the Analog Devices AD77xx series).

To further complicate the issue, mixed-signal ICs have both analog and digital ports, and because of this, much confusion has resulted with respect to proper grounding techniques. In addition, some mixed-signal ICs have relatively low digital currents, while others have high digital currents. In many cases, these two types must be treated differently with respect to optimum grounding.

Digital and analog design engineers tend to view mixed-signal devices from different perspectives, and the purpose of this section is to develop a general grounding philosophy that will work for most mixed-signal devices, without having to know the specific details of their internal circuits.

From the previous discussion it should be clear that the issue of grounding cannot be handled in a "cookbook" approach. Unfortunately we cannot give a list of things to do that will guarantee success. We can say that there are certain things that if they are not done will probably lead to difficulties. And, what works in one frequency range may not necessarily work in another frequency range. And, often, there are competing requirements. The best way to handle grounding is to understand how the currents flow.

## Star Ground

The "star" ground philosophy builds on the theory that there is one single ground point in a circuit to which all voltages are referred. This is known as the *star ground* point. It can be better understood by a visual analogy—the multiple conductors extending radially from the common schematic ground resemble a star. Note that the star point need not look like a star—it may be a point on a ground plane—but the key feature of the star ground system is that all voltages are measured with respect to a particular point in the ground network, not just to an undefined "ground" (i.e., wherever one can clip a probe).

This star grounding philosophy is reasonable theoretically, but is difficult to implement practically. For example, if we design a star ground system, drawing out all signal paths to minimize signal interaction and the effects of high impedance signal or ground paths, we often find implementation problems. When the power supplies are added to the circuit diagram, they either add unwanted ground paths, or their supply currents flowing in the existing ground paths are sufficiently large or noisy (or both) to corrupt the signal transmission. This particular problem can often be avoided by having separate power supplies (and thus separate ground returns) for the various circuit portions. For example, separate analog and digital supplies with separate analog and digital grounds, joined at the star point, are common in mixed-signal applications.

## Separate Analog and Digital Grounds

As a fact of life, digital circuitry is noisy. Saturating logic, such as TTL and CMOS, draws large, fast current spikes from its supply during switching. However, logic stages, with hundreds of millivolts (or more) of noise immunity, usually have little need for high levels of supply decoupling.

On the other hand, analog circuitry is quite vulnerable to noise on both power supply rails and grounds. So, it is very sensible to separate analog and digital circuitry and to prevent digital noise from corrupting analog performance. Such separation involves separation of both ground returns *and* power rails, which is inconvenient in a mixed-signal system.

Nevertheless, if a mixed-signal system is to deliver full performance capability, it is often essential to have separate analog and digital grounds, and separate power supplies. The fact that some analog circuitry will "operate" (i.e., function) from a single +5 V supply does *not* mean that it may optimally be operated from the same noisy +5 V supply as the microprocessor and dynamic RAM, the electric fan, and other high current devices! What is required is that the analog portion *operate with full performance from such a low voltage supply*, not just be functional. This distinction will by necessity require quite careful attention to both the supply rails and the ground interfacing.

Note that analog and digital ground in a system must be joined at some point (the star ground concept), to allow signals to be referred to a common potential. This star point, or analog/digital common point, is chosen so that it does not introduce digital currents into the ground of the analog part of the system—it is often convenient to make the connection at the power supplies.

Note also that many ADCs and DACs have separate *analog ground* (AGND) and *digital ground* (DGND) pins. On the device data sheets, users are often advised to connect these pins together at the package. This seems to conflict with the advice to connect analog and digital ground at the power supplies, and, in systems with more than one converter, with the advice to join the analog and digital ground at a single point.

There is, in fact, no conflict. The labels "analog ground" and "digital ground" on these pins refer to the internal parts of the converter to which the pins are connected, and not to the system grounds to which they must go. For example, with an ADC, generally these two pins should be joined together and to the *analog* ground of the system. It is not possible to join the two pins within the IC package, because the analog part of the converter cannot tolerate the voltage drop resulting from the digital current flowing in the bond wire to the chip. But they can be so tied, *externally*.

Figure 12-49 illustrates this concept of ground connections for an ADC. If these pins are connected in this way, the digital noise immunity of the converter is diminished somewhat, by the amount of common-mode noise between the digital and analog system grounds. However, since digital noise immunity is of the order of hundreds or thousands of millivolts, this factor is unlikely to be important.

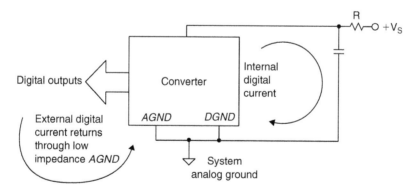

**Figure 12-49: Analog (AGND) and digital ground (DGND) pins of a data converter should be returned to system analog ground**

The analog noise immunity is diminished only by the external digital currents of the converter itself flowing in the analog ground. These currents should be kept quite small, and this can be minimized by ensuring that the converter outputs do not see heavy loads. A good solution toward this is to use a low input current buffer at the ADC output, such as a CMOS buffer-register IC.

If the logic supply to the converter is isolated with a small resistance and decoupled to analog ground with a local 0.1 µF capacitor, all the fast-edge digital currents of the converter will return to ground through the capacitor, and will not appear in the external ground circuit. If the analog ground impedance is maintained low, as it should be for adequate analog performance, additional noise due to the external digital ground current should rarely present a problem.

## Ground Planes

Related to the star ground system discussed earlier is the use of a *ground plane*. To implement a ground plane, one side of a double-sided PCB (or one layer of a multi-layer one) is made of continuous copper and used as ground. The theory behind this is that the large amount of metal will have as low a resistance as is possible. It will, because of the large flattened conductor pattern, also have as low an inductance as possible. It then offers the best possible conduction, in terms of minimizing spurious ground difference voltages across the conducting plane.

Note that ground plane concept can also be extended to include *voltage planes*. A voltage plane offers advantages similar to a ground plane, i.e., a very low impedance conductor, but is dedicated to one (or more) of the system supply voltages. Thus a system can have more than one voltage plane, as well as a ground plane.

While ground planes solve many ground impedance problems, it should still be understood they are not a panacea. Even a continuous sheet of copper foil has residual resistance and inductance, and in some circumstances, these can be enough to prevent proper circuit function. Designers should be wary of injecting very high currents in a ground plane, as they can produce voltage drops that interfere with sensitive circuitry.

The importance of maintaining a low impedance large area ground plane is critical to all analog circuits today. The ground plane not only acts as a low impedance return path for decoupling high frequency currents (caused by fast digital logic) but also minimizes EMI/RFI emissions. Because of the shielding action of the ground plane, the circuit's susceptibility to external EMI/RFI is also reduced.

Ground planes also allow the transmission of high speed digital or analog signals using transmission line techniques (microstrip or stripline) where controlled impedances are required.

The use of "buss wire" is totally unacceptable as a "ground" because of its impedance at the equivalent frequency of most logic transitions. For instance, #22 gauge wire has about 20 nH/inch inductance. A transient current having a slew rate of 10 mA/ns created by a logic signal would develop an unwanted voltage drop of 200 mV at this frequency flowing through 1 inch of this wire:

$$\Delta v = L \frac{\Delta i}{\Delta t} = 20\,\text{nH} \times \frac{10\,\text{mA}}{\text{ns}} = 200\,\text{mV} \tag{12-12}$$

For a signal having a 2 V peak-to-peak range, this translates into an error of about 200 mV, or 10% (approximate 3.5-bit accuracy). Even in all-digital circuits, this error would result in considerable degradation of logic noise margins.

Figure 12-50 shows an illustration of a situation where the digital return current modulates the analog return current (A). The ground return wire inductance and resistance is shared between the analog and digital circuits, and this is what causes the interaction and resulting error. A possible solution is to make the digital return current path flow directly to the GND REF as shown in (B). This is the fundamental concept of a "star," or single-point ground system. Implementing the true single-point ground in a system which contains multiple high frequency return paths is difficult because the physical length of the individual return current wires will introduce parasitic resistance and inductance which can make obtaining a low impedance high frequency ground difficult. In practice, the current returns must consist of large area ground planes for low impedance to high frequency currents. Without a low impedance ground plane, it is therefore almost impossible to avoid these shared impedances, especially at high frequencies.

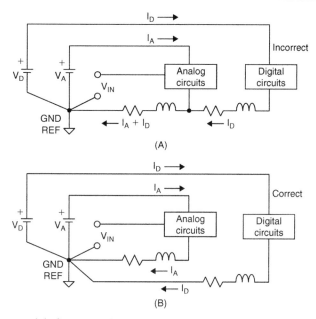

**Figure 12-50: Digital currents flowing in analog return path create error voltages**

All integrated circuit ground pins should be soldered directly to the low impedance ground plane to minimize series inductance and resistance. The use of traditional IC sockets is not recommended with high speed devices. The extra inductance and capacitance of even "low profile" sockets may corrupt the device performance by introducing unwanted shared paths. If sockets must be used with DIP packages, as in prototyping, individual "pin sockets" or "cage jacks" may be acceptable. Both capped and uncapped versions of these pin sockets are available (AMP part numbers 5-330808-3 and 5-330808-6). They have spring-loaded gold contacts which make good electrical and mechanical connection to the IC pins. Multiple insertions, however, may degrade their performance.

Power supply pins should be decoupled directly to the ground plane using low inductance ceramic surface mount capacitors. If through-hole mounted ceramic capacitors must be used, their leads should be less than 1 mm. The ceramic capacitors should be located as close as possible to the IC power pins. Ferrite beads may also be required for additional decoupling.

So, the more ground the better—right? Ground planes solve many ground impedance problems, but not all. Even a continuous sheet of copper foil has residual resistance and inductance, and in some circumstances, they can be enough to prevent proper circuit function. Figure 12-51 shows such a problem—and a possible solution.

Consider the application in Figure 12-51. Due to the realities of the mechanical design, the connector, which has power input is on one side and the power output section, which needs to be near the heat sinking, which, in turn, needs to be on the other side of the board. The board has a ground plane 100 mm wide and a power amplifier draws 15 A. If the ground plane is 0.038 mm thick and 15 A flows in it, there will be a voltage drop of 68 μV/mm. This voltage drop would cause quite serious problems to any ground-referenced precision circuitry sharing the PCB. We can slit the ground plane so that high current does not flow in the region of the precision circuitry, instead forcing it to flow around the slit. This can possibly solve the problem (which in this case it did)—even though the voltage gradient will increase in those parts of the ground plane where the current does flow.

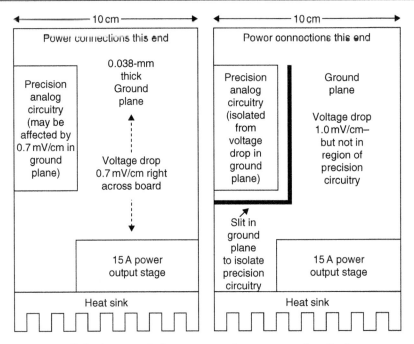

**Figure 12-51: A slit in the ground plane can reconfigure current flow for better accuracy**

## Grounding and Decoupling Mixed-Signal ICs with Low Digital Currents

Sensitive analog components such as amplifiers and voltage references are always referenced and decoupled to the analog ground plane. *The ADCs and DACs (and other mixed-signal ICs) with low digital currents should generally be treated as analog components and also grounded and decoupled to the analog ground plane.* At first glance, this may seem somewhat contradictory, since a converter has an analog and digital interface and usually has pins designated as *analog ground* (AGND) and *digital ground* (DGND). The diagram shown in Figure 12-52 will help to explain this seeming dilemma.

Inside an IC that has both analog and digital circuits, such as an ADC or a DAC, the grounds are usually kept separate to avoid coupling digital signals into the analog circuits. Figure 12-52 shows a simple model of a converter. There is nothing the IC designer can do about the wirebond inductance and resistance associated with connecting the bond pads on the chip to the package pins except to realize they are there. The rapidly changing digital currents produce a voltage at point B which will inevitably couple into point A of the analog circuits through the stray capacitance, $C_{STRAY}$. In addition, there is approximately 0.2 pF unavoidable stray capacitance between every pin of the IC package! It is the IC designer's job to make the chip work in spite of this. However, in order to prevent further coupling, the AGND and DGND pins should be joined together externally to the *analog* ground plane with minimum lead lengths. Any extra impedance in the DGND connection will cause more digital noise to be developed at point B; it will, in turn, couple more digital noise into the analog circuit through the stray capacitance. *Note that connecting DGND to the digital ground plane applies $V_{NOISE}$ across the AGND and DGND pins and invites disaster!*

> *The name "DGND" on an IC tells us that this pin connects to the digital ground of the IC. This does not imply that this pin must be connected to the digital ground of the system. It could correctly be referred to as "Digital Return."*

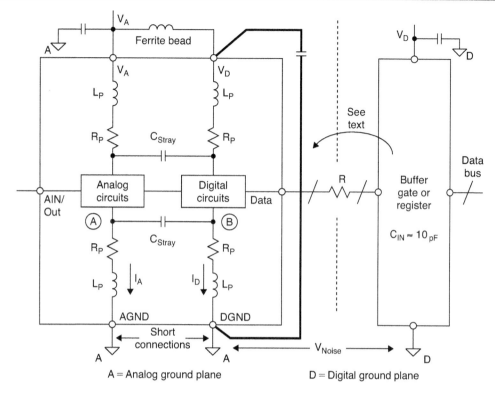

**Figure 12-52: Proper grounding of mixed-signal ICs with low internal diginal currents**

It is true that this arrangement may inject a small amount of digital noise onto the analog ground plane. These currents should be quite small, and can be minimized by ensuring that the converter output does not drive a large fanout (they normally cannot, by design). Minimizing the fanout (which, in turn, means lower currents) on the converter's digital port will also keep the converter logic transitions relatively free from ringing and minimize digital switching currents, and thereby reduce any potential coupling into the analog port of the converter. The logic supply pin ($V_D$) can be further isolated from the analog supply by the insertion of a small lossy ferrite bead as shown in Figure 12-52. The internal transient digital currents of the converter will flow in the small loop from $V_D$ through the decoupling capacitor and to DGND (this path is shown with a heavy line on the diagram). The transient digital currents will therefore not appear on the external analog ground plane but are confined to the loop. The $V_D$ pin decoupling capacitor should be mounted as close to the converter as possible to minimize parasitic inductance. These decoupling capacitors should be low inductance ceramic types, typically between 0.01 and 0.1 μF.

Again, not one grounding scheme will be appropriate for all applications. But by understanding the options and planning ahead problems will be minimized.

## Treat the ADC Digital Outputs with Care

It is always a good idea (as shown in Figure 12-52) to place a buffer register adjacent to the converter to isolate the converter's digital lines from noise on the data bus. The register also serves to minimize loading on the digital outputs of the converter and acts as a Faraday shield between the digital outputs and the data bus (see Figure 12-53). Even though many converters have three-state outputs/inputs, these registers are on the die and still allow the signals on the data pins to couple into sensitive areas. This isolation register still represents good design practice. In some cases it may be desirable to add an additional buffer register on the analog ground plane next to the converter output to provide greater isolation.

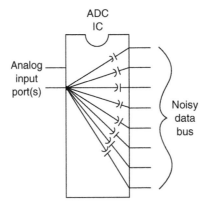

**Figure 12-53: A high speed ADC IC using a buffer/latch at the output shows enhanced immunity to digital data bus noise**

The series resistors (labeled "R" in Figure 12-54) between the ADC output and the buffer-register input help to minimize the digital transient currents which may affect converter performance. The resistors isolate the digital output drivers from the capacitance of the buffer-register inputs. In addition, the RC network formed by the series resistor and the buffer-register input capacitance acts as a lowpass filter to slow down the fast edges.

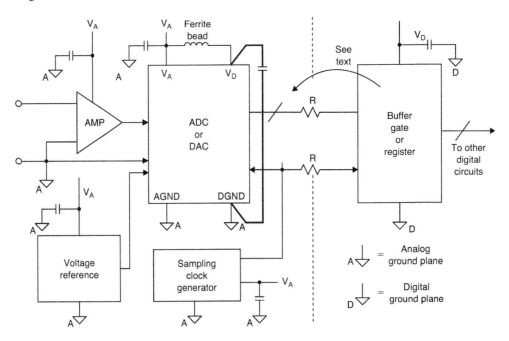

**Figure 12-54: Grounding and decoupling points**

A typical CMOS gate combined with PCB trace and a through-hole will create a load of approximately 10 pF. A logic output slew rate of 1 V/ns will produce 10 mA of dynamic current if there is no isolation resistor:

$$\Delta I = C \frac{\Delta v}{\Delta t} = 10 \, pF \times \frac{1 \, V}{ns} = 10 \, mA \tag{12-13}$$

A 500 Ω series resistors will minimize this output current and result in a rise and fall time of approximately 11 ns when driving the 10 pF input capacitance of the register:

$$t_r = 22 \times t = 22 \times R \times C = 22 \times 500 \, \Omega \times 10 \, pF = 11 \, ns \tag{12-14}$$

TTL registers should be avoided, since they can appreciably add to the dynamic switching currents because of their higher input capacitance.

The buffer register and other digital circuits should be grounded and decoupled to the *digital* ground plane of the PC board. Notice that any noise between the analog and digital ground plane reduces the noise margin at the converter digital interface. Since digital noise immunity is of the orders of hundreds or thousands of millivolts, this is unlikely to matter. The analog ground plane will generally not be very noisy, but if the noise on the digital ground plane (relative to the analog ground plane) exceeds a few hundred millivolts, then steps should be taken to reduce the digital ground plane impedance, thereby maintaining the digital noise margins at an acceptable level. Under no circumstances should the voltage between the two ground planes exceed 300 mV, or the ICs may be damaged.

Separate power supplies for analog and digital circuits are also highly desirable. The analog supply should be used to power the converter. If the converter has a pin designated as a digital supply pin ($V_D$), it should either be powered from a separate analog supply or filtered as shown in the diagram. All converter power pins should be decoupled to the analog ground plane, and all logic circuit power pins should be decoupled to the digital ground plane as shown in Figure 12-54. If the digital power supply is relatively quiet, it may be possible to use it to supply analog circuits as well, but be very cautious.

In some cases it may not be possible to connect $V_D$ to the analog supply. Some of the newer, high speed ICs may have their analog circuits powered by $+5V$, but the digital interface powered by $+3V$ to interface to 3 V logic. In this case, the $+3V$ pin of the IC should be decoupled directly to the analog ground plane. It is also advisable to connect a ferrite bead in series with the power trace that connects the pin to the $+3V$ digital logic supply.

The sampling clock generation circuitry should be treated like analog circuitry and also be grounded and heavily decoupled to the analog ground plane. Phase noise on the sampling clock produces degradation in system SNR as will be discussed shortly.

## Sampling Clock Considerations

In a high performance sampled data system a low phase noise crystal oscillator should be used to generate the ADC (or DAC) sampling clock because sampling clock jitter modulates the analog input/output signal and raises the noise and distortion floor. The sampling clock generator should be isolated from noisy digital circuits and grounded and decoupled to the analog ground plane, as is true for the op amp and the ADC.

The effect of sampling clock jitter on ADC SNR is given approximately by the equation:

$$SNR = 20 \log_{10} \left[ \frac{1}{2\pi ft_j} \right] \tag{12-15}$$

where SNR is the SNR of a perfect ADC of infinite resolution where the only source of noise is that caused by the RMS sampling clock jitter, $t_j$. Note that f in the above equation is the analog input frequency. Just working through a simple example, if $t_j = 50\,ps$ RMS, $f = 100\,kHz$, then SNR $= 90\,dB$, equivalent to about 15-bits dynamic range.

It should be noted that $t_j$ in the above example is the root-sum-square (RSS) value of the external clock jitter *and* the internal ADC clock jitter (called aperture jitter). However, in most high performance ADCs, the internal aperture jitter is negligible compared to the jitter on the sampling clock.

Since degradation in SNR is primarily due to external clock jitter, steps must be taken to ensure the sampling clock is as noise-free as possible and has the lowest possible phase jitter. This requires that a crystal oscillator be used. There are several manufacturers of small crystal oscillators with low jitter (less than 5\,ps RMS) CMOS compatible outputs (e.g., MF Electronics, 10 Commerce Dr., New Rochelle, NY 10801, Tel. 914-576-6570 and Wenzel Associates, Inc., 2215 Kramer Lane, Austin, TX 78758, Tel. 512-835-2038).

Ideally, the sampling clock crystal oscillator should be referenced to the analog ground plane in a split ground system. However, this is not always possible because of system constraints. In many cases, the sampling clock must be derived from a higher frequency multi-purpose system clock which is generated on the digital ground plane. It must then pass from its origin on the digital ground plane to the ADC on the analog ground plane. Ground noise between the two planes adds directly to the clock signal and will produce excess jitter. The jitter can cause degradation in the SNR and also produce unwanted harmonics.

This can be remedied somewhat by transmitting the sampling clock signal as a differential signal using either a small RF transformer as shown in Figure 12-55 or a high speed differential driver and receiver IC. If an active differential driver and receiver are used, they should be ECL to minimize phase jitter. In a single $+5\,V$ supply system, ECL logic can be connected between ground and $+5\,V$ (PECL), and the outputs AC coupled into the ADC sampling clock input. In either case, the original master system clock must be generated from a low phase noise crystal oscillator.

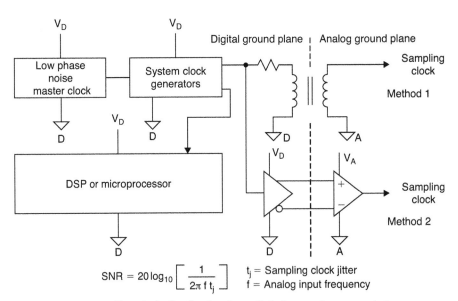

$$SNR = 20\log_{10}\left[\frac{1}{2\pi f\, t_j}\right] \qquad \begin{array}{l} t_j = \text{Sampling clock jitter} \\ f = \text{Analog input frequency} \end{array}$$

**Figure 12-55: Sampling clock distribution from digital-to-analog ground planes**

## *The Origins of the Confusion about Mixed-Signal Grounding*

Most ADC, DAC, and other mixed-signal device data sheets discuss grounding relative to a single PCB, usually the manufacturer's own evaluation board. This has been a source of confusion when trying to apply these principles to multicard or multi-ADC/DAC systems. The recommendation is usually to split the PCB ground plane into an analog plane and a digital plane. It is then further recommended that the AGND and DGND pins of a converter be tied together and that the analog ground plane and digital ground plane be connected at that same point as shown in Figure 12-56. This essentially creates the system "star" ground at the mixed-signal device.

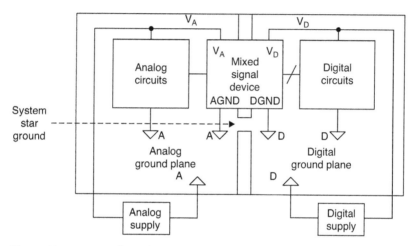

**Figure 12-56: Grounding mixed-signal ICs: single PC board (typical evaluation/test board)**

All noisy digital currents flow through the digital power supply to the digital ground plane and back to the digital supply; they are isolated from the sensitive analog portion of the board. The system star ground occurs where the analog and digital ground planes are joined together at the mixed-signal device. While this approach will generally work in a simple system with a single PCB and single ADC/DAC, it is not optimum for multicard mixed-signal systems. In systems having several ADCs or DACs on different PCBs (or on the same PCB, for that matter), the analog and digital ground planes become connected at several points, creating the possibility of ground loops and making a single-point "star" ground system impossible. For these reasons, this grounding approach is not recommended for multicard systems, and the approach previously discussed should be used for mixed-signal ICs with low digital currents.

## *Summary: Grounding Mixed-Signal Devices with Low Digital Currents in a Multicard System*

Figure 12-57 summarizes the approach previously described for grounding a mixed-signal device which has low digital currents. The analog ground plane is not corrupted because the small digital transient currents flow in the small loop between $V_D$, the decoupling capacitor, and DGND (shown as a heavy line). The mixed-signal device is for all intents and purposes treated as an analog component. The noise $V_N$ between the ground planes reduces the noise margin at the digital interface but is generally not harmful if kept less than 300 mV by using a low impedance digital ground plane all the way back to the system star ground.

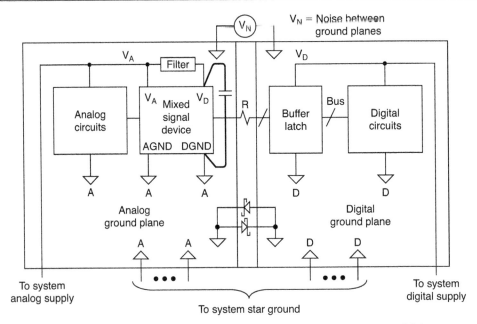

**Figure 12-57: Grounding mixed-signal ICs with low internal digital currents: multiple PC boards**

However, mixed-signal devices such as sigma–delta ADCs, codecs, and DSPs with on-chip analog functions are becoming more and more digitally intensive. Along with the additional digital circuitry come larger digital currents and noise. For example, a sigma–delta ADC or DAC contains a complex digital filter which adds considerably to the digital current in the device. The method previously discussed depends on the decoupling capacitor between VD and DGND to keep the digital transient currents isolated in a small loop. However, if the digital currents are significant enough and have components at DC or low frequencies, the decoupling capacitor may have to be so large that it is impractical. Any digital current which flows outside the loop between VD and DGND must flow through the analog ground plane. This may degrade performance, especially in high resolution systems.

It is difficult to predict what level of digital current flowing into the analog ground plane will become unacceptable in a system. All we can do at this point is to suggest an alternative grounding method which may yield better performance.

### Summary: Grounding Mixed-Signal Devices with High Digital Currents in a Multicard System

An alternative grounding method for a mixed-signal device with high levels of digital currents is shown in Figure 12-58. The AGND of the mixed-signal device is connected to the analog ground plane, and the DGND of the device is connected to the digital ground plane. The digital currents are isolated from the analog ground plane, but the noise between the two ground planes is applied directly between the AGND and DGND pins of the device. For this method to be successful, the analog and digital circuits within the mixed-signal device must be well isolated. The noise between AGND and DGND pins must not be large enough to reduce internal noise margins or cause corruption of the internal analog circuits.

874

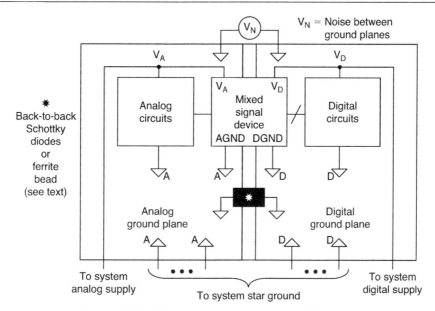

**Figure 12-58: High digital currents: multiple PC boards**

Figure 12-58 shows optional Schottky diodes (back-to-back) or a ferrite bead connecting the analog and digital ground planes. The Schottky diodes prevent large DC voltages or low frequency voltage spikes from developing across the two planes. These voltages can potentially damage the mixed-signal IC if they exceed 300 mV because they appear directly between the AGND and DGND pins. As an alternative to the back-to-back Schottky diodes, a ferrite bead provides a DC connection between the two planes but isolates them at frequencies above a few megahertz where the ferrite bead becomes resistive. This protects the IC from DC voltages between AGND and DGND, but the DC connection provided by the ferrite bead can introduce unwanted DC ground loops and may not be suitable for high resolution systems.

## Grounding DSPs with Internal Phase-Locked Loops

As if dealing with mixed-signal ICs with AGND and DGNDs was not enough, DSPs such as the ADSP-21160 SHARC with internal phase-locked-loops (PLLs) raise issues with respect to proper grounding. The ADSP-21160 PLL allows the internal core clock (which determines the instruction cycle time) to operate at a user-selectable ratio of 2, 3, or 4 times the external clock frequency, CLKIN. The CLKIN rate is the rate at which the synchronous external ports operate. Although this allows using a lower frequency external clock, care must be taken with the power and ground connections to the internal PLL as shown in Figure 12-59.

In order to prevent internal coupling between digital currents and the PLL, the power and ground connections to the PLL are brought out separately on pins labeled $AV_{DD}$ and AGND, respectively. The $AV_{DD}$ +2.5 V supply should be derived from the $V_{DD\,INT}$ +2.5 V supply using the filter network as shown. This ensures a relatively noise-free supply for the internal PLL. The AGND pin of the PLL should be connected to the digital ground plane of the PC board using a short trace. The decoupling capacitors should be routed between the $AV_{DD}$ pin and AGND pin using short traces.

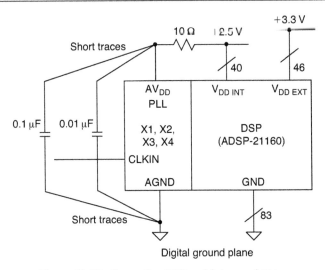

**Figure 12-59: Grounding DSPs with internal PLLs**

## Grounding Summary

There is no single grounding method which will guarantee optimum performance 100% of the time! This section has presented a number of possible options depending on the characteristics of the particular mixed-signal devices in question. It is helpful, however, to provide for as many options as possible when laying out the initial PC board.

It is mandatory that at least one layer of the PC board be dedicated to ground plane! The initial board layout should provide for non-overlapping analog and digital ground planes, but pads and vias should be provided at several locations for the installation of back-to-back Schottky diodes or ferrite beads, if required. Pads and vias should also be provided so that the analog and digital ground planes can be connected together with jumpers if required.

The AGND pins of mixed-signal devices should in general always be connected to the analog ground plane. An exception to this are DSPs which have internal PLLs, such as the ADSP-21160 SHARC. The ground pin for the PLL is labeled AGND, but should be connected directly to the digital ground plane for the DSP.

## Grounding for High Frequency Operation

The "ground plane" layer is often advocated as the best return for power and signal currents, while providing a reference node for converters, references, and other subcircuits. However, even extensive use of a ground plane does not ensure a high quality ground reference for an AC circuit.

The simple circuit of Figure 12-60, built on a two-layer PCB, has an AC and DC current source on the top layer connected to a via (via 1) at one end and to a single U-shaped copper trace connected to via 2. Both vias go through the circuit board and connect to the ground plane. Ideally, the impedance is zero and the voltage appearing across the current source is also zero.

This simple schematic hardly begins to show the actual subtleties. But an understanding of how the current flows in the ground plane from via 1 to via 2 makes the realities apparent and shows how ground noise in high frequency layouts can be avoided.

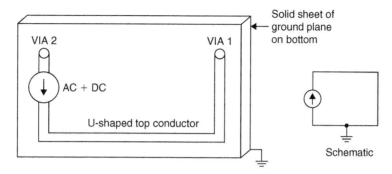

**Figure 12-60: Schematic and layout of current source with U-shaped trace on PC board and return through ground plane**

The DC current flows in the manner shown in Figure 12-61, as one might surmise, taking the path of least resistance from via 1 to via 2. Some current spreading occurs, but little current flows a substantial distance from this path. In contrast, the AC current does not take the path of least resistance; it takes the path of least impedance, which, in turn, depends on inductance.

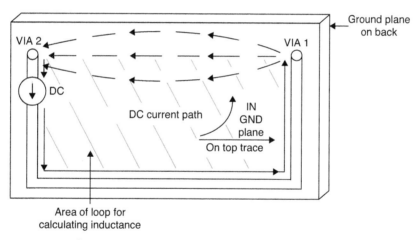

**Figure 12-61: DC current flow for Figure 12-60**

Inductance is proportional to the area of the loop made by the current flow; the relationship can be illustrated by the right-hand rule and the magnetic field shown in Figure 12-62. Inside the loop, current along all parts of the loop produces magnetic field lines that add constructively. Away from the loop, however, field lines from different parts add destructively; thus, the field is confined principally within the loop. A larger loop has greater inductance. This means that, for a given current level, it has more stored magnetic energy ($Li^2$), greater impedance ($X_L = jL$), and hence will develop more voltage for a given frequency.

Which path will the current choose in the ground plane? Naturally, the lower impedance path. Considering the loop formed by the U-shaped surface lead and the ground plane and neglecting resistance, high frequency AC current will follow the path with the least inductance, hence the least area.

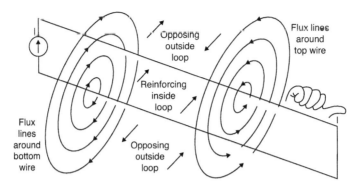

**Figure 12-62: Magnetic field lines and inductive loop (right hand rule)**

In the example shown, the loop with the least area is quite evidently formed by the U-shaped top trace and the portion of the ground plane directly underneath it. So while Figure 12-61 shows the DC current path, Figure 12-63 shows the path that most of the AC current takes in the ground plane, where it finds minimum area, directly under the U-shaped top trace. In practice, the resistance in the ground plane causes the current to flow at low and mid-frequencies to somewhere between straight back and directly under the top conductor. However, the return path is nearly under the top trace as low as 1 or 2 MHz.

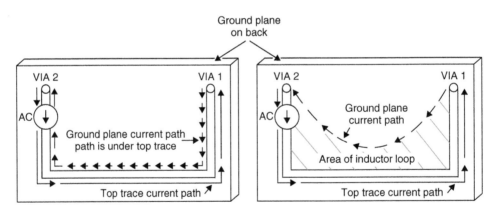

**Figure 12-63: AC current path without (left) and with (right) resistance in the ground plane**

## Be Careful with Ground Plane Breaks

Wherever there is a break in the ground plane beneath a conductor, the ground plane return current must by necessity flow *around* the break. As a result, both the inductance and the vulnerability of the circuit to external fields are increased. This situation is diagrammed in Figure 12-64, where conductors A and B must cross one another.

Where such a break is made to allow a crossover of two perpendicular conductors, it would be far better if the second signal were carried across both the first and the ground planes by means of a piece of wire. The ground plane then acts as a shield between the two signal conductors, and the two ground return currents, flowing in opposite sides of the ground plane as a result of skin effects, do not interact.

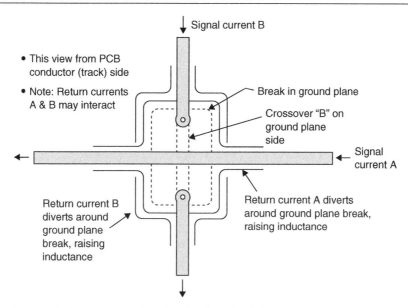

**Figure 12-64: A ground plane break raises circuit inductance and increases vulnerability to external fields**

With a multi-layer board, both the crossover and the continuous ground plane can be accommodated without the need for a wire link. Multi-layer PCBs are expensive and harder to troubleshoot than simpler double-sided boards, but do offer even better shielding and signal routing. The principles involved remain unchanged but the range of layout options is increased.

The use of double-sided or multi-layer PCBs with at least one continuous ground plane is undoubtedly one of the most successful design approaches for high performance mixed-signal circuitry. Often the impedance of such a ground plane is sufficiently low to permit the use of a single ground plane for both analog and digital parts of the system. However, whether or not this is possible does depend on the resolution and bandwidth required, and the amount of digital noise present in the system.

### References: Grounding

1.  A. Rich, "Shielding and Guarding," **Analog Dialogue**, Vol. 17, No. 1, 1983, p. 8.

2.  R. Morrison, **Grounding and Shielding Techniques, 4th Edition**, John Wiley, Inc., New York, 1998, ISBN: 0471245186.

3.  H. W. Ott, **Noise Reduction Techniques in Electronic Systems, 2nd Edition,** John Wiley, Inc., New York, 1988, ISBN: 0-471-85068-3.

4.  P. Brokaw, "An IC Amplifier User's Guide to Decoupling, Grounding and Making Things Go Right for a Change," **Analog Devices AN202**.

5.  P. Brokaw and J. Barrow, "Grounding for Low- and High-Frequency Circuits," **Analog Devices AN345**.

6.  J. Barrow, "Avoiding Ground Problems in High Speed Circuits," **RF Design**, July 1989.

7.  B. I. Bleaney and B. Bleaney, **Electricity and Magnetism**, Oxford at the Clarendon Press, 1957, pp 23, 24, and 52.

8.  W. C. Rempfer, "Get all the Fast ADC Bits You Pay For," **Electronic Design, Special Analog Issue**, June 24, 1996, p. 44.

9.  M. Sauerwald, "Keeping Analog Signals Pure in a Hostile Digital World," **Electronic Design, Special Analog Issue**, June 24, 1996, p. 57.

10. W. Kester and James Bryant, "Grounding in High Speed Systems," **High Speed Design Techniques**, Analog Devices, 1996, Chapter 7, p. 7–27.

11. H. W. Johnson and M. Graham, **High-Speed Digital Design**, PTR Prentice Hall, 1993, ISBN: 0133957241.

12. W. Kester, "A Grounding Philosophy for Mixed-Signal Systems," **Electronic Design Analog Applications Issue**, June 23, 1997, p. 29.

13. R. Morrison, **Solving Interference Problems in Electronics**, John Wiley, New York, 1995.

14. C. D. Motchenbacher and J. A. Connelly, **Low Noise Electronic System Design**, John Wiley, New York, 1993.

15. M. Montrose, **EMC and the Printed Circuit Board**, IEEE Press, New York, 1999, (IEEE Order Number PC5756).

# *Decoupling*

It is imperative to properly decouple all ICs in a high speed and/or high precision application. This decoupling should include a small (typically 0.01–0.1 µF) capacitor. This capacitor should have good high frequency characteristics. Surface mount multi-layer ceramics are ideal. The purpose of this capacitor is to shunt any high frequency noise away for the IC. This is because the power supply rejection ratio drops with frequency, as shown in Figure 12-65. While this plot is for an op amp, all linear circuits and converters have the same general shape, rejection falling with increasing frequency. Keeping the high frequency noise out of the IC helps keep it from getting to the output (of a linear circuit) or affecting the noise (of a converter).

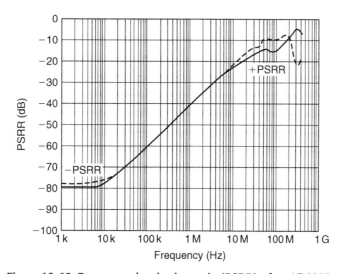

Figure 12-65: Power supply rejection ratio (PSRR) of an AD8029

In addition to the high frequency cap there should be liberal use of larger electrolytic capacitors (10–100 µF). These capacitors are not required at every chip. The purpose of these capacitors is to provide a local reservoir of charge so that instantaneous current demands can be provided from a local source, instead of having to come from a power supply which may be relatively far away and subject to the inductance and resistance of the PCB traces.

## *Local High Frequency Bypass/Decoupling*

As we have stated, each individual analog stage requires local, high frequency decoupling. *These stages are provided directly at the power pins, of* all *individual analog stages.* Figure 12-66 shows the preferred technique, in both correct (A) and incorrect example implementations (B). In (A), a typical 0.1 µF chip ceramic capacitor goes directly to the opposite PCB side ground plane, by virtue of the via, and on to the

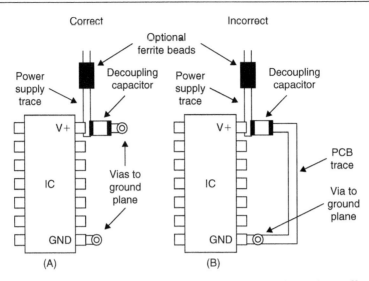

**Figure 12-66: Localized high frequency supply filter(s) provides optimum filtering and decoupling via short low inductance path (ground plane)**

IC's GND pin by a second via. In contrast, the less desirable arrangement (B) adds additional PCB trace inductance in the ground path of the decoupling cap, reducing effectiveness.

The general technique shown here is suitable for single-rail power supply, but the concept obviously extends to dual-rail systems. Note that if the decoupled IC in question is an op amp, the GND pin shown is the $-V_S$ pin. For dual-supply op amp uses, there is no op amp GND pin per se, so the dual decoupling networks should go directly to the ground plane when used, or other local ground.

*All* high frequency (i.e., ≤10 MHz) ICs should use a bypassing scheme similar to Figure 12-66 for best performance. Trying to operate op amps and other high performance ICs without local bypassing is almost always folly. It *may* be possible in a few circumstances, *if* the circuitry is strictly micropower in nature, and the gain bandwidth in the kilohertz range. To put things into an overall perspective, however, note that a pair of 0.1 µF ceramic bypass caps costs less than 25 cents. Hardly a worthy savings compared to the potential grief and lost time of troubleshooting a system without bypassing!

In contrast, the ferrite beads are not 100% necessary, but they will add extra HF noise isolation and decoupling, which is often desirable. Possible caveats here would be to verify that the beads never saturate, when the op amps are handling high currents.

Note that with some ferrites, even before full saturation occurs, some beads can be nonlinear, so if a power stage is required to operate with a low distortion output, this should also be lab checked.

The effects of inadequate decoupling on harmonic distortion performance are dramatically illustrated in Figure 12-67. Figure 12-67(A) shows the spectral output of the AD9631 op amp driving a 100 Ω load with proper decoupling (output signal is 20 MHz, $2V_{p-p}$). Notice that the second harmonic distortion at 40 MHz is approximately −70 dBc. If the decoupling is removed, the distortion is increased, as shown in the Figure 12-67(B). Figure 12-67(A) also shows stray RF pickup in the wiring connecting the power supply to the op amp test fixture. Unlike lower frequency amplifiers, the power supply rejection ratio of many high frequency amplifiers is generally fairly poor at high frequencies. For example, at 20 MHz, the power supply rejection ratio of the AD9631 is less than 25 dB. This is the primary reason for the degradation in

Proper decoupling        No decoupling

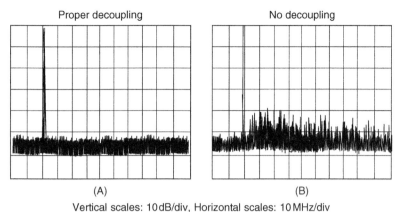

(A)                (B)

Vertical scales: 10 dB/div, Horizontal scales: 10 MHz/div

**Figure 12-67: Effects of inadequate decoupling on harmonic distortion performance of the AD9611 Op Amp**

performance with inadequate decoupling. The change in output signal produces a corresponding signal dependent load current change. The corresponding change in power supply voltage due to inadequate decoupling produces a signal dependent error in the output which manifests itself as an increase in distortion.

Inadequate decoupling can also severely affect the pulse response of high speed amplifiers such as the AD9631. Figure 12-68, shows normal operation and the effects of removing all decoupling capacitors on the AD9631 in its evaluation board. Notice the severe ringing on the pulse response for the poorly decoupled condition.

Proper decoupling        No decoupling

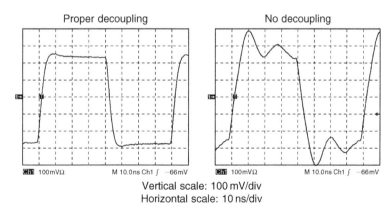

Vertical scale: 100 mV/div
Horizontal scale: 10 ns/div

**Figure 12-68: Effects of inadequate decoupling on the phase response of the AD9631 op amp**

## Ringing

An inductor in series or parallel with a capacitor forms a resonant, or "tuned," circuit, whose key feature is that it shows marked change in impedance over a small range of frequency. Just how sharp the effect is depends on the relative Q of the tuned circuit. The effect is widely used to define the frequency response of narrowband circuitry, but can also be a potential problem source.

If stray inductance and capacitance (which may or may not be stray) in a circuit should form a tuned circuit, then that tuned circuit may be excited by signals in the circuit, and ring at its resonant frequency.

An example is shown in Figure 12-69, where the resonant circuit formed by an inductive power line and its decoupling capacitor may possibly be excited by fast pulse currents drawn by the powered IC.

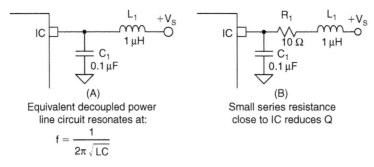

(A)
Equivalent decoupled power
line circuit resonates at:

$$f = \frac{1}{2\pi \sqrt{LC}}$$

(B)
Small series resistance
close to IC reduces Q

**Figure 12-69: Resonant circuit formed by power line decoupling**

While normal trace inductance and typical decoupling capacitances of 0.01–0.1 µF will resonate well above a few megahertz, an example 0.1 µF capacitor and 1 µH of inductance resonate at 500 kHz. Left unchecked, this could present a resonance problem, as shown in the case (A). Should an undesired power line resonance be present, the effect may be minimized by lowering the Q of the inductance. This is most easily done by inserting a small resistance (~10 Ω) in the power line close to the IC, as shown in the case (B).

## References: Decoupling

1. P. Brokaw, "An IC Amplifier User's Guide to Decoupling, Grounding and Making Things Go Right for a Change," **Analog Devices AN202**.

2. H. W. Ott, **Noise Reduction Techniques in Electronic Systems, 2nd Edition**, John Wiley, Inc., New York, 1988, ISBN: 0-471-85068-3.

3. M. Sauerwald, "Keeping Analog Signals Pure in a Hostile Digital World," **Electronic Design, Special Analog Issue**, June 24, 1996, p. 57.

4. J. Grame and B. Baker, "Design Equations Help Optimize Supply Bypassing for Op Amps," **Electronic Design, Special Analog Issue**, June 24, 1996, p. 9.

5. J. Grame and B. Baker, "Fast Op Amps Demand More Than a Single-Capacitor Bypass," **Electronic Design, Special Analog Issue**, November 18, 1996, p. 9.

6. J. S. Pattavina, "Bypassing PC Boards: Thumb Your Nose at Rules of Thumb," **EDN**, October 22, 1998, p. 149.

7. C. D. Motchenbacher and J. A. Connelly, **Low Noise Electronic System Design**, John Wiley, New York, 1993.

8. W. Jung, W. Kester and B. Chesnut, "Power Supply Noise Reduction and Filtering," portion of Section 8 within W. Kester, Editor, **Practical Design Techniques for Power and Thermal Management**, Analog Devices, Inc., 1998, ISBN 0-916550-19-2.

# *Thermal Management*

For reliability reasons, systems with appreciable power dissipation are increasingly called upon to observe *thermal management*. All semiconductors have some specified safe upper limit for junction temperature ($T_J$), usually on the order of 150°C (sometimes 175°C). Like maximum power supply voltages, maximum junction temperature is a worst-case limitation which should not be exceeded. In conservative designs an ample safety margin should be included. Note that this is critical, since semiconductor lifetime is inversely related to operating junction temperature. Simply put, the cooler ICs are, the longer their lifetimes will be.

This limitation of power and temperature is basic, and is illustrated by a typical data sheet statement as in Figure 12-70. In this case it is for the AD8017AR, an 8-pin SOIC device.

> The maximum power that can be safely dissipated by the AD8017 is limited by the associated rise in junction temperature. The maximum safe junction temperature for plastic encapsulated device is determined by the glass transition temperature of the plastic, approximately +150°C. Temporarily exceeding this limit may cause a shift in parametric performance due to a change in the stresses exerted on the die by the package. Exceeding a junction temperature of +175°C for an extended period can result in device failure.

**Figure 12-70: Maximum power dissipation data sheet statement for the AD8017AR, an ADI thermally enhanced SOIC packaged device**

Tied to these statements are certain conditions of operation, such as the power dissipated by the device, and the package mounting to the PCB. In the case of the AD8017AR, the part is rated for 1.3 W of power at an ambient of 25°C. This assumes operation of the 8-lead SOIC package on a two-layer PCB with about 4 in$^2$ (~2,500 mm$^2$) of 2 ounce copper for heat sinking purposes. Predicting safe operation for the device under other conditions is covered below.

## *Thermal Basics*

The symbol θ is generally used to denote *thermal resistance*. Thermal resistance is in units of °C/watt (°C/W). Unless otherwise specified, it defines the resistance heat encounters transferring from a hot IC junction to the ambient air. It might also be expressed more specifically as $\theta_{JA}$, for *thermal resistance, junction-to-ambient*. $\theta_{JC}$ and $\theta_{CA}$ are two additional θ forms used, and are further explained below.

In general, a device with a thermal resistance θ equal to 100°C/W will exhibit a temperature differential of 100°C for a power dissipation of 1 W, as measured between two reference points. Note that this is a linear relationship, so 1 W of dissipation in this part will produce a 100°C differential (and so on, for other powers). For the AD8017AR example, θ is about 95°C/W, so 1.3 W of dissipation produces about a 124°C junction-to-ambient temperature differential. It is of course this rise in temperature that is used to predict the internal temperature, in order to judge the thermal reliability of a design. With the ambient at 25°C, this allows an internal junction temperature of about 150°C. In practice most ambient temperatures are above 25°C, so less power can then be handled.

For any power dissipation P (in watts), one can calculate the effective temperature differential ($\Delta T$) in °C as:

$$\Delta T = P \times \theta \qquad\qquad (12\text{-}16)$$

where $\theta$ is the total applicable thermal resistance.

Figure 12-71 summarizes a number of basic thermal relationships.

- $\theta$ = Thermal resistance (°C/W)
- P = Total device power dissipation (W)
- T = Temperature (°C)
- $\Delta T$ = Temperature differential = P × $\theta$
- $\theta_{JA}$ = Junction-ambient thermal resistance
- $\theta_{JC}$ = Junction-case thermal resistance
- $\theta_{CA}$ = Case-ambient thermal resistance
- $\theta_{JA} = \theta_{JC} + \theta_{CA}$
- $T_J = T_A + (P \times \theta_{JA})$
- Note: $T_{J(Max)}$ = 150°C (sometimes 175°C)

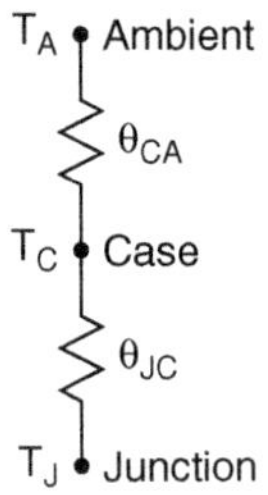

$T_A$ • Ambient

$\theta_{CA}$

$T_C$ • Case

$\theta_{JC}$

$T_J$ • Junction

**Figure 12-71: Basic thermal relationships**

Note that series thermal resistances, such as the two shown at the right, model the total thermal resistance path a device may see. Therefore the total $\theta$ for calculation purposes is the sum, i.e., $\theta_{JA} = \theta_{JC} + \theta_{CA}$. Given the ambient temperature $T_A$, P, and $\theta$, then $T_J$ can be calculated. As the relationships signify, to maintain a low $T_J$, either $\theta$ or the power being dissipated (or both) must be kept low. A low $\Delta T$ is the key to extending semiconductor lifetimes, as it leads to lower maximum junction temperatures.

In ICs, one temperature reference point is always the device junction, taken to mean the hottest spot inside the chip operating within a given package. The other relevant reference point will be either $T_C$, the case of the device, or $T_A$, that of the surrounding air. This then leads in turn to the above-mentioned individual thermal resistances, $\theta_{JC}$ and $\theta_{JA}$.

Taking the most simple case first, $\theta_{JA}$ is the thermal resistance of a given device measured between its *junction* and the *ambient* air. This thermal resistance is most often used with small, relatively low power ICs such as op amps, which often dissipate 1 W or less. Generally, $\theta_{JA}$ figures typical of op amps and other small devices are on the order of 90–100°C/W for a plastic 8-pin DIP package, as well as the better SOIC packages.

It should be clearly understood that these thermal resistances are *highly* package dependent, as different materials have different degrees of thermal conductivity. As a general rule-of-thumb, thermal resistance of conductors is analogous to electrical resistances, i.e., copper is the best, followed by aluminum, steel, and so on. Thus copper lead-frame packages offer the highest performance, i.e., the lowest $\theta$.

## Heat Sinking

By definition, a *heat sink* is an added low thermal resistance device attached to an IC to aid heat removal. A heat sink has additional thermal resistance of its own, $\theta_{CA}$, rated in °C/W. However, most current IC packages do not easily lend themselves to heat sink attachment (exceptions are older TO-99 metal can types). Devices meant for heat sink attachment will often be noted by a $\theta_{JC}$ dramatically lower than the $\theta_{JA}$. In this case $\theta$ will be composed of more than one component. Thermal impedances add, making a net calculation relatively simple. For example, to compute a net $\theta_{JA}$ given a relevant $\theta_{JC}$, the thermal resistance of the heat sink, $\theta_{CA}$, or *case* to *ambient* is added to the $\theta_{JC}$ as:

$$\theta_{JA} = \theta_{JC} + \theta_{CA} \qquad (12\text{-}17)$$

and the result is the $\theta_{JA}$ for that specific circumstance.

More generally however, modern op amps *do not* use commercially available heat sinks. Instead, when significant power needs to be dissipated, such as $\geq 1\,W$, low thermal resistance copper PCB traces are used as the heat sink. In such cases, the most useful form of manufacturer data for this heat sinking is the boundary conditions of a sample PCB layout, and the resulting $\theta_{JA}$ for those conditions. This is in fact the type of specific information supplied for the AD8017AR, as mentioned earlier. Applying this approach, example data illustrating thermal relationships for such conditions are shown by Figure 12-72. These data apply for an AD8017AR mounted to a heat sink with an area of about 4 square inches on a two-layer, 2 ounce copper PCB.

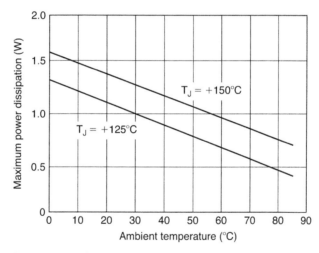

**Figure 12-72: Thermal rating curves for AD8017AR op amp**

These curves indicate the maximum power dissipation versus temperature characteristic for the AD8017, for maximum junction temperatures of both 150°C and 125°C. Such curves are often referred to as *derating* curves, since allowable power decreases with ambient temperature.

With the AD8017AR, the proprietary ADI *Thermal Coastline* IC package is used, which allows additional power to be dissipated with no increase in the SO-8 package size. For a $T_{J(MAX)}$ of 150°C, the upper curve shows the allowable power in this package, which is 1.3 W at an ambient of 25°C. If a more conservative $T_{J(MAX)}$ of 125°C is used, the lower of the two curves applies.

A performance comparison for an 8-pin standard SOIC and the ADI Thermal Coastline version is shown in Figure 12-73. Note that the Thermal Coastline provides an allowable dissipation at 25°C of 1.3 W, whereas a standard package allows only 0.8 W. In the Thermal Coastline heat transferal is increased, accounting for the package's lower $\theta_{JA}$.

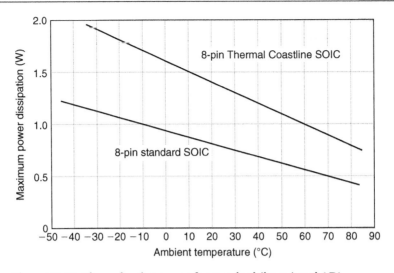

**Figure 12-73: Thermal rating curves for standard (lower) and ADI Thermal Coastline (upper) 8-Pin SOIC packages**

Even higher power dissipation is possible, with the use of IC packages better able to transfer heat from chip to PCB. An example is the AD8016 device, available with two package options rated for 5.5 and 3.5 W at 25°C, respectively, as shown in Figure 12-73.

Taking the higher rated power option, the AD8016ARP PSOP3 package, when used with a 10 inch$^2$ of 1 ounce heat sink plane, the combination is able to handle up to 3 W of power at an ambient of 70°C, as noted by the upper curve. This corresponds to a $\theta_{JA}$ of 18°C/W, which in this case applies for a maximum junction temperature of 125°C (Figure 12-74).

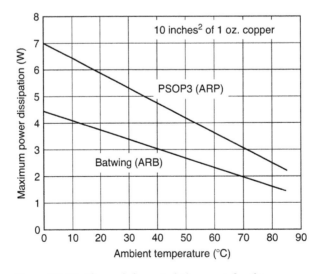

**Figure 12-74: Thermal characteristic curves for the AD8016 BATWING (lower) and PSOP3 (upper) packages, for TJ(max) equal to 125°C**

The reason the PSOP3 version of the AD8016 is so better able to handle power lies with the use of a large area copper slug. Internally, the IC die rests directly on this slug, with the bottom surface exposed as shown in Figure 12-75. The intent is that this surface be soldered directly to a copper plane of the PCB, thereby extending the heat sinking.

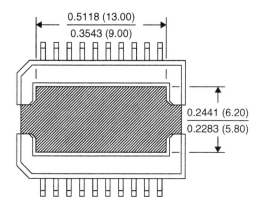

**Figure 12-75: Bottom view of AD8016 20-lead PSOP3 package showing copper slug for aid in heat transfer (central grayed area)**

For reliable, low thermal resistance designs with op amps, several design *Do's and Don't's* are listed below. Consider all of these points that may be practical.

1. *Do use as large an area of copper as possible for a PCB heat sink, up to the point of diminishing returns.*
2. *In conjunction with (1), do use multiple (outside) PCB layers, connected together with multiple vias.*
3. *Do use as heavy copper as is practical (2 ounce or more preferred).*
4. *Do provide sufficient natural ventilation inlets and outlets within the system, to allow heat to freely move away from hot PCB surfaces.*
5. *Do orient power-dissipating PCB planes vertically, for convection-aided airflow across heat sink areas.*
6. *Do consider the use of external power buffer stages, for precision op amp applications.*
7. *Do consider the use of forced air for situations where several watts must be dissipated in a confined space.*
8. *Do not use solder mask planes over heat-dissipating traces.*
9. *Do not use excessive supply voltages on ICs delivering power.*

Both of the AD8016 package options are characterized for both still and moving air, but the thermal information given above applies *without* the use of directed airflow. Therefore, adding additional airflow lowers thermal resistance further (see Reference 2).

For the most part, these points are obvious. However, one that could use some elaboration is number 9. Whenever an application requires only modest *voltage* swings (such as, e.g., standard video, $2V_{p-p}$) a wide supply voltage range can often be used. But operation of an op amp driver on higher supply voltages produces a large IC dissipation, even though the load power is constant (Figure 12-76).

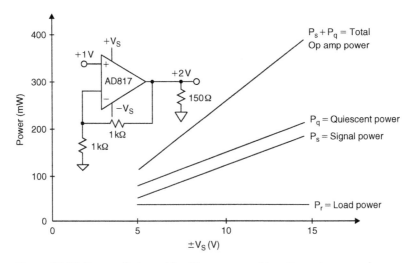

**Figure 12-76: Power dissipated in video op amp driver for various supply voltages with low voltage output swing**

In such cases, as long as the distortion performance of the application does not suffer, it can be advantageous to operate the IC on lower supplies, say $\pm 5V$, as opposed to $\pm 15V$. The above example data were calculated on a DC basis, which will generally tax the driver more in terms of power than a sinewave or a noise-like waveform, such as a DMT signal (see Reference 2). The general principles still hold for these AC waveforms, i.e., the op amp power dissipation is high when load current is high and the voltage is low.

While there is ample opportunity for high power handling with the thermally enhanced packages described above for the AD8016 and AD8107, the increasingly popular smaller IC packages actually move in an opposite direction. Without question, it is true that today's smaller packages do noticeably sacrifice thermal performance. But, it must be understood that this is done in the interest of realizing a smaller size for the packaged op amp, and, ultimately, a much greater final PCB density for the overall system.

These points are illustrated by the thermal ratings for the AD8057 and AD8058 family of single and dual op amp devices, as is shown in Figure 12-77. The AD8057 and AD8058 op amps are available in three different packages. These are the SOT-23-5, and the 8-pin µSOIC, along with standard SOIC.

As the data show, as the package size becomes smaller and smaller, much less power is capable of being removed. Since the lead frame is the only heat sinking possible with such tiny packages, their thermal performance is thus reduced. The $\theta_{JA}$ for the packages mentioned is 240, 200, and 160°C/W, respectively. Note that this is more of a *package* than *device* limitation. Other ICs with the same packages have similar characteristics.

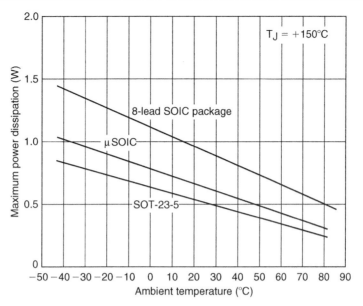

**Figure 12-77: Comparative thermal performance for several AD8057/58 op amp package options**

## Data Converter Thermal Considerations

At first glance, one might assume that the power dissipation of an ADC or a DAC will remain constant for a given power supply voltage. However, many data converters, especially CMOS ones, have power dissipations that are highly dependent on not only output data loading but also the sampling clock frequency. Since many of the newer high speed converters can dissipate between 1.5 and 2 W maximum power under the worst-case operating conditions, this point must be well understood in order to ensure that the package is mounted in such a way as to maintain the junction temperature within acceptable limits at the highest expected operating temperature.

The previous discussion in this chapter on grounding emphasized that the digital outputs of high performance ADCs, especially those with parallel outputs, should be lightly loaded (5–10 pF) in order to prevent digital transient currents from corrupting the SNR and SFDR. Even under light output loading, however, most CMOS and BiCMOS ADCs have power dissipations which are a function of sampling clock frequency and in some cases, the analog input frequency and amplitude.

For example, Figure 12-78 shows the AD9245 14-bit, 80-MSPS, 3 V CMOS ADC power dissipation versus frequency for a 2.5 MHz analog input and 5 pF output loading of the data lines. The graphs show the digital and analog power supply currents separately as well as the total power dissipation. Note that total power dissipation can vary between approximately 310 and 380 mW as the sampling frequency is varied between 10 and 80 MSPS.

The AD9245 is packaged in a 32-pin leadless chip-scale package as shown in Figure 12-79. The bottom view of the package shows the exposed paddle which should be soldered to the PC board ground plane for best thermal transfer. The worst-case package junction-to-ambient resistance, $\theta_{JA}$, is specified as 32.5°C/W, which places the junction $32.5° \times 0.38 = 12.3°C$ above the ambient for a power dissipation of 380 mW. For a maximum operating temperature of $+85°C$, this places the junction at a modest $85 + 12.3 = 97.3°C$.

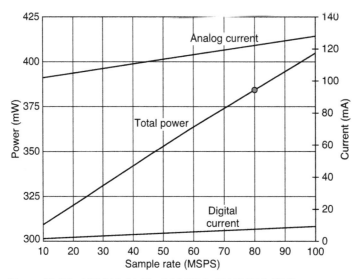

Figure 12-78: AD9245 14-Bit, 80-MSPS, 3 V CMOS ADC power
dissipation versus sample rate for 2.5-MHz input, 5-pF output loads

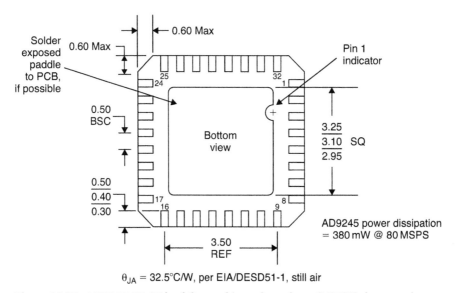

Figure 12-79: AD9245 CP-32 lead-frame chip-scale package (LFCSP), bottom view

The AD9430 is a high performance 12-bit, 170/210-MSPS 3.3 V BiCMOS ADC. Two output modes are available: dual 105-MSPS demultiplexed CMOS outputs, or 210-MSPS LVDS outputs. Power dissipation as a function of sampling frequency is shown in Figure 12-80. Analog and digital supply currents are shown for CMOS and LVDS modes for an analog input frequency of 10.3 MHz. Note that in the LVDS mode and a sampling frequency of 210 MSPS, total supply current is approximately 455 mA—yielding a total power dissipation of 1.5 W.

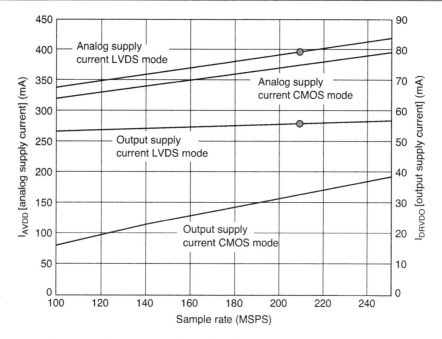

Total current @ 210 MSPS, LVDS mode = 55 mA + 400 mA = 455 mA
Total power dissipation = 3.3 V × 455 mA = 1.5 W

**Figure 12-80: AD9430 12-bit 170/210-MSPS ADC supply current versus sample rate for a 10.3 MHz input**

The AD9430 is available in a 100-lead thin plastic quad flat package with an exposed pad (TQFP/EP) as shown in Figure 12-81. The conductive pad is connected to chip ground and should be soldered to the PC board ground plane. The $\theta_{JA}$ of the package when soldered to the ground plane is 25°C/W in still air. This places the junction $25 \times 1.5 = 37.5$°C above the ambient temperature for 1.5°W of power dissipation. For a maximum operating temperature of +85°C, this places the junction at $85 + 37.5 = 122.5$°C.

The AD6645 is a high performance 14-bit, 80/105-MSPS ADC fabricated on a high speed complementary bipolar process (XFCB, and offers the highest SFDR (89 dBc) and SNR (75 dB)). Although there is little variation in power as a function of sampling frequency, the maximum power dissipation of the device is 1.75 W. The package is a thermally enhanced 52-lead PowerQuad 4® with an exposed pad as shown in Figure 12-82.

It is recommended that the exposed center heat sink be soldered to the PC board ground plane to reduce the package $\theta_{JA}$ to 23°C/W in still air. For 1.75 W of power dissipation, this places the junction temperature $23 \times 1.75 = 40.3$°C above the ambient temperature. For a maximum operating temperature of +85°C, this places the junction at $85 + 40.3 = 125.3$°C. The thermal resistance of the package can be reduced to 17°C/W with 200 LFPM airflow, thereby reducing the junction temperature to 30°C above the ambient, or 115°C for an operating ambient temperature of +85°C.

High speed CMOS DACs (such as the TxDAC® series) and DDS ICs (such as the AD985x series) also have clock-rate dependent power dissipation. For example, in the case of the AD9777 16-bit, 160-MSPS dual interpolating DAC, power dissipation is a function of clock rate, output frequency, and the enabling

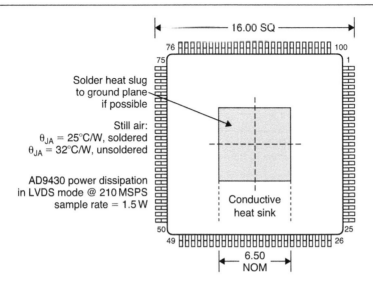

Solder heat slug to ground plane if possible

Still air:
$\theta_{JA} = 25°C/W$, soldered
$\theta_{JA} = 32°C/W$, unsoldered

AD9430 power dissipation in LVDS mode @ 210 MSPS sample rate = 1.5 W

16.00 SQ

Conductive heat sink

6.50 NOM

Figure 12-81: AD9430 100-lead e-PAD TQFP

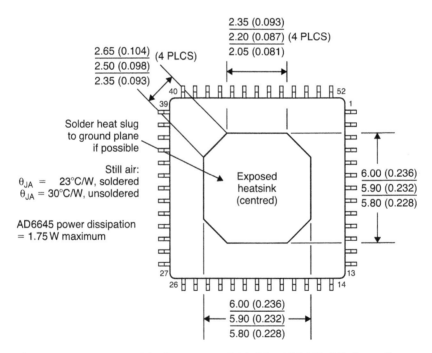

2.65 (0.104)
2.50 (0.098) (4 PLCS)
2.35 (0.093)

2.35 (0.093)
2.20 (0.087) (4 PLCS)
2.05 (0.081)

Solder heat slug to ground plane if possible

Still air:
$\theta_{JA} = 23°C/W$, soldered
$\theta_{JA} = 30°C/W$, unsoldered

AD6645 power dissipation = 1.75 W maximum

Exposed heatsink (centred)

6.00 (0.236)
5.90 (0.232)
5.80 (0.228)

6.00 (0.236)
5.90 (0.232)
5.80 (0.228)

Figure 12-82: AD6645 52-lead power-quad 4 (LQFP_ED) (SQ-52) thermally enhanced package, bottom view

of the PLL and the modulation functions. Power dissipation on 3.3 V supplies can range from 380 mW ($f_{DAC}$ = 100 MSPS, $f_{OUT}$ = 1 MHz, no interpolation, no modulation) to 1.75 W ($f_{DAC}$ = 400 MSPS, $f_{DATA}$ = 50 MHz, $f_s/2$ modulation, PLL enabled). These and similar parts in the family are also offered in thermally enhanced packages with exposed pads for soldering to the PC board ground plane.

These discussions on the thermal application issues of op amps and data converters have not dealt with the classic techniques of using clip-on (or bolt-on) type heat sinks. They also have not addressed the use of forced air cooling, generally considered only when tens of watts must be handled. These omissions are mainly because these approaches are seldom possible or practical with today's op amp and data converter packages.

The more general discussions within References 4–7 can be consulted for this and other supplementary information.

## References: Thermal Management

1. Data sheet for **AD8017 Dual High Output Current, High Speed Amplifier**, Analog Devices, Inc., http://www.analog.com.

2. Data sheet for **AD8016 Low Power, High Output Current, xDSL Line Driver**, Analog Devices, Inc., http://www.analog.com/.

3. "Power Consideration Discussions," data sheet for **AD815 High Output Current Differential Driver**, Analog Devices, Inc., http://www.analog.com/.

4. W. Jung and W. Kester, "Thermal Management," portion of Section 8 within W. Kester, Editor, **Practical Design Techniques for Power and Thermal Management**, Analog Devices, Inc., 1998, ISBN 0-916550-19-2.

5. General Catalog, **Aavid Thermal Technologies, Inc.**, One Kool Path, Laconia, NH, 03246, (603) 528-3400.

6. Seri Lee, "How to Select a Heat Sink," **Aavid Thermal Technologies**, http://www.aavid.com.

7. Seri Lee, "Optimum Design and Selection of Heat Sinks," **11th IEEE SEMI-THERM™ Symposium**, 1995, http://www.aavid.com.

# INDEX

- Subject Index
- Analog Devices' Parts Index

Printed and bound by CPI Group (UK) Ltd, Croydon, CR0 4YY

03/10/2024

01040339-0004